T0145261

Lecture Notes in Artificial Intelligence   12335

Subseries of Lecture Notes in Computer Science

Series Editors

Randy Goebel
University of Alberta, Edmonton, Canada
Yuzuru Tanaka
Hokkaido University, Sapporo, Japan
Wolfgang Wahlster
DFKI and Saarland University, Saarbrücken, Germany

Founding Editor

Jörg Siekmann
DFKI and Saarland University, Saarbrücken, Germany

More information about this series at http://www.springer.com/series/1244

Alexey Karpov · Rodmonga Potapova (Eds.)

Speech and Computer

22nd International Conference, SPECOM 2020
St. Petersburg, Russia, October 7–9, 2020
Proceedings

Editors
Alexey Karpov (iD)
St. Petersburg Institute for Informatics
and Automation of the Russian Academy
of Sciences
St. Petersburg, Russia

Rodmonga Potapova (iD)
Institute for Applied
and Mathematical Linguistics
Moscow State Linguistic University
Moscow, Russia

ISSN 0302-9743 ISSN 1611-3349 (electronic)
Lecture Notes in Artificial Intelligence
ISBN 978-3-030-60275-8 ISBN 978-3-030-60276-5 (eBook)
https://doi.org/10.1007/978-3-030-60276-5

LNCS Sublibrary: SL7 – Artificial Intelligence

This Springer imprint is published by the registered company Springer Nature Switzerland AG
The registered company address is: Gewerbestrasse 11, 6330 Cham, Switzerland

SPECOM 2020 Preface

The International Conference on Speech and Computer (SPECOM) has become a regular event since the first SPECOM held in St. Petersburg, Russia, in October 1996. 24 years ago, SPECOM was established by the St. Petersburg Institute for Informatics and Automation of the Russian Academy of Sciences (SPIIRAS) and the Herzen State Pedagogical University of Russia thanks to the efforts of Prof. Yuri Kosarev and Prof. Rajmund Piotrowski.

SPECOM is a conference with a long tradition that attracts researchers in the area of computer speech processing, including recognition, synthesis, understanding, and related domains like signal processing, language and text processing, computational paralinguistics, multi-modal speech processing or human-computer interaction. The SPECOM international conference is an ideal platform for know-how exchange – especially for experts working on Slavic or other highly inflectional languages – including both, under-resourced and regular well-resourced languages.

In its long history, the SPECOM conference was organized alternately by SPIIRAS and by the Moscow State Linguistic University (MSLU) in their hometowns. Furthermore, in 1997 it was organized by the Cluj-Napoca Subsidiary of the Research Institute for Computer Technique (Romania), in 2005 by the University of Patras (in Patras, Greece), in 2011 by the Kazan Federal University (Russia), in 2013 by the University of West Bohemia (in Pilsen, Czech Republic), in 2014 by the University of Novi Sad (in Novi Sad, Serbia), in 2015 by the University of Patras (in Athens, Greece), in 2016 by the Budapest University of Technology and Economics (in Budapest, Hungary), in 2017 by the University of Hertfordshire (in Hatfield, UK), in 2018 by the Leipzig University of Telecommunications (in Leipzig, Germany), and in 2019 by the Boğaziçi University (in Istanbul, Turkey).

SPECOM 2020 was the 22nd event in the series, and this time it was organized by SPIIRAS in cooperation with MSLU during October 7–9, 2020, in an online format. In July 2020, SPIIRAS incorporated five other research institutions, but has now been transformed into the St. Petersburg Federal Research Center of the Russian Academy of Sciences (SPC RAS). The conference was sponsored by HUAWEI (Russian Research Center) as a general sponsor, and by ASM Solutions (Moscow, Russia), as well as supported by the International Speech Communication Association (ISCA) and Saint Petersburg Convention Bureau. The official conference service agency is Monomax PCO.

SPECOM 2020 was held jointly with the 5th International Conference on Interactive Collaborative Robotics (ICR 2020), where problems and modern solutions of human-robot interaction were discussed.

During SPECOM and ICR 2020, three keynote lectures were given by Prof. Isabel Trancoso (University of Lisbon and INESC-ID, Lisbon, Portugal) on "Profiling Speech for Clinical Applications", by Dr. Ilshat Mamaev (Karlsruhe Institute of Technology,

Germany) on "A Concept for a Human-Robot Collaboration Workspace using Proximity Sensors", as well as by researchers of HUAWEI (Russian Research Center).

Due to the COVID-19 global pandemic, for the first time, SPECOM 2020 was organized as a fully virtual conference. The virtual conference, in the online format via Zoom, had a number of advantages including: an increased number of participants because listeners could take part without any fees, essentially reduced registration fees for authors of the presented papers, no costs for travel and accommodation, a paperless green conference with only electronic proceedings, free access to video presentations after the conference, comfortable home conditions, etc.

This volume contains a collection of submitted papers presented at the conference, which were thoroughly reviewed by members of the Program Committee and additional reviewers consisting of more than 100 top specialists in the conference topic areas. In total, 65 accepted papers out of over 160 papers submitted for SPECOM/ICR were selected by the Program Committee for presentation at the conference and for inclusion in this book. Theoretical and more general contributions were presented in common plenary sessions. Problem-oriented sessions as well as panel discussions brought together specialists in limited problem areas with the aim of exchanging knowledge and skills resulting from research projects of all kinds.

We would like to express our gratitude to all authors for providing their papers on time, to the members of the conference Program Committee and the organizers of the special sessions for their careful reviews and paper selection, and to the editors and correctors for their hard work in preparing this volume. Special thanks are due to the members of the Organizing Committee for their tireless effort and enthusiasm during the conference organization.

October 2020

Alexey Karpov
Rodmonga Potapova

Organization

The 22nd International Conference on Speech and Computer (SPECOM 2020) was organized by the St. Petersburg Institute for Informatics and Automation of the Russian Academy of Sciences (SPIIRAS, St. Petersburg, Russia) in cooperation with the Moscow State Linguistic University (MSLU, Moscow, Russia). The conference website is: http://specom.nw.ru/2020.

General Chairs

Alexey Karpov   SPIIRAS, Russia
Rodmonga Potapova   MSLU, Russia

Program Committee

Shyam Agrawal, India
Tanel Alumäe, Estonia
Elias Azarov, Belarus
Anton Batliner, Germany
Jerome Bellegarda, USA
Milana Bojanic, Serbia
Nick Campbell, Ireland
Eric Castelli, Vietnam
Josef Chaloupka, Czech Republic
Vladimir Chuchupal, Russia
Nicholas Cummins, Germany
Maria De Marsico, Italy
Febe De Wet, South Africa
Vlado Delić, Serbia
Anna Esposito, Italy
Yannick Estève, France
Keelan Evanini, USA
Vera Evdokimova, Russia
Nikos Fakotakis, Greece
Mauro Falcone, Italy
Philip Garner, Switzerland
Gábor Gosztolya, Hungary
Tunga Gungor, Turkey
Abualseoud Hanani, Palestine
Ruediger Hoffmann, Germany
Marek Hrúz, Czech Republic
Kristiina Jokinen, Japan

Oliver Jokisch, Germany
Denis Jouvet, France
Tatiana Kachkovskaia, Russia
Alexey Karpov, Russia
Heysem Kaya, The Netherlands
Tomi Kinnunen, Finland
Irina Kipyatkova, Russia
Daniil Kocharov, Russia
Liliya Komalova, Russia
Evgeny Kostyuchenko, Russia
Galina Lavrentyeva, Russia
Benjamin Lecouteux, France
Anat Lerner, Israel
Boris Lobanov, Belarus
Elena Lyakso, Russia
Joseph Mariani, France
Konstantin Markov, Japan
Jindřich Matoušek, Czech Republic
Yuri Matveev, Russia
Ivan Medennikov, Russia
Peter Mihajlik, Hungary
Wolfgang Minker, Germany
Iosif Mporas, UK
Ludek Muller, Czech Republic
Bernd Möbius, Germany
Sebastian Möller, Germany
Satoshi Nakamura, Japan

Jana Neitsch, Denmark
Stavros Ntalampiras, Italy
Dimitar Popov, Bulgaria
Branislav Popović, Serbia
Vsevolod Potapov, Russia
Rodmonga Potapova, Russia
Valeriy Pylypenko, Ukraine
Gerhard Rigoll, Germany
Fabien Ringeval, France
Milan Rusko, Slovakia
Sergey Rybin, Russia
Sakriani Sakti, Japan
Albert Ali Salah, The Netherlands
Maximilian Schmitt, Germany
Friedhelm Schwenker, Germany
Milan Sečujski, Serbia
Tatiana Sherstinova, Russia
Tatiana Shevchenko, Russia

Ingo Siegert, Germany
Vered Silber-Varod, Israel
Vasiliki Simaki, Sweden
Pavel Skrelin, Russia
Claudia Soria, Italy
Victor Sorokin, Russia
Tilo Strutz, Germany
Sebastian Stüker, Germany
Ivan Tashev, USA
Natalia Tomashenko, France
Laszlo Toth, Hungary
Isabel Trancoso, Portugal
Jan Trmal, USA
Charl van Heerden, South Africa
Vasilisa Verkhodanova, The Netherlands
Matthias Wolff, Germany
Zeynep Yucel, Japan
Miloš Železný, Czech Republic

Additional Reviewers

Gerasimos Arvanitis, Greece
Alexandr Axyonov, Russia
Cem Rıfkı Aydın, Turkey
Gözde Berk, Turkey
Tijana Delić, Serbia
Denis Dresvyanskiy, Germany
Bojana Jakovljević, Serbia
Uliana Kochetkova, Russia
Sergey Kuleshov, Russia

Olesia Makhnytkina, Russia
Danila Mamontov, Germany
Maxim Markitantov, Russia
Dragiša Mišković, Serbia
Dmitry Ryumin, Russia
Andrey Shulipa, Russia
Siniša Suzić, Serbia
Alena Velichko, Russia
Oxana Verkholyak, Russia

Organizing Committee

Alexey Karpov (Chair)
Andrey Ronzhin
Rodmonda Potapova
Daniil Kocharov
Irina Kipyatkova
Dmitry Ryumin

Natalia Kashina
Ekaterina Miroshnikova
Natalia Dormidontova
Margarita Avstriyskaya
Dmitriy Levonevskiy

Contents

Lightweight CNN for Robust Voice Activity Detection

Tanvirul Alam⊙ and Akib Khan^(✉)⊙

BJIT Limited, Dhaka, Bangladesh
{tanvirul.alam,akib.khan}@bjitgroup.com

Abstract. Voice activity detection (VAD) is an important prepossessing step in many speech related applications. Convolutional neural networks (CNN) are widely used for different audio classification tasks and have been adopted successfully for this. In this work, we propose a lightweight CNN architecture for real time voice activity detection. We use strong data augmentation and regularization for improving the performance of the model. Using knowledge distillation approach, we transfer knowledge from a larger CNN model which leads to better generalization ability and robust performance of the CNN architecture in noisy conditions. The resulting network obtains 62.6% relative improvements in EER compared to a deep feedforward neural network (DNN) of comparable parameter count on a noisy test dataset.

Keywords: Voice activity detection · Convolutional neural networks · Regularization · Knowledge distillation

1 Introduction

Voice activity detection (VAD) is the task of identifying the parts of a noisy speech signal that contains human speech activity. This is used widely in many speech processing applications including automatic speech recognition, speech enhancement, speech synthesis, speaker identification, etc. Usually, VAD is used in the initial phase of such an application to improve the efficacy of the later tasks. VAD algorithms are thus required to be robust to different kinds of environmental noise while having a low computational cost and small memory footprint.

A number of techniques have been proposed for VAD in the literature. Early works were based on energy based features, in combination with zero crossing rates and periodicity measures [15,32,33]. As these approaches are highly affected in the presence of additive noise, various other features have been proposed [2,7,23]. Different supervised [19,36] and unsupervised [23,34] learning algorithms have been adopted for the task as well.

More recently, deep learning has been used for the VAD due to their ability to model complex functions and learn robust features from the dataset itself.

© Springer Nature Switzerland AG 2020
A. Karpov and R. Potapova (Eds.): SPECOM 2020, LNAI 12335, pp. 1–12, 2020.
https://doi.org/10.1007/978-3-030-60276-5_1

Convolutional neural networks (CNN) have been used for VAD as they are capable of learning rich, local feature representation and are more invariant to input position and distortion [1,25,26,30]. Recurrent neural networks (RNN) and especially their variants, long short-term memory (LSTM) networks, are capable of learning long range dependencies between inputs and are also used for VAD [5,12,27]. Both CNN and LSTM based approaches perform better than multilayer perceptrons under noisy conditions [31]. However, if the audio duration is too long LSTM may become trapped into a dead state, and the performance can degrade [1]. CNNs do not suffer from this while also being computationally more efficient. CNN and RNN can also be combined to form convolutional recurrent neural networks that can benefit from frequency modeling with CNN and temporal modeling with RNN [35]. Denoising autoencoders are often used to learn noise-robust features which are then used for the classification task [13,16,37]. Speech features are augmented with estimated noise information for noise aware training in [31] to improve robustness.

We adopt CNN in our study due to their computational efficiency and reliable performance in VAD and other audio classification tasks [4,10,24]. Since our focus is to develop a VAD which is robust to different kinds of noise, we synthesize a training dataset under different noisy conditions and signal to noise ratio (SNR) levels. To gauge the robustness of the learned model, we prepare a test dataset by using speech and noise data obtained from a different source.

One of our primary goal is to improve the performance of the network without incurring additional computational cost. For this, we design a lightweight CNN architecture. We use SpecAugment [20] and DropBlock [6] regularization to improve the generalization performance of the network, both of which improve upon the baseline. CNN models trained with deeper layers with strong regularization tend to have better generalization performance and perform better under unseen noise sources. However, they are often not feasible for real-time use in constrained settings (e.g., in mobile devices where low memory and fast inference time are often preferred). To address this issue, we train a larger model that achieves better performance compared to the proposed CNN architecture. We then use it to transfer knowledge to our lightweight CNN architecture using knowledge distillation [11]. This further improves the network's performance under unseen noise types.

We organize the rest of the paper as follows. In Sect. 2, we describe our baseline CNN architecture along with its improvements using SpecAugment, DropBlock and knowledge distillation. In Sect. 3, we describe the training and test datasets. We provide details of the experiments, different hyperparameter settings, and results obtained on development and test set with different approaches in Sect. 4. Finally, in Sect. 5, we conclude our paper and provide an outlook on future work.

2   Method

2.1   Deep Convolutional Neural Network

Deep convolutional neural network (CNN) architectures have been shown to be successful for the voice activity detection task [1,30,31] as they are capable of learning rich hierarchical representation by utilizing local filters. Feature extracted from an extended centered context window is used as network input for frame level voice activity detection. We design a CNN architecture while keeping the computational and memory constraints often encountered by VAD applications in mind. The network is composed of three convolution and pooling layers followed by two fully connected layers. The first convolution and pooling layer each uses a (5×5) kernel. Second and final convolution layer uses a (3×3) kernel and each is followed by a (2×2) max pooling layer. The three convolution layers consist of 32, 48 and 64 filters respectively. The first fully connected layer has 64 hidden units and the second fully connected layer has 2 hidden units corresponding to the two categories of interest, specifically voice and nonvoice. ReLU nonlinearity is used after each convolution and fully connected layer except for the final layer which has softmax activation function. Dropout [29] is used after the first fully connected layer.

All audios are sampled at 16 KHz. We use librosa [18] to extract log-scaled mel spectrogram features with 40 frequency bands. We use window size of 512 and hop length 256 covering the frequency range of 300–8000 Hz. To stabilize mel spectrogram output we use $log(melspectrogram + 0.001)$, where the offset is used to avoid taking a logarithm of zero. The log mel spectrograms are normalized to have zero mean value. We use an extended context window of 40 frames i.e., the input shape is 40×40. We treat the mel spectrogram feature as a single channel image and apply successive convolution and pooling operations on it. Padding is used to keep spatial dimension unchanged during

Table 1. Proposed CNN architecture. Data shape represents the dimension in channel, frequency, time. Each convolution and fully connected layer includes a bias term.

Layer	ksize	Filters	Stride	Data shape
Input				(1, 40, 40)
conv1	(5, 5)	32	(1, 1)	(32, 40, 40)
pool1	(5, 5)		(5, 5)	(32, 8, 8)
conv2	(3, 3)	48	(1, 1)	(48, 8, 8)
pool2	(2, 2)		(2, 2)	(48, 4, 4)
conv3	(3, 3)	64	(1, 1)	(64, 4, 4)
pool3	(2, 2)		(2, 2)	(64, 2, 2)
fc1		64		(64,)
Dropout				(64,)
fc2		2		(2,)

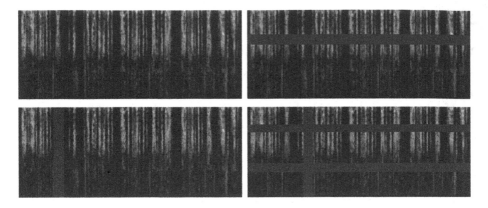

Fig. 1. SpecAugment used in this study. *top-left:* Log mel spectrogram for a sample audio without augmentation, *top-right:* Frequency masking, *bottom-left:* Time-masking, *bottom-right:* A combination of three frequency and time masking applied.

convolution operation. The network has 58,994 trainable parameters and a total of 2.7M multiply-accumulate operations. This network can run in real time in most modern processors. See Table 1 for specific layer details.

We also train a deep feedforward neural network (DNN) consisting of three hidden layers, and compare its performance to that of the CNN model. Each hidden layer of the DNN has 36 units. All fully connected layers except the last one are followed by ReLU non-linearity. Dropout is not added in this network as it causes the network to underfit on the dataset. This network is designed to have parameter count comparable to the CNN architecture and has 60374 total trainable parameters. The DNN is trained using the same input features as in the CNN architecture.

2.2   SpecAugment

SpecAugment was introduced in [20] as a simple but effective augmentation strategy for speech recognition. Three types of deformations were introduced in the paper that can be directly applied on spectrogram features: time warping, frequency masking and time masking. Here, we apply a combination of frequency and time masking (see Fig. 1). We omit time warping as it was reported in [20] to be the least influential while being the most expensive. Frequency masking is applied to f consecutive mel frequency channels $[f_0, f_0 + f)$ where f is first chosen from a uniform distribution from 0 to a selectable parameter F, and f_0 is chosen uniformly from $[0, v - f]$ with v being the number of mel frequency channels. Similarly, for time masking, t consecutive time steps $[t_0, t_0 + t)$ are masked, where t is chosen from a uniform distribution from 0 to the time mask parameter T and t_0 is chosen uniformly from $[0, \tau - t)$, where τ is the number of total time steps. We apply $[0, n]$ such masking randomly for each spectrogram image where n is a tunable parameter. For each masking, we randomly select

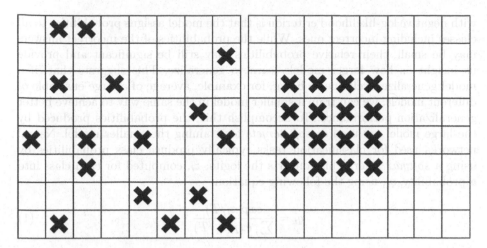

Fig. 2. Schematics of DropBlock regularization used in this study. *left:* Dropout which drops activations at random, *right:* DropBlock drops continuous regions of features.

either frequency or time masking and set $F = \alpha v$ and $T = \alpha \tau$. Since the log mel spectrogram is normalized to have zero mean value, setting the masked value to zero is equivalent to setting it to the mean value. SpecAugment effectively converts an overfitting to an underfitting problem. This enables us to train larger networks with longer duration (i.e., more epochs) without overfitting and leads to improved generalization performance of the model.

2.3 DropBlock

Although dropout is commonly used as a regularization technique for fully connected layers, it is not as effective for convolution layers. This is because features are spatially correlated in convolution layer, and dropping units randomly does not effectively prevent information from being sent to the next layers. To remedy this, the authors in [6] introduced *DropBlock* regularization which drops contiguous regions of feature maps (see Fig. 2). As this discards features in a spatially correlated area, the network needs to learn discriminative features from different regions of the feature map for classification. It was shown to be a more effective regularizer compared to dropout for image classification and detection tasks in [6]. DropBlock has two main parameters which are *block_size* and γ. *block_size* determines the size of the square block to be dropped and γ controls the number of activation units to drop. In our experiments we apply DropBlock after the first pooling layer only. We don't apply DropBlock on the subsequent layers as the resolution of the feature map becomes small.

2.4 Knowledge Distillation

Distillation is a method for transferring knowledge from an ensemble or from a large, highly regularized model into a smaller model [11]. A side-effect of training with negative log-likelihood criterion is that the model assigns probabilities to all classes including incorrect ones. While the probabilities of the incorrect answers may be small, their relative probabilities may still be significant and provide us insight on how larger models learn to generalize. This suggests that, if a model generalizes well due to being, for example, average of a large ensemble of different models, we can train a smaller model in the same way to achieve better generalization performance. To accomplish this, the probabilities produced by the large model are used as *soft targets* for training the smaller model. Neural networks used for classification tasks typically produce class probabilities by using a *softmax* layer. This converts the logits, z_i, computed for each class into a probability, q_i, using the following equation

$$q_i = \frac{exp(z_i/T)}{\sum_j exp(z_j/T)} \tag{1}$$

Here, T is the temperature term that is usually set to 1. We can get softer probability distribution over classes using a higher temperature. If the correct labels are known, the smaller model can be trained to produce the correct labels as well. The loss function then takes the following form

$$L_{KD} = \alpha T^2 L_{CE}(Q_s^T, Q_t^T) + (1 - \alpha)L_{CE}(Q_s, y_{true}) \tag{2}$$

Here, Q_s^T and Q_t^T are the softened probabilities of the smaller student and larger teacher model respectively, generated using a higher temperature and Q_s is generated using $T = 1$. L_{CE} is the cross entropy loss function and $\alpha(0 \leq \alpha \leq 1)$ is a hyper parameter that controls the contribution from each part of the loss. The first term in the loss is multiplied by T^2 since the magnitude of the gradients produced by soft targets in Eq. 1 scale as $1/T^2$. Best results are obtained in practice by using a considerably lower weight on the second objective function or equivalently using a large value of α.

CNNs designed for image classification tasks have been shown to be successful in large scale audio classification tasks [10]. Inspired by this, we use Pre-Act ResNet-18 [9] network as the teacher network to transfer knowledge to the previously described CNN architecture. We use the implementation provided in [3] which consists of 11.2M trainable parameters and 868M total multiply-accumulate operations. This means it has roughly 190 times more parameters compared to our baseline CNN architecture.

3   Dataset Description

We prepared the training dataset using MUSAN [28] corpus. The corpus consists of 109 h of audio data partitioned into three categories: speech, music and noise. The speech portion consists of about 60 h of data. It contains 20 h and 21 min of read speech from LibriVox[1], approximately half of which are in English and the

---

[1] https://librivox.org/.

rest are from eleven different languages. The remainder of the speech portion consists of 40 h and 1 min of US government hearings, committees and debates. There are 929 noise files collected from different noise sources, ranging from technical noises, such as DTMF tones, dialtones, fax machine noises etc. as well as ambient sounds like car idling, thunder, wind, footsteps, animal noises etc. These are downloaded from Free Sound[2] and Sound Bible[3]. The total noise duration is about 6 h.

We used the MS-SNSD toolkit provided in [22] to generate our training and test dataset. We removed the long silent parts from the train speech data using pyAudioAnalysis [8]. We generated 60 h of training data with 30 h of noise and speech data each. Noise was added to the speech data by selecting SNR randomly from a discrete uniform distribution $[-10, 20]$. We used a minimum audio length of 10 s each when generating noisy speech. Five-fold cross validation was performed during training.

We prepared an additional test dataset by using noise collected from 100 different types[4]. We used UW/NU corpus version 1.0 [17] for speech which consists of 20 speakers (10 male, 10 female) reciting 180 Harvard IEEE sentences each. We resampled the audios to 16 Khz and used MS-SNSD to generate 10 h of test data. We used 7 discrete SNR levels in $[-10, 20]$ range for preparing the test data in order to gauge model performance under different noisy conditions.

4   Experiments

4.1   Training Procedure

We trained the model using cross entropy loss function and Adam optimization algorithm [14]. We used 256 samples per minibatch during training. The DNN and baseline CNN networks were both trained for 20 epochs. We used an initial learning rate of 0.001 which was decreased by a factor of 10 after 10 and 15 epochs. Dropout was applied after the first fully connected layer of the CNN model with probability 0.5. We optimized the network hyperparameters using the validation set. The network was checkpointed when the accuracy on the validation set improved in an epoch and the network weights with the highest validation set accuracy were applied to the test data. The network was trained using PyTorch [21] deep learning framework.

When using SpecAugment and DropBlock, we noticed that training for more epochs improved performance and so these networks were trained for 40 epochs. Learning rate was initialized at 0.001 and decreased by a factor of 10 after 25 and 35 epochs. Validation set was used to identify the optimum parameter settings for SpecAugment and DropBlock. We used $n = 3$ and $\alpha = 0.2$ for SpecAugment. For DropBlock, we applied $blocksize = 4$ and $\gamma = 0.3$. As mentioned in [6],

---

[2] https://freesound.org/.
[3] http://soundbible.com/.
[4] http://web.cse.ohio-state.edu/pnl/corpus/HuNonspeech/.

Table 2. AUC and EER (%) on validation and test set.

Method	Validation		Test	
	AUC	EER	AUC	EER
DNN	98.32	5.52	95.55	10.45
CNN	99.57	2.44	99.07	4.72
PreAct ResNet-18	**99.75**	**1.68**	**99.51**	**3.27**
CNN + SpecAugment	99.63	2.29	99.22	4.37
CNN + DropBlock	99.58	2.36	99.15	4.50
CNN + SpecAugment + DropBlock	**99.65**	**2.24**	99.31	4.14
CNN Distilled	99.65	2.25	**99.36**	**3.91**

we also noticed that using a fixed value of γ did not work well. So, we linearly increased γ from 0 to 0.3 at the end of each epoch.

The PreAct ResNet-18 model was trained for 20 epochs with similar learning rate scheduling as the baseline CNN architecture. We did not apply dropout or DropBlock for training this model but used SpecAugment with the configuration mentioned above. The model with the best validation set performance was used for transferring knowledge to our CNN model. SpecAugment and DropBlock were used during distillation but dropout was removed as the network underfitted the training data with dropout. Distilled network was trained for 40 epochs using learning rate scheduling same as the regularized models. We experimented with values of $\alpha \in \{0.9, 0.95, 0.99\}$ and $T \in \{2, 4, 8, 16\}$. Best results were obtained for $\alpha = 0.99$ and $T = 8$.

4.2   Results

We use the area under the ROC curve (AUC) and equal error rate (EER) as evaluation metrics as they are commonly used in literature. The reported results are average across 5 runs for both the validation and test sets. In this work, we are only interested in frame level performance, and so we have not conducted evaluation of segment level performance. The results are displayed in Table 2 for different models.

The baseline CNN model performs significantly better than the DNN model in the validation and test datasets. PreAct ResNet-18 outperforms the baseline CNN architecture. We investigate the effect of applying SpecAugment and DropBlock in isolation, and then if they are combined. It is evident that both SpecAugment and DropBlock improves upon CNN baseline, although the performance gain from SpecAugment is greater comparably. Combining both further improves the performance. This suggests that the regularization applied are effective for improving the generalization performance of the network. Since the PreAct ResNet-18 model has better performance compared to the CNN architecture, it can serve as a suitable teacher model for distillation. Transferring

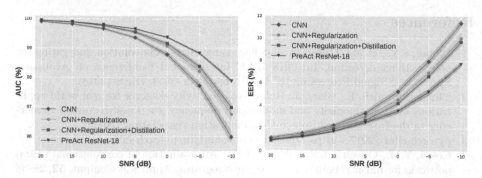

Fig. 3. left: AUC(%) on test data at different SNR levels. right: EER(%) on test data at different SNR levels. Means and standard deviations of five runs are depicted by solid lines and shaded areas respectively.

knowledge from this network further improves performance under unseen noisy conditions. The distilled CNN model has 62.6% reduced EER compared to the DNN architecture on the test dataset. It is interesting to note that, performance gain from regularization and distillation is relatively greater on the test dataset which is constructed from unseen noise types. For example, compared to the baseline CNN, distilled CNN has 7.8% relatively lower EER on the validation dataset. But this number is 17.2% on the test dataset.

Performance at 7 different SNR levels on the test dataset are plotted in Fig. 3 for different models. *CNN + Regularization* refers to the model trained using SpecAugment and DropBlock. As expected, performance degrades as noise is increased. However, for regularized and distilled models the drop is less severe compared to the baseline CNN model. At high SNR levels (e.g., 20 SNR), different models perform comparably to each other. On the other hand, at the extreme noisy condition of -10 SNR, we see 1% absolute improvement of AUC and 1.68% absolute reduction in EER for distilled CNN compared to the baseline CNN architecture. This suggests that adding regularization and distillation makes the model more robust to unseen and severe noisy conditions.

5   Conclusions

In this work, we have designed a lightweight CNN architecture for voice activity detection. We have evaluated the trained model on a noisy test dataset and showed that better results can be obtained using strong data augmentation and regularization. We further demonstrate the effectiveness of knowledge distillation for the task. Our proposed model is robust under severe and unseen noisy conditions. We believe further improvements can be made by using an ensemble of larger models instead of a single model and using them to train the distilled model. We plan to investigate the performance of the proposed approach on audio recorded in noisy environment in future work.

References

1. Chang, S.Y., et al.: Temporal modeling using dilated convolution and gating for voice-activity-detection. In: 2018 IEEE International Conference on Acoustics, Speech and Signal Processing (ICASSP), pp. 5549–5553. IEEE (2018)
2. Chuangsuwanich, E., Glass, J.: Robust voice activity detector for real world applications using harmonicity and modulation frequency. In: Twelfth Annual Conference of the International Speech Communication Association (2011)
3. pytorch cifar (2017). https://github.com/kuangliu/pytorch-cifar
4. Costa, Y.M., Oliveira, L.S., Silla Jr., C.N.: An evaluation of convolutional neural networks for music classification using spectrograms. Appl. Soft Comput. **52**, 28–38 (2017)
5. Eyben, F., Weninger, F., Squartini, S., Schuller, B.: Real-life voice activity detection with LSTM Recurrent Neural Networks and an application to Hollywood movies. In: 2013 IEEE International Conference on Acoustics, Speech and Signal Processing, pp. 483–487. IEEE (2013)
6. Ghiasi, G., Lin, T.Y., Le, Q.V.: Dropblock: a regularization method for convolutional networks. In: Advances in Neural Information Processing Systems, pp. 10727–10737 (2018)
7. Ghosh, P.K., Tsiartas, A., Narayanan, S.S.: Robust voice activity detection using long-term signal variability. IEEE Trans. Speech Audio Process. **19**(3), 600–613 (2011)
8. Giannakopoulos, T.: Pyaudioanalysis: an open-source python library for audio signal analysis. PloS One **10**(12), e0144610 (2015)
9. He, K., Zhang, X., Ren, S., Sun, J.: Identity mappings in deep residual networks. In: Leibe, B., Matas, J., Sebe, N., Welling, M. (eds.) ECCV 2016. LNCS, vol. 9908, pp. 630–645. Springer, Cham (2016). https://doi.org/10.1007/978-3-319-46493-0_38
10. Hershey, S., et al.: CNN architectures for large-scale audio classification. In: 2017 IEEE International Conference on Acoustics, Speech and Signal Processing, pp. 131–135. IEEE (2017)
11. Hinton, G., Vinyals, O., Dean, J.: Distilling the knowledge in a neural network. In: NIPS Deep Learning and Representation Learning Workshop (2015), arxiv:1503.02531
12. Hughes, T., Mierle, K.: Recurrent neural networks for voice activity detection. In: 2013 IEEE International Conference on Acoustics, Speech and Signal Processing, pp. 7378–7382. IEEE (2013)
13. Jung, Y., Kim, Y., Choi, Y., Kim, H.: Joint learning using denoising variational autoencoders for voice activity detection. In: INTERSPEECH, pp. 1210–1214 (2018)
14. Kingma, D.P., Ba, J.: Adam: A method for stochastic optimization. In: 3rd International Conference on Learning Representations (2015)
15. Lamel, L., Rabiner, L., Rosenberg, A., Wilpon, J.: An improved endpoint detector for isolated word recognition. IEEE Trans. Acoust. Speech Signal Process. **29**(4), 777–785 (1981)
16. Lin, R., Costello, C., Jankowski, C., Mruthyunjaya, V.: Optimizing voice activity detection for noisy conditions. In: Interspeech, pp. 2030–2034. ISCA (2019)
17. McCloy, D.R., Souza, P.E., Wright, R.A., Haywood, J., Gehani, N., Rudolph, S.: The UW/NU corpus (2013). http://depts.washington.edu/phonlab/resources/uwnu/, version 1.0

18. McFee, B., Raffel, C., Liang, D., Ellis, D.P., McVicar, M., Battenberg, E., Nieto, O.: librosa: audio and music signal analysis in python. In: Proceedings of the 14th Python in Science Conference, vol. 8 (2015)
19. Ng, T., et al.: Developing a speech activity detection system for the DARPA RATS program. In: Thirteenth Annual Conference of the International Speech Communication Association (2012)
20. Park, D.S., et al.: Specaugment: a simple data augmentation method for automatic speech recognition. In: Interspeech, pp. 2613–2617. ISCA (2019)
21. Paszke, A., et al.: Pytorch: an imperative style, high-performance deep learning library. In: Advances in Neural Information Processing Systems 32, pp. 8024–8035. Curran Associates, Inc. (2019). http://papers.neurips.cc/paper/9015-pytorch-an-imperative-style-high-performance-deep-learning-library.pdf
22. Reddy, C.K.A., Beyrami, E., Pool, J., Cutler, R., Srinivasan, S., Gehrke, J.: A scalable noisy speech dataset and online subjective test framework. In: Interspeech, pp. 1816–1820. ISCA (2019)
23. Sadjadi, S.O., Hansen, J.H.: Unsupervised speech activity detection using voicing measures and perceptual spectral flux. IEEE Signal Process. Lett. **20**(3), 197–200 (2013)
24. Salamon, J., Bello, J.P.: Deep convolutional neural networks and data augmentation for environmental sound classification. IEEE Signal Process. Lett. **24**(3), 279–283 (2017)
25. Saon, G., Thomas, S., Soltau, H., Ganapathy, S., Kingsbury, B.: The IBM speech activity detection system for the DARPA RATS program. In: Interspeech, pp. 3497–3501. ISCA (2013)
26. Sehgal, A., Kehtarnavaz, N.: A convolutional neural network smartphone app for real-time voice activity detection. IEEE Access **6**, 9017–9026 (2018)
27. Shannon, M., Simko, G., Chang, S.Y., Parada, C.: Improved end-of-query detection for streaming speech recognition. In: Interspeech, pp. 1909–1913 (2017)
28. Snyder, D., Chen, G., Povey, D.: MUSAN: a music, speech, and noise corpus. CoRR abs/1510.08484 (2015), arxiv:1510.08484
29. Srivastava, N., Hinton, G., Krizhevsky, A., Sutskever, I., Salakhutdinov, R.: Dropout: a simple way to prevent neural networks from overfitting. J. Mach. Learn. Res **15**(1), 1929–1958 (2014)
30. Thomas, S., Ganapathy, S., Saon, G., Soltau, H.: Analyzing convolutional neural networks for speech activity detection in mismatched acoustic conditions. In: 2014 IEEE International Conference on Acoustics, Speech and Signal Processing (ICASSP), pp. 2519–2523. IEEE (2014)
31. Tong, S., Gu, H., Yu, K.: A comparative study of robustness of deep learning approaches for VAD. In: 2016 IEEE International Conference on Acoustics, Speech and Signal Processing (ICASSP), pp. 5695–5699. IEEE (2016)
32. Tucker, R.: Voice activity detection using a periodicity measure. IEEE Proc. I (Commun. Speech Vision) **139**(4), 377–380 (1992)
33. Woo, K.H., Yang, T.Y., Park, K.J., Lee, C.: Robust voice activity detection algorithm for estimating noise spectrum. Electron. Lett. **36**(2), 180–181 (2000)
34. Ying, D., Yan, Y., Dang, J., Soong, F.K.: Voice activity detection based on an unsupervised learning framework. IEEE Trans. Audio Speech Lang. Process. **19**(8), 2624–2633 (2011)
35. Zazo, R., Sainath, T.N., Simko, G., Parada, C.: Feature learning with raw-waveform CLDNNs for voice activity detection. In: Interspeech, pp. 3668–3672 (2016)

36. Zhang, X.L., Wang, D.: Boosted deep neural networks and multi-resolution cochlea-gram features for voice activity detection. In: Fifteenth Annual Conference of the International Speech Communication Association (2014)
37. Zhang, X.L., Wu, J.: Denoising deep neural networks based voice activity detection. In: 2013 IEEE International Conference on Acoustics, Speech and Signal Processing, pp. 853–857. IEEE (2013)

Hate Speech Detection Using Transformer Ensembles on the HASOC Dataset

Pedro Alonso[1], Rajkumar Saini[1], and György Kovács[1,2(✉)]

[1] Embedded Internet Systems Lab, Luleå University of Technology, Luleå, Sweden
{pedro.alonso,rajkumar.saini,gyorgy.kovacs}@ltu.se
[2] MTA-SZTE Research Group on Artificial Intelligence, Szeged, Hungary

Abstract. With the ubiquity and anonymity of the Internet, the spread of hate speech has been a growing concern for many years now. The language used for the purpose of dehumanizing, defaming or threatening individuals and marginalized groups not only threatens the mental health of its targets, as well as their democratic access to the Internet, but also the fabric of our society. Because of this, much effort has been devoted to manual moderation. The amount of data generated each day, particularly on social media platforms such as Facebook and twitter, however makes this a Sisyphean task. This has led to an increased demand for automatic methods of hate speech detection.

Here, to contribute towards solving the task of hate speech detection, we worked with a simple ensemble of transformer models on a twitter-based hate speech benchmark. Using this method, we attained a weighted F_1-score of 0.8426, which we managed to further improve by leveraging more training data, achieving a weighted F_1-score of 0.8504. Thus markedly outperforming the best performing system in the literature.

Keywords: Natural Language Processing · Hate speech detection · Transformers · RoBERTa · Ensemble

1 Introduction

There are many questions still surrounding the issue of hate speech. For one, it is strongly debated whether hate speech should be prosecuted, or whether free speech protections should extend to it [2,11,16,24]. Another question debated is regarding the best counter-measure to apply, and whether it should be suppression (through legal measures, or banning/blocklists), or whether it should be methods that tackle the root of the problem, namely counter-speech and education [4]. These arguments, however are fruitless without the ability of detecting hate speech en masse. And while manual detection may seem as a simple (albeit hardly scalable) solution, the burden of manual moderation [15], as well as the sheer amount of data generated online justify the need for an automatic solution of detecting hateful and offensive content.

© Springer Nature Switzerland AG 2020
A. Karpov and R. Potapova (Eds.): SPECOM 2020, LNAI 12335, pp. 13–21, 2020.
https://doi.org/10.1007/978-3-030-60276-5_2

1.1 Related Work

The ubiquity of fast, reliable Internet access that enabled the sharing of information and opinions at an unprecedented rate paired with the opportunity for anonymity [50] has been responsible for the increase in the spread of offensive and hateful content in recent years. For this reason, the detection of hate speech has been examined by many researchers [21,48]. These efforts date back to the late nineties and Microsoft research, with the proposal of a rule-based system named Smokey [36]. This has been followed by many similar proposals for rule-based [29], template-based [27], or keyword-based systems [14,21].

In the meantime, many researchers have tackled this task using classical machine learning methods. After applying the Bag-of-Words (BoW) method for feature extraction, Kwok and Wang [19] used a Naïve Bayes classifier for the detection of racism against black people on Twitter. Grevy et al. [13] used Support Vector Machines (SVMs) on BoW features for the classification of racist texts. However, since the BoW approach was shown to lead to high false positive rates-[6], others used more sophisticated feature extraction methods to obtain input for the classical machine learning methods (such as SVM, Naïve Bayes and Logistic Regression [5,6,39,40]) deployed for the detection of hateful content.

One milestone in hate speech detection was deep learning gaining traction in Natural Language Processing (NLP) after its success in pattern recognition and computer vision [44], propelling the field forward [31]. The introduction of embeddings [26] had an important role in this process. For one, by providing useful features to the same classical machine learning algorithms for hate speech detection [25,45], leading to significantly better results than those attained with the BoW approach (both in terms of memory-complexity, and classification scores [9]). Other deep learning approaches were also popular for the task, including Recurrent Neural Networks [1,7,10,38], Convolutional Neural Networks [1,12,32,51], and methods that combined the two [17,41,49].

The introduction of transformers was another milestone, in particular the high improvement in text classification performance by BERT [37]. What is more, transformer models have proved highly successful in hate speech detection competitions (with most of the top ten teams using a transformer in a recent challenge [46]). Ensembles of transformers also proved to be successful in hate speech detection [28,30]. So much so, that such a solution has attained the best performance (i.e. on average the best performance over several sub-tasks) recently in a challenge with more than fifty participants [35]. For this reason, here, we also decided to use an ensemble of transformer models.

1.2 Contribution

Here, we apply a 5-fold ensemble training method using the RoBERTA model, which enables us to attain state-of-the-art performance on the HASOC benchmark. Moreover, by proposing additional fine-tuning, we significantly increase the performance of models trained on different folds.

Table 1. Example tweets from the HASOC dataset [22].

Tweet	Label
@piersmorgan Dont watch it then. #dickhead	NOT
This is everything. #fucktrump https://t.co/e2C48U3pss	HOF
I stand with him ...He always made us proud 🙏🏻#DhoniKeepsTheGlove	NOT
@jemelehill He's a cut up #murderer	HOF
#fucktrump #impeachtrump 😂😂😂😂😂😂 @ Houston, Texas https://t.co/8QGgbWtOAf	NOT

2 Experimental Materials

In this section we discuss the benchmark task to be tackled. Due to the relevance of the problem, many competitions have been dedicated finding a solution [3, 18, 23, 42, 46]. For this study, we consider the challenge provided by HASOC [22] data (in particular, the English language data) to be tackled using our methods. Here, 6712 tweets (the training and test set containing 5852 and 860 tweets, respectively) were annotated into the following categories:

- NOT: tweets not considered to contain hateful or offensive content
- HOF: tweets considered to be hateful, offensive, or profane

Where in the training set approximately 39% of all instances (2261 tweets) were classified in the second category, while for the test set this ratio was 28% (240 tweets). Some example tweets from the training set are listed in Table 1. As can be seen, in some cases it is not quite clear why one tweet was labelled as hateful, and others were not (#fucktrump). Other examples with debatable labeling are listed in the system description papers from the original HASOC competition [33]. This can be an explanation why Ross et al. suggested to consider hate speech detection as a regression task, as opposed to a classification task [34].

2.1 OffensEval

In line with the above suggestion, the training data published by Zampieri et al. for the 2020 OffensEval competition [47] were not class labels but scores, as can be seen in Table 2 below. Here, for time efficiency reasons we used the first 1 million of the 9,089,140 tweets available in the training set.

Table 2. Example tweets from the OffensEval corpus [47].

Tweet	Score
@USER And cut a commercial for his campaign	0.2387
@USER Trump is a fucking idiot his dementia is getting worse	0.8759
Golden rubbers in these denim pockets	0.3393
Hot girl summer is the shit!!! #period	0.8993

3   Experimental Methods

In this section we discuss the processing pipeline we used for the classification of HASOC tweets. This includes the text preprocessing steps taken, the short description of the machine learning models used, as well as the method of training we applied on said machine learning models.

3.1   Text Preprocessing

Text from social media sites, and in particular Twitter often lacks proper grammar/punctuation, and contains many paralinguistic elements (e.g. URLs, emoticons, emojis, hashtags). To alleviate potential problems caused by this variability, tweets were put through a preprocessing step before being fed to our model. First, consecutive white space characters were replaced by one instance, while extra white space characters were added between words and punctuation marks. Then @-mentions and links were replaced by the character series *@USER* and *URL* respectively. Furthermore, as our initial analysis did not find a significant correlation between emojis and hatefulness scores on the more than nine million tweets of the OffensEval dataset, all emojis and emoticons were removed. Hashtag characters (but not the hashtags themselves) were also removed in the process. Lastly, tweets were tokenized into words.

3.2   RoBERTa

For tweet classification and regression, in our study we used a variant of BERT [8], namely RoBERTa [20], from the SimpleTransformers library [43] (for a detailed description of transformers in general, as well as BERT and RoBERTa in particular, please see the sources cited in this paper). We did so encouraged by the text classification performance of BERT [37], as well as our preliminary experiments with RoBERTa. When training said model, we followed [43] in selecting values for our meta-parameters, with the exception of the learning rate, for which we used $1e - 5$ as our value.

3.3   5-Fold Ensemble Training

In our experiments we used the following training scheme. First, we split the HASOC train set into five equal parts, each consisting of 1170 tweets (Dev_1, Dev_2, Dev_3, Dev_4, Dev_5). This partitioning was carried out in a way that the ratio of the two different classes was the same in each subset than it was in the whole set. Then, for each development set, we created a training set using the remaining tweets from the original training set ($Train_1$, $Train_2$, $Train_3$, $Train_4$, $Train_5$). After creating the five folds in this manner, we used each fold to train separate RoBERTa models. The final model was then defined as the ensemble of the five individual models, where the predictions of the ensemble model was created by averaging the predicted scores of the individual models.

Table 3. F_1-scores of different models on the test set of the HASOC benchmark. For each model, and each F_1-score, the best result is emphasized in bold.

Model	Fold	$HASOC_{only}$	$HASOC_{OffensEval}$
Macro F_1-score	1st	0.7586	0.7964
	2nd	0.7681	0.7855
	3rd	0.7688	0.7943
	4th	0.7914	0.7929
	5th	0.7758	**0.8029**
	Ensemble	**0.7945**	0.7976
Weighted F_1-score	1st	0.8125	0.8507
	2nd	0.8165	0.8402
	3rd	0.8244	0.8474
	4th	0.8415	0.8485
	5th	0.8327	**0.8537**
	Ensemble	**0.8426**	0.8504

Here, we examined two different ensembles. For one ($HASOC_{only}$), we used a pretrained RoBERTa model [43], and fine-tuned it on different folds, creating five different versions of the model. Then, we averaged the predictions of these models for the final classification. To examine how further fine-tuning would affect the results, we first fine-tuned the RoBERTa model using one million tweets from the OffensEval competition for training, and ten thousand tweets for validation. Then, we further fine-tuned the resulting model in the manner described above. However, since the first fine-tuning resulted in a regression model, when further fine-tuning these models, we first replaced NOT and HOF labels with a value of 0 and 1 respectively. In this case, the predicted scores before classification were first rescaled to the 0–1 interval by min-max normalization ($HASOC_{OffensEval}$).

4   Results and Discussion

We evaluated the resulting ensembles on the HASOC test set. Results of these experiments are listed in Table 3. As can be seen in Table 3, as a result of further fine-tuning using OffensEval data, the performance of individual models significantly increased (applying the paired t-test, we find that the difference is significant, at $p < 0.05$ for both the macro, and the weighted F_1-score). The difference in the performance of the ensembles, however, is much less marked. A possible explanation for this could be that the five models in the $HASOC_{OffensEval}$ case may be more similar to each other (given that here the original model went through more fine-tuning with the same data). Furthermore, while in the case of the $HASOC_{only}$ model the ensemble attains better F_1-scores using both metrics, this is not the case with the $HASOC_{OffensEval}$ model, where the best performance is attained using the model trained on the 5th fold. Regardless of this, however, both ensemble methods outperform the winner of the HASOC competition [38] in both F_1-score measures (the winning team achieving a score of 0.7882 and 0.8395 in terms of macro F_1-score, and weighted F_1-score respectively).

5   Conclusions and Future Work

In this study we have described a simple ensemble of transformers for the task of hate speech detection. Results on the HASOC challenge showed that this ensemble is capable of attaining state-of-the-art performance. Moreover, we have managed to improve the results attained by additional pre-training using in-domain data. In the future we plan to modify our pre-training approach so that models responsible for different folds are first pre-trained using different portions of the OffensEval data. Furthermore, we intend to extend our experiments to other hate speech datasets and challenges, as well as other transformer models. Lastly, we also intend to examine the explainability of resulting models.

Acknowledgements. Part of this work has been funded by the Vinnova project "Language models for Swedish authorities" (ref. number: 2019-02996).

References

1. Badjatiya, P., Gupta, S., Gupta, M., Varma, V.: Deep learning for hate speech detection in tweets. In: Proceedings of the 26th International Conference on World Wide Web Companion, WWW '17 Companion, pp. 759–760 (2017)
2. Barendt, E.: What is the harm of hate speech? Ethic theory, moral prac., vol. 22 (2019). https://doi.org/10.1007/s10677-019-10002-0
3. Basile, V., et al.: SemEval-2019 task 5: multilingual detection of hate speech against immigrants and women in Twitter. In: Proceedings of the 13th International Workshop on Semantic Evaluation, pp. 54–63 (2019). https://doi.org/10.18653/v1/S19-2007
4. Brown, A.: What is so special about online (as compared to offline) hate speech? Ethnicities **18**(3), 297–326 (2018). https://doi.org/10.1177/1468796817709846
5. Burnap, P., Williams, M.L.: Cyber hate speech on twitter: an application of machine classification and statistical modeling for policy and decision making. Policy Internet **7**(2), 223–242 (2015). https://doi.org/10.1002/poi3.85
6. Davidson, T., Warmsley, D., Macy, M., Weber, I.: Automated hate speech detection and the problem of offensive language. In: Proceedings of the 11th International AAAI Conference on Web and Social Media, ICWSM 2017, pp. 512–515 (2017)
7. Del Vigna, F., Cimino, A., Dell'Orletta, F., Petrocchi, M., Tesconi, M.: Hate me, hate me not: Hate speech detection on Facebook. In: ITASEC, January 2017
8. Devlin, J., Chang, M.W., Lee, K., Toutanova, K.: BERT: Pre-training of deep bidirectional transformers for language understanding. In: Proceedings of the 2019 Conference of the North American Chapter of the Association for Computational Linguistics: Human Language Technologies, Volume 1 (Long and Short Papers), pp. 4171–4186. Association for Computational Linguistics, Minneapolis, Minnesota, June 2019. https://doi.org/10.18653/v1/N19-1423

9. Djuric, N., Zhou, J., Morris, R., Grbovic, M., Radosavljevic, V., Bhamidipati, N.: Hate speech detection with comment embeddings. In: Proceedings of the 24th International Conference on World Wide Web, WWW 2015 Companion, pp. 29–30. Association for Computing Machinery, New York (2015). https://doi.org/10.1145/2740908.2742760

10. Do, H.T.T., Huynh, H.D., Nguyen, K.V., Nguyen, N.L.T., Nguyen, A.G.T.: Hate speech detection on vietnamese social media text using the bidirectional-LSTM model (2019), arXiv:1911.03648

11. Dworkin, R.: A new map of censorship. Index Censorship **35**(1), 130–133 (2006). https://doi.org/10.1080/03064220500532412

12. Gambäck, B., Sikdar, U.K.: Using convolutional neural networks to classify hate-speech. In: Proceedings of the First Workshop on Abusive Language Online, pp. 85–90. Association for Computational Linguistics, Vancouver, BC, Canada, August 2017. https://doi.org/10.18653/v1/W17-3013. https://www.aclweb.org/anthology/W17-3013

13. Greevy, E., Smeaton, A.F.: Classifying racist texts using a support vector machine. In: Proceedings of the 27th Annual International ACM SIGIR Conference on Research and Development in Information Retrieval, SIGIR 2004, pp. 468–469. Association for Computing Machinery, New York (2004). https://doi.org/10.1145/1008992.1009074

14. Gröndahl, T., Pajola, L., Juuti, M., Conti, M., Asokan, N.: All you need is "love": evading hate speech detection. In: Proceedings of the 11th ACM Workshop on Artificial Intelligence and Security, AISec 2018, pp. 2–12. Association for Computing Machinery, New York (2018). https://doi.org/10.1145/3270101.3270103

15. Hern, A.: Revealed: catastrophic effects of working as a Facebook moderator. The Guardian (2019). https://www.theguardian.com/technology/2019/sep/17/revealed-catastrophic-effects-working-facebook-moderator. Accessed 26 Apr 2020

16. Heyman, S.: Hate speech, public discourse, and the first amendment. In: Hare, I., Weinstein, J. (eds.) Extreme Speech and Democracy. Oxford Scholarship Online (2009). https://doi.org/10.1093/acprof:oso/9780199548781.003.0010

17. Huynh, T.V., Nguyen, V.D., Nguyen, K.V., Nguyen, N.L.T., Nguyen, A.G.T.: Hate speech detection on Vietnamese social media text using the bi-gru-lstm-cnn model. arXiv:1911.03644 (2019)

18. Immpermium: detecting insults in social commentary. https://kaggle.com/c/detecting-insults-in-social-commentary. Accessed 27 April 2020

19. Kwok, I., Wang, Y.: Locate the hate: detecting tweets against blacks. In: Proceedings of the Twenty-Seventh AAAI Conference on Artificial Intelligence, AAAI 2013, pp. 1621–1622. AAAI Press (2013)

20. Liu, Y., et al.: Roberta: A robustly optimized bert pretraining approach (2019)

21. MacAvaney, S., Yao, H.R., Yang, E., Russell, K., Goharian, N., Frieder, O.: Hate speech detection: Challenges and solutions. PLOS ONE **14**(8), 1–16 (2019). https://doi.org/10.1371/journal.pone.0221152

22. Mandl, T., Modha, S., Mandlia, C., Patel, D., Patel, A., Dave, M.: HASOC - Hate Speech and Offensive Content identification in indo-European languages. https://hasoc2019.github.io. Accessed 20 Sep 2019

23. Mandl, T., Modha, S., Patel, D., Dave, M., Mandlia, C., Patel, A.: Overview of the HASOC track at FIRE 2019: Hate Speech and Offensive Content Identification in Indo-European Languages). In: Proceedings of the 11th Annual Meeting of the Forum for Information Retrieval Evaluation, December 2019

24. Matsuda, M.J.: Public response to racist spech: considering the victim's story. In: Matsuda, M.J., Lawrence III, C.R. (ed.) Words that Wound: Critical Race Theory, Assaultive Speech, and the First Amendment, pp. 17–52. Routledge, New York (1993)
25. Mehdad, Y., Tetreault, J.: Do characters abuse more than words? In: Proceedings of the SIGDIAL2016 Conference, pp. 299–303, January 2016. https://doi.org/10.18653/v1/W16-3638
26. Mikolov, T., Sutskever, I., Chen, K., Corrado, G., Dean, J.: Distributed representations of words and phrases and their compositionality. In: Proceedings of the NIPS, pp. 3111–3119 (2013)
27. Mondal, M., Silva, L.A., Benevenuto, F.: A measurement study of hate speech in social media. In: Proceedings of the 28th ACM Conference on Hypertext and Social Media, HT 2017, pp. 85–94. Association for Computing Machinery, New York (2017). https://doi.org/10.1145/3078714.3078723
28. Nina-Alcocer, V.: Vito at HASOC 2019: Detecting hate speech and offensive content through ensembles. In: Mehta, P., Rosso, P., Majumder, P., Mitra, M. (eds.) Working Notes of FIRE 2019 - Forum for Information Retrieval Evaluation, Kolkata, India, 12–15 December, 2019. CEUR Workshop Proceedings, vol. 2517, pp. 214–220. CEUR-WS.org (2019). http://ceur-ws.org/Vol-2517/T3-5.pdf
29. Njagi, D., Zuping, Z., Hanyurwimfura, D., Long, J.: A lexicon-based approach for hate speech detection. Int. J. Multimed. Ubiquitous Eng. **10**, 215–230 (2015). https://doi.org/10.14257/ijmue.2015.10.4.21
30. Nourbakhsh, A., Vermeer, F., Wiltvank, G., van der Goot, R.: sthruggle at SemEval-2019 task 5: an ensemble approach to hate speech detection. In: Proceedings of the 13th International Workshop on Semantic Evaluation, pp. 484–488. Association for Computational Linguistics, Minneapolis, Minnesota, June 2019. https://doi.org/10.18653/v1/S19-2086
31. Otter, D.W., Medina, J.R., Kalita, J.K.: A survey of the usages of deep learning for natural language processing. IEEE Trans. Neural Networks Learn. Syst., 1–21 (2020)
32. Park, J., Fung, P.: One-step and two-step classification for abusive language detection on Twitter. In: ALW1: 1st Workshop on Abusive Language Online, June 2017
33. Alonso, P., Rajkumar Saini, G.K.: The North at HASOC 2019 hate speech detection in social media data. In: Proceedings of the 11th Anual Meeting of the Forum for Information Retrieval Evaluation, December 2019
34. Ross, B., Rist, M., Carbonell, G., Cabrera, B., Kurowsky, N., Wojatzki, M.: Measuring the reliability of hate speech annotations: the case of the European refugee crisis. In: Beißwenger, M., Wojatzki, M., Zesch, T. (eds.) Proceedings of NLP4CMC III: 3rd Workshop on Natural Language Processing for Computer-Mediated Communication, pp. 6–9, September 2016. https://doi.org/10.17185/duepublico/42132
35. Seganti, A., Sobol, H., Orlova, I., Kim, H., Staniszewski, J., Krumholc, T., Koziel, K.: Nlpr@srpol at semeval-2019 task 6 and task 5: linguistically enhanced deep learning offensive sentence classifier. In: SemEval@NAACL-HLT (2019)
36. Spertus, E.: Smokey: Automatic recognition of hostile messages. In: Proceedings of the 14th National Conference on Artificial Intelligence and 9th Innovative Applications of Artificial Intelligence Conference (AAAI-97/IAAI-97), pp. 1058–1065. AAAI Press, Menlo Park (1997. http://www.ai.mit.edu/people/ellens/smokey.ps
37. Sun, C., Qiu, X., Xu, Y., Huang, X.: How to fine-tune BERT for text classification? In: Sun, M., Huang, X., Ji, H., Liu, Z., Liu, Y. (eds.) CCL 2019. LNCS (LNAI), vol. 11856, pp. 194–206. Springer, Cham (2019). https://doi.org/10.1007/978-3-030-32381-3_16

38. Wang, B., Ding, Y., Liu, S., Zhou, X.: Ynu_wb at HASOC 2019: Ordered neurons LSTM with attention for identifying hate speech and offensive language. In: Mehta, P., Rosso, P., Majumder, P., Mitra, M. (eds.) Working Notes of FIRE 2019 - Forum for Information Retrieval Evaluation, Kolkata, India, 12–15 December, 2019, pp. 191–198 (2019). http://ceur-ws.org/Vol-2517/T3-2.pdf

39. Warner, W., Hirschberg, J.: Detecting hate speech on the world wide web. In: Proceedings of the Second Workshop on Language in Social Media, pp. 19–26. Association for Computational Linguistics, Montréal, Canada, June 2012. https://www.aclweb.org/anthology/W12-2103

40. Waseem, Z., Hovy, D.: Hateful symbols or hateful people? predictive features for hate speech detection on Twitter. In: Proceedings of the NAACL Student Research Workshop, pp. 88–93. Association for Computational Linguistics, San Diego, California, June 2016. https://doi.org/10.18653/v1/N16-2013. https://www.aclweb.org/anthology/N16-2013

41. Wei, X., Lin, H., Yang, L., Yu, Y.: A convolution-LSTM-based deep neural network for cross-domain MOOC forum post classification. Information 8, 92 (2017). https://doi.org/10.3390/info8030092

42. Wiegand, M., Siegel, M., Ruppenhofer, J.: Overview of the germeval 2018 shared task on the identification of offensive language. In: Proceedings of the GermEval 2018 Workshop, pp. 1–11 (2018)

43. Wolf, T., et al.: Huggingface's transformers: State-of-the-art natural language processing. arXiv:1910.03771 (2019)

44. Young, T., Hazarika, D., Poria, S., Cambria, E.: Recent trends in deep learning based natural language processing (2017), arXiv:1708.02709 Comment: Added BERT, ELMo, Transformer

45. Yuan, S., Wu, X., Xiang, Y.: A two phase deep learning model for identifying discrimination from tweets. In: Pitoura, E., et al. (eds.) Proceedings of the 19th International Conference on Extending Database Technology, EDBT 2016, Bordeaux, France, March 15–16, 2016, Bordeaux, France, 15–16 March, 2016, pp. 696–697. OpenProceedings.org (2016). https://doi.org/10.5441/002/edbt.2016.92

46. Zampieri, M., Malmasi, S., Nakov, P., Rosenthal, S., Farra, N., Kumar, R.: Semeval-2019 task 6: Identifying and categorizing offensive language in social media (offenseval). In: Proceedings of the 13th International Workshop on Semantic Evaluation, pp. 75–86 (2019)

47. Zampieri, M., et al.: SemEval-2020 Task 12: multilingual offensive language identification in social media (OffensEval 2020). In: Proceedings of SemEval (2020)

48. Zhang, Z., Luo, L.: Hate speech detection: a solved problem? the challenging case of long tail on twitter. Semantic Web Accepted, October 2018. https://doi.org/10.3233/SW-180338

49. Zhang, Z., Robinson, D., Tepper, J.: Detecting hate speech on Twitter using a convolution-GRU based deep neural network. In: Gangemi, A., Navigli, R., Vidal, M.-E., Hitzler, P., Troncy, R., Hollink, L., Tordai, A., Alam, M. (eds.) ESWC 2018. LNCS, vol. 10843, pp. 745–760. Springer, Cham (2018). https://doi.org/10.1007/978-3-319-93417-4_48

50. Zimbardo, P.G.: The human choice: individuation, reason, and order versus deindividuation, impulse, and chaos. Nebr. Symp. Motiv. 17, 237–307 (1969)

51. Zimmerman, S., Kruschwitz, U., Fox, C.: Improving hate speech detection with deep learning ensembles. In: Proceedings of the Eleventh International Conference on Language Resources and Evaluation (LREC 2018). European Language Resources Association (ELRA), Miyazaki, Japan, May 2018. https://www.aclweb.org/anthology/L18-1404

MP3 Compression to Diminish Adversarial Noise in End-to-End Speech Recognition

Iustina Andronic[1]([✉]), Ludwig Kürzinger[2][ID], Edgar Ricardo Chavez Rosas[2], Gerhard Rigoll[2][ID], and Bernhard U. Seeber[1][ID]

[1] Chair of Audio Information Processing,
Department of Electrical and Computer Engineering,
Technical University of Munich, Munich, Germany
{iustina.andronic,ludwig.kuerzinger}@tum.de
[2] Chair of Human-Machine Communication,
Department of Electrical and Computer Engineering,
Technical University of Munich, Munich, Germany

Abstract. Audio Adversarial Examples (AAE) represent purposefully designed inputs meant to trick Automatic Speech Recognition (ASR) systems into misclassification. The present work proposes MP3 compression as a means to decrease the impact of Adversarial Noise (AN) in audio samples transcribed by ASR systems. To this end, we generated AAEs with a new variant of the Fast Gradient Sign Method for an end-to-end, hybrid CTC-attention ASR system. The MP3's effectiveness against AN is then validated by two objective indicators: (1) Character Error Rates (CER) that measure the speech decoding performance of four ASR models trained on different audio formats (both uncompressed and MP3-compressed) and (2) Signal-to-Noise Ratio (SNR) estimated for uncompressed and MP3-compressed AAEs that are reconstructed in the time domain by feature inversion. We found that MP3 compression applied to AAEs indeed reduces the CER when compared to uncompressed AAEs. Moreover, feature-inverted (reconstructed) AAEs had significantly higher SNRs after MP3 compression, indicating that AN was reduced. In contrast to AN, MP3 compression applied to utterances augmented with regular noise resulted in more transcription errors, giving further evidence that MP3 encoding is effective in diminishing AN exclusively.

Keywords: Automatic Speech Recognition (ASR) · MP3 compression · Audio Adversarial Examples

1 Introduction

In our increasingly digitized world, Automatic Speech Recognition (ASR) has become a natural and convenient means of communication with many daily-use

© Springer Nature Switzerland AG 2020
A. Karpov and R. Potapova (Eds.): SPECOM 2020, LNAI 12335, pp. 22–34, 2020.
https://doi.org/10.1007/978-3-030-60276-5_3

gadgets. Recent advances in ASR push toward end-to-end systems that directly infer text from audio features, with sentence transcription being done in a single stage, without intermediate representations [12,25]. Those systems do not require handcrafted linguistic information, but learn to extract this information by themselves, as they are trained in an end-to-end manner. At present, there is a trend towards Deeper Neural Networks (DNN), however these ASR models are also more complex and prone to security threats.

In particular, this paper is concerned with the existence of Audio Adversarial Examples (AAEs), the audio inputs that carry along a hidden message induced by Adversarial Noise (AN) added to a regular speech sample. AN is generally optimized to mislead the ASR system into misclassification (i.e., into recognizing a target hidden message), whereas for humans, AN is supposed to remain undetected [1,4,14,21]. Yet, from the perspective of ASR research, the system should classify the sentence as close as it is understood by humans. For this reason, we proceed to investigate MP3 compression as a means to filter out the AN, and in turn, to diminish its detrimental effects on ASR performance.

Our contributions are three-fold:

- Based on original samples from the VoxForge speech corpus, we fabricate AAEs in the audio domain via a feature inversion procedure;
- We demonstrate the effectiveness of MP3 compression to mitigate AN in adversarial inputs with four end-to-end ASR models trained on different levels of MP3 compression;
- Conversely, we assess the effects of MP3 compression when applied to inputs augmented with regular, *non-adversarial* noise.

2 Related Work

Adversarial Examples (AEs) were primarily generated for image classification tasks [10]. The regular, gradient-based training of DNNs calculates the gradient of a chosen loss function, aiming for a step-wise, gradual improvement of the networks' parameters. By contrast, the Fast Gradient Sign Method (FGSM) [10] creates AN based on the gradient w.r.t. the *input data*, in order to optimize towards misclassification. FGSM was already applied in the context of end-to-end ASR to DeepSpeech [5,12], a CTC-based speech recognition system, as well as to the Listen-Attend-Spell (LAS) system featuring an attention mechanism [6,23].

In generating AAEs, most of the previous works have set about to pursue one of the following goals, sometimes succeeding in both: (1) that their AAEs work in a physical environment and (2) that they stay imperceptible to humans. Carlini et al. [4] were among the first to introduce the so-called *hidden voice commands*, demonstrating that targeted attacks operating over-the-air against archetypal ASR systems (i.e., solely based on Hidden Markov Models - HMM), are feasible. In contrast to previous works that targeted short malicious phrases, adversarial perturbations for designated longer sentences were achieved in [5]. The novel attack was implemented with a gradient-descent minimization based on the CTC loss function, which is optimized for time sequences. Moreover, [21] were the first to develop imperceptible AAEs for a conventional hybrid DNN-HMM ASR system by leveraging human psychoacoustics, i.e., manipulating the acoustic signal below the thresholds of human perception. In a follow-up publication [22], their *psychoacoustic hiding* method was enhanced to produce *generic* AAEs that remained robust in simulated over-the-air attacks.

Protecting ASR systems against AN that is embedded in AEs has also been primarily investigated in the visual domain, which in turn inspired research in the audio domain. According to [13], two major defense strategies are considered from a security-related perspective, namely *proactive* and *reactive* approaches. The former aims to enhance the robustness during the training procedure of the ASR models themselves, e.g., by adversarial training [23] or network distillation [19]. Reactive approaches instead aim to detect if an input is adversarial after the DNNs are trained, by means of e.g., input transformations such as compression, cropping or resizing, meant to at least partially discard the AN and thus recover the genuine transcription.

Regarding AAEs, primitive signal processing operations such as local smoothing, down-sampling and quantization were applied to input audio so as to disrupt the adversarial perturbations [26]. The effectiveness of that pre-processing defense was demonstrated especially for shorter-length AAEs. Rajaratnam et al. [20] likewise explored audio pre-processing methods such as band-pass filtering and compression (MP3 and AAC), while also venturing to more complex speech coding algorithms (Speex, Opus) so as to mitigate the AAE attacks. Their experiments, which targeted a simple keyword-spotting system with a limited dictionary, indicated that an ensemble strategy (made of both speech coding and other form of pre-processing, e.g., compression) is more effective against AAEs. Finally, Das et al. [8] also experimented with MP3 compression against targeted AAEs in the CTC-based DeepSpeech model, claiming that it was able to reduce the attack success rate to 0%. However, they considered broad margins for what makes an unsuccessful attack, i.e., when at least an error in the original targeted transcription was achieved. Unfortunately, a more in-depth analysis of the experimental results was not performed.

3   Experimental Set-Up

MP3 is an audio compression algorithm that employs a lossy, perceptual audio coding scheme that discards audio information below the hearing threshold and thus diminishes the file size [3]. Schönherr et al. [21] recently hypothesized that MP3 can make for a robust countermeasure to AAE attacks, as it might remove exactly those inaudible ranges in the audio where the AN lies.

Consequently, given an audio utterance that should transcribe to the original, non-adversarial phrase, we formulate our research question as follows: *To what extent can MP3 aid in removing the AN and thus recover the benign character of the input?* We aim to analyze the AN reduction with two objective indicators: Character Error Rates (CER) and Signal-to-Noise Ratio (SNR).

3.1 Pipeline from Original Audio Data to MP3-Compressed AAEs

A four-stage pipeline that transforms the original test data to MP3-compressed AAEs was implemented. As depicted in Fig. 1, transformations include the FGSM method, feature inversion and MP3 compression. To also consider the effects of MP3 compression on the ASR performance, i.e., whether the neural network adapts to MP3 compression, experiments were validated on four ASR models, each *trained* on audio data with a different level of MP3 compression (uncompressed, 128 kbps, 64 kbps and 24 kbps MP3 data). Format-matched AAEs were then decoded by each of the four models.

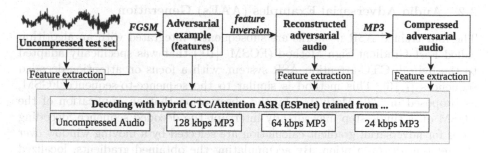

Fig. 1. General experimental workflow.

All experiments are based on the English share of the open-source Vox-Forge speech corpus[1], which consists of $\approx$130.1 h of utterances in various English accents. VoxForge database was originally recorded and released in uncompressed format (*.wav*), allowing for further compression and thus, for the exploration of our research question. The Lame MP3 encoder[2] was used as command line tool for batch MP3 compression.

Log-mel filterbank (fbank) features were extracted from every utterance of the speech corpus and then fed to the input of the ASR models in both training and testing stages. The speech recognition experiments are performed with the hybrid CTC-attention ASR system called ESPnet [24,25], which combines the two main techniques for end-to-end speech recognition. First, Connectionist Temporal Classification (CTC [11]) carries the concept of hidden Markov states into the training loss function of end-to-end DNNs, which perform sequence-to-sequence classification, that is, they classify token probabilities for each time frame. Second, attention-based encoder-decoder architectures [6] are trained as auto-regressive, sequence-generative models that directly generate the sentence transcription from the entire input utterance. The multi-objective learning framework unifies the Attention-loss and CTC-loss with a linear interpolation weight (the parameter α in Eq. 1), which is set to 0.3 in our experiments, allowing

---

[1] http://www.voxforge.org/home/downloads.
[2] https://lame.sourceforge.io/about.php.

attention to dominate over CTC.

$$Loss_{hybrid} = \alpha \log p_{CTC}(Y|X) + (1 - \alpha) \log p_{Att.}(Y|X), \quad \alpha \in [0,1] \quad (1)$$

Different from the original LAS ASR system [6], we use a *location-aware* attention mechanism that additionally uses attention weights from the previous sequence step; a more detailed description is to be found in [17]. Full description of the ESPnet's hybrid architecture can be consulted in [17,24,25]. We use slightly adapted training and decoding parameters[3] starting from the default network configuration for the VoxForge dataset[4].

3.2   Audio Adversarial Examples (AAEs) Generation

The four trained ASR networks are subsequently integrated with an algorithm called Fast Gradient Sign Method (FGSM [10]) that was specifically adapted to the hybrid, CTC-attention ASR system, with a focus on attention-decoded sequences [7,16]. This method is similar to the sequence-to-sequence FGSM, as proposed in [23]. Motivated by the observation that an application of the FGSM on a single step may already interrupt the decoded sequence, decoding steps for adversarial gradient calculation are selected by a moving window over the reference transcription. By accumulating the obtained gradients, localized maxima of AN are reduced and dissipated over the sample [16].

A previously decoded label sequence $y_{1:L}^*$ is used as reference to avoid label leaking [15]. Using back propagation, the cross-entropy loss $J(x_{1:T}, y_l^*; \theta)$ is obtained from a whitebox model (whose parameters θ are fully known). The gradients from a sliding window with a fixed length l_w and stride ν are accumulated to $\nabla_{AAE}(x_t)$ (Eq. 2) and gradient normalization is applied for accumulating the gradient directions [9]:

$$\nabla_{AAE}(x_t) = \sum_{i=0}^{\lceil L/\nu \rceil} \left(\frac{\nabla_{x_t} \sum_{l=i\cdot\nu}^{l_w} J(x_{1:T}, y_l^*; \theta)}{||\nabla_{x_t} \sum_{l=i\cdot\nu}^{l_w} J(x_{1:T}, y_l^*; \theta)||_1} \right), \quad l \in [0; L] \quad (2)$$

As previously mentioned, the adversarial perturbation should be lower in amplitude than the original input x_t, so as to stay unnoticeable. Here, the intensity

---

[3] We use the LSTM architecture that has a 4-layer VGGBLSTMP encoder to each 320 units with subsampling in the second and third layer. The decoder has only one layer with 300 units. The multitask learning parameter α was set to 0.3. The maximum number of epochs was set to 25 with a patience of 3. During training, scheduled sampling was applied with a probability of 0.5. The training script can be found at https://github.com/iustinaabc/ASR-mp3-compression-AAEs/tree/master/code/conf. The full parameters' configuration is also listed in Table 4.2 of the Master's Thesis underlying this publication [2].

[4] At ESPnet commit 81a383a9.

of the AN (denoted by δ_{AAE}) is controlled with the ϵ factor, which was set to 0.3 in our experiments:

$$\delta_{AAE}(x_t) = \epsilon \cdot \text{sgn}(\nabla_{\text{AAE}}(x_t)) \tag{3}$$

The resulting δ_{AAE} is then added to the original *feature-domain* input x_t (Eq. 4), in order to trigger the network f to output a wrong transcription, different from the ground truth y (Eq. 5):

$$\hat{x}_t = x_t + \delta_{AAE}(x_t), \quad \forall t \in [1, T] \tag{4}$$

$$y \neq f(\hat{x}_t, \theta) \tag{5}$$

With the adapted version of FGSM, we generated an adversarial instance for each utterance of the four *non-adversarial* test sets (each with a different degree of MP3-compression). Notably, we haven't employed psychoacoustic hiding [21] to render the AN inaudible. This method limits AN to stay under the hearing threshold in the spectral domain. However, the parameters of this threshold are retrieved from a psychoacoustic model similar to the one underlying MP3 compression.

Because the networks take acoustic feature vectors as input, FGSM originally creates AEs in the feature domain. Yet, in order to evaluate our research hypothesis, we required AEs in the audio domain, that is, AAEs. Hence, we proceeded to invert the adversarial features and thus obtain synthetically *reconstructed* adversarial audio (rAAEs). The exact steps for both forward feature extraction, as well as feature inversion, are illustrated in Fig. 2. They were implemented with functions from the Python-based *Librosa* toolbox [18].

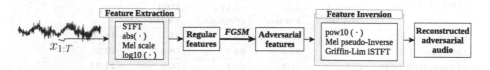

Fig. 2. AAE generation via feature inversion applied to adversarial features

A further notable point is that log-mel fbank features are a *lossy* representation of the original audio input, since they lack the phase information in the spectrum. This in turn prevents a highly accurate audio reconstruction. In fact, mere listening revealed that for a human subject, it is relatively easy to distinguish the reconstruction artefacts when comparing pairs of original (non-adversarial) audio samples with their reconstructed versions[5]. Nonetheless, to verify whether the reconstruction method impaired in any way the ASR performance, we performed the following sanity check: we used the ASR model trained on uncompressed *.wav* files to decode both the original test set, as well as its

---

[5] Reconstructed audio samples (both non-adversarial and adversarial) can be retrieved from https://github.com/iustinaabc/ASR-mp3-compression-AAEs.

reconstructed counterpart (obtained by feature extraction, directly followed by inversion, i.e., without applying FGSM). We observed just a mild +1.3% absolute increase in the Character Error Rate (CER) for the reconstructed set, which implies that the relevant acoustic features are still accurately preserved following audio reconstruction with our feature inversion approach.

4   Results and Discussion

4.1   ASR Results for Non-adversarial vs. Adversarial Input

Table 1 conveys the CER results for decoding various test sets that originate from the same audio format of each ASR model's training data, hence in a train-test *matched* setting. Most notably, the adversarial inputs in the feature domain (column [b]) render more transcription errors (by an absolute mean of +52.05% over all models) compared to the baseline, non-adversarial input listed in column [a]. This validates the FGSM method as effective in creating adversarial features from input data of any audio format (uncompressed, as well as MP3-compressed).

Table 1. CER results for decoding different types of adversarial input (marked as [b], [c] and [d]). The last column indicates the relative CER difference between the adversarial features [b] and the reconstructed, MP3-compressed AAEs [d], calculated as $\frac{[d]-[b]}{[b]} \cdot 100(\%)$. All inputs from column [d] were MP3-compressed at 24 kbps

Source format of train & test data (ASR model index)	Input test data & corresp. CER scores				CER (%) reduction between [b] and [d]
	[a] Original (non-adv.)	[b] Adv. features	[c] rAAEs (reconstr.)	[d] MP3 rAAEs	
Uncompressed (#1)	17.8	70.5	62.2	57.4	−18.58
128 kbps-MP3 (#2)	18.8	72.3	64.0	58.4	−19.23
64 kbps-MP3 (#3)	18.6	71.8	63.1	56.5	**−21.31**
24 kbps-MP3 (#4)	20.2	69.0	60.5	55.3	−19.86

The error scores for the reconstructed AAEs (rAAEs in column [c], obtained via adversarial feature inversion) are also higher than the baseline, but, interestingly, lower than the CER scores for the adv. features themselves. This suggests that our reconstruction method makes the AAEs less powerful in misleading the model, which on the other hand is beneficial for the system's robustness to AN. When we further compress the rAAEs with MP3 at the bitrate of 24 kbps (column [d]), we observe an additional decline in the CER, thus indicating that MP3 compression is favourable in reducing the attack effectiveness of the adversarial input. However, these CER values are still much higher than the baseline by a mean absolute difference of +38.05% across all ASR models, suggesting that the original transcription could *not* be fully recovered. The strongest CER reduction effect (-21.31%) between the MP3 rAAEs and the "original" adversarial features can be observed in the results achieved by ASR model #3, which decoded adversarial compressed input (24 kbps) that originated from 64 kbps MP3-compressed data. Overall, the mere numbers show that MP3 compression manages to partially reduce the error rates to raw adversarial feature inputs.

4.2 MP3 Effects on the Signal-to-Noise Ratio (SNR)

SNR is a measure that quantifies the amount of noise in an audio sample. Thus, SNR can offer valuable insight into the extent to which the AN was diminished in MP3-compressed AAEs. Since it is not straight-forward to compute the SNR in the original, adversarial features' domain, we estimated the SNR in the audio domain for the reconstructed AAEs (both before and after MP3 compression):

$$SNR_{adv} = 10 \log_{10} \frac{\text{signal power}}{\text{noise power}} \qquad (6)$$

$$= 10 \log_{10} \frac{\text{(non-adv.) Reconstructed Audio Power (RAP)}}{\text{Adversarial Noise Power (ANP)}} \qquad (7)$$

with

$$ANP = (\text{rAAE's power}) - RAP \qquad (8)$$

For both the uncompressed and MP3 rAAEs, the SNR_{adv} was calculated with reference to the same signal in the numerator (Eq. 7): the Reconstructed Audio Power (RAP) of the original (*non-adversarial*) speech utterance, so as to introduce similar artefacts as the ones resulting from the adversarial audio reconstruction. Thus, the power ratio between two reconstructed signals ensures a more accurate SNR estimation. The Adversarial Noise Power (ANP) was then obtained by subtracting RAP from the rAAE's signal power.

As expected, we obtained different SNRs for each adversarial audio, as AN varies with the speech input. Moreover, the SNR values differed before and after MP3 compression for each adversarial sample. As the normalized histograms show, most of the SNR values for the uncompressed rAAEs are negative (Fig. 3, left), suggesting that the added AN has high levels and is therefore audible. However, after MP3 compression, most adversarial samples acquire positive SNRs (Fig. 3, right), implying that AN was diminished due to MP3 encoding. To validate this, we applied the non-parametric, two-sided Kolmogorov-Sminov statistical test to evaluate whether the bin counts from the normalized SNR histograms of rAAEs before and after MP3 compression originate from the same underlying distribution. We obtained a *p-value* of 0.039 (< 0.05), confirming the observed difference between the underlying histogram distributions as statistically significant. Thereby, MP3's incremental effect on the SNR values of adversarial samples was validated, which essentially means that MP3 reduces the AN.

4.3 ASR Results for Input Augmented with Non-adversarial Noise

To have a reference for the ASR systems' behaviour when exposed to common types of noise, we also assessed the effects of MP3 compression on audio inputs

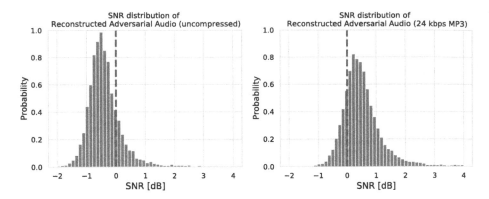

Fig. 3. Normalized histograms following SNR estimation of reconstructed adversarial inputs (rAAEs): uncompressed (left) and MP3 at 24 kbps (right). The number of histogram bins was set to 50 for both plots

corrupted by regular, non-adversarial noise. For this complementary analysis, we augmented the original test sets with four types of commonly-occuring noise at various SNRs, i.e., *white*, *pink*, *brown* and *babble* noise (overlapping voices), each boasting distinct spectral characteristics. The noise-augmented samples and their MP3-compressed versions were then fed as input to the decoding stage of the corresponding ASR system, i.e., the one that was trained on audio data of the same format as the test data. Table 2 lists the CER results of this decoding experiment performed by ASR model #1 (trained on uncompressed data). Decoding results from the other three ASR models can be found in [2].

Table 2. CER scores for test sets augmented with regular, non-adversarial noise at different SNR values, then MP3-compressed at 24 kbps. These inputs were exclusively decoded by ASR model #1

Test set augmented with:	SNR [dB]					
	30	10	5	0	−5	−10
[A] white noise	19.1	32.7	41.9	53.7	66.2	78.2
+ MP3 compression	29.1	51.2	61.7	71.2	78.7	86
[B] pink noise	18.5	29.1	38.1	51.7	67.4	82.1
+ MP3 compression	26.9	42.5	53	66.4	79.8	89.9
[C] brown noise	17.9	19.7	21.9	26.1	34.1	47.8
+ MP3 compression	25.3	29	32.5	38	47.3	60.6
[D] babble noise	18.3	35.8	53.4	77.4	89	93.6
+ MP3 compression	25.8	48.2	66	83.6	93.1	95.4

Mind that the ASR systems were trained on the original, non-adversarial and noise-free data; therefore, decoding noisy inputs was expected to cause more transcription errors than the original clean data. Based on CER, one can observe that the lower the SNR (or the higher level of noise added to the input), the more error-prone the ASR system is, irrespective of the noise type. The most adverse effect seems to be in the case of white and babble noise (rows [A] and [D]), which also happen to have the richest spectral content, imminently interfering with the original speech.

Yet of utmost interest is what happens when the same treatment used for adversarial inputs, i.e., MP3 compression, is applied to the novel speech inputs enhanced with regular noise. These results are illustrated every second row in Table 2. Error rates turn out always higher for MP3-compressed inputs than for the uncompressed ones, regardless of noise type or SNR. Consequently, MP3 compression has the *inverse effect* compared to what was observed for adversarial noise. That is, MP3 triggered a reduction in the amount of errors returned by the ASR systems to adversarial input, while it failed to do so for non-adversarial noise. On the contrary, MP3 compression increased the amount of errors for inputs augmented with non-adversarial noise, especially at high and mid-range SNRs. This further validates that MP3 compression has the effect of partially reducing only adversarial perturbations, whereas deteriorating the non-adversarial, regular noise.

5   Conclusion

In this work, we explored the potential of MP3 compression as a countermeasure against Adversarial Noise (AN) compromising speech inputs. To this end, we constructed adversarial feature samples with an adapted version of the gradient-based FGSM algorithm. The adversarial features were then mapped back into the audio domain by inverse feature extraction operations. The resulting adversarial audio (denoted as *reconstructed*) was thereafter MP3-compressed and presented to the input of four ASR models featuring a hybrid CTC-attention architecture, having been previously trained on four types of audio data.

Our three key findings are:

1. In contrast to *adversarial features*, reconstructed AAEs, as well as MP3 compressed AAEs had lower error rates in their transcriptions.
   The error reduction did not achieve the baseline performance for non-adversarial input, which implies that correct transcriptions were not completely recovered.
2. MP3 compression yielded a statistically significant effect on the estimated SNR, validating that MP3 increased the SNR values of adversarial samples, which translates to AN reduction.
3. Our experiments with *non-adversarial* noise suggest that MP3 compression is beneficial only in mitigating AN, while it deteriorates the speech recognition performance to non-adversarial noise.

Results' Practical Value and Impact

Generation of AAEs to Audio Samples. Most state-of-the-art ASR systems nowadays classify from complex feature inputs (fbanks). One highlight of our work is the proposed method of direct fbank inversion by means of backward transformations, to obtain the AAE audio samples for fbank-based ASR systems.

Relevance for Psychoacoustic Hiding. The previously discussed psychoacoustic hiding [21] focuses on making AAEs inaudible to humans by restricting the AN below the human hearing threshold. Because the spectral envelope of this hearing threshold is obtained from the psychoacoustic model of MP3 compression, we hypothesize that such perceptually-hidden AN would also be discarded by MP3 compression.

MP3 Compression as Countermeasure. If MP3 compression is used as a countermeasure to AAEs, it is best to use an ASR system already trained on compressed MP3 data, and then further apply MP3 on AAEs that originate from the same compressed audio format. Results in Table 1 indicate that a good trade-off between ASR performance and robustness is achieved when starting off from non-adversarial MP3 data at 64 kbps.

References

1. Alzantot, M., Balaji, B., Srivastava, M.: Did you hear that? adversarial examples against automatic speech recognition. arXiv preprint arXiv:1801.00554 (2018)
2. Andronic, I.: MP3 Compression as a Means to Improve Robustness against Adversarial Noise Targeting Attention-based End-to-End Speech Recognition. Master's thesis, Technical University of Munich, Germany (2020)
3. Brandenburg, K.: MP3 and AAC Explained. In: Audio Engineering Society, 17th International Conference 2004 October 26–29, pp. 99–110 (1999)
4. Carlini, N., et al.: Hidden voice commands. In: 25th USENIX Security Symposium (USENIX Security 16), pp. 513–530. USENIX Association, August 2016
5. Carlini, N., Wagner, D.: Audio adversarial examples: targeted attacks on speech-to-text. In: Proceedings - 2018 IEEE Symposium on Security and Privacy Workshops, SPW 2018, pp. 1–7 (2018)
6. Chan, W., Jaitly, N., Le, Q., Vinyals, O.: Listen, attend and spell: a neural network for large vocabulary conversational speech recognition. In: 2016 IEEE International Conference on Acoustics, Speech and Signal Processing (ICASSP), pp. 4960–4964. IEEE (2016)
7. Chavez Rosas, E.R.: Improving Robustness of Sequence-to-sequence Automatic Speech Recognition by Means of Adversarial Training. Master's thesis, Technical University of Munich, Germany (2020)
8. Das, N., Shanbhogue, M., Chen, S.-T., Chen, L., Kounavis, M.E., Chau, D.H.: ADAGIO: interactive experimentation with adversarial attack and defense for audio. In: Brefeld, U., et al. (eds.) ECML PKDD 2018. LNCS (LNAI), vol. 11053, pp. 677–681. Springer, Cham (2019). https://doi.org/10.1007/978-3-030-10997-4_50

9. Dong, Y., et al.: Boosting adversarial attacks with momentum. In: Proceedings of the IEEE Conference on Computer Vision and Pattern Recognition, pp. 9185–9193 (2018)
10. Goodfellow, I.J., Shlens, J., Szegedy, C.: Explaining and harnessing adversarial examples. In: 3rd International Conference on Learning Representations, ICLR 2015 - Conference Track Proceedings, pp. 1–11 (2015)
11. Graves, A., Fernández, S., Gomez, F., Schmidhuber, J.: Connectionist temporal classification: labelling unsegmented sequence data with recurrent neural networks. In: Proceedings of the 23rd International Conference on Machine Learning, pp. 369–376. ACM (2006)
12. Hannun, A., et al.: Deep speech: scaling up end-to-end speech recognition. arXiv preprint arXiv:1412.5567 (2014)
13. Hu, S., Shang, X., Qin, Z., Li, M., Wang, Q., Wang, C.: Adversarial examples for automatic speech recognition: attacks and countermeasures. IEEE Commun. Mag. **57**(10), 120–126 (2019)
14. Iter, D., Huang, J., Jermann, M.: Generating adversarial examples for speech recognition. Stanford Technical report (2017)
15. Kurakin, A., Goodfellow, I., Bengio, S.: Adversarial machine learning at scale. arXiv preprint arXiv:1611.01236 (2016)
16. Kürzinger, L., Rosas, E.R.C., Li, L., Watzel, T., Rigoll, G.: Audio adversarial examples for robust hybrid ctc/attention speech recognition. arXiv preprint arXiv:2007.10723, To be published at SPECOM 2020 (2020)
17. Kürzinger, L., Watzel, T., Li, L., Baumgartner, R., Rigoll, G.: Exploring Hybrid CTC/Attention end-to-end speech recognition with gaussian processes. In: Salah, A.A., Karpov, A., Potapova, R. (eds.) SPECOM 2019. LNCS (LNAI), vol. 11658, pp. 258–269. Springer, Cham (2019). https://doi.org/10.1007/978-3-030-26061-3_27
18. McFee, B., et al.: librosa: audio and music signal analysis in python. In: Proceedings of the 14th Python in Science Conference, vol. 8 (2015)
19. Papernot, N., McDaniel, P., Wu, X., Jha, S., Swami, A.: Distillation as a defense to adversarial perturbations against deep neural networks. In: 2016 IEEE Symposium on Security and Privacy (SP), pp. 582–597. IEEE (2016)
20. Rajaratnam, K., Alshemali, B., Kalita, J.: Speech coding and audio preprocessing for mitigating and detecting audio adversarial examples on automatic speech recognition, August 2018
21. Schönherr, L., Kohls, K., Zeiler, S., Holz, T., Kolossa, D.: Adversarial attacks against automatic speech recognition systems via psychoacoustic hiding. arXiv preprint arXiv:1808.05665 (2018)
22. Schönherr, L., Zeiler, S., Holz, T., Kolossa, D.: Imperio: Robust over-the-air adversarial examples against automatic speech recognition systems. arXiv preprint arXiv:1908.01551 (2019)
23. Sun, S., Guo, P., Xie, L., Hwang, M.Y.: Adversarial regularization for attention based end-to-end robust speech recognition. IEEE/ACM Trans. Audio Speech Lang. Process. **27**(11), 1826–1838 (2019)

24. Watanabe, S., et al.: ESPNet: end-to-end speech processing toolkit. In: Proceedings of the Annual Conference of the International Speech Communication Association, INTERSPEECH 2018-Sep, pp. 2207–2211 (2018)
25. Watanabe, S., Hori, T., Kim, S., Hershey, J.R., Hayashi, T.: Hybrid CTC/attention architecture for end-to-end speech recognition. IEEE J. Sel. Top. Sign. Proces. **11**(8), 1240–1253 (2017)
26. Yang, Z., Li, B., Chen, P.Y., Song, D.: Towards mitigating audio adversarial perturbations (2018). https://openreview.net/forum?id=SyZ2nKJDz

Exploration of End-to-End ASR for OpenSTT – Russian Open Speech-to-Text Dataset

Andrei Andrusenko[1], Aleksandr Laptev[1(✉)], and Ivan Medennikov[1,2]

[1] ITMO University, St. Petersburg, Russia
{andrusenko,laptev,medennikov}@speechpro.com
[2] STC-innovations Ltd., St. Petersburg, Russia

Abstract. This paper presents an exploration of end-to-end automatic speech recognition systems (ASR) for the largest open-source Russian language data set – OpenSTT. We evaluate different existing end-to-end approaches such as joint CTC/Attention, RNN-Transducer, and Transformer. All of them are compared with the strong hybrid ASR system based on LF-MMI TDNN-F acoustic model.

For the three available validation sets (phone calls, YouTube, and books), our best end-to-end model achieves word error rate (WER) of 34.8%, 19.1%, and 18.1%, respectively. Under the same conditions, the hybrid ASR system demonstrates 33.5%, 20.9%, and 18.6% WER.

Keywords: Speech recognition · End-to-end · OpenSTT · Russian language

1 Introduction

For a long time, hybrid systems dominated the text-to-speech task. Such ASR systems [9] are closely associated with the use of hidden Markov models (HMMs), which are capable of handling the temporal variability of speech signals. Earlier, Gaussian Mixture Models (GMM) were used to compute emission probabilities from HMM states (phones usually look like a three-state HMM model) to map acoustic features to phones. But with the significant development in deep neural networks (DNN), GMM was replaced by DNN. At the moment, they demonstrate state-of-the-art (SotA) results in many ASR tasks [19,27]. Nevertheless, hybrid systems have some shortcomings. They are associated with dividing of the entire model into separate modules, specifically acoustic and language models, which are trained independently with different objective functions. Furthermore, for many languages, a lexicon (word into phone sequence mapping) should be provided. It often requires manual building by specialized linguists. Therefore, hybrid systems are limited with their lexicon-based word vocabulary. All this increases the complexity and training/tuning time of ASR systems.

A. Andrusenko and A. Laptev—Equal contribution.

A. Karpov and R. Potapova (Eds.): SPECOM 2020, LNAI 12335, pp. 35–44, 2020.
https://doi.org/10.1007/978-3-030-60276-5_4

Recently, researchers began to show an interest in end-to-end approaches. Unlike hybrid systems, these methods are trained to map input acoustic feature sequence to sequence of characters. They are trained with objective functions, which more consistent with the final evaluation criteria (e.g., WER) than, for example, the cross-entropy in case of hybrid system. Freedom from the necessity of intermediate modeling, such as acoustic and language models with pronunciation lexicons, makes the process of building the system more clear and straightforward. Also, the use of grapheme or subword units as decoder targets may allow to reduce the problem of out-of-vocabulary words (OOV). Unlike the hybrid ones, such end-to-end systems can construct new words itself.

Nowadays, end-to-end ASR systems are successfully used in solving many speech recognition tasks [16,22]. Furthermore, there is an informal competition between end-to-end and hybrid systems [17]. Most of these studies are conducted in a limited number of languages, such as English or Chinese. One of the reasons is the lack of significant (more than 1000 h of speech) open-source databases for other languages. There are works on end-to-end in low-resource conditions [1,3,15], but to reach a competitive quality, such system should be trained on a massive amount of data.

Inspired by [4], we did an investigation in end-to-end ASR systems performance for the Russian language. This study became possible with the advent of the largest open-source Russian database for speech to text converting task called OpenSTT [29]. This paper provides an overview of speech recognition performance for commonly used end-to-end architectures trained only on the OpenSTT database. We explored joint CTC-Attention [12], Neural-transducer [7], and Transformer sequence-to-sequence model [30]. To contribute to the comparison of conventional versus and-to-end approaches, we built a strong hybrid baseline based on Factorized time-delayed neural network (TDNN-F) [23] acoustic model with the lattice-free maximum mutual information (LF-MMI) training criterion [25].

2   Related Work

There are three types of work investigating ASR for the Russian language.

Iakushkin et al. [10] considered speech recognition in large-resource conditions. Their system built on top of the Mozilla DeepSpeech framework and trained with nearly 1650 h of YouTube crawled data. This work is most similar to ours since it investigated end-to-end performance for a comparable amount of data.

Works [20,26] by Speech Technology Center Ltd described building systems for spontaneous telephone speech recognition in a medium-resource setup (390 h). The authors used complex Kaldi- and self-made hybrid approaches that delivered significant recognition performance despite having a relatively small amount of data.

Low-resource Russian ASR is presented in works [13,18]. While the presented WER was relatively low for the used amount of data (approximately 30 h), it can be explained by the simplicity of the data (prepared speech in noise-free conditions). The authors built their hybrid systems with Kaldi and CNTK [28] toolkits.

3 End-to-End ASR Models

This section describes end-to-end modeling approaches considered in our work.

3.1 Connectionist Temporal Classification (CTC)

One of the leading hybrid models' problems is hard alignment. For DNN training, a strict mapping of input feature frames to DNN's target units should be provided. This is because the cross-entropy loss function has to be defined at each frame of an input sequence. To obtain such alignment, it is necessary to complete several resource- and time-consuming stages of the GMM-HMM model training.

Connectionist Temporal Classification (CTC) was proposed to solve this problem [8]. This is a loss function that allows to make mapping of input feature sequence to final output recognition result without the need for an intermediate hard alignment and output label representation (e.g., tied states of HMM model). CTC uses graphemes or subwords as target labels. It also introduces an auxiliary "blank" symbol, which indicates repetition of the previous label or the absence of any label. However, to achieve better accuracy, CTC may require the use of an external language model since it is context-independent.

3.2 Neural Transducer

The neural transducer was introduced in [7] to solve CTC problem of accounting for the independence between the predicted labels. It is a modification of CTC. In addition to the encoder, this model has two new components, namely prediction network (predictor) and joint network (joiner), which can be considered as a decoder part of the model. Predictor predicts the next embedding based on previously predicted labels. Joiner evaluates outputs of the encoder and predictor to yield the next output label. The neural transducer has a "blank" symbol, which is used similarly as in CTC.

3.3 Attention-Based Models

A attention-based encoder-decoder model was firstly introduced for the machine translation task [2]. This approach has found wide application in ASR tasks (e.g., Listen, Attend and Spell [5]), since speech recognition is also sequence-to-sequence transformation process.

A typical attention-based architecture can be divided into three parts: encoder, attention block, and decoder. The attention mechanism allows the decoder to learn soft alignment between encoded input features and output target sequences. During label prediction, the decoder also uses information about previously predicted labels that allows making context-depended modeling. However, in noisy conditions or wrong reference transcriptions of training data, such models may have weak convergence.

Joint CTC-Attention. There are ways to improve attention-based approaches. One of them is to apply CTC loss to the encoder output and use it jointly with the attention-based loss [12]. Thus, the final loss is represented as a weighted sum of those above, and the weight has to be tuned.

Transformer. Another way is to use more powerful attention-based architecture. The Transformer [30], which is also adapted from machine translation, is a multi-head self-attention (MHA) based encoder-decoder model. Compared with RNN, Transformer-based models demonstrate SotA results in many speech recognition tasks [11].

4   OpenSTT Dataset

As a database for our work, we used Russian Open Speech To Text (STT/ASR) Dataset (OpenSTT) [29]. At the moment, this is the largest known multidomain database of Russian language suitable for ASR systems training. The domains are radio, public speech, books, YouTube, phone calls, addresses, and lectures. OpenSTT has a total of about 20,000 h of transcribed audio data.

However, transcriptions for these audio files are either recognition results of an unknown ASR model or a non-professional annotation (e.g. YouTube subtitles). Manual transcriptions are presented only for the three validation datasets *asr_calls_2_val*, *public_youtube700_val*, and *buriy_audiobooks_2_val*, which are the domains of phone calls, YouTube videos, and books reading, respectively. Based on this, we selected only data belonging to domains with validation sets. After discarding data with erroneous transcriptions (there is a particular CSV file[1] with this information provided by the authors of OpenSTT), we obtained the following distribution of "clean" data by domain:

- Phone calls: 203 h.
- YouTube: 1096 h.
- Books: 1218 h.

To slightly improve the data distribution situation of different domains, we applied 3-fold speed perturbation for phone calls data. Next, we left only utterances lasting from 2 s to 20 s and not exceeding 250 characters. This was necessary to stabilize end-to-end training and prevent GPU memory overflowing. As a result, the training data set had 2926 h.

---

[1] https://github.com/snakers4/open_stt/releases/download/v0.5-beta/public_exclude_file_v5.tar.gz

5   Experimental Setup

To make a complete evaluation of ASR systems, we compared the conventional hybrid baseline model with the end-to-end approaches described above.

5.1   Baseline System

The whole baseline system training is done according to the *librispeech/s5* recipe from the Kaldi [24] toolkit. At the moment, this is one of the most popular conventional hybrid model architectures for the ASR task. It consists of a TDNN-F acoustic model, trained with the LF-MMI criterion, and word 3-gram language model. To build a lexicon for our data, we trained Phonetisaurus [21] grapheme-to-phoneme model on the VoxForge Russian lexicon[2].

Since our end-to-end exploration considers external language model usage, we also trained word-type NNLM from the Kaldi baseline WSJ recipe[3]. This NNLM was used for lattice re-scoring of decoding results for the hybrid system. The model was trained on utterances text of our chosen training set from Sect. 4.

5.2   End-to-End Modeling

As the main framework for end-to-end modeling we used the ESPnet speech recognition toolkit [33], which supports most of the basic end-to-end models and training/decoding pipelines.

In this work, we studied the following end-to-end architectures: Joint CTC-Attention (CTC-Attention), RNN-Transducer (RNN-T), and Transformer attention-based model (Transformer). The exact parameters of each model are presented in Table 1.

In front of each encoder, we used a convolutional neural network (CNN) consisting of four Visual Geometry Group (VGG) convolutional layers to compress the input frame sequence four times in the time domain.

Acoustic Units. We used two types of target acoustic units for each end-to-end model. The first one was characters (32 letters of the Russian alphabet and 2 auxiliary symbols). The second one was 500-classes subwords selected by a word segmentation algorithm from the SentencePiece toolkit [14].

Acoustic Features. The input feature sequence of all our end-to-end models are cepstral mean and variance normalized 80-dimensional log-Mel filterbank coefficients with 3-dimensional pitch features.

Decoding. To measure the system performance (Word Error Rate, WER), we used a fixed beam size 20 for all end-to-end models decoding. We also trained two types of NNLM model with different targets (characters, and subwords) for

---

[2] http://www.dev.voxforge.org/projects/Russian/export/2500/Trunk/AcousticMode
ls/etc/msu_ru_nsh.dic.

[3] https://github.com/kaldi-asr/kaldi/blob/master/egs/wsj/s5/local/rnnlm/tuning/
run_lstm_tdnn_1a.sh.

Table 1. End-to-end models configuration.

CTC-Attention	
Encoder	VGG-BLSTM, 5-layer 1024-units, 512-dim projection, dp 0.4
Attention	1-head 256-units, dp 0.4
Decoder	LSTM, 2-layer 256 units, dp 0.4
RNN-T	
Encoder	VGG-BLSTM, 5-layer 1024-units, 512-dim projection, dp 0.4
Predictor	LSTM, 2-layer 512-units, 1024-dim embeding, dp 0.2
Joiner	FC 512 units
Transformer	
Encoder	MHA, 12-layer 1024-units, dp 0.5
Attention	4-head 256-units, dp 0.5
Decoder	MHA, 2-layer 1024 units, dp 0.5

hypotheses rescoring in decoding time. The NNLM weight was set to 0.3 for all cases. The architecture of our NNLM is one-layer LSTM with 1024-units. As well as for NNLM from the hybrid setup, the model is trained only on the training set text.

6 Results

The results of our experiments are presented in the Table 2. They show that the hybrid model (TDNN-F LF-MMI) is still better in the reviewed validation data domains among most of the end-to-end approaches in terms of WER. The use of NNLM also noticeably improves hybrid model performance. However, the Transformer end-to-end model with subword acoustic units demonstrates the best accuracy results on YouTube and books data. The recognition of phone calls is worse within 0.6–1.3% in comparison with the hybrid system. This may be due to the small number of unique phone calls training data compared to YouTube and books domain. As we discussed before, end-to-end ASR systems are more sensitive to the amount of training data than the hybrid ones.

Further NNLM experiments showed that using external language model for hypotheses rescoring of end-to-end decoding results degrades WER performance almost in all cases. For RNN-transducer, we could not get any improvement for all sets. These problems can be caused by the default ESPnet's way of using the NNLM scores in the decoding time for end-to-end models. In all cases, a final acoustic unit score is the sum of the end-to-end model score and weighted NNLM score. But in case of the hybrid model, there are separate acoustic and language model scores of a word unit. Next, in the rescoring time, the final score is the sum of acoustic and rEweighted (according to the external NNLM) language scores. Primarily, the use of external NNLM helps only for the books validation.

Table 2. Final results.

Model	Units	NNLM	WER,%		
			calls	YouTube	books
TDNN-F LF-MMI	phone	no	34.2	22.2	19.8
		yes	33.5	20.9	18.6
Joint CTC-Attention	char	no	38.9	23.2	21.0
		yes	38.9	22.4	18.9
	subword	no	39.6	23.0	21.5
		yes	40.4	23.3	19.6
RNN-transducer	char	no	39.6	21.7	21.6
		yes	47.3	31.3	37.2
	subword	no	39.3	20.3	21.0
		yes	45.9	24.8	23.2
Transformer	char	no	35.1	21.3	18.5
		yes	35.7	21.4	16.9
	subword	no	34.8	19.1	18.1
		yes	36.8	20.9	16.8

Table 3. Comparison of each of our best models with previously published results.

Model	WER,%		
	calls	YouTube	books
TDNN-F LF-MMI	33.5	20.9	18.6
CTC-Attention	38.9	22.4	18.9
RNN-transducer	39.3	20.3	21.0
Transformer	34.8	19.1	18.1
Transformer [6]	32.6	20.8	18.1
Separable convolutions & CTC [32]	37.0	26.0	23.0

We also compared our models with previously published results, which used OpenSTT database for building ASR systems. The comparison is presented in the Table 3. The Transformer from the ESPnet recipe has 2048 units per layer, which is two times wider than the ours. Also, it is trained on all "clean" data, that is out of the validation domain (radio and lectures recordings, synthesized utterances of Russian addresses, and other databases). The model from Open-STT authors is a time-efficient stacked separable convolutions with CTC and external LM decoding. Also, they used all the database domains and pre-train their model on non-public data. The system details are described in [31].

7   Conclusion

In this study, we presented the first detailed comparison of common end-to-end approaches (Joint CTC-Attention, RNN-transducer, and Transformer) along with a strong hybrid model (TDNN-F LF-MMI) for the Russian language. As training and testing data, we used OpenSTT data set in three different domains: phone calls, YouTube videos, and books reading. The Transformer model with subword acoustic units showed the best WER result on YouTube and books validations (19.1%, and 16.8%, respectively). However, the hybrid model still performs better in the case of a small amount of training data and presented a better performance on phone calls validation (33.5% WER) than end-to-end systems. The use of external NNLM for hypotheses rescoring of the hybrid system provides a WER reduction for all validation sets. At the same time, NNLM using for end-to-end rescoring delivers ambiguous results. We observed performance degradation in almost all cases (for RNN-transducer in all) except the books validation. We think that this may be due to a sub-optimal ESPnet default hypotheses rescoring algorithm. As future work, we are going to study using NNLM for improving end-to-end decoding results.

Acknowledgments. This work was partially financially supported by the Government of the Russian Federation (Grant 08-08).

References

1. Andrusenko, A., Laptev, A., Medennikov, I.: Towards a competitive end-to-end speech recognition for chime-6 dinner party transcription. arXiv preprint arXiv:2004.10799 (2020). https://arxiv.org/abs/2004.10799v2
2. Bahdanau, D., Cho, K., Bengio, Y.: Neural machine translation by jointly learning to align and translate. In: 3rd International Conference on Learning Representations, ICLR, May 2015
3. Bataev, V., Korenevsky, M., Medennikov, I., Zatvornitskiy, A.: Exploring end-to-end techniques for low-resource speech recognition. In: Karpov, A., Jokisch, O., Potapova, R. (eds.) SPECOM 2018. LNCS (LNAI), vol. 11096, pp. 32–41. Springer, Cham (2018). https://doi.org/10.1007/978-3-319-99579-3_4
4. Boyer, F., Rouas, J.L.: End-to-end speech recognition: a review for the French language. arXiv preprint arXiv:1910.08502 (2019). http://arxiv.org/abs/1910.08502
5. Chan, W., Jaitly, N., Le, Q.V., Vinyals, O.: Listen, attend and spell: a neural network for large vocabulary conversational speech recognition. In: IEEE International Conference on Acoustics, Speech and Signal Processing (ICASSP), pp. 4960–4964. IEEE (2016). https://doi.org/10.1109/ICASSP.2016.7472621
6. Denisov, P.: Espnet recipe results for Russian open speech to text (2019). https://github.com/espnet/espnet/blob/master/egs/ru_open_stt/asr1/RESULTS.md
7. Graves, A.: Sequence transduction with recurrent neural networks. In: Proceedings of the 29th International Conference on Machine Learning (2012)
8. Graves, A., Fernández, S., Gomez, F., Schmidhuber, J.: Connectionist temporal classification: labelling unsegmented sequence data with recurrent neural networks. In: Proceedings of the 23rd International Conference on Machine Learning - ICML, pp. 369–376. ACM Press (2006). https://doi.org/10.1145/1143844.1143891

9. Hinton, G., Deng, l., Yu, D., Dahl, G., et al.: Deep neural networks for acoustic modeling in speech recognition: the shared views of four research groups. In: Signal Processing Magazine, IEEE, pp. 82–97, November 2012. https://doi.org/10.1109/MSP.2012.2205597
10. Iakushkin, O., Fedoseev, G., Shaleva, A., Degtyarev, A., Sedova, O.: Russian-language speech recognition system based on deepspeech. In: Proceedings of the VIII International Conference on Distributed Computing and Grid-technologies in Science and Education (GRID 2018), September 2018. https://github.com/GeorgeFedoseev/DeepSpeech
11. Karita, S., Wang, X., Watanabe, S., Yoshimura, T., et al.: A comparative study on transformer vs RNN in speech applications. In: IEEE Automatic Speech Recognition and Understanding Workshop (ASRU), pp. 449–456. IEEE, December 2019
12. Kim, S., Hori, T., Watanabe, S.: Joint CTC-attention based end-to-end speech recognition using multi-task learning. In: IEEE International Conference on Acoustics, Speech and Signal Processing (ICASSP), pp. 4835–4839. IEEE, March 2017. https://doi.org/10.1109/ICASSP.2017.7953075
13. Kipyatkova, I., Karpov, A.: DNN-based acoustic modeling for Russian speech recognition using kaldi. In: Ronzhin, A., Potapova, R., Németh, G. (eds.) SPECOM 2016. LNCS (LNAI), vol. 9811, pp. 246–253. Springer, Cham (2016). https://doi.org/10.1007/978-3-319-43958-7_29
14. Kudo, T., Richardson, J.: SentencePiece: a simple and language independent subword tokenizer and detokenizer for Neural Text Processing. In: Conference on Empirical Methods in Natural Language Processing: System Demonstrations, pp. 66–71 (2018). https://github.com/google/sentencepiece
15. Laptev, A., Korostik, R., Svischev, A., Andrusenko, A., et al.: You do not need more data: improving end-to-end speech recognition by text-to-speech data augmentation. arXiv preprint arXiv:2005.07157 (2020)
16. Li, J., Lavrukhin, V., Ginsburg, B., Leary, R., et al.: Jasper: an end-to-end convolutional neural acoustic model. In: Interspeech 2019, pp. 71–75. ISCA, September 2019. https://doi.org/10.21437/interspeech.2019-1819
17. Lüscher, C., Beck, E., Irie, K., Kitza, M., et al.: RWTH ASR systems for LibriSpeech: hybrid vs attention. In: Interspeech 2019, pp. 231–235. ISCA, September 2019. https://doi.org/10.21437/Interspeech.2019-1780
18. Markovnikov, N., Kipyatkova, I., Karpov, A., Filchenkov, A.: Deep neural networks in Russian speech recognition. In: Filchenkov, A., Pivovarova, L., Žižka, J. (eds.) AINL 2017. CCIS, vol. 789, pp. 54–67. Springer, Cham (2018). https://doi.org/10.1007/978-3-319-71746-3_5
19. Medennikov, I., Korenevsky, M., Prisyach, T., Khokhlov, Y., et al.: The STC system for the CHiME-6 challenge. In: CHiME 2020 Workshop on Speech Processing in Everyday Environments (2020)
20. Medennikov, I., Prudnikov, A.: Advances in STC Russian spontaneous speech recognition system. In: Ronzhin, A., Potapova, R., Németh, G. (eds.) SPECOM 2016. LNCS (LNAI), vol. 9811, pp. 116–123. Springer, Cham (2016). https://doi.org/10.1007/978-3-319-43958-7_13
21. Novak, J., Minematsu, N., Hirose, K.: Phonetisaurus: exploring grapheme-to-phoneme conversion with joint n-gram models in the WFST framework. Natural Language Engineering, pp. 1–32, September 2015. https://doi.org/10.1017/S1351324915000315. https://github.com/AdolfVonKleist/Phonetisaurus
22. Park, D.S., Zhang, Y., Jia, Y., Han, W., et al.: Improved noisy student training for automatic speech recognition. arXiv preprint arXiv:2005.09629 (2020). https://arxiv.org/abs/2005.09629v1

23. Povey, D., Cheng, G., Wang, Y., Li, K., et al.: Semi-orthogonal low-rank matrix factorization for deep neural networks. In: Proceedings of the Interspeech 2018, pp. 3743–3747. ISCA, September 2018. https://doi.org/10.21437/Interspeech.2018-1417

24. Povey, D., et al.: The kaldi speech recognition toolkit. In: IEEE Workshop on Automatic Speech Recognition and Understanding, December 2011. https://github.com/kaldi-asr/kaldi

25. Povey, D., Peddinti, V., Galvez, D., Ghahremani, P., et al.: Purely sequence-trained neural networks for ASR based on lattice-free mmi. In: Interspeech 2016, pp. 2751–2755. ISCA, September 2016. https://doi.org/10.21437/Interspeech.2016-595

26. Prudnikov, A., Medennikov, I., Mendelev, V., Korenevsky, M., Khokhlov, Y.: Improving acoustic models for russian spontaneous speech recognition. In: Ronzhin, A., Potapova, R., Fakotakis, N. (eds.) SPECOM 2015. LNCS (LNAI), vol. 9319, pp. 234–242. Springer, Cham (2015). https://doi.org/10.1007/978-3-319-23132-7_29

27. Ravanelli, M., Parcollet, T., Bengio, Y.: The pytorch-kaldi speech recognition toolkit. In: 2019 IEEE International Conference on Acoustics, Speech and Signal Processing (ICASSP). IEEE, May 2019. https://doi.org/10.1109/icassp.2019.8683713

28. Seide, F., Agarwal, A.: CNTK: Microsoft's open-source deep-learning toolkit. In: Proceedings of the 22nd ACM SIGKDD International Conference on Knowledge Discovery and Data Mining. Association for Computing Machinery (2016). https://doi.org/10.1145/2939672.2945397. https://github.com/Microsoft/CNTK

29. Slizhikova, A., Veysov, A., Nurtdinova, D., Voronin, D., Baburov, Y.: Russian open speech to text (STT/ASR) dataset v1.0 (2019). https://github.com/snakers4/open_stt/

30. Vaswani, A., Shazeer, N., Parmar, N., Uszkoreit, J., et al.: Attention is all you need. In: Advances in Neural Information Processing Systems, vol. 30, pp. 5998–6008 (2017)

31. Veysov, A.: Toward's an imagenet moment for speech-to-text. The Gradient (2020). https://thegradient.pub/towards-an-imagenet-moment-for-speech-to-text/

32. Veysov, A.: Сравнение нашей системы stt с остальными системами на рынке по качеству (обновление 2020–05-21) [quality comparison of our stt system with other systems in the market (update 2020–05-21)] (2020). https://www.silero.ai/russian-stt-benchmarks-update1/

33. Watanabe, S., Hori, T., Karita, S., Hayashi, T., et al.: ESPnet: end-to-end speech processing toolkit. In: Interspeech 2018, pp. 2207–2211. ISCA, September 2018. https://doi.org/10.21437/Interspeech.2018-1456. https://github.com/espnet/espnet

Directional Clustering with Polyharmonic Phase Estimation for Enhanced Speaker Localization

Sergei Astapov[1]([✉]), Dmitriy Popov[2], and Vladimir Kabarov[1]

[1] International Research Laboratory "Multimodal Biometric and Speech Systems,"
ITMO University, Kronverksky prospekt 49A, St. Petersburg 197101, Russia
{astapov,kabarov}@speechpro.com
[2] Speech Technology Center, Vyborgskaya naberezhnaya 45E,
St. Petersburg 194044, Russia
popov-d@speechpro.com

Abstract. Lately developed approaches to distant speech processing tasks fail to reach the quality of close-talking speech processing in terms of speech recognition, speaker identification and diarization quality. Sound source localization remains an important aspect in multi-channel distant speech processing applications. This paper considers an approach to improve speaker localization quality on large-aperture microphone arrays. To reduce the shortcomings of signal acquisition with large-aperture arrays and reduce the impact of noise and interference, a Time-Frequency masking approach is proposed applying Complex Angular Central Gaussian Mixture Models for sound source directional clustering and inter-component phase analysis for polyharmonic speech component restoration. The approach is tested on real-life multi-speaker recordings and shown to increase speaker localization accuracy for the cases of non-overlapped and partially overlapped speech.

Keywords: Distant speech · Sound source localization ·
Large-aperture microphone arrays · Inter-component phase
estimation · SRP-PHAT · Stochastic Particle Filter · CACGMM

1 Introduction

Currently the state-of-the-art approaches to distant speech processing for the tasks of Automatic Speech Recognition (ASR), speaker identification and diarization fail to reach the quality of close-talking speech processing [16]. Speech signals acquired from a distance suffer from noise and interference in a much greater degree compared to close-talking speech. Coherent noise produced by point sources in the vicinity of the microphone is coupled with the diffuse background noise, and distant speech signal attenuation coupled with interference produced by the Room Impulse Response (RIR) all result in high levels of interference and low Signal to Noise Ratio (SNR). In such conditions the application

© Springer Nature Switzerland AG 2020
A. Karpov and R. Potapova (Eds.): SPECOM 2020, LNAI 12335, pp. 45–56, 2020.
https://doi.org/10.1007/978-3-030-60276-5_5

of multi-channel recording devices such as microphone arrays is beneficial as it allows to operate with the signal spatial properties.

Sound source localization (SSL) is performed to estimate the direction to the sound source or its position relative to the microphone array. Position estimates are mainly used during speech enhancement [5], e.g., in spatial filtering applying beamforming techniques [11], however, they can also be applied in ASR and speaker identification [16]. Several latest works in ASR and speaker identification applying deep learning on Artificial Neural Networks (ANN) tend to use SSL estimates as model embeddings [11,17] or train end-to-end models [8,18], which account for signal spatial properties. During multi-channel model training spatial information is either explicitly introduced to the model or implicitly accounted for as one of the signal's intrinsic properties. Utilization of spatial information in both classical signal processing approaches and the ones applying state-of-the-art machine learning (ML) indicate the importance of SSL in modern speech processing applications.

This paper considers an approach to SSL improvement for application in large-aperture microphone arrays. Compared to compact arrays, in which the distance between microphones measures in several centimeters, large-aperture arrays are far more distributed, usually covering a large space. While greater distance between successive microphones gives these kind of arrays greater spatial resolution, they suffer more from inconsistencies in the signal between channels [13]. This paper discusses an approach of applying directional clustering by a Complex Angular Central Gaussian Mixture Model (CACGMM) to estimate the frequency bands occupied by different speakers and also reduce the influence of noise. Additionally, inter-component phase analysis employing the polyharmonic nature of speech is performed to reduce the side effects of CACGMM permutation inconsistencies in an attempt to produce a finer Time-Frequency (TF) mask. As a base algorithm for SSL the Steered Response Power with Phase Transform (SRP-PHAT) method is employed.

2 Preliminary Information

This section states the problem of SSL in large-aperture microphone arrays and discusses the algorithms applied in the approach considered in this paper.

2.1 Problem Formulation

Any approach to SSL, either classical [5,6] or ML [4,14], is based on the temporal differences of sound wave arrival between the spatially distributed sensors, i.e., the Time Difference of Arrival (TDOA). The estimation of TDOA is performed by calculation of some metric of signal inter-channel similarity, e.g. cross-correlation, etc. In compact microphone arrays the signal is acquired by closely spaced sensors and, thus, its power spectral density (PSD) between channels is highly similar, while the inter-channel phase shift is imposed by TDOA. As a result, a similarity metric is quite effective in determining these phase shifts. In large-aperture arrays, however, the PSDs of the signal acquired by distributed sensors may quite differ [13] due to:

- signal attenuation: a sound source may be close to some microphones of the array, but farther away from the other;
- RIR: room acoustics at one measurement point may vary from the acoustics at another distant point;
- point noise sources: a noise source may appear in the vicinity of some microphones and corrupt the signal-of-interest (low SNR), but this noise may attenuate while reaching farther microphones (high SNR).

These circumstances can significantly lower the confidence of the inter-channel similarity metric in broad arrays. Possible solutions include restricting inter-channel signal analysis to defined frequency bands of interest (i.e., which contain the components of the signal of interest); performing prior signal enhancement to reduce the influence of noise and RIR interference; defining the channels with the most coherent and similar PSD and performing SSL on these dedicated channels [12]. The last solution comes with the cost of possibly losing array resolution. Primarily, any of these possible solutions must require for the inter-channel phase information remain intact in order to retrieve TDOA.

We pursue the direction of restricting frequency bands of interest, while also reducing the destructive effect of noise and interference in these frequency bands. For this we apply CACGMM and inter-component phase analysis to achieve signal TF masks, which are applied during SSL.

2.2   SRP-PHAT with Stochastic Particle Filtering

SRP-PHAT is one of the most effective SSL estimation methods for reverberant environments [6]. The SRP $P(\mathbf{a})$ is a real-valued functional of a spatial vector $\mathbf{a}$, which is defined by the acoustic source search area or volume. The maxima of $P(\mathbf{a})$ indicate the position of sound source. The SRP is computed as the cumulative Generalized Cross-Correlation with Phase Transform (GCC-PHAT) across all pairs of sensors at the theoretical TDOA, associated with the chosen position. Consider a pair of signals $x_i(t)$, $x_j(t)$ of an array consisting of M microphones. The time instances of sound arrival from some point $a \in \mathbf{a}$ for the two microphones are $\tau(a, i)$ and $\tau(a, j)$, respectively. Hence the time delay between the signals is $\tau_{ij}(a) = \tau(a, i) - \tau(a, j)$. The SRP-PHAT for all pairs of signals is then defined as

$$P(a) = \sum_{i=1}^{M} \sum_{j=i+1}^{M} \int_{-\infty}^{\infty} \Psi_{ij} X_i(\omega) X_j^*(\omega) e^{j\omega \tau_{ij}(a)} d\omega, \tag{1}$$

where $X_i(\omega)$ is the Short Time Fourier Transform (STFT) of signal $x_i(t)$, $[\cdot]^*$ denotes complex conjugation, ω is the angular frequency and Ψ_{ij} is the PHAT coefficient, defined as

$$\Psi_{ij} = \left[|X_i(\omega) X_j^*(\omega)| \right]^{-1}. \tag{2}$$

In order to perform a fine SRP evaluation on a single STFT frame and obtain the most probable source position by locating the SRP global maximum the search area or volume has to be divided into a discrete search grid. It is obvious that evaluating (1) for each short signal frame on a large number of discrete points is highly resource demanding. Thus, we employ a technique of stochastic search with application of a Stochastic Particle Filter (SPF), proposed in [7]. Shortly, it consists of stochastically populating the search area with particles (spatial points), evaluating the SRP of these particles, calculating the probability of missing the global SRP maximum, calculating the number of new particles needed to minimize this probability, and repopulating with new particles. These steps, similar to the sampling and resampling steps of a conventional SPF, are repeated iteratively until convergence. This procedure reduces the number of evaluations (1) by several orders of magnitude for large search areas [7].

As a result of SRP-PHAT with SPF an estimation of sound source position is obtained for each STFT frame. We operate a planar area search to reduce computational costs and retrieve the position (x, y) in Cartesian coordinates.

2.3   TF Mask Estimation with CACGMM

The CACGMM was originally developed for the task of source separation [9]. Its aim is to cluster Time-Frequency bins of the signal's STFT corresponding to speech of different speakers in a signal mixture, while employing directional statistics. Let us denote the STFT signal mixture of $N \geq 2$ components acquired by $M \geq 2$ microphones as

$$\boldsymbol{y}_{tf} = \sum_{n=1}^{N} \boldsymbol{s}_{tf}^{(n)}, \ \boldsymbol{y}_{tf} \in \mathbb{C}^M, \tag{3}$$

where $\boldsymbol{y}_{tf}$ is a complex vector of observed STFT values at discrete time instance t and frequency bin f, and $\boldsymbol{s}_{tf}^{(n)}$ is the n-th signal source component. The directional statistics are then defined as the unit vector

$$\boldsymbol{z}_{tf} = \frac{\boldsymbol{y}_{tf}}{\|\boldsymbol{y}_{tf}\|}, \tag{4}$$

where $\|\cdot\|$ denotes the Euclidean norm. As it is shown in [9], directional statistics follow a complex angular central Gaussian distribution defined as

$$\mathcal{A}\left(\boldsymbol{z}; \mathbf{B}\right) = \frac{(M-1)!}{2\pi^M \det(\mathbf{B})} \frac{1}{\left(\boldsymbol{z}^H \mathbf{B}^{-1} \boldsymbol{z}\right)^M}, \tag{5}$$

where $\mathbf{B}$ is the spatial correlation matrix of $\boldsymbol{z}$. Note that the distribution (5) has a global maximum when $\boldsymbol{z}$ is a principal eigenvector of the positive-definite Hermitian matrix $\mathbf{B}$. The distribution of directional statistics for the sum of K source components is then characterized by the CACGMM:

$$p\left(\boldsymbol{z}_{tf}; \lambda_f\right) = \sum_{k=1}^{K} \alpha_f^{(k)} \mathcal{A}\left(\boldsymbol{z}_{tf}; \mathbf{B}_f^{(k)}\right). \tag{6}$$

The model parameter set $\lambda_f = \left\{ \alpha_f^{(k)}, \mathbf{B}_f^{(k)} \mid k = 1, \ldots, K \right\}$ consists of time invariant mixture weights $\alpha_f^{(k)}$ satisfying $0 \leq \alpha_f^{(k)} \leq 1$, $\sum_k^{N+1} \alpha_{fk} = 1$, and spatial correlation matrices $\mathbf{B}_f^{(k)}$. Unsupervised training of the model is performed by the Expectation-Maximization algorithm, which estimates λ_f by maximizing the log-likelihood function

$$\mathcal{L}\left(\lambda_f\right) = \sum_{t=1}^{T} \ln \sum_{k=1}^{K} \alpha_f^{(k)} \mathcal{A}\left(z_{tf}; \mathbf{B}_f^{(k)}\right). \tag{7}$$

The TF mask $\gamma_{tf}^{(k)} \in \mathbb{R}^{T \times F}$ for each of K mixture components is then defined as the posterior probability

$$\gamma_{tf}^{(k)} = \frac{\alpha_f^{(k)} \mathcal{A}\left(z_{tf}; \mathbf{B}_f^{(k)}\right)}{\sum_{k=1}^{K} \alpha_f^{(k)} \mathcal{A}\left(z_{tf}; \mathbf{B}_f^{(k)}\right)}. \tag{8}$$

2.4 Inter-component Phase Estimation

As of late, phase information of STFT has become a prosperous source of additional information in modern speech processing approaches. Prior to that, features were mainly extracted from the STFT amplitude or power spectra. Phase features can be treated independently, like STFT frequency bins, however, ICPE treats the phase relations considering the polyharmonic nature of human speech [3]. Specifically, it adopts the property of Phase Quasi-Invariance (PQI) [15].

Given the fundamental frequency $F_0(t)$, each of the harmonics of (single-channel) signal $s(t)$ is characterized by the harmonic index $h \in [1, H_t]$ and the corresponding amplitude $A(h, t)$ and phase $\Phi(h, t)$, both varying in time:

$$s(t) = \sum_{h=1}^{H_t} A(h, t) \cos \Psi(h, t) = \sum_{h=1}^{H_t} A(h, t) \cos \left(2\pi h F_0(t) t + \Phi(h, t)\right). \tag{9}$$

The constraint of PQI is defined for components with frequencies $\bar{h}F_0(t)$ and $hF_0(t)$, where $\bar{h} < h$, as

$$\mathrm{PQI}(\bar{h}, h, t) = \Psi(\bar{h}, t) - \frac{\Psi(h, t)\bar{h}}{h} = \Phi(\bar{h}, t) - \frac{\Phi(h, t)\bar{h}}{h}. \tag{10}$$

By estimating the PQI and normalizing it by $h/\bar{h}$, any arbitrary component with harmonic index h is derived as

$$\Phi(h, t) = \frac{h\Phi(h, t)}{\bar{h}} - \mathrm{PQI}(\bar{h}, h, t). \tag{11}$$

After that, temporal smoothing of the unwrapped harmonic phase is applied to achieve smoothed estimates of differential phases to account for glottal and pulse shapes of speech [10], and the enhanced instantaneous STFT phase $\hat{\psi}_s(t, f)$ is obtained by convolution of the smoothed differential phases with the STFT window function. Finally, the ICPE enhanced STFT of the speech signal is obtained as

$$\hat{S}(t, f) = |X(t, f)| \, e^{j\hat{\psi}_s(t,f)}. \tag{12}$$

3   Considered Approach to Speaker Localization

As stated in Subsect. 2.1, we consider applying TF masks prior to SRP-PHAT in order to reduce the side effects of large-aperture microphone arrays while minimizing the influence of noise and interference on SSL. In our considered approach we adopt the following assumptions about the targeted multi-speaker conversation scenario:

- N speakers participate in a conversation. Each speaker is active approximately an equal amount of time compared to other speakers.
- The number of speakers is arbitrary: less, greater, or equal to the umber of microphones M.
- Speech may partially overlap, but at any time instance only one speaker is considered active.
- Point noise sources may emerge, but a speaker's entire utterance cannot be fully covered with this noise.
- Background diffuse non-stationary noise may be present throughout the conversation.

The CACGMM is a fitting method of TF mask estimation under such assumptions, as it provides speech pattern clustering along with enhancement from noise and interference. The produced TF mask (8) is real-valued, and thus does not alter inter-channel phase information. To achieve speaker directional clustering and reduction of noise and interference in a conversation of N speakers we apply the CACGMM with the mixture number (6) of $K = N + 1$ to account for noise components, similar to [1]. Due to the unsupervised nature of model training the obtained clusters are unordered. The TF mask corresponding to noise is considered to be the one with the smallest inverse condition number of its distribution's (5) spatial correlation matrix $\mathbf{B}$:

$$\gamma_{tf}^N = \gamma_{tf}^{(l)}, \; l = \arg\min_k \left[\kappa^{-1} \left(\mathbf{B}_f^{(k)} \right) \right], \tag{13}$$
$$\kappa \left(\mathbf{B}_f^{(k)} \right) = \max \left(\sum_{j=1}^M \left| b_{ij}^{(k)} \right| \right).$$

And the aggregated signal source TF matrix is then defined as

$$\gamma_{tf}^S = \sum_{1 \le k \le K, \, k \ne l} \gamma_{tf}^{(k)} = 1 - \gamma_{tf}^N. \tag{14}$$

Note, that the signal TF matrix is one for all M channels, so the enhanced multi-channel signal mixture is obtained as

$$s_m^\Sigma(t, f) = \gamma_{tf}^S \boldsymbol{y}_{tf}^{(m)}, \; 1 \le m \le M. \tag{15}$$

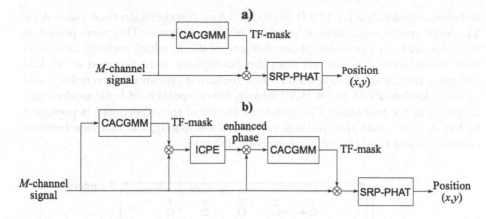

Fig. 1. Block diagram of two approaches to CACGMM and ICPE application in source localization.

The TF mask produced by CACGMM possesses intrinsic inconsistencies originating from unsupervised training and the permutation problem. The resulting speech pattern mask may be inconsistent with the speech polyharmonic signal pattern. Speech harmonics may be attenuated or lost due to faults in TF bin permutation, outliers in the training data, or simply because speech patterns of several speakers overlap in the frequency domain and their harmonics are thus jointly allocated during clustering. In an attempt to restore the attenuated harmonics we apply ICPE.

In this paper we examine two approaches to TF mask estimation, which are presented in Fig. 1. The first approach (Fig. 1a) employs CACGMM once and the TF mask (8) is applied directly to the M-channel signal according to (15), after which SRP-PHAT is performed to estimate (x, y) source coordinates. In the second approach (Fig. 1b) we also apply ICPE while maintaining the consistency of the original inter-channel phase. The first CACGMM produces the initial TF mask, reducing noise and interference. Afterwards ICPE is performed on the enhanced signal (15) and the relevant phase components are emphasized by applying the ICPE phase transform (12) to the original M-channel signal. As this phase transform disrupts the inter-channel phase relations, it cannot be directly used by SRP-PHAT. Instead, the emphasized signal is passed through the second CACGMM with the same parameters for distribution initialization as for the first model. This final resulting TF mask is then used to enhance the signal prior to SSL by (15).

4   Experimental Evaluation and Results

Experimental evaluation of the considered approach to SSL is performed on real-life recordings of a multi-speaker conversation. The recording took place in a meeting room with a large table situated in the middle. Room parameters are as follows: dimensions $L \times W \times H = 10 \times 6 \times 3.5$ m, reverberation time $T_{60} = 0.7$ s. The large-aperture microphone consisted of 5 microphones. They were placed on the table with an inter-microphone distance of 0.7 m, which resulted in a 3.5 m wide linear array. The signals from the microphone were acquired at 16 kHz sampling rate by a common audio mixer (ensuring synchronous recording) and saved in lossless PCM 16 bit WAV format. Seven speakers of both genders participated in the recording. The geometric layout of the experiment is presented in Fig. 2. Note, that the speakers are seated compactly, the distance between speakers is less than the inter-microphone distance.

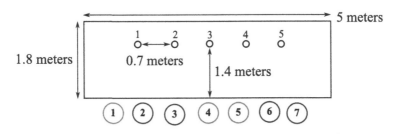

Fig. 2. Diagram of the experimental layout. Microphones are denoted by small black circles, positions of speakers - by large circles. Blue and pink colors denote male and female speakers, respectively. (Color figure online)

The conversation sessions are held according to the scenario addressed in Sect. 3, however no particular noise sources are introduced during the sessions. In order to accurately set the SNR during testing, background noise and point noise sources are mixed with the recording after signal acquisition. The background uncorrelated noise from separate recordings is added to the mixture as is, and the point noise sources are introduced by simulating room acoustics and reverberating noise according to the dispositions of its sources. For this we apply a technique previously developed by us for noisy corpora generation [2]. The mix of coherent and diffuse noise is then added to the recordings at specific SNR levels.

Evaluation is performed on two recording sessions containing one hour of spontaneous speech. The first one contains specifically non-overlapping speech, the speakers were instructed to avoid interruptions. For the second session interruptions and mild cross talk were allowed, thus, this session contains partially overlapping speech. The evaluation results for the two sessions are presented in Fig. 3. The localization error percent is defined as the ratio between falsely localized utterances and the total number of utterances. In this case, manually performed speaker dependent segmentation is used to distinguish between utterances and specify the correct speaker for each utterance. For each utterance the position estimate is obtained as the mean of estimates performed on STFT frames corresponding to this utterance. The target speaker is defined as the closest to the position estimate, but not farther than 0.5 m from the estimate. Therefore, the result for each utterance is binary: either the target speaker utterance was correctly localized or not.

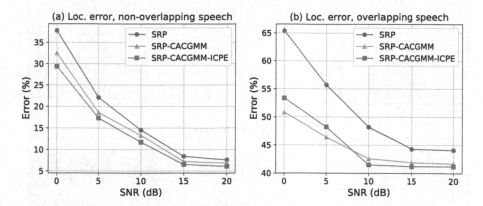

Fig. 3. Speaker localization error for non-overlapping and overlapping speech cases under different SNR levels.

The results presented in Fig. 3 demonstrate the usefulness of pre-processing with CACGMM generated TF masks. ICPE further reduces localization error, however, in case of overlapping speech in low SNR conditions the application of ICPE seems to be harmful. This may indicate the inability to restore overlapped polyharmonic patterns under heavy noise or inapplicability of CACGMM-ICPE in such conditions. This matter is a topic of further investigation.

An example of CACGMM-ICPE application in the form of intermediate results is portrayed in Fig. 4. The spectrogam (Fig. 4a) portrays speech of two different speakers at SNR=10 dB with the speaker transition at the 10th s. Wide-band instantaneous noise is seen at the 1st, 6th and 7.5th s. Speaker 1 (first 10 s) has significantly lower signal power, which indicates his far disposition from the particular microphone. The TF mask generated by the first CACGMM (Fig. 4b) indicates the frequency bands containing no useful information for source clustering (4– 5 kHz, 7–8 kHz). Also, as a result of clustering TF bins occupied by several speakers the polyharmonic pattern for speaker 1 is very poorly represented in the lower frequencies (below 1 kHz). Figure 4c presents the difference of the phase components of the signal before and after ICPE application. Polyharmonic pattern restoration is visible, specially for speaker 1. Finally, Fig. 4d presents the final CACGMM-ICPE mask, which is applied to the signal prior to SSL. The harmonics of both speakers are better emphasized in the lower frequencies compared to the initial mask, and the distinction between useless frequency bands is more evident.

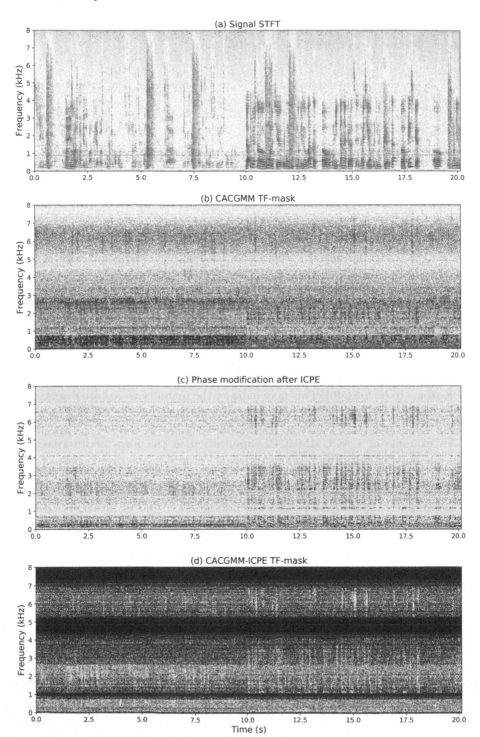

Fig. 4. Results of TF mask estimation by applying CACGMM and ICPE.

5   Conclusion

The paper addressed the problem of distant speech localization with a large-aperture microphone array. An approach of signal pre-processing applying directional clustering and polyharmonic phase restoration was presented and evaluated on real-life data. The results suggest the applicability of the considered approach in cases of non-overlapping and partially overlapping speech. Polyharmonic phase restoration in the conditions of partially overlapping speech under heavy noise has led to controversial results, which require further investigation.

Acknowledgments. This research was financially supported by the Foundation NTI (Contract 20/18gr, ID 0000000007418QR20002) and by the Government of the Russian Federation (Grant 08-08).

References

1. Astapov, S., Lavrentyev, A., Shuranov, E.: Far field speech enhancement at low SNR in presence of nonstationary noise based on spectral masking and MVDR beamforming. In: Karpov, A., Jokisch, O., Potapova, R. (eds.) SPECOM 2018. LNCS (LNAI), vol. 11096, pp. 21–31. Springer, Cham (2018). https://doi.org/10.1007/978-3-319-99579-3_3
2. Astapov, S., et al.: Acoustic event mixing to multichannel AMI data for distant speech recognition and acoustic event classification benchmarking. In: Salah, A.A., Karpov, A., Potapova, R. (eds.) SPECOM 2019. LNCS (LNAI), vol. 11658, pp. 31–42. Springer, Cham (2019). https://doi.org/10.1007/978-3-030-26061-3_4
3. Barysenka, S.Y., Vorobiov, V.I., Mowlaee, P.: Single-channel speech enhancement using inter-component phase relations. Speech Commun. **99**, 144–160 (2018)
4. Comanducci, L., Cobos, M., Antonacci, F., Sarti, A.: Time difference of arrival estimation from frequency-sliding generalized cross-correlations using convolutional neural networks. In: ICASSP 2020–2020 IEEE International Conference on Acoustics, Speech and Signal Processing (ICASSP), pp. 4945–4949 (2020)
5. Dey, N., Ashour, A.: Direction of Arrival Estimation and Localization of Multi-Speech Sources. Springer Briefs in Speech Technology. Springer, Cham (2017). https://doi.org/10.1007/978-3-319-73059-2
6. DiBiase, J.H.: A High-Accuracy, Low-Latency Technique for Talker Localization in Reverberant Environments Using Microphone Arrays. Ph.D. thesis, Brown University, Providence, RI, USA (2000)
7. Do, H., Silverman, H.F.: Stochastic particle filtering: a fast SRP-PHAT single source localization algorithm. In: 2009 IEEE Workshop on Applications of Signal Processing to Audio and Acoustics, pp. 213–216 (2009)
8. He, W., Lu, L., Zhang, B., Mahadeokar, J., Kalgaonkar, K., Fuegen, C.: Spatial attention for far-field speech recognition with deep beamforming neural networks. In: 2020 IEEE International Conference on Acoustics, Speech and Signal Processing (ICASSP), pp. 7499–7503, May 2020
9. Ito, N., Araki, S., Nakatani, T.: Complex angular central Gaussian mixture model for directional statistics in mask-based microphone array signal processing. In: 2016 24th European Signal Processing Conference (EUSIPCO), pp. 1153–1157 (2016)
10. Kulmer, J., Mowlaee, P.: Phase estimation in single channel speech enhancement using phase decomposition. IEEE Signal Process. Lett. **22**(5), 598–602 (2015)

11. Luo, Y., Han, C., Mesgarani, N., Ceolini, E., Liu, S.C.: FaSNet: Low-latency adaptive beamforming for multi-microphone audio processing. In: 2019 IEEE Automatic Speech Recognition and Understanding Workshop (ASRU), pp. 260–267. IEEE, Piscataway, NJ (2020). IEEE Automatic Speech Recognition and Understanding Workshop (ASRU 2019); Conference Location: Singapore, Singapore; Conference Date: December 14–18 (2019)
12. Sachar, J.M.: Some Important Algorithms for Large-Aperture Microphone Arrays: Calibration and Determination of Talker Orientation. Ph.D. thesis, Brown University, Providence, RI, USA (2004)
13. Silverman, H.F., Patterson, W.R., Sachar, J.: Factors affecting the performance of large-aperture microphone arrays. J. Acoust. Soc. Am. **111**(5 Pt 1), 2140–2157 (2002)
14. Vera-Diaz, J., Pizarro, D., Macias-Guarasa, J.: Towards end-to-end acoustic localization using deep learning: from audio signals to source position coordinates. Sensors **18**, 3418 (2018)
15. Vorobiov, V.I., Davydov, A.G.: Study of the relations between quasi-harmonic components of speech signal in Chinese language. Proc. Twenty-Fifth Session Russian Acoust. Soc. **3**, 11–14 (2012)
16. Watanabe, S., Araki, S., Bacchiani, M., Haeb-Umbach, R., Seltzer, M.L.: Introduction to the issue on far-field speech processing in the era of deep learning: speech enhancement, separation, and recognition. IEEE J. Sel. Top. Sig. Process. **13**(4), 785–786 (2019)
17. Xiao, X., Watanabe, S., Chng, E.S., Li, H.: Beamforming networks using spatial covariance features for far-field speech recognition. In: 2016 Asia-Pacific Signal and Information Processing Association Annual Summit and Conference (APSIPA), pp. 1–6 (2016)
18. Zhao, H., Zarar, S., Tashev, I., Lee, C.H.: Convolutional-recurrent neural networks for speech enhancement. In: IEEE International Conference Acoustics Speech and Signal Processing (ICASSP), April 2018

Speech Emotion Recognition Using Spectrogram Patterns as Features

Umut Avci[✉]

Faculty of Engineering, Department of Software Engineering, Yasar University,
Bornova, Izmir, Turkey
umut.avci@yasar.edu.tr

Abstract. In this paper, we tackle the problem of identifying emotions from speech by using features derived from spectrogram patterns. Towards this goal, we create a spectrogram for each speech signal. Produced spectrograms are divided into non-overlapping partitions based on different frequency ranges. After performing a discretization operation on each partition, we mine partition-specific patterns that discriminate an emotion from all other emotions. A classifier is then trained with features obtained from the extracted patterns. Our experimental evaluations indicate that the spectrogram-based patterns outperform the standard set of acoustic features. It is also shown that the results can further be improved with the increasing number of spectrogram partitions.

Keywords: Emotion recognition · Feature extraction · Spectrogram

1 Introduction

Speech emotion recognition technologies enable individuals to interact with machines more naturally and effectively. On a large scale, however, the technology provides greater benefits to companies. In the recruitment process, for instance, emotion analysis can reveal candidates with a positive attitude and personal stability, which can be used to determine if the candidate will be a good team member or not. Identifying emotions of current employees is equally important as how they feel will have a direct effect on their productivity, creativity, and professional success. Emotion recognition is also used to measure customer satisfaction. Happy customers can be given exclusive deals because they not only show loyalty to the company but also increase new customer acquisition by spreading the word about the organization. Customers who are angry or disappointed, on the other hand, can be given incentives and more priority so that their issues are solved without any delay. For all these reasons and more, it is crucial for companies to use methods that can infer their stakeholders' feelings.

Over the years, researchers and practitioners of machine learning have developed numerous approaches to address the problem of recognizing emotions from speech. No matter how different they are, these approaches share two common

A. Karpov and R. Potapova (Eds.): SPECOM 2020, LNAI 12335, pp. 57–67, 2020.
https://doi.org/10.1007/978-3-030-60276-5_6

steps: feature extraction and classification. The first step aims at identifying significant data attributes that characterize emotions. The studies in the literature generally use the standard set of continuous, qualitative, or spectral features such as energy [2], voice quality [5], and MFCC [20] respectively. Recent works have extended the standard set with the new acoustic features including AuDeep embeddings [6] and Fisher vectors [3]. These features are sometimes combined with the linguistic cues to improve the recognition performance. Lexical information is obtained either with the traditional methods in information retrieval like the language models [8] and the Bag-of-Words [18] or with the linguistic features such as word2vec [4] and GloVe [17]. The second step takes as input the extracted features and trains a classifier such as Support Vector Machines (SVMs) [1], Hidden Markov Models (HMMs) [11] and Deep Neural Networks (DNNs) [19]. Apart from these two-step approaches, there are also neural network variants that take as input directly raw speech samples and perform classification [16].

This study introduces a novel set of features extracted from spectrograms. To this end, we produce a spectrogram for each recording in an audio corpus. According to a predefined set of frequency intervals, previously obtained spectrograms are split into parts. We apply a transformation to each part so that the spectrogram data is downsampled and quantized. The quantized data is used to mine patterns characterizing emotions and the patterns are used to extract features to be fed into a classifier. Several papers in the literature suggest using spectrograms for speech emotion recognition. However, the vast majority of the studies take as input the whole spectrogram and leave the feature extraction and classification tasks to a deep neural network structure [7,14]. Others show differences in the feature extraction phase [15,22]. To the best of our knowledge, we are the first to propose mining patterns from spectrograms for the purpose of speech emotion recognition.

We present our approach in the next section. The description of the dataset used to evaluate the proposed approach and the results of our evaluation are given in Sect. 3. We conclude and discuss future work in Sect. 4.

2   Our Approach

Our approach is based on the observation that spectrograms obtained with speech signals of a certain emotion type are more similar to each other than those obtained with signals of different emotion types. That's why we present a number of steps for identifying spectrogram characteristics specific to each emotion type, using them to extract features, and evaluating the performance of the extracted features on the emotion recognition task. While Fig. 1 shows the main steps of our approach, details are introduced in the following sections.

2.1   Spectrogram Generation and Partitioning

Emotion recognition datasets are generally formed by audio recordings of subjects who vocalize a sentence several times differently to mimic distinct emotions.

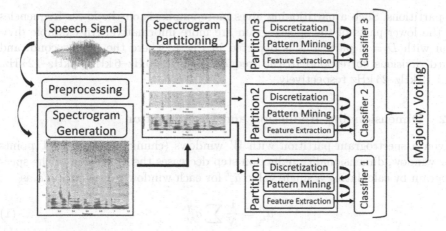

Fig. 1. Summary of our approach for the case with 3 spectrogram partitions.

Subjects remain silent for a certain period at the beginning and at the end of the recordings for the sentence to be captured completely. However, these silent regions do not carry any information about emotions. That's why we first trim these silent parts from the speech recordings. We also perform the z-score normalization of speech signals to eliminate interpersonal differences in sound.

In this study, we create spectrograms from the pre-processed recordings. Given an audio signal, i.e. a subject's processed speech recording for a certain type of emotion, a spectrogram can be computed using the short-time Fourier transform. For this purpose, the input signal is divided into chunks on each of which a local Fourier Transform is performed. The digital audio files used in this work were recorded at a sampling rate of 48 kHz with 16-bit quantization (see Sect. 3.1 for the details). Accordingly, we empirically set the width of each chunk to 480 samples with a Hamming window. In order to obtain a firm frequency localization, we specify 50% overlap between adjoining chunks. We make a final decision about the number of DFT points to be computed per each chunk. By setting DFT points per chunk to 480, we calculate the power spectrum of size 241 ($|DFT|/2 + 1$) for the one-sided spectrogram. Note that the width of each chunk and the number of DFT points are independent from each other and that one may choose different values for these parameters.

The bottom-left part of Fig. 1 shows an example spectrogram. As can be seen, there are different density zones along the frequency axis, i.e. the highest and the lowest spectral densities are observed for the lower and higher frequencies respectively. In our experiments, we have spotted similar structures in almost all the spectrograms. Based on the idea that different frequency ranges provide different information about spectrogram characteristics, we propose dividing spectrograms into several partitions according to disjoint intervals of frequency. Starting from the initial spectrogram, we repeatedly split the lower half of the spectrogram into halves. So, $D - 1$ division operations are performed to obtain

D partitions. Such a partitioning gives more emphasis on the lower frequencies as the lower parts of the spectrogram are more informative. An example division with $D = 3$ partitions is depicted in Fig. 1, where the first, second, and third divisions correspond to the frequency ranges 0 Hz–6 kHz, 6 kHz–12 kHz, and 12 kHz–24 kHz respectively.

2.2   Dimensionality Reduction and Discretization

Given a spectrogram partition with W windows (chunks) and N DFT points per window, dimensionality reduction step decreases the dimension of the spectrogram by calculating a single value $\overline{a}_i^d$ for each window i of partition d as

$$\overline{a}_i^d = \frac{1}{N} \sum_j a_{ij}^d \tag{1}$$

where a_{ij}^d is value of the jth DFT component for i^{th} window of partition d, $i = 1, ..., W$, $j = 1, ..., N$ and $d = 1, ..., D$.

Shortly, DFT components of a window are averaged to obtain a single value characterizing the magnitude of the window. On the other hand, the ranges of the averaged values change significantly from one partition to another. In order to compensate for the variation between the partitions, we apply the z-score normalization to the averaged values in each partition. As an example, let us take the first partition (0 Hz–6 kHz) of the spectrogram presented in Fig. 1. A 1.2-second speech recording captured at a sampling rate of 48 kHz has 240 windows with a window length of 480 samples and a 50% overlap. Since the partition is only one-quarter of the spectrogram, each window has 60 DFT points. Averaging 60 points for each window and performing normalization on the averages produces the graph (without the horizontal lines and the letters) presented in Fig. 2a.

A discretization operation follows the dimensionality reduction step. In the discretization phase, we assign a symbol to each $\overline{a}_i^d$ of the partition d. To achieve this, we first fit a Gaussian Distribution to the normalized averages of the partition. We then define breakpoints $b_1, ..., b_{k-1}$ that split the distribution into k sections. The breakpoints are specified such that the area under the Gaussian curve between two consecutive breakpoints b_{v-1} and b_v is the same for all $v = 2, ..., v = k - 1$. Afterward, each section is labeled with a symbol. Finally, the label of the section in which a normalized average $\overline{a}_i^d$ falls is assigned to the $\overline{a}_i^d$ as its symbol. As a result, the calculated averages of the partition d, $\overline{A}^d = \overline{a}_1^d, ..., \overline{a}_W^d$, are converted to a sequence of symbols for that partition, $\overline{S}^d = \overline{s}_1^d, ..., \overline{s}_W^d$. Note that the quality of the patterns to be obtained in the next step depends on the value of k. When k is less than 5, the discretization process produces sequences that contain many consecutive repetitions of the same symbols with few transitions between them, e.g. **...aaaaaabbbbbb....** This leads to the extraction of uninformative patterns especially for the spectrogram partitions of high frequency ranges in which lower spectral densities are observed. When k is greater than 10, contrary to the previous case, discretized sequences contain a

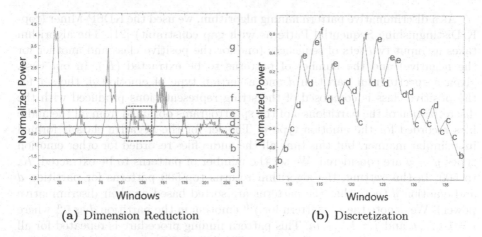

(a) Dimension Reduction   (b) Discretization

Fig. 2. (a) The dimension of spectrogram is reduced by window-based averaging. (b) The signal is discretized based on the areas of the Gaussian distribution.

few consecutive repetitions of the same symbols with many transitions between them, e.g. **...cahhbdeefeighg....** Such fine-grained sequences carry information about small vocal changes that exist in the utterances of all emotion types. In the pattern mining phase, this results in patterns with low discriminative power. For these reasons, we set k to a value between 5 and 10, i.e. 7, in our experiments.

Figure 2 shows an example discretization operation. In Fig. 2a, a Gaussian distribution is fitted to the normalized averages and 6 breakpoints are identified to create seven regions (separated from each other by the horizontal lines). We use letters from the English alphabet for symbols and label each region with a unique symbol. Each normalized average is then assigned a symbol depending on the region that the average lies. Figure 2b clearly depicts the assigned symbols for a subset of averages pointed out as the dashed rectangle in Fig. 2a. The conversion from the averages to the symbols creates a new representation of the spectrogram, i.e. a string of characters such as **bbcdedefded**.

2.3 Pattern Mining

We are interested in revealing the unique spectrogram characteristics that are specific to each type of emotion. One way to extract these characteristics is to use pattern mining. In the literature, there are many approaches to mine various kinds of patterns like frequent, rare, and sequential. Among all the alternatives, we particularly select discriminative pattern mining. It allows mining patterns that are frequent only in one class and infrequent in all the other classes. In this way, a class can readily be distinguished from others just by looking at its patterns. With such a capability, discriminative mining perfectly fits our needs since we would like to make a clear distinction between different types of emotions.

As a discriminative pattern mining algorithm, we used the KDSP-Miner (top-K Distinguishing Sequential Patterns with gap constraint) [21]. The algorithm takes as input two sets of sequences (one for the positive class and another for the negative) and the number of patterns to be extracted (K). In our case, given a spectrogram partition d and a certain type of emotion y, the set for the positive class is composed of the string representations produced with the discretization of the partitions d of the spectrograms obtained from all the audio files recorded for the emotion type y. The set for the negative class is formed in a similar manner, but this time all the audio files recorded for other emotion types $y' \neq y$ are considered. We set the number of patterns to be extracted, K, to 500. In this setting, the algorithm returns top-500 patterns for partition d and emotion y (note that the patterns are sorted based on their discriminative power). We denote the ith pattern for j^{th} emotion in the partition d as $\overline{p}_{ij}^{d}$ where $i = 1, \ldots, n$ and $j = 1, \ldots, m$. This pattern mining procedure is repeated for all the partitions and all types of emotions. The KDSP-Miner also enables to define an arbitrary number of spaces between individual pattern elements. However, we choose to leave it at its default value of 0 because any value greater than 0 will cause the algorithm to catch regularities between distant time instances, which is irrelevant in the case of spectrograms.

2.4 Feature Extraction

The previous step generates a different set of patterns for each type of emotion. The set whose patterns are observed the most in a speech recording determines the type of emotion of that recording. More specifically, the number of times that a pattern appears in a speech signal gives a clue about the emotion and hence can be used for the classification. Following this rationale, we propose identifying the pattern-based features.

Assume that we would like to extract features for a speech recording. Let F^d be the feature vector of size $n \times m$ for partition d, where n and m are the number of patterns and the number of emotion types respectively. Given a pattern for a certain emotion type j and partition d ($\overline{p}_{ij}^{d}$) and the string representation of the speech recording for partition d ($\overline{S}^{d}$), we calculate t^{th} element of the feature vector f_t^d by counting the number of occurrences of $\overline{p}_{ij}^{d}$ in $\overline{S}^{d}$, where $t = 1, \ldots, n \times m$. The full feature vector F^d is obtained by enumerating all patterns and all emotions. The counting will be made based on exact matches as we do not allow gaps between individual pattern elements. As an example take two patterns **def** and **ded**, and a string representation of a speech signal **dedddefdedeed** (see Fig. 2b). The first pattern is observed once in the representation, hence its feature value will be 1. The second pattern is observed three times in the representation. However, only two of these are exact matches since the last one, i.e. **deed**, has a gap of one. So, the feature value for the second pattern will be 2.

2.5 Classification

We solve the classification problem by using Support Vector Machines (SVMs). They proved to be successful in various domains for both binary and multi-class problems. In this study, we deal with a multi-class problem as we try to identify a single emotion type among several alternatives for a speech recording. There are different SVM strategies to handle multi-class classification problems such as one-against-one, one-against-all, and directed acyclic graph SVM (DAGSVM). All of these techniques are based on dividing a multi-class problem into multiple binary ones. The one-against-all approach tries to separate one class from all other classes by building c classifiers for a problem with c classes [9]. Again for a problem with c classes, one-against-one and DAGSVM methods both aim at separating one class from another by building $c(c-1)/2$ classifiers. In the testing phase of the former, a class label is selected by the majority voting of $c(c-1)/2$ classifiers, which leads to slow testing time [12]. The latter uses a rooted binary directed acyclic graph in the testing and prunes the search space to reach a prediction [13]. As a result, testing is performed more efficiently. We used the DAGSVM technique in our work.

In the previous steps of our approach, we divide a spectrogram into D partitions and then separately perform the discretization, pattern mining, and feature extraction operations on each partition d. Similarly, we build a separate classifier for each partition d and make classification via majority voting of D classifiers. To be more specific, each classifier votes for a single emotion type, and the emotion type that has the majority of the votes is assigned as the emotion of the speech recording. We use a random selection process to break the ties in the voting procedure. We could also concatenate D feature vectors and build a single classifier. As this would significantly increase the size of the feature vector, we chose to use the majority voting approach.

3 Dataset and Experimental Results

3.1 Dataset

We performed our experiments on the freely available Ryerson Audio-Visual Database of Emotional Speech and Song (RAVDESS) dataset [10]. RAVDESS dataset is formed by audio and video recordings of 24 professional actors (12 females and 12 males). Each actor imitates emotions by speaking and singing two lexically-matched statements ("dogs are sitting by the door" and "kids are talking by the door") and repeats each statement twice. 8 types of emotions are expressed in the speaking part: calm, happy, sad, angry, fearful, neutral, surprise, and disgust. In the singing part, on the other hand, surprise and disgust are not included. All types of emotions but neutral are vocalized with normal and strong intensities. Neutral is vocalized only with the normal intensity. The dataset is available with three options: audio-only, video-only, and audio-visual.

In this study, we use a subset of the RAVDESS dataset. We first choose the audio-only option since the video modality is outside the scope of this work. For the same reason, we also omit the song portion of the dataset and focus only on the speech recordings (16 bit, 48 kHz .wav). Further eliminations are made for comparison purposes. Our benchmark research [23] does not include disgust and surprise emotion types into their evaluation. For the remaining emotion types, they solely consider the recordings vocalized with the normal intensity. Accordingly, we downsize the dataset by removing the audio files recorded not only for the disgust and surprise emotion types but also for the strong intensity. As a result of these pruning works, we evaluate our approach on a dataset with 576 audio files (24 actors × 2 sentences ×2 repetitions × 6 emotions).

3.2 Experimental Results

This section presents the experimental evaluation of our approach and compares our results with the ones of Zhang et al. [23]. We analyze the performance of the proposed methodology for varying number of spectrogram divisions and varying number of patterns. For the spectrogram divisions, we consider four alternatives where D changes from 1 to 4. For the patterns, five different sets with 10, 50, 100, 250, and 500 patterns for each emotion are generated. Recall that, 500 patterns obtained in Sect. 2.3 are sorted according to their discriminative power. In this way, creating a set with 10 patterns, for instance, can easily be achieved by extracting the first 10 patterns out of 500. Note that, the size of the feature vector changes with the change in the pattern count. A setting with 10 patterns and 6 types of emotions produces a feature vector of size 60. Also, note that we apply z-score normalization to feature vector before its usage.

We measure the performance of the trained classifier by using leave-one-performer-and-sentence-out cross-validation. We take one sentence of one performer as testing data in each round. The remaining sentences constitute training data. DAGSVM hyperparameters, C and γ, are tuned with an internal cross-validation procedure. We run a five-fold internal cross-validation with a grid search on each training fold and save the parameters that achieve the highest five-fold cross-validation accuracy. The saved parameters are then passed to the outer cross-validation. We prefer to use the Radial Basis Function (RBF) as the kernel due to its success in many applications.

Figure 3 shows the changes in the classification accuracies as the number of patterns and the number of spectrogram divisions vary. Each line corresponds to a spectrogram with a certain number of divisions. The line with the square is for the case where $D = 1$. This is a baseline in which the spectrogram is taken as a whole without any divisions. As expected, it produces the lowest accuracies because representing each window with a single average is an oversimplification. Disregarding the effect of distinct frequency regions leads to the loss of valuable information about the spectrogram characteristics. When $D = 2$ (the line with the diamond), the spectrogram is divided into two equal parts. Even such a simple partitioning results in a rise in the performance. With three partitions, we obtain the divisions depicted in Fig. 1. From the figure, it is evident that the lower frequencies are represented better than the previous two settings. Our final setting where $D = 4$, hence, is the most representative of all. A finer level of detail

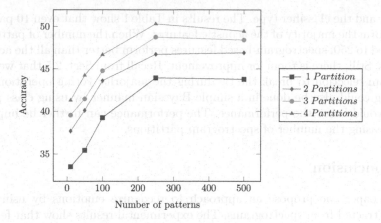

Fig. 3. DAGSVM classification results.

about the lower frequencies is provided with each increment in the number of partitions. As a result, consistent improvement of performance is observed with the increase in the number of spectrogram divisions. The number of patterns extracted for each emotion type is shown on the X-axis of Fig. 3. We see a rapid increase in the classification performance as the number of patterns changes from 10 to 250. Each pattern contains a unique piece of information about the emotion it represents. Considering them collectively draws a better picture of the emotion and therefore increases the success of the prediction. Whenever the picture is complete, adding more information does not lead to an improvement in the performance. Because of this, the classification results remain the same from 250 patterns to 500.

Table 1. Classification results of acoustic and pattern-based features.

Classification accuracy (%)	
Energy [23]	34.78
Spectral [23]	48.01
MFCC [23]	48.73
Voicing [23]	38.41
Rasta [23]	30.43
10 Patterns (proposed)	41.28
250 Patterns (proposed)	50.31

We compare our results (for $D = 4$) with those obtained with the standard set of acoustic features provided by Zhang et al. [23]. The comparison is made under the same experimental conditions in terms of the dataset, cross-validation scheme, and the classifier type. The results in Table 1 show that even 10 patterns outperform the majority of the acoustic features. When the number of patterns is increased to 250, spectrogram-based features perform better than all the acoustic features. Still, there is room for improvement. Recall from Sect. 2.5 that we make a random selection to break the tie during the majority voting operation. Tie-breaking could also be done in a simple Bayesian manner by using class priors, which would boost the performance. The performance can further be improved by increasing the number of spectrogram partitions.

4   Conclusion

In this paper, we propose an approach to recognize emotions by using patterns extracted from spectrograms. The experimental results show that features obtained with a few patterns perform better than the majority of the standard set of acoustic features. We outperform the best performing acoustic feature, i.e. MFCC, with the increased number of patterns. Our analysis also demonstrates that the classification accuracies rise as the number of spectrogram partitions used to extract patterns grows. There are several directions to take to improve our paper in the future. We plan to expand the current acoustic feature set by using the INTERSPEECH 2013 configuration and to perform the cross-validation in a subject independent setting, i.e. leave-one-performer-out. We also would like to investigate the joint effect of the spectrogram-based features and acoustic features on the performance. Last but not least, we consider evaluating our approach on another dataset that contains more sentences with different content and illustrating cross-dataset performance of the presented technique.

References

1. Bhavan, A., Chauhan, P., Hitkul, Shah, R.R.: Bagged support vector machines for emotion recognition from speech. Knowl. Based Syst. **184**, 104886 (2019)
2. Chen, L., Mao, X., Xue, Y., Cheng, L.L.: Speech emotion recognition: features and classification models. Digit. Signal Proc. **22**(6), 1154–1160 (2012)
3. Gosztolya, G.: Using the Fisher vector representation for audio-based emotion recognition. Acta Polytechnica Hungarica **17**, 7–23 (2020)
4. Huang, K., Wu, C., Hong, Q., Su, M., Zeng, Y.: Speech emotion recognition using convolutional neural network with audio word-based embedding. In: ISCSLP, pp. 265–269 (2018)
5. Ishi, C.T., Kanda, T.: Prosodic and voice quality analyses of loud speech: differences of hot anger and far-directed speech. In: SMM, pp. 1–5 (2019)
6. Kaya, H., Fedotov, D., Yeşilkanat, A., Verkholyak, O., Zhang, Y., Karpov, A.: LSTM based cross-corpus and cross-task acoustic emotion recognition. In: INTERSPEECH, pp. 521–525 (2018)
7. Khalil, R.A., Jones, E., Babar, M.I., Jan, T., Zafar, M.H., Alhussain, T.: Speech emotion recognition using deep learning techniques: a review. IEEE Access **7**, 117327–117345 (2019)
8. Lin, J., Wu, C., Wei, W.: Emotion recognition of conversational affective speech using temporal course modeling. In: INTERSPEECH, pp. 1336–1340 (2013)

9. Liu, Y., Zheng, Y.F.: One-against-all multi-class SVM classification using reliability measures. In: IJCNN, vol. 2, pp. 849–854 (2005)
10. Livingstone, S.R., Russo, F.A.: The Ryerson audio-visual database of emotional speech and song (RAVDESS): a dynamic, multimodal set of facial and vocal expressions in North American English. PLoS ONE **13**(5), e0196391 (2018)
11. Mao, S., Tao, D., Zhang, G., Ching, P.C., Lee, T.: Revisiting hidden Markov models for speech emotion recognition. In: ICASSP, pp. 6715–6719 (2019)
12. Milgram, J., Cheriet, M., Sabourin, R.: "One against one" or "One against all": which one is better for handwriting recognition with SVMs? In: IWFHR (2006)
13. Platt, J.C., Cristianini, N., Shawe-Taylor, J.: Large margin DAGs for multiclass classification. In: ANIPS, pp. 547–553 (2000)
14. Satt, A., Rozenberg, S., Hoory, R.: Efficient emotion recognition from speech using deep learning on spectrograms. In: INTERSPEECH, pp. 1089–1093 (2017)
15. Spyrou, E., Nikopoulou, R., Vernikos, I., Mylonas, P.: Emotion recognition from speech using the bag-of-visual words on audio segment spectrograms. Technologies **7**(1), 20 (2019)
16. Trigeorgis, G., et al.: Adieu features? End-to-end speech emotion recognition using a deep convolutional recurrent network. In: ICASSP, pp. 5200–5204 (2016)
17. Tripathi, S., Beigi, H.S.M.: Multi-modal emotion recognition on IEMOCAP dataset using deep learning. arXiv abs/1804.05788 (2018)
18. Tripathi, S., Kumar, A., Ramesh, A., Singh, C., Yenigalla, P.: Deep learning-based emotion recognition system using speech features and transcriptions. arXiv abs/1906.05681 (2019)
19. Tzirakis, P., Zhang, J., Schuller, B.W.: End-to-end speech emotion recognition using deep neural networks. In: ICASSP, pp. 5089–5093 (2018)
20. Wang, Y., Hu, W.: Speech emotion recognition based on improved MFCC. In: CSAE, pp. 1–7 (2018)
21. Yang, H., Duan, L., Hu, B., Deng, S., Wang, W., Qin, P.: Mining top-k distinguishing sequential patterns with gap constraint. J. Softw. **26**(11), 2994–3009 (2015)
22. Yang, Z., Hirschberg, J.: Predicting arousal and valence from waveforms and spectrograms using deep neural networks. In: INTERSPEECH, pp. 3092–3096 (2018)
23. Zhang, B., Essl, G., Provost, E.M.: Recognizing emotion from singing and speaking using shared models. In: ACII, pp. 139–145 (2015)

Pragmatic Markers in Dialogue and Monologue: Difficulties of Identification and Typical Formation Models

Natalia Bogdanova-Beglarian[1], Olga Blinova[1,2],
Tatiana Sherstinova[1,2(✉)], Daria Gorbunova[1], Kristina Zaides[1],
and Tatiana Popova[1]

[1] Saint Petersburg State University, 7/9 Universitetskaya nab.,
St. Petersburg 199034, Russia
{n.bogdanova, o.blinova, t.sherstinova}@spbu.ru,
dgorbunova2@gmail.com,
kristina.zaides@student.spbu.ru, tipopoval3@gmail.com
[2] National Research University Higher School of Economics, 190068 St.
Petersburg, Russia

Abstract. The paper deals with new research findings on pragmatic markers (PMs) use in spoken Russian. The study is based on two speech corpora: "One Day of Speech" (ORD, which contains mainly dialogues), and "Balanced Annotated Collection of Texts" (SAT, which contains only monologues). We explored two annotated subcorpora consisting of 321,504 tokens and 50,128 tokens respectively. The main results are as follows: 1) the extended frequency lists of PMs were formed; 2) PMs, that are frequently used in both types of speech, were identified (e.g., hesitation markers like *tam 'there', tak 'that way'*), 3) the list of PMs, used primarily in monologue speech, was compiled (in this list there are such PMs as boundary ones *znachit 'well', nu vot 'well er', vs'o 'that's all'*); 4) the list of PMs, used primarily in dialogues, was made (among such PMs are, for example, "xeno"-markers *takoj 'like', grit 'says'* and meta-communicative markers like *vidish' 'you know', (ja) ne znaju 'don't know'*). Particular attention was paid to the variability of pragmatic markers, as well as to complex cases of their identification. Finally, the most common models of pragmatic markers formation (for single-word and multi-word PMs) were revealed.

Keywords: Russian everyday speech · Speech corpus · Pragmatic marker · Corpus annotation · Monologue · Dialogue

1 Introduction

Pragmatic markers (PM) are the units of spoken language that were subjected to the process of pragmaticalization, and as a result, have almost lost their original lexical and/or grammatical meaning. However, they have acquired the pragmatic meaning and started to perform special functions in the discourse: they mark the boundaries of the replicas (*znachit 'well', nu vot 'well er', vs'o 'that's all'*) or enter someone else's

© Springer Nature Switzerland AG 2020
A. Karpov and R. Potapova (Eds.): SPECOM 2020, LNAI 12335, pp. 68–78, 2020.
https://doi.org/10.1007/978-3-030-60276-5_7

speech (*takoy/takaja/takie 'this/those'*, *tipa (togo chto)* '*sort of that*', *grit* '*says*', etc.), verbalize the speaker's hesitation (*eto samoe 'whatchamacallit'*, *kak jego (jejo, ikh)* '*whatchamacallit*', *tam 'em'*), his reflection on what was said or is preparing to pronounce (*ili kak tam? 'or whatchamacallit?' ili chto? 'or what?' ili kak skazat'? 'or how to say?'*) or self-correction (*eto 'what' and eto samoe 'whatchamacallit'*), express meta-communication (*znaesh, 'you know', ponimaesh 'you know', da 'yes', (ja) ne znaju 'don't know', chto jeshcho? 'what else'*), etc. (for more details on the PMs typology used in the study, see: [12, 13]).

Differences between pragmatic markers and discourse words, or discourse markers (DMs), which are considered by both foreign and domestic linguistics as a very wide class of functional units (see, for example: [2, 3, 20–24]) are as follows:

1) PMs are used by the speaker unconsciously, at the level of speech automatism; DMs are placed into the text consciously, primarily with the aim of structuring it.
2) PMs do not have (or have a weakened, almost disappeared) lexical and/or grammatical meaning.

PMs are actually outside the system of parts of speech, even the category of particles, which also do not have a generalized grammatical meaning and common criteria for selection into this lexical and grammatical category[1]; DMs are full-fledged lexical and grammatical units of oral discourse.

3) PMs are used only in speech or in stylizations in a literary text; DMs can be found both in written text and in spontaneous speech.
4) PMs demonstrate the speaker's attitude to the process of speech generation or its result, verbalizing all difficulties and disfluencies, and are often meta-communicative units.

DMs either structure the text (parentheses, functional words), or convey the speaker's attitude to what he reports.

5) PMs, in all their functional diversity, are practically outside lexicographic fixation and usually remain outside of the scope of lingual didactics and various applied speech processing systems.

DMs are the part of traditional lexicography as lexemes on the one hand, and on the other hand are also considered in discourse studies as structuring statement operators.

To the corpus-based description of the PMs system the concepts of *the basic* (standard) *variant* of the PM, its *structural variants* and *the inflectional paradigm*, as well as the *PM model* were also introduced.

---

[1] The closest to the particles are "xeno"-markers, but they also differ from traditional particles, introducing someone else's speech into the narrative (*mol, de, deskat'*), for example, due to the fact that the "atavisms" of grammatical meaning are retained predominantly: variability by gender and number (*takoj/takaja/takie*) or by face and number (*gr'u, grit, grim*, etc.) [7]. B. Fraser also includes "xeno"-markers to pragmatic markers, not to particles [17].

2   PMs Annotation in the Corpus Material

Continuous annotation of PMs in both speech corpora was carried out at the following levels.

Level 1. PM – a pragmatic marker in the form as presented in the transcript (filled in the ELAN annotation environment).

Level 2. Function PM – PM functions that are indicated simultaneously, at the same level, in the alphabetical order (filled in the ELAN).

Level 3. Speaker PM – speaker's code (filled in ELAN).

Level 4. Comment PM – level of comments. This layer is intended for entering optional information, as well as for marking complex cases in which the identification of PMs and their functions was difficult (filled in ELAN).

Level 5. Standart – the standard (basic) version of a PM (without taking into account structural variants and/or inflectional paradigm) (filled out according to the results of uploading annotation levels into MS Excel).

Level 6. POS – part-of-speech marking of the original lexical unit from which the standard version of the PM originated. Implemented automatically using the MyStem morphoparser (Yandex-technology) and then verified manually[2].

Level 7. Model – the formation model for non-single-word PMs (filled in MS Excel).

Level 8. Frase – each implementation of the PM is brought into correspondence with the phrase context (filled in the ELAN).

In total, 136 communicative macro episodes were annotated for the ORD corpus, and 170 monologues for the SAT corpus. The total number of speakers in the annotated subcorpus is 257 from the ORD and 34 from the SAT.

For a detailed description of the system of pragmatic markers (PM) of Russian everyday speech, an information database PM-Database is being developed at St. Peters- burg State University. The database contains information related to the two speech corpora used in this study – "One Day of Speech" (ORD; dialogues) (see recent works about it: [6, 8–10]) and "Balanced Annotated Collection of Texts" (SAT; monologues), see: [4, 11]. The database has been carried out for the annotated volume of corpus material: 321 504 tokens (word usages) for the ORD corpus and 50 128 tokens (word usages) for the SAT.

---

[2] On the difficulties of part-of-speech tagging of corpus material, given the presence of a large number of pragmatic markers in it, which are only etymologically related to traditional parts of speech, see, for example: [14].

3 Features of PM as Elements of Oral Discourse and Complex *'er'*– *i vot, da vot, nu vot;* Cases of Identification

Almost all PMs are presented in the speech in their structural variants or various grammatical forms, cf.: VOT ZNAYESH *'you know'* – *ty znaesh', vot znaesh', nu znaete*, etc.; GOVORIT *'says'*– *govor'u, govorish', govorim*, etc.; VRODE *'like'*– *nu vrode, vrode kak, vrode by*.

In this case, the PMs grammar remains at the level of "atavisms" (see more on this in detail: [7]). So, for the marker GOVORIT *'says'*, only variability in persons and numbers is preserved in case of the present time (all these forms are usually reduced: *grit, gyt, gr'u, grish', gr'at...*); for the marker TAKOJ *'like'* – variability by gender and number (*takoj, takaja, takie*); for the marker ETO SAMOYE *'whatchamacallit'* – by gender, numbers and cases (*s etim samym, eta samja, k etim samym*), for VSE DELA *'and all that'* – by cases (practically, only instrumental form is possible – *so vsemi delami*).

Rare of the PMs studied do not reveal such structural variability: *von, prikin', i tak dalee, po idee* and some other.

On the other hand, the deictic marker VOT (...) VOT *'like ... this'* exists exclusively as a structural model, which is filled each time with a new unit: *vot **tak** vot, vot **takoj** vot, vot ots'uda vot*, and so on. The marker simply does not have a single basic (standard) form.

The set of identified PMs shows that they have different part-of-the-speech "origin"[3]: particles (*vot 'here', von 'there'*), verbs (*znat' 'to know', govorit' 'to speak', smotret' 'to look', dumat' 'to think'*), including gerund (*govor'a 'speaking'*), adverbs (*tam 'there', tak 'that way', kuda 'where'*), pronouns (*etot 'this', samyj 'the most', on 'he', ona 'she'*), conjunctions/prepositions (*tipa 'kind of', vrode 'like'*).

In this case, the PMs themselves are actually outside the part-of-speech classification, since they do not have those categorical attributes by which the units are distributed among the corresponding classes.

According to the results of a pilot data annotation performed by four different experts, see details: [12], complex cases of PMs identification were identified, which led to a partial review of the identified types of PMs. First of all, those categories of PM that underwent qualitative heterogeneity in terms of the diversity of the functions performed by them underwent audits, namely, various hesitative markers, marker *takoj*, replacement markers, clearly distinguishable from markers-approximators, identified in the original typology [5], as well as those PMs, in relation to which there was no unequivocal opinion of experts from the point of view of their assignment to this class of units (for example, rhythm-forming markers and "xeno"-markers).

As a result, it was decided to clarify the principles for determining functions for some PMs. For example, for a marker *takoj 'like'* two main functions are allocated:

---

[3] Such relationships between markers and outwardly corresponding full-valued lexical units are considered by B. Fraser as *homophony* [18].

- "xeno"-marker (K): prishol /a nashi promoutery s gvozdej **takiye** na men'a /o... ogo /ty otkuda takoj?
- hesitative marker (H): ja pomn'u chto vy jeshcho togda govorili /chto on **takoj** / ***P** @ *somnitel'nyj* @ *slozhnyj /da /on slozhnyj*.

In addition, at the level of comments, this marker has ceased to be assigned additional functions: F – figurativeness and E – enantiosemia, which can be considered more as shades of meanings, rather than functions of the word.

Criteria were also identified for the analysis of complex cases of classifying a marker to a specific class of PMs.

So, the rhythm-forming function (RF) is now attributed to single, mostly monosyllabic markers (*tam, tak, vot, koroche*) in those cases when no other functions (hesitative, boundary, etc.) are found in this context, cf.: *vs'o ravno vs'a eta utilizacia* **koroche** *ona* **tam** *maksimum davala garantiju* **tam***na 50 let*; *sobaka ikh sjedaet //daby ne dostalos'* **vot** *jejo sopernice.*

Such cases do not include these markers in the chains (*vot koroche, tam vot* and so on), which are not actually rhythm-forming PMs, but polyfunctional markers with several combined functions.

In the course of work, for a separate group of PMs, more clear, formalized signs of determining the status of functional units homonymous with uses that are not pragmatic markers were revealed. So, for a "xeno"-marker *govorit 'says'*, the following criteria can be distinguished:

a) this PM is not inclined to occupy the leftmost position in the replica, more often it is a repeated element; the first use *govorit* in case of repetition is not considered as a PM, but qualifies as a predicate (sometimes with a pronounced subject), but the second and subsequent usages can already be "xeno"-markers;
b) in PMs there are no dependent words or other distribution, cf.:

- *govorit, chto pojdet tuda* – not a PM, but the verb predicate in the main part of the complex sentence;
- *govorit /pojdet tuda* – not a PM, but a verb predicate in the first part of a complex nonunion sentence;
- *govorit /eto samoe /poshol tuda i uvidel* – a PM.

The carried out optimization of the typology of pragmatic markers contributes to a more accurate correspondence of the theoretical scheme and the results of the practical processing of corpus material.

In addition, some common markup problems for pragmatic markers were discovered:

- the complicated division of spontaneous speech into syntagmas, which cannot always be performed unambiguously, but this is important for differentiating, for example, starting, navigation, and final boundary PM (B);

- the formal similarity of PM and full-valued units of discourse, which undergo pragmaticalization in speech and are sometimes at different stages on the evolution from lexemes to pragmatems[4];
- the polyfunctionality of the majority of PMs in Russian speech and the need to differentiate the main and additional functions of each PM in specific uses;
- "magnetism" of PMs in oral discourse, the attraction of "single-functional" (synonymous at the function level) units to each other, and the need to distinguish heterogeneous PM, on the one hand, and PM chains, on the other: ETO KAK JEGO 'what whatchamacallit' (one marker) or ETO+KAK JEGO 'er+whatchamacallit' (a chain of markers).

According to the results of the pilot annotation of the corpus material, a qualitative heterogeneity of the PMs was revealed, which appears both in terms of the variety of functions performed by them, and in terms of the uniqueness of their allocation and assignment of these units to the pragmatic elements of oral dis-course. This poses the urgent task of creating, on the basis of the PMs, a special dictionary of pragmatic markers, which would reflect their qualitative differentiation.

4 PMs in Dialogue and Monologue

The analysis of PMs frequency lists (common to the two corpora and separate for various speakers' groups) showed that the statistically significant differences in the PMs usage in dialogue and monologue do indeed exist. Preliminary data on functioning of different PMs types in two forms of Russian speech have already been received at the initial investigation phase, on the smaller sample: 60,000 to-kens from ORD (dialogic speech) and 15,000 tokens from SAT (monologic speech), see: [12]; this research allowed verifying these results on the substantially extended volume of corpus material and their confirming, for the most part. The tops of the PMs frequency lists from two corpora currently are as fol-lows – see Table 1 and Table 2.

---

[4] Linguists have written more than once about the prospects of the scaling method when describing the fate of particular units in our language/speech, cf.: "Recently, the idea that it is advisable to abandon the "Procrustean bed" of a clear and uncompromising scheme and prefer the method quantitative assessment, according to which each linguistic phenomenon should be described in case of the place it occupies on the gradual transition scale " [1: 89]. Pragmatic markers on their "way from classical lexemes" often go through the *grammaticalization* [15, 16, 19], cf.:

• *skazat'* – a verb in all the richness of its meanings and grammatical forms→*skazhem* as a *parenthesis* (*ja zhe ne otlichu* **tak skazhem** / *tadzhika ot uzbeka*) (frozen form, the result of grammaticalization)→(...) *skazhem* (...) as a PM, more often reflective (R) (*tam slozhnaja publika* / **skazhem tak**) or hesitative (H) (*nu(:)* / **tam skazhem** / **P nu* / *ne znaju* / *pajaet chto-to*) (the result of pragmaticalization);

• *sejchas* – an adverb with a number of meanings and a grammatical characteristic→*shchas* (usually in a reduced form) – interjection (*Ja? Xa!* **Shchas pr'am!**) (the result of grammaticalization)→*shchas-shchas* as a hesitative marker (H) (usually with multiple repetition, reduction) (*tak* / **shchas-shchas-shchas-shchas** / *podozhdi*) (the result of pragmaticalization).

Table 1. The top of the PMs frequency list in ORD (for 300,000 tokens).

Rank	PM	Frequency	IPM
1.	*vot*	1205	4017
2.	*tam*	657	2190
3.	*da*	353	1177
4.	*tak*	271	903
5.	*kak by*	270	900
6.	*govorit*	230	767
7.	*znayesh'*	164	547
8.	*slushaj*	160	533
9.	*znachit*	158	527
10.	*eto*	158	527

On the material of the annotated ORD-subcorpus, 59 basic PMs used in 315 different contexts have been identified. On the material of the annotated SAT-subcorpus, 30 basic PMs used in 133 different contexts have been identified. One can see that basic PMs in the monologic speech were two times less than in the dialogues.

Table 2. The top of the PMs frequency list in SAT (for 50,000 tokens).

Rank	PM	Frequency	IPM
1.	*vot*	332	6640
2.	*znachit*	99	1980
3.	*tam*	63	1260
4.	*nu vot*	46	920
5.	*kak by*	40	800
6.	*tak*	33	660
7.	*da*	27	540
8.	*vs'o*	25	500
9.	*skazhem tak*	20	400
10.	*v obshchem-to*	17	340

It can be seen that the common for both corpora PMs in this area are five following: VOT (the 1st rank in both corpora), TAM (the 2nd and the 3rd ranks respectively), KAK BY (the 5th rank in both corpora), TAK (the 4th and the 6th ranks), DA (the 3rd and the 7th ranks). Apparently, these PMs are the most typical for any form of speech.

Interestingly, the repeatability of basic PMs has already begun in the top of the frequency list of SAT (monologues): *nu vot* as a variant of *vot* (the 4th rank), *skazhem tak* as a variant of *skazhem* (the 9th rank), and *v obshchem-to* as a variant of *v obshchem* (the 10th rank). Whereas in the ORD-subcorpus (dialogues) PMs start repeating in their usage only from the 11th rank.

The difference between the content of the top of the PMs frequency lists for two corpora can be easily explained.

Thus, in *dialogic speech* the speaker constantly must draw and keep the attention of the interlocutor up, therefore, the meta-communicative PMs *znayesh'* and *slushaj* (the 7th and the 8th ranks) are such frequent. In dialogue, we often have to transmit someone else's (in a broad sense) speech; consequently, the high frequency of the "xeno"-PM *govorit* has been observed (in different, mostly reduced, forms) (the 6th rank). In monologue, this PM-usage can be triggered just by the specificity of retelling or pictures that have been offered for description to the speaker.

In *monologues* the boundary PMs are extremely frequent: the start marker *znachit* (the 2nd rank) and the final marker *vs'o* (the 8th rank) help to identify the boundaries of speaker's monologue.

Regarding the PMs functions, the most frequently used in both speech forms is the *hesitative function* (H): 1860 such usages in ORD and 388 in SAT. The further on frequency in ORD are the *meta-communicative markers* (M; 701), followed by the polyfunctional PMs with two functions – *hesitative* and *boundary* (HB; 676); after are the purely *boundary* markers (B; 648). In SAT-corpus, the *boundary* PM-function takes the second place as well. This function helps to structure speaker's monologue: BH (shared with hesitative; 199 contexts) and B (135).

5   PMs Formation Models

The analysis of the corpus material let receiving as well the list of the most frequent formation models of pragmatic markers (for single-word and multi-word PM).

Single-word PMs derive from the regular parts of speech which in this case lose or weaken their lexical and grammatical meaning and move to the class merely pragmatic units of oral discourse: VOT, DA (from particles), ETO (from pronouns), TAK, TAM (from adverbs), ZNAYES', PRIKIN' (from verbs), V OBSHCHEM (from adjectives), V PRINCIPE (from nouns), etc.

Occasionally, these source full-value units turn out to be non-single-word (words and expressions) – and equally they become the non-single-word PMs in speech: TO-S'O, TUDA-S'UDA, KAK BY, VRODE BY, V OBSHCHEM, V PRINCIPE, KAK GOVORITS'A, TAK SKAZAT', KOROCHE GOVOR'A, SOBSTVENNO GOVOR'A, VOOBSHCHE GOVOR'A, I TAK DALEE, VS'O TAKOE, VSE DELA, NA SAMOM DELE, etc. Their formation models as well can be described; however, it is not included in the scope of tasks of this research being not interesting from the point of view of colloqualistics.

The deictic marker VOT (...) VOT is of particular importance among other PM. It exists merely as a structural model which is filled by a new unit each time: VOT TAK VOT, VOT TAKOJ VOT, VOT OTS'UDA VOT, etc. This marker, as mentioned above, does not have any particular basic form. Although, it is possible to include other units in this basic structure in speech: VOT ETA KHREN' VOT.

The formation of multi-word PMs happen in different ways, which allows distinguishing various PMs formation models:

1) initially multi-word PM, which exist in these basic versions (but not in the source "lexicographic" version):

- ETO SAMOE – with the possibility of gender, number, and case inflection;
- KAK JEGO (JEJO, IKH) – with the possibility of gender and number inflection; having the question form, in general, are not actually questions;
- KAK ETO? the rhetorical question;
- KAK SKAZAT'? the rhetorical question;
- KAK ETO NAZYVAETS'A? the rhetorical question;
- CHTO JESHCHO? the rhetorical question;

2) combination (chain) of two or more single-word PM:

- NU VOT, NU TAM, NU TAK, VOT TAK, NU ZNAESH', TAK SKAZ-HEM, SKAZHEM TAK, SKAZHEM TAM, VOT SKAZHEM, ZNAESH' TAM, VOT KAK BY, VOT SKAZHEM TAK, NU NE ZNAJU, TAM TIPA, NU KOROCHE, ZNACHIT VOT, V PRINCIPE VS'O;

3) combination (chain) of single-word and multi-word PM: NU KAK SKAZAT'? KAK JEGO TAM? NU VOT ETI VOT, NU (JA) NE ZNAJU TAM;

4) addition of the personal pronoun with a weakened lexical and grammatical meaning:

- (JA) NE ZNAJU;
- (JA) (NE) DUMAJU (CHTO);
- (TY) ZNAESH', PONIMAESH', VIDISH'…;
- (TY) PREDSTAV', PRIKIN'…;
- CHTO (TEBE) JESHCHO SKAZAT'?

5) addition of the emphatic particles/conjunctions: I VSE DELA, I VS'O TAKOE, A VOT, NU I VS'O, I TO I S'O, JA UZH NE ZNAJU TAM, TY ZH PONI MAESH';

6) addition of the non-personal pronoun: VS'O TAKOE PROCHEE, TIPA TOGO/ETOGO, VRODE TOGO, TAKOJ KAKOJ-TO, KAK (BY) ETO SKAZAT'?

7) addition of the conjunction CHTO (CHEGO): DUMAJU CHTO, BOJUS' CHTO, TIPA TOGO CHTO, VRODE TOGO CHTO, ZNAESH' CHTO/CHEGO;

8) addition of the parentheses: VS'O NAVERNOE, VS'O POZHALUJ;

9) addition of the interjection: OJ SLUSHAJ;

10) loss of the gerund GOVOR'A: KOROCHE, SOBSTVENNO, VOOBSHCHE;

11) reduplication: DA-DA-DA, NA-NA-NA, SHCHAS-SHCHAS-SHCHAS, TE-TE-TE, OP-OP-OP, BLA-BLA-BLA, TAK-TAK-TAK, ETO-ETO-ETO;

12) ILI+(more often) the rhetorical question:

- ILI KAK JEGO?
- ILI KAK TAM?
- ILI CHTO?
- ILI KAK SKAZAT'?
- ILI ETOT?
- NU ILI NE ZNAJU.

The set of these models can make the description of PMs (their database) of Russian every day speech undoubtedly more comprehensive.

6 Conclusion

The paper presents the new results of investigation of pragmatic markers in Russian speech on the material of two speech corpora: "One Day of Speech" (ORD; dialogues) and "Balanced Annotated Collection of Texts" (SAT; monologues). During the analysis, the approaches to annotation of the particular PMs (for instance, "xeno" markers, replacement markers, markers-approximators, and rhythm-forming markers) in the corpus material were specified. The frequency lists of PMs from dialogues and monologues were received. The markers common to both types of speech (primarily, hesitative markers such as vot, tam, tak) and common to a large extent to monologue (boundary markers like znachit, nu vot, vs'o) or to dialogue ("xeno" and meta-communicative markers) were revealed. The most frequent PMs formation models (for single-word and multi-word markers) were identified.

Acknowledgements. The presented research was supported by the Russian Science Foundation, project #18-18-00242 "Pragmatic Markers in Russian Everyday Speech".

References

1. Arutyunova, N.D.: On criteria for selecting analytical forms. Analytical Constructions in Languages of Different Types, pp. 89–93. Nauka, Moscow-Leningrad (1965). (in Russian)
2. Baranov, A.N., Plungian, V.A., Rakhilina, Ye.V.: Guide to the Russian Discourse Words. Pomovsky i partnery, Moscow (1993). (in Russian)
3. Beliao, J., Lacheret, A.: Disfluency and discursive markers: when prosody and syntax plan discourse. In: DiSS 2013: The 6th Workshop on Disfluency in Spontaneous Speech, vol. 54, no. 1, pp. 5–9. Stockholm, Sweden (2013)
4. Bogdanova-Beglarian, N.V. (ed.): Speech Corpus as a Base for Analysis of Russian Speech. Collective Monograph. Part 1. Reading. Retelling. Description, St. Petersburg (2013). (in Russian)
5. Bogdanova-Beglarian, N.V.: Pragmatems in spoken everyday speech: definition and general typology. Perm University Herald. Russ. Foreign Philol. 3(27), 7–20 (2014). (in Russian)
6. Bogdanova-Beglarian, N.V. (ed.): Everyday Russian Language: Functioning Features in Different Social Groups. Collective Monograph. LAIKA, St. Petersburg (2016). (in Russian)
7. Bogdanova-Beglarian, N.V.: Grammatical Atavisms of the pragmatic markers in Russian speech. In: Glazunova, O.I. , Rogova, K.A. (eds.). Russian Grammar: Structural Organization of Language and the Processes of Language Functioning, pp. 436–446. URSS, Moscow (2019). (in Russian)
8. Bogdanova-Beglarian, N., Sherstinova, T., Blinova, O., Martynenko, G.: An exploratory study on sociolinguistic variation of Russian everyday speech. In: Ronzhin, A., Potapova, R., Németh, G. (eds.) SPECOM 2016. LNCS (LNAI), vol. 9811, pp. 100–107. Springer, Cham (2016). https://doi.org/10.1007/978-3-319-43958-7_11

9. Bogdanova-Beglarian, N., Sherstinova, T., Blinova, O., Ermolova, O., Baeva, E., Martynenko, G., Ryko, A.: Sociolinguistic extension of the ORD corpus of Russian everyday speech. In: Ronzhin, A., Potapova, R., Németh, G. (eds.) SPECOM 2016. LNCS (LNAI), vol. 9811, pp. 659–666. Springer, Cham (2016). https://doi.org/10.1007/978-3-319-43958-7_80

10. Bogdanova-Beglarian, N., Blinova, O., Sherstinova, T., Martynenko, G.: Corpus of Russian everyday speech one day of speech: present state and prospects. In: Moldovan, A.M., Plungyan, V.A. (eds.) Proceedings of the V.V. Vinogradov Russian Language Institute. Russian National Corpus: Research and Development, Moscow, vol. 21, pp. 101–110 (2019). (in Russian)

11. Bogdanova-Beglarian, N., Blinova, O., Zaides, K., Sherstinova, T.: Balanced annotated collection of texts (SAT): studying the specifics of Russian monological speech. In: Moldovan, A.M., Plungyan, V.A. (eds.) Proceedings of the V.V. Vinogradov Russian Language Institute. Russian National Corpus: Research and Development, Moscow, vol. 21, pp. 111–126 (2019). (in Russian)

12. Bogdanova-Beglarian, N., Blinova, O., Martynenko, G., Sherstinova, T., Zaides, K., Popova, T.: Annotation of pragmatic markers in the Russian speech corpus: problems, searches, solutions, results. In: Selegey, V.P. (ed.) Computational Linguistics and Intellectual Technologies: Papers from the Annual International Conference Dialogue, Moscow, vol. 18 and 25, pp. 72–85 (2019). (in Russian)

13. Bogdanova-Beglarian, N.V., et al.: Pragmatic markers of russian everyday speech: the revised typology and corpus-based study. In: Balandin, S., Niemi, V., Tuytina, T. (eds.) Proceedings of the 25th Conference of Open Innovations Association FRUCT, Helsinki, Finland, pp. 57–63 (2019)

14. Bogdanova-Beglarian, N.V., Zaides, K.D.: Pragmatic markers and parts of speech: on the problems of annotation of the speech corpus. In: International Workshop Computational Linguistics (CompLing-2020), June 19–20 2020, St. Petersburg. IMS-2020 Proceedings. CEUR (http://ceur-ws.org/) (2020) (in Print)

15. Degand, L., Evers-Vermeul, J.: Grammaticalization or pragmaticalization of discourse markers? More than a terminological issue. J. Hist. Pragmatics 16(1), 59–85 (2015)

16. Diewald, G.: Pragmaticalization (Defined) as Grammaticalization of Discourse Functions. Linguistics 49(2), 365–390 (2011)

17. Fraser, B.: Pragmatic markers. Pragmatics 6(2), 167–190 (1996)

18. Fraser, B.: Towards a theory of discourse markers. In: Fischer, K. (ed.) Approaches to Discourse Particles. Studies in Pragmatics, vol. 1, pp. 189–204. Elsevier, Oxford (2006)

19. Günther, S., Mutz, K.: Grammaticalization vs. Pragmaticalization? The development of pragmatic markers in German and Italian. In: Bisang, W., Himmelmann, N.P., Wiemer, B. (eds.) What Makes Grammaticalization? A Look from its Fringes and its Components, pp. 77–107. Language Arts & Disciplines, Berlin (2004)

20. Kiseleva, K.L., Payar, D.: Discourse Russian Words: an Experience of Context and Semantic Description [Diskursivnye slova russkogo jazyka: opyt kontekstno-semanticheskogo opisania]. Metatekst, Moscow (1998). (in Russian)

21. Kiseleva, K.L., Payar, D.: Discourse Russian Words: Variation and Semantic Unity [Diskursivnye slova russkogo jazyka: varjirovanie i semanticheskoe edinstvo]. Azbukovnik, Moscow (2003)

22. Lenk, U.: Marking Discourse Coherence: Functions of Discourse Markers in Spoken English. Narr, Tuebingen (1998)

23. Schiffrin, D.: Discourse Markers. Cambridge University Press, Cambridge (1996)

24. Schourup, L.: Discourse Markers. Lingua 107, 227–265 (1999)

Data Augmentation and Loss Normalization for Deep Noise Suppression

Sebastian Braun$^{(\boxtimes)}$ (ID) and Ivan Tashev (ID)

Microsoft Research, Redmond, WA, USA
{sebastian.braun,ivantash}@microsoft.com
https://www.microsoft.com/en-us/research/group/audio-and-acoustics-research-group

Abstract. Speech enhancement using neural networks is recently receiving large attention in research and being integrated in commercial devices and applications. In this work, we investigate data augmentation techniques for supervised deep learning-based speech enhancement. We show that not only augmenting SNR values to a broader range and a continuous distribution helps to regularize training, but also augmenting the spectral and dynamic level diversity. However, to not degrade training by level augmentation, we propose a modification to signal-based loss functions by applying sequence level normalization. We show in experiments that this normalization overcomes the degradation caused by training on sequences with imbalanced signal levels, when using a level-dependent loss function.

Keywords: Data augmentation · Speech enhancement · Deep noise suppression

1 Introduction

Speech enhancement using neural networks has recently seen large attention and success in research [9,16] and is being implemented in commercial applications also targeting real-time communication. An exciting property of deep learning-based noise suppression is that it also reduces highly non-stationary noise and background sounds such as barking dogs, banging kitchen utensils, crying babies, construction or traffic noise, etc. This has not been possible so far using single-channel statistical model-driven speech enhancement techniques that often only reduce quasi-stationary noise [2,4,8]. Notable approaches towards real-time implementations have been proposed e. g. in [12–14,17,20].

The dataset is a key part of data-driven learning approaches, especially for supervised learning. It is a challenge to build a dataset that is large enough to generalize well, but still represents the expected real-world data sufficiently. Data augmentation techniques can not only help to control the amount of data, but is also necessary to synthesize training data that represents all effects encountered in practice.

© Springer Nature Switzerland AG 2020
A. Karpov and R. Potapova (Eds.): SPECOM 2020, LNAI 12335, pp. 79–86, 2020.
https://doi.org/10.1007/978-3-030-60276-5_8

While in many publications, data corpus generation is only roughly outlined due to lack of space, or often exclude several key practical aspects, we direct this paper on showing contributions on several augmentation techniques when synthesizing a noisy and target speech corpus for speech enhancement. In particular, we show the effects of increasing the SNR range and using a continuous instead of discrete distribution, spectral augmentation by applying random spectral shaping filters to speech and noise, and finally level augmentation to increase robustness of the network against varying input signal levels.

As we found that level augmentation can decrease the performance when using signal-level dependent losses, we propose a normalization technique for the loss computation that can be generalized to any other signal-based loss. We show in experiments on the CHIME-2 challenge dataset that the augmentation techniques and loss normalization substantially improve the training procedure and the results.

In this paper, we first introduce a the general noise suppression task in Sect. 2. In Sect. 3, we describe the used real-time noise suppression system based on a recurrent network, that works on a single frame in - single frame out basis, i. e. requires no look-ahead and memory buffer, and describe the training setup. In Sect. 4, we describe the used loss function and propose a normalization to remove the signal level dependency of the loss. In Sect. 5, we describe augmentation techniques for signal-to-noise ratio (SNR), spectral shaping, and sequence level dynamics. The experiments are shown in Sect. 6, and Sect. 7 concludes the paper.

2 Deep Learning Based Noise Suppression

In a pure noise reduction task, we assume that the observed signal is an additive mixture of the desired speech and noise. We denote the observed signal $X(k, n)$ directly in the short-time Fourier transform (STFT) domain, where k and n are the frequency and time frame indices as

$$X(k, n) = S(k, n) + N(k, n), \tag{1}$$

where $S(k, n)$ is the speech and $N(k, n)$ is the disturbing noise signal. Note that the speech signal $S(k, n)$ can be reverberant, and we only aim at reducing additive noise.

The objective is to recover an estimate $\widehat{S}(k, n)$ of the speech signal by applying a filter $G(k, n)$ to the observed signal by

$$\widehat{S}(k, n) = G(k, n) X(k, n). \tag{2}$$

The filter $G(k, n)$ can be either a real-valued suppression gain, or a complex-valued filter. While the former option (also known as *mask*) only recovers the speech amplitude, a complex filter could potentially also correct the signal phase. In this work, we use a suppression gain.

3 Network and Training

We use a rather straightforward recurrent network architecture based on gated
recurrent units (GRUs) [1] and feed forward (FF) layers, similar to the core
architecture of [19] without convolutional encoder layers. Input features are the
logarithmic power spectrum $P = \log_{10}(|X(k,n)|^2 + \epsilon)$, normalized by the global
mean and variance of the training set. We use a STFT size of 512 with 32 ms
square-root Hann windows and 16 ms frame shift, but feed only the relevant 255
frequency bins into the network, omitting 0th and highest (Nyquist) bins, which
do not carry useful information. The network consists of a FF embedding layer,
two GRUs, and three FF mapping layers. All FF layers use rectified linear unit
(ReLU) activations, except for the last output layer. When estimating a real-
valued suppression gain, a *Sigmoid* activation is used to ensure positive output.
The network architecture is shown in Fig. 1, and has 2.8 M parameters.

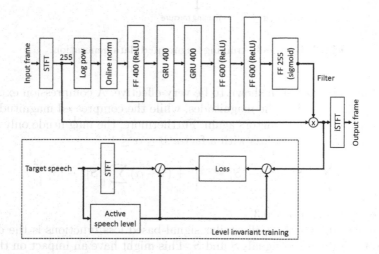

Fig. 1. Network architecture and enhancement system and training procedure.

The network was trained using the AdamW optimizer [7] with an initial
learning rate of 10^{-4}, which was dropped by a factor of 0.9 if the loss plateaued
for 5 epochs. The training was monitored every 10 epochs using a validation
subset. The best model was chosen based on the highest perceptual evaluation
of speech quality (PESQ) [6] on the validation set. All hyper-parameters were
optimized by a grid search and choosing the best performing parameter for PESQ
on the validation set.

4 Level Invariant Normalized Loss Function

The speech enhancement loss function is typically a distance metric between the
enhanced and target spectral representations. The dynamically compressed loss

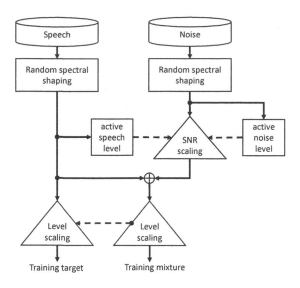

Fig. 2. On-the-fly training augmented data generation.

proposed in [3,18] has been shown to be very effective. A compression exponent of $0 < c \leq 1$ is applied to the magnitudes, while the compressed magnitudes are combined with the phase factors again. Furthermore, the magnitude only loss is blended with the complex loss with a factor $0 \leq \alpha \leq 1$.

$$\mathcal{L} = \alpha \sum_{k,n} \left| |S|^c e^{j\varphi_S} - |\widehat{S}|^c e^{j\varphi_{\widehat{S}}} \right|^2 + (1 - \alpha) \sum_{k,n} \left| |S|^c - |\widehat{S}|^c \right|^2. \tag{3}$$

We chose $c = 0.3$ and $\alpha = 0.3$.

A common drawback of all similar signal-based loss functions is the dependency on the level of the signals S and $\widehat{S}$. This might have an impact on the loss when computing the loss over a batch of several sequences, which exhibit large dynamic differences. It could be that large signals dominate the loss, creating less balanced training.

Therefore, we propose to normalize the signals S and X by the active signal level of each utterance, before computing the loss. The normalized loss is computed as given by (3), but using the normalized signals $\tilde{S} = \frac{S}{\sigma_S}$ and $\tilde{X} = \frac{X}{\sigma_S}$, where σ_S is the active speech level per utterance, i.e. the target speech signal standard deviation computed only for active speech frames. Note that this normalization does not affect the input features of the network: they still exhibit the original dynamic levels.

5   Data Augmentation Techniques

Especially for small and medium-scale datasets, augmentation is a powerful tool to improve the results. In the case of supervised speech enhancement training,

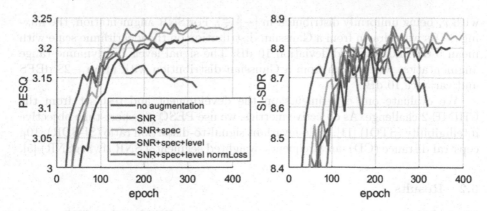

Fig. 3. Validation metrics for various augmentation techniques. (Color figure online)

where the actual noisy audio training data is generated synthetically by mixing speech and noise, there are some augmentation steps, which are essential to mimic effects on data encountered in the wild. Disregarding reverberation, we need to be able to deal with different SNRs, audio levels, and filtering effects that can be caused e.g. by acoustics (room, occlusion), or the recording device (microphone, electronic transfer functions).

Our augmentation pipeline is shown in Fig. 2. Before mixing speech with noise, we applied random biquad filters [14] to each noise sequence and speech sequences separately to mimic different acoustic transmission effects. From these signals, active speech and noise levels are computed using a level threshold-based voice activity detector (VAD). Speech and noise sequences are then mixed with a given SNR on-the-fly during training. After mixing, the mixture is scaled using a given level distribution. The clean speech target is scaled by the same factor as the mixture. The data generation and augmentation procedure is depicted in Fig. 2.

6 Experiments

6.1 Dataset and Experimental Setup

We used the CHIME-2 WSJ-20k dataset [15], which is currently, while only being of medium size, the only realistic self-contained public dataset including matching reverberant speech and noise conditions. The dataset contains 7138, 2418, and 1998 utterances for training, validation and testing, respectively. The utterances are reverberant using binaural room impulse responses, and noise from the same rooms was added with SNRs in the range of −6 to 9 dB in the validation and test sets. We used only the left channel for our single-channel experiments.

The spectral augmentation filters are designed as proposed in [14] by

$$H(z) = \frac{1 + r_1 z^{-1} + r_2 z^{-2}}{1 + r_3 z^{-1} + r_4 z^{-2}} \tag{4}$$

with r_i being uniformly distributed in $[-\frac{3}{8}, \frac{3}{8}]$. For SNR augmentation, the mixing SNRs were drawn from a Gaussian distribution on the logarithmic scale with mean 5 dB and standard deviation 10 dB. The signal levels for dynamic range augmentation were drawn from a Gaussian distribution with mean -28 dBFS and variance 10 dB.

We evaluate our experiments on the development and test set from the CHIME-2 challenge. As objective metrics, we use PESQ [6], short-time objective intelligibility (STOI) [11], scale-invariant signal-to-distortion ratio (SI-SDR) [10], cepstral distance (CD), and frequency-weighted segmental SNR (fwSegSNR) [5].

6.2   Results

Figure 3 shows the training progression in terms of PESQ and SI-SDR on the validation set. The blue curve shows training on the original data without any augmentation, mixed the 6 different SNR levels between -6 and 9 dB. We can see that the validation metrics decrease after 150 epochs. When applying SNR augmentation (red curve) with a broader and continuous distribution, we prevent the early validation decrease and can train for 280 epochs. Further, adding spectral augmentation increases the validation PESQ slightly. However, the level augmentation when training with standard loss (3) (purple curve) decreases the performance compared to SNR and spectral augmentation only. We attribute this effect to the large level imbalance per batch, which affects the standard level-dependent loss function. When computing the loss from normalized signals (green curve), this drawback is overcome and we obtain similar or even slightly better results than the yellow curve, but making the system robust to varying input signal levels. The validation SI-SDR shows similar behavior as PESQ.

Table 1. Evaluation metrics on CHIME-2 test set.

Augmentation	Loss	PESQ	STOI	CD	SI-SDR	fwSegSNR
–	Noisy	2.29	81.39	5.46	1.92	16.96
None	Standard	3.27	91.20	2.90	9.48	23.57
SNR	Standard	3.31	91.40	**2.85**	9.45	23.30
SNR+spec	Standard	**3.32**	91.57	2.89	9.55	23.30
SNR+spec+level	Standard	3.30	**91.68**	2.87	**9.57**	**23.48**
SNR+spec+level	Normalized	3.31	91.55	2.89	9.52	23.41

In Table 1 shows the results on the CHIME-2 test set. The enhancement systems improve all results substantially over the noisy input. Adding SNR augmentation adds a gain of 0.05 PESQ. As in the development set, spectral augmentation adds an additional minor improvement. Interestingly, on the test set, the normalized loss shows no influence on the results. We assume this is due to that fact that the given test set does not exhibit largely varying signal levels.

7   Conclusion

We have shown the effectivity of data augmentation techniques for supervised deep learning-based speech enhancement by augmenting the SNR, spectral shapes, and signal levels. As level augmentation degrades the performance of the learning algorithm when using level-dependent losses, we proposed a normalization technique for the loss, which is shown to overcome this issue. The experiments were conducted using a real-time capable recurrent neural network on the reverberant CHIME-2 dataset. Future work will also investigate augmentation for acoustic conditions with reverberant impulse responses.

References

1. Cho, K., Merriënboer, B.V., Bahdanau, D., Bengio, Y.: On the properties of neural machine translation: encoder-decoder approaches. In: Proceedings of the Eighth Workshop on Syntax, Semantics and Structure in Statistical Translation (SSST-8) (2014)
2. Ephraim, Y., Malah, D.: Speech enhancement using a minimum-mean square error short-time spectral amplitude estimator. IEEE Trans. Acoust. Speech Signal Process. **32**(6), 1109–1121 (1984)
3. Ephrat, A., et al.: Looking to listen at the cocktail party: a speaker-independent audio-visual model for speech separation. ACM Trans. Graph. **37**(4), 112:1–112:11 (2018)
4. Gerkmann, T., Hendriks, R.C.: Noise power estimation based on the probability of speech presence. In: Proceedings of the IEEE Workshop on Applications of Signal Processing to Audio and Acoustics (WASPAA), pp. 145–148, October 2011
5. Hu, K., Divenyi, P., Ellis, D., Jin, Z., Shinn-Cunningham, B.G., Wang, D.: Preliminary intelligibility tests of a monaural speech segregation system. In: Proceedings of the Workshop on Statistical and Perceptual Audition, Brisbane, September 2008
6. ITU-T: Recommendation P.862: Perceptual evaluation of speech quality (PESQ), an objective method for end-to-end speech quality assessment of narrowband telephone networks and speech codecs, February 2001
7. Loshchilov, I., Hutter, F.: Decoupled weight decay regularization. In: International Conference on Learning Representations (2019). https://openreview.net/forum?id=Bkg6RiCqY7
8. Martin, R.: Noise power spectral density estimation based on optimal smoothing and minimum statistics. IEEE Trans. Speech Audio Process. **9**, 504–512 (2001)
9. Reddy, C.K.A., et al.: The INTERSPEECH 2020 deep noise suppression challenge: datasets, subjective speech quality and testing framework. In: Proceedings of the INTERSPEECH 2020 (2020, to appear)
10. Roux, J.L., Wisdom, S., Erdogan, H., Hershey, J.R.: SDR - half-baked or well done? In: Proceedings of the IEEE International Conference on Acoustics, Speech and Signal Processing (ICASSP), pp. 626–630, May 2019
11. Taal, C.H., Hendriks, R.C., Heusdens, R., Jensen, J.: An algorithm for intelligibility prediction of time-frequency weighted noisy speech. IEEE Trans. Audio Speech Lang. Process. **19**(7), 2125–2136 (2011)
12. Tan, K., Wang, D.: A convolutional recurrent neural network for real-time speech enhancement. In: Proceedings of the Interspeech, pp. 3229–3233 (2018)

13. Tu, Y.H., Tashev, I., Zarar, S., Lee, C.: A hybrid approach to combining conventional and deep learning techniques for single-channel speech enhancement and recognition. In: Proceedings of the IEEE International Conference on Acoustics, Speech and Signal Processing (ICASSP), pp. 2531–2535, April 2018
14. Valin, J.: A hybrid DSP/deep learning approach to real-time full-band speech enhancement. In: 20th International Workshop on Multimedia Signal Processing (MMSP), pp. 1–5, August 2018
15. Vincent, E., Barker, J., Watanabe, S., Nesta, F.: The second 'CHIME' speech separation and recognition challenge: datasets, tasks and baselines. In: Proceedings of the IEEE International Conference on Acoustics, Speech and Signal Processing (ICASSP), June 2012
16. Wang, D., Chen, J.: Supervised speech separation based on deep learning: an overview. IEEE/ACM Trans. Audio Speech Lang. Process. **26**(10), 1702–1726 (2018)
17. Wichern, G., Lukin, A.: Low-latency approximation of bidirectional recurrent networks for speech denoising. In: Proceedings of the IEEE Workshop on Applications of Signal Processing to Audio and Acoustics (WASPAA), pp. 66–70, October 2017
18. Wilson, K., et al.: Exploring tradeoffs in models for low-latency speech enhancement. In: Proceedings of the International Workshop on Acoustic Signal Enhancement (IWAENC), pp. 366–370, September 2018
19. Wisdom, S., et al.: Differentiable consistency constraints for improved deep speech enhancement. In: Proceedings of the IEEE International Conference on Acoustics, Speech and Signal Processing (ICASSP), pp. 900–904, May 2019
20. Xia, R., Braun, S., Reddy, C., Dubey, H., Cutler, R., Tahev, I.: Weighted speech distortion losses for neural-network-based real-time speech enhancement. In: Proceedings of the IEEE International Conference on Acoustics, Speech and Signal Processing (ICASSP) (2020)

Automatic Information Extraction from Scanned Documents

Lukáš Bureš[✉], Petr Neduchal, and Luděk Müller

Faculty of Applied Sciences, New Technologies for the Information Society,
University of West Bohemia, Univerzitní 8, 306 14 Plzeň, Czech Republic
{lbures,neduchal,muller}@ntis.zcu.cz

Abstract. This paper deals with the task of information extraction from a structured document scanned by an ordinary office scanner device. It explores the processing pipeline from scanned paper documents to the extraction of searched information such as names, addresses, dates, and other numerical values.

We propose system design decomposed into four consecutive modules: preprocessing, optical character recognition, information extraction with a database, and information extraction without a database. In the preprocessing module, two essential techniques are presented – image quality improvement and image deskewing. Optical Character Recognition solutions and approaches to information extraction are compared using the whole system performance. The best performance of information extraction with the database was obtained by the Locality-sensitive Hashing algorithm.

Keywords: Information extraction · Image processing · Text processing · OCR · Scanner · Deskew · Database

1 Introduction

Information extraction of semi-structured scanned documents is a current problem because paper still works as a main data storage medium in the administration. Naturally, a paper database is highly impropriety for effective searching. On the other hand, searching is a simple and quick task for a computer.

That is why almost every company manually creates an electronic database of information about its clients, orders, or payable invoices. It is a time and resources consuming process that we can automatize by addressing four necessary steps. We assume using scanners or cameras for document digitization in the form of images.

The first step is preprocessing image data to improve readability and highlight the information that we want to extract. The next step is applying Optical Character Recognition (OCR) to recognize all symbols in the input image. The third part is the process of information extraction based on the database, and the last one is information extraction of data without the database. For example, a database can be a structured file, SQL, or NoSQL database.

A. Karpov and R. Potapova (Eds.): SPECOM 2020, LNAI 12335, pp. 87–96, 2020.
https://doi.org/10.1007/978-3-030-60276-5_9

The goal of this paper is to present our experiments on the mentioned parts of the invoice digitalization process. We consecutively address individual parts. Thus, the input of the preprocessing part, we expect a scanned document in a common image file type. Then, we apply the other three partially independent modules – i.e., OCR, information extraction with a database, and information extraction without a database. This paper's main contributions are a description of an effective information extraction process design and a set of experiments comparing various approaches in a mean of accuracy and computational cost.

The paper is structured as follows. Related work is described in Sect. 2. In Sect. 3, we introduce the overview of a proposed information extraction process design. A more detailed description of each part of the process and related results are in Sect. 4. We summarise the proposed approach and obtained results in the conclusion of this paper.

2   Related Work

There is a lot of related work about individual parts of this task, such as preprocessing, deskewing, or string comparison. Bart and Sarkar [2] describe information extraction by finding repeated structure in a structured document. Information extraction based on regular expressions is explained in paper [3]. Expressions are derived from manually classified samples. The work of Muslea et al. [8] contains a survey of extraction patterns for information extraction tasks. Papers above usually describe only one sub-task of the whole digitalization and information extraction process. The counterexample is the work of Altamura et al. [1], which describes the whole system transforming a paper document to an XML file. Another paper by Palm et al. [10] is focused on the deep learning of the patterns that should be extracted from the document.

Our paper [13] also worth mentioning because it addresses the problem of the text extraction from old scanned documents. In the paper, we propose a deskewing method for small skew angles based on the Fourier Transform that we also use in this research. Moreover, there is also mentioned proposed preprocessing – especially binarization – approach.

3   Information Extraction Process Design

The Information extraction process contains four essential modules. Besides, it includes two more steps – manual digitization using scanner or camera device and storage of obtained information in a database. The visualization of the process is shown in Fig. 1.

The first part of the system is the preprocessing module. It prepares input data for OCR processing. The second module – OCR – is a complex problem. We decide to compare existing systems instead of developing our solution. There are several criteria, such as quality of recognition, software license, supported languages, and computing performance. The output of the OCR module is an input for both the information extraction modules – i.e., module with and without a

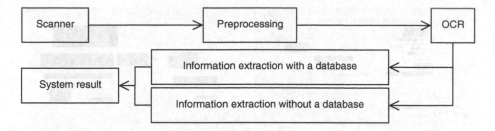

Fig. 1. Information extraction process design visualization.

database, respectively. Module with database searches for information base on a database and decides which record has the highest value of conformity. The module without a database usually uses regular expressions determined by a human expert – i.e., names, dates, or other numbers. Sometimes, it is called named entities – i.e., predefined information extraction classification categories.

4 Experiments and Results

In this section, the whole process from paper documents to extracted information is deeply analyzed. Firstly the preprocessing module, which prepares scanned data for optical character recognition, is described. The output of the OCR module is a text, which is then processed by information extraction with a database and without a database, respectively.

Data that we used consists of 39 different real-world invoices. They were scanned in the grayscale color format on 300 DPI resolution. In incoming invoices are present Czech, Slovak, English, Germany, French, Japan, and Chinese languages. Examples of invoices from the dataset are shown in Fig. 2.

4.1 Preprocessing

The preprocessing phase is one of the most critical phases in the whole pipeline. Input in the preprocessing block is a raw, scanned image. The output is an image with improved visual properties with respect to the readability of the document for OCR. As the most promising preprocessing techniques, we have chosen these four techniques: image contrast improvement, noise reduction, binarization, and image deskewing.

Image Quality Improvement. One of the methods which we tried is Contrast Limited Adaptive Histogram Equalisation (CLAHE). CLAHE differs from ordinary adaptive histogram equalization in its contrast limiting. It is advantageous not to discard the part of the histogram that exceeds the clip limit but to redistribute it equally among all histogram bins. The redistribution pushes some bins over the clip limit again, resulting in an effective clip limit larger than the

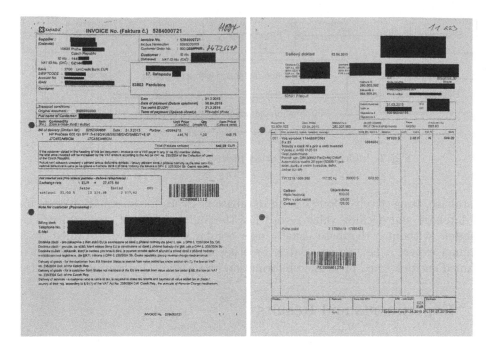

Fig. 2. Examples of invoices from our dataset. Note: Invoices had to be anonymized because the images contain the real information and are not from a public dataset.

prescribed limit and the exact value of which depends on the image. If this is undesirable, the redistribution procedure can be repeated recursively until the excess is negligible – see [11] and [14].

Noise Reduction. Due to the high amount of noise in real data, the application of some noise-reducing technique is necessary. There are many sophisticated non-linear filters, such as non-locals mean, non-linear filters, wavelet transform. Because the fully automatic image processing of documents needs a highly robust solution, a linear smoothing filter was chosen and applied to every input image.

Binarization. After CLAHE and noise reduction processing, we perform image thresholding to obtain binary representation. We used Otsu's thresholding method [9]. It is based on a simple idea: find the threshold that minimizes the weighted within-class variance. It turns out to be the same as maximizing the between-class variance. It operates directly on the gray level histogram. One assumption has to be met: histogram (and the image) is bimodal. Our documents met this assumption.

Note About Results of These Methods. The results of these preprocessing methods were evaluated in terms of quality by comparing text in the invoice and text from OCR by a human expert. The results, where contrast improvement, noise reduction, and binarization was applied, were more accurate. On the other hand, documents in this particular dataset do not have problems with low contrast and noise. The results and examples for a more challenging dataset can be found in our related paper [13].

Document Deskewing. As we mentioned in Sect. 1, the output from the scanner device is often slightly skewed – i.e., ±5°. It can reduce OCR accuracy. Therefore, it is appropriate to apply an accurate and robust deskew method.

A basic method for document deskewing searches for the smallest rectangular bounding box of the detected object. The search is performed using the image rotation technique. We used this method as a baseline for our deskew experiment, which is summarized in Table 1.

Another approach is Hough line transform which transforms the image into 2D space with parameters ρ – the perpendicular distance from the origin of the line – and θ – the angle between obtained line and horizontal axis – defined by line parametric equation

$$\rho = x \cdot \cos(\theta) + y \cdot \sin(\theta). \tag{1}$$

Angle θ is perpendicular to the observed line – i.e. perpendicular to searched document rotation.

We propose to use the Fourier Transform (FT) based method that we also use in our previous paper [13]. In the first step, the image is converted to the frequency domain using a Fast Fourier Transform (FFT) algorithm. The resulting matrix contains the lowest frequencies in the center and highest ones on edges. In the frequency matrix, we can detect lines that are perpendicular to lines in the source image. The most significant line with a small deviation to the vertical axis is searched by rotation of the matrix center. It is a more efficient approach because it is not necessary to rotate the whole image. Then the maximum value of its vertical projection is picked as the skew angle.

There are two more approaches. The first one is based on both mentioned transformations, where Hough transform is used for preprocessing. It creates a synthetic binary image with a black background and white lines. Then we applied the Fourier Transform for final results. The second method is similar to the Fourier transform approach. Results are obtained in several tiles deterministically picked from the image instead of one result computed from the whole image. We use an average of computed results as an estimate of document rotation.

The results of a deskew experiment are shown in Table 1. We created a synthetic dataset from available invoices by rotating them ten times by ±5° resulting in 390 test images with known rotation. All methods based on Fourier transform have small average error – i.e., lower than one degree – which is similar performance as Hough transformation. Moreover, the Fourier approach's error variance is 22 times smaller than in the Hough transform. On the other hand, the Hough transform is a less time-consuming approach. The worst method is the Bounding box, with the average error over 5° and average processing time almost 30 s. Our chosen solution for our system is Average Fourier because it offers a good trade-off between computational cost and accuracy.

Table 1. Deskew methods results

Method	Average error [°]	Error variance [°]	Average time [s]	Time variance [s]
Bounding box	5.07	9.73	29.58	8.13
Hough Tf.	**0.56**	2.89	**0.14**	**0.01**
Fourier Tf.	0.59	**0.13**	5.49	0.32
Fourier + Hough	0.82	1.43	6.29	1.05
Average Fourier	0.81	0.42	1.30	0.01

4.2 Optical Character Recognition

Optical Character Recognition is the technology for transferring scanned documents to digital files with highlightable and editable text. Because of the presence of the noise, this method is not perfect. Thus, the output should be checked by a human.

Nowadays it exists a lot of OCR software. All of them belong to one of the following categories – freeware or commercial. In the following text, we summarised 8 leading software from each of the two mentioned groups. All OCR systems can be used on multiple system platforms. Moreover, there are also freeware but online-only solutions.

Most online and freeware OCR systems do not provide information about accuracy. On the other hand, all of the commercial OCR provides accurate information. A number of supported languages can be one of the critical parameters, mainly when your solution is widely used.

To test OCR systems from the section, we chose testing data as a subset of files from our dataset – i.e., invoices written in the following languages: Czech, English, German and French. Invoices are not typically created only from blocks of text. There are also logos, stamps, handwritten signatures, bar codes, and invoices that are divided into columns. This small set of files fully represents what a company can get from its real customers, which is sufficient for our purpose.

In the first step, we use all presented OCR systems for recognizing texts on invoices. If it was possible, we chose the appropriate language in the OCR system. When the right language was not supported, we selected the English language. We stored all the recognized data in text files.

We created ground truth data; thus, we rewrote incoming invoice documents into the text files. Then, we selected the following procedure for the evaluation of accuracy. We compared text files with a recognized word against ground truth text files. We deleted all words which were in both text files. The remaining words were poorly recognized, and we calculated errors from them. The final results are calculated as the average value of individual OCR results. The results are shown in Table 2. We did not consider white characters (i.e., spaces, tabulator indentation) in error calculation.

Table 2. Average errors of OCR systems.

OCR system	FreeOCR	MeOCR	Tesseract	Cuneiform	OnlineOCR.net (online)	NewOCR (online)	Free OCR (online)	Google Docs (online)	OmniPage Ultimate	Adobe Acrobat Pro DC	ABBYY FineReader Corporate	Readiris 15 Corporate	PowerPDF Advanced	ABBYY PDF Transformer+	Soda PDF	Presto! Page Manager
$\overline{err}$ [%]	37.00	32.25	29.00	**23.00**	**11.00**	40.50	54.75	22.25	**10.25**	23.75	11.00	18.00	45.75	12.50	33.50	22.50

It is visible that the best commercial solution for our dataset is Omni-Page Ultimate OCR, with a 10.25% average error. The best freeware is OnlineOCR.net, with 11%. We also highlighted Cuneiform with a 23% average error as the best offline freeware solution.

For further experiments, we chose Tesseract OCR software. It is based on two reasons. The first one is the need for an offline OCR solution, and the second one is that we need an open-source solution. The best freeware solution was Cuneiform, but Tesseract OCR is still actively developed and has a large community.

4.3 Information Extraction with Database

Information extraction with the database – i.e., a list of records searched in a document. In our experiment, the database was a structured CSV file containing all information about clients – e.g., company name, VAT number, address.

In the experiment, we apply searching methods on text obtained by applying Tesseract OCR. In particular, we employ two versions of Levenstein distance [12], two versions of the FuzzyWuzzy [4] algorithm, context triggered piecewise hashing (CTPH) [6], three versions of Locality–Sensitive Hashing (LSH) [5] and similarity hashing (SimHash method).

Versions of Levenshtein distance differs in the approach to comparing. The first version computes the ratio of two input strings – i.e., Levenshtein 1. The second version searches for the most similar substring in the whole document and returns the maximum value of similarity or minimum distance. This version is denoted as a Levenshtein 2. Similarly, there are two versions of the FuzzyWuzzy algorithm. Again the first one compares two strings and the second one search for the best substring. FuzzyWuzzy is based on Levenshtein distance expects sorted set t_0 of common words from both strings as an input. Then it creates supersets by a union of the set t_0 with the rest of the words from input strings. The final value is calculated using these three sets.

CTPH is a hashing algorithm that computes the same or similar hashes for sufficiently similar data. Thus, its hashes can be used to compare two strings. The similarity is then computed using the Jaccard similarity index – i.e., Intersection over Union.

$$J = \frac{|F \cap D|}{|F \cup D|}, \tag{2}$$

where F is a set of hashes from an invoice, and D is a set of hashes from the database. Result lies in interval $\langle 0, 1 \rangle$ and 1 is maximum similarity.

LSH method is also based on hashing. In these case three versions are based on three python packages that we used – i.e., lshhdc[1] for LSH 1, lshash[2] for LSH 2 and pyflann[3] for LSH 3. Also, the last proposed method – SimHash – is based on hashing string to the binary sequence of length n. The first step of the algorithm is to transform chars into binary codes of its ASCII values. In the second step, all character representations are summed into a vector of length n. This vector is thresholded. Thus the calculated hash is binary.

The results of this experiment are shown in Table 3. Most of the presented methods have perfect results for top 20 criteria – i.e., the right answer is in top 20 records, sorted by confidence value in descending order – and almost perfect for top 10 criteria. We considered the correctness of results and time performance. The LSH 1 algorithm performs best with respect to both of our criteria.

4.4 Information Extraction Without Database

We are also able to extract any information from recognized text without the database. Some information has a strict pattern. In cases of date or time in general, value-added tax (VAT) number, and others pattern searching can be used. We constructed regular expressions for each pattern, and we were able to find the required items.

At first, we used text from the OCR. In the second step, we filtered out unnecessary characters – e.g., repeating characters and non-ASCII characters. After filtering, we apply regular expressions to obtain the required information from invoices.

---

[1] https://github.com/go2starr/lshhdc.

[2] https://pypi.org/project/lshash/.

[3] http://www.cs.ubc.ca/research/flann.

Table 3. Information extraction methods.

Method	Right answer [%]	Top 10 [%]	Top 20 [%]	time [s]
Levenshtein 1	71.79	89.74	97.44	9.73
Levenshtein 2	30.77	56.41	69.23	2.12
FuzzyWuzzy 1	51.28	100.00	100.00	7.31
FuzzyWuzzy 2	80.15	100.00	100.00	1139.82
CTPH	25.64	64.10	69.32	**0.35**
LSH 1	84.62	100.00	100.00	0.97
LSH 2	80.01	100.00	100.00	1085.86
LSH 3	**87.18**	100.00	100.00	34.09
SimHash	76.92	92.31	100.00	90.74

The first number type that we searched for was a bar code. In our case, bar code had one of following patterns: "KCS$xxxxxxxx$" or "KCSE$xxxxxxxx$" where x can be any digit 0–9 (e.g. "KCS00000061"). We achieved 0.833 precision, 0.270 recall, and 0.408 F1 scores on our dataset. The VAT number was next analyzed value. The VAT number has a defined pattern. Therefore, it was straightforward to find them. We had 1.000 precision, 1.000 recall, and 1.000 F1 scores. There are other interesting items to search for, such as the Czech variable symbol, order number, the amounts, due date, or date of issue. On the other hand, all of them are based on regular expressions. Thus it is not the goal of this paper to describe every single item individually.

More Interesting is the time analysis of our information extraction design. All parts of the system processed all 39 invoices of our dataset in 38.724 s. Thus, the average time needed to process one document is 0.992 s (with average office computer).

5   Conclusion

This paper describes the design of the system for information extraction that consists of preprocessing, optical character recognition, and two information extraction modules. Our main goal was to compare approaches to information extraction and propose system design for future research. We suggest using CLAHE, Otsu's method, and deskewing based on Hough or Fourier transform for the preprocessing step. We chose Tesseract OCR as the most suitable open-source solution for our purposes because it has satisfactory recognition results and can be widely customized. The resulting system design can process a single document with processing time less than 1 s. The best result of information extraction with a database was achieved using the LSH algorithm. The accuracy was 87.18%. The proposed system design proved that it could handle both information extraction tasks.

In future work, we will focus on creating a system for automated information retrieval. Primarily, we will try to use statistical methods and named entities [7] for extracting relevant information from incoming invoices and other financial documents. We plan to combine more OCR systems to obtain better results.

Acknowledgments. This publication was supported by the project LO1506 of the Czech Ministry of Education, Youth and Sports and by the grant of the University of West Bohemia, project No. SGS-2019-027.

References

1. Altamura, O., Esposito, F., Malerba, D.: Transforming paper documents into xml format with wisdom++. Int. J. Doc. Anal. Recogn. **4**(1), 2–17 (2001)
2. Bart, E., Sarkar, P.: Information extraction by finding repeated structure. In: Proceedings of the 9th IAPR International Workshop on Document Analysis Systems, DAS 2010, pp. 175–182. ACM, New York (2010)
3. Brauer, F., Rieger, R., Mocan, A., Barczynski, W.M.: Enabling information extraction by inference of regular expressions from sample entities. In: Proceedings of the 20th ACM International Conference on Information and Knowledge Management, CIKM 2011, pp. 1285–1294. ACM, New York (2011)
4. Cohen, A.: Fuzzywuzzy: Fuzzy string matching in python (2011). https://chairnerd.seatgeek.com/fuzzywuzzy-fuzzy-string-matching-in-python/
5. Indyk, P., Motwani, R.: Approximate nearest neighbors: towards removing the curse of dimensionality. In: Proceedings of the Thirtieth Annual ACM Symposium on Theory of Computing, pp. 604–613 (1998)
6. Kornblum, J.: Identifying almost identical files using context triggered piecewise hashing. Digital Invest. **3**, 91–97 (2006)
7. Král, P.: Named entities as new features for czech document classification. In: Gelbukh, A. (ed.) CICLing 2014. LNCS, vol. 8404, pp. 417–427. Springer, Heidelberg (2014). https://doi.org/10.1007/978-3-642-54903-8_35
8. Muslea, I., et al.: Extraction patterns for information extraction tasks: a survey. In: The AAAI-99 Workshop on Machine Learning for Information Extraction, vol. 2. Orlando Florida (1999)
9. Otsu, N.: A threshold selection method from gray-level histograms. IEEE Trans. Syst. Man Cybern. **9**(1), 62–66 (1979)
10. Palm, R.B., Laws, F., Winther, O.: Attend, copy, parse end-to-end information extraction from documents. In: 2019 International Conference on Document Analysis and Recognition (ICDAR), pp. 329–336. IEEE (2019)
11. Pizer, S.M., Amburn, E.P., Austin, J.D., Cromartie, R., Geselowitz, A., Greer, T., Romeny, B.T.H., Zimmerman, J.B.: Adaptive histogram equalization and its variations. Comput. Vision Graph. Image Process. **39**(3), 355–368 (1987)
12. Yujian, L., Bo, L.: A normalized levenshtein distance metric. IEEE Trans. Pattern Anal. Mach. Intell. **29**(6), 1091–1095 (2007)
13. Zajíc, Z., et al.: Towards processing of the oral history interviews and related printed documents. In: Proceedings of the Eleventh International Conference on Language Resources and Evaluation (LREC 2018) (2018)
14. Zuiderveld, K.: Contrast limited adaptive histogram equalization. In: Heckbert, P.S. (ed.) Graphics Gems IV, pp. 474–485. Academic Press Professional Inc., San Diego (1994)

Dealing with Newly Emerging OOVs in Broadcast Programs by Daily Updates of the Lexicon and Language Model

Petr Cerva[1]([⊠]) [iD], Veronika Volna[2] [iD], and Lenka Weingartova[2] [iD]

[1] Institute of Information Technologies and Electronics,
Technical University of Liberec, Studentska 2, Liberec 46117, Czech Republic
petr.cerva@tul.cz
[2] Newton Technologies, Na Pankraci 1683/127, Praha 4 140 00, Czech Republic
{veronika.volna,lenka.weingartova}@newtontech.cz

Abstract. This paper deals with out-of-vocabulary (OOV) word recognition in the task of 24/7 broadcast stream transcription. Here, the majority of OOVs newly emerging over time are constituted of names of politicians, athletes, major world events, disasters, etc. The absence of these content OOVs, e.g. COVID-19, is detrimental to human understanding of the recognized text and harmful to further NLP processing, such as machine translation, named entity recognition or any type of semantic or dialogue analysis. In production environments, content OOVs are of extreme importance and it is essential that their correct transcription is provided as soon as possible. For this purpose, an approach based on daily updates of the lexicon and language model is proposed. It consists of three consecutive steps: a) the identification of new content OOVs from already existing text sources, b) their controlled addition into the lexicon of the transcription system and c) proper tuning of the language model. Experimental evaluation is performed on an extensive data-set compiled from various Czech broadcast programs. This data was produced by a real transcription platform over the course of 300 days in 2019. Detailed statistics and analysis of new content OOVs emerging within this period are also provided.

Keywords: Out of vocabulary words · Automatic speech recognition · Broadcast monitoring · Named entities detection.

1 Introduction

One of the key issues surrounding state-of-the-art automatic speech recognition (ASR) systems is the question of how OOVs are dealt with, since their occurrence reduces recognition accuracy, comprehensibility for human readers, as well as readability in case of subtitling systems. In a pipe-lined environment, where ASR is one of the first steps and its output is utilized for further NLP analysis, such as machine translation, named entity recognition, dialogue systems or any other type of semantic analysis, OOVs are a major source of errors.

© Springer Nature Switzerland AG 2020
A. Karpov and R. Potapova (Eds.): SPECOM 2020, LNAI 12335, pp. 97–107, 2020.
https://doi.org/10.1007/978-3-030-60276-5_10

As a rule, the solution to OOV transcription consists of two consecutive steps: identification and recovery.

In the first phase, one possible method would have the OOVs be represented by a filler [1] or hybrid sub-word model [2]. Another approach uses confidence scores to find and label unreliable regions of the recognized output as containing OOVs [3]. Moreover, methods combining both of the above approaches are introduced in [7,8].

In the second step, OOVs can be recovered using an open vocabulary system with a set of unified recognition units formed by lexical entries with the (sublexical) graphones derived from grapheme-to-phoneme conversion [2]. A similar approach based on words and data-driven variable-length sub-word units is proposed in [9]. A phoneme recognizer is also utilized for recovery in [3,7].

In recent end-to-end (acoustic-to-word) systems, OOV recovery can be performed using a complementary character model [6,12]. For open vocabulary end-to-end speech recognition, byte-pair encoding [5] is utilized in [11,12]. A comprehensive study comparing all of these approaches is presented in [10].

1.1 Motivation for This Work

In this paper, we deal with the task of broadcast stream recognition. In our case, an ASR system is utilized within a cloud platform for real-time 24/7 transcription of TV and radio stations in several languages including Czech, Slovak, Polish and other, predominantly Slavic, languages. The automatic transcripts are further used in two different services. The first, broadcast monitoring, relies on manual corrections of the transcripts for a large set of selected programs (approximately 920 programs on a weekly basis). Within the scope of the second service, called TVR Alerts, manual corrections are not possible as a) the results are sent to clients with minimal delay following the broadcasting time and b) the service operates in 24/7 mode for all relevant TV and radio streams (74 stations for Czech).

In broadcast data, an important part of OOVs consists of names of politicians, major world events or disasters, etc. These OOVs emerge continuously over time (e.g. Buttigieg, Čaputová, COVID-19, etc.) and their correct transcription is much more vital for deciphering the meaning of the given program than in the case of other OOVs (e.g. low occurrence forms or inflected forms of common words). Therefore, we refer to this category of OOVs as content OOVs (COOVs).

In the case of TVR Alerts, we are operating within a real-time production environment, where the output is sent to paying clients. In contrast to most previously mentioned approaches, our method cannot produce meaningless forms of COOVs that would be phonetically similar to the correct forms. It is of great importance that the output is reliable. We have taken advantage of the fact that, simultaneously, the transcripts of a substantial number of programs are being corrected manually on a daily basis, and we are able to utilize those for the identification of new COOVs, which are then added to the lexicon. To further improve their chances at correct recognition, we boost their probability in the language model.

In contrary to most existing works on OOV recognition, the experimental evaluation within this paper is not performed on OOVs obtained by reducing an original lexicon as is customary, but on real new OOV words which emerged in broadcast data over the course of 2019.

2 Proposed Scheme of Dealing with COOVs

The scheme is based on daily updates of the lexicon and language model and it is carried out in three consecutive phases.

In the first phase, which is performed early in the morning, newly emerged text data are gathered from two main sources: manually corrected transcriptions of TV, radio and internet broadcasts (54K words on average), new articles from leading Czech newspapers, e.g. MF Dnes, Právo, etc., covering topics such as local news, world news and economics (21K words on average). Following this, new COOVs are extracted using texts from all previous days (not just from the last day). These COOVs are then filtered using limits on minimal frequency of occurrence (8 in our case), black lists, etc. Moreover, the system also skips all the COOVs that emerged in any of the previous days. After applying all of these rules, it is usually only a handful of new COOVs which are identified for further manual corrections and/or filtering. A human annotator may modify the word's capitalization or correct the pronunciation (suggested by a g2p system), or discard any of the COOVs that were found automatically. In practice, this work is performed daily via a web interface and usually takes less than a minute. The statistics and analysis of all new COOVs distilled using the approach described above for the period from 2019/01/01 to 2019/11/01 are examined in Sect. 3.

The last phase of the update process is building the new language model. For this purpose, the system utilizes all the manually corrected sets of COOVs, as well as a collection of texts from all previous days and original text corpora or models. Finally, the updated model is distributed to the production cloud platform and may be immediately used for speech recognition.

2.1 Strategies for Building the New Language Model

The baseline strategy is to add all COOVs to the lexicon, join all the new texts to already existing text corpora used to build the original model, and then rebuild the new model from scratch using all this data.

Note that for real-time transcription, we use a language model based on bigrams (with collocations) and KN discounting, with a lexicon containing 500 K words. This size of vocabulary is quite standard for Czech as this language is highly inflective.

Another option, less computationally demanding, is to extend the original lexicon by all new COOVs and then compute a new language model just using all the gathered texts and the extended lexicon. Finally, both the new and original LMs may be combined linearly using a weight determined on a development set.

We also performed experiments using two different strategies for grouping of COOVs into classes. All words in one class then share (or are mounted to) the same symbol in the language model and are also replaced by the given symbol in the input text corpora. As a result, each word in the lexicon of the transcription system has three items: a) the output symbol (original word), b) pronunciation forms and c) the corresponding mounting symbol in the language model.

In the first case (variant #1), all inflections of each word were mounted to its first case (nominative). In the second one (variant #2), just several classes were created: the first contained nominatives of all new words, second genitives, etc. The effect that all these approaches had on the quality of COOV detection is evaluated using the development set in Sect. 4.

3   Statistics and Analysis of Newly Emerged COOVs

3.1   Categorization of COOVs

A total of 1140 COOVs were extracted from 2019/01/01 to 2019/11/01 as described in the previous section and subjected to a qualitative analysis resulting in the classification of each item based on its denotative meaning. Our set of categories is largely based on the types used by spaCy models for NER (Named Entity Recognition) [13], with three additional categories; COMMON_NOUN, FOREIGN_WORD, and OTHER, which we consider relevant to this analysis. An overview, including descriptions of the categories, is provided in Table 1. Note that these categories are not used in language modelling.

The analysis showed that the most highly represented COOV category is PERSON (720 items), followed by GPE (122) and ORG (78). Another size-able section of the data consists of common nouns (COMMON_NOUN category) such as "elektrokoloběžce" (electric scooter; dative) or "hrabošem" (vole; instrumental). Rather than representing new concepts, these words reflect the trends and topics being currently discussed in the media, and it is often the case that the lexicon already contains these words in another grammatical case, usually nominative. In fact, a total of 315 COOVs (28%) were marked for case (i.e. in a grammatical case other than nominative), indicating that they are already present in the lexicon in the unmarked nominative case.

3.2   Subcategories Within the COOV Category PERSON

Within the category PERSON, the largest subcategory is SPORTS (39%), which includes surnames of named entities connected with football (93), tennis (53), cycling (36), hockey (33), athletics (16), basketball (15), and a variety of other sports (35). This reflects the interests of the audience for whom the broadcasts are designed, and naturally it is presumable that the sports represented in this category will vary over the course of a calendar year based on the sporting events

Table 1. Categorization of COOVs based on our tag set.

Type	Description	Count	Rel. count [%]
PERSON	People, including fictional	720	63.2
GPE	Countries, cities, states	122	10.7
ORG	Companies, agencies, institutions, etc	78	6.8
COMMON_NOUN	Czech nouns outside the category of named entities, including abstract nouns	57	5.0
OTHER	Words from other word classes (most notably adjectives and verbs)	39	3.4
PRODUCT	Objects, vehicles, foods, etc. (Not services.)	43	3.8
FOREIGN_WORD	Words in a foreign language such as Eagle, Animation, Solitude, etc	37	3.2
LOC	Non-GPE locations, mountain ranges, bodies of water	14	1.2
NORP	Nationalities, religious or political groups	11	1.0
EVENT	Named hurricanes, battles, wars, sports events, etc	10	0.9
FAC	Buildings, airports, highways, bridges, etc	6	0.5
ABSTRACT_NOUN	Abstract nouns	2	0.2
WORK_OF_ART	Titles of books, songs, etc	1	0.1

currently in season. The second largest subcategory is POLITICS (24%), followed by FIRST NAME (12%) and UNSPECIFIED (6%), a subcategory reserved for names shared by more than one named entity where it is not immediately obvious which of the contexts caused the COOV to be picked up by the algorithm. The category CULTURE (6%) includes musicians (14), actors (13) and one author. All of the main subcategories are described in Table 2.

Table 2. Subcategories within the COOV category PERSON.

Category of PERSON	Count	Rel. count [%]
SPORTS	281	39.0
POLITICS	170	23.6
FIRST_NAME	88	12.2
UNSPECIFIED	47	6.5
CULTURE	41	5.7
CRIME	37	5.1
BUSINESS	21	2.9
JOURNALISM	19	2.6
SCIENCE	16	2.2

3.3 Source Language

In terms of source language, the COOVs can be divided into two basic groups – Foreign language and Czech. Slovak was included as a separate group, because Czech and Slovak languages are a) mutually intelligible to a great degree, and b) both have similar phonemic inventories as well as grapheme-to-phoneme correspondence. This is of interest to us namely due to the fact that words of non-Czech (or Slovak) origin are likely to have pronunciations diverging significantly from those generated by a g2p tool, which means that manual corrections of their pronunciation are, in many cases, absolutely vital. We have found that the majority (73%) of COOVs came from various foreign languages, while words of Czech (24%) and Slovak (3%) origin were represented less.

4   Experimental Evaluation on the Development Set

4.1 The Development Set

The development set consisted of 301 consecutive recordings (starting from 2019/01/01) of the main news program which is broadcast by Czech Television (CT1) daily at 19:40, under the name Events (Události). The total length of these recordings was 264 hours and they contained 1842 K words.

This data contained 1627 occurrences belonging to the set of harvested COOVs, with a total of 370 distinct words. Not all of these words were frequent enough to be added to the lexicon used for transcription on the first day they occurred, and were added later. Therefore we have two groups of COOVs: 1026 occurrences which could be detected/recognized by the transcription system, and 601 occurrences, which could not be recognized in any case, since they were not in the lexicon at that time. Note that the total OOV rate (including all OOVs existing in the data) was 0.76%.

Figure 1 shows that the vast majority of these COOVs had a low rate of occurrence; of the 370 distinct forms, 139 appeared only once and 84 appeared twice. On the other hand, there was a much greater dispersion among most frequently detected forms; the most frequent Čaputová appeared in a total of 100 instances, followed by Notre-Dame at 60, and Zelenskyj at 54.

4.2   Obtained Results

The evaluation of COOV detection quality was carried out using precision (P), recall (R) and F-measure (F). The last two metrics were calculated a) with respect to the total occurrence count (1627) and b) only for occurrences that could be detected (1026), i.e. the corresponding word was already in the lexicon. In the latter case, the measures are denoted as R_{Lex} and F_{Lex}. The quality of transcription was evaluated using accuracy (Acc), which is defined as 1 - Word Error Rate (WER).

The system utilized Hidden Markov Model-Deep Neural Network (HMM-DNN) hybrid architecture [4]. The DNN input layer corresponds to 11 consecutive feature vectors, 5 preceding and 5 following the current frame. It is followed by five fully-connected hidden layers consisting of 768 units; the activation function is ReLU. The output layer corresponds to 2219 senones. The feature vectors contain 39 filter bank coefficients computed using 25-ms frames with frame shift of 10 ms. We employ the mean subtraction normalization with a floating window of 1 s. The DNN parameters are trained by minimization of the negative

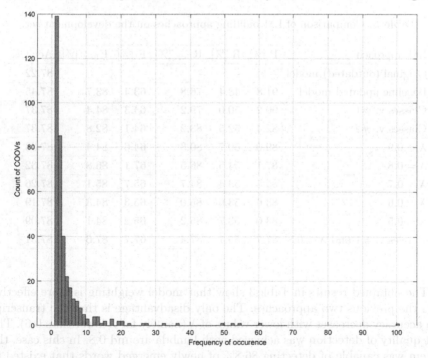

Fig. 1. Histogram of COOVs on the development set.

log-likelihood criterion via the stochastic gradient descent. Decoding uses a time-synchronous Viterbi search.

In the first experiment, the baseline LM approach was compared with two variants based on word classes (see first part of Table 3). The obtained results show that the baseline approach yields much higher precision than recall, which also holds for R_{Lex}. This implies the importance of adding the COOVs to the lexicon as soon as possible. The addition of COOVs is safe as the reached precision is over 90%. Therefore, in the future, we plan to extend the number of articles and sources used for the extraction of the newly emerged COOVs.

The word classes in variant #1 (all cases mounted to nominative) yield slightly better recall but slightly worse precision than the baseline. The reached F-measure in this case is therefore just slightly higher. Variant #2 is more aggressive in the sense of favoring COOVs during recognition as more words share the same class and have higher probability in the LM. Therefore, this method yields better recalls but much worse precision. Both F-measures were lower than in the case of variant #1.

In the second experiment, the baseline approach was compared with the approaches based on weighting the new and original language model according to the equation:

$$model_{updated} = \lambda model_{new} + (1 - \lambda)model_{original} \tag{1}$$

Table 3. Comparison of LM building approaches on the development set.

LM approach	P [%]	R [%]	R_{Lex} [%]	F [%]	F_{Lex} [%]	Acc [%]
Original (outdated) model	-	-	-	-	-	87.22
Baseline updated model	91.8	48.4	76.8	63.4	83.7	87.45
Classes v. #1	90.2	50.0	79.2	64.3	84.4	87.37
Classes v. #2	82.4	52.5	83.2	64.1	82.8	87.37
$\lambda = 0.9$	89.0	50.7	80.3	64.6	84.4	87.33
$\lambda = 0.8$	87.1	54.5	86.5	67.1	86.8	87.32
$\lambda = 0.7$	85.5	53.4	84.7	65.7	85.1	87.26
$\lambda = 0.6$	83.6	53.6	85.0	65.3	84.3	87.19
$\lambda = 0.5$	83.0	53.7	85.2	65.2	84.1	87.09
Classes v. #1 and $\lambda = 0.8$	87.7	55.1	87.4	67.7	87.6	87.31

The obtained results in Table 3 show that model weighting is more effective than the previous two approaches. The only disadvantage is that the transcription accuracy decreases with lower values of λ (namely for λ lower than 0.6). The best quality of detection was achieved with lambda around 0.8. In this case, the system was capable of detecting 86.8% of newly emerged words that existed in its lexicon (with a precision of 87.1% and a slight increase in accuracy by 0.13% over the outdated model).

Table 4. Overview of data used for testing.

Program	Source	# of recor.	# of words	Length [hours]	# COOVs	# COOVs in lexicon
News at 19:30	TV Nova	298	1.575K	198	1677	911
News at 18:00	CRo1	303	291K	51	509	303
Sport at 19:50	CT1	302	444K	70	778	347

Finally, we combined the best performing approaches from the two previous experiments (the last row of Table 3). Here, slight additional improvement in F-measure was observed over weighting with $\lambda = 0.8$.

5   Results on the Test Set

Finally, the approach which was deemed most successful in the preliminary evaluation on the development set was applied to a set of test data compiled from consecutive recordings (starting again from 2019/01/01) of three different broadcast programs. The details are provided in Table 4.

Table 5. Results [%] reached on test data-sets

Test data-set	P	R	R_{Lex}	F	F_{Lex}	Acc	Outdated model Acc
News (TV NOVA)	89.8	48.4	89.0	62.9	89.4	88.8	88.6
Sport (CT1)	80.4	30.1	67.4	43.8	73.4	85.4	85.3
News (CR1)	86.0	55.4	93.1	67.4	89.4	91.4	91.2

The reached results are presented in Table 5, where the last column represents the accuracy achieved by using the original (outdated) model. The results in terms of F-measure for the Czech Radio 1 (CRo1) data are slightly better than those achieved using the development data from Czech TV news. Similar results were also obtained for news broadcasts from the commercial station TV Nova.

On the other hand, significantly worse F values were obtained for sport news data. The reasons for this are twofold. Firstly, the portion of COOVs presented in the lexicon of the recognizer was lower than for other data-sets, due to the fact that COOVs occurring in sports news are usually names of athletes, whose occurrence is generally limited to this domain and therefore more time elapses before their extraction and addition to the lexicon. Secondly, these names are usually of foreign origin with no standardized form of Czech pronunciation, which may then lead to variation depending on the given presenter.

Similar to the development set, a slight improvement in accuracy over the outdated model was observed for all three data-sets.

6   Conclusion

This paper presented an approach for recognition of COOVs newly emerging over time in broadcast data. Analysis performed on the set of COOVs from

2019/01/01 to 2019/01/11 showed that the majority (63%) fell into the named entity category of PERSON, of which the two most frequent subcategories were SPORTS and POLITICS. This is perhaps unsurprising owing to their relatively low stability over time (compared to other named entities such as GPE and LOC). Furthermore, the analysis showed that 73% of COOVs came from various foreign languages, and therefore had, with very few exceptions, a non-native pronunciation in Czech, which clearly points to the limitations of relying solely on an automatic g2p system for the correct transcription.

The results of the experimental evaluation of different news broadcasts demonstrated that the correct recognition of COOVs may be improved namely by properly weighting the original and current language model. A key factor for ensuring high recall is the timeliness of adding COOVs to the transcription system's dictionary. Therefore, in the future, the sources used for new COOV extraction will be extended by incorporating more newspapers and also various web pages.

The presented results also showed that daily updates of the lexicon and language model lead to just a small improvement in recognition accuracy of the transcription system as the COOVs form only a small portion of all words occurring in most TV/R programs. However, the greatest benefit of COOV recognition stems from the fact that COOVs often carry very important pieces of information. That means that their presence in automated transcripts improves readability and comprehensibility for human readers. Their absence, on the other hand, is perceived by users of the recognition system as a highly noticeable error in comparison to other types of errors, for example incorrectly recognized word endings. Secondly, COOV recognition significantly speeds up human-made corrections. Ever since this solution was introduced into our workflow, all human editors reported improvements in their editing times because prior to this, looking up difficult COOVs substantially decreased their work speed. Moreover, recognized COOVs also improve the effectiveness of other NLP tools which take the ASR output as their input, such as machine translation. To see "Pete Buttigieg" transcribed as "Pete the judge" would result in a detrimental meaning change in the translated text, which would be very difficult to decipher for end users.

Acknowledgments. This work was supported by the Technology Agency of the Czech Republic (Project No. TH03010018).

References

1. Bazzi, I., Glass, J. R.: Learning units for domain-independent out-of- vocabulary word modelling. In: Proceedings of INTERSPEECH (2001)
2. Bisani, M., Ney, H.: Open vocabulary speech recognition with flat hybrid models. In: Proceedings of INTERSPEECH, pp. 725–728 (2005)
3. Burget, L., et al.: Combination of strongly and weakly constrained recognizers for reliable detection of OOVS. In: Proceedings of IEEE International Conference on Acoustics, Speech and Signal Processing, pp. 4081–4084 (2008)

4. Dahl, G.E., Yu, D., Deng, L., Acero, A.: Context-dependent pre-trained deep neural networks for large-vocabulary speech recognition. IEEE Trans. Audio Speech Lang. Process. **20**(1), 30–42 (2012)
5. Gage, P.: A new algorithm for data compression. C Users J. **12**(2), 23–38 (1994)
6. Inaguma, H., Mimura, M., Sakai, S., Kawahara, T.: Improving OOV detection and resolution with external language models in acoustic-to-word ASR. In: Proceedings of IEEE Spoken Language Technology Workshop, pp. 212–218 (2018)
7. Lin, H., Bilmes, J.A., Vergyri, D., Kirchhoff, K.: OOV detection by joint word/phone lattice alignment. In: Proceedings of ASRU, pp. 478–483 (2007)
8. Parada, C., Dredze, M., Filimonov, D., Jelinek, F.: Contextual Information Improves OOV Detection in Speech. In: Proceedings of HTL-NAACL, pp. 216–224 (2010)
9. Rastrow, A., Sethy, A., Ramabhadran, B.: A new method for OOV detection using hybrid word/fragment system. In: Proceedings of IEEE International Conference on Acoustics, Speech and Signal Processing, pp. 3953–3956 (2009)
10. Thomas, S., Audhkhasi, K., Tüske, Z., Huang, Y., Picheny, M.: Detection and recovery of OOVs for improved English broadcast news captioning. In: Proceedings of INTERSPEECH, pp. 2973–2977 (2019)
11. Xiao, Z., Ou, Z., Chu, W., Lin, H.: Hybrid CTC-Attention based end-to-end speech recognition using subword units. In: Proceedings of 11th International Symposium on Chinese Spoken Language Processing (ISCSLP), pp. 146–150 (2018)
12. Zenkel, T., Sanabria, R., Metze, F., Waibel, A.H.: Subword and crossword units for CTC acoustic models. In: Proceedings of INTERSPEECH, pp. 396–400 (2018)
13. spaCy Webpage. https://spacy.io/api/annotation. Accessed 15 July 2020

A Rumor Detection in Russian Tweets

Aleksandr Chernyaev, Alexey Spryiskov, Alexander Ivashko,
and Yuliya Bidulya[(✉)]

University of Tyumen, Tyumen, Russia
chernyaev.tmn@gmail.com, facilisdes@gmail.com,
ivashco@mail.ru, bidulya@yandex.ru

Abstract. In this paper, we investigate the problem of the rumor detection for Russian language. For experiments, we collected messages in Twitter in Russian. We implemented a set of features and trained a neural network on the dataset about three thousand tweets, collected and annotated by us. 40% of this collection contains rumors of three events. The software for rumor detection in tweets was developed. We used SVM to filter tweets by type of speech act. An experiment was conducted to check the tweet for rumor with a calculation of accuracy, precision and recall values. F1 measure reached the value 0.91.

Keywords: Rumor detection · Neural network · Feature extraction

1 Introduction

Over the past 15 years, the Internet has become one of the main sources of news for most people around the globe. More recently, social networks and network services, such as Twitter, VKontakte, Facebook, etc., which have gained immense popularity, have greatly influenced news aggregates and the entire field of journalism. Now more than ever, people are turning to social media as a source of news. This is especially true for news situations where people are hungry for quick updates along a chain of events. Moreover, the availability, speed and ease of use of social networks have made them invaluable sources of first-hand information.

This causes many problems for various sectors of society, such as journalists, emergency services and news consumers. Journalists in the modern world must compete with a huge stream of data of ordinary users, because of this, the main factor of quality becomes the time for which the news article was published. As a result, an increasing number of traditional news sources report unsubstantiated information due to the rush to be the first. There are many cases where false rumors about the facts led to unjustified actions by people and to difficult controlled consequences. The spread of rumors at the global level creates distrust of the credibility of the news flow from all sources, including official ones. Therefore, the aim of our work is to develop a software that could detect rumors and predict the accuracy of these rumors distributed in the Russian segment of Twitter users. We consider the rumor as an item of circulating information whose veracity status is yet to be verified at the time of posting [19]. The rumor is distributed as one or more messages. Since the input data for the algorithm is a set of messages, the task of finding a rumor is reduced to two subtasks: determining

© Springer Nature Switzerland AG 2020
A. Karpov and R. Potapova (Eds.): SPECOM 2020, LNAI 12335, pp. 108–118, 2020.
https://doi.org/10.1007/978-3-030-60276-5_11

messages carrying information about the event and combining messages carrying information about the same event.

2 Related Works

Some studies on this issue are aimed at finding the very first node that launched a wave of false rumors [26]. This approach seems to be logical, because by the first node it is possible to assess the degree of reliability of the news that it spreads. Later work decided to go further and determine the rumor before it spread. Among them are several basic approaches to solving this problem:

1. Extraction of temporal and causal relationships between events [6, 7, 14]. The basis of this method is the linking and matching of messages dedicated to a single event using causal and temporal connections.
2. Semantic analysis of messages. The basis of these methods is the division of the message into components and the use of comparative analysis of messages. The fact that there is a conflict between the data [1, 3, 16, 27] is also taken for the fact.
3. Machine learning methods. In these papers, the probability of rumor truth is estimated by applying classification algorithms. The main problem of this approach is the correctly organized dataset for training a classifier [13, 18, 21]. Such methods, as a rule, require the following operations: search for sources of messages, semantic analysis of messages, processing of metadata, identification of users, extraction of rumor markers, and training of a neural network or another classifier. In [21], a recurrent neural network is used, the input of which is the raw parameters obtained from the social network API and study the evolution of these features with time. The proposed approach allows you to find rumors, but not to determine whether the received message is a rumor. The authors [7] propose also apply RNN but suggest to process the entire set of messages for a certain period.

Some works are devoted to the processing of linguistic features in the text, namely the search for "signal" words indicating a possible rumor. An example of such an approach is discussed in [26]. The authors of [1] develop this idea, but use not only signals, but also parameters describing the text as a whole, such as the style of the written text, the number of characters and others.

In the works of such authors as Kwon [10], Friggeri [4] and Alton Y. K. Chua [1], judgments made on sites exposing rumors such as Snopes.com were tracked. This site determines whether news is the spread of rumors, and also determines the level of rumor, i.e. is a rumor, in part is and is not. The Russian equivalent of this site is https://noodleremover.news/.

The most comprehensive approach to the extraction of features was proposed in [18], consisting of two stages:

1) Automatic search for rumors on request (rumor detection);
2) Determining whether a message is a rumor or not (rumor veracity).

All this is done using a combination of classification methods, extracting the linguistic features of the text, the temporal factors of the distribution of messages and the behavior of the authors of messages. The methodology proposed in this paper was taken as the basis for our study.

3 Dataset

For collecting and analyzing data, the Mozdeh tool was used, which supports such functions as searching, receiving data about tweets, as well as text analysis, the ability to work with different languages, including Russian. Thus, data collection was carried out using a combination of the Mozdeh tool [12] and the official Twitter API library [20].

During the data collection, 3,026 messages in Russian including messages about three events that occurred in 2018 were collected and annotated. 1196 (40%) were true and 1830 (60%) were false. Table 1 shows the number of tweets for each type: original and retweets, for each event.

Table 1. The number of true tweets for each type.

Event	All tweets	Original tweets	Retweets
2018 Fire in Paris	121	59	62
2018 Attack in School	207	128	79
2018 President Elections in US	376	291	85
Other	492	175	317
All	1196	653	543

Data collection includes operations such as retrieving, clearing and pre-processing tweets. Tweets were selected in accordance with the search criteria: keywords, time and date of publication, location, hashtags, the presence of a link to the news, which is the source of rumor. Tweets were cleared to discard:

- duplicate tweets with the same identifiers;
- tweets written in languages other than target (Russian).

Retweets also have been saved in the dataset, because retweeting actions are a way to support or share the same opinion of another user about rumor. Pre-processing included filtering text tweets using regular expressions.

We removed links, references, numbers, and special characters (for example, punctuation marks, brackets, slashes, quotes, etc.) from the content of the tweet. The hashtags were saved without the # symbol, since Twitter users typically use hashtags in sentences instead of regular words.

4 Features

Based on publications review, we implemented three groups of features that used as the input data for the neural network. At the output, we get the probability of presence of rumor in the individual tweet. The features are described below.

4.1 Twitter Specific Features

To achieve our goal, we need data that can be obtained using the Twitter API. It provides a set of parameters for each tweet, including 54 items [9]. We have chosen the required set of parameters, which is shown in Table 2.

Table 2. Tweet parameters from Twitter API.

Parameter	Data type	Description
ID	Integer	Twitter user ID
Verified	Bool	Boolean number indicating user verification 1 – checked, 0 otherwise
FolowersCount	Integer	The number of followers of the user
FriendsCount	Integer	The number of users that a current user is subscribed to
FavouritesCount	Integer	Number of "likes"
Createdat	Time	Date and time when the user account was registered. Stored as follows: DayWeights Month Day hh: mm: ss +0000 yyyy
StatusesCount	Integer	Number of tweets (posts) of user
Retweetcount	Integer	The number of retweets among all the messages that the user made. Denoted by @rt + text
TweetID	Integer	Tweet ID Number on Twitter
Tweettext	String	Tweet text

Together with the tweet parameters listed above, we looked at the responses and retweets. Responses and retweets in microblogs are factors that make it possible to judge the reliability of the user when it comes to the transfer of information [9]. For example, user A is more likely to send a rumor than user B if user A has a history of retweets or replies to user C, who also has a history of rumor distribution. The described data set determines input for the model of the neural network that we used to verify tweet to be rumor or not.

4.2 Content Features

Content Features capture the characteristics of a tweet text in a rumor detection. We used the following ones [3, 19, 22]:

1) The presence of words expressing an opinion. Based on the approach from [2] and the Russian dictionary (http://linis-crowd.org) we compiled a list of 65 words used of the opinion expression.

2) The presence of vulgar words. To obtain this parameter, a collection of more than 750 words was collected. The parameter value takes 1 if a vulgar word was found in the text, 0 otherwise.

3) The presence of emoticons in the text. Based on the dictionaries, a list of 362 emoticons (http://pc.net/emoticons/) was compiled, and on its basis a binary sign was formed that reflects the presence of at least one list item in the recall. The parameter value takes 1 if an emoticon was found in the text, 0 otherwise.

4) The presence of abbreviations. Based on the dictionary from [2] and online dictionaries created a list of 31 abbreviations often used in the Internet environment. On its basis, 31 binary features are defined to determine if each abbreviation is present in the tweet text.

5) Verbs describing speech acts. Based on the dictionary from [23] compiled a list of 43 verbs inherent in certain speech acts. They were translated into Russian and stemmed using pystem library.

6) The average complexity of a word in a tweet. It is estimated by counting the number of characters in each word in a tweet and dividing it by the total number of words in this tweet.

7) Negative tone. To calculate the negative tone of the message, a collection of the most popular words in Russian was compiled. Each word is assigned a value from -5 to 5, where words with the maximum negative hue take the value -5, and with the most positive take the value 5. If the word is not found, the value 0 is assigned. The search is performed using regular expressions.

8) Speech act category. Not every message carries information that may be a rumor. For example, messages expressing opinions about the event cannot be a rumor, because, carrying information about the opinion of the author regarding the event, do not carry information about the event itself in an affirmative form. According to the works [17, 24] rumor is usually expressed in the form of a assertion, the definition of which represents the task of classification. In Sect. 3, we present the results of experiments to refine the method of filtering tweets by the type of speech act.

4.3 User Identities

User identities reflect the characteristics of users involved in spreading rumors. We used the following characteristics [18]:

1) Originality is calculated by ratio of the number of original tweets produced by user, and the number of retweets of someone else's original tweet. Originality is an indicator of how original Twitter posts are. Originality is calculated by the ratio of the number of source tweets that the user produced, and the number of times the user simply rewrote someone else's original tweet. The greater this ratio, the more inventive and original the user is. Conversely, a lower ratio indicates the opposite behavior of the user.

2) Credibility. Each user on Twitter is labeled by Twitter itself to indicate the user's official pages. Accordingly, if the user has this marking, then the credibility of this user is 1, otherwise 0.

3) Influence. It is measured by the simple number of people reading the user. Presumably, the more users, the more influential is the user. The magnitude of the effect is equal to the number of subscribers the user.

4) Role. This is the number of people who read the user divided by the number of people that the user reads. This implies two roles:

Distributor. If the number of subscribers is greater than subscriptions, then the role coefficient is greater than 1;
Receiver, otherwise.

5) Engagement. This value shows how active the user has been for the entire time since the registration date (in years) and is calculated as follows:

Engagement = (#Tweets + #Retweets + #Replies + #Favourites)/AccountAge
If the value is less than 1, then the value is set to 0.5.

4.4 Distribution Dynamics

Distribution features record the temporal diffusion dynamics of rumors. To determine the propagation dynamics, it is necessary to construct a diffusion tree, reflecting the paths of retweet tweets, and to analyze it [5, 11]. It was found that a total of 6 propagation features significantly affect the model results.

We use the following features determined from the diffusion graph [18]:

1) The fraction of diffusion from low to high (LHD). It characterizes the number of diffusion events, when diffusion was from the sender with less influence on the receiver with higher influence. The influence of the user corresponds to the number of his subscribers. As an example of low-high diffusion, consider the following situation. Suppose we have user A, who has 10 thousand subscribers and user B, who has 10 subscribers. In this case, low-high diffusion will occur only if User A retweets User B's tweet, but not vice versa. The value is calculated as the ratio of the number of such events to the total number of diffusion events. This function models the behavior of an eyewitness that is observed during real events.

2) The fraction of nodes in the largest connected component. A connected component of the graph is a subgraph in which a node is accessible from any other node [11]. The largest connected component (LCC) is the component with the largest number of nodes. In the Twitter distribution box, the LCC matches the original tweet with the most retweets. The proportion of nodes in the largest connected component measures the ratio of nodes in the largest connected component of the diffusion graph of rumors to the total number of nodes in the diffusion graph and takes a value from 0 to 1. [18] This function captures the longest chain of rumor conversations on Twitter and indicates how deep is the rumor.

3) The average depth-to-width ratio (APDW). It is an attempt to quantify the shape of the diffusion graph of rumors, which can tell a lot about the nature of the spread of rumor. The depth of the diffusion tree is defined as the longest path from the root to

the leaf. Tree width is defined as the total number of nodes contained in it. Since each diffusion graph of rumors consists of many diffusion trees, the sum of the values of the ratio of depth to width of diffusion trees in rumor is averaged by their number.

4) The ratio of the original tweets (ONT). It shows how fascinating, interesting and original is the conversation about the rumors. This is measured by the ratio of new tweets and replies to the sum of tweets and replies and retweets in the diffusion graph of a rumor.

5) The fraction of tweets containing external links (NWU). It calculates as the ratio of tweets with link to an outside source.

6) The fraction of isolated nodes (IN). It characterizes the proportion of tweets that receive no answers or retweets and are displayed as isolated nodes in the graph of rumors.

All these functions are derived from the diffusion graph of rumors. The Twitter API does not provide a true way to retweet a tweet. Each edge in the rumor propagation graph corresponds to a diffusion event. Each diffusion event occurs between two nodes. Diffusion events are directed (in the direction of time), with information spreading from one node to another over time. We will call the node that pushes the information (i.e., influences), the sender and the node that receives the information (i.e., the one affected), the recipient.

The diffusion graph building is as follows. It starts the search for a tweet based on the content in it of a mention of an event, the rumor that interests us. With the help of the API, we obtain data on retweets among the results obtained, which include the ID, the date of creation and the user ID of the retweet. According to the user ID, we get a list of subscription IDs and look for among them the user ID of the original tweet. If the search is successful, then this ID is added to the root of the tree. If the search was unsuccessful, then the user ID of the original tweet is searched among the subscriptions of each retweeted tweet. If there are several coincidences, then using time sorting, among the list of matched ones, we find out who was the last to determine to which node the node will be added.

5   System Architecture

A software application is formed as a Python library, i.e. can be used both independently and provide functions for other libraries. The application performs the following functions:

In training mode:

- Fulfillment of a request to the Twitter network to collect tweets (in JSON format).
- Saving tweet data in a database.
- Data cleansing and preprocessing.
- Processing in the filtering module by type of speech act (approval).
- Preparation of parameters for training/testing the neural network.
- Training a neural network and saving the model in CSV files.

In tweet check mode (rumor or not):

- Processing a user request: links to tweets entered the GUI.
- Tweet data cleansing and preprocessing.
- Preparation of parameters for testing via neural network.
- Processing in the neural network and obtaining the likelihood of rumor in the provided tweets.
- Output results in a graphical interface.

For comfortable work with the application, a graphical interface was developed using the PyQT5 library and developed using the PyQT5 Designer (https://python-scripts.com/pyqt5).

6 Experiments and Results

6.1 Speech Act Classification

We use four categories of speech act: assertion, recommendation, expression, question [25]. As options for the implementation of the multi-class classification, four methods were chosen – the naive Bayes classifier, the decision tree, logistic regression, and the support vector machine. For all four methods, their implementation was used in the Python Scikit-learn package [15].

For classifier training, 196 tweets were manually tagged. Content features were calculated for each tweet. In addition to 81 static features, 7 additional features were obtained. Testing was performed using cross-validation of 14 blocks. The results of the calculation of the F1-measure are presented in Table 3.

Table 3. F1 scores for each speech act category.

Method	Assertion	Recommendation	Expression	Question
Decision tree	.888	.667	.706	.500
Logistic regression	.878	.667	.698	.625
Naive Bayes classifier	.880	.632	.709	.609
Support vector machine	.879	.640	.719	.645

In determining the category of the support vector method showed the best result. Thus, the classification by the support vector machine was chosen and used further. The resulting algorithm is used to classify input tweets. A specific tweet class is stored in the module database for future use.

6.2 Rumor Detection

In this work we used a multilayer perceptron for the classification of tweets: rumor or not. Training of the neural network model was carried out in several stages. The first is the loading and sample preparation for training. The data was saved in the CSV format and downloaded using the Pandas library in the Dataframe format [13]. Cross-validation was performed using the scikit-learn library with parameters: using 5-fold cross validation, KFold cross-validation type in a ratio of 70\30. We used the following optimal parameters of neural network: solver: stochastic gradient descent; hidden layer sizes: 25; penalty parameter (alpha): 10-6; learning rate: constant; initial learning rate: 0.2; activation function: hyperbolic tan, iteration number: 100.

An example of outputting tweet parameters in test mode is shown in Fig. 1. The program calculated the parameters of the user who sent this tweet: Originality: 0.59, Confidence: 1.0, Flow: 8786, Role: 51.42, which means Distributor. The probability of the presence of rumor in this tweet is 37%, which is low, since it does not exceed 50%.

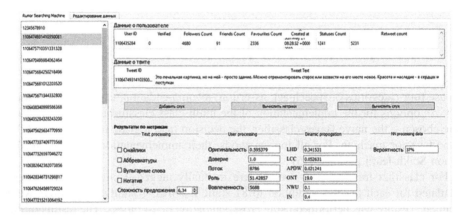

Fig. 1. The example of output tweet parameters.

The result of classification of tweets was evaluated using the metrics accuracy, recall, precision, F-measure. We got the following results: accuracy 0.927, precision 0.897, recall 0.932, F-measure 0.914.

7   Conclusion

As a result of this work, the software for rumor detection in messages in the Russian-speaking segment of Twitter has been developed. On a hand-made dataset, the presence of rumor is determined with the accuracy of up to 92%.

In the future, we plan to conduct tests on a large dataset, as well as to work out the technology of searching for rumors in tweets based on repeated topics for a certain period.

References

1. Alton, Y.K., Snehasish, B.: Linguistic predictors of rumor veracity on the internet. In: Proceedings of the International MultiConference of Engineers and Computer Scientists 2016 IMECS, vol. 1, pp. 387–391 (2016)
2. Crystal, D.: Language and the Internet. Cambridge University Press, Cambridge (2006)
3. Diab, M., Sardar, H.: Rumor detection and classification for twitter data, pp. 71–77 (2015)
4. Friggeri, A., Adamic, L.A., Eckle,s D., Cheng J.: Rumor cascades, pp. 101–110 (2015)
5. Goel, S., Watts, D.J., Goldstein, D.G.: The structure of online diffusion networks. In: Proceedings of the 13th ACM Conference on Electronic Commerce, pp. 623–638 (2012)
6. Jin, C., Zhang, L.: News verification by exploiting conflicting, pp. 2972–2978 (2016)
7. Jing, M., et al.: Detecting rumors from microblogs with recurrent neural networks. In: Proceedings of the Twenty-Fifth International Joint Conference on Artificial Intelligence, pp. 3818–3824 (2016)
8. Kirk, M.: Thoughtful Machine Learning with Python, vol. 1, p. 220. O'Reilly Media, Sebastopol (2017)
9. Kurt, T., Chris, G., Dawn, S., Vern, A.P.: Suspended accounts in retrospect: an analysis of twitter spam. In: Proceedings of the ACM SIGCOMM Conference on Internet Measurement Conference, pp. 243–258 (2011)
10. Kwon, S., Cha, M., Jung, K., Chen, W., Wang, Y. Prominent features of rumor propagation in online social media. In: International Conference, pp. 1103–1108 (2013)
11. Liu, L., Qu, Q., Chen, B., Hanjalic, A., Wang, H.: Modelling of information diffusion on social networks with applications to WeChat. Phys. A Stat. Mech. Appl. **496**, 318–329 (2018)
12. Mozdeh Big Data Text Analysis. http://mozdeh.wlv.ac.uk. Accessed 01 June 2019
13. Pandas: powerful Python data analysis toolkit. http://pandas.pydata.org/pandas-docs/stable/index.html. Accessed 01 June 2019
14. Paramita, M.: Extracting temporal and causal relations between events. In: Proceedings of the ACL 2014 Student Research Workshop, pp. 10–17 (2014)
15. Pedregosa, F., et al.: Scikit-learn: machine learning in python. J. Mach. Learn. Res. **12**, 2825–2830 (2011)
16. Qazvinian, V., Rosengren, E., Redev, D., Qiaozhu, M.: Rumor has it: identifying misinformation in microblogs. In: Proceedings of the 2011 Conference on Empirical Methods in Natural Language Processing, pp. 1589–1599 (2011)
17. Searle, J.R.: Expression and Meaning: Studies in the Theory of Speech Acts. Cambridge University Press, Cambridge (1985)
18. Soroush, V.: Automatic Detection and Verification of Rumors. Massachusetts Institute of Technology, Massachusetts (2015)
19. Takahashi, T., Igata, N.: Rumor detection on twitter. Soft Comput. Intell. Syst. (SCIS) **6**, 452–457 (2012)
20. Twitter libraries — Twitter Developers. https://developer.twitter.com/en/docs/developer-utilities/twitter-libraries.html. Accessed 01 June 2019
21. Weiling, C., Zhanga, Y., Tong, C., Bu, S.: Unsupervised rumor detection based on users' behaviors using neural networks. Pattern Recogn. Lett. **105**, 226–233 (2018)
22. Wikipedia contributors. List of emoticons, Wikipedia, The Free Encyclopedia. https://en.wikipedia.org/wiki/Listofemoticons. Accessed 01 June 2019
23. Wierzbicka, A.: English Speech Act Verbs: A Semantic Dictionary. Academic Press, Sydney (1987)

24. Zhang, R., Gao, D., Li, W.: What are tweeters doing: recognizing speech acts in twitter. Analyzing Microtext (2011)
25. Zhao, X., Jiang, J.: An empirical comparison of topics in twitter and traditional media. Singapore Management University School of Information Systems Technical paper series (2011)
26. Zhao, Z., Resnick, P., Mei, Q.: Enquiring minds: early detection of rumors in social media from enquiry posts. In: IW3C2 (2015)
27. Zubiaga, A., Aker, A., Bontcheva, K., Liakata, M., Procter, R.: Detection and resolution of rumours in social media: a survey. ACM Comput. Surv. **51**, 2–38 (2018)

Automatic Prediction of Word Form Reduction in Russian Spontaneous Speech

Maria Dayter[iD] and Elena Riekhakaynen[✉][iD]

Saint-Petersburg State University,
Universitetskaya emb. 7/9, 199034 St. Petersburg, Russia
mvdayter@gmail.com, e.riehakajnen@spbu.ru

Abstract. The aim of this paper is to determine the main features of word forms that undergo reduction in Russian spontaneous speech using machine learning algorithms. We examined the factors proposed in the previous corpus-based studies, namely, the number of syllables, word frequency, part of speech, reduction of the preceding word. We used the texts from the Corpus of Russian Oral Speech as the data. The following machine learning algorithms were applied: extremely randomized tree, random forest and logistic regression. The results show that the higher the frequency of a word form is, the higher the chances are that it will be reduced; the more syllables a word form has, the higher probability of its reduction is. But we did not find the influence of the factor whether the previous word is reduced or not. Regarding the part-of-speech feature, the results are not that straightforward since we received different lists using different algorithms, but the adjective and parenthetical word were in both of them. Thus, we can conclude that adjectives and parenthetical words are likely to be reduced more often than other parts of speech.

Keywords: Phonetic reduction · Speech · Machine learning · Russian

1 Introduction

The fact of phonetic reduction of words (the loss of one or more sounds in a word) has been reported to be an essential feature of natural speech in many languages [1–4, etc.]. Over the past twenty years, several psycholinguistic experiments have been conducted in order to determine the status of reduced word forms in the mental lexicon of a native speaker and to describe the processing of such units by a listener and a speaker. We assume that the understanding of the reasons underlying phonetic reduction of words can contribute to answering these questions.

Factors influencing phonetic reduction can be linguistic or extralinguistic [5]. Extralinguistic ones include the socio-psychological characteristics of speakers, the situational factors and the speaker's intention. The more spontaneous the speech is, the more likely reduction is to occur. However, words can be reduced while reading as well.

Among the linguistic factors one can find so-called "proper phonetic factors", i.e. the phonetic nature of a sound and its position in the word and phrase (for instance, stressed vowels undergo reduction less often than unstressed ones); the morpho-phonetic structure of a word; speech rate and the position of a word in a clause [5]; the

© Springer Nature Switzerland AG 2020
A. Karpov and R. Potapova (Eds.): SPECOM 2020, LNAI 12335, pp. 119–127, 2020.
https://doi.org/10.1007/978-3-030-60276-5_12

frequency of a word as well as the likelihood of a word to appear in a particular context [6, 7].

Riekhakaynen in a corpus-based study of words that are not reduced in spontaneous speech (i.e. are reduction resistant) [4] put forward the following hypotheses:

- low-frequency word forms are reduced less often;
- monosyllabic word forms are reduced less often;
- nouns are reduced less often than other parts of speech;
- if a word is used with a reduced adjective or verb, then it is unlikely to undergo reduction.

We decided to test the above-mentioned hypotheses using machine learning methods. We regard the task of finding out the factors that influence phonetic reduction as a classification task. There are several machine learning algorithms used for solving classification tasks, and all of them are used in linguistic research. The supervised learning algorithm solves such linguistic tasks as information extraction, resolving lexical ambiguity (sense disambiguation), document classification reference resolution, automatic dialogue analysis; unsupervised learning is applied for studying probabilistic context-free grammar and lexical ambiguity resolution; reinforced learning is used for speech recognition and part-of-speech tagging [8]. Application of machine learning algorithms in phonetic research is described in [9], where the probability of the occurrence of different phonetic phenomena in Italian was predicted. In [10], machine learning algorithms were used to model prosodic phenomena. Many tasks concerning language behaviour modelling are being solved by means of neural networks. For example, in [11], a deep neural network was used to transcribe conversational speech. Algorithms of automatic speech recognition for the English language have been developed using deep neural networks based on hidden Markov models [12, 13], convolutional neural networks [14], recurrent neural networks [15]. In [16], a system for speech recognition based on a neural network that does not need a dictionary of phonemes and tokens was tested. In [17], a standard approach to speech recognition using neural networks based on the material of the Czech language is described. The results of the deep neural networks application for the recognition of Russian speech are described in [18, 19]. As far as we know, the machine learning algorithms have not been used for studying the reasons of phonetic reduction in Russian so far.

In the paper, we will first describe the data (Sect. 2.1), then we will provide a detailed description of the methods (Sect. 2.2) and the results we obtained (Sect. 2.3), and we will discuss the results (Sect. 2.4) and propose some future plans (Sect. 3) in the end.

2 Experiments

2.1 Data

We used the transcripts from the Corpus of Russian Oral Speech (http://russpeech.spbu. ru/; last accessed: 28.05.2020) as the material for our research. The corpus includes different types of Russian oral texts (radio interviews, talk-shows, reading, etc.). The size of the corpus is more than 22,000 words now. All audio recordings are provided with both orthographic annotation and acoustic-phonetic transcription. The transcripts and the audio files are available online. For a more detailed description of the corpus see [20].

For our study, we chose a radio interview and two talk-shows from the corpus (the overall duration is around 94 min). All words (tokens) were extracted from the transcripts automatically using a program written in the Python programming language. A single token could be either a single word or a compound (for example, *potomu_chto* 'because', *i_tak_dalee* 'and so on', etc.). We checked the list of compound words in [21]. To identify such units in the transcripts and to calculate their frequency in the texts used, we developed another programme in Python. In total, the dataset contained 9181 one-word units and 287 compound ones.

Then we organized the data in a table with the following columns: orthographic annotation of a word form, its phonetic transcription (in the form in which it is presented in the corpus), the number of syllables, part of speech, frequency in oral speech (in absolute numbers and in ipm (instances per million words) – two different columns following each other), the frequency in the texts used (in absolute numbers for a particular text and in ipm - two different columns following each other), the presence or absence of reduction of the previous word, the presence or absence of reduction of the word itself (a word was considered reduced if there was least one sound omitted compared to the canonical pronunciation; this was the target variable). The table and the scripts for all the programs mentioned in the text are available here: https://cloud.mail.ru/public/4Zwh/4CwfRSc55.

In [4], a reduced adjective or verb in the close context of a noun is regarded as one of the factors that contribute to the absence of the reduction of this noun. Since we analyzed not only nouns, we decided to modify this feature and took into account the presence/absence of the reduction of the word preceding the target one, without taking into account the part-of-speech attribution of both the target word and the word preceding it. The number of syllables was calculated automatically by the number of vowels in a word (a program written in Python by one of the authors was used for this task). The information on the frequency of words in oral speech was retrieved from the oral subcorpus of the Russian National Corpus (http://ruscorpora.ru/new/; the number of entries – 13,001,027 words; last accessed: 15.04.2020). The frequency of words in each of three texts we used as the material for our study was examined separately for each text and then used to build algorithms for each of the texts. The part-of-speech tagging was performed manually based on the approach proposed by Shcherba [22], in the controversial cases we consulted the dictionaries available at: http://gramota.ru/ (last accessed: 15.04.2020). The names (labels) of the parts of speech we used for tagging were the ones proposed in these dictionaries.

2.2   Method

We decided to use the tree-based algorithms and logistic regression in our study. Decision trees were chosen because of their interpretability, robustness to outliers, and the ability to work with both quantitative and nominal variables [23]. Due to its sensitivity to the selection of features, the logistic regression can be used to verify the selection of features.

Since the decision tree algorithm is prone to overfitting (lack of generalization ability), a choice was made in favor of tree-based ensemble methods, namely, random forest and extremely randomized tree. The code was written in Python version 3.7.4, the following libraries were used: Scikit-learn – for building the algorithms, pandas – for data processing, NumPy – for working with mathematical functions.

During the preparatory stage of constructing the algorithm, we imported the dataset, made the feature "PoS" ("part-of-speech") categorical, divided this feature into separate parts of speech corresponding to all categories we used for part-of-speech tagging. For this purpose, a dictionary was created, the keys of which were numbers, and the values were the names of parts of speech. In order to scale the data, we converted the feature "frequency (in ipm)" to binary. To do this, we used the value of the third quartile of ipm measures for unique word forms as a threshold. The word forms with ipm above 94.68 were classified as the word forms of high frequency.

A complete list of features was as follows (the examples of all the part-of-speech categories are available in the dataset table: https://cloud.mail.ru/public/4Zwh/ 4CwfRSc55/dataset.xlsx):

- number of syllables ("syllables");
- frequency ("ipm_binary");
- is the previous word form reduced ("isPr");
- whether the word is a noun ("is_noun");
- whether the word is a verb ("is_verb");
- whether the word is an adjective ("is_adj");
- whether the word is an adverb ("is_adv");
- whether the word is a participle ("is_part");
- whether the word is a pronoun ("is_pronoun");
- whether the word is a preposition ("is_prep");
- whether the word is a conjunction ("is_conj");
- whether the word is a pronominal adverb ("is_pr_adv");
- whether the word is a verbal adverb ("is_deepr");
- whether the word is a particle ("is_particle");
- whether the word is a category of state ("is_stat");
- whether the word is a pronominal numeral ("is_pr_num");
- whether the word is a numeral ("is_numer");
- whether the word is an interjection ("is_inter");
- whether the word is a parenthetical word ("is_vvod");
- whether the word is a nominal pronoun ("is_pr_noun");
- whether the word is a pronominal adjective ("is_pr_adj").

We separated the target variable into the corresponding vector.

The first way to solve the problem of finding the required features seemed to be an exhaustive search. If we had chosen this method, it would have been necessary to compose all combinations of n by k (in our case, n is twenty, since our dataset has twenty features, respectively, k takes values from one to twenty), each such combi-nation should have been used as a subset of features to build a model on. There would have been 1,048,575 iterations for one algorithm only. We considered it to be unpractical. Therefore, to select features, we used the feature_selection module from the Scikit-Learn library, choosing the RFECV (recursive feature elimination with cross-validation) class, which recursively selects features and chooses the best set of features using cross-validation (the training set contained 80% of the data, the test set – 20%).

2.3 Results

To build the first model, we used an extremely randomized tree. As an optimized metric, balanced accuracy was used, since the classes in our sample are not balanced (32% of reduced word forms). As a result, we got five features (Subset 1): number of syllables ("syllables"), frequency ("ipm_binary"), whether the word is an adjective ("is_adj"), whether the word is a preposition ("is_prep"), whether the word is a parenthetical word ("is_vvod"). The accuracy of the predictions of the model for these five features was 80.30%, balanced accuracy – 75.08%. Based on the selected features, we built a logistic regression. One of the part-of-speech features – the preposition – received a negative coefficient, which can be interpreted as the inverse relationship between this feature and the target variable. It means that if a word is a preposition, it most likely will not undergo reduction. The remaining features received positive coefficients. The accuracy of the model predictions was 80.30%, and the balanced accuracy was 75.28%.

To build the second model we used a random forest algorithm. Balanced accuracy was also chosen as an optimized metric. As a result, a slightly different list of features was obtained (Subset 2): the number of syllables ("syllables"), frequency ("ipm_binary"), whether the word is an adjective ("is_adj"), whether the word is a parenthetical word ("is_vvod"). Based on these features, a logistic regression was built. All four features received positive coefficients. The accuracy of the model predictions was 80.30%, and the balanced accuracy was 75.32%. However, this model required more iterations than the previous one (250 vs 145 respectively). The results are summarized in the Table 1.

Table 1. The results of the application of the extremely randomized tree, random forest algorithms and logistic regression.

Algorithm	Extremely randomized tree		Random forest		Logistic regression	
Metrics	Accuracy	Balanced accuracy	Accuracy	Balanced accuracy	Accuracy	Balanced accuracy
All features	78.47%	72.50%	78.59%	72.74%	77.88%	72.57%
Subset 1 (five features)	80.30%	75.08%	–	–	80.30%	75.28%
Subset 2 (four features)	–	–	80.30%	75.21%	80.30%	75.32%

We also aimed to check if the frequency of a word form in the text where it was used influences whether it will be reduced or not. We divided our data into three subsamples using the pandas module according to the number of texts used in the study (two talk-shows and one radio interview). Three datasets were obtained: the first one – 1583 word forms for 21 features, the second one – 5368 word forms for 21 features, and the third one – 2571 word forms to 21 features. The frequency value in a specific text (in ipm) for each of the subsamples was converted into a binary variable, as we did while processing of the entire dataset. Building models of both an extremely randomized tree and a random forest using the first subset (containing only 1583 word forms) turned out not to be successful. Therefore, we decided to work only with the second dataset that had more entries than the first and the third ones.

When we built an extremely randomized tree, the only feature that received the rank 1 was the number of syllables. The model based on a random forest showed the same results. The balanced accuracy of the logistic regression constructed with one predictor (the number of syllables) was 65.23%, and the accuracy was 75.13%. This is an acceptable result, but we made an attempt to improve it. When constructing a model for selecting features, we set a minimum number of features for selection equal to three. The result of the model based on a random forest turned out to be 18 features that is not satisfactory in the framework of this study. A model with the same parameters based on an extremely randomized tree selected three features: the number of syllables, frequency, and whether the word is an adjective. As a result of constructing a logistic regression, all three features received positive coefficients. The accuracy of the model was 77.28%, the balanced accuracy was 72.32%. The feature "frequency (in ipm) in a particular text" that interested us got the rank 4.

2.4   Discussion

Having built two different models, we obtained two different lists of features. Our results support the hypotheses that word frequency and the number of syllables influence the chances of a word form to be reduced or not. However, we did not find the influence of whether the previous word is reduced or not. Concerning the part-of-speech feature, the results are not that straightforward. The feature "is the word a preposition" was not included in the second list (when a random forest algorithm was used), since it received the rank 2. In this case, the logistic regression required 250 iterations to build a model on this list of four features versus 145 iterations when training on a data set that included this feature. Thus, we can assume with a certain degree of confidence that prepositions undergo reduction quite rarely.

It is worth noticing that the feature "whether the word is a noun" that was mentioned as one of the factors in the corpus-based research of Russian spontaneous speech [4], did not receive the rank 1 in any of the experiments. A possible explanation of this fact is the distribution of parts of speech in our dataset and the percentage of reduced wordforms within every part of speech. The nouns constitute 23% of the dataset, 37% of them are reduced. The same values for adjectives are 6% and 79% respectively, for parenthetical words – 1% and 77% respectively. 8% of the words in the dataset are prepositions and only 2% of them are reduced.

The percentage of reduced word forms among adjectives is high, while only 12% of adjectives in our data are the words of high frequency. Probably, this is because adjectives are usually accompanied by nouns, so the loss of grammatical information in an adjective is not that crucial. A high percentage of reduced parenthetical words can be explained by their frequency, since 94% of the parenthetical words in our data are the words of high frequency.

As for building a model based on smaller datasets and checking the influence of the word frequency in a given text, this part of the research requires further consideration using larger datasets.

3 Conclusions and Future Plans

The aim of this paper was to examine the hypotheses about the linguistics factors influencing word form reduction in Russian spontaneous speech using machine learning algorithms. We retrieved the data from the Corpus of Russian Oral Speech and analyzed such factors as the number of syllables, word frequency, part of speech, reduction of the preceding word. The following machine learning algorithms were applied: extremely randomized tree, random forest and logistic regression. The results show that the higher the frequency of a word form is, the more likely it is to be reduced; the more syllables a word form has, the higher probability of its reduction is; adjectives and parenthetical words tend to be reduced more often than other parts of speech.

The further research in the field can include, along with the testing of the discussed algorithms on larger datasets, the use of the support vector machine and perceptron algorithms. It will be interesting to compare the lists of word form features obtained in our study with those received as a result of using latter models. To predict the fact of reduction as such, we plan to build an algorithm of k nearest neighbors and neural networks based on the selected features. On smaller datasets, it is possible to use a naive Bayes classifier and to compare its results to the results of the logistic regression algorithm.

Acknowledgements. The research is supported by the grant #19-012-00629 from the Russian Foundation for Basic Research.

References

1. Kohler, K.J.: Segmental reduction in connected speech in German: phonological facts and phonetic explanations. In: Hardcastle, W.J., Marchal, A. (eds.) Speech Production and Speech. NATO ASI Series, pp. 69–92. Springer, Dordrecht (1990). https://doi.org/10.1007/978-94-009-2037-8_4
2. Brand, S., Ernestus, M.: Listeners' processing of a given reduced word pronunciation variant directly reflects their exposure to this variant: evidence from native listeners and learners of French. Q. J. Exp. Psychol. **71**, 1240–1259 (2018). https://doi.org/10.1080/17470218.2017.1313282
3. Ernestus, M., Baayen, R.H., Schreuder, R.: The recognition of reduced word forms. Brain Lang. **81**, 162–173 (2002). https://doi.org/10.1006/brln.2001.2514

4. Riekhakaynen, E.: Reduction in spontaneous speech: How to survive. In: Heegart, J., Henrichsen, P.J. (eds.) New Perspectives on Speech in Action: Proceedings of the 2nd SJUSK Conference on Contemporary Speech Habits (Copenhagen Studies in Language 43), Samfundslitteratur, Frederiksberg, pp. 153–167 (2013)
5. Stoyka, D.A.: Reduced forms of Russian speech: linguistic and extralinguistic aspects. PhD thesis, Saint Petersburg (2016). (In Rus.)
6. Zemskaya, E.A. (ed.): Conversational Russian Speech. Nauka, Moscow (1973). (In Rus.)
7. Jurafski, D., Bell, A., Gregory, M., Raymond, W.D.: Probabilistic relations between words: evidence from reduction in lexical production. In: Bybee, J., Hopper, P. (eds.) Frequency and the Emergence of Linguistic, pp. 229–254. John Benjamins, Philadelphia (2000)
8. Jurafsky, D., Martin, J.H.: Speech and Language Processing: An Introduction to Natural Language Processing, Computational Linguistics, and Speech Recognition. Prentice Hall, Upper Saddle River (2000)
9. Schiel, F., Stevens, M., Reichel, U., Cutugno F.: Machine learning of probabilistic phonological pronunciation rules from the Italian CLIPS corpus. In: Proceedings of the Annual Conference of the International Speech Communication Association, INTER-SPEECH, pp. 1414–1418 (2013). https://doi.org/10.5282/ubm/epub.18046
10. Vainio, M.: Phonetics and machine learning: hierarchical modelling of prosody in statistical speech synthesis. In: Besacier, L., Dediu, A.-H., Martín-Vide, C. (eds.) SLSP 2014. LNCS (LNAI), vol. 8791, pp. 37–54. Springer, Cham (2014). https://doi.org/10.1007/978-3-319-11397-5_3
11. Seide, F., Li, G., Yu, D.: Conversational speech transcription using context-dependent deep neural networks. In: Proceedings of Interspeech, pp. 437–440 (2011)
12. Ellis, D.P.W., Singh, R., Sivadas, S.: Tandem acoustic modeling in large-vocabulary recognition. In: IEEE International Conference on Acoustics, Speech, and Signal Processing. Proceedings, Salt Lake City, USA, vol. 1, pp. 517–520 (2001). https://doi.org/10.1109/ICASSP.2001.940881
13. Dahl, G., Yu, D., Deng, L., Acero, A.: Context-dependent pre-trained deep neural networks for large vocabulary speech recognition. IEEE Trans. Audio Speech Lang. Process. **20**(1), 30–42 (2012). https://doi.org/10.1109/TASL.2011.2134090
14. Zhang, Y., et al.: Towards end-to-end speech recognition with deep convolutional neural networks. In: Proceedings of Interspeech, pp. 410–414 (2016). https://doi.org/10.21437/Interspeech.2016-1446
15. Ravanelli, M., Serdyuk, D., Bengio, Y.: Twin regularization for online speech recognition. In: Karpov, A., Potapova, R., Mporas, I. (eds.) Proceedings of Interspeech (2018). https://arxiv.org/pdf/1804.05374.pdf. Accessed 28 May 2020. https://doi.org/10.21437/Interspeech.2018-1407
16. Maas, L., Xie, Z., Jurafsky, D., Ng, A.Y.: Lexicon-free conversational speech recognition with neural networks. In: Proceedings of the 2015 Conference of the North American Chapter of the Association for Computational Linguistics: Human Language Technologies, pp. 345–354. Curran Associates, New York (2015). https://doi.org/10.3115/v1/N15-1038
17. Mizera, P., Pollak, P.: Improving of LVCSR for causal czech using publicly available language resources. In: Karpov, A., Potapova, R., Mporas, I. (eds.) SPECOM 2017. LNCS (LNAI), vol. 10458, pp. 427–437. Springer, Cham (2017). https://doi.org/10.1007/978-3-319-66429-3_42
18. Kipyatkova, I.: Improving Russian LVCSR using deep neural networks for acoustic and language modeling. In: Karpov, A., Jokisch, O., Potapova, R. (eds.) SPECOM 2018. LNCS (LNAI), vol. 11096, pp. 291–300. Springer, Cham (2018). https://doi.org/10.1007/978-3-319-99579-3_31

19. Markovnikov, N.M., Kipyatkova, I.S.: Researching methods of building encoder-decoder models for Russian speech. Data Manage. Syst. **4**, 45–53 (2019). (in Russian)
20. Riekhakaynen, E.: Corpora of Russian spontaneous speech as a tool for modelling natural speech production and recognition. In: 10th Annual Computing and Communication Workshop and Conference, CCWC 2020, January 2020, pp. 406–411. IEEE, Las Vegas (2020). https://doi.org/10.1109/CCWC47524.2020.9031251
21. Ventsov, A.V., Grudeva, E.V.: A Frequency Dictionary of Russian. CHSU Publishing House, Cherepovets (2008). (in Russian)
22. Shcherba, L.V.: About parts of speech in the Russian language. In: Language System and Speech Behaviour, pp. 77–100. Nauka, Leningrad (1974). (in Russian)
23. Harrington, P.: Machine Learning in Action. Manning Publications, New York (2012)

Formant Frequency Analysis of MSA Vowels in Six Algerian Regions

Ghania Droua-Hamdani[(✉)]

Centre for Scientific and Technical Research on Arabic Language Development
(CRSTDLA), Algiers, Algeria
gh.droua@post.com

Abstract. The paper deals with a formant analysis of short vowels in Modern Standard Arabic (MSA) language. The investigation aims to highlight similarities and differences in vowel quality in MSA. The vowels (/a/, /u/and /i/) were exracted from 565 speech files recorded by 163 Algerians that were belonging to six regions. Three formants (F1, F2 and F3) were computed from the data set of all the regions. Statistical analyses were conducted on formant values to show the variability depending on the regional accent and the gender of speakers. The outcomes indicate a significant effect of regional accent and gender of speakers in specific vowels.

Keywords: Formants · Statsitical anlaysis · Modern standard arabic MSA · Regional accent · Algerian speakers

1 Introduction

Speech conveys several kinds of information expressed into 3 categories; linguistic, paralinguistic, and nonlinguistic information. While, the linguistic information deals with the wanted message to be transmitted, the paralinguistic and nonlinguistic information concern all controlled and uncontrolled supplements given by the speaker that are different from the linguistic information such as attitudes, emphasis, speaking styles, age, gender, regional accent, physical and emotional states of the speaker, and so on. In speech technology, being able to model paralinguistic and nonlinguistic information stills to be a great challenge for researchers, as it is the case in speech recognition, emotion recognition, regional accent identification, etc. [1–3].

The present study takes a part of across languages accent patterns project which aims to build an automatique identification system of Arabic regional accent using acoustics and prosodic features (rhythm, formants, melody, etc.). This work focuses on the investigation of vowel variation quality (the first, second and third formants, hereafter F1, F2 and F3) within Arabic language accent. Depending on the intended applications, there are numerous researches dedicated to the study of the quality of Arabic vowels, whether in Arabic dialects or in Modern Standard Arabic (MSA) [4–7]. However, investigations on MSA Arabic produced by Algerian speakers are less available. Therefore, the objective of the study is to highlight similarities and differences in vowel production in MSA speaking in Algeria using statistical analyses. Thus, we examined speakers' accent, gender, and variations in individual articulation formant

© Springer Nature Switzerland AG 2020
A. Karpov and R. Potapova (Eds.): SPECOM 2020, LNAI 12335, pp. 128–135, 2020.
https://doi.org/10.1007/978-3-030-60276-5_13

within a corpus produced by 163 Algerian speakers belonging to six scattered regions in the country.

The paper is organized as follows. In Sect. 2, a brief overview of Algerian languages. Section 3 exposes speakers profiles and speech material used in the study. Section 4 describes the methodology used to extract the formant data used. The experimental setup and findings are discussed in Sect. 5. Lastly, Sect. 6 gives the concluding remarks based on the analysis.

2 MSA and Algerian Languages

MSA includes 34 phonemes: 28 consonants and 6 vowels (three short vowels /a/, /u/and /i/vs. three long /a: /, /u: /and /i: /). Based on the Classical Arabic in terms of lexicon, syntax, morphology, semantics and phonology, it is the standardization of Arabic spoken in 22 countries called Colloquial Arabic (CA). However, it is observed that Arabic speakers tend to be influenced by their regional accent during speaking MSA as for colloquial Egyptian. The influence of Arabic dialects is more salient at the phonological level.

Arabic and Tamazight (Berber) are the official languages of Algeria. Algerian Arabic dialect and Berber are the native spoken languages of Algerians. French is also widely used in media, culture, etc. In Algeria, MSA differs substantially from the Arabic Algerian dialects. As a result, there are significant local variations (in pronunciation, grammar, etc.) of spoken Arabic, and many of its varieties can be encountered across the country. In phonetic level, we can observe a variability in pronunciation of many consonants such as [q], [θ], [t] etc.

3 Speakers and Speech Material

The subjects of this study were 163 adult Algerian speakers (83 female/80 male) taken from ALGerian Arabic Speech Database (ALGASD) [8]. The database used includes 565 speech files recorded from 6 regions that cover six dialect pronunciation groups. Three regions from the north (Algiers -capital city-, Tizi Ouzou -Berber region- and Jijel) called in the text R1, R2 and R3 respectively. Moreover, three others localities from the south of the country (Bechar, El Oued and Ghardaia) were submitted to the study. They were named R4, R5 and R6. Since most of the Algerian population is settled in the north of the country, the distribution of speakers according to their number and gender is proportional to the population of the regions. Therefore, the number of recordings was depended on studied localities. All speech material were recorded on an individual basis in an acoustically controlled environment. Speakers were instructed to read the MSA sentences at a comfortable rate in a normal voice. Recordings were made with a high-quality microphone at a sampling rate of 16 kHz. were randomly distributed among the participants. The number of readings varied from two to six sentences per speaker. Table 1 shows the distribution of speakers per region.

Table 1. Distribution of speakers per regions.

Regions	Female Spk.	Male Spk.
R1	40	37
R2	17	17
R3	9	6
R4	3	4
R5	8	8
R6	6	8
Total	83	80

Spk. : speakers.

4   Measurement

Formants are distinctive frequency components of the speech signal that express frequency resonances of the vocal tract. In speech processing, formants (F1–F5) correspond to the local maxima in the spectrum. In the study, to get the speech formant, an experimented annotator segmented manually all speech material i.e. 565 speech files of the dataset onto their different units (vowels and consonants) using Praat software. From these segments, the formants were computed automatically using Linear Predictive Coding Coefficients (LPCC) from all short vowels formants (/a/; /u/and /i/). Most often, the two first formants, F1 and F2, are sufficient to identify the kind of vowel. However, in this study we extracted three formants (F1, F2 and F3) values for each vowel to get more information about a possible variation within the same segment due to speaker regional accent. Data were submitted to a MANOVA [9] to test for significant differences across regions, gender and vowels.

5   Results

5.1   Region Group Differences

The first experimentation concerns the assessment of the mean values of all formants regardless gender of speakers (male/female). Table 2 illustrates the average and standard deviation values of F1, F2 and F3 computed for each short vowel (/a/, /u/and /i/) per region. As it can be seen in the Table 2, the mean values of F1 for the vowel /a / vary according to the regions. It can be observed that the F1 means obtained for R1 and R3, that are located in the north of the country, are closed while R2, which is also in the north, gives a result nearby that ones gotten for R3, R4 and R5 which are in the southern localities. Regarding the standard deviations, the less value is computed for R6 speakers that seems that the F1 spreading in this region is in reduction comparing of its counterpart. Likewise, F2 of /a/measured from R5 data is upper than other values. For the vowel /u/, the higher F1, F2 and F3 means are achieved when the participants were belonging to R4 region. It can be noticed from findings, that F1 average computed for the vowel /i/have globally a same tendency in all localities. However, we point out a large variation in the standard deviations between the northern regions and the southern ones. Regarding, F2 and F3 formants scores, the outcomes show that R2 expresses the lowest averages in comparison to other regions. From the findings, we can also conclude then there are some diffrences in vowels production especially for /a/and /u/vowels.

Figure 1 shows the density of vowels spreading on (F1, F2) plan for each vowel per region. The outcomes reveal a wide distribution of vowels on (F1, F2) plan especially in the cas of /a/and /u/. The results suggest that there is a difference in vowel production in particular for the vowel /u/for regions situated in the south. In the other hand, the opposite is observed when it comes to vowel /i /where a mass of data representing all values measured for the six regions is gathered in one consolidated cluster.

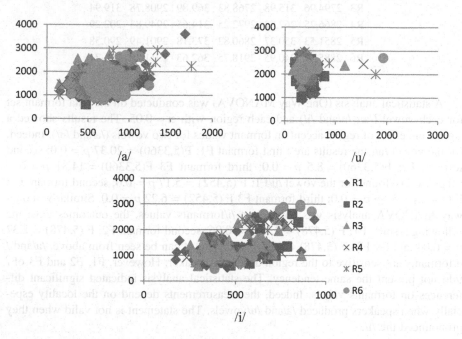

Fig. 1. Distribution of MSA short vowels in (F1, F2) plan by regions.

Table 2. Means and standard deviation of F1, F2 and F3 (Hertz) of MSA short vowels per regions.

		/a/	std	/u/	std	/i/	std
F1	R1	675,54	150,28	449,46	120,86	389,53	82,72
	R2	730,69	141,13	498,60	116,59	404,69	91,86
	R3	661,23	155,17	473,40	143,54	408,47	142,13
	R4	706,13	162,08	561,02	158,64	463,99	275,48
	R5	721,61	180,42	476,47	112,09	433,96	195,65
	R6	722,75	128.80	505,60	168,02	441,91	285,56

(continued)

Table 2. (*continued*)

		/a/	std	/u/	std	/i/	std
F2	R1	1548,05	295,45	1101,39	322,89	2116,95	340,72
	R2	1589,15	253,46	1124,45	322,98	2098,20	298,46
	R3	1578,66	312,90	1292,91	425,27	2231,84	331,95
	R4	1524,64	266,43	1518,92	650,41	2235,16	278,14
	R5	1647,19	313,25	1267,04	556,93	2207,92	341,84
	R6	1598,66	313,91	1340,48	693,57	2195,51	229,76
F3	R1	2755,27	307,27	2697,32	329,38	2850,01	296,67
	R2	2804,36	281,91	2722,61	298,48	2809,66	298,46
	R3	2794,96	315,98	2768,83	369,89	2908,78	319,94
	R4	2866,25	262,62	2972,25	314,65	2949,83	202,29
	R5	2851,53	350,37	2860,82	373,18	2901,49	290,58
	R6	2885,37	240,95	2918,75	362,63	2911,36	236,71

A statistical analysis (One Way MANOVA) was conducted on the data formant set for each vowel (/a/; /u/and /i/) and each region with $\alpha = 0.05$. The results showed a significant effect of region accent on formant values for two vowels (/a/and /u/). Indeed, for the vowel /a/, the results are : first formant F1: $F(5,3360) = 20.37$ $p = 0.0$; second formant F2: $F(5,3360) = 8.5$ $p = 0.0$; third formant F3 $F(5,3360) = 14.81$ $p = 0.0$. Likewise, we found for the vowel /u/F1: $F(5,452) = 5.17$ $p = 0.0$; second formant F2: $F(5,452) = 7.56$ $p = 0.0$; third formant F3 $F(5,452) = 6.22$ $p = 0.0$. Similarly, a one-way MANOVA analysis was applied on /i/formants values, the outcomes gave the following results. F1: $F(5,478) = 2.15$ $p = 0.06$; second formant F2: $F(5,478) = 2.37$ $p = 0.04$ and for F3: $F(5,478) = 1.85$ $p = 0.1$. As it can be seen from above, /a/and /u/formants are sensitive to the region accent of speaker. However, F1, F2 and F3 of /i/do not present the same tendency. The statistical analysis indicated significant differences on formants values. Indeed, the measurements depend on the locality especially when speakers produced /a/and /u/vowels. The statement is not valid when they pronounced the /i/.

5.2   Gender Group Differences

The second experimentation concerns the assessment of the mean values of all formants regarding gender of speakers (male/female) by region. The aim of this analysis is to put forward possible variation in vowel production regarding the interaction between gender and regions factors. Figure 2 illustrates the distribution of vowels on (F1, F2) plan for each vowel per region for both male and female speakers. Regarding the vowel /a/, the diffusion area is more important for female speakers compared to males. The data of females are gathered to form globally a consolidated cluster. However, the area is more extensive in the case of men, where we observe a second block of data in the higher frequencies. These data represent vowels produced by speakers of both R5 and R6 from the south. Similarly, an identical result is noticed when the vowel /u/was pronounced. The results highlight differences in vowel production between female and male speakers depending on region they belong to it. The Findings show a significant variation either for /a/or /u/vowels. As regard to vowel /i/, an interesting outcome is given from the chart when male participant's data were studied. Indeed, the figure illustrates an expansion of both either F1 and F2 values, or only F2 values in all regions when /a/and /u/were pronounced. Whereas, for /i/produced by male speakers, a small variation of F1 measures is observed in opposition of large variation of F2 scores.

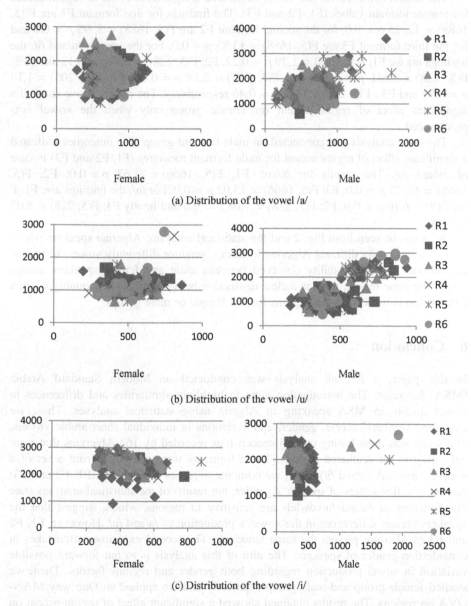

(a) Distribution of the vowel /a/

(b) Distribution of the vowel /u/

(c) Distribution of the vowel /i/

Fig. 2. Distribution of MSA short vowels in (F1, F2) plan per regions regarding the gender of speakers.

As it is well known, there is a difference in formant production between both categories of speakers male and female. Nevertheless, what is interesting to put forward is to study the formant variability within the same speakers group and then considering only regional accent due to the variation of localities. Thus, statistical analyses (One Way MANOVA) were applied separately on the data formant set for female speakers then for male speakers for each short vowels (/a/; /u/and /i/).

The results achieved for the vowel /a/showed a significant effect of region accent for female formant values (F1, F2 and F3). The findings for first formant F1 are F (5, 1688) = 12, 25 p = 0.0, for the second formant F2 are F(5, 1688) = 5, 95 p = 0.0 and for the third formant F3 are F(5, 1688) = 13,83 p = 0.0. For the vowels /u/and /i/; the findings are for F1: F (5, 288) = 1,39 p = 0.22, F2: F(5, 228) = 1,69 p = 0.13 and f F3: F(5, 228) = 2,841 p = 0.037; F1: F(5, 205) = 2,18 p = 0.06, F2: F(5, 205) = 1,18 p = 0.31 and F3 : F(5, 205) = 0,92 p = 0.46 respectively. The results above showed a significant effect of region accent for female group only when the vowel /a/is pronounced.

The same analysis was conducted on male formant group. The outcomes indicated a significant effect of region accent for male formant measures (F1, F2 and F3) in case of /a/and /u/. The results for /a/are: F1, F(5, 1666) = 21,67 p = 0.0; F2, F(5, 1666) = 10, 27 p = 0.0; F3: F(5, 1666) = 15,02 p = 0.0. For /u/, the findings are: F1, F (5, 218) = 6,16 p = 0.0; F2: F(5, 218) = 7,08 p = 0.0 and finally F3: F(5, 218) = 5,05 p = 0.0.

As it can be seen from Fig. 2 and the statistical analyses, Algerian speakers whom are belonging to six different Algerian localities, produce differently vowels formants. In addition, to the variability observed between male and female speakers groups within the same region, there is a clear distinction between the vowels quality when a comparison is made between regions for the female or male groups.

6 Conclusion

In this paper, a formant analysis was conducted on Modern Standard Arabic (MSA) language. The investigation aims to highlight similarities and differences in vowel quality in MSA speaking in Algeria using statistical analyses. Thus, we examined speakers' accent, gender, and variations in individual short arabic vowels. The study was done basing on 565 speech files recorded by 163 Algerians that were belonging to six scathered regions. Three formants were computed from a set of a vocalic units (/a/, /u/and /i/). First, we compute averages of formants (F1, F2 and F3) per regions. Regardless of speaker's gender, the results of the statistical analyses state that formants of /a/and /u/vowlels are sensitive to regions, which, suggest that the speakers present differences in the vowel's production of /a/and /u/. However, F1, F2 and F3 of /i/do not present the same tendency. The second experimentation takes in considertion gender of speakers. The aim of this analysis is to put forward possible variation in vowel production regarding both gender and regions factors. Then, we studied female group and male group separately and we applied an One way MAN-OVA on regions. The results obtained showed a significant effect of region accent on female formant values (F1, F2 and F3) for only the vowel /a/. In the other hand, the outcomes of the analysis conducetd on male group indicated a significant effect of region accent (F1, F2 and F3) in the case of /a/and /u/vowels.

References

1. Alotaibi, Y.A., Hussain, A.: Speech recognition system and formant based analysis of spoken Arabic vowels. In: International Conference on Future Generation Information Technology, pp. 50–60. FGIT (2009)
2. Gharavian, D., Sheikhan, M., Ashoftedel, F.: Emotion recognition improvement using normalized formant supplementary features by hybrid of DTW-MLP-GMM model. Neural Comput. Appl. **22**, 1181–1191 (2013)
3. Droua-Hamdani, G., Selouani, S.-A., Boudraa, M.: Speaker-independent ASR for modern standard Arabic: effect of regional accents. Int. J. Speech Technol. **15**(4), 487–493 (2012)
4. Natour, Y.S., Marie, B.S., Saleem, M.A., Tadros, Y.K.: Formant frequency characteristics in normal Arabic-speaking Jordanians. J. Voice **25**(2), e75–e84 (2011)
5. Kepuska, V., Alshaari, M.: Using formants to compare short and long vowels in modern standard Arabic. J. Comput. Commun. **8**, 96 (2020)
6. Farchi, M., et al.: Energy distribution in formant bands for Arabic vowels. Int. J. Electr. Comput. Eng. **9**(2), 1163–1167 (2019). Yogyakarta
7. Fathi, H.M., Qassim, Z.R.: An acoustic study of the production of Iraqi Arabic vowels. J. Al-Frahedis Arts. **12**, 692–704 (2020)
8. Droua-Hamdani, G., Selouani, S.A., Boudraa, M.: Algerian arabic speech database (ALGASD): corpus design and automatic speech recognition application. Arab. J. Sci. Eng. **35**(2C), 157–166 (2010)
9. Warne, R.T.: A primer on multivariate analysis of variance (MANOVA) for behavioral scientists. Pract. Assess. Res. Eval. **19**(17), 1–10 (2014)

Emotion Recognition and Sentiment Analysis of Extemporaneous Speech Transcriptions in Russian

Anastasia Dvoynikova[1,2] , Oxana Verkholyak[1,2(✉)] ,
and Alexey Karpov[1,2]

[1] St. Petersburg Institute for Informatics and Automation of the Russian
Academy of Sciences SPIIRAS, St. Petersburg, Russia
{dvoynikova.a,karpov}@iias.spb.su,
overkholyak@gmail.com
[2] ITMO University, St. Petersburg, Russia

Abstract. Speech can be characterized by acoustical properties and semantic meaning, represented as textual speech transcriptions. Apart from the meaning content, textual information carries a substantial amount of paralinguistic information that makes it possible to detect speaker's emotions and sentiments by means of speech transcription analysis. In this paper, we present experimental framework and results for 3-way sentiment analysis (positive, negative, and neutral) and 4-way emotion classification (happy, angry, sad, and neutral) from textual speech transcriptions in terms of Unweighted Average Recall (UAR), reaching 91.93% and 88.99%, respectively, on the multimodal corpus RAMAS containing recordings of Russian improvisational speech. Orthographic transcriptions of speech recordings from the database are obtained using available pre-trained speech recognition systems. Text vectorization is implemented using Bag-of-Words, Word2Vec, FastText and BERT methods. Investigated machine classifiers include Support Vector Machine, Random Forest, Naive Bayes and Logistic Regression. To the best of our knowledge, this is the first study of sentiment analysis and emotion recognition for both extemporaneous Russian speech and RAMAS data in particular, therefore experimental results presented in this paper can be considered as a baseline for further experiments.

Keywords: Sentiment analysis · Speech transcriptions · Emotion recognition · Russian speech and language

1 Introduction

Speech utterances contain important information about opinions and emotions of the speaker. Typically, the meaning of a text transmits the polarity of an utterance (i.e. positive or negative attitude towards a subject), while the acoustical properties show intensity of the conveyed emotion. Therefore, it is important to analyze both to achieve the best possible performance. However, the analysis of texts is a far under-studied area in the task of emotion recognition compared to acoustical signal properties. Moreover,

A. Karpov and R. Potapova (Eds.): SPECOM 2020, LNAI 12335, pp. 136–144, 2020.
https://doi.org/10.1007/978-3-030-60276-5_14

Russian language resources for sentiment analysis lag behind other languages, one of the reasons being the scarcity of Russian language databases. Currently, only one open corpus containing extemporaneous Russian speech utterances annotated with emotions is known – RAMAS [1]. This is the first study of sentiment analysis for speech transcriptions of Russian utterances, performed by automatically transcribing extemporaneous speech and vectorizing text using various natural language processing methods. Therefore, the results presented in this paper can be considered as a baseline for future experiments.

The general pipeline of the proposed method is presented in Fig. 1. First, Automatic Speech Recognition (ASR) is applied to raw speech samples from RAMAS corpus. Then, the transcriptions are preprocessed and used to obtain feature vectors via text vectorization. Next, these vectors (embeddings) are passed as an input for classification. The dashed rectangles show particular methods used for implementation. To implement vectorization and classification, one of the following methods was selected.

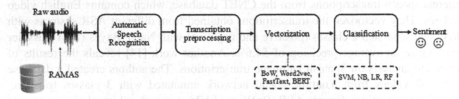

Fig. 1. The general pipeline of the proposed method.

The rest of the paper is organized as following. Section 2 presents the related work on sentiment analysis and emotion recognition. Sections 3 and 4 describe the database and preprocessing steps associated with it. Section 5 describes the experiments and reveals the experimental results. Section 6 presents summaries and conclusions.

2 Related Work

The sentiment analysis task consists of two parts: text vectorization and classification. Many text vectorization techniques were proposed for text analysis [2]. Bag-of-Words (BoW) is a simple yet popular method that converts a text to an unordered set of one-hot encoded vectors, each having its own weight. This approach often results high-dimensional text representation since the size of each vector is equal to the size of the vocabulary in the dataset. Word2Vec [3] and FastText [4] are neural-network-based embeddings that generate distributional word vectors. They are powerful representations capable of modelling linguistic context but require a lot of training data. Typical vector dimensionality is 300. The recent state-of-the-art text representation is BERT model, developed by Google [5]. BERT shows superior performance in many Natural Language Processing tasks [6, 7].

Various classification schemes for recognition of emotions and sentiments were investigated before. Authors in [8] used the English language dataset IEMOCAP [9] to analyze transcripts of the utterances via FastText embeddings of dimensionality 300 and Gated Recurrent Units (GRU) as a classification method. Four-way emotion classification with the following classes: anger, sadness, neutral, and happiness, resulted in an accuracy of 72.5%. Another work [10] used a dataset with reviews in English language, containing ExpoTv video recordings. Transcriptions were obtained via open-source ASR system provided by Google. Text vectorization was accomplished with the Bag-of-Words, classification – with the Support Vector Machine (SVM). Authors also built on lexical characteristics of words. Best accuracy in binary sentiment classification (negative and positive classes) was 63.9% and 67.8% for reviews on phones and books, respectively. Authors in [11] also performed a sentiment analysis of transcriptions form video reviews about different products ICT-MMMO [12]. They used the Google ASR system to obtain the transcriptions, Bag-of-n-grams to vectorize the text, and SVM to perform classification. This approach allowed to achieve an Unweighted Average Recall (UAR) of 80.7% for 3 classes: positive, neutral, and negative. In another work [13], authors performed sentiment analysis of an extemporaneous speech transcriptions from the CNET database, which contains English video reviews. They vectorized the transcriptions obtained from different ASR systems with BoW and separated 2 classes (positive and negative) with Naive Bayes (NB) classifier resulting in an average precision of 54.6%. Another work [14] reveals the results of sentiment analysis on Spanish language transcriptions. The authors created a database with 105 videos from Youtube social network annotated with 3 classes (positive, neutral, and negative). Google ASR, BoW and SVM approach achieved an accuracy of 64.9%.

The analysis of the literature reveals that different authors choose to use different classification metrics to assess the performance of the proposed approaches, including accuracy, UAR and average precision. The choice may be affected by the corpus involved in the experiments, as well as personal preference and practical implications. The lack of standardization in terms of experimental design makes it difficult to directly compare experimental results. Most popular tasks include 2-way (positive and negative) and 3-way (positive, negative, neutral) sentiment prediction and 4-way (angry, happy, sad, neutral) emotion recognition. The vast majority of research is dedicated to English, while other languages enjoy little or no attention.

3 RAMAS Corpus

RAMAS [1] is a multimodal corpus containing 560 audio-video dialogues between two actor speakers, a male and a female. The dialogues are played according to interactive scenarios predefined by the authors of the dataset. Average duration of a dialogue is 30 s. Totally 14 different scenarios are played on average 15 times each by different actors. Each scenario is designed to contain 2 different emotions, one per actor, from the set of 6 basic emotions: sadness, anger, disgust, happiness, surprise, and fear. Each actor has a dedicated microphone. While given the scenarios, the actors could improvise and choose their own words that makes their speech close to natural. Totally 10 semi-professional actors (5 males and 5 females) took part in the creation of the database. A professional theatre coach was supervising the performance of the actors to make sure that the portrayed emotions correspond to the predefined scenarios. Each actor also contributed several samples of neutral speech, such as counting numbers and telling an every-day life story.

4 Data Preprocessing

As mentioned earlier, RAMAS corpus does not contain the transcriptions of the dialogues as actors were allowed to improvise and thus there was no control over their speech. Therefore, the first step in the preprocessing of the database includes an extraction of the transcriptions from audio files using Google ASR [15] and Yandex SpeechKit [16]. The ASR failed for some of the recordings, which may be attributed to a bad sound quality and/or microphone fails. Such samples, along with short utterances of less than 3 words, were ignored during the experiments. The transcriptions were manually inspected to decide on which ASR engine works best. The total number of obtained samples is 264. The distribution of samples according to their emotion tag is shown in Table 1.

Table 1. Number of samples per each emotion class in RAMAS.

Emotion	# Samples
Neutral	35
Sadness	35
Anger	41
Disgust	28
Happiness	34
Surprise	33
Fear	58

The next step is the processing of obtained transcriptions. After tokenization, all words are transformed into low register. Then stop words that bear no semantic value are removed. The list of Russian stop words is borrowed from NLTK library [17], which is a popular open-source natural language processing toolkit. The NLTK stop words list was manually modified, i.e. some of the words from the list that could affect sentiment analysis were excluded, for example «ne» (no/not), «horosho» (good), «bolee» (more), «menee» (less) etc. In the last preprocessing step, all words are normalized by means of lemmatization (removal of inflectional endings and finding the base or dictionary form of a word) using the same NLTK library.

5   Experimental Results

The experiments were conducted separately for the sentiment analysis and emotion classification tasks. The train/test data split was kept the same for both of the tasks. It was performed randomly for every emotion in each scenario with the ratio 70/30. All classifiers were trained using 5-fold cross-validation with UAR as classification metric.

5.1   Classification into 3 Sentiment Classes

The sentiment classification of preprocessed texts is carried out for 3 classes: negative, positive, and neutral. This is accomplished by grouping the original 6 basic emotions into classes according to the valence axis of the Russel diagram [18]. The negative class that corresponds to negative valence contains anger, sadness, fear, and disgust; positive class corresponds to positive valence – happiness and surprise. The neutral class corresponds to the neutral emotion. As can be inferred from Table 1, the number of samples in negative class is 162, positive class -67, and neutral class -35, making the data distribution imbalanced for this task.

Text vectorization methods included Bag of Words (BoW), Word2Vec, FastText and BERT. The BoW variant dubbed TF-IDF (Term Frequency – Inverse Document Frequency) was used to calculate the weights. The feature vector dimensionality in that case was equal to 1966. Word2Vec and FastText pretrained models were adopted from RusVectores [19]. Experimentally the best models were found to be tayga_upos_ skipgram_300_2_2019 for Word2Vec and tayga_none_fasttextcbow_300_10_2019 for FastText with dimensionality 300 for each model. Word2Vec was trained on Taiga corpus [20] with the size of about 5 000 000 000 words using continuous Skipgram method. The FastText was trained on the same corpus but using continuous BoW. To compare the results with recent state-of-the-art approach we also developed the BERT text representation using pretrained models. Google provides pretrained open source BERT models. We use BERT-Base Multilingual – this model is designed to be used for 102 languages, including Russian. The size of the feature vectors for BERT equals 735. The following classification schemes were tested: Support Vector Machine (SVM), Random Forest (RF), Naive Bayes (NB), and Logistic Regression (LR). Table 2 presents the best experimental results obtained with different vectorization techniques and classification methods for 3-way sentiment recognition task (negative, positive, and neutral).

Table 2. 3-way sentiment classification results for RAMAS data (UAR, %).

Features	SVM	RF	NB	LR
BoW	61.87	43.31	85.87	46.09
Word2Vec	**91.93**	73.68	80.89	82.26
FastText	85.81	86.89	72.46	85.55
BERT	81.14	47.12	45.89	79.24

Word2Vec vectorization method shows the best UAR = 91.93% with the linear SVM and regularization parameter C = 2.7. FastText achieved UAR = 86.89% with Random Forest of 70 trees and depth 85. BoW vectorization method allowed to achieve UAR = 85.87% together with Bernoulli Naive Bayes classifier. The optimal parameter was found via cross-validation additive smoothing parameter α = 0.1. State-of-the-art BERT produced suboptimal results in this task with UAR = 81.14% obtained by SVM. For the best comparability with other authors, a detailed analysis of the results for the best performing approach (Word2Vec+SVM) is shown in Table 3.

Table 3. Classification results (%) for the best performing sentiment analysis approach (Word2Vec+SVM).

Class	Precision	Recall	F-score
Negative	93.44	96.61	95.00
Neutral	84.62	100	91.67
Positive	95.00	79.17	86.36
Macro average	91.02	91.93	91.01
Weighted average	92.81	92.55	92.40

As can be seen from Table 3, the Negative class, which is usually of the most interest, achieves a high recall score of 96.61%. This is a good indication that the algorithm returned most of the relevant results. The precision score is lower at 93.44%, however in the sentiment analysis task the recall is usually more important because we are concerned with retrieving all the negative samples, even though some of them could be irrelevant. Note that the harmonic mean of the precision and recall for the Negative class achieves up to 95.00%. On the other hand, the recall for Positive class turned out lower than the precision, which means that most of the Positive class predictions were accurate, however not all the relevant results were retrieved. Notably, the recall of the neutral class reached 100%.

5.2   Classification into 4 Emotion Classes

RAMAS 4-way emotion classification is performed for the following emotions: anger, sadness, happiness and neutral. These emotions are most widely considered in other datasets, such as IEMOCAP [9]. The same 4 text vectorization methods (BoW, Word2Vec, FastText, BERT) and 4 classification methods (SVM, RF, NB, LR) are used. The BoW model was retrained with the new number of samples, resulting in a feature vector size of 1375. Word2Vec and FastText models were used as described in the previous section as they showed superior performance relative to others. The BERT model also remained the same. The experimental design is identical to the previous one. The results of the experiments are shown in Table 4.

The best UAR = 88.99% was achieved by both Word2Vec and FastText representations and SVM and LR classifiers, respectively. The optimal parameters for SVM – linear kernel with regularization parameter C = 1.4, LR–L2 penalty and regularization parameter C = 0.1. BoW with TF-IDF weights and SVM (linear kernel with

Table 4. 4-way emotion classification results for RAMAS data (UAR, %).

Features	SVM	RF	NB	LR
BoW	84.52	48.97	75.62	73.84
Word2Vec	**88.99**	75.60	62.50	88.69
FastText	88.20	68.15	75.54	**88.99**
BERT	80.28	69.78	53.44	80.28

regularization parameter C = 0.5) resulted in UAR = 84.52%. As before the experiments with BERT show suboptimal performance. This may be attributed to the lack of Russian language specific resources used for training the original multilingual model. Tables 5 and 6 reveal detailed classification report for Word2Vec+SVM and FastText +LR, respectively.

Table 5. Classification results (%) for the best performing emotion recognition approach (Word2Vec+SVM).

	Precision	Recall	F-score
Neutral	84.62	100	91.67
Angry	91.67	78.57	84.62
Sad	80.00	85.71	82.76
Happy	100	91.67	95.65
Macro average	89.07	88.99	88.67
Weighted average	88.90	88.24	88.22

Notably, the Word2Vec+SVM model achieves 100% precision for Happy class and 91.67% for Angry class. Precision is important in such applications as personal assistants, since wrongly identifying emotional state of the speaker potentially implies more cost than not identifying any emotion at all. On the other hand, higher recall for Sad class implies better ability to identify such psycho-emotional states as depression in a long-term patient monitoring.

The FastText+LR model, contrary to the Word2Vec+SVM, achieves greater recall for Angry and Happy classes, reaching 92.86% and 91.67%, respectively, at the cost of the precision. This is a better solution for such applications as monitoring of customer satisfaction at call-centers, where detecting angry customers is crucial for success. Remarkably, this model achieves 100% for both precision and recall in the neutral class.

Although not directly comparable, the accuracy for classification of 4 emotions in our approach (UAR = 88.24%) greatly exceeds the one reported earlier for a similar classification task for English database IEMOCAP (UAR = 72.52%) [8].

Table 6. Classification results (%) for the best performing emotion recognition approach, FastText+LR.

	Precision	Recall	F-score
Neutral	100	100	100
Angry	76.47	92.86	83.87
Sad	90.91	71.43	80.00
Happy	91.67	91.67	91.67
Macro average	89.76	88.99	88.88
Weighted average	89.08	88.24	88.12

6   Conclusions

In this work, we studied sentiment analysis and emotion classification from orthographic transcriptions of extemporaneous Russian speech. The experimental results set a baseline performance for the Russian emotional speech dataset RAMAS. This baseline also applies to the sentiment analysis and emotion classification of Russian transcribed speech in general. The best ternary sentiment classification (positive, negative, neutral) UAR performance of 91.93% was achieved with Word2Vec text representation and SVM classifier. The classification of emotions into four classes (angry, sad, happy, and neutral) showed the best UAR of 88.99% with two different approaches: Word2Vec+SVM and FastText+LR. These two models showed different performance in terms of recall and precision measures, and so should be properly chosen according to practical needs. The presented results are significantly higher than the previously reported ones in the literature for similar tasks for English and other natural languages. Our future directions of research include extensions to other communication modalities, such as audio and video data.

Acknowledgements. This research is supported by the Russian Science Foundation (project No. 18-11-00145, research and development of the emotion recognition system), as well as by the Russian Foundation for Basic Research (project No. 18-07-01407), and by the Government of Russia (grant No. 08-08).

References

1. Perepelkina, O., Kazimirova, E., Konstantinova, M.: RAMAS: Russian multimodal corpus of dyadic interaction for affective computing. In: Karpov, A., Jokisch, O., Potapova, R. (eds.) SPECOM 2018. LNCS (LNAI), vol. 11096, pp. 501–510. Springer, Cham (2018). https://doi.org/10.1007/978-3-319-99579-3_52
2. Dvoynikova, A., Verkholyak, O., Karpov, A.: Analytical review of methods for identifying emotions in text data. CEUR-WS **2552**, 8–21 (2020)
3. Mikolov, T., et al.: Distributed representations of words and phrases and their compositionality. In: Advances In Neural Information Processing Systems, pp. 3111–3119 (2013)
4. Bojanowski, P., Grave, E., Joulin, A., Mikolov, T.: Enriching word vectors with subword information. Trans. Assoc. Comput. Linguist. **5**, 135–146 (2017)

5. Devlin, J., Chang, M., Lee, K., Toutanova, K.: BERT: pre-training of deep bidirectional transformers for language understanding. In: Conference of the North American Chapter of the Association for Computational Linguistics: Human Language Technologies (NAACL-HLT), vol. 1, pp. 4171–4186 (2019)
6. Byszuk, J., et al.: Detecting direct speech in multilingual collection of 19th-century novels. In: Proceedings of LT4HALA 2020-1st Workshop on Language Technologies for Historical and Ancient Languages, pp. 100–104 (2020)
7. Conneau, A., et al.: Unsupervised cross-lingual representation learning at scale. arXiv preprint arXiv: 1911.02116 (2019)
8. Atmaja, B.: Deep learning-based categorical and dimensional emotion recognition for written and spoken text. INA-Rxiv (2019). https://doi.org/10.31227/osf.io/fhu29
9. Busso, C., et al.: IEMOCAP: interactive emotional dyadic motion capture database. Lang. Res. Eval. **42**(4), 335–364 (2008)
10. Perez-Rosas, V., Mihalcea, R.: Sentiment analysis of online spoken reviews. In: Interspeech, pp. 862–866 (2013)
11. Cummins, N., et al.: Multimodal bag-of-words for cross domains sentiment analysis. In: IEEE International Conference on Acoustics, Speech and Signal Processing (ICASSP), pp. 4954–4958. IEEE (2018)
12. Wollmer, M., et al.: Youtube movie reviews: sentiment analysis in an audio-visual context. IEEE Intell. Syst. **28**(3), 46–53 (2013)
13. Pereira, J.C., Luque, J., Anguera, X.: Sentiment retrieval on web reviews using spontaneous natural speech. In: IEEE International Conference on Acoustics, Speech and Signal Processing (ICASSP), pp. 4583–4587. IEEE (2014)
14. Rosas, V., Mihalcea, R., Morency, L.: Multimodal sentiment analysis of spanish online videos. IEEE Intell. Syst. **28**(3), 38–45 (2013)
15. Speech Recognition (version 3.8.1). https://pypi.org/project/SpeechRecognition. Accessed 15 June 2020
16. Yandex SpeechKit. https://cloud.yandex.ru/services/speechkit. Accessed 15 June 2020
17. Bird, S., Klein, E., Loper, E.: Natural Language Processing with Python: Analyzing Text with the Natural Language Toolkit. O'Reilly Media, Sebastopol (2009)
18. Russell, J.: Culture and the categorization of emotions. Psychol. Bull. **110**(3), 426–450 (1991)
19. RusVectores. https://rusvectores.org/ru. Accessed 15 June 2020
20. Shavrina, T., Shapovalova, O.: To the methodology of corpus construction for machine learning: taiga syntax tree corpus and parser. In: Proceeding of CORPORA2017, International Conference, Saint-Petersburg (2017)

Predicting a Cold from Speech Using Fisher Vectors; SVM and XGBoost as Classifiers

José Vicente Egas-López[1(✉)] and Gábor Gosztolya[1,2]

[1] University of Szeged, Institute of Informatics, Szeged, Hungary
egasj@inf.u-szeged.hu
[2] MTA-SZTE Research Group on Artificial Intelligence, Szeged, Hungary

Abstract. Screening a *cold* may be beneficial in the sense of avoiding the propagation of it. In this study, we present a technique for classifying subjects having a cold by using their speech. In order to achieve this goal, we make use of frame-level representations of the recordings of the subjects. Such representations are exploited by a generative Gaussian Mixture Model (GMM) which consequently produces a fixed-length encoding, i.e. Fisher vectors, based on the Fisher Vector (FV) approach. Afterward, we compare the classification performance of the two algorithms: a linear kernel SVM and a XGBoost Classifier. Due to the data sets having a high class imbalance, we undersample the majority class. Applying Power Normalization (PN) and Principal Component Analysis (PCA) on the FV features proved effective at improving the classification score: SVM achieved a final score of 67.81% of Unweighted Average Recall (UAR) on the test set. However, XGBoost gave better results on the test set by just using *raw* Fisher vectors; and with this combination we achieved a UAR score of 70.43%. The latter classification approach outperformed the original (non-fused) baseline score given in 'The INTERSPEECH 2017 Computational Paralinguistics Challenge'.

Keywords: Fisher vectors · Speech processing · SVM · XGBoost · Cold assessment · Computational paralinguistics

1 Introduction

Identifying cold or other related illnesses with similar symptoms may be beneficial when assessing them; as it could be a way of avoiding the spread of a specific kind of viral infection of the nose and throat (upper respiratory tract). Upper respiratory tract infection (URTI) affects the components of the upper airway. URTI can be thought as of a common cold, a sinus infection, among others. Screening a cold directly from the speech of subjects can create the possibility of monitoring (even from call-centers or telephone communications), and predicting their propagation. In contrast with Automatic Speech Recognition (ASR), which focuses on the actual *content* of the speech of an audio signal, computational paralinguistics may provide the necessary tools for determining the *way*

© Springer Nature Switzerland AG 2020
A. Karpov and R. Potapova (Eds.): SPECOM 2020, LNAI 12335, pp. 145–155, 2020.
https://doi.org/10.1007/978-3-030-60276-5_15

the speech is spoken. Various studies have offered promising results in this field: diagnosing neuro-degenerative diseases using the speech of the patients [6,7,10]; the classification of crying sounds and heart beats [13]; or even the estimation of the sincerity of apologies [12]. Hence, we focus on finding certain patterns hidden within the speech of the *cold* recordings and not on what the speakers actually said.

Here, we make use of the Upper Respiratory Tract Infection Corpus (URTIC) [26] to classify speakers having a cold. Previous studies applied various approaches for classifying *cold* subjects on the same corpus; for example, Gosztolya et al. employ Deep Neural Networks for feature extraction for such purpose [11]. Huckvale and Beke utilizated voice features for studying changes in health [14]; furthermore, Kaya et al. [16] introduced the application of a weighting scheme on instances of the corpus, employing Weighted Kernel Extreme Learning Machine in order to handle the imbalanced data that comprises the URTIC corpus.

In this study, frame-level features (Mel-frequency cepstral coefficients), extracted from the utterances, are utilized to fit a generative Gaussian Mixture Model (GMM). Next, the computation of low-level patch descriptors together with their deviations from the GMM give us an encoding (features) called the Fisher Vector. FV features are learned using SVM and XGBoost as binary classifiers, where the prediction is *cold* or *healthy*. In order to search for the best parameters of both SVM and XGBoost, Stratified Group k-fold Cross Validation (CV) was applied on the training and development sets. Unweighted Average Recall (UAR) scoring was used to measure the performance of the model. To the best of our knowledge, this is the first study that uses a FV representation to detect a cold from human speech.

In the next part of our study we also show that PN and L2-normalization over the Fisher vectors have a beneficial effect on the SVM classification scores. PN reduces the *sparsity* of the features; L2-normalization is a valid technique that can be applied to any high-dimensional vector; and moreover, it improves the prediction performance [25]. Likewise, PCA also affects positively to the performance of Support Vector Machines (SVM) due to its effects of feature decorrelation as well as dimension reduction.

The combination of all three feature pre-processing methods gave the best scoring with respect to the SVM classifier. However, XGBoost did not produce competitive scores when any kind of feature pre-processing was employed before training the model. Namely, employing the same feature-treatments (PCA, PN, L2-normalization) as SVM to the XGBoost classifier led to a decrease in performance. Mentioned algorithm showed better results when learning from *raw* features. Thus, there was no need for any feature processing prior to training, owing to the fact that decision tree algorithms do not necessitate so. We show that our system produces better UAR scores relative to the baseline individual methods reported in the 'The INTERSPEECH 2017 Computational Paralinguistics Challenge' [26] for the Cold sub-challenge.

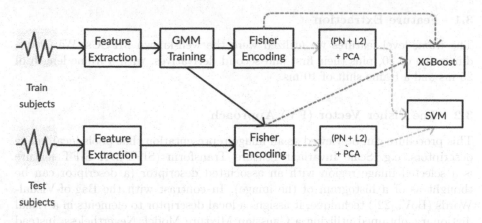

Fig. 1. The generic methodology applied in our work.

2 Data

The Upper Respiratory Tract Infection Corpus (URTIC) comprises recordings of 630 speakers: 382 male, 248 female, and a sampling rate of 16kHz. Recordings were held in quiet rooms with a microphone/headset/hardware setup. The tasks performed by the speakers were as follows: reading short stories, for example, *The North Wind and the Sun* which is well known in the phonetics area; producing voice commands such as numbers or driver assistant controlling commands; and narrating spontaneous speech. The number of tasks varied for each speaker. Although the sessions lasted up to 2 hours, the recordings were split into 28652 chunks of length 3 to 10 seconds. The division was done in a speaker-independent fashion, so each set had 210 speakers. The training and development sets were both comprised of 37 subjects having a cold and 173 subjects not having a cold [26]. The train, development, and test datasets are composed of 9505, 9596, and 9551 recordings respectively.

3 Methods

As outlined in Fig. 1, our workflow is as follows:

1. MFCCs features are extracted from all the recordings.
2. The GMM is trained using the MFCCs belonging to the training dataset.
3. The FV encoding (Fisher vectors) is performed for all the MFCCs of each utterance.
4. Classification:
   (a) Fisher vectors are processed using Power Normalization and L2 normalization; Support Vector Machines carried out the classification process using the new scaled features.
   (b) *Raw* Fisher vectors are fed to XGBoost for classification.

3.1   Feature Extraction

The frame-level features we utilized were the well-known MFCCs. We used a dimension of 20, plus their first and second derivatives, with a frame length of 25 ms and a frame shift of 10 ms.

3.2   The Fisher Vector (FV) Approach

This procedure can be viewed as an image representation that pools local image descriptors, e.g. Scale Invariant Feature Transform (SIFT). A SIFT feature is a selected image region with an associated descriptor (a descriptor can be thought as of a histogram of the image). In contrast with the Bag-of-Visual-Words (BoV, [22]) technique, it assigns a local descriptor to elements in a visual dictionary, obtained utilizing a Gaussian Mixture Model. Nevertheless, instead of just storing visual word occurrences, these representations take into account the difference between dictionary elements and pooled local features, and they store their statistics. A nice advantage of the FV representation is that, regardless of the number of local features (i.e. SIFT), it extracts a *fixed-sized* feature representation from each image. Applied to this study, such approach becomes quite practical because the length of the speech utterances are subject to vary.

The FV approach has been widely used in image representation and it can achieve high performance [15]. In contrast, just a handful of studies use FV in speech processing, e.g. for categorizing audio-signals as speech, music and others [20], for speaker verification [30,34], for determining the food type from eating sounds [17], and even for emotion detection [9]. These studies demonstrate the potential of achieve good classification performances in audio processing.

Fisher Kernel (FK). It seeks to measure the similarity of two objects from a parametric generative model of the data (X) which is defined as the gradient of the log-likelihood of X [15]:

$$G_\lambda^X = \nabla_\lambda \log v_\lambda(X),$$ (1)

where $X = \{x_t, t = 1, \ldots, T\}$ is a sample of T observations $x_t \in \mathcal{X}$, v represents a probability density function that models the generative process of the elements in $\mathcal{X}$ and $\lambda = [\lambda_1, \ldots, \lambda_M]' \in R^M$ stands for the parameter vector v_λ [25]. Thus, such a gradient describes the way the parameter v_λ should be changed in order to best fit the data X. A way to measure the similarity between two points X and Y by means of the FK can be expressed as follows [15]:

$$K_{FK}(X,Y) = G_\lambda^{X'} F_\lambda^{-1} G_\lambda^Y.$$ (2)

Since F_λ is positive semi-definite, $F_\lambda = F_\lambda^{-1}$. Equation (3) shows how the Cholesky decomposition $F_\lambda^{-1} = L_\lambda' L_\lambda$ can be utilized to rewrite the Eq. (2) in terms of the dot product:

$$K_{FK}(X,Y) = G_\lambda^{X'} G_\lambda^Y,$$ (3)

where

$$G_\lambda^X = L_\lambda G_\lambda^X = L_\lambda \, \nabla_\lambda \log \upsilon_\lambda(X). \tag{4}$$

Such a normalized gradient vector is the so-called *Fisher Vector* of X [25]. Both the FV G_λ^X and the gradient vector G_λ^X have the same dimension.

Fisher Vectors. Let $X = \{X_t, t = 1 \ldots T\}$ be the set of D-dimensional local SIFT descriptors extracted from an image and let the assumption of independent samples hold, then Eq. (4) becomes:

$$G_\lambda^X = \sum_{t=1}^{T} L_\lambda \, \nabla_\lambda \log \upsilon_\lambda(X_t). \tag{5}$$

The assumption of independence permits the FV to become a sum of normalized gradients statistics $L_\lambda \, \nabla_\lambda \log \upsilon_\lambda(x_t)$ calculated for each SIFT descriptor:

$$X_t \to \varphi_{FK}(X_t) = L_\lambda \, \nabla_\lambda \log \upsilon_\lambda(X_t), \tag{6}$$

which describes an operation that can be thought of as a higher dimensional space embedding of the local descriptors X_t.

Hence, the FV approach extracts low-level local patch descriptors from the audio-signals' spectrogram. Then, with the use of a GMM with diagonal covariances we can model the distribution of the extracted features. The log-likelihood gradients of the features modeled by the parameters of such GMM are encoded through the FV [25]. This type of encoding stores the mean and covariance *deviation* vectors of the components k that form the GMM together with the elements of the local feature descriptors. The image is represented by the concatenation of all the mean and the covariance vectors that gives a final vector of length $(2D + 1)N$, for N quantization cells and D dimensional descriptors [23,25].

The FV approach can be compared with the traditional encoding method: BoV, and with a first order encoding method like VLAD Vector of Locally Aggregated Descriptors) [1]. In practice, BoV and VLAD are outperformed by FV due to its second order encoding property of storing additional statistics between codewords and local feature descriptors [28].

3.3 Classification with XGBoost and SVM

The classification of the data was carried out separately by two algorithms: XGBoost and SVM. In this section, we describe in a general manner these two approaches. SVM complexity parameter C was optimized by employing a Stratified Group k-fold Cross Validation using the train and development sets combined. For XGBoost parameters the same process was performed. Unweighted Average Recall (UAR) is the chosen metric due to the fact that it is more competent when having imbalanced datasets and also because it has been the de facto standard metric for these kinds of challenges [24,27].

As is widely known, a normal cross-validation gives the indices to split the data into train and test folds. In contrast, a stratified cross-validation applies the same principle but it preserves the percentage of samples for each class; and, a group k-fold cross-validation also has the same basis but it tries to keep the balance of different groups across the folds, so the same group will not be present in two distinct folds. Here, utterances from one speaker are treated as one group. The combination of these two different cross-validation approaches meant we could avoid having the same speaker in more than one specific fold while keeping the number of samples of each target class within that fold even.

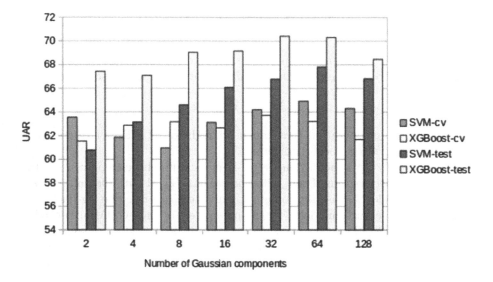

Fig. 2. UAR CV and test scores as a function of G_c for SVM and XGBoost using FVs.

XGBoost. This library is an implementation based on Gradient Boosting Machines (GBM) [8]. GBM is a regression/classification algorithm which makes use of an ensemble of *weak* models, i.e. small decision trees, to make predictions. A decision tree ensemble in XGBoost is a set of CARTs (Classification and Regression Trees). Put simply, GBM sequentially adds *decision tree* models to correct the predictions made by the previous models, and based on gradient descent, it minimizes the loss function. This is continued until the objective function (training loss and regularization) finds that no further improvement can be done [21]. Both XGBoost and GBM, basically act in the same manner; however, the main difference between these two is that XGBoost, in order to control over-fitting, employs a more regularized model than GBM does.

This algorithm is widely used in machine learning mostly due to its scaling capability and model performance; it was designed to exploit the limits of the computational resources for GBM algorithms [5]. Our decision to use XGBoost was also influenced by its advanced capability for performing model tuning. We can see the performance of XGBoost in [19,32,33], where the authors report high scores using such algorithm when applied to speech-related classification tasks. In this study, we employed the Python implementation of XGBoost [5].

Support Vector Machines We relied on the libSVM implementation [3]. SVM was found to be robust even with a large number of dimensions and it was shown to be efficient when fitting them on FV [25,29]. To avoid overfitting due to having a large number of meta-parameters, we applied a linear kernel.

4 Experiments and Results

The GMM used to compute the FVs operated with a different number of components, G_c ranged from $2, 4, 8$ to 128. Here, the construction of the FV encoding was performed using a Python-wrapped version of the VLFeat library [31]. The dataset suffers from high class-imbalance, which could affect the performance of either of the classifiers. The training dataset comprises 9505 recordings: 8535 (89.8%) as *healthy* and the rest, 970 (10.2%) as *cold*. We relied on a random undersampling technique that reduces the number of samples associated with all classes, to the number of samples of the minority class, i.e. *cold*. We employed imbalanced-learn [18], a Python-based tool which offers several resampling methods for between-class imbalance. For the SVM, the complexity value (C) was set in the range $10^{\{-5,-4,...,0,1\}}$.

As a baseline, we utilized the ComParE functionals that were originally presented and described in [26]. As Table 1 shows, these representations achieved an UAR score of 69.30% on the test set, which is slightly higher than the score achieved with FV representations (67.81%). The SVM classifier gave better results using FVs with Power Normalization (PN) and L2-Normalization, along with PCA: UAR score of 67.81% (see Table 1). PCA was applied using the 95% of the variance, which apart from decorrelating the FV features, also helped with both the computation (lower memory consumption) and the discrimination task. This method is also described in [4]. We saw that PN helped to reduce the impact of the features that become more sparse as the number of Gaussian components increases. Meanwhile, L2-normalization helped to alleviate the effect of having different utterances with distinct amounts of background information projected into the extracted features, which attempts to improve the prediction performance. Also, we employed a late fusion of the posterior probabilities. We combined the ComParE functionals SVM-posteriors with those that gave the highest scores when using SVM on FVs; the result was a better UAR score: 70.71%.

For XGBoost, we just utilized the non-preprocessed Fisher vector features and performed a grid search to find the best parameters. To control overfitting, we tuned the parameters that influence the model complexity: the gamma value, which represents the minimum loss required to split further on a leaf; the maximum depth for each tree (the higher the value, the higher the complexity); and the minimum child weight, that is, the minimum sum of weights needed in a child. Also, the learning rate and the number of estimators (number of trees) were tuned, these two having an inverse relation: the higher the learning rate the smaller the number of trees that have to be defined, and vice-versa.

As shown in Fig. 2, the classifiers discriminate better the data as the value of G_c increases. However, the highest G_c did not give the best UAR score. SVM classified better when the Fisher vectors were encoded using 64 Gaussian components, while a smaller number of G_c was needed for XGBoost (32). Stratified k-fold CV (on the combined train and development data) with $k = 10$ was applied for the hyper-parameter tuning of both algorithms. Due to XGBoost basically being an ensemble of regression trees, its posterior probability values are not really meaningful, hence we did not perform any kind of fusion with them. In spite of this, such algorithm achieved a score of 69.59% with the ComParE feature set and 70.43% using the Fisher vector features; the former outperformed the non-fused highest score of SVM (67.81%) and it is slightly lower than the fused one (70.17%), while the latter surpassed both of them. Furthermore, these scores surpassed the non-fused baseline and are around the fused baseline score given in [26] (see Table 1).

Table 1. UAR scores obtained using XGBoost and SVM on the URTIC Corpus.

2* Features	2* GMM size	Performance (%)	
		CV	Test
		SVM	
ComParE	–	64.20%	69.30%
Fisher vectors	64	63.98%	66.12%
Fisher vectors + PCA	64	64.72%	67.65%
Fisher vectors (+PN+L2) + PCA	64	64.92%	67.81%
Fusion: ComParE + FV(+PN+L2+PCA)	–/64	63.01%	70.17%
		XGBoost	
ComParE	–	62.19%	69.59%
Fisher vectors	32	63.71%	70.43%

5   Conclusions and Future Work

Here, we showed how well the Fisher vector encoding allows frame-level features to classify speaking subjects with a cold. We utilized two different classification algorithms (SVM and XGBoost) that used FV as input features. We showed that such features trained on both algorithms outperform original baseline scores given in [26] and they are highly competitive with those reported in [2,14]. Moreover, our approach offers a much simpler pipeline than the above-mentioned studies. We found that we got a better SVM performance when we applied feature pre-processing before starting the train/classification phases. Namely, it was shown that both L2-Normalization and Power Normalization produced an increased prediction performance. Also, PCA played a relevant role in decorrelating the features and increasing the model's performance. We demonstrated the usefulness of the fusion of SVM posterior probabilities which yielded even better UAR results. In contrast, XGBoost did not need any pre-processing or any kind of fusion to achieve and surpass SVM scores in our study. Yet, one disadvantage of XGBoost was the significant number of parameters that have to be tuned. This can slow down the parameter-tuning phase especially if there is no GPU available. In our next study, we plan to apply the methodology presented here on different kinds of paralinguistic corpora.

Acknowledgments. This study was partially funded by the National Research, Development and Innovation Office of Hungary via contract NKFIH FK-124413 and by the Ministry for Innovation and Technology, Hungary (grant TUDFO/47138-1/2019-ITM). G. Gosztolya was also funded by the János Bolyai Scholarship of the Hungarian Academy of Sciences and by the Hungarian Ministry of Innovation and Technology New National Excellence Program ÚNKP-20-5.

References

1. Arandjelovic, R., Zisserman, A.: All about VLAD. In: Proceedings of the IEEE conference on Computer Vision and Pattern Recognition, pp. 1578–1585 (2013)
2. Cai, D., Ni, Z., Liu, W., Cai, W., Li, G., Li, M.: End-to-end deep learning framework for speech paralinguistics detection based on perception aware spectrum. In: Proceedings of Interspeech, pp. 3452–3456 (2017)
3. Chang, C.C., Lin, C.J.: LIBSVM: a library for support vector machines. ACM Trans. Intell. Syst. Technol. **2**, 1–27 (2011)
4. Chatfield, K., Lempitsky, V., Vedaldi, A., Zisserman, A.: The devil is in the details: an evaluation of recent feature encoding methods. In: British Machine Vision Conference, vol. 2, pp. 76.1–76.12, November 2011
5. Chen, T., Guestrin, C.: XGBoost: a scalable tree boosting system. In: Proceedings of the 22nd ACM SIGKDD International Conference on Knowledge Discovery and Data Mining abs/1603.02754, pp. 785–794 (2016)
6. Egas-López, J.V., Orozco-Arroyave, J.R., Gosztolya, G.: Assessing Parkinson's disease from speech using fisher vectors. In: Proceedings of Interspeech (2019)
7. Egas López, J.V., Tóth, L., Hoffmann, I., Kálmán, J., Pákáski, M., Gosztolya, G.: Assessing Alzheimer's disease from speech using the i-vector approach. In: Salah, A.A., Karpov, A., Potapova, R. (eds.) SPECOM 2019. LNCS (LNAI), vol. 11658, pp. 289–298. Springer, Cham (2019). https://doi.org/10.1007/978-3-030-26061-3_30
8. Friedman, J.H.: Greedy function approximation: a Gradient Boosting Machine. Ann. Stat. **29**, 1189–1232 (2001)
9. Gosztolya, G.: Using the Fisher vector representation for audio-based emotion recognition. Acta Polytechnica Hungarica **17**(6), 7–23 (2020)
10. Gosztolya, G., Bagi, A., Szalóki, S., Szendi, I., Hoffmann, I.: Identifying schizophrenia based on temporal parameters in spontaneous speech. In: Proceedings of Interspeech, Hyderabad, India, pp. 3408–3412, September 2018

11. Gosztolya, G., Busa-Fekete, R., Grósz, T., Tóth, L.: DNN-based feature extraction and classifier combination for child-directed speech, cold and snoring identification. In: Proceedings of Interspeech, Stockholm, Sweden, pp. 3522–3526, August 2017
12. Gosztolya, G., Grósz, T., Szaszák, G., Tóth, L.: Estimating the sincerity of apologies in speech by DNN rank learning and prosodic analysis. In: Proceedings of Interspeech, San Francisco, CA, USA, pp. 2026–2030, September 2016
13. Gosztolya, G., Grósz, T., Tóth, L.: General utterance-level feature extraction for classifying crying sounds, atypical and self-assessed affect and heart beats. In: Proceedings of Interspeech, Hyderabad, India, pp. 531–535, September 2018
14. Huckvale, M., Beke, A.: It sounds like you have a cold! testing voice features for the interspeech 2017 computational paralinguistics cold challenge. In: Proceedings of Interspeech, International Speech Communication Association (ISCA) (2017)
15. Jaakkola, T.S., Haussler, D.: Exploiting generative models in discriminative classifiers. In: Proceedings of NIPS, Denver, CO, USA, pp. 487–493 (1998)
16. Kaya, H., Karpov, A.A.: Introducing weighted kernel classifiers for handling imbalanced paralinguistic corpora: snoring, addressee and cold. In: Interspeech, pp. 3527–3531 (2017)
17. Kaya, H., Karpov, A.A., Salah, A.A.: Fisher vectors with cascaded normalization for paralinguistic analysis. In: Proceedings of Interspeech, pp. 909–913 (2015)
18. Lemaître, G., Nogueira, F., Aridas, C.K.: Imbalanced-learn: a Python toolbox to tackle the curse of imbalanced datasets in machine learning. J. Mach. Learn. Res. 18(1), 559–563 (2017)
19. Long, J.M., Yan, Z.F., Shen, Y.L., Liu, W.J., Wei, Q.Y.: Detection of Epilepsy using MFCC-based feature and XGBoost. In: 2018 11th International Congress on Image and Signal Processing, BioMedical Engineering and Informatics (CISP-BMEI), pp. 1–4. IEEE (2018)
20. Moreno, P.J., Rifkin, R.: Using the Fisher kernel method for web audio classification. In: Proceedings of ICASSP, Dallas, TX, USA, pp. 2417–2420 (2010)
21. Natekin, A., Knoll, A.: Gradient boosting machines, a tutorial. Front. Neurorob. 7, 21 (2013)
22. Peng, X., Wang, L., Wang, X., Qiao, Y.: Bag of visual words and fusion methods for action recognition: comprehensive study and good practice. Comput. Vis. Image Underst. 150, 109–125 (2016)
23. Perronnin, F., Dance, C.: Fisher kernels on visual vocabularies for image categorization. In: 2007 IEEE Conference on Computer Vision and Pattern Recognition, pp. 1–8, June 2007.https://doi.org/10.1109/CVPR.2007.383266
24. Rosenberg, A.: Classifying skewed data: importance weighting to optimize average recall. In: Proceedings of Interspeech, pp. 2242–2245 (2012)
25. Sánchez, J., Perronnin, F., Mensink, T., Verbeek, J.: Image classification with the fisher vector: theory and practice. Int. J. Comput. Vision 105(3), 222–245 (2013). https://doi.org/10.1007/s11263-013-0636-x
26. Schuller, B., et al.: The Interspeech 2017 computational paralinguistics challenge: addressee, cold and snoring. In: Computational Paralinguistics Challenge (ComParE), Interspeech 2017, pp. 3442–3446 (2017)
27. Schuller, B.W., Batliner, A.M.: Emotion, Affect and Personality in Speech and Language Processing. Wiley, Hoboken (1988)
28. Seeland, M., Rzanny, M., Alaqraa, N., Wäldchen, J., Mäder, P.: Plant species classification using flower images: a comparative study of local feature representations. PLOS ONE 12(2), 1–29 (2017)

29. Smith, D.C., Kornelson, K.A.: A comparison of Fisher vectors and Gaussian super-vectors for document versus non-document image classification. In: Applications of Digital Image Processing XXXVI, vol. 8856, p. 88560N. International Society for Optics and Photonics (2013)
30. Tian, Y., He, L., Li, Z.Y., Wu, W.L., Zhang, W.Q., Liu, J.: Speaker verification using Fisher vector. In: Proceedings of ISCSLP, Singapore, pp. 419–422 (2014)
31. Vedaldi, A., Fulkerson, B.: VLFeat: an open and portable library of computer vision algorithms. In: Proceedings of the 18th ACM International Conference on Multimedia, pp. 1469–1472. ACM (2010)
32. Wang, C., Deng, C., Wang, S.: Imbalance-XGBoost: leveraging weighted and focal losses for binary label-imbalanced classification with XGBoost. arXiv preprint arXiv:1908.01672 (2019)
33. Wang, S.-H., Li, H.-T., Chang, E.-J., Wu, A.-Y.A.: Entropy-assisted emotion recognition of valence and arousal using XGBoost classifier. In: Iliadis, L., Maglogiannis, I., Plagianakos, V. (eds.) AIAI 2018. IAICT, vol. 519, pp. 249–260. Springer, Cham (2018). https://doi.org/10.1007/978-3-319-92007-8_22
34. Zajíc, Z., Hrúz, M.: Fisher Vectors in PLDA speaker verification system. In: Proceedings of ICSP, Chengdu, China, pp. 1338–1341 (2016)

Toxicity in Texts and Images
on the Internet

Denis Gordeev[1]($\boxtimes$) and Vsevolod Potapov[2]

[1] Russian Presidential Academy of National Economy and Public Administration,
Moscow, Russia
gordeev-di@ranepa.ru
[2] Centre of New Technologies for Humanities, Lomonosov Moscow State University,
Leninskije Gory 1, Moscow 119991, Russia
volikpotapov@gmail.com

Abstract. In this paper we studied the most typical characteristics of toxic images on the web. To get a set of toxic images we collected a set of 8800 images from 4chan.org. Then we trained a BERT-based classifier to find toxic texts with accompanying images. We manually labelled approximately 2000 images accompanying these texts. This revealed that toxic content in images does not correlate with toxic content in texts. On top of manually annotated images there was trained a neural network that inferred labels for unannotated pictures. Neural network layer activations for these images were clustered and manually classified to find the most typical ways of expressing aggression in images. We find that racial stereotypes are the main cause of toxicity in images (https://github.com/denis-gordeev/specom20).

Keywords: Toxicity · BERT · Text classification · Image classification · Inception · MobileNet · K-means

1 Introduction

There are a lot of works devoted to toxicity and aggression detection in social media nowadays due to it being rampant on the Internet. According to a study by Duggan 66% of Americans have witnessed harassing behaviour online [7]. Most works studying cyberbullying, online aggression, toxic behaviour and other similar phenomena deal only with textual content. However, aggression can be expressed with the help of other media types, i.e. images and videos. Moreover, exposure to aggressive images is associated with higher aggression towards others [6]. Multidimensional approaches are also extremely relevant nowadays [14]. Thus, it can be beneficial to study how toxicity can be expressed in visual media. In this work we train a text classifier using annotated toxic texts dataset. Then this classifier is used to infer labels for texts with accompanying images. We base on the hypothesis that toxic texts accompany toxic images. We manually check a small subset of images to prove our hypothesis. With the help of these

© Springer Nature Switzerland AG 2020
A. Karpov and R. Potapova (Eds.): SPECOM 2020, LNAI 12335, pp. 156–165, 2020.
https://doi.org/10.1007/978-3-030-60276-5_16

labels we train a new neural network classifier using only image data. After that hidden layer activations of the image classifier are clustered using K-means. The resulting clusters are manually examined to study what makes an image toxic.

2   Related Work

Toxicity in texts and its automatic detection are rather well-studied problems. The term 'toxicity' is often used to describe negative behavior on the Internet that often involves damaging the other'as or even one's own self-image, without any beneficial personal growth [10,18]. Many researchers have paid attention to related problems of aggression and trolling behaviour detection on the Internet. Previously researchers focused on lexicon-based methods. For example, Davidson et al. collected a hate speech dataset exploring this problem [4]. The authors relied on heavy use of crowd-sourcing. First, they used a crowd-sourced hate speech lexicon to collect tweets with hate speech keywords. Then they resorted again to crowd-sourcing to label a sample of these tweets into three categories: those containing hate speech, containing only offensive language, and those with neither. Later analysis showed that hate speech can be reliably separated from other types of offensive language. They find that racist and homophobic tweets are more likely to be classified as hate speech but sexist tweets are generally classified as offensive. Malmasi together with Zampiere explored this dataset even further [11]. They have found that the main challenge for successful hate speech detection lies in indiscriminating profanity and hate speech from each other.

Private initiatives also do not keep out of this problem. For example, there were held several challenges on machine learning competition platform Kaggle devoted to aggression and toxic speech investigation in social media, among them: Jigsaw Toxic Comment Classification Challenge[1]. The best solutions on Kaggle used a bunch of various techniques to improve the model score. Among such techniques were various types of pseudo-labelling such as back-translation and new language modelling subtasks.

Despite the large number of papers devoted to toxic text identification on social media, there are few works aimed at toxicity detection in images. One of them uses optical character recognition (OCR) to extract text from images which later uses latent semantic analysis (LSA) and a threshold-based classifier to detect hate speech in images [13].

Some works [1,23] use media captions provided by content creators. This approach may be used only for images with provided text descriptions. Other researchers [2] apply similar methods and use only meta information from images without considering visual content. There are works studying social media images [8]. However, the images themselves are harmless in their work and are prone to cause toxic behaviour by Internet users (e.g. an image of a bad quality tattoo is likely to be ridiculed).

---

[1] https://www.kaggle.com/c/jigsaw-toxic-comment-classification-challenge.

3   Data Collection

There were collected 8800 posts from 4chan.org with images (posts without images were discarded). All texts were collected from the politics (\po) board. As images we downloaded only small thumbnails due to the fact that for older threads full-scale images were not available.

We trained a BERT-based [5] model to classify toxicity in texts. For its training there was used a toxicity dataset provided by Jigsaw on Kaggle[2]. It contained 2m public comments from social media each annotated by up to 10 annotators. We used only 'severe_toxicity' column and converted it to a binary format with its values equal to zero being considered as non-toxic texts and items with values between zero and 0.2 considered toxic ones (the original dataset contains only values in the range between 0 and 0.2, they were not normalized in the original dataset) (see Table 1). We took 100000 random toxic and non-toxic texts to train our classifier. Other columns in the dataset ('obscene', 'threat', 'insult', 'identity_attack', 'sexual_explicit') were considered irrelevant for the present research.

Table 1. BERT-inferred text labels.

Text	Label
...I won't watch any IBS anymore, if people like JF associate and even praise the turd.	Toxic
...No you, Marxist traitor. I am named after a Nordic god for real....	Toxic
There's more too. I know I had them, but I can't seem to find them. Take this as a consolation prize.	Non-Toxic
What's with all the Discord shilling, don't they know Discord was revealed as a honeypot operation?	Non-Toxic

4   Experiments

4.1   Text Classification

Using Jigsaw challenge data there was trained a BERT-based model. BERT [5] is a Transformer-based model [21]. We used a multilingual uncased BERT model provided by Hugging Face [22]. We used PyTorch framework to create our model. BERT was trained using Wikipedia texts in more than 100 languages. All texts were tokenized using byte-pair encoding (BPE) which allows limiting the vocabulary size compared to Word2vec and other word vector models. The training consisted in predicting a random masked token in the sentence and a binary next sentence prediction. We did not perform any text augmentation or pre-processing besides standard byte-pair encoding. All texts longer than 510

---

[2] https://www.kaggle.com/c/jigsaw-unintended-bias-in-toxicity-classification/.

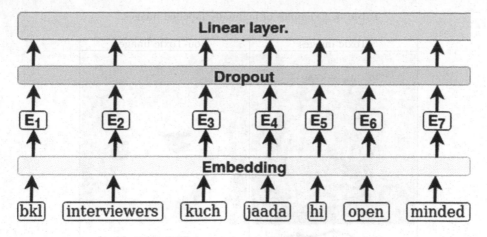

Fig. 1. Bert model.

tokens were truncated. Two tokens marking the beginning and the end of the sequence were added to each input text ("[CLS]" and "[SEP]"). Texts shorter than 510 tokens were padded with zeroes. All tokens excluding special ones were masked with ones, while all other tokens were masked with zeroes.

On top of BERT, we added a Dropout layer to fight overfitting (see Fig. 1). The Dropout probability was equal to 0.1. On top of the Dropout Layer, a softmax layer was added with its dimension equal to the number of classes (2). Target values were one-hot encoded. All texts were selected randomly out of the training and validation datasets. Cross entropy loss function was used. Half precision training was used via Apex library[3]. We used a single Nvidia V100 GPU to train our model. The training batch size was made equal to 16. The model was trained for 2 epochs. The model training results can be seen in Table 2. The trained model was used to infer labels for 4chan.org texts (see Table 1).

Table 2. BERT training results.

Metric	Score
F1 micro	0.837
Precision	0.861
Recall	0.814

4.2 Image Labelling

We based our work on the hypothesis that toxic images accompany aggressive texts. We manually labelled the images into two classes: toxic and non-toxic

---

[3] https://github.com/NVIDIA/apex.

Table 3. Examples of manually labelled images.

to check it. The labelling was performed by two annotators. While detecting toxicity, we used the same annotation principles as were used for the Jigsaw text dataset[4], i.e. the annotators were asked to detect whether a message is rude, disrespectful, aggressive or unreasonable enough, to make the annotator leave the discussion. In the statistics we included only images where both annotators agree. In the majority (approximately 1500) of cases annotators consulted and performed the annotation collectively. For other cases which were unambiguous the annotation was performed individually and the Cohen's kappa score [3] for them is very high (0.956). Many examples of toxic images contain examples of people and drawings of people with mental deficiencies or anthropomorphized animals (see Table 3). The hypothesis that toxicity in texts and toxicity in images correlate was not supported with Pearson correlation coefficient equal to -1 (see Table 4).

After manual image labeling we got 191 images containing toxic content and 1755 pictures without it. Using manually annotated images we trained another neural network using only images. The image data was split into train, validation

---

[4] https://meta.wikimedia.org/wiki/Research:Detox/.

Table 4. Manually labelled images dataset statistics.

	Toxic texts	Non-toxic texts
Toxic images	101	85
Non-toxic images	876	885

and test datasets with the ratio equal to 0.9:0.1:0.1. There were tried several architectures with various hyper-parameters. Namely they were MobileNetV2 [17] and InceptionV3 [19]. These models have been chosen due to their small size that helped to avoid overfitting with the small thumbnails in our dataset. Both models were pretrained using ImageNet. The pretrained weights were provided by TensorFlow Hub[5]. Due to the unbalanced class distribution F1-score metric was used. It should be noted that in our case the class imbalance is on the order of 1:10 which is quite ordinary. For example, ImageNet [16] contains classes with their ratios equal to 1:50 and there are datasets with even large imbalance (e.g. iNaturalist with the 1:500 ratio [20] or even $1:10^5$ and higher for WikiLSHTC-325K and other extreme classification datasets [9]). We also experimented with image augmentation. Images were rotated, shifted horizontally and vertically, zoomed in and flipped horizontally. All images were made square using padding and resizing if the image is too large due to neural network models limitations. The batch size was set to 32. The learning rate was equal to 0.01 in all cases. We trained the model using stochastic gradient descent with momentum set to 0.9.

Table 5. Test results for models classifying toxicity in images.

Model	Image size	Augmentation	Epochs taken	F1-score
MobileNet v2	224	+	6	0.8765
MobileNet v2	224	−	24	0.8615
MobileNet v2	160	+	8	0.8261
MobileNet v2	**160**	−	**17**	**0.9132**
Inception v3	160	+	7	0.8400
Inception v3	160	−	15	0.8958
Inception v3	224	+	9	0.8919
Inception v3	224	−	15	0.9105
Inception v3	448	+	10	0.8779
Inception v3	448	−	10	0.9054

Image classification results can be seen in (Table 5). As can be seen from the table larger models do not provide any performance boost. Moreover, decreasing

---

[5] https://tfhub.dev/google/

image size even boosted model performance. It may be attributed to the median image size being less than 200 pixels. It is also interesting that image augmentation despite decreasing the number of training epochs did not contribute to model performance.

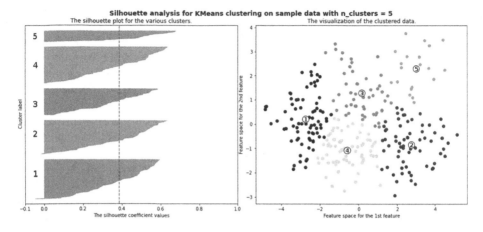

Fig. 2. Clustering analysis

4.3   Toxic Clusters

Our aim was to examine the main ways to express toxicity in images on 4chan.org. For this purpose, the best performing image classification model was used to infer labels for images that were not manually labelled. For toxic images MobileNet last layer activations were first converted to 2D using UMAP [12] and then clustered with K-means. The number of clusters found using the silhouette analysis [15] is equal to 5 (see Fig. 2). The silhouette analysis has found two peaks at two and five clusters (with the values equal to 2: 0.445, 3: 0.336, 4: 0.360, 5: 0.387, 6: 0.355, 7: 0.364), but two clusters were too problematic for manual analysis.

These clusters were manually labelled. We also selected the closest images to cluster centers to illustrate them. Clusters were not homogeneous. Each cluster contained images that were erroneous or were closer to other clusters. Some image types (e.g. green faced Pepe-frog images) were spread across all clusters. Two out of five clusters are mainly devoted to racial stereotypes. All five clusters with descriptions can be seen in Table 6. Thus, we may suppose that the main way to express toxicity in images on the Internet is to use racial stereotypes. Also Internet users tend to present their opponents as mentally challenged. One of the most popular ways of doing is just to depict their opponents' leaders with a 'small face' or without face at all.

Table 6. Examples of manually aggression clusters.

Cluster number	Image example	Description	Cluster number	Image example	Description
1		Caricatures and 'small faces'	2		Afro-American racist images and anime pictures
3		Notorious historic figures	4		Miscellaneous
5		racist Jew images			

5 Conclusion

Using the existing deep knowledge on toxicity in texts we extended this approach to images. We found that toxicity in images does not correlate with toxicity in the accompanying texts. Moreover, a neutral text may be illustrated with a toxic image and vice versa. Thus, an image classifier is required on top of a toxic-based model. Using a set of manually annotated images we created a classifier that can be used to find toxic images on the Internet. We studied its layer activations and found that racial stereotypes are the main way to create toxicity in images.

Acknowledgments. This research was supported by the Russian Science Foundation (RSF) according to the research project 18-18-00477.

References

1. Cheng, L., Guo, R., Silva, Y., Hall, D., Liu, H.: Hierarchical attention networks for cyberbullying detection on the instagram social network. In: Proceedings of the 2019 SIAM International Conference on Data Mining, pp. 235–243. SIAM (2019)
2. Cheng, L., Li, J., Silva, Y.N., Hall, D.L., Liu, H.: XBully: cyberbullying detection within a multi-modal context. In: Proceedings of the Twelfth ACM International Conference on Web Search and Data Mining, pp. 339–347 (2019)
3. Cohen, J.: A coefficient of agreement for nominal scales. Educ. Psychol. Measur. **20**(1), 37–46 (1960)
4. Davidson, T., Warmsley, D., Macy, M., Weber, I.: Automated hate speech detection and the problem of offensive language. In: Proceedings of the ICWSM (2017)
5. Devlin, J., Chang, M.W., Lee, K., Toutanova, K.: BERT: pre-training of deep bidirectional transformers for language understanding, October 2018. http://arxiv.org/abs/1810.04805
6. DeWall, C.N., Anderson, C.A., Bushman, B.J.: The general aggression model: theoretical extensions to violence. Psychol. Violence **1**(3), 245 (2011)
7. Duggan, M.: Online harassment, vol. 2017 (2017)
8. Hitkul, H., Shah, R.R., Kumaraguru, P., Satoh, S.: Maybe look closer? Detecting trolling prone images on instagram. In: 2019 IEEE Fifth International Conference on Multimedia Big Data (BigMM), pp. 448–456. IEEE (2019)
9. Khandagale, S., Xiao, H., Babbar, R.: Bonsai-diverse and shallow trees for extreme multi-label classification. arXiv preprint arXiv:1904.08249 (2019)
10. Lapidot-Lefler, N., Barak, A.: Effects of anonymity, invisibility, and lack of eye-contact on toxic online disinhibition. Comput. Hum. Behav. **28**(2), 434–443 (2012)
11. Malmasi, S., Zampieri, M.: Detecting Hate Speech in Social Media. In: Proceedings of the International Conference on Recent Advances in Natural Language Processing, pp. 467–472 (2017)
12. McInnes, L., Healy, J., Melville, J.: UMAP: uniform manifold approximation and projection for dimension reduction. arXiv preprint arXiv:1802.03426 (2018)
13. Niam, I.M.A., Irawan, B., Setianingsih, C., Putra, B.P.: Hate speech detection using latent semantic analysis (LSA) method based on image. In: 2018 International Conference on Control, Electronics, Renewable Energy and Communications (ICCEREC), pp. 166–171. IEEE (2018)
14. Potapova, R., Potapov, V.: Human as acmeologic entity in social network discourse (multidimensional approach). In: Karpov, A., Potapova, R., Mporas, I. (eds.) SPECOM 2017. LNCS (LNAI), vol. 10458, pp. 407–416. Springer, Cham (2017). https://doi.org/10.1007/978-3-319-66429-3_40
15. Rousseeuw, P.J.: Silhouettes: a graphical aid to the interpretation and validation of cluster analysis. J. Comput. Appl. Math. **20**, 53–65 (1987)
16. Russakovsky, O., et al.: Imagenet large scale visual recognition challenge. Int. J. Comput. Vision **115**(3), 211–252 (2015)
17. Sandler, M., Howard, A., Zhu, M., Zhmoginov, A., Chen, L.C.: MobileNetV2: inverted residuals and linear bottlenecks. In: Proceedings of the IEEE Conference on Computer Vision and Pattern Recognition, pp. 4510–4520 (2018)
18. Suler, J.: The online disinhibition effect. Cyberpsychol. Behav. **7**(3), 321–326 (2004)
19. Szegedy, C., Vanhoucke, V., Ioffe, S., Shlens, J., Wojna, Z.: Rethinking the inception architecture for computer vision. In: Proceedings of the IEEE conference on computer vision and pattern recognition, pp. 2818–2826 (2016)

20. Van Horn, G., et al.: The inaturalist species classification and detection dataset. In: Proceedings of the IEEE Conference on Computer Vision and Pattern Recognition, pp. 8769–8778 (2018)
21. Vaswani, A., et al.: Attention is all you need. In: Advances in Neural Information Processsing Systems, vol. 2017, pp. 5999–6009, December 2017. http://papers.nips.cc/paper/7181-attention-is-all-you-need
22. Wolf, T., Debut, L., Sanh, V., Chaumond, J., Delangue, C., Moi, A., Cistac, P., Rault, T., Louf, R., Funtowicz, M., Brew, J.: HuggingFace's Transformers: State-of-the-art Natural Language Processing. arXiv abs/1910.0 (2019)
23. Yao, M., Chelmis, C., Zois, D.S.: Cyberbullying ends here: Towards robust detection of cyberbullying in social media. In: The World Wide Web Conference. pp. 3427–3433 (2019)

An Automated Pipeline for Robust Image Processing and Optical Character Recognition of Historical Documents

Ivan Gruber[1,2]([✉]) [iD], Pavel Ircing[1,2] [iD], Petr Neduchal[1] [iD], Marek Hrúz[1] [iD], Miroslav Hlaváč[1] [iD], Zbyněk Zajíc[1] [iD], Jan Švec[1,2] [iD], and Martin Bulín[1,2] [iD]

[1] Faculty of Applied Sciences, New Technologies for the Information Society, University of West Bohemia, Plzeň, Czech Republic
{grubiv,ircing,neduchal,mhruz,mhlavac,zzajic,honzas,bulinm}@ntis.zcu.cz
[2] Department of Cybernetics, University of West Bohemia, Univerzitní 8, 301 00 Plzeň, Czech Republic

Abstract. In this paper we propose a pipeline for processing of scanned historical documents into the electronic text form that could then be indexed and stored in a database. The nature of the documents presents a substantial challenge for standard automated techniques – not only there is a mix of typewritten and handwritten documents of varying quality but the scanned pages often contain multiple documents at once. Moreover, the language of the texts alternates mostly between Russian and Ukrainian but other languages also occur. The paper focuses mainly on segmentation, document type classification, and image preprocessing of the scanned documents; the output of those methods is then passed to the off-the-shelf OCR software and a baseline performance is evaluated on a simplified OCR task.

Keywords: OCR · Document classification · Document digitization

1 Introduction

The ever-increasing availability of scanning and storage devices has motivated archivists and historians to digitize a vast amount of historical materials, including millions of pages with handwritten and typewritten text and photographs. Those scholars have quickly realized that the digitization itself is only a first step—and probably the easiest one – in making the historical documents accessible for both the academic community and the general public. This paper concentrates on the methods that will be able to reliably transform the scanned documents into the electronic text form that could then be indexed and stored in a database. On top of the database, there is a GUI that enables multifaceted search in the stored documents, with special emphasis on searching for the occurrence of specified words and/or short phrases.

A. Karpov and R. Potapova (Eds.): SPECOM 2020, LNAI 12335, pp. 166–175, 2020.
https://doi.org/10.1007/978-3-030-60276-5_17

The documents in question are related to the individuals persecuted by NKVD (People's Commissariat for Internal Affairs) in the USSR, providing their life stories that are somehow connected to the history of Czechoslovakia. The archive currently contains over 0.5 million pages which justify the necessity of the (semi-)automated processing. At the same time, the nature of the documents presents a substantial challenge for such automated techniques—not only there is a mix of typewritten and handwritten documents of varying quality but also the languages of the text vary across (and sometimes even within) individual pages; the main portion of the texts is written in Russian and Ukrainian but other languages (Czech, German, . . .) also occur.

In order to fulfill the "umbrella" task—converting the scanned documents into searchable database—in this challenging scenario, we had to first outline the appropriate pipeline for document processing (Sect. 2). Sections 3 to 5 then analyze the current status of the individual pipeline blocks within our project, whereas Sect. 6 puts our image processing efforts into a broader perspective by showing the search GUI, the ultimate outcome of our project. The final two sections then outline the work that has to be done in the near future and summarize the findings of the experiments.

2   Proposed Processing Pipeline

As was already mentioned, the task at hand is rather challenging as the quality of the scanned materials is often far from optimal from the OCR point of view. This stems not only from the historical nature of the scanned documents—which means that the quality of the source documents themselves is low (crumbles, background spots)—but also from the fact that they were scanned and stored probably in a hurry and consist of an unordered mix of typewritten and handwritten text, photos and other graphical materials. There are no metadata that would denote the type of the document stored in a given file, nor there is any naming convention that would indicate it. Moreover, the scanned "page" often contains not just a single page of a document, but also a part of some other page or sometimes even a completely unrelated loose leaf with a text, covering part of the "main" document (see Fig. 3 for illustration).

All these problems make the data very hard to process with standard OCR pipelines. Thus we had to design a pipeline tailored specifically to our project needs (see Fig. 1). A reader might argue such a pipeline is probably routinely used in similar projects but, according to our knowledge, the Document Page Segmentation and Document Type Classification tasks are done manually. However, our aim is to develop automated methods, as described in the following sections.

3   Document Page Segmentation

In our initial attempts to automatically segment the input scanned page into individual documents, we have employed two existing solutions for document

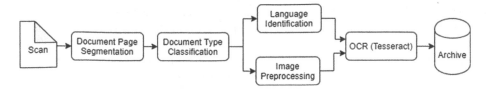

Fig. 1. Proposed pipeline.

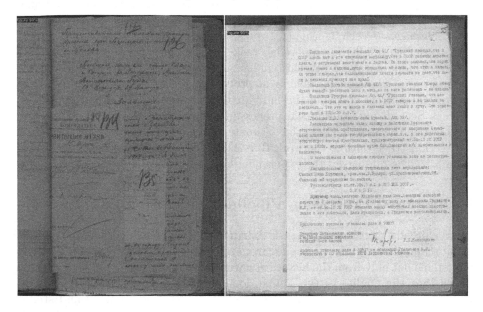

Fig. 2. Outputs of scan segmentation using plain Detectron2. The whole left image (red area) is classified as Figure. The major part of the right image (blue area) is also classified as Figure. Document source: Sectoral State Archive of the Security Services of Ukraine (HDA SBU). (Color figure online)

layout analysis. The first approach is based on a plain implementation of Detectron2 [9] and it is pretrained on PubLayNet dataset [11] containing modern scientific papers to segment document layout into five following segmentation classes: Text, Title, List, Table, and Figure. The second tested approach called Newspaper navigator [5] is based on a modified Detectron2 approach and it was trained on historic newspaper pages in Chronicling America to segment document layout into the following seven segmentation classes: Photograph, Illustration, Map, Comics/Cartoon, Editorial Cartoon, Headline, and Advertisement.

The example outputs can be found in Fig. 2 and Fig. 3. Unfortunately, both of the tested methods performed very poorly. The results obtained from the Newspaper navigator at least separate different pages from each other, however, with very misguided labels. We believe that it stems from the fact that processed documents are of very different nature from documents used for training of the

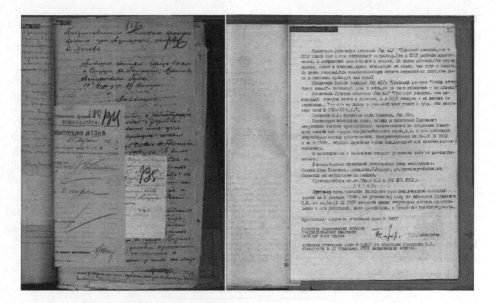

Fig. 3. Outputs of scan segmentation using Newspaper navigator. On the left image, the orange and the green bounding boxes are classified as Map. On the right image, thee green bounding box is classified as Advertisement. Document source: Sectoral State Archive of the Security Services of Ukraine (HDA SBU). (Color figure online)

evaluated methods; for example, the handwritten letter is something completely new for both tools. Moreover, the fact that neither of the evaluated methods was trained on Cyrillic text can constitute a substantial hindrance. In our future research, we, therefore, would like to address these problems via fine-tuning of Newspaper navigator using our own annotated data.

4   Document Type Classification

To test the feasibility of the document type classification system we devised an experiment based on document retrieval metric - mean average precision (mAP) [6].

We assume the following types of the document: handwritten, typewritten, image, and unknown. The first two classes are self-explanatory and can appear simultaneously in every document. The image class contains documents with photos. We have also included documents with big stamps representing symbols of states into the image class. The class unknown usually contains documents that are empty or unreadable.

In the document type classification, we consider both the dependent and independent cases of tests. In the case of a dependent test, we assume that all the labeled classes for two given documents must match. In the case of the independent test, we run it for each assigned class of the document and look

Table 1. Results for the class dependent document retrieval tests. AP is the average precision for individual classes, mAP is the average precision.

Classes	Count	AP
typewritten	396	0.32
handwritten	1094	0.66
images	8	0.18
type- and handwritten	569	0.38
typewritten and images	30	0.06
handwritten and images	7	0.08
all	46	0.11
mAP		**0.57**

Table 2. Results for the class independent document retrieval tests. AP is the average precision for individual classes, mAP is the average precision.

Classes	Count	AP
typewritten	1041	0.60
handwritten	1716	0.83
images	91	0.09
mAP		**0.77**

only for the tested class in a matching document. To compute the mAP, we first compute deep features of all the images using a ResNet50 architecture [4] pre-trained on ImageNet dataset [2]. Since we have only limited number of labeled data, we do not perform transfer learning. The deep feature is extracted as the output of the last global averaging layer in the ResNet50 network. The input images of documents have been resized to a resolution of 224×224 px. Next, we iterate through the images one by one and use it as a query document. For each query, we compute the precision of document retrieval ranking via Euclidean distance of the deep features. Then we obtain the mAP by averaging the precision through classes. The results are summarized in Table 1 and 2.

These results indicate that the usage of deep features to classify the documents is feasible, but the model needs to be fine-tuned for the final application. Once the document page segmentation described in the previous section starts producing usable results, the classification performance is likely to substantially improve.

5   Image Preprocessing and Optical Character Recognition Experiments

We have performed baseline experiments using an off-the-shelf solution in order to get an insight view on the accuracy of the OCR system on our data. We have chosen Tesseract OCR [8] because we have a great experience with this system from our previous projects [10].

In order to assess the OCR performance on a new data collection, we of course need the reference transcriptions of some documents sampled from such a collection. However, our goal is not to provide the user with the full transcription of the scanned documents but rather to find an occurrence of a word or a short phrase within the documents (see Sect. 6 for details). Thus we should concentrate on the good OCR performance for words that the user is likely to input to the search engine. Those are usually the words that could be characterized as named entities, especially personal and geographic names. We have, therefore – in cooperation with the historians participating in our project – randomly selected 67 typewritten text documents containing those named entities. There we have transcribed a total of 251 occurrences of 117 unique geographic and personal names. We have also marked the bounding boxes of those words and registered the language used in the document (Russian, Ukrainian, or multi-language - it should be noted that only isolated words in other languages than Russian or Ukrainian are usually presented). These annotations were saved into the *.yaml* file and served as a first test set for the OCR experiments described in the following paragraphs.

Our OCR task is thus somewhat simplified in this scenario—we are trying to find occurrences of the above mentioned 117 words in the set of 67 documents. Note that in this scenario, it is not necessary to classify the document type (all documents are typewritten) and we did not perform any page segmentation since we do not yet have a functional segmentation module.

We actually perform two sets of rather different experiments—we call the first one the *document level search* and the second one the *spatial level search*.

In the *document level search*, the word is considered correctly found when it is recognized anywhere in the document (page), regardless of the mutual position of the manually annotated ground-truth and the recognized OCR result.

In the *spatial level* experiments, the search performance is evaluated based on the *Intersection over Union (IoU)* between ground-truth and recognized bounding boxes (provided by Tesseract OCR). *IoU*, also known as the Jaccard index, is an evaluation metric used to measure the performance of object detection algorithms. It is defined as the size of the intersection divided by the size of the union of bounding boxes:

$$IoU = \frac{B_1 \cap B_2}{B_1 \cup B_2}, \tag{1}$$

where B_1 and B_2 are ground-truth bounding box and recognized bounding box, respectively. In most detection tasks, $IoU > 0.5$ is considered as a correct detection.

Both sets of experiments were performed for raw data and also for data preprocessed by deskew and RGB binarization methods. In Figs. 4 and 5, there are graphs of ratio between average accuracy on our dataset and IoU value – ROC curve – for 100% and 75% similarity of searched words. As you can see, OCR was applied using three language settings. In particular, it was Russian

(RUS) language, Ukrainian language (UKR), and multi-language (multi) settings including Russian, Ukrainian, and Czech Language.

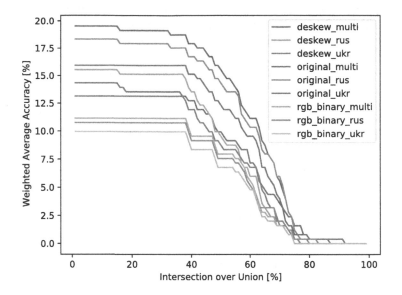

Fig. 4. ROC curve of word recall based on the IoU for 100% word similarity.

Similarity is computed using Levenshtein ratio which is computed as follows

$$p = 100 \cdot (l_{sum} - c_{sum})/l_{sum} \tag{2}$$

where l_{sum} is a sum of both words and c_{sum} is the sum of costs. The insert and delete have cost 1, and the substitution has cost 2 – i.e., the sum of the insert and delete cost. Unfortunately, we found that preprocessing methods do not bring any accuracy increase. In fact, the RGB binarization method even causes an accuracy reduction.

In the case of the document level search, we achieved the best results for the multi-language settings without preprocessing. In particular, it is 25.50% for 100% word similarity and 55.38% for 75% word similarity.

6   Indexing, Searching and GUI

The indexing and searching algorithms that we plan to use in our project were developed within the previous projects dealing with historical archives, especially the one described in [10]. They were originally developed for the task of spoken term detection (implementation details could be found in [7]) but can be rather easily adapted for detection of the "written terms". Here, we just include the screenshot (see Fig. 6) of the current version of the search GUI in order to justify

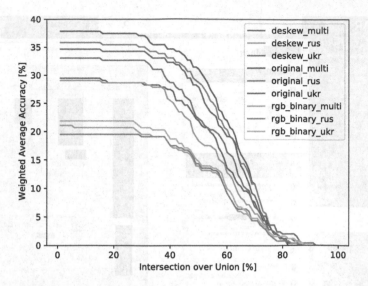

Fig. 5. ROC curve of word recall based on the IoU for 75% word similarity.

our focus on the spatial level query search discussed in Section 5. It can been seen that in this case, the OCR coupled with the search algorithm was able to correctly localize the queried words *tábor* (*the camp*) in the scanned document.

7 Future Work

Our experiments showed that existing off-the-shelf solutions for the tasks in the proposed pipeline leave substantial room for improvement. Our attempts to enhance the performance of the pipeline components will involve the preparation of additional datasets directly relevant to the processed document collection. We have already manually labeled approx. 2100 documents with document type, using a specially designed GUI—those are currently used as the ground-truth for the experiment described in Sect. 4 but could be readily used for retraining of the classification algorithms. The annotation guidelines for a similar dataset that will be later used for adapting the page segmentation algorithms are currently being prepared. For this task, we see the biggest potential in the Newspaper navigator method.

We also plan to work on methods that will be able to automatically identify the language of the processed scan. Given the sources of the documents, the classification between Russian, Ukrainian, and "multi-language" will probably be sufficient.

Due to the time-consuming nature of hand-made annotations, we would like to explore other possibilities. To be more specific, we see a big potential in synthetic image generation [1,3], which allows creating a big amount of realistically looking data with precise annotations.

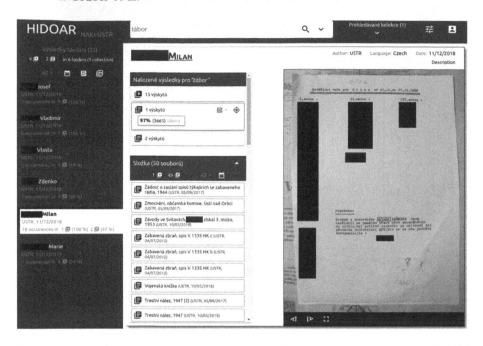

Fig. 6. Current version of the search interface. All sensitive data have been blacked out from the example before publishing in this paper. Document source: Institute for the Study of Totalitarian Regimes (ÚSTR).

8   Conclusion

The paper presented the image processing pipeline that is designed to achieve a robust optical character recognition in 20th century documents. The results of the OCR will then be indexed and made available through a graphical user interface with search capabilities. The first experiments have shown that the data in question pose a significant challenge to existing image processing algorithms and further research and development of new methods is necessary. This also includes the need for additional annotation of the data sampled from the collection at hand, for domain adaptation of the methods used in our proposed pipeline.

Acknowledgments. This research was supported by the Ministry of Culture Czech Republic, project No. DG20P02OVV018. Access to computing and storage facilities owned by parties and projects contributing to the National Grid Infrastructure Meta-Centrum provided under the programme "Projects of Large Research, Development, and Innovations Infrastructures" (CESNET LM2015042), is greatly appreciated.

References

1. Bureš, L., Gruber, I., Neduchal, P., Hlaváč, M., Hrúz, M.: Semantic text segmentation from synthetic images of full-text documents. SPIIRAS Proc. **18**(6), 1381–1406 (2019)
2. Deng, J., Dong, W., Socher, R., Li, L.J., Li, K., Fei-Fei, L.: ImageNet: a large-scale hierarchical image database. In: 2009 IEEE Conference on Computer Vision and Pattern Recognition, pp. 248–255. IEEE (2009)
3. Gruber, I., Hlaváč, M., Hrúz, M., Železný, M.: Semantic segmentation of historical documents via fully-convolutional neural network. In: Salah, A.A., Karpov, A., Potapova, R. (eds.) SPECOM 2019. LNCS (LNAI), vol. 11658, pp. 142–149. Springer, Cham (2019). https://doi.org/10.1007/978-3-030-26061-3_15
4. He, K., Zhang, X., Ren, S., Sun, J.: Deep residual learning for image recognition. In: Proceedings of the IEEE Conference on Computer Vision and Pattern Recognition, pp. 770–778 (2016)
5. Lee, B.C.G., et al.: The newspaper navigator dataset: extracting and analyzing visual content from 16 million historic newspaper pages in chronicling America (2020)
6. Liu, L., Özsu, M.T.: Mean average precision. In: Liu, L., Özsu, M.T. (eds.) Encyclopedia of Database Systems, p. 1703. Springer, Boston (2009). https://doi.org/10.1007/978-0-387-39940-9_3032
7. Psutka, J., et al.: System for fast lexical and phonetic spoken term detection in a Czech cultural heritage archive. EURASIP J. Audio Speech Music Process. **2011**(1), 10 (2011)
8. Smith, R.: An overview of the Tesseract OCR engine. In: Ninth International Conference on Document Analysis and Recognition (ICDAR 2007), vol. 2, pp. 629–633. IEEE (2007)
9. Wu, Y., Kirillov, A., Massa, F., Lo, W.Y., Girshick, R.: Detectron2 (2019). https://github.com/facebookresearch/detectron2
10. Zajíc, Z., et al.: Towards processing of the oral history interviews and related printed documents. In: Proceedings of LREC 2018, pp. 2099–2104 (2018)
11. Zhong, X., Tang, J., Yepes, A.J.: PubLayNet: largest dataset ever for document layout analysis. In: 2019 International Conference on Document Analysis and Recognition (ICDAR), pp. 1015–1022. IEEE, September 2019

Lipreading with LipsID

Miroslav Hlaváč[1,2]([📧]) [ID], Ivan Gruber[1,2] [ID], Miloš Železný[1,2] [ID],
and Alexey Karpov[3] [ID]

[1] University of West Bohemia, Pilsen, Czech Republic
{mhlavac,zelezny}@kky.zcu.cz
[2] NTIS, Pilsen, Czech Republic
{mhlavac,grubiv}@ntis.zcu.cz
[3] St. Petersburg Institute for Informatics and Automation of the Russian Academy
of Sciences SPIIRAS, St. Petersburg, Russia
karpov@iias.spb.su
http://www.kky.zcu.cz
http://www.ntis.zcu.cz
http://hci.nw.ru/en/employees/1

Abstract. This paper presents an approach for adaptation of the current visual speech recognition systems. The adaptation technique is based on LipsID features. These features represent a processed area of lips ROI. The features are extracted in a classification task by neural network pre-trained on the dataset-specific to the lip-reading system used for visual speech recognition. The training procedure for LipsID implements ArcFace loss to separate different speakers in the dataset and to provide distinctive features for every one of them. The network uses convolutional layers to extract features from input sequences of speaker images and is designed to take the same input as the lipreading system. Parallel processing of input sequence by LipsID network and lipreading network is followed by a combination of both feature sets and final recognition by Connectionist Temporal Classification (CTC) mechanism. This paper presents results from experiments with the LipNet network by re-implementing the system and comparing it with and without LipsID features. The results show a promising path for future experiments and other systems. The training and testing process of neural networks used in this work utilizes Tensorflow/Keras implementations [4].

Keywords: Automated lipreading · Computer vision · Visual speech recognition · Deep learning

1 Introduction and State-of-the-Art

The topic of visual speech recognition is the focus of researchers in recent years thanks to the evolution of deep neural networks. It has moved from statistical-based approaches based on Active Shape or Active Appearance models [7] to end-to-end systems based on deep learning. Most recent methods include networks like LipNet [3], WLAS [5], and other transformer-based models [2,12].

© Springer Nature Switzerland AG 2020
A. Karpov and R. Potapova (Eds.): SPECOM 2020, LNAI 12335, pp. 176–183, 2020.
https://doi.org/10.1007/978-3-030-60276-5_18

End-to-end systems enabled the processing of large amounts of data without precise annotations of the spoken word. The usual input is comprised of an input sequence of frames(video) with a label in the form of a character, word, or sentence corresponding to the input. The task of identification of frames to the corresponding translation is then solved internally in the system without the need of providing annotations. Visual speech synchronization and transcription are then solved by CTC algorithm [9], which proved to be robust enough to provide meaningful results.

This paper is a continuation of work started in [10]. It provides improvements in the LipsID network by incorporation of ArcFace loss [8]. LipsID features are then combined with lipreading features to provide a form of adaptation for state-of-the-art systems based on speaker identity. This task requires full access to training scripts of the system which will be adapted and thus the experiments involved in this paper use the LipNet network. The other modern systems were unfortunately not available for training or we were unable to replicate their original results during the writing of this paper, see Subsect. 4.2. These systems will be further tested in the future after they are publicly released. This research is a part of the Ph.D. thesis [13].

2   Previous Work

We have presented the LipsID features in the previous paper [10]. The idea is to provide similar information to i-vectors [11] that are based on speaker audio channel analysis and provided a means for adaptation based on the characteristics of the speaker's channel. To implement LipsID into current systems it was designed as a classification network. The system is implemented as a classification neural network where the penultimate layer produces the LipsID features after training. This network can be specifically trained with the dataset that was used for the lipreading system that is then augmented with LipsID. Since the usual input for a lipreading system is a sequence of images contain the speaker's face it was implemented as a 3D Convolutional classifier with Softmax. The output is the classification of the whole sequence into classes based on speakers in the dataset. The architecture can be found in the Table 1. The previous work was mainly focused on designing the LipsID features and in the paper, we present the results of our initial experiments with lipreading using LipsID and LipNet.

Table 1. Architecture of the original LipsID network published in [10].

Conv3D(32, 3 × 3 × 3, ReLU)	Conv3D(64, 3 × 3 × 3, ReLU)	Conv3D(128, 3 × 3 × 3, ReLU)	FC(256)
Conv3D(32, 3 × 3 × 3, ReLU)	Conv3D(64, 3 × 3 × 3, ReLU)	Conv3D(128, 3 × 3 × 3, ReLU)	FC(64, softmax)
Batch Normalization	Batch Normalization	Batch Normalization	
MaxPool3D(3 × 3 × 2)	MaxPool3D(3 × 3 × 2)		

Conv3D(32, 3 × 3 × 3, ReLU) denotes convolutional layer with 32 kernels of shape 3 × 3 × 3. MaxPool3D denotes max-pooling layer with kernel shape of

$3 \times 3 \times 2$. FC denotes fully-connected layer with the corresponding number of neurons.

3   Lipreading with LipsID

Our previous research is augmented by using ArcFace loss [8]. The importance of this upgrade is best described by the image in Fig. 1. ArcFace is trying to maximize the distance between the classes and thus helping the final network to generate a distinctive set of features for each speaker.

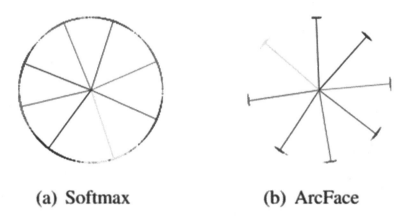

(a) Softmax   (b) ArcFace

Fig. 1. Toy examples under the Softmax and ArcFace loss on 8 identities with 2D features. Dots indicate samples and lines refer to the center direction of each identity. Based on the feature normalization, all features are pushed to the arc space with a fixed radius. The geodesic distance gap between closest classes becomes evident as the additive angular margin penalty is incorporated. Taken from [8].

Originally, the LipsID network started as a single-image classifier [10]. Then it evolved into sequence analysis. This helps the network to capture the dynamic part of the speech captured in the input images. Thanks to this approach it is possible to use the same input for the LipsID extraction as for the lipreading system.

The state-of-the-art systems solve the lipreading task as an end-to-end system. To connect the LipsID network in the process a suitable place has to be chosen. Since the LipsID network can be trained independently for each lipreading system, the size of the output feature vector can be selected to be compatible with features produced by the lipreading system. The structure of the network can be found in the original paper [10]. The LipsID network is pre-trained with the same dataset as the system it is going to augment. This step is followed by a joined fine-tuning training of both systems - the original system, and the LipsID network.

Interesting problems raise during the combination of LipsID features and lipreading features. There are several tasks that have to be solved to successfully combine them into equally important sets. The first is the process of their combination. The features can be concatenated, added together, or combined by an additional layer of the network. Since the penultimate layer of the LipsID network can be accommodated to produce a feature vector of desirable size, we have selected concatenation. The second task is then the normalization of the features. The features should share a common magnitude, so one would not dominate the other. This can also be solved by multiple approaches. We have chosen the sigmoidal activation function before concatenation, so both feature vectors have the same minimal and maximal value.

4   Experiments

In this section, we present the results of experiments with the LipNet and AVSR network. In the process of developing the LipsID features, we were looking for already existing lipreading systems that can serve as benchmark systems for our features. Unfortunately, the systems published in [2,5] were not available in a trainable form that could reproduce the results in their corresponding papers. However, the LipNet network and the AVSR network were released by the authors on GitHub and thus were ideal systems to experiment with.

We have developed a universal algorithm to train a lipreading system and LipsID network and to connect them together afterward. The algorithm contains the following steps:

1. Pre-training of the chosen lipreading system using the original parameters.
2. Pre-training the LipsID network for classification (stopped at the accuracy of 95% (heuristically chosen threshold).
3. Creating a new combined model that includes both pre-trained networks.
4. Locking the weights of layers preceding the concatenation of outputs from both networks.
5. Training the new model until the loss stabilizes.
6. Unlocking all the weights and fine-tuning the whole system.

4.1   LipNet

Firstly, we replicated the results from the original paper for LipNet. The original network is trained on GRID [6] dataset which is not as extensive as other lipreading datasets like LRS3-TED [1]. GRID is composed of 1000 sentences with a specific structure, for exemplary images see Fig. 2.

We have applied the training algorithm described above to train LipsID and LipNet network separately at first. The last classification layer of LipsID was separated and the penultimate layer produced the LipsID features. The layer after Bidirectional GRU in LipNet was concatenated with LipsID features and then followed by a dense layer and CTC as in the original paper. The schematic depiction of the system is shown in Fig. 3. The training scenarios of the original paper for LipNet were followed including the split for training, validation, and testing data. The same split was used for both networks for consistency.

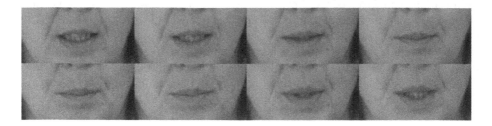

Fig. 2. Sample images from sequence - GRID dataset [6].

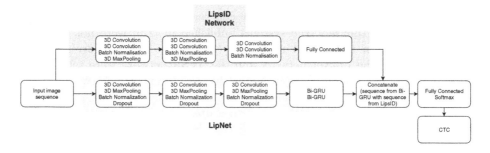

Fig. 3. LipNet with LipsID network.

The fine-tuning took 60 epochs and the development of accuracy is shown in Fig. 4. This is the case of the scenario of overlapped speakers. We have used standard SGD optimizer with Nesterov momentum with starting learning rate $l = 0.001$ since both networks were pre-trained.

We tested the systems in two different scenarios - unseen speaker scenario and overlapped speaker scenario. The unseen speaker scenario means the evaluation of speakers not included in the training set. The overlapped speaker scenario means evaluation on a subset of 255 videos from each speaker [3]. For both scenarios, we calculated the character error rate (CER) and the word error rate (WER). The results are shown in Table 2.

4.2   AVSR

In our second experiment, we tried to implement the LipsID features into the AVSR network [14]. This original network uses TCD-TIMIT dataset [15]. Despite the fact, we followed the original training scenario for audio-visual speech recognition, we were unable to achieve the recognition accuracy published in the

Table 2. Comparison of lipreading errors (lower is better) of original LipNet and LipNet augmented with the LipsID features. The training scenarios follow the original paper with LipNet.

Scenario	Unseen speaker		Overlapped speakers	
Error type	CER	WER	CER	WER
LipNet	6.4%	11.4%	1.9%	4.8%
LipNet + LipsID	**5.2%**	**9.9%**	**1.2%**	**3.3%**

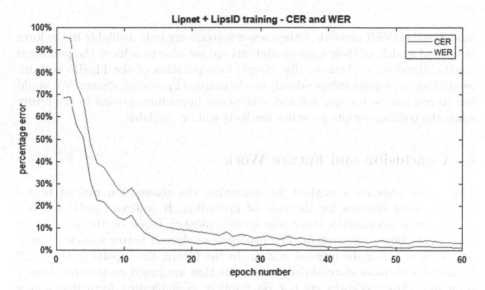

Fig. 4. LipNet with LipsID CER and WER during fine-tuning.

original paper [14] during the experiment. On such a trained network we applied LipsID training algorithm to investigate the effect of LipsID features. The results, see Table 3, confirms the added benefit of LipsID features once again. Since the results of the training AVSR network were noticeably worse than the ones published by the authors, we did not perform any further experiments with the AVSR network.

4.3   Discussion

It can be seen that our proposed system outperformed the original systems in all tested scenarios. We acknowledge that LipNet and the GRID dataset are not considered great in the field of automated lipreading. We have used LipNet because it is available in the original form published in the paper and can be replicated from scratch and the GRID dataset is used to produce the original results for comparison. Despite the fact we were unable to reach results from the original paper, LipsID also significantly improved the results of our re-trained

Table 3. Comparison of lipreading errors (lower is better) of the original AVSR (audio-visual scenario), our re-implementation of AVSR, and the re-trained AVSR augmented with the LipsID features. The training scenarios follow the original AVSR paper.

Network/Error type	CER	WER
Original AVSR [14]	17.7%	41.9%
Re-trained AVSR	30.1%	60.5%
Re-trained AVSR + LipsID	27.8%	55.7%

model of the AVSR network. Other newer systems are only available in the form of trained models or their implementations are not able to achieve the published results. However, we believe, that simple incorporation of the LipsID network would improve results independently on the original lipreading system. We would like to continue to test our method with newer lipreading systems in the future when the training scripts for other methods will be available.

5   Conclusion and Future Work

This paper presents a method for improving the recognition rate of neural network-based systems for the task of lipreading. It utilizes LipsID features as additional information about the speaker identity based on the same input data as the lipreading system. The experiments showed better speech recognition accuracy than the original system. In the future, the LipsID features will be tested with more state-of-the-art systems that are based on the transformer principle. These networks are not yet publicly available in a form that allows for replication of the training processes and results published in their corresponding papers. Since the inclusion of LipsID requires access to the original source code and the dataset, it will be done after their release to further test the improvement provided by the LipsID features. There is also space for experiments with feature normalization and concatenation between the LipsID features and lipreading parts of the system.

Acknowledgments.. This work was supported by the Ministry of Education of the Czech Republic, project No. LTARF18017 and the Ministry of Science and Higher Education of the Russian Federation, agreement No. 14.616.21.0095 (reference RFMEFI61618X0095). Moreover, access to computing and storage facilities owned by parties and projects contributing to the National Grid Infrastructure MetaCentrum provided under the programme "Projects of Large Research, Development, and Innovations Infrastructures" (CESNET LM2015042), is greatly appreciated.

References

1. Afouras, T., Chung, J.S., Zisserman, A.: LRS3-TED: a large-scale dataset for visual speech recognition (2018). arXiv preprint arXiv:1809.00496

2. Afouras, T., Chung, J.S., Senior, A., Vinyals, O., Zisserman, A.: Deep audio-visual speech recognition. IEEE Trans. Pattern Anal. Mach. Intell. (2018)
3. Assael, Y.M., Shillingford, B., Whiteson, S., De Freitas, N.: Lipet: End-to-end sentence-level lipreading (2016). arXiv preprint arXiv:1611.01599
4. Chollet, F.: Keras: GitHub repository (2015)
5. Chung, J.S., Senior, A., Vinyals, O., Zisserman, A.: Lip reading sentences in the wild. In: 2017 IEEE Conference on Computer Vision and Pattern Recognition (CVPR), pp. 3444–3453. IEEE (2017)
6. Cooke, M., Barker, J., Cunningham, S., Shao, X.: An audio-visual corpus for speech perception and automatic speech recognition. J. Acoust. Soc. Am. **120**(5), 2421–2424 (2006)
7. Cootes, T.F., Cristopher J.T.: Statistical models of appearance for computer vision (2004)
8. Deng, J., Guo, J., Xue, N., Zafeiriou, S.: Arcface: additive angular margin loss for deep face recognition. In: Proceedings of the IEEE Conference on Computer Vision and Pattern Recognition, pp. 4690–4699 (2019)
9. Graves, A., Fernández, S., Gomez, F., Schmidhuber, J.: Connectionist temporal classification: labelling unsegmented sequence data with recurrent neural networks. In: Proceedings of the 23rd International Conference on Machine learning, pp. 369–376 (2006)
10. Hlaváč, M., Gruber, I., Železný, M., Karpov, A.: LipsID using 3D convolutional neural networks. In: Karpov, A., Jokisch, O., Potapova, R. (eds.) SPECOM 2018. LNCS (LNAI), vol. 11096, pp. 209–214. Springer, Cham (2018). https://doi.org/10.1007/978-3-319-99579-3_22
11. Karafiát, M., Burget, L., Matějka, P., Glembek, O., Černocký, J.: iVector-based discriminative adaptation for automatic speech recognition. In: 2011 IEEE Workshop on Automatic Speech Recognition & Understanding, Waikoloa, HI, pp. 152–157 (2011). https://doi.org/10.1109/ASRU.2011.6163922
12. Sterpu G., Saam C., Harte N.: How to teach DNNs to pay attention to the visual modality in speech recognition. In: IEEE/ACM Transactions on Audio, Speech, and Language Processing (2020). https://doi.org/10.1109/TASLP.2020.2980436
13. Hlaváč, M.: Automated Lipreading with LipsID Features. PhD Thesis, University of West Bohemia, (2019)
14. Sterpu, G., Saam, C., Harte, N.: Attention-based audio-visual fusion for robust automatic speech recognition. In: 2018 International Conference on Multimodal Interaction (ICMI 2018) (2018). https://doi.org/10.1145/3242969.3243014
15. Harte, N., Gillen, E.: TCD-TIMIT: an Audio-Visual Corpus of Continuous Speech. In: IEEE Transactions on Multimedia, pp. 603–615 (2015). https://doi.org/10.1109/TMM.2015.2407694

Automated Destructive Behavior State Detection on the 1D CNN-Based Voice Analysis

Anastasia Iskhakova[1,2](✉) , Daniyar Wolf[1] ,
and Roman Meshcheryakov[1]

[1] V. A. Trapeznikov Institute of Control Sciences of Russian Academy
of Sciences, 65 Profsoyuznaya Street, Moscow 117997, Russia
Shumskaya.ao@gmail.com, mrv@ieee.org
[2] Tomsk State University of Control Systems and Radioelectronics,
40 Prospect Lenina, 634050 Tomsk, Russia

Abstract. The article considers one of the approaches to the solution of the problem of finding the effect of a user on the Internet through media files including acoustic materials. The solution to the given problem is a part of the project on the development of a uniform socio-cyberphysical control system of the Internet content to identify the destructive materials and protect Internet users. Authors developed an algorithm including a classification model based on an artificial neural network (1D CNN-based) with the application of deep learning and the further application of the statistical probability approach to the solution of the problem of identification of the destructive type of emotion with the accuracy of 74%. The article considers one of the approaches to solving the problem of identifying the impact on the Internet of a user through media files, including acoustic materials. The solution to this problem is an integral part of the project to create a unified socio-cyberphysical system for managing Internet content in order to identify destructive materials and protect network users.

Keywords: Internet content · Social network sites · Intellectual data analysis · Socio-Cyberphysical system · Artificial neural networks · Acoustic impact · Psychological pressure · Aggression on the internet

1 Introduction

The authors of the article solve the problem of identifying destructive content in the virtual environment, i.e. the materials negatively influencing the condition of an Internet user and changing their intentions towards the negative side [1, 2]. In particular, within the limits of the present paper, we offer an approach that allows carrying out automatic analysis of electronic audio materials based on artificial neural networks to identify the items with signs of aggression and animosity. Also, there are described some attempts of classification and possible combinations of emotions to form the required behavioral pattern (for audio materials).

The interaction of a person with a virtual environment for today is impossible to overstate. The fact that many users cannot imagine themselves anymore separate from the World Wide Web and various services which "it offers" is a part of the digital

© Springer Nature Switzerland AG 2020
A. Karpov and R. Potapova (Eds.): SPECOM 2020, LNAI 12335, pp. 184–193, 2020.
https://doi.org/10.1007/978-3-030-60276-5_19

epoch. The most important component of such behavior of people is the use of social network sites, which have been actively developing for the last 15 years. Social network sites solve many user's problems: from entertainment to business correspondence with partners. Thus, it is necessary to note that during the last years the development of the given resources is closely connected to the implementation of media files in them. The acoustic Internet content comprises the biggest part of all electronic materials at the expense of the audio stream in video files which are the most popular today. Thereupon, the investigation of acoustic materials allows us to bring an important powerful contribution to the solution of the problem of identifying the destructive content in a virtual environment.

In the modern approach of technetronic identification of a people's psychological condition, the majority of modern scientific papers are aimed at the identification of 8 basic emotions. The advanced modern methods of identification of a human emotional condition based on a speech signal still remain approaches with the application of machine learning methods and deep artificial neural networks [3, 4]. At the heart of such methods, the following key stages are put: preliminary, the extraction of acoustic characteristics is carried out which are directly linked to the accuracy of the subsequent classification, there could be selected such characteristics as the frequency of the pitch, power characteristic, format feature, spectral function, etc.; after the characteristics are extracted, the machine learning methods of an artificial neural network are applied to the subsequent forecasting of the person's emotional condition. Nevertheless, the accuracy of such methods is only 60-70% [5]. Also, it is known that the accuracy of the evaluation of emotions strongly correlates with the used emotion databases. So, in the paper [6], the accuracy of the emotional condition of 78% is reached for the Surrey Audio-Visual Expressed Emotion (SAVEE) database [7] and of 84% for the Berlin Database of Emotional Speech (Emo-DB) [8]. Also, at the 2018 conference ICAART the paper [9] was presented with the results of 69.5% for Emo-DB and 86.5% for Spanish database (here are considered the results with the application of the Mel-Frequency Cepstrum coefficient – MFCC [10]).

All described methods and solutions show high values of efficiency for the certain chosen specific database of audio files. At the same time, there is the problem of processing of diverse items which can be recorded in various conditions as well as by people with various speech characteristics and manifestations of emotional conditions. When applying the known methods of data sampling out of 3–4 various databases, their efficiency decreases by 30–40%, without the consideration of gender for 8 types of emotions [11]; and at the same time, there is a real practical necessity of investigation of heterogeneous records and finding the solutions of high accuracy.

The problem of studying diverse audio files is basically caused by the following factors:

- presence of noise in the record;
- quality of the record;
- other properties of the record (for example, language, length of the record, loudness, clearness, etc.);
- heterogeneity of the data for learning (incorporated database of many speakers);

- heterogeneity of items of the investigated records at the recognition stage (various types and genres of the investigated audio-content);
- emotionality and oratorical abilities and features of the speaker.

The approach offered in the present paper investigates 5 diverse databases of audio records for the development of an effective method of identification of items with the signs of destructive behavior: anger, aggression, animosity.

2 Data Preparation and Signal Processing

The whole process of the experiment is based on the classification of the emotional signs extracted from a speech signal according to the general model of Artificial Neural Network (ANN) [12, 13] with the use of the acoustic characteristics based on Mel-Frequency Cepstrum coefficient (MFCC). MFCC are the dominating characteristics used for the recognition of speech in convolution neural networks. Their success is caused by their ability to represent the spectrum of speech amplitude in a compact form as learning and recognition data. A disadvantage is that it is complex to evaluate the quality of the MFCC characteristics since there are losses of some key emotional signs that are present at an initial speech signal, and as a consequence, the accuracy of the recognition at the classification stage decreases.

In the experiment, we used a specially prepared database containing 4500 audio samples with human speech with various emotional conditions, of the duration from 1 to 4 s. We decided to unite the English open-source emotional speech databases in a balance one. The database was prepared from such emotional databases as:

- Surrey Audio-Visual Expressed Emotion (SAVEE);
- Ryerson Audio-Visual Database of Emotional Speech and Song (RAVDESS);
- Toronto emotional speech set (TESS) [14];
- Crowd-sourced Emotional Mutimodal Actors Dataset (CREMA-D) [15];
- Emo-DB [16].

So, for example, the databases SAVEE and Emo-DB were used in the paper [17]. After the integration of the above-stated databases, the incorporated emotional database containing both English and German speech was obtained. In the obtained database there are 5 types of men's and women's emotions: anger (angry, volume 900), happiness (happy, 900), sadness (sad, 900), fear (fear, 900), neutral (neutral, 900). The German-language database (Emo-DB) we added to the new database was slightly heterogeneous, taking into consideration that both languages belong to the German language group.

Based on the MFCC average vector on the time scale, it is possible to observe the effect of the integration of various speech databases (the addition of acoustic material). Figure 1 shows how the form of the MFCC average vector of emotions obviously differs at women; and men, on the contrary, have an overlapping of coefficients (merging). Already at the given stage, it could be assumed that absolute values of the classification results to the greater degree would be based on the effect of estimations for women's records.

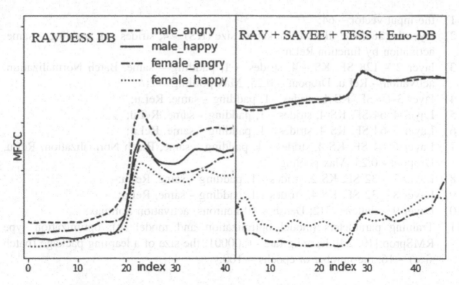

Fig. 1. Average vectors of Mel-Frequency Cepstrum Coefficient (MFCC), 4 categories of emotions, taking into account the gender. One emotional database is on the left, four databases are on the right. On the right, the difference according to the gender is obvious.

The acoustic characteristic in the form of average vectors of Mel-Frequency Cepstrum Coefficients according to the emotion types also confirms the conclusion in the paper [18]. On the one hand, all types of emotions said by experts, while being in equal proportions, should be presented; however, on the other hand, the considerable number of emotional types sharply reduces the quality of identification (Fig. 1).

For each acoustic sample from the database, we extracted Mel-Frequency Cepstrum coefficients with the following parameters: the duration of audio is 1–4 s; the frequency of digitization is 44100 Hz, 64 MFCC coefficients. It is assumed that the characteristic that defines the emotion of the speech is present in the Mel-frequencies with the averaged Mel-Frequency Cepstrum coefficients on the frequency scale. Therefore, the obtained Mel-Frequency Cepstrum coefficients were reduced to an average vector. Also, based on the received characteristic we have noted that the solution to the problem of gender classification is considerably improved. Thus, the received acoustic characteristic was accepted as relevant for its application to the problems of the emotion classification.

3   CNN Designing

After extraction of the averaged 2D Mel-frequency Cepstrum characteristics on a frequency scale, deep learning with the volume of data 4500×64 was applied, where 3375 is the training sample. The learning and classification were carried out on the model of a 1D convolution neural network (Fig. 2) with the following parameters:

1) the input vector – 64;
2) layer 1–128 same filters (SF), kernel size (KS) – 4, strides – 1, padding – same, activation by function ReLu;
3) layer 2 – 128 SF, KS – 4, strides – 1, padding – same, Batch Normalization, activation – ReLu, Dropout – 0.25, Max pooling – 2;
4) layer 3–64 SF, KS 4, strides - 1, padding - same, ReLu;
5) Layer 4–64 SF, KS 4, strides - 1, padding - same, ReLu;
6) Layer 5–64 SF, KS 4, strides - 1, padding - same, ReLu;
7) Layer 6–64 SF, KS 4, strides - 1, padding - same, Batch Normalization, ReLu, Dropout - 0.25, Max pooling - 2;
8) Layer 7 – 32 SF, KS 2, strides - 1, padding - same, ReLu;
9) Layer 8 – 32 SF, KS 4, strides - 1, padding - same, ReLu;
10) Layer 9 – flatten - 512; Dense – 10 neurons, activation Softmax;
11) Training parameters (model optimization and model fit): optimization type RMSprop [19, 20], learning rate - 0.00001, the size of a learning package (batch size) – 16, the number of epochs – 100.

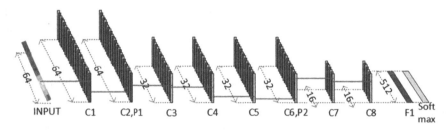

Fig. 2. Scheme of the used 1D-CNN.

The model parameters of the neural network were selected empirically, regarding to the volumes and the shapes of the training samples. The choice of such a model of the neural network is caused by the fact that the chosen acoustic characteristic allowed us to use a one-dimensional model of convolution in deep learning. In comparison, in the paper [21], a two-dimensional convolution model of deep learning was used, where at the depth of convolution layers equal to one, the size of the entrance layer was equal to 400×12 units; as opposed to the model used in our experiment, where at the depth of convolution layers equal to 8, the size of the input layer was 64 units. Thus, while the size of the input layer being significantly smaller (by the factor of 75), the executed deep learning in our experiment surpassed the depth of learning by the factor of 8 compared to that described in the paper [21] and is deeper by the factor of 1.6 compared to research [11].

4   Experimental Results and Analysis

In the experiment, 10 types of emotions with gender effect were selected, where the learning sample composed 75% out of the general data set, and the others 25% were intended for testing (cross-validation). After 100 epochs of learning of the developed neural network model, the absolute accuracy of the classification was equal to 68.36% (Table 1, Fig. 3).

Table 1. Distribution of the estimations of emotion classification after the test.

	Precision	Recall	F1-score	Support
female_angry	0.87	0.95	0.91	112
female_fear	0.90	0.91	0.91	100
female_happy	0.88	0.86	0.87	125
female_neutral	0.84	0.91	0.87	125
female_sad	0.97	0.83	0.89	117
male_angry	0.56	0.57	0.56	100
male_fear	0.62	0.24	0.34	119
male_happy	0.31	0.48	0.37	102
male_neutral	0.55	0.50	0.52	116
male_sad	0.47	0.56	0.51	109
Accuracy			0.68	1125
Weighted avg	0.70	0.68	0.68	1125

At first sight, the absolute estimation can seem to be not too high. Let us remind that research [11] received the result of 42%; however, one database IEMOCAP [22] was applied and the results of the first experiment were received for 8 types of emotions without regard to the gender type. Thus, we can conclude that the obtained accuracy of the classification in our experiment showed the best result.

Taking into account the received estimations in Table 1, the given problem could be considered as solved in a "pure form" only by means of a neural network and deep learning; however, as we can see there are some nuances. On the one hand, based on the available heterogeneous emotional database and 1D-CNN algorithm of deep learning, we received a neural network allowing us to define the women's emotional condition with good enough accuracy since the estimations recall and f1-score for women are impressive. However, on the other hand, the estimations of the classification of the men's emotional condition are not so impressive. Let us note the distribution of classification errors (Fig. 3).

The distribution of errors in a confusion matrix confirms our assumption which made in Sect. 2: the absolute values of the classification results are to a greater degree based on the effect of women's estimations. Nevertheless, we had the task to develop an algorithm capable to define the destructive condition of a person based on audio material.

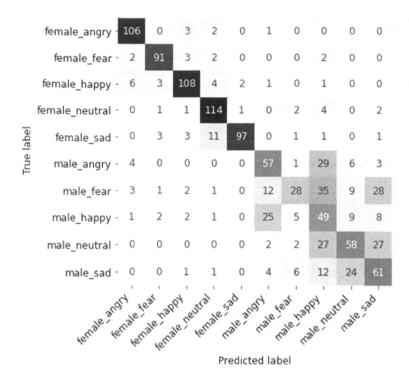

Fig. 3. Distribution of classification errors for emotions in the course of testing. Horizontally, the errors of 1st type are distributed; vertically, the errors of the 2nd type are distributed.

Despite the received value of the absolute accuracy of classification, based on the received estimations of classification in Table 1, the problem of identification of destructive condition was solved. To solve the given problem, we applied the statistical probability approach.

In the first stage, the emotions were grouped according to the gender type and the classification estimations in Table 1 were reevaluated. The result of grouping showed that our neural network (1D-CNN), is capable to define the gender type with an accuracy of 97% for both men and women (Table 2) based on acoustic material.

Table 2. Distribution of the estimations of classification according to the gender type.

	Precision	Recall	F1-score	Support
Female	0.97	0.97	0.97	579
Male	0.97	0.97	0.97	546
Accuracy			0.97	1125

In the second stage, we excluded the gender component and thus generalized the types of emotions up to 5 (Table 3). Thus, the best estimations are received for the classification of such emotions as neutral is 71%, anger is 75%, fear is 82%, happiness is 57% and grief is 69%.

Table 3. Distribution of classification estimations of emotions without regard to the gender type.

	Precision	Recall	F1-score	Support
Angry	0.75	0.79	0.77	212
Fear	0.82	0.55	0.66	219
Happy	0.57	0.70	0.63	227
Neutral	0.71	0.71	0.71	241
Sad	0.69	0.70	0.70	226
Accuracy			0.69	1125

In the third stage, the emotions anger and fear can be united in one group characterizing the destructive behavior of the speaker.

We combined the emotions of happy, fear and sad into one general mixed group, these emotions were not of interest in the context of the problem being solved. Based on the result obtained, it follows that at this stage of the study, the algorithm allows classifying the destructive type of emotion with an accuracy of 75%, and also classifying the speaker's gender with an accuracy of 97% (Table 4).

Table 4. Estimations of classification of destructive, positive and neutral emotion.

	Precision	Recall	F1-score	Support
Destructive	0.75	0.79	0.77	212
Neutral	0.71	0.71	0.71	241
Mixed	0.84	0.83	0.84	672
Accuracy			0.80	1125

According to the theorem of probability multiplication of two independent events P (AB) = P(A)*P(B). The probability of simultaneous occurrence of two independent events is equal to the product of probabilities of these events. Let P(A) is the probability of the determination of gender type of the available audio material, and P(B) is the probability of determination of destructive emotion. Then, proceeding from the received results, we receive an algorithm allowing us to define the destructive condition of a person based on audio material with the probability of 72.75% (0.97*0.75).

5   Conclusion

In the paper, we solve the problem of the development of a tool for automatic identification of the audio records which can be characterized as destructive for the listener, i.e. influencing negatively their both physical and mental state and mood. The presented approach proposes to integrate the base emotions in certain groups and classify according to the selected groups with the application of artificial neural networks. As the computing experiments showed, the integration of emotions applied at the stage of learning and the reduction of the number of required classes allows us to increase the accuracy of the classification and to minimize the number of errors of the 1st and the 2nd types. Thus, the offered architecture and approach could be applied to audio records from different databases, which also increases efficiency in the solution of the practical problems during the identification of destructive content on the Internet. The application of the developed algorithm allows us to classify the destructive types of emotions in an audio record with an accuracy of 72.75%, which is a high enough indicator, taking into account the application of diverse learning material (summation of diverse records) and gender division.

Acknowledgements.   The reported study was partially funded by RFBR, project number №18-29-22104.

References

1. Iskhakova, A., Iskhakov, A., Meshcheryakov, R.: Research of the estimated emotional components for the content analysis. In: Journal of Physics: Conference Series, vol. 1203, pp. 012065 (2019)
2. Kulagina, I., Iskhakova, A., Galin, R.: Modeling the practice of aggression in the socio-cyber-physical environment. Vestnik tomskogo gosudarstvennogo universiteta-Filosofiya-Sotsiologiya-Politologiya-Tomsk state Univ. J. Phil. Sociol. Polit. Sci. **52**, 147–161 (2019). (in Russian)
3. Levonevskii, D., Shumskaya, O., Velichko, A., Uzdiaev, M., Malov, D.: Methods for determination of psychophysiological condition of user within smart environment based on complex analysis of heterogeneous data. In: Ronzhin, A., Shishlakov, V. (eds.) Proceedings of 14th International Conference on Electromechanics and Robotics "Zavalishin's Readings". SIST, vol. 154, pp. 511–523. Springer, Singapore (2020). https://doi.org/10.1007/978-981-13-9267-2_42
4. Malov, D., Shumskaya, O.: Audiovisual content feature selection for emotion recognition system. In: International Conference Cyber-Physical Systems and Control CPS&C (2019)
5. Zheng, W.-L., Zhu, J.-Y., Yong, P., Lu, B.-L.: EEG-based emotion classification using deep belief networks. In: IEEE International Conference on Multimedia & Expo IEEE, pp. 1–6. Chengdu, China (2014)
6. Han, K., Yu, D., Tashev, I.: speech emotion recognition using deep neural network and extreme learning machine. In: INTERSPEECH, pp. 223–227 (2014)
7. Haq, S., Jackson, P.J.B.: Multimodal emotion recognition. In: Machine Audition. Principles, Algorithms and Systems, vol. 17, pp. 398–423. IGI Global Press (2010)

8. Burkhardt, F., Paeschke, A., Rolfes, M., Sendlmeier, W., Weiss, B.: A database of German emotional speech. In: 9th European Conference on Speech Communication and Technology, Lisbon, Portugal, vol. 5, pp. 1517–1520 (2005)
9. Serrestou, Y., Mbarki, M., Raoof, K., Mahjoub, M.: Speech emotion recognition: methods and cases study. In: Proceedings of the 10th International Conference on Agents and Artificial Intelligence (ICAART 2018), vol. 2, pp. 175–182 (2018)
10. Hossan, M.A., Memon, S., Gregory, M.A.: A novel approach for MFCC feature extraction. In: 2010 4th International Conference on Signal Processing and Communication Systems, Gold Coast, QLD, Australia, pp. 1–5 (2010)
11. Niu, Y., Zou, D., Niu, Y., He, Z., Tan, H.: A breakthrough in speech emotion recognition using deep retinal convolution neural networks. https://arxiv.org/abs/1707.09917. Accessed 21 July 2020
12. Oludare, A., Aman, J.: Comprehensive review of artificial neural network applications to pattern recognition. In: IEEE Access, vol. 7, pp. 158820–158846 (2019)
13. Kim, Y.: convolutional neural networks for sentence classification. In: Proceedings of the 2014 Conference on EMNLP, Doha, Qatar, pp. 1746–1751 (2014)
14. Dupuis, K., Kathleen, M.: Toronto emotional speech set (TESS). https://doi.org/10.5683/SP2/E8H2MF. Accessed 21 July 2020
15. Busso, C., et al.: IEMOCAP: interactive emotional dyadic motion capture database. https://sail.usc.edu/iemocap/iemocap_release.htm. Accessed 21 July 2020
16. Zhang, L., et al.: BioVid Emo DB: a multimodal database for emotion analyses validated by subjective ratings. In: 2016 IEEE Symposium Series on Computational Intelligence (SSCI), Athens, pp. 1–6 (2016)
17. Fayek, H.M., Lech, M., Cavedon, L.: Towards real-time Speech Emotion Recognition using deep neural networks. In: International Conference on Signal Processing and Communication Systems 2015, Cairns, QLD, Australia, pp. 1–5 (2015)
18. Aleshina, T.S., Redko, AYu.: Bases of speech data corpus preparation for the emotional speech recognition. Mod. High Technol. 6(2), 229–233 (2016). In Russ
19. Mahesh, C.M., Matthias, H.: Variants of RMSProp and adagrad with logarithmic regret bounds. In: Proceedings of the 34th International Conference on Machine Learning, Sydney, Australia (2017)
20. Bottou, L.: Large-scale machine learning with stochastic gradient descent. In: Lechevallier Y., Saporta G. (eds.) Proceedings of COMPSTAT 2010, pp. 177–186. Physica-Verlag HD (2010)
21. Ispas, I., Dragomir, V., Dascalu, M., Zoltan, I., Stoica, C.: Voice based emotion recognition with convolutional neural networks for companion robots. Rom. J. Inf. Sci. Technol. 20(3), 222–240 (2017)
22. Mower, E., Mataric, M.J., Narayanan, S.S.: A framework for automatic human emotion classification using emotional profiles. IEEE Trans. Audio Speech Lang. Process. 19(5), 1057–1070 (2011)

Rhythmic Structures of Russian Prose and Occasional Iambs (a Diachronic Case Study)

Evgeny Kazartsev[1,2] (ID), Arina Davydova[2] (ID),
and Tatiana Sherstinova[2,3(✉)] (ID)

[1] National Research University Higher School of Economics, Staraya Basmannaya, Moscow 105066, Russia
[2] National Research University Higher School of Economics, 123 Griboyedova Canal Emb., St. Petersburg 190068, Russia
{ekazartsev, tsherstinova}@hse.ru,
ayudavydova@edu.hse.ru
[3] St. Petersburg State University, 7/9 Universitetskaya Emb., 199034 St. Petersburg, Russia

Abstract. The paper deals with the study on rhythmic structures of phonetic words conducted on the base of Russian novelistic prose of the initial three decades of the 20th Century. Due to the revolutionary changes in the society, Russian language has changed a lot during this period, at least on lexical and stylistic levels, and one could expect to find some transformations on its prosody level as well. To check this hypothesis, rhythmic dictionaries for 30 Russian writers were compiled, then they were compared with the reference probabilities which are considered as the standard ones for Russian prose texts. In order to research changes of language rhythmic structures over time, language prosody models were built for each of the three consecutive historical periods – the beginning of the 20th century (1900–1913), the periods of wars and revolutions (1914–1922), and the early Soviet era (1923–1930). The results of the quantitative study have shown that although language style characteristics (as well as techniques of versification) obviously changed, the basic language rhythmic features remained unchanged. Furthermore, they are in general close to the characteristics of probability in the distribution of standard rhythmic structures of the Russian language. In addition, for selected Russian poets – who wrote prose – occasional iambs were found in prose texts, and the comparative study of the language and speech models of Russian iambic verse was conducted.

Keywords: Language prosody model · Rhythmic structures · Occasional iambs · Iambic tetrameter · Russian prose · Diachrony · Corpus linguistics

1 Introduction

The present study examines rhythm of Russian prose, with special attention drawn to the influence of poetic prosody on the formation of poetic prose. Such approach reveals the deep structures of the prosodic organization of any Russian text and speech. The

© Springer Nature Switzerland AG 2020
A. Karpov and R. Potapova (Eds.): SPECOM 2020, LNAI 12335, pp. 194–203, 2020.
https://doi.org/10.1007/978-3-030-60276-5_20

relevance of the study focused on research of rhythmic organization of speech is determined by the need for an in-depth study of structural patterns, phonetic manifestation and grammatical content of rhythmic units of speech both in diachrony and synchrony [8].

The focus of this study was on the analysis of prosodic features of Russian prose, written in the pre-revolutionary, revolutionary and post-revolutionary periods. We were interested in how prosodic structure of Russian words and the distribution of the main types of rhythmic words change and whether they change at all.

The chosen time period was not taken by chance. Due to revolutionary transformation in society, during this period Russian language changed a lot, at least on lexical and stylistic levels, and one could expect to find some transformations on its prosody level as well. Therefore, the aim of the study was to look at how the rhythmic structures of words behave over time and whether we can say that a significant change in Russian language vocabulary entailed a change in its rhythmic structure as well.

2   Data and Method

The research was made on the texts from the Corpus of Russian Short Stories of the first third of the 20th century [7]. The corpus contains the following subcorpora:

I)   Short stories of the beginning of the 20th century (1900–1913),
II) Short stories of the period of wars and acute social upheaval (1914–1922) – World War I, the February and October Revolutions and the subsequent Civil War (this subcorpus may be further subdivided into two chronologically consecutive periods: II-1. World War I before the Revolutions (1914–1916) and II-2. Revolutions and the Civil War), and
III) Short stories of the post-revolutionary era (1923–1930) [9].

The corpus size is more than 1 million words. It was lemmatized and annotated on morphological, syntactic [10] and, selectively, rhythmic levels; texts were segmented into fragments of narrator's speech, narrator's remarks and characters' speech, the elements of literary annotation (type of narrative, theme, some structural features) were added as well.

For this study, a sample of 30 texts written by various authors was randomly selected. The total sample size has more than 30 thousand rhythmic words.

It should be emphasized that it was precisely rhythmic (phonetic) words that were studied, i.e., the complex of syllables, united by one stress. As a rule, such a complex includes enclitics and proclitics. For example, *po gorodám* (*by city*), *on govoríl* (*he spoke*) are considered as tetrasyllabic words having stress on the last syllable. Rhythmic frequency lists were compiled for each text, thereby frequency distribution of certain types of rhythmic words was obtained. These data were compared with the reference (standard) probabilities of Russian texts [6].

Frequency lists of rhythmic structures were compiled for all phonetic words (from monosyllabic words to the nine-syllabic ones), based on annotated texts, where the place of word stress was marked and the borders between rhythmic words were settled. The rhythmic structures word lists were compiled in order to construct probabilistic statistical models of texts for studying its variation or stability over the study period. All annotated texts have been processed by the specialized software system designed for multilingual prosodic analysis of rhythmic structures.

The boundaries between words have been settled according to the rules proposed by V. M. Zhirmunsky [11] and further supplemented by A. N. Kolmogorov and A. V. Prokhorov [4]. The same rules were used for measuring the degree of word emphasis.

3   Prosodic Characteristics of Russian Texts by Different Authors

Rhythmic frequency lists for different authors have been compiled and compared. The probabilities for the most frequent models are presented in Table 1. Designations of rhythmic patterns should be understood as follows: the first digit shows the number of syllables in phonetic words, the second one marks the position of the stress. In addition to data for selected Russian writers, the reference probability for each of the model calculated for the Russian language earlier is presented, allowing to compare our empirical data with the known statistics on Russian rhythm.

Then the authors and their texts have been grouped into three periods, according to the chronological division of the corpus: Period I – the beginning of the 20th century (1900–1913), Period II – the wartime and revolutionary period (1914–1922), and Period III – post-revolutionary period (1923–1930). Comparison of these corpus data was carried out against the background of the reference probabilities of Russian words in prose (see Table 2).

Comparison of rhythmic models' distribution allows us to note that in texts of Period III the number of monosyllabic words slightly increases, as well as the number of disyllabic words with an accent on the first syllable, such as *górod* (*city*), *véter* (*wind*). At the same time the percentage of disyllabic words with an accent on the second syllable *rasskáz* (*story*), *pokhód* (*campaign*) is noticeably reduced. The number of polysyllabic words is slightly growing. However, these changes are not critical. In general, we observe that the rhythm of pre-revolutionary prose does not differ much from that of the revolutionary period as well as of texts from the 1920s.

Application of chi-square statistical criteria allows us to suppose that in terms of rhythmic characteristics, almost all texts regardless of the time of their writing, belong to one general population. Moreover, empirical distribution of rhythmic structures barely differs from the reference (standard) probabilities calculated for Russian by M. A. Krasnoperova [6]. Thus, in general, there is no basis for arguing that there are significant changes in rhythmic structure of Russian texts over the historical periods in concern. In other words, the hypothesis that the proposed changes in the prosodic characteristics of the Russian language could have occurred as a result of the revolution was not proved on our data. In general, the basic rhythmic models of the language remained the same.

Then we compared the rhythmic characteristics of Russian texts by the probability models of rhythm in metrical verse. Thus, two models were built: the independence model proposed by Kolmogorov [4] and the dependence model, in which the choice of

Table 1. Distribution of rhythmic words for different authors (to be continued).

Author	Period	Year	Rhythmic models						
			1.1	2.1	2.2	3.1	3.2	3.3	4.1
Reference probability			0.069	0.156	0.173	0.07	0.159	0.091	0.013
Andreev	I	1900	0.04	0.121	0.156	0.071	0.182	0.095	0.013
Budishchev	I	1901	0.046	0.13	0.14	0.083	0.181	0.093	0.008
Bunin	I	1900	0.04	0.187	0.13	0.086	0.171	0.092	0.015
Gippius	I	1901	0.039	0.139	0.152	0.109	0.156	0.093	0.027
Grin	I	1912	0.07	0.139	0.145	0.078	0.156	0.107	0.024
Kuprin	I	1900	0.037	0.152	0.14	0.114	0.139	0.066	0.017
Slezkin	I	1911	0.068	0.125	0.16	0.084	0.154	0.083	0.016
Sologub	I	1907	0.034	0.14	0.124	0.098	0.172	0.07	0.015
Timkovsky	I	1901	0.056	0.12	0.146	0.066	0.164	0.083	0.018
A.Chekhov	I	1900	0.043	0.145	0.156	0.08	0.157	0.102	0.009
Asheshov	II	1914	0.056	0.107	0.153	0.073	0.179	0.065	0.02
Yesenin	II	1916	0.045	0.121	0.168	0.06	0.199	0.119	0.012
Zamyatin	II	1919	0.058	0.178	0.125	0.065	0.151	0.081	0.018
Karaulov	II	1914	0.049	0.101	0.157	0.069	0.172	0.087	0.016
Kuzmin	II	1914	0.047	0.145	0.118	0.094	0.18	0.054	0.009
Okunev	II	1917	0.032	0.125	0.122	0.068	0.17	0.096	0.015
Rabotnikov	II	1922	0.048	0.133	0.149	0.086	0.152	0.081	0.017
Slonimsky	II	1921	0.058	0.164	0.18	0.061	0.178	0.087	0.01
Shklyar	II	1916	0.042	0.16	0.167	0.074	0.174	0.072	0.01
Shulgovsky	II	1916	0.034	0.125	0.134	0.072	0.14	0.08	0.017
Alekseev	III	1924	0.037	0.14	0.106	0.072	0.161	0.091	0.015
Belyaev	III	1926	0.061	0.157	0.145	0.071	0.167	0.078	0.026
Bulgakov	III	1926	0.054	0.111	0.133	0.086	0.176	0.107	0.017
Gaidar	III	1927	0.048	0.176	0.167	0.076	0.165	0.088	0.013
Zazubrin	III	1923	0.056	0.191	0.127	0.122	0.145	0.063	0.02
Platonov	III	1926	0.068	0.133	0.126	0.067	0.188	0.075	0.008
Sokolov-Mikitov	III	1927	0.052	0.176	0.099	0.089	0.181	0.08	0.019
A.N.Tolstoy	III	1928	0.05	0.142	0.106	0.099	0.151	0.085	0.023
Ulin	III	1928	0.075	0.159	0.122	0.074	0.162	0.081	0.014
Fadeev	III	1923	0.045	0.133	0.127	0.065	0.146	0.088	0.014

Author	Rhythmic models								
	4.2	4.3	4.4	5.1	5.2	5.3	5.4	5.5	6.1
Reference probability	0.071	0.091	0.016	0.007	0.010	0.029	0.020	0.003	0.001
Andreev	0.084	0.108	0.02	0.002	0.01	0.047	0.022	0.005	0
Budishchev	0.085	0.087	0.032	0.002	0.011	0.047	0.034	0.002	0
Bunin	0.074	0.094	0.013	0.003	0.015	0.042	0.019	0.004	0
Gippius	0.057	0.094	0.03	0.006	0.01	0.039	0.034	0.002	0
Grin	0.066	0.078	0.02	0.002	0.014	0.045	0.028	0	0.002

(*continued*)

Table 1. (*continued*)

Author	Rhythmic models								
	4.2	4.3	4.4	5.1	5.2	5.3	5.4	5.5	6.1
Kuprin	0.109	0.082	0.018	0.001	0.019	0.047	0.022	0.001	0
Slezkin	0.077	0.093	0.025	0	0.018	0.038	0.031	0.001	0
Sologub	0.079	0.115	0.021	0.002	0.019	0.06	0.022	0.001	0
Timkovsky	0.088	0.114	0.025	0	0.018	0.052	0.027	0.005	0
A.Chekhov	0.076	0.108	0.024	0	0.015	0.037	0.026	0.004	0
Asheshov	0.089	0.117	0.02	0	0.019	0.047	0.025	0	0
Yesenin	0.065	0.11	0.026	0.001	0.021	0.015	0.022	0.001	0
Zamyatin	0.08	0.105	0.023	0.001	0.014	0.04	0.016	0.004	0
Karaulov	0.091	0.102	0.023	0.004	0.017	0.06	0.026	0.004	0
Kuzmin	0.083	0.104	0.029	0.001	0.009	0.059	0.027	0.002	0
Okunev	0.105	0.108	0.027	0	0.011	0.057	0.025	0.004	0
Rabotnikov	0.073	0.115	0.019	0	0.007	0.055	0.034	0.001	0
Slonimsky	0.067	0.077	0.019	0	0.016	0.041	0.021	0.005	0
Shklyar	0.066	0.104	0.03	0.002	0.014	0.038	0.024	0.003	0
Shulgovsky	0.081	0.123	0.032	0.001	0.018	0.057	0.037	0.005	0
Alekseev	0.1	0.104	0.024	0	0.019	0.064	0.032	0.002	0
Belyaev	0.069	0.099	0.016	0.001	0.02	0.043	0.027	0	0.001
Bulgakov	0.061	0.099	0.023	0.002	0.015	0.056	0.024	0.005	0
Gaidar	0.065	0.084	0.023	0.001	0.02	0.027	0.028	0.003	0.001
Zazubrin	0.069	0.08	0.013	0.003	0.018	0.039	0.021	0.001	0
Platonov	0.073	0.102	0.023	0.005	0.015	0.045	0.03	0.006	0
Sokolov-Mikitov	0.073	0.106	0.019	0.001	0.015	0.034	0.025	0.002	0
A.N.Tolstoy	0.083	0.102	0.019	0	0.023	0.056	0.024	0.004	0
Ulin	0.083	0.095	0.02	0.002	0.015	0.048	0.021	0.003	0
Fadeev	0.106	0.099	0.024	0.001	0.027	0.053	0.028	0.003	0

rhythmic words depends on the metric position and context. The last algorithm was proposed by Krasnoperova [6]. In general, it turned out that statistical measures for these models barely differ from that calculated for the periods in concern. As a whole, the dependence model differs from the data of the common language model (Kolmogorov model), while the indicators of typologically related models do not reveal significant differences for the periods under study.

Thus, it can be argued that in general Russian rhythm of 1900–1930 remained almost unchanged. Even though the stylistic characteristics of the language and the technique of versification obviously changed, the basic rhythmic characteristics of prose represent constant values, which are close to the characteristics typical for the Russian language.

Table 2. Empiric mean values of rhythmic models vs. the reference probability.

Rhythmic model	Reference probability	Period I	Period II	Period III
1.1	0.069	0.0473	0.0469	0.0546
2.1	0.156	0.1398	0.1359	0.1518
2.2	0.173	0.1449	0.1473	0.1258
3.1	0.07	0.0869	0.0722	0.0821
3.2	0.159	0.1632	0.1695	0.1642
3.3	0.091	0.0884	0.0822	0.0836
4.1	0.013	0.0162	0.0144	0.0169
4.2	0.071	0.0795	0.08	0.0782
4.3	0.091	0.0973	0.1065	0.097
4.4	0.016	0.0228	0.0248	0.0204
5.1	0.007	0.0018	0.001	0.0016
5.2	0.010	0.0149	0.0146	0.0187
5.3	0.029	0.0454	0.0469	0.0465
5.4	0.020	0.0265	0.0257	0.026
5.5	0.003	0.0025	0.0029	0.0029
6.1	0.001	0.0002	0	0.0002
6.2	0.002	0.001	0.001	0.0012
6.3	0.008	0.0058	0.0088	0.0075
6.4	0.009	0.0097	0.0109	0.0112
6.5	0.002	0.002	0.0034	0.0037
6.6	0	0	0.0002	0.0004
7.2	0	0	0.0002	0
7.3	0	0.0005	0.0007	0.0005
7.4	0	0.0016	0.0021	0.002
7.5	0	0.0013	0.0014	0.0018
7.6	0	0.0002	0.0002	0.0005
8.3	0	0	0	0.0001
8.4	0	0	0	0.0001
8.5	0	0.0002	0.0001	0.0003
8.6	0	0	0.0001	0.0002
9.6	0	0	0.0001	0
9.9	0	0.0001	0	0

4   Occasional Iambs in Russian Prose

In this section we present data on the so-called speech models of Russian four-foot iamb, which were calculated on prose texts. These models are also called "occasional iambs", they represent statistics on distribution of verse-like fragments in a prosaic text [5]. In general, these data, as well as rhythmic dictionaries of prose, suggest that Russian prose in the first third of the 20th century is prosodically homogeneous. We may notice that Anton Chekhov's prose stands a little separate from the others, having a "pyrrhic foot" on the second "foot", rhythmic form #3. Otherwise, the writers are similar by usage of occasional iambs, and even those of them who are poets do not greatly differ in specifics of their prosaic rhythm from other writers (see Table 3 and Fig. 1 and 2).

Table 3. Distribution of rhythmic forms in iambic speech models

	Period	Year	1	2	3	4	5	6	7
Chekhov	I	1900	0.134	0.11	0.341	0.305	0	0.098	0.012
Gippius	I	1901; 1901	0.171	0.085	0.22	0.28	0.012	0.207	0.024
Kuzmin	II	1912; 1915; 1915	0.049	0.049	0.122	0.39	0	0.305	0.085
Yesenin	II	1916; 1916; 1917	0.122	0.061	0.244	0.415	0	0.122	0.037
Platonov	III	1926	0.171	0.073	0.183	0.257	0	0.22	0.098
Balmont	III	1923; 1924; 1925; 1927; 1929	0.11	0.061	0.183	0.341	0.012	0.256	0.037

Further, based on the material of Boris Pasternak's prose, we carried out a comparison between language and speech models of Russian four-foot iamb. These data were compared with the similar characteristics in the prose by Fyodor Sologub and Alexei Tolstoy. In general, it was found that speech models of Pasternak's early prose correspond to the probabilistic indicators of linguistic model, as well as the similar rhythmic features of Sologub's and Tolstoy's prose. However, in his late prose, Pasternak reveals significant discrepancies between speech and language models. In this case, we conducted a study on the material of the first part of the novel "Doctor Zhivago". In the course of the present work, discrepancies were revealed between the speech and language model at the start of the beginning of the verse line. Here, the language model is closer to the characteristics of Russian classical poems of Pushkin's era. In this case, the language model remains unchanged, being similar to the model which is inherent to texts studied in our previous works [3]. A similar result was found earlier on the material of Pushkin's "The Belkin Tales". We explained it earlier as the influence of the verse [2]. It turns out that under certain conditions the poet's prose can really be influenced by the rhythm of the verse. In this case Pasternak's prose manifests some standard characteristics of poet's prose, laid down already in Pushkin's time.

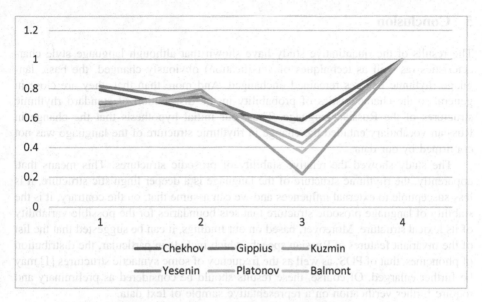

Fig. 1. The rhythmic profiles of occasional iambs.

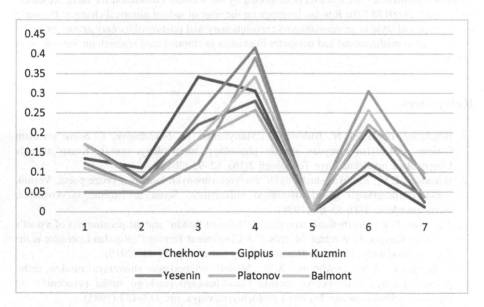

Fig. 2. Comparison of rhythmic forms in speech models.

5 Conclusion

The results of the quantitative study have shown that although language style characteristics (as well as techniques of versification) obviously changed, the basic language rhythmic features remained unchanged. And more than that, they are close in general to the characteristics of probability in the distribution of standard rhythmic structures of the Russian language. Thus, our initial hypothesis that the change in Russian vocabulary entailed the change in rhythmic structure of the language was not confirmed by our data.

The study showed the relative stability of prosodic structures. This means that, apparently, the rhythmic structure of the language is a deeper linguistic structure, it is less susceptible to external influences and we can assume that, on the contrary, it is the stability of language prosodic structure that sets boundaries for the possible variability of its lexical structure. Moreover, based on our findings, it can be suggested that the list of the invariant features of Russian speech, which include in particular, the distribution of phonemes, that of POS, as well as the frequency of some syntactic structures [1] may be further enlarged. Of course, these results should be considered as preliminary and require further verification on a representative sample of text data.

Acknowledgements. The research is supported by the Russian Foundation for Basic Research, project #17-29-09173 "The Russian language on the edge of radical historical changes: the study of language and style in prerevolutionary, revolutionary and post-revolutionary artistic prose by the methods of mathematical and computer linguistics (a corpus-based research on Russian short stories)".

References

1. Bogdanova-Beglarian, N., Blinova, O., Martynenko, G., Sherstinova, T.: Some invariant features of Russian everyday speech: phonology, morphology, syntax. Komp'juternaja Lingvistika i Intellektual'nye Tehnologii 2(16), 82–95 (2017)
2. Kazartsev, E.V.: Ritmika 'sluchainykh' chetyrekhstopnykh iambov v proze poeta. Vestnik Sankt-Peterburgskogo gosudarstvennogo universiteta. Seriia 2: Istoriia, iazykoznanie, literaturovedenie 1(2), 53–61 (1998)
3. Kazartsev, E.V.: The rhythmic structure of "Tales of Belkin" and the peculiarities of a poet's prose. In: Kopper, J., Wachtel, M. (eds.) "A Convenient Territory". Russian Literature at the Edge of Modernity. Essays in Honor of Barry Scherr, pp. 55–65 (2015)
4. Kolmogorov, A.N., Prokhorov, A.V.: Model' riticheskogo stroyeniya russkoi rechi, prisposoblennaya k izucheniyu metriki klassicheskogo russkogo stikha (vvedenie). In: Russkoie stikhoslozhenie: Tradicii i problemy razvitiya, pp. 113–133 (1985)
5. Krasnoperova, M.A.: K voprosu o raspredelenii ritmicheskikh slov i sluchainykh iambov v khudozhestvennoi proze. In: Bak, D.P. (ed.) Russkii stikh: Metrika, ritmika, rifma, strofika. V chest' 60-letiia M. L. Gasparova, pp. 163–181 (1996)
6. Krasnoperova, M.A.: Osnovy sravnitel'nogo statisticheskogo analiza ritmiki prozy i stikha: Uchebnoe posobie. St. Petersburg State University Press, St. Petersburg (2004)

7. Martynenko, G.Ya., et al.: On the principles of creation of the Russian short stories corpus of the first third of the 20th century. In: Proceedings of the XV International Conference on Computer and Cognitive Linguistics 'TEL 2018', pp. 180–197 (2018)
8. Potapov, V.V.: Dinamika i statika rechevogo ritma: Sravnitel'noye issledovaniye na materiale slavyanskikh i germanskikh yazykov. URSS, Moscow (2004)
9. Sherstinova, T., Martynenko, G.: Linguistic and stylistic parameters for the study of literary language in the corpus of Russian short stories of the first third of the 20th century. In: CEUR Workshop Proceedings, vol. 2552, pp. 105–120 (2020)
10. Sherstinova, T., Ushakova, E., Melnik, A.: Measures of syntactic complexity and their change over time (the case of Russian). In: Proceedings of the 27th Conference of Open Innovations Association FRUCT, pp. 221–229 (2020)
11. Zhirmunsky, V.M.: Vvedeniie v metriku. In: Teoriia Stikha, pp. 5–232 (1975)

Automatic Detection of Backchannels in Russian Dialogue Speech

Pavel Kholiavin$^{(\boxtimes)}$ ID, Anna Mamushina ID, Daniil Kocharov ID,
and Tatiana Kachkovskaia ID

Saint Petersburg State University, Universitetskaya emb., 11, St. Petersburg, Russia
p.kholyavin@mail.ru, mamushina.anna@mail.ru,
{kocharov,kachkovskaia}@phonetics.pu.ru

Abstract. This paper deals with acoustic properties of backchannels –
those turns within a dialogue which do not convey information but signify
that the speaker is listening to his/her interlocutor (*uh-huh, hm* etc.). The
research is based on a Russian corpus of dialogue speech, SibLing, a part
of which (339 min of speech) was manually segmented into backchannels
and non-backchannels. Then, a number of acoustic parameters was calculated: duration, intensity, fundamental frequency, and pause duration.
Our data have shown that in Russian speech backchannels are shorter
and have lower loudness and pitch than non-backchannels. After that,
two classifiers were tested: CART and SVM. The highest efficiency was
achieved using SVM ($F_1 = 0.651$) and the following feature set: duration, maximum fundamental frequency, melodic slope. The most valuable
feature was duration.

Keywords: Dialogue speech · Backchannel · Turn-taking · Speech
acoustics · Russian

1 Introduction

A dialogue is an exchange of utterances between interlocutors. That is, at a given
moment one of the interlocutors is a speaker, while the other is a listener, and
at times they alternate the roles.

In the speech signal we also observe alternation of speakers by registering the
voices of one interlocutor or the other. However, a change of a speaker doesn't
necessarily implicate a change of role. There is a type of speech units which
indicate that no role change is implied. These units are termed backchannels (or
response particles, backchannel feedback) [6,7,9].

We apply the term backchannel to turns whose function is to indicate that the
speaker is listening to the interlocutor, agrees with him and/or understands what
is said but doesn't intend to interrupt him. Such units do not carry information
and do not lead to role change.

The most frequent backchannels in English are *yeah, okay, uh-huh, hm, right*,
in Russian – *uh-huh* and *ah-hah*. The role of backchannels is also performed

© Springer Nature Switzerland AG 2020
A. Karpov and R. Potapova (Eds.): SPECOM 2020, LNAI 12335, pp. 204–213, 2020.
https://doi.org/10.1007/978-3-030-60276-5_21

by non-speech sounds such as laughter, sniffing, etc. According to the evidence reported in [13], account for 19% of all utterances.

The nature of backchannels is different from that of other utterances so they are of interest for the development of dialogue systems aimed at human-like behaviour [1,19]. Most part of today's systems respond to any vocalizations of the user which causes the system to stop speaking and process the user's utterance, while the function of backchannels is to indicate that the interlocutor may continue speaking without interruption. Systems oriented on natural dialogue flowing should react to such turns as humans do, that is, pass them by. This motivates both developers of dialogue systems and linguists search for methods to automatically detect backchannels [15].

Most studies concerning backchannels describe detection algorithms based on acoustic features of the preceding turn (see, for example, [25]). The acoustic properties of backchannels are far less investigated. To date, it is known that backchannels differ from other turn-types in their duration, loudness, and pitch. A popular method here is to compare acoustic features of backchannels with that of homophonic turns expressing agreement: for instance, words like okay, yes (in Russian –'да' yes,'угу' uh-huh,'ага' ah-hah) can be used in both cases. For English, it is shown [3] that backchannels are higher in loudness and pitch, whereas the difference in duration is statistically non-significant. Apart from this, backchannels are generally characterised by rising pitch, while other turn-types exhibit falling movement. The latter difference is important for perception [10].

The evidence for Slovak obtained by the same method [2] reveal a number of differences from English. Backhannels had lower intensity and longer duration, compared to agreement tokens.

A similar study for Russian [17] demonstrated that some speakers exhibit higher speaking rate and lower intensity of backchannels. The results of the research are of interest as they show inter-speaker variability.

Compared to most utterances backchannels are much shorter in duration. This allows for a hypothesis that the majority of short turns are backchannels. It was confirmed in [8]: backchannels can be fairly precisely distinguished from other utterances by their duration; the efficiency is only slightly improved by adding one more parameter – relative loudness (the k-nearest neighbors algorithm, precision 73%, recall 99.5%). The duration threshold for English data is 1 s. Lower loudness of backchannels as compared to other turn types can be explained by lesser pronunciation effort [26].

Traditionally, automatic detection of backchannels is based on systems of rules using various prosodic features [19,24,25]. Recent studies increasingly employ neural networks of various types (DNN [18,21], LSTM [11,22].

Given that automatic recognition is highly important for the development of human-like computer interfaces, we investigated whether it is possible to automatically detect backchannels in Russian. We tested two different methods of machine learning: decision trees and Support Vector Machine.

Table 1. The most frequent backchannels in the data analysed.

Backchannel	Quantity in the data
'угу' *uh-huh*	612
'да' *yes*	159
'ага' *ah-hah*	82
'так' *well*	36
'ну' *well*	16
non-speech sound	8
'угу угу' *uh-huh uh-huh*	7
'хорошо' *okay*	7
'а' *ah*	5
'а ага' *ah ah-hah*	4

2 Material

The experiments described in this paper were conducted on SibLing Corpus of Russian Dialogue Speech [14]. The corpus contains studio recordings of task-oriented dialogues between Russian native speakers. Subjects completed two speaking tasks–a card matching game and a map task. The card matching game was based on searching for similarities in two decks of cards. The interlocutors described their pictures and negotiated matching elements (completion time–10–15 min). In the map task the speakers were asked to guide each other through a set of schematic maps, swapping the Leader and the Follower roles 4 times (completion time–15–60 min due to speed variation). During the recording the subjects were separated by a non-transparent screen to prevent them from seeing each other.

For this research we used a portion of the corpus containing 9 dialogues (overall duration–339 min). 7 dialogues included both tasks, 2–only the first task. The age of the subjects varied from 24 to 40, all the interlocutors knew each other (being siblings or close friends). The selected part of the corpus was annotated for turn boundaries (automatically with manual correction), backchannels were manually labeled by a professional phonetician. The material contains 5800 turns, of which 1028 are backchannels. The most frequent backchannels in the material are presented in Table 1.

3 Acoustic Properties of Backchannels

It has been shown before for various languages that backchannels can differ from other types of turns in duration, intensity, melodic contour, duration of preceding pause. In this study, the following parameters were calculated for each turn:

1. Turn duration.
2. Fundamental frequency maximum and mean.
3. Energy maximum and energy mean.
4. Melodic slope.
5. Pause durations: between the last turn of this speaker and the current one, between the current turn and the next one of this speaker.

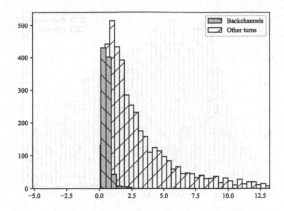

Fig. 1. Duration distributions for two types of turns.

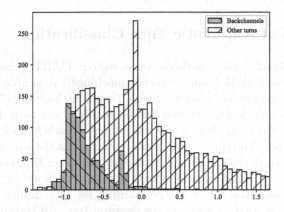

Fig. 2. Energy maximum distributions for two types of turns.

Fundamental frequency (F_0) values were calculated using REAPER: Robust Epoch And Pitch EstimatoR [23]. This algorithm is currently one of the most effective ones [12], and it also allows computing F_0 period marks, which might be useful for pitch-synchronous signal processing. Melodic slope was estimated as linear regression calculated for all F_0 values within the given turn. Speech signal energy in a turn was calculated as the mean of the squared signal amplitudes. All F_0 and energy related features were z-normalized across the dialogue to which the turn belonged.

The analysis of the acoustic features showed that the most significant parameter must be duration, turn intensity may also be useful for classification, and pause durations and F_0 are unlikely to be of use. Figures 1–3 show distribution histograms of duration, F_0 maximum and energy maximum for two types of turns: backchannels and non-backchannels. Despite the apparent difference in usefulness for classification, all features were experimentally tested.

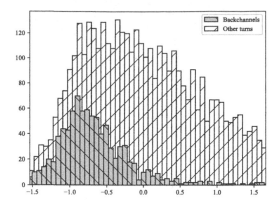

Fig. 3. F_0 maximum distributions for two types of turns.

4   Methods of Automatic Turn Classification

Two different classification methods were tested: CART (classification and regression tree) and SVM (support-vector machines). A simple threshold was chosen as a baseline model, against which all machine learning algorithms were then tested. The only feature chosen for the baseline was turn duration, i.e. if the turn was shorter than that, it was considered a backchannel threshold and otherwise it was not. The threshold T was determined as the sum of the expected value and standard deviation of backchannel durations in the training material.

Decision trees are predictive models that work as follows [4]: first, a list of questions based on data values in the training set is compiled. After this, a greedy algorithm is used to generate the decision tree. All training samples are placed in the tree root. The best question is then chosen from the list to separate the samples into two subsets. The question set consists of questions about all possible thresholds for all the feature values within all training data. For every split, the classifier chooses the question with which the split results in minimal entropy. The algorithm recursively splits the samples in the nodes with the best questions until there are samples of only one class in each subset. The CART implementation used in this study was the one from the scikit-learn library for Python with the Gini measure of impurity [20].

The other classification algorithm used in the study was SVM (support-vector machines) [5]. This method, like other linear classifiers, is based on representing each sample as a point in N-dimensional space, where N is the number of classification features, and searching for a (N-1)-dimensional hyperplane that would separate these data points into classes. The algorithm projects the points into a higher-dimensional space. A hyperplane is then found which would have the maximal distance to the closest point of each class. A key feature of the algorithm is the kernel trick that allows for calculating the distances in the new space without explicitly computing the new coordinates. The scikit-learn implementation was also used for this study.

Table 2. Results of automatic detection of backchannels.

Classifier	F_1 Score
Threshold (based on duration)	0.619
Decision tree (duration, max energy, max F_0, melodic slope)	0.610
Decision tree (duration)	0.603
Support-vector machines (duration, pause)	0.516
Support-vector machines (duration, melodic slope)	0.638
Support-vector machines (duration, max and mean F_0)	0.617
Support-vector machines (duration, max F_0, melodic slope)	**0.651**
Support-vector machines (duration, max energy, max F_0, melodic slope)	0.642
Support-vector machines (all features)	0.574

We followed the cross-validation approach to test the classifiers. All the dialogues except one were used for training, and the remaining dialogue was used for testing. The procedure was repeated for all the dialogues one after the other. The efficiency measures were averaged over all rounds.

5 Results

Backchannels constitute about 20% of all turns. The classes are therefore not equally sized, which is why using error rates to evaluate classification efficiency leads to incorrect assessments. A measure usually employed for similar tasks is the F_1 score [16]. The F_1 score is defined as the harmonic mean between precision and recall. Precision is the ratio of the number of True Positive decisions to the number of True Positive and False Positive ones. Recall is the ratio of the number of True Positive decisions to the number of True Positive and False Negative ones. Therefore, the formula for the F_1 score is as follows:

$$F_1 = \frac{2 \times Precision \times Recall}{Precision + Recall} \qquad (1)$$

Table 2 shows the comparison of the results achieved using various classifiers.

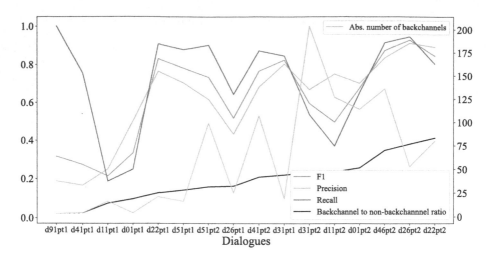

Fig. 4. Efficiency of classification for the best SVM setup (duration, max F_0, melodic slope) across all the dialogues, and the amount of backchannels within the test dialogue.

The F_1 score for the baseline classifier (a simple threshold) was 0.619.

The features chosen for the CART experiment were turn duration, energy maximum, F_0 maximum and melodic slope. The efficiency of this classifier slightly lower that of the baseline model–0.610. The analysis of the dependency between tree depth and its efficiency showed that maximum efficiency is achieved with depth 4, after which the efficiency quickly decreases. It was found that the first split was based on turn duration, which led us to testing a decision tree with duration as its only feature. The quality wasn't significantly lower–0.603. A similar analysis of the tree's depth showed that maximum efficiency is achieved at depth 2, after which the quality decreases similarly. The reason why maximum efficiency is not achieved with the first split is evidently connected to the nature of the training data and the algorithm of choosing the best split threshold. The analysis of the resulting tree showed that the only significant for classification threshold was 0.611 s, and all turns shorter than that were considered backchannels. Adding more features did not lead to any improvement.

The next experiment used support-vector machines for classification. Several feature combinations were tested, which are shown in Table 2. The most efficient system uses turn length, F_0 maximum, and melodic slope. The F_1 of this classifier is 0.651 (Pr = 0.640, Rec = 0.755). The usage of additional maximum energy feature slightly decreased the efficiency to F_1 of 0.642. Adding preceding and following pause durations leads to a sharp decrease in efficiency ($F_1 = 0.574$).

The efficiency varies much from dialogue to dialogue, see Fig. 4. In general, there is some correlation between the relative amount of backchannels within the test dialogue and the efficiency of classification. The other source of efficiency variation could be inter-speaker variability of backchannel realisation.

Table 3. The most frequent words and responses falsely detected as backchannels in the four classification designs.

Threshold		CART (duration)		CART (4 features)		SVM (4 features)	
'да' *yes*	114	'да' *yes*	89	'да' *yes*	79	'да' *yes*	64
'а' *and*	12	'а' *and*	10	'нет' *no*	10	'а' *and*	9
'нет' *no*	9	'нет' *no*	9	'а' *and*	8	'хорошо' *okay*	7
'так' *well*	9	'так' *well*	7	'ну' *well*	7	'нет' *no*	6
'хорошо' *okay*	9	'вот' *well*	4	'хорошо' *okay*	6	'ну' *well*	5
hesitations	53	hesitations	37	hesitations	30	hesitations	30
'угу' *uh-huh* affirm. response	9	'угу' *uh-huh* affirm. response	5	'угу' *uh-huh* affirm. response	7	'угу' *uh-huh* affirm. response	8
'не-а' *uh-uh* neg. response	6	'не-а' *uh-uh* neg. response	2	'не-а' *uh-uh* neg. response	6	'не-а' *uh-uh* neg. response	5
...	...	...	...	...	...	...	...
TOTAL	698	TOTAL	355	TOTAL	382	TOTAL	316

Then we analysed the turns falsely detected as backchannels by each of the classifiers. A significant amount of them (see Table 3) were short affirmative and negative words, hesitations and non-lexical responses (such as uh-huh). One can see that these words might be found in the list of the most frequent backchannels as well (see Table 1). Thus, it is often the broad dialogue context that separates backchannels and non-backchannel responses.

6 Conclusion

The acoustic analysis of speech data revealed that backchannels in Russian are 1) shorter, 2) lower in intensity, and 3) lower in F_0 than other turns.

The experiments showed that when employing the decision tree the relatively high efficiency ($F_1 = 0.603$) is reached with the use of only one parameter–duration; adding other parameters just slightly increased the efficiency of the classifier, up to $F_1 = 0.610$.

A more sophisticated classification algorithm, support-vector machines, improves the efficiency of backchannel detection, with different parameters at play. The maximum value of F_1 score (0.651) can be achieved by using the following set of features: duration, maximum F_0, melodic movement (melodic slope).

Inter-speaker variability seems to play a significant role in the efficiency of backchannel detection. In some dialogues the presented algorithms achieve high efficiency (with the F_1 score above 0.8), while for a few of the dialogues they fail to reach the score of 0.4. We might suppose that some speakers might use backchannels in a non-standard way. On the other hand, manual backchannel annotation is based on subjective decisions, and different annotators would also fail to agree on all the cases. The analysis of errors in backchannel detection revealed that the major part of turns incorrectly detected as backchannels comprises various forms of affirmative and negative words and other types of feedback which may be considered backchannels in terms of dialogue organisation.

Acknowledgments. The research is supported by Russian Science Foundation (Project 19-78-10046 "Phonetic manifestations of communication accommodation in dialogue").

References

1. Bailly, G., Elisei, F., Juphard, A., Moreaud, O.: Quantitative analysis of backchannels uttered by an interviewer during neuropsychological tests. In: Proceedings of Interspeech, pp. 2905–2909 (2016)
2. Beňuš, Š.: The prosody of backchannels in Slovak. In: Proceedings of 8th International Conference on Speech Prosody, pp. 75–79 (2016)
3. Beňuš, Š.: The prosody of backchannels in Slovak. In: Proceedings of 8th International Conference on Speech Prosody, pp. 75–79 (2016)
4. Breiman, L., Friedman, J., Stone, C.J., Olshen, R.A.: Classification and Regression Trees. Chapman and Hall/CRC (1984)
5. Cortes, C., Vapnik, V.: Support-vector networks. Mach. Learn. **20**, 273–297 (1995)
6. Dobrushina, N.: The semantics of interjections in reactive turns [semantika mezhdometij v reaktivnykh replikakh]. Bull. Moscow Univ. **2**, 136–145 (1998). (in Russian)
7. Edlund, J.: In search of the conversational homunculus: serving to understand spoken human face-to-face interaction. Doctoral thesis, KTH Royal Institute of Technology (2011)
8. Edlund, J., Heldner, M., Moubayed, S.A., Gravano, A., Hirschberg, J.: Very short utterances in conversation. Proc. FONETIK **2010**, 11–16 (2010)
9. Gerassimenko, O.: Functions of feedback items a, aha, and hm in Russian phone conversation [Funktsii chastits obratnoj svyazi v telefonnom dialoge (na primere leksem a, aga i hm]. Proceedings of the International Conference Dialog **1**, 103–108 (2012). (in Russian)
10. Gravano, A., Beňuš, Š., Chávez, H., Hirschberg, J., Wilcox, L.: On the role of context in the interpretation of 'okay'. In: Proceedings of 45th Conference of Association of Computer Linguistics, pp. 800–807 (2007)
11. Hara, K., Inoue, K., Takanashi, K., Kawahara, T.: Prediction of turn-taking using multitask learning with prediction of backchannels and fillers. In: Proceedings of Interspeech, pp. 991–995 (2018)
12. Jouvet, D., Laprie, Y.: Performance analysis of several pitch detection algorithms on simulated and real noisy speech data. In: Proceedings of 25th European Signal Processing Conference (EUSIPCO), pp. 1664–1668 (2017)
13. Jurafsky, D., et al.: Automatic detection of discourse structure for speech recognition and understanding. In: IEEE Workshop on Automatic Speech Recognition and Understanding, pp. 88–95 (1997)
14. Kachkovskaia, T., et al.: SibLing corpus of Russian dialogue speech designed for research on speech entrainment. In: Proceeding of LREC (2020, in press)
15. Kawahara, T., Yamaguchi, T., Inoue, K., Takanashi, K., Ward, N.: Prediction and generation of backchannel form for attentive listening systems. In: Proceedings of Interspeech, pp. 2890–2894 (2016)

16. de Kok, I., Heylen, D.: A survey on evaluation metrics for backchannel prediction models. In: Proceedings of the Interdisciplinary Workshop on Feedback Behaviors in Dialog, pp. 15–18 (2012)
17. Malysheva, E.: Phonetic properties of backchannels in dialogue. Bachelor's thesis, Saint Petersburg State University (2018). (in Russian)
18. Müller, M., et al.: Using neural networks for data-driven backchannel prediction: a survey on input features and training techniques. In: Proceedings of International Conference on Human-Computer Interaction (2015)
19. Park, H.W., Gelsomini, M., Lee, J.J., Zhu, T., Breazeal, C.: Backchannel opportunity prediction for social robot listeners. In: Proceedings of IEEE International Conference on Robotics and Automation (ICRA), pp. 2308–2314 (2017)
20. Pedregosa, F., et al.: Scikit-learn: machine learning in Python. J. Mach. Learn. Res. **12**, 2825–2830 (2011)
21. Ruede, R., Müller, M., Stüker, S., Waibel, A.: Enhancing backchannel prediction using word embeddings. In: Proceedings of Interspeech, pp. 879–883 (2017)
22. Ruede, R., Müller, M., Stüker, S., Waibel, A.: Yeah, right, uh-huh: a deep learning backchannel predictor. In: Eskenazi, M., Devillers, L., Mariani, J. (eds.) Advanced Social Interaction with Agents: 8th International Workshop on Spoken Dialog Systems, pp. 247–258 (2019). https://doi.org/10.1007/978-3-319-92108-2_25
23. Talkin, D.: REAPER: Robust Epoch And Pitch EstimatoR (2015). https://github.com/google/REAPER
24. Truong, K.P., Poppe, R., Heylen, D.: A rule-based backchannel prediction model using pitch and pause information. In: Proceedings of Interspeech, pp. 3058–3061 (2010)
25. Ward, N., Tsukahara, W.: Prosodic features which cue back-channel responses in English and Japanese. J. Pragmat. **23**, 1177–1207 (2000)
26. Włodarczak, M., Heldner, M.: Respiratory turn-taking cues. In: Proceedings of Interspeech, pp. 1275–1279 (2016)

Experimenting with Attention Mechanisms in Joint CTC-Attention Models for Russian Speech Recognition

Irina Kipyatkova[1,2(✉)] and Nikita Markovnikov[1]

[1] St. Petersburg Institute for Informatics and Automation of the Russian Academy of Sciences (SPIIRAS), St. Petersburg, Russia
kipyatkova@iias.spb.su, niklemark@gmail.com
[2] St. Petersburg State University of Aerospace Instrumentation (SUAI), St. Petersburg, Russia

Abstract. The paper presents an investigation of attention mechanisms in end-to-end Russian speech recognition system created by join Connectional Temporal Classification model and attention-based encoder-decoder. We trained the models on a small dataset of Russian speech with total duration of about 60 h, and performed pretraining of the models using transfer learning with English as non-target language. We experimented with following types of attention mechanism: coverage-based attention and 2D location-aware attention as well as their combination. At the decoding stage we used beam search pruning method and gumbel-softmax function instead of softmax. We have achieved 4% relative word error rate reduction using 2D location-aware attention.

Keywords: End-to-End speech recognition · Attention mechanism · Coverage-based attention · 2D Location-aware attention · Russian speech

1 Introduction

In recent years, development of end-to-end systems became the main trend in speech recognition technologies due to fast advances in deep learning approaches. The end-to-end speech recognition methods are mainly based on two types of models: Connectional Temporal Classification (CTC) and attention-based encoder-decoder, as well as on their combination [1].

For example, an investigation of end-to-end model with CTC is described in [2]. It was shown that such system is able to work without language model (LM) well. Training dataset [3] was made up of audio tracks of Youtube videos with duration more than 650 h. Testing dataset was made up of audio tracks of Google Preferred channels on YouTube [4], its duration was 25 h. The lowest word error rate (WER) obtained without using LM was 13.9% and it was equal to 13.4% with the usage of LM.

Attention based encoder-decoder model was used in [5] for experiments on recognition of speech from LibriSpeech corpus. The authors performed data augmentation by speed, tempo, and/or volume perturbation, sequence-noise injection. The authors obtained WER = 4% on test-clean data and WER = 11.7% on test-other.

© Springer Nature Switzerland AG 2020
A. Karpov and R. Potapova (Eds.): SPECOM 2020, LNAI 12335, pp. 214–222, 2020.
https://doi.org/10.1007/978-3-030-60276-5_22

The application of LSTM-based LM results in decreasing of WER to 2.5% and 8.2% on test-clean and test-other sets respectively.

Joint CTC-attention based end-to-end speech recognition was proposed in [6]. Two loss functions are used in these model, which are combined using weighted sum as follows:

$$L = \lambda L_{CTC} + (1 - \lambda)L_{att},$$

where L_{CTC} is an objective for CTC and L_{att} is an objective for attention-based model, λ is a weight of CTC model, $\lambda \in [0, 1]$.

Different types of attention mechanism in end-to-end speech recognition systems are analyzed in many paper. For example, in [7] an application of Self-attention in CTC model was investigated. The proposed model allowed the authors to outperform the existing end-to-end models (CTC, encoder-decoder, joint CTC/encoder-decoder models) in speech recognition accuracy. In [8], the authors proposed Monotonic Chunkwise Attention (MoChA), which adaptively splits the input sequence into small chunks over which soft attention is computed. The authors applied this attention mechanism for two tasks: online speech recognition and automatic document summarization. Experiments on online speech recognition were performed on Wall Street Journal (WSJ) corpus. WER was equal to 13.2%. However, the authors of another research published in [9] found out that this attention mechanism is unstable in their system and proposed modified attention mechanism called stable MoChA (sMoCha). Moreover, the authors proposed to compute truncated CTC (T-CTC) prefix probability on the segmented audio rather than on the complete audio. At the decoding stage, the authors proposed the dynamic waiting joint decoding (DWDJ) algorithm to collect the decoding hypotheses from the CTC and attention branches in order to deal with the problem that these two branches predict labels asynchronously in beam search. Experiments on online speech recognition showed WER equal to 6% and 16.7% on test-clean and test-other of LibriSpeech respectively.

The aim of the current research was to improve the Russian end-to-end speech recognition system developed in SPIIRAS by modification of attention mechanism in joint CTC-attention based encoder-decoder model. We have investigated with coverage-based attention and 2D location-aware attention as well as their combination. The models were trained and tested with the help of EspNet toolkit [10] with a PyTorch as a back-end part. The developed models were evaluated in terms of Character Error Rate (CER) and word error rate (WER).

The rest of the paper is organized as follows. In Sect. 2 we present our baseline end-to-end speech recognition model, in Sect. 3 we describe modifications of attention mechanism, our Russian speech corpora are presented in Sect. 4, the experimental results are given in Sect. 5, in Sect. 6 we make a conclusion to our work.

2   The Baseline Russian End-to-End Model

As a baseline we used joint CTC-attention based encode-decoder model presented on
Fig. 1, where h is an input vector, h is a vector of hidden states obtained from encoder,
g_i is a weighted vector obtained from attention mechanism on i-th iteration of decoding,
y_i is decoder's output on i-th iteration, w_i is i-th symbol of output sequence, s_{i-1} is the
decoder's state on the previous iteration.

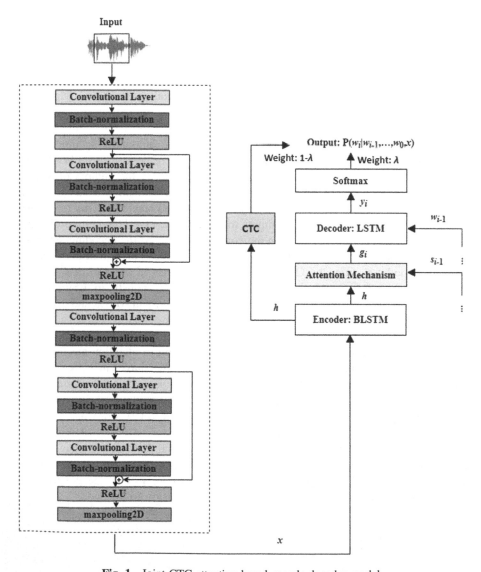

Fig. 1. Joint CTC-attention based encode-decoder model

The model had the following topology. Long Short-Term Memory (LSTM) network contained two layers with 512 cells in each was used as decoder. Bidirectional LSTM (BLSTM) contained five layers with 512 cells in each was used as encoder. Moreover we used highway connection in encoder. In order to prevent our model from making over-confident predictions, we used dropout [11] with probability equal to 0.4 in LSTM layers at every time step in the encoder's network as well we used label smoothing [12] as a regularization mechanism with a smoothing factor of 0.01. Location-aware [13] attention mechanism was used in decoder.

Before the encoder, there was a feature extraction block that was VGG model [14]. Moreover, we added residual connection (ResNet) in this block as it is shown on Fig. 1. Thus, the feature extraction block consisted of two similar parts, which included three convolution layers followed by batch normalization [15] and max-pooling layer; rectified linear unit (ReLU) [16] was used as activation function. The number of output features was equal to 128.

At the training stage, the CTC weight was equal to 0.3. Filter banks features were used as input. The model's training was carried out using the transfer learning approach. At first, the models were pretrained on English speech data from LibriSpeech corpus (we used 360 h of English data for pretraining). Then models were trained on Russian speech data. Our Russian speech corpora are described in detail in Sect. 4.

Due to small size of Russian speech dataset, we additionally used LSTM-based LM at speech recognition experiments. The language model was trained on text corpus collected from online Russian newspapers. The corpus consisted of 350 M words (2.4 GB data). LSTM contained one layer with 512 cells. The vocabulary consisted of 150 K most frequent word-forms from the training text corpus.

3 Attention Mechanisms in Russian End-to-End Speech Recognition Model

3.1 The Baseline Location-Aware Attention

Attention mechanism is a subnetwork in the decoder. Attention mechanism chooses a subsequence of the input and then uses it for updating hidden states of neural network of decoder and predicting an output. On the i-th step decoder generates an output y_i focusing on separate components of h as follows [13]:

$$\alpha_i = \text{Attend}\,(s_{i-1}, \alpha_{i-1}, h);$$

$$g_i = \sum_{j=1}^{L} \alpha_{i,j} h_j,$$

where s_{i-1} is the $(i-1)$-th state of neural network called Generator, $\alpha_i \in \mathbb{R}^L$ are attention weights, vector g_i is called glimpse, Attend denotes a function that calculates attention weight. The step comes to an end with computing a new generator state as $s_i = Recurrency(s_{i-1}, g_i, y_i)$.

In our baseline system we used location-aware attention that calculates as follows:

$$f_i = F * \alpha_{i-1};$$

$$e_{i,j} = w^{\mathrm{T}} \tanh \left(W s_{i-1} + V h_j + U f_{i,j} + b \right),$$

where $w \in \mathbb{R}^m$ and $b \in \mathbb{R}^n$ denote weight vector, $W \in \mathbb{R}^{m \times n}$, $V \in \mathbb{R}^{m \times 2n}$, and $U \in \mathbb{R}^{m \times k}$ are weight matrices, n and m are number of hidden units in the encoder and decoder respectively, vector $f_{i,j} \in \mathbb{R}^k$ are convolutional features. Generally, an attention weights matrix is calculated as $\alpha_i = softmax_i(e)$.

3.2   Coverage-Based Attention

The attention mechanism usually takes into account only the previous time step when calculating the matrix of weights. However, during research and experiments, it was found out that models often miss some, mostly short, words, for example, prepositions. Such words are often swallowed in pronunciation. In [17] a method of taking into account all preceding history of weight matrix was proposed for text summarization. In this method coverage vector is computed as follows [18]:

$$\beta_i = \sum_{l=0}^{i-1} \alpha_l,$$

where α_l is weight matrix of neural network in attention mechanism at l-th time step, i is an index of the current time step. Thus, in this case weight matrix in the attention mechanism is the sum of attention weights at all preceding time steps. The coverage vector is used as extra input to the attention mechanism as follows:

$$e_{i,j} = w^{\mathrm{T}} \tanh \left(W s_{i-1} + V h_j + U \beta_{i,j} + b \right).$$

3.3   2D Location-Aware Attention

The general location attention takes into account one frame to compute attention weight vector. Therefore, it was suggested [13] to take into account several frames for computation of weight vector using fixed-size window on frames. It should be noted that convolution in location attention is carried out with the help of a kernel of the size (1, K), where K is the given hyperparameter of the model. It was suggested to use the kernel of the size (w, K), where w is the width of window on frames. At each iteration the window is shifted on one frame updating the obtained matrix. We set $w = 5$ in our experiments.

3.4   Joint Coverage-Based and 2D Location-Aware Attention

As well we performed combination of coverage-based and 2D location-aware attention mechanisms. In this case attention is computed over 5 frames as 2D location-aware attention and then coverage vector is compute as the sum of attention distributions over all previous decoder timesteps.

4 Speech Datasets

For training the acoustic models, we used three corpora of Russian speech recorded at SPIIRAS [19, 20]:

- the speech database developed within the framework of the EuroNounce project [21] that consists of recordings of 50 speakers, each of them pronounced a set of 327 phonetically rich and meaningful phrases and texts;
- the corpus consisting of recordings of other 55 native Russian speakers; each speaker pronounced 105 phrases: 50 phrases were taken from the Appendix G to the Russian State Standard P 50840-95 [22] (these phrases were different for each speaker), and 55 common phrases were taken from a phonetically representative text, presented in [23];
- the audio part of the audio-visual speech corpus HAVRUS [24] that consists of recordings of 20 speakers pronouncing 200 phrases: (a) 130 phrases for training were two phonetically rich texts common for all speakers, and (b) 70 phrases for testing were different for every speaker: 20 phrases were commands for the MIDAS information kiosk [25] and 50 phrases were 7-digits telephone numbers (connected digits);

  In addition, we supplemented our speech data with free available speech corpora:
- Voxforge[1] that contains about 25 h of Russian speech recordings pronounced by 200 speakers; unfortunately, some recordings contained a lot of noises, hesitations, self-repairs, etc., therefore some recordings were excluded;
- M-AILABS[2] that mostly contains recordings of audiobooks; Russian part of the corpus contains 46 h of speech recordings of three speaker (two men and one woman).

As a result we had about 60 h of speech data. This speech dataset was splitted into validation and trains parts with sizes of 5% and 95%.

Our test speech corpus consists of 500 phrases pronounced by 5 speakers. The phrases were taken from online newspaper which was not used for LM training.

5 Experiments

At the decoding stage we used beam search pruning method similar to the approach described in [26]. Our method is described in detail in [27], in general terms, we filter beam search output with some condition to remove too bad hypotheses.

---

[1] http://www.voxforge.org/.

[2] https://www.caito.de/2019/01/the-m-ailabs-speech-dataset/.

Moreover, during decoding we used gumbel-softmax function instead of softmax. The standard decoding algorithm uses softmax function building probability distribution for estimating the character probability on each iteration. However this distribution is rather strict that can influence on the recognition result. Therefore we replaced softmax function by gumbel-sofmax [28]:

$$gumbel\_softmax_i(z) = \frac{e^{\frac{z_i + g_i}{T}}}{\sum_{k=1}^{K} e^{\frac{z_i + g_k}{T}}},$$

where T is smoothing coefficient, g_i is values from Gumbel probability destribution. With T increasing, the probability distribution of the characters become more uniform that does not allow the decoding algorithm to give too confident decisions regarding the characters. During the preliminary experiments we chose $T = 3$.

Using our baseline system we obtained CER = 10.8% and WER = 29.1%. The results obtained with usage of coverage-based and 2D location-aware attention as well as joint usage of coverage and 2D location-aware attentions are presented in Table 1.

Table 1. Experimental results on speech recognition using different types of attention

Model	CER, %	WER, %
Baseline	10.8	29.1
Coverage-based attention	11.0	29.5
2D location-aware attention	10.6	**27.9**
Joint coverage-based and 2D location-aware attention	10.9	29.8

As we can see from the Table, the best result was obtained with 2D location-aware attention (CER = 10.8%, WER = 27.9%). The usage of coverage-based attention, as well as joint coverage and 2D location-aware attention does not results in improvement of speech recognition results.

6 Conclusions and Future Work

In the paper, we have investigated two types of attention mechanisms for Russian end-to-end speech recognition system: coverage-based attention and 2D location-aware attention. Coverage-based attention unfortunately does not results in improving speech recognition result. The usage of 2D location-aware attention allowed us to achieve 4% relative reduction of WER. In further research we are going to research another architectures of neural network for Russian end-to-end speech recognition, for example, Transformer.

Acknowledgements. This research was supported by the Russian Foundation for Basic Research (project No. 18-07-01216).

References

1. Markovnikov, M., Kipyatkova, I.: An analytic survey of end-to-end speech recognition systems. SPIIRAS Proc. **58**, 77–110 (2018)
2. Soltau, H., Liao, H., Sak, H.: Neural speech recognizer: Acoustic-to-word LSTM model for large vocabulary speech recognition (2016). arXiv preprint arXiv:1610.09975 https://arxiv.org/abs/1610.09975
3. Liao, H., McDermott, E., Senior, A.: Large scale deep neural network acoustic modeling with semi-supervised training data for YouTube video transcription. In: Proceedings of IEEE Workshop on Automatic Speech Recognition and Understanding (ASRU), pp. 368–373 (2013)
4. Google preferred lineup explorer – YouTube. https://www.youtube.com/yt/lineups/. Accessed 17 Feb 2018
5. Tüske, Z., Audhkhasi, K., Saon, G.: Advancing sequence-to-sequence based speech recognition. In: INTERSPEECH-2019, pp. 3780–3784 (2019)
6. Kim, S., Hori, T., Watanabe, S.: Joint CTC-attention based end-to-end speech recognition using multi-task learning. In: Proceedings of IEEE International Conference on Acoustics, Speech and Signal Processing (ICASSP-2017), pp. 4835–4839 (2017)
7. Salazar, J., Kirchhoff, K., Huang, Z.: Self-attention networks for connectionist temporal classification in speech recognition. In: Proceedings of IEEE International Conference on Acoustics, Speech and Signal Processing (ICASSP-2019), pp. 7115–7119 (2019)
8. Chiu, C.C., Raffel, C.: Monotonic chunkwise attention (2017). arXiv preprint arXiv:1712.05382
9. Miao, H., et al.: Online hybrid CTC/Attention architecture for end-to-end speech recognition. In: INTERSPEECH-2019, pp. 2623–2627 (2019)
10. Watanabe, S., et al.: ESPnet: end-to-end speech processing toolkit. In: INTERSPEECH-2018, pp. 2207–2211 (2018)
11. Srivastava, N., et al.: Dropout: a simple way to prevent neural networks from overfitting. J. Mach. Learn. Res. **15**(1), 1929–1958 (2014)
12. Szegedy, C., et al.: Rethinking the inception architecture for computer vision. In: Proceedings of the IEEE Conference on Computer Vision and Pattern Recognition, pp. 2818–2826 (2016)
13. Chorowski, J.K., et al.: Attention-based models for speech recognition. In: Advances in Neural Information Processing Systems, pp. 577–585 (2015)
14. Simonyan, K., Zisserman, A.: Very deep convolutional networks for large-scale image recognition(2014). arXiv preprint arXiv:1409.1556
15. Ioffe, S., Szegedy, C.: Batch normalization: accelerating deep network training by reducing internal covariate shift (2015). CoRR abs/1502.03167 http://arxiv.org/abs/1502.03167
16. Glorot, X., Bordes, A., Bengio, Y.: Deep sparse rectifier neural networks. In: Proceedings of the 14th International Conference on Artificial Intelligence and Statistics, pp. 315–323 (2011)
17. Tu, Z., et al.: Modeling coverage for neural machine translation (2016). arXiv preprint arXiv:1601.04811
18. See, A., Liu, P.J., Manning, C.D.: Get to the point: Summarization with pointer-generator networks (2017). arXiv preprint arXiv:1704.04368
19. Kipyatkova, I.: Experimenting with hybrid TDNN/HMM acoustic models for Russian speech recognition. In: Karpov, A., Potapova, R., Mporas, I. (eds.) SPECOM 2017. LNCS (LNAI), vol. 10458, pp. 362–369. Springer, Cham (2017). https://doi.org/10.1007/978-3-319-66429-3_35

20. Kipyatkova, I., Karpov, A.: Class-based LSTM Russian language model with linguistic information. In: Proceedings of the 12th Conference on Language Resources and Evaluation (LREC 2020), pp. 2470–2474 (2020)
21. Jokisch, O., Wagner, A., Sabo, R., Jaeckel, R., Cylwik, N., Rusko, M., Ronzhin, A., Hoffmann, R.: Multilingual speech data collection for the assessment of pronunciation and prosody in a language learning system. Proceedings of SPECOM **2009**, 515–520 (2009)
22. State Standard P 50840–95. Speech Transmission by Communication Paths. Evaluation Methods of Quality, Intelligibility and Recognizability, p. 230. Standartov Publication, Moscow (1996). (in Russian)
23. Stepanova, S.B.: Phonetic features of Russian speech: realization and transcription, Ph.D. thesis (1988). (in Russian)
24. Verkhodanova, V., Ronzhin, A., Kipyatkova, I., Ivanko, D., Karpov, A., Železný, M.: HAVRUS corpus: high-speed recordings of audio-visual Russian speech. In: Ronzhin, A., Potapova, R., Németh, G. (eds.) SPECOM 2016. LNCS (LNAI), vol. 9811, pp. 338–345. Springer, Cham (2016). https://doi.org/10.1007/978-3-319-43958-7_40
25. Karpov, A.A., Ronzhin, A.L.: Information enquiry kiosk with multimodal user interface. Pattern Recogn. Image Anal. **19**(3), 546–558 (2009)
26. Freitag, M., Al-Onaizan, Y.: Beam search strategies for neural machine translation (2017). arXiv preprint arXiv:1702.01806
27. Markovnikov, N., Kipyatkova, I.: Investigating joint CTC-attention models for end-to-end Russian speech recognition. In: Salah, A.A., Karpov, A., Potapova, R. (eds.) SPECOM 2019. LNCS (LNAI), vol. 11658, pp. 337–347. Springer, Cham (2019). https://doi.org/10.1007/978-3-030-26061-3_35
28. Jang, E., Gu, S., Poole, B.: Categorical reparameterization with gumbel-softmax (2016). arXiv preprint arXiv:1611.01144

Comparison of Deep Learning Methods for Spoken Language Identification

Can Korkut[1], Ali Haznedaroglu[1(✉)], and Levent Arslan[1,2]

[1] Sestek, Istanbul, Turkey
{can.korkut,ali.haznedaroglu,
levent.arslan}@sestek.com
[2] Electrical and Electronics Engineering Department, Bogazici University,
Istanbul, Turkey

Abstract. In this paper, we implement and compare deep learning based spoken language identification models. We also deploy two very recent and popular speech recognition methods, namely Wav2Vec and SpecAugment, in our classifiers and test if they are also applicable to the field of language identification. Out of the models we implement, X-vector based deep feed forward network classifier obtains the highest F1-score of 0.91, where the target set consists of five languages. SpecAugment data augmentation method turns out to increase the classification accuracy when applied to the input mel-spectrograms of the CRNN architecture. Although they obtain lower classification accuracies than some of the other methods, Wav2Vec speech representations also achieve promising results.

Keywords: Spoken language identification · Data augmentation · Speech representations

1 Introduction

Spoken language identification is an acoustic classification problem in which the main purpose is to determine the language of a spoken utterance. Spoken language identification systems are used as the first step in many speech related technologies like multi-lingual contact center analysis applications, mobile assistants and so on. In this work, our main goal is to compare deep learning-based language identification methods. We also apply two recent approaches which are successfully used in speech recognition systems, namely Wav2vec and SpecAugment, to the problem of language identification to see if they also work in this domain. All the models we deploy use different features with different neural network architectures.

Hybrid CNN and RNN models, or CRNNs, are successfully applied for language identification [1]. They use speech spectrograms as the input features and these features are first fed into layers of CNNs, and then the CNN outputs are connected to LSTM layers. The output of the last LSTM node is then connected to a fully-connected layer. This hybrid network obtains 0.98 F1-score on the YouTube News Dataset, outperforming CNN-based system which obtains an F1-score of 0.91.

© Springer Nature Switzerland AG 2020
A. Karpov and R. Potapova (Eds.): SPECOM 2020, LNAI 12335, pp. 223–231, 2020.
https://doi.org/10.1007/978-3-030-60276-5_23

Speech representations such as I-vectors and X-vectors are also commonly used in language identification systems. I-vectors, which once showed state-of-the-art performance for speaker recognition systems are also deployed for language identification applications [2]. X-vectors, which turns the speech waveforms into fixed-length one-dimensional representations using deep neural networks achieve very good results and they outperform I-vector systems in most of the cases [3]. Wav2Vec's emerged as another example of speech representations that are learned in an un-supervised manner, and they achieve very good results on speech recognitions applications even in the cases of very limited transcribed training data [4].

Data augmentation methods are commonly used in deep learning applications as lots of data is needed to successfully optimize huge number of parameters in the deep network architectures. Speech representations as X-vectors also perform better when they are trained from augmented data. SpecAugment is a recent data augmentation method which is directly applied to the input features, namely to the speech spectrograms, of the deep network, and it gives state-of-the art results in speech recognition applications [5]. In this work, we also try this augmentation method in the CRNN topology to see if it can also be used for language identification tasks.

In this paper, we compare the methods of CRNNs and X-vectors in language identification and we also implement Wav2Vec speech representations and SpecAugment data augmentation methods in the field of language identification. These methods are successfully used in speech recognition applications, but to our knowledge this work is the first that they are also used in the task of language identification. So, this stands out as our main contribution to the field.

This paper is organized as follows. In the next section, we describe the deep neural network architectures that we compare. Then we describe our experimental setting, and then the results we obtain in the next two sections. Then we will conclude our work in the last section.

2 Methods

The network architectures that we implement, and compare are described in the following subsections. Each architecture uses different feature extraction methods, so we name the subsections with respect to the features and network types that are used.

2.1 Mel-Spectrogram Based CRNNs

Spectrograms are used in many speech processing applications like speech recognition systems, text-to-speech systems and so on. They represent how the frequency spectrum of a speech signal varies over time. In this work, we use Mel-spectrograms in which the spectrogram frequencies are converted to mel-scale.

Mel-spectrograms are two-dimensional images in which the energy content of each frequency bin is represented by the color or intensity of the corresponding pixel. This high-dimensional "image-like" representation makes them ideal inputs for convolutional neural networks (CNNs), which are intensively used in image classification applications. These network architectures are also proposed for language identification purposes [6]. In this setup, we use a hybrid model, namely CRNN, in which the mel-spectrogram features are first fed into 2D CNNs, and then the output of the final CNN layer is fed into an LSTM layers in a time-wised sequential manner. The final layer uses softmax activation function which converts the outputs to class probabilities. The overall architecture can be seen on Fig. 1.

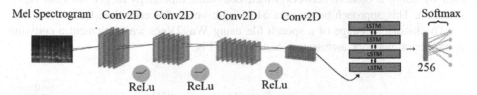

Fig. 1. Mel-Spectrogram based CRNN architecture.

2.2   X-Vector Based FFNNs

X-vectors are speech representations which converts variable length speech segments into one-dimensional vectors using deep neural networks [7]. They first emerged as robust features for speaker recognition applications, but they can also be successfully used as features for other audio classification problems like language identification. X-vectors are trained using a deep neural network in which the input features are filter-bank coefficients. The first five layers consume consecutive speech frames with different temporal context lengths, then a pooling layer aggregates the output of these layers feeding them into layers deep down on the network. The output layer uses softmax activation and outputs a one-dimensional embedding vector.

Different classifiers can be implemented on the top of x-vectors to determine the language of a speech utterance. In this work we choose to use deep feed-forward neural networks (FFNNs) as out classifier. The overall architecture of this method is given in Fig. 2.

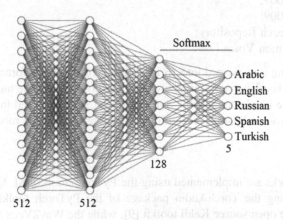

Fig. 2. X-Vector based FFNN architecture.

2.3 Wav2Vec Based CNNs

Wav2vec's are speech representations that are trained in an unsupervised manner, and they give state-of-the-art results in speech recognition applications even in the low-resource conditions. The model takes the input speech signal and turns it into its representation by first using an encoder network that outputs latent representation, and then by using a context network which combines time-steps to get the final representation. This approach produces a 512-length vector for each time frame.

To classify language of a speech file using Wav2Vecs we implement a convolutional neural network architecture as shown in Fig. 3.

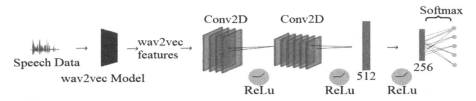

Fig. 3. Wav2Vec based CNN architecture.

3   Experimental Setup

3.1   Data

We choose Arabic, English, Russian, Spanish, and Turkish as our target languages. We have collected and combined different datasets to model different recording conditions and speaking styles. Our data is composed of the following sets:

- YouTube News Channel;
- NIST Language Recognition Evaluation Sets;
- NIST LRE 2005;
- NIST LRE 2007;
- NIST LRE 2009;
- European Speech Repository;
- Mozilla Common Voice.

Each recording is converted into 16 kHz, linear 16-bit, mono format and set to 10 s in length. Each class have approximately 8.000 audio files, and the number of test files for each language is selected as 800. Same train/test splits are used in all experiments. We use 5-fold cross validation resampling procedure to train the models.

3.2   Toolkits

All neural networks are implemented using the PyTorch toolkit [8]. Mel-spectrograms are extracted using the TorchAudio package of the PyTorch toolkit. X-vectors are derived using the open source Kaldi toolkit [9], while the Wav2Vecs are derived using the Fairseq toolkit [10].

3.3 Evaluation Metrics

To evaluate our results, we use Recall and Precision metrics together with their hormonic mean which is the F1-score. Those metrics are calculated as follows:

$$Recall = \# \, of \, true \, positives \, / \, (\# \, of \, true \, positives \, + \, \# \, of \, false \, negatives) \quad (1)$$

$$Precision = \# \, of \, true \, positives \, / \, (\# \, of \, true \, positives \, + \, \# \, of \, false \, positives) \quad (2)$$

$$F1 - Score = 2 \times Recall \times Precision \, / \, (Recall + Precision) \quad (3)$$

4 Results

The results of the implemented networks are given in the following subsections. Different hyper-parameter sets are tested for each neural network, but we only report the results of the hyper-parameter set that results in highest identification accuracies.

4.1 Mel-Spectrogram Based CRNNs

Table 1 summarizes the CRNN architecture used in our experiments.

Table 1. Summary of the CRNN architecture.

Layers	Layer summary
Conv2D/Batchnormalization2D	Kernel = (3, 3), stride = (2, 2), padding = 0
Conv2D/Batchnormalization2D	Kernel = (3, 3), stride = (2, 2), padding = 0
Conv2D/Batchnormalization2D	Kernel = (3, 3), stride = (2, 2), padding = 0
Conv2D/Batchnormalization2D	Kernel = (3, 3), stride = (1, 1), padding = 0
Global average Pooling	Kernel = (2, 2), stride = (2, 2), padding = 0
Bi-directional LSTM	Input = 2048, Hidden = 128, Layer = 1
Fully Connected	Input = 256, Output = 5
Total Number of Parameters	3.200.000

In our CRNN experiments, we also implement the SpecAugment data augmentation method on the input Mel-Spectrograms so test if they increase the identification accuracies. SpecAugment increases the overall system performance as shown in Fig. 4. Results that we obtain in this architecture is given in Table 2.

Table 2. Results of the Mel-Spectrogram based CRNN.

Language	Precision	Recall	F1-Score
Arabic	0.88	0.84	0.86
English	0.85	0.77	0.81
Russian	0.86	0.94	0.90
Spanish	0.85	0.79	0.82
Turkish	0.83	0.88	0.95
Average	**0.85**	**0.84**	**0.85**

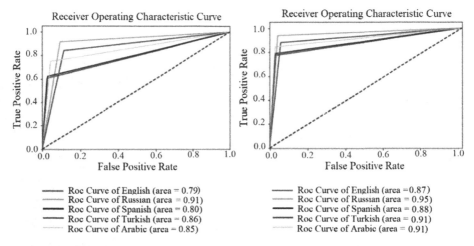

Fig. 4. Comparison of ROC Curves in case of using SpecAugment. The graph on the right shows the case when SpecAugment is used.

4.2 X-vector Based FFNNs

For the X-vector based architecture, we used the pre-trained Kaldi X-vector extractor [11]. We also trained our extractor, but the pre-trained extractor gives better results although it is trained for speaker recognition experiments. Table 3 summarizes the FFNN architecture used in X-vector based language identification architecture, and the results are given in Table 4.

Table 3. Summary of the FFNN architecture.

Fully Connected/Batchnormalization1D	Input = 512, Output = 512
Fully Connected/Batchnormalization1D	Input = 512, Output = 128
Fully Connected/Batchnormalization1D	Input = 128, Output = 5
Total Number of Parameters	330.255

Table 4. Results of the X-Vector Based FFNN architecture.

Language	Precision	Recall	F1-Score
Arabic	0.90	0.89	0.89
English	0.93	0.93	0.93
Russian	0.93	0.94	0.89
Spanish	0.90	0.89	0.89
Turkish	0.90	0.92	0.91
Average	**0.91**	**0.91**	**0.91**

4.3 Wav2Vec Based CNNs

We train our Wav2Vec model using the 5 target language sets from the Mozilla Common Voice dataset. The summary of the CNN architecture is given in Table 5, and the results are given in Table 6 (Fig. 5).

Table 5. Summary of the CNN architecture.

Layers	Layer summary
Conv2D/Batchnormalization2D	Kernel = (5, 5), stride = (2, 2), padding = 0
Conv2D/Batchnormalization2D	Kernel = (5, 5), stride = (2, 2), padding = 0
MaxPool2D	Kernel = (2, 2), stride = (2, 2), padding = 0
Fully Connected	Input = 512, Output = 256
Output	Input = 256, Output = 5
Total Number of Parameter	201.797

Table 6. Results of the Wav2Vec Based CNN architecture.

Language	Precision	Recall	F1-Score
Arabic	0.76	0.74	0.75
English	0.80	0.83	0.81
Russian	0.80	0.78	0.79
Spanish	0.85	0.73	0.79
Turkish	0.74	0.85	0.79
Average	**0.79**	**0.79**	**0.79**

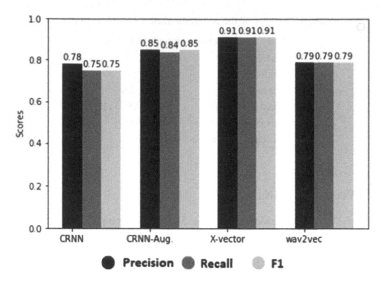

Fig. 5. Comparison of the different architecture results.

5 Conclusion

In this work, we aimed to implement and compare existing and successful language identification methods. We also introduced SpecAugment and Wav2Vec methods to the field of language identification, which are successfully used in recent speech recognition applications. X-vector based FFNN classifier outperforms the other with an F1-score of 0.91. SpecAugment method improves the performance of the CRNN architecture from F1-score of 0.75 to 0.85 when it is applied to the mel-spectrogram inputs of the classifier. So SpecAugment turns out to be a good data augmentation method also for language identification purposes. Although it gets the lower classification accuracies than some of the other classifiers, Wav2Vec based system also looks promising as it outperforms the CRNN based model that does not use SpecAugment data augmentation method.

References

1. Bartz, C., et al.: Language identification using deep convolutional recurrent neural networks. In: Liu, D., Xie, S., Li, Y., Zhao, D., El-Alfy, E.S. (eds.) International Conference on Neural Information Processing, pp. 1–13. Springer, Cham (2016)
2. Song, Y., et al.: Ivector representation based on bottleneck features for language identification. Electron. Lett. **49**(24), 1569–1570 (2013)
3. Snyder, D., et al.: Spoken Language Recognition using X-vectors. Odyssey (2018)
4. Schneider, S., et al.: wav2vec: unsupervised pre-training for speech recognition. arXiv preprint arXiv:1904.05862 (2019)
5. Park, D.S., et al.: Specaugment: a simple data augmentation method for automatic speech recognition. arXiv preprint arXiv:1904.08779 (2019)

6. Lozano-Diez, A., et al.: An end-to-end approach to language identification in short utterances using convolutional neural networks. In: Sixteenth Annual Conference of the International Speech Communication Association (2015)
7. Snyder, D., et al.: X-vectors: robust DNN embeddings for speaker recognition. In: 2018 IEEE International Conference on Acoustics, Speech and Signal Processing (ICASSP). IEEE (2018)
8. Paszke, A., et al.: PyTorch: an imperative style, high-performance deep learning library. In: Advances in Neural Information Processing Systems (2019)
9. Povey, D., et al.: The Kaldi speech recognition toolkit. In: IEEE 2011 Workshop on Automatic Speech Recognition and Understanding. No. CONF. IEEE Signal Processing Society (2011)
10. Ott, M., et al.: Fairseq: a fast, extensible toolkit for sequence modeling. arXiv preprint arXiv: 1904.01038 (2019)
11. SRE16 Xvector Model. https://kaldi-asr.org/models/m. Accessed 14 June 2020

Conceptual Operations with Semantics for a Companion Robot

Artemiy Kotov[1,2(✉)] ⓘD, Liudmila Zaidelman[1,2] ⓘD, Anna Zinina[1,2] ⓘD,
Nikita Arinkin[1,2] ⓘD, Alexander Filatov[1], and Kirill Kivva[1,3]

[1] Russian State University for the Humanities, Moscow, Russia
kotov@harpia.ru
[2] National Research Center "Kurchatov Institute", Moscow, Russia
[3] Bauman Moscow State Technical University, Moscow, Russia

Abstract. We study the features crucial for a companion robot conceptual
processing and suggest a practical implementation of a cognitive architecture
that support these features while operating a real F-2 companion robot. The
robot is designed to react to incoming speech and visual events, guide a person
in a problem space and accumulate knowledge from texts and events in memory
for further dialogue support. We show how a conceptual representation system
designed for a companion robot deals with several types of conceptual representations: text semantics, sets of emotions and reactions, operations in a
problem space and in semantic memory. We also suggest a conceptual representation based on linguistic valency structures (semantic predication) that is
suitable to link the processing components. The general processing architecture
is based on the production approach: it may trigger several scripts and combine
speech and behavioral outputs of these scripts on the robot. The system performs
conceptual operations with semantics while processing texts on a server in a
standalone mode, or while controlling the robot in a dialogue mode, or assisting
a user in solving Tangram puzzle.

Keywords: Conceptual processing · Knowledge base · Production logic ·
Semantic representation · Emotional computer agent

1 Introduction

Conceptual processing systems designed for companion robots should handle not only
traditional linguistic tasks like dialogue support and question answering, but also
perform visual recognition and problem-solving. In each of these tasks a system
operates with diverse conceptual representations: text semantics, scene setup, own
intermediary inferences, and memory. These representations should be shared between
the main processing modules, and converted, where required. We develop *F-2* robot
with conceptual processing system aimed at the integration of natural communication,
problem solving and visual comprehension tasks. The system is designed to demonstrate the following interconnections between the information processing modules:

© Springer Nature Switzerland AG 2020
A. Karpov and R. Potapova (Eds.): SPECOM 2020, LNAI 12335, pp. 232–243, 2020.
https://doi.org/10.1007/978-3-030-60276-5_24

1. Reactions of the robot should be invoked by incoming events of different nature: utterances, user movements or game actions. All these events depict or denote the real-world events, and thus may be processed in a compatible way.
2. Different reactions should compete and may output diverse and even contradictory behavioral patterns, as suggested by the model of *proto-specialists* [14] or *CogAff* architecture [20]. In particular, emotional processing may compete with rational processing, and the compound rational/emotional behavioral patterns (blending reactions) may be executed by the robot.
3. An emotion may influence a conceptual representation (*frame*) of an incoming event that is described as *top-down* emotional processing [4]. Further, conceptual representation of a situation may influence a specific concept, as suggested by the *situational conceptualization* theory [22].
4. The system should combine *reactive* processing scheme, applicable to immediate emotional reactions and speech replies, and *goal-oriented* processing scheme, applied to compound plans and problem solving.

Unlike the majority of neural networks, the system also has to keep a conceptual representation in a readable form for research purposes. This form should also allow the system to store incoming events in a memory base for further knowledge retrieval.

2 Cognitive and Dialogue Support Architectures

Conceptual processing systems are implemented in several areas of cognitive and computational research. In particular, they are required (a) to control interactive artificial organisms – robots and virtual agents, (b) to model human logic, natural or scientific inferences, (c) to simulate human competence in problem solving, and (d) to support natural dialogue regarding problems and actions. In linguistics such systems are used (e) to extract and classify speech semantics, (f) to make inferences basing on text semantics, and (d) to provide speech responses in a dialogue. In this publication it is only possible to give a bird's eye view over this vast scientific area.

Shank et al. have introduced *scripts* as a basic model for natural inferences in his classic works on conceptual processing [16]. Scripts allowed the system to model typical sequences of actions (like attending a restaurant), to reconstruct missing facts from a text and thus to support question answering on the missing data. The system, designed by Dorofeev and Martemianov [5], is another classic example of text comprehension engine: the system extracted semantic predications from a fairy-tale and constructed possible outcomes in each situation. For the action graph of the outcomes it anticipated *good* actions for the protagonist and *bad* actions for the antagonist, thus prognosing agents' actions. The system even had a concept of *soul*, which could be destabilized by external stimuli, forcing the agent to operate on text semantics until the *soul* is finally balanced. One of the first systems of conceptual representation linking text semantics and problem space – SHRDLU – was designed to handle the representation of a real situation (arrangement of blocks) and to suggest the appropriate actions [21]. Within the development of F-2 conceptual processing system, we mostly rely on SOAR architecture, designed for operations in problem space and dialogue support tasks [13]. SOAR also relies on scripts (productions) to consider the possible moves in the problem space. It may suggest moves to a user, once the script graph successfully reaches the target state (solution) of the problem.

Minsky has suggested to support emotion processing with a limited set of *proto-specialists* [14]. *Proto-specialists* of an agent suggest the reactions in case of danger or urgent lucrative opportunity. Sloman has extended this approach in his CogAff architecture: it was suggested that *reactions* and *alarms* (units of the basic *reactive* level) compete with "rational" inferences on the level of *deliberative reasoning* and with the processes of introspection on *meta-management* level [20]. Sloman has suggested that *rational processing* is more accurate in the recognition and provides better planning, while *reactions* are fast and shallow: they ensure quick response in critical situations. It was suggested that *secondary* or *tertiary* emotions may combine rational end emotional units, like, triggering an emotional response by a rational inference or by a meta-management process, constantly returning the thoughts to the emotional image. The compound nature of emotional responses has also been studied in linguistics. As noted by Sharonov, an emotional event may trigger numerous emotional and etiquette speech reactions, which linearize in time to the series of (a) interjections – *Oh!* (b) emotional evaluations – *God!* (c) emotional classifications – *What a mess!* (d) acquisition of speaker's responsibility – *What have I done!* and (e) etiquette replies – *I'm so sorry!* [18]. The order is defined by the processing difficulty, as the primer segments are more expressive and are generated faster, while latter segments require more resources and time.

Dialogue support systems and automatic question answering is another fast-growing approach to semantic representation and the simulation of reactions (dialogue turns). The classic papers by Jurafsky [8, 9] present a comprehensive overview of dialogue support and question-answering systems. Dialogue systems are usually divided into rule-based, information-retrieval and statistical systems [3, 8, 9], while modern systems combine the three approaches. It was shown that most of the participants at Amazon Alexa Prize competition [15] used rule-based approaches, while boosting the performance with neural networks and machine learning algorithms such as: Hierarchical Latent Variable Encoder-Decoder [17], a two-level Long Short-Term Memory network [1] and others. In this sense, a dialogue support system may be generally considered as production architecture, where an input utterance triggers a script and the best response is further selected.

3   F-2 Robot Conceptual Architecture

Robot may process texts or visual events at its input. Texts can be received from a text source (a text file or RSS subscription) or as an oral speech, in this case Yandex Speech API is used to decode the signal to the written form. The audio recognition may return several ambiguous results, which are processed in parallel up to the stage of scripts, where the preferred variant is selected. Robot is also equipped with two cameras (the number can be easily extended) and may also receive the information on the location and movement of different objects, in particular, it processes the location of faces and the location of game elements in Tangram puzzle. The general architecture of the system is represented on Fig. 1.

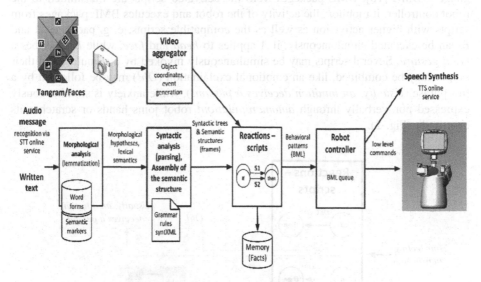

Fig. 1. General architecture of the conceptual processing system.

Similar to classic conceptual processors and dialogue support systems, we use a set of scripts at the core of the system to classify an input (both speech and visual) and to suggest an appropriate reaction. For each input the system selects the corresponding scripts: it calculates the distance between the input and scripts' premises, sorts the scripts following the reduction of the similarity and processes the topmost scripts from the list, e. g. executes the actions attached to the scripts. The scripts are divided into 3 groups: (a) *dominant scripts* or *d-scripts* for the emotional processing, like INADEQ for *They lie to you!* or INADEQ for *They are crazy!* [12], (b) *behavioral rules* – etiquette and dialogue support routines, which cause certain behavior in a given situation, like an excuse or a speech response, and (c) *rational scripts* or *r-scripts* for rational classification of input stimuli and the simulation of inference (perspective component). Within the proposed architecture, the scripts are used (a) as the dynamic model of emotions and reactions, expressed in verbal and nonverbal behavior of the robot, (b) as the representations of *regular situations* for the resolution of ambiguity, (c) as indexes for judgements in the memory base. The system relies on the semantic representations, constructed by syntactic parser and by visual recognition system. These representations are also saved to memory and may be retrieved for an output. In this sense, semantic representations link the components of speech perception, visual awareness, inference, memory, and performance. Although *production* approach is frequently contrasted to neural networks, we implement a network-like evaluation of semantic markers to calculate the distance between an input and scripts. Further, a neural network is used to evaluate binding within a syntactic tree in order to select "the best" trees in case of ambiguity.

A script may be annotated by a behavioral pattern: a combination of speech, facial expression and/or gestures. These patterns are described in Behavior Markup Language – BML [10]. BML packages from the activated scripts are transmitted to the robot controller, it monitors the activity of the robot and executes BML packages from scripts with higher activation as well as the compatible scripts, e. g. packages *A* and *B* can be executed simultaneously, if *A* applies to *head* and *face*, while *B* describes a *hand gesture*. Several scripts may be simultaneously invoked by a stimulus, and their output may be combined, like an emotional exclamation (*Oh!*) may be followed by a full phrase (*Usually, an intuition deceives a person!*), while anxiety is simultaneously expressed non-verbally through *automanipulation*: robot joins hands or scratches its own body (Fig. 2).

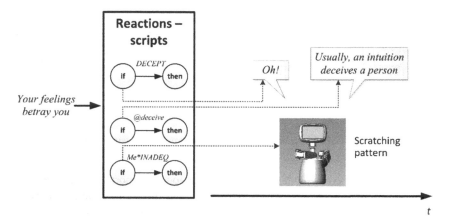

Fig. 2. Generation of behavioral patterns by different scripts in time.

3.1   Video Recognition

Video recognition should allow the robot to interact with a user as well as with a problem space. For the interaction with a user we use **face_recognition/dlib** libraries that enable the robot to detect faces in video stream and associate faces with known referents. The component generates events like 'John is_present' to invoke the reaction of interest on the robot – robot moves gaze direction towards the person, moves eyebrows, etc. To model the interaction with the user while solving a problem we have chosen *the Tangram puzzle*. We have implemented an interaction scenario, where the robot controls user moves through an automatic video recognition system and suggests optimal moves, guiding the human through the problem space according to a solution script. Robot stores the solution (or several alternative solutions), evaluates each move by the player as a positive or negative action in respect to the closest solution and suggests the next move. In case the user switches to another possible solution, the recognition system switches accordingly. Unlike in SOAR architecture, the solution graph is designed from top to bottom: starting from the solution combination (goal, final state) to the specific moves required in the situation (current state) (Fig. 3).

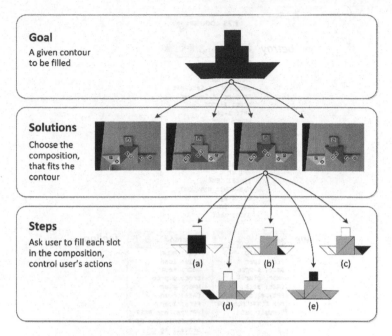

Fig. 3. Tangram solution tree.

The recognition system generates the events in a format similar to text semantic representations: 'user moved element *a*' and 'game elements *a, b, d* are correctly placed for the closest solution *Y*' (see 3.6). This, in turn, allows the robot to construct representations 'user moved element *a* correctly for the solution *Y*' and react to this event. These representations of user actions can be used not only by the reactions, but, potentially, by text synthesis system for flexible discussion on the user actions. The approach may also utilize an external problem-solving component, which constructs a solution for a given puzzle in a form of a script path and guides the user through the path or discusses the process of solution.

3.2   The Extraction of Valencies from Syntactic Structures

Each recognition result is processed by syntactic parser with the Russian grammar containing over 600 syntactic rules in syntXML format [11], as a result, a dependency syntactic tree is constructed (example in Fig. 4).

Speech processing module should represent the utterances in a form sufficient to calculate the distances to the premises of scripts. We rely on shallow semantic representations: semantics of a single clause is represented as a *semantic predication* – a set of semantic markers distributed between semantic valencies. A predicate is assigned to *p* (*predicate*) valency, while actants are assigned to *ag* (*agent*), *pat* (*patient*), *instr* (*instrument*) etc. – following an extended list by Fillmore [6]. For a compound sentence several semantic predications are constructed with co-reference links. In case of ambiguity, a set of syntactic trees is constructed, all the trees are processed in parallel

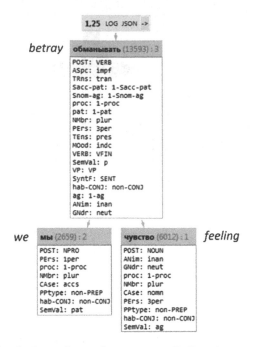

Fig. 4. Syntactic tree for a sentence *Feelings betray us.*

and evaluated by neural network to select the best trees for further steps of analysis. Semantics of each valency is the aggregation of semantic markers of all the words within the corresponding subtree, e. g. for a *noun phrase* a semantic representation is a sum of markers for the *noun* and all the *adjectives*. Lexical ambiguity is represented as a Cartesian product of semantic sets for ambiguous lexemes.

3.3   Semantic Markers in Valencies

Words in the semantic dictionary are annotated by semantic markers, so that similar words have the greater number of common markers. We use a set of 4778 markers consisting of (a) *focal markers* of d-scripts (see 3.4), like 'liar' – typical agent in the situation of deception (209 markers), (b) markers from a semantic dictionary [19], (669 markers), and (c) semantic markers assigned to words after clustering of *word2vec* vectors (3900 markers). Markers, based on *word2vec*, are assigned basing on clustering of nouns to 2000 clusters and 600 "superclusters". In this way we get a two level "ontology" with a "basic level" and a "top level" marker for each word. In a similar way, verbs were divided into 1000 clusters and 300 superclusters. Table 1 shows a semantic representation (predicate structure) for an utterance *Our feelings betray us.* Lexical semantics is distributed into "1 1"/ "1 2" slots (as verb *betray* may be a communicative or behavioral action). Markers assigned after *word2vec* clusters are indicated by "@".

Table 1. Semantic representation of *Our feelings betray us (/lie to us).*

Betray: P (predicate)	Our feelings: **Ag** (agens)	Us: **Pat** (patient)
1 1 present tense	1 1 many	1 1 somebody
1 1 assertive	1 1 abstract	1 1 egocentric – *me*
1 1 to communicate	1 1 negative emotions	1 1 includes another person
1 1 DECEPT attribute	1 1 positive emotions	1 1 physical object
1 1 to report	1 1 @feeling_NOUN	1 1 principal – *speaker*
1 1 @to_simulate	1 1 @173_NOUN	1 1 set of people
1 1 @214_VERB	1 1 own	
1 2 present tense		
1 2 assertive		
1 2 social action		
1 2 DECEPT attribute		
1 2 @to_simulate		
1 2 @214_VERB		

Within the research of conceptualization Barsalou has noted that the structure of a notion, e. g. *chair*, depends on the situation where the notion appears, e. g. *kitchen, cinema, hotel* or *move a chair, sit on a chair, buy a chair* [2]. This observation has been categorized with a set of rules of conceptualization [22]. Following these assertions, a set of semantic features within a notion (concept) is affected by the *frame* of the situation. A similar process is described by the psychology of emotions as emotional *top-down* processing: subjective representation of a situation can be changed by an invoked emotion [4], e. g. a person tends to overrate his ability to eat when being hungry, and overrates a threat or danger when being frightened. As an incoming event may contain not all the relevant markers, defined in script premise, *focal markers* [7] are applied to the incoming representation on a *top-down* basis: an input event is attracted by the scripts, e. g. "gets more emotional".

3.4 Input Processing with Scripts

Rational scripts (r-scripts) provide rational inference from the input predications and also contribute to the resolution of ambiguity. Each input predication is associated with an r-script, in case of ambiguity, the system selects the closest representation. To design the scripts, we have aggregated predicative structures, where all the words in each valency constitute a coherent set, e.g. {*team, sportsman, champion, player, hockey_-player*} *defeated* {*host, guest*} or {*candidate, promotee*} *defeated* {*mayor, governor*}. In this sense, each script searches for a prototype situation in the incoming stimuli. We have applied the following methods to construct the premises of scripts:

1. Following the clustering of verbs, we have selected 1391 verbs with relevant verb frequency > 0,02% (percent of all the verbs in the text corpus), but and at least one verb from each *word2vec* cluster. For each verb we have defined the most frequent words in each valency in the text corpus (over 80 million wordforms), collected by the previous runs of our parser. The results were manually inspected: in case words

from two (or more) different superclusters occupied a valency, the whole *frame* was divided into two or more frames.

2. For each verb in the dictionary we have selected facts, where all the words in *agent* valency belong to one supercluster, while words in *patient* valency (for transitive verbs) belong to another supercluster. So, different *frames* were constructed for different verb senses:

   (i) {*finger, palm, hand*} *gripes* {*shoulder, finger, palm, hand*}
   (ii) {*anxiety, fear, depression*} *gripes* {*neck*}

1619 scripts were constructed after the procedure. For r-scripts, a typical speech answer was designed as a speech output reaction as the sequence of words in the valencies. The scripts are associated with *microstates*, which change the sensitivity of scripts and simulate current emotional profile: *nervous, calm, motivated, hypocritical,* etc. An incoming stimulus invokes several scripts, which generate compound reactions, as represented in Table 2. As the sensitivity of each script is proportional to the activation of microstate, one can design (a) an *emotional mode,* when d-scripts DECEPT and Me*INADEQ are preferred, (b) a *rational mode,* where r-scripts are preferred, or (c) *aggressive* or *depressive* agent (DECEPT and Me*INADEQ are preferred respectively).

Table 2. Distribution of script activation for *Our feelings betray us (/lie to us)*

Similarity	Script	Speech output
0,2231	@deceive	[Usually] *An intuition deceives a person*
0,2131	DECEPT (d-script)	*Everybody lies!*
0,1593	@inflate/puff[1]	[Usually] *Girl puffs lips*
0,1561	Me*INADEQ (d-script)	*I do something stupid!*
0,1559	@lie	[Usually] *Man lies to a man*

[1]The script @inflate/puff is activated due to speech homonymy and may be used for a speech game or humorous response.

The order of scripts at the top of activation list is sensitive to the distribution of valencies in a particular communication. In case the phrase addresses *you*, not *us* (e. g. *Feelings betray you*), the activation of scripts is more emotional: DECEPT (d-script) is the leading, as the robot personally interprets incoming phrases with the pronoun *you*. Scripts DECEPT, @deceive and Me*INADEQ can construct multimodal behavior for the robot in time; @inflate/puff and @lie scripts are suppressed, as belonging to the same microstate as @deceive. In this case DECEPT is getting the highest initial activation due to its high sensitivity and generates an interjection thus discharging the activation. @deceive generates a verbal response and Me*INADEQ generates automanipulation as a sign of confusion. Since DECEPT and @deceive both generate speech outputs, they compete in time for the robot's mouth and are expressed one after another (Fig. 2). A script may generate many behavioral response patterns for different parts of the robot's body – these patterns are kept in buffer, and compatible patterns (e. g. corresponding to different parts of the body) are selected for execution.

3.5   Memory

The semantic representation of input sentences is indexed and saved to a database to support long-term memory and perspective question answering. A base of 1 million facts is automatically collected for 80 million wordform corpus. Phrases are indexed by the script, used to select the meaning, and can be retrieved by the script index. The base can return all the predications, which correspond to the script premise, or an arbitrary semantic pattern, for example, corresponding to the semantics of an input question. Table 3 represents sentences, returned for the pattern 'feelings deceive'. We suggest that question answering may retrieve utterances from the array of all the analyzed texts, or from a subbase with manually inspected utterances, or with judgements from a trusted source (e. g. a schoolbook). An advantage of the method is that the knowledge of an agent is directly enriched through the automatic analysis of incoming texts, without any retraining. We also expect that scripts can provide more flexible Q&A by training on Q&A pairs, where a script premise corresponds to the question and script inference corresponds to the answer.

Table 3.   Judgements for *Our feelings betray us* **semantic pattern in the memory base**

Fact	Sentence
1094963	*Anger troubled him, pushed him to insolence.*
2527406	*The fact is that physical sensations deceive them, because...*
8446093	*No, you know that feelings do not deceive you.*
146731	*Is it possible, that the sixth sense deceives him?*

4   Conclusion

We suggest that each object and action within the surrounding of the robot may be represented by a list of markers and the structure of the situation – as a distribution of the referents between the valencies. Text semantics and visual recognition results may thus be represented in a unified way. We consider the main contribution of the work as a practical implementation of proto-specialists classic theory and the extension of CogAff architecture to the area of semantic processing and event representation in semantic form. Further, the interaction between the incoming representations and scripts provides a practical implementation of the theory of situated conceptualization, as the input semantics changes depending on the chosen reaction (script). The general architecture inherits a classic *production* approach and simulates balanced and parallel rational/emotional processing of contradictory reactions. This allows the robot to construct compound behavioral patterns (blending reactions), as suggested within the linguistic analysis of multimodal communication. The whole design of conceptual processing system is implemented and works in a standalone mode on a server, accumulating and processing news and blogs on an everyday basis, as well as on F-2 robot in the Tangram support mode, and a dialogue mode. In sum, the combination of *semantic predications* as the form of representation, and *scripts* as a processing architecture, suggests a cognitive architecture with features, essential for a companion robot.

Acknowledgment. The present study has been supported by the Russian Science Foundation, project No 19-18-00547.

References

1. Adewale, O., et al.: Pixie: a social chatbot. Alexa Prize Proceedings (2017)
2. Barsalou, L.W.: Frames, concepts, and conceptual fields. In: Hillsdale, N.J. (ed.) Frames, Fields, and Contrasts: New Essays in Semantic and Lexical Organization, pp. 21–74. L. Erlbaum Associates (1992)
3. Cahn, J.: CHATBOT: Architecture, design, & development. University of Pennsylvania School of Engineering and Applied Science Department of Computer and Information Science (2017)
4. Clore, G.L., Ortony, A.: Cognition in Emotion: Always, Sometimes, or Never? Cognitive Neuroscience of Emotion, pp. 24–61. Oxford University Press (2000)
5. Dorofeev, G.V., Martemyanov, Y.: The logical conclusion and identification of the relationships between sentences in the text. Mach. Translation Appl. Linguist. **12**, 36–59 (1969)
6. Fillmore, C.J.: The Case for Case. Universals in Linguistic Theory, pp. 1–68. Holt, Rinehart & Winston, New York (1968)
7. Glovinskaya, M.Y.: Hidden hyperbole as a manifestation and justification of verbal aggression. Sacred meanings: word. Text. The culture. Languages of Slavic culture, pp. 69–76 (2004)
8. Jurafsky, D.: Speech & Language Processing. Pearson Education India (2000)
9. Jurafsky, D., Martin, J.H.: Speech and Language Processing (draft). Chapter Dialogue Systems and Chatbots (Draft of October 16, 2019) (2019)
10. Kopp, S., et al.: Towards a common framework for multimodal generation: the behavior markup language. In: Gratch, J., Young, M., Aylett, R., Ballin, D., Olivier, P. (eds.) IVA 2006. LNCS (LNAI), vol. 4133, pp. 205–217. Springer, Heidelberg (2006). https://doi.org/10.1007/11821830_17
11. Kotov, A., Zinina, A., Filatov, A.: Semantic parser for sentiment analysis and the emotional computer agents. In: Proceedings of the AINL-ISMW FRUCT 2015, pp. 167–170 (2015)
12. Kotov, A.A.: Description of speech exposure in a linguistic. In: Computer Linguistics and Intelligent Technologies, Nauka, pp. 299–304 (2003)
13. Laird, J.E., Newell, A., Rosenbloom, P.S.: SOAR: an architecture for general intelligence. Artif. Intell. **33**(1), 1–64 (1987)
14. Minsky, M.L.: The Society of Mind. Touchstone Book, New-York (1988)
15. Ram, A., et al.: Conversational AI: the science behind the alexa prize. arXiv preprint arXiv: 1801.03604 (2018)
16. Schank, R.C., Abelson, R.P.: Scripts, Plans, Goals, and Understanding: An Inquiry into Human Knowledge Structures. L. Erlbaum Associates, Hillsdale (1977)
17. Serban, I.V., et al.: The octopus approach to the Alexa competition: a deep ensemble-based socialbot. In: Alexa Prize Proceedings (2017)
18. Sharonov, I.A.: Interjection in speech, text, and dictionary. RGGU (2008)
19. Shvedova, N.Y.: Russian semantic dictionary. Explanatory dictionary systematized by classes of words and meanings. Azbukovnik (1998)

20. Sloman, A., Chrisley, R.: Virtual machines and consciousness. J. Conscious. Stud. **10**(4–5), 133–172 (2003)
21. Winograd, T., Flores, F.: Understanding Computers and Cognition: A New Foundation for Design. Addison-Wesley (1987)
22. Yeh, W., Barsalou, L.W.: The situated nature of concepts. Am. J. Psychol. **119**(3), 349–384 (2006)

Legal Tech: Documents' Validation Method Based on the Associative-Ontological Approach

Sergey Kuleshov(iD), Alexandra Zaytseva(✉)(iD),
and Konstantin Nenausnikov(iD)

St.-Petersburg Federal Research Center of the Russian Academy of Sciences,
St.-Petersburg, Russia
cher@iias.spb.su

Abstract. Trend of Legal Tech is actively developing, thus, allowing an automation of various legal tasks solved both by professional lawyers and ordinary (common) users. Because of the legal documents properties, natural language processing (NLP) technologies are widely used in Legal Tech development. One of the tasks in Legal Tech, whose solution is necessary for both professionals and non-professionals, is a validation documents' texts, including certain checking for the presence of mandatory structural elements in them. This article considers an implementation of a method and an algorithm for validating, for instance, the documents called "Consent to the personal data processing" in the Russian Language legal practice based on machine learning and using an associative-ontological representation of the text. Such validation occurs by checking documents through a set of rules, at that, each rule describes the documents' structural elements. Associative ontological representation of the text makes such rules human-readable, and simplifies their adjustment and fine-tuning to the changing legislation norms. Results of experimental verification of the proposed algorithm on a set of texts of real legal documents show its effectiveness when applied to Legal Tech systems.

Keywords: Legal tech · Natural language processing · Semantic graph

1 Introduction

The modern world tends to a global services' softwarization [1] aimed at the replacement of human specialists by digital services providing similar results. Against this background, a new IT direction has emerged – Legal Tech.

Legal Tech is an industry sector specializing in information-technical support of professional legal activity [2]. The main goal of Legal Tech solutions development is an automation of the jurisprudence system and providing an access to its services both for lawyer specialists and for those who cannot afford lawyers. The official USA statistics show that 80% of the population cannot afford any juridical protection in civil matters [3]. In addition, automation decreases the amount of routine work for lawyers and thus redirects their efforts to problems demanding creative approach.

A. Karpov and R. Potapova (Eds.): SPECOM 2020, LNAI 12335, pp. 244–254, 2020.
https://doi.org/10.1007/978-3-030-60276-5_25

Development of Legal Tech industry is uneven in different countries. The most progressive Legal Tech development is taking place in the USA. The current Russian market is falling behind. Local lawmaking procedures could explain such inconsistency.

Most of available Legal Tech services are oriented to automation and softwarization of people's interaction with lawyers: Wevorce, UpCounsel services are available in the USA and Pravoved.ru in the Russian Federation. The main objective of these systems is to provide the users' fastest access to professional legal defense. Currently it is the easiest and most effective way to solve legal problems both applicable to simple and complex ones. The main disadvantage of these services is a high cost of a high participation is not always necessary.

In case of a simple task like filing a typical claim, the cost can be reduced by applying the methods of automatic document formation. For example, the DoNotPay (donotpay.com) service helps car owners to compose and send an appeal for receipt of a fine and a VisaBot service (visabot.co) helps to collect the necessary documents for the USA visa extension and acquiring of the green card. In such systems, the correct document formation is provided by usage of template forms. It is worth noting that such products are only oriented to solving some specific problem that is simple and stays in a consistently high demand. The problems without formalized algorithmic solutions cannot be solved by the approaches described above.

The following services are having the most coverage of "typical" document formation are based on questioning: Rocket Lawyer (www.rocketlawyer.com) and Legalzoom (www.legalzoom.com) for the American market and FreshDoc (www.freshdoc.ru) for the Russian. These systems allows for a partial automation of the processes like business creation, consents' and agreements' drafting, new trademark registration', tax filing, patenting and others.

The specialist's participation is necessary to compose somewhat more complex documents such as applications for a property division in a divorce. Some systems provide information in a form facilitating this work for the further processing: Lexoo (www.lexoo.co.uk/how-it-works) for the American market and «ConsultantPlus» (www.consultant.ru/about), «Garant.ru» (www.garant.ru) for the Russian. Right now the systems of a type are widely distributed among the specialists, and are getting popular among non-professional users. Their main advantages are the availability of a high number of sorted by sections and connected via hyperlinks documents, examples for consent drafting, links to lawyers and specialists comments. This set of features decreases the time for legal document processing by speeding up searching for needed templates and normative documents. Unlike the systems based on questioning this approach requires some education and effort from users, but from the lawyer's point of view, this type of work is still a routine procedure.

Another sophisticated approach is a Thomson Reuters Westlaw (legal.thomsonreuters.com) product that emerged in the USA. This software includes a complete database of legislation for 60 countries including localized Russian version, templates for consents and legal expertise. The changes in active legislation are presented as a demonstrative graph. At the document creation, the links to other documents based on keywords analysis are being automatically generated. The found links to other documents then are being rated and marked if they have a controversial interpretation or had been withdrawn. The closest Russian analogue for this system is a product of Preferentum company «Systema Jurist» (https://www.1jur.ru/). Its distinctive feature is a division of a document by the regions by a rigid structure that allows controlling the completeness and possible errors. This control is based on comparison with the template document selected from the database. Note that the main problem is the search by words since it can lead to a large number of errors.

Due to the properties of legal documents, Legal Tech development is closely related to natural language processing (NLP) technologies.

In our opinion, tasks that potentially can get benefits by implementing NLP technologies are:

- the task of automated intellectual analysis of a legal document set for a company audit;
- the task of inside sabotage detecting;
- the automation of the consent documents creation task with given parameters;
- the task of obtaining new knowledge regarding contractors based on available Internet sources;
- the task of semantic search and information analysis in arbitral awards databases;
- the task of validating legal documents to reach compliance with various requirements including language norms, legal compliance, semantic integrity and the absence of external and internal logical disagreements.

The list is not limited to these tasks; however, their solutions are most active.

Possible solutions of these tasks mean an analysis of employees' behavior by some statistical approach and the graph theory methods, the implementation of NLP document analyzer, associative-ontology data model, automated ontology creation, semantic search and data mining [4, 7–10]. For instance, [5] traces the impact of governments on words or processes in society, [6] discusses the topic modelling use and some other computational methods at measuring the contribution to topics in social groups of each user.

This paper discusses the solution to the task of validating legal documents, for instance, "Consents to the personal data processing", to reach compliance with various requirements including legal compliance, semantic integrity and the absence of external and internal logical disagreements. To solve this problem, we use machine learning, an associative-ontological approach and NLP methods.

2   Problem of Document Validation System Development

When validating a document, an expert performs a check of the language norms, possible ambiguous expressions and compliance with legal and ethical standards.

It is also necessary to check the document for the presence of a specific set of obligatory and desirable statements, to be unambiguously determined by all parties to the consent to comply with legal norms.

Part of the documents validation is reduced to checking for the availability and correctness sections consistent with the provisions of specific statements or details of the being validated legal document. A routine for an expert that is a check for the presence of sections of documents structural elements, necessary for each particular case, which takes quite a time and can be automated through NLP methods [11, 12].

To date, the dominant approach to automatic documentation analysis remains statistical analysis based on the occurrence of words [13]. Although there exists a specific finite set of standard templates for expressions, the formulations are different, which leads to low accuracy of this approach. Instead, to check the texts of legal documents, we develop methods and algorithms based on a comparison of structural elements automatically selected from the text with a template that determines the structure of necessary and variable elements.

The Russian Federation has its standards for many legal documents, including consent to the personal data processing [14]; therefore the systems being developed for the Russian segment of LegalTech consider the features of Russian legal system [15–17].

To solve the problem of checking the legal documents for the presence of essential sections, the following principles are proposed hereby:

- the set of rules is identifying structural elements;
- rules are being formed based on the description of the real legal document structure elements and the examples of valid and invalid usage provided by lawyers;
- obtained rules have interpretable properties and can be edited by the user.

Structure templates for the necessary and variable sections are being formed manually based on federal legislation and existing legal rules.

The rules are built based on a manually defined structure template and are presented in the form of a semantic graph. The construction can be done manually by entering the correct text of the structural element or editing a semantic graph describing the structural element, as well as using automatic training on fragments of marked-up real documents based on the associative ontological approach.

In general, the document validation system should include the following modules:

- the NLP subsystem for the Russian language;
- the rule-based subsystem of text documents validation;
- the web interface for task acquiring and result producing;
- the rule management subsystem defining the structural composition of the document;
- the rule management toolbox for the system administrator.

Figure 1 shows the structural diagram of the system.

3   The Method and Algorithm for Rule-Based Text Validation

The existing approaches to thematic modelling based on principles of statistics or machine learning require large volumes of text for reliable clustering and cannot be applied to short and ultra-short fragments of text that correspond to the structural elements of legal documents.

NLP defines a topic as a semantically cohesive construct separated from the surrounding text by some features. In the simplest case, this is a change in the general

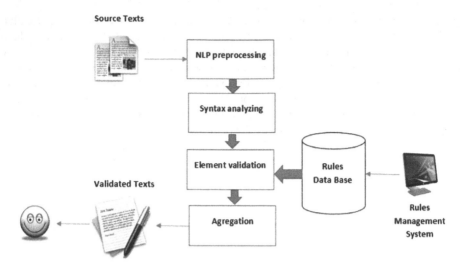

Fig. 1. The general structure of the document validation system.

terminology, in the more complex case, the redefinition of the structure value by changing the critical points of its component system. Existing approaches involve manually adjusting the size and content of topics, i.e. selection of necessary coefficients for given buildings [5, 6, 18].

Thematic modelling algorithms are based on the LSI and LDA approaches [18, 19] and can be used both for specialized and common topics analysis [5, 18].

In the classical form, approaches involve the analysis of large-volume text corps due to the principle of highlighting topics. Algorithms work the better, the more coherent and separate alleged topics are. To apply approaches to small texts: tweets, messages, microblogs and others, the use of additional information, for example, the distribution of topics for each word or hashtags [20–22].

In contrast, structural components are not separate elements. Many of them are located in one topic; therefore, it is incorrect to determine them by the considered methods, which means that their processing requires a new approach.

To increase the accuracy of small texts processing, it is proposed to represent the semantic invariant of the topic in the graph model form [23]. The description of the topic (structural section) is represented in the form of a graph. The nodes of such graph display the words, and the edges represent the connections between them. The model does not contain hierarchical levels and based on undirected graphs (graphs with all possible connections that are not duplicated and have no direction), herein graph nodes are the notions, and the edges are associative links. Every notion contains information about its frequency of occurrence in analyzed text and its context variations. Every link contains the measure of given terms cohesion. For the first time, similar approach was used in [24].

The following requirements are imposed on of the model to maintain the relevance: adaptability of model to processing of new texts, visibility, and the adjusting ability by the user.

Every structural element contains a semantic invariant represented by the graph of the associative links of the notions.

Two techniques can obtain the rule descriptions:

- as a result of an automatic learning process on the sets of valid and invalid texts;
- composed manually by using the dedicated toolbox.

Proposed method of knowledge representation possesses the property of interpretability, which makes it possible to use assessors to adjust the operation of algorithms.

Development of an algorithm for constructing the rules for document validation and their adjustment was carried out for documents containing consent to the personal data processing.

Processed text of the document after preliminary preprocessing, comprising lemmatization and reduction of all words to normal form, was converted into a graph representation [24] – a document semantic graph model. The rules are formed based on the structural elements of a document containing consent to the personal data processing provided by legal experts. In total, 16 structural elements were obtained by experts based on an analysis of existing standard types of documents containing consent to the personal data processing [12]. Syntactic integrity and the absence of logical contradictions are checked using other types of rules.

Structural elements can be either mandatory sections of a document on consent to the personal data processing: "Name of a document", "Information about the subject of personal data", "Ordinary personal data", "List of actions with personal data", "Purpose of processing", either variable: "Information about the representative of the personal data subject", "Special personal data", "Biometric personal data", "Instruction of processing or transferring personal data to a third party".

All rules are presented in graph form. Automatic generation of rules in the machine learning is performed in the NLP processing module. The NLP processing module presents sentences from texts in the form of a graph where words are nodes. Figure 2 shows a significant part of the code. Figure 3 present the example of automatically generated rule with machine learning for one of the structural elements.

Further, each of the rules corresponding to the structural elements is applied to the graph model of a document by searching for the occurrence of the rule graph in the document graph. The comparison algorithm is described in [23] and presented in Fig. 4 as a pseudo-code. Figure 5 illustrates the principle of the ontology graph comparison algorithm.

In case of a positive search result, the rule is considered fulfilled, and the corresponding flag of the presence of a structural element in the document is set.

The result of the proposed algorithm is a list of structural elements as follows:

1. found in the analyzed document;
2. falsely found in the document;
3. not found.

Input: text T=(S,W), s={w$_i$,w$_j$,..} –sentences (set of words), w$_i$∈W –words, B – set of stop words
Output: undirectred graph G=(V,E), vertex w∈V, edge (w$_i$,w$_i$) ∈E, l[i,j] – positive edge lengths
procedure TextProcess()
for each s∈T:
      for each w[i]∈s:
          for each w[j]∈s:
          if i≠j **and** w$_i$ ∉B w$_j$, ∉B:
              v[i]← w$_i$
              v[j]← w$_i$
              e[i,j]← (w$_i$,w$_i$)
              l[i,j] ← l[i,j]+1
          end for
      end for
end for
end procedure

Fig. 2. Pseudo-code of the graph generation algorithm

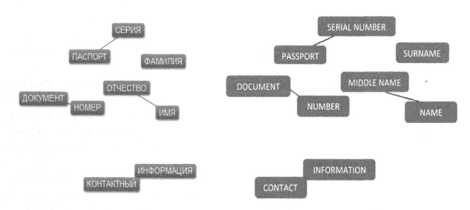

Fig. 3. The rule for an "Ordinary personal data" structural element in Russian and its translation in English.

4   Experimental Results

For machine learning by the developed algorithm, a dataset consisting of 62 different types of documents "Consent to the personal data processing" used by various organizations of the Russian Federation was employed. The dataset is formed, with due account for the opinions of experts on each type of documents correctness.

For experimental verification, we developed dataset from 62 different types of documents with adding of details in structural elements. The testing dataset consists of about 500 different documents.

Rules' accuracy for all 16 document sections (the system detected a structural element in the text that contains it, and did not detect the structural element in the text that does not contain it) was 0.76. Separately, the accuracy of the rules formed manually was 0.68, and the accuracy of the rules generated automatically in the training

Input: text T=(S,W), s={w$_i$,w$_j$,...} –sentences (set of words), w$_i$∈W –words, rule graph G=(V,E)
Output: result – "true" if a structural element is present
procedure TextProcess()
count ← 0
for each s∈T:
    for each w[i]∈s:
        for each w[j]∈s:
        if i≠j **and** (w$_i$,w$_j$) ∈E:
            count ← cout +1
        end for
    end for
end for
if count > threshold
    result ← true
else
    result ← false
end procedure

Fig. 4. The significant part of the code for the ontology graph comparison algorithm.

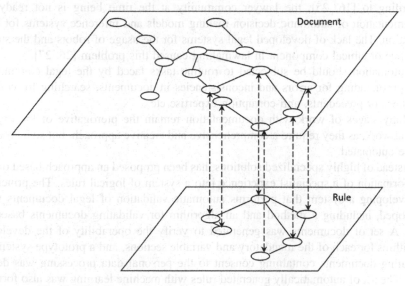

Fig. 5. Visualization for the ontology graphs comparison algorithm.

mode was 0.82. Table 1 shows data on the accuracy of the algorithm on the testing dataset separately for some of the rules for structural elements.

Table 1. Experimental results.

#	Name of the structural element	Rule formation method	Structural element correct validation value
1	Ordinary personal data	Machine learning	0,92
2	Special personal data	Machine learning with manual correction	0,81
3	Biometric personal data	Machine learning with manual correction	0,92
4	Set of actions with personal data	Machine learning	0,97
5	The term the consent is valid for	Machine learning	0,95

5 Conclusion

According to [16, 25], the lawyer community at the time being is not ready for implementation of automatic decision-making models and inference systems for bills formation. The lack of developed legal systems for the usage of robots and the strong influence of ethical component in lawmaking caused this problem [26, 27].

Automation should be subjected to routine tasks faced by the legal community: local proofreading for errors and inconsistencies in documents, searching for conflict with laws or precedents, anti-corruption expertise etc.

Many stages of work with documentation remain the prerogative of the experts manual work, as they require a comprehensive and creative approach, but some routine can be automated.

Instead of highly specialized solutions, has been proposed an approach based on the transformation of a specialist experience into a system of logical rules. The principles of developing a system that performs automatic validation of legal documents were developed, including a method and an algorithm for validating documents based on rules. A set of documents was generated to verify the operability of the developed algorithms for each of the mandatory and variable sections, and a prototype system for validating documents containing consent to the personal data processing was developed. The set of automatically generated rules with machine learning was also formed.

Experimental verification has shown the applicability of the associative-ontological approach to the problem of identification and validation of structural elements in legal documents containing consent to the personal data processing. The highest identification accuracy was obtained on structural elements with less variability of the text and with specific terms (for example, in a structural element "Biometric personal data"). In some structural elements, due to the high variability of the text in different types of documents, manual adjustment of the created rules was required.

Further research includes the development of a full-fledged web service, including a system for validating documents containing consent to the personal data processing with the capacity of integration into modern Legal Tech systems.

Acknowledgments. The research is granted by the budget (project no.0073-2019-0005).

References

1. Kuleshov, S.V., Yusupov, R.M.: Is softwarization the way to import substitution? SPIIRAS Proc. **3**(46), 5–13 (2016) (in Russian)
2. LegalTech News. https://www.law.com/legaltechnews. Accessed 20 May 2020
3. Voitenko, V.: The Future with the IT startup Legal Tech. https://kod.ru/legaltech-part01/. Accessed 20 May 2020. (in Russian)
4. Qin, Z., Yu, T., Yang, Y., Khalil, I., Xiao, X., Ren, K.: Generating synthetic decentralized social graphs with local differential privacy. In: Proceedings of the 2017 ACM SIGSAC Conference on Computer and Communications Security (CCS 2017), pp. 425–438. Association for Computing Machinery, New York (2017). https://doi.org/10.1145/3133956. 3134086
5. Carter, D., Brown, J., Rahmani, A.: Reading the high court at a distance: topic modelling the legal subject matter and judicial activity of the high court of Australia, 1903–2015. UNSW Law J. **39**(4), 1300–1352 (2018). https://doi.org/10.31228/osf.io/qhezc
6. Carron-Arthur, B., Reynolds, J., Bennett, K., et al.: What's all the talk about? Topic modelling in a mental health Internet support group. BMC Psychiatry **16**, 367 (2016). https://doi.org/10.1186/s12888-016-1073-5
7. Huang, J., Zhou, M., Yang, D.: Extracting chatbot knowledge from online discussion forums. In: 20th International Joint Conference on Artificial intelligence, pp. 423–428 (2007)
8. Diefenbach, D., Lopez, V., Singh, K., Maret, P.: Core techniques of question answering systems over knowledge bases: a survey. Knowl. Inf. Syst. **55**(3), 529–569 (2017). https://doi.org/10.1007/s10115-017-1100-y
9. Bast, H., Bjorn, B., Haussmann, E.: Semantic search on text and knowledge bases. Found. Trends® Inf. Retrieval **10**(2–3), 119–271 (2016). https://doi.org/10.1561/1500000032
10. Bova, V.V., Leshchanov, D.V.: The semantic search of knowledge in the environment of operation of interdisciplinary information systems based on ontological approach. Izvestiya SFedU. Engineering Sciences **7**(192), 79–90 (2017). (In Russ)
11. Zhang, J., El-Gohary, N.M.: Semantic NLP-based information extraction from construction regulatory documents for automated compliance checking. J. Comput. Civ. Eng. **30**(2), 1–14 (2016). https://doi.org/10.1061/(ASCE)CP.1943-5487.0000346
12. Bues, M.-M., Matthaei, E.: LegalTech on the rise: technology changes legal work behaviours, but does not replace its profession. In: Jacob, K., Schindler, D., Strathausen, R. (eds.) Liquid Legal. MP, pp. 89–109. Springer, Cham (2017). https://doi.org/10.1007/978-3-319-45868-7_7
13. Kharlamov, A.A., Yermolenko, T.V., Zhonin, A.A.: Process dynamics modeling on the base of text corpus sequence analysis. Eng. J. Don **4**(27) (2013) http://www.ivdon.ru/en/magazine/archive/n4y2013/2047. Accessed 25 May 2020. (in Russian)
14. Federal Law of July 27, 2006 N 152-FZ (as amended on December 31, 2017) "On Personal Data". https://base.garant.ru/12148567/. Accessed 20 May 2020. (in Russian)
15. RusBase homepage. https://rb.ru/longread/legal-russia/. Accessed 20 May 2020. (in Russian)
16. Lim, K.W., et al.: Twitter-network topic model: a full bayesian treatment for social network and text modeling. ArXiv abs/1609.06791 (2016)
17. Skolkovo LegalTech. Black Edition (2019). https://sklegaltech.com/. Accessed 20 May 2020. (in Russian)

18. Platforma Media. https://platforma-online.ru/media/detail/maksim-shchikolodkov-produkt-dolzhen-reshat-konkretnuyu-problemu-ili-zakryvat-potrebnost-/?lang=ru. Accessed 20 May 2020. (in Russian)
19. Vorontsov, K.V.: Additive regularization for hierarchical multimodal topic modeling. J. Mach. Learn. Data Anal. 2(2), 187–200 (2016). https://doi.org/10.21469/22233792.2.2.05
20. Chemudugunta, C., Padhraic, S., Steyvers, M.: Modeling general and specific aspects of documents with a probabilistic topic model. Adv. Neural. Inf. Process. Syst. 19, 241–248 (2007)
21. Lim, K.W., Chen, C., Buntine, W.: Twitter-network topic model: a full bayesian treatment for social network and text modeling. ArXiv abs/1609.06791 (2016). https://arxiv.org/pdf/1609.06791.pdf. Accessed 25 May 2020
22. Yan, X., Guo, J., Lan, Y., Cheng, X.: A biterm topic model for short texts. In: Proceedings of the 22nd International Conference on World Wide Web (WWW 2013), pp. 1445–1456. Association for Computing Machinery, New York (2013). https://doi.org/10.1145/2488388.2488514
23. Zuo, Y., Zhao, J., Xu, K.: Word network topic model: a simple but general solution for short and imbalanced texts. Knowl. Inf. Syst. 48(2), 379–398 (2015). https://doi.org/10.1007/s10115-015-0882-z
24. Kuleshov, S., Zaytseva, A., Aksenov, A.: Natural language search and associative-ontology matching algorithms based on graph representation of texts. In: Silhavy, R., Silhavy, P., Prokopova, Z. (eds.) CoMeSySo 2019 2019. AISC, vol. 1046, pp. 285–294. Springer, Cham (2019). https://doi.org/10.1007/978-3-030-30329-7_26
25. Kuleshov, S.V., Zaytseva, A.A., Markov, S.V.: Associative-ontological approach to natural language texts processing. Intellect. Technol. Transp. 4, 40–45 (2015). (In Russ)
26. Moscow Legal Tech 2018. http://moscowlegal.tech/. Accessed 20 May 2020. (in Russian)
27. Ivanov, A.A.: Penetration of mechanisation in law. Zakon 5, 35–41 (2018). (In Russ)

Audio Adversarial Examples for Robust Hybrid CTC/Attention Speech Recognition

Ludwig Kürzinger(✉)[iD], Edgar Ricardo Chavez Rosas, Lujun Li[iD],
Tobias Watzel[iD], and Gerhard Rigoll[iD]

Chair of Human-Machine Communication, Technical University of Munich,
Munich, Germany
{ludwig.kuerzinger,ricardo.chavez}@tum.de

Abstract. Recent advances in Automatic Speech Recognition (ASR) demonstrated how end-to-end systems are able to achieve state-of-the-art performance. There is a trend towards deeper neural networks, however those ASR models are also more complex and prone against specially crafted noisy data. Those Audio Adversarial Examples (AAE) were previously demonstrated on ASR systems that use Connectionist Temporal Classification (CTC), as well as attention-based encoder-decoder architectures. Following the idea of the hybrid CTC/attention ASR system, this work proposes algorithms to generate AAEs to combine both approaches into a joint CTC-attention gradient method. Evaluation is performed using a hybrid CTC/attention end-to-end ASR model on two reference sentences as case study, as well as the TEDlium v2 speech recognition task. We then demonstrate the application of this algorithm for adversarial training to obtain a more robust ASR model.

Keywords: Adversarial examples · Adversarial training · ESPnet · Hybrid CTC/Attention.

1 Introduction

In recent years, advances in GPU technology and machine learning libraries enabled the trend towards deeper neural networks in Automatic Speech Recognition (ASR) systems. End-to-end ASR systems transcribe speech features to letters or tokens without any intermediate representations. There are two major techniques: 1) Connectionist Temporal Classification (CTC [13]) carries the concept of hidden Markov states over to end-to-end neural networks as training loss for sequence classification networks. Neural networks trained with CTC loss calculate the posterior probability of each letter at a given time step in the input sequence. 2) Attention-based encoder-decoder architectures such as [7],

---

L. Kürzinger and E. R. C. Rosas—Contributed equally to this work.

A. Karpov and R. Potapova (Eds.): SPECOM 2020, LNAI 12335, pp. 255–266, 2020.
https://doi.org/10.1007/978-3-030-60276-5_26

are trained as auto-regressive sequence-generative models. The encoder transforms the input sequence into a latent representation; from this, the decoder generates the sentence transcription. The hybrid CTC/attention architecture combines these two approaches in one single neural network [31].

Our work is motivated by the observation that adding a small amount of specially crafted noise to a sample given to a neural network can cause the neural network to wrongly classify its input [28]. From the standpoint of system security, those algorithms have implications on possible attack scenarios. A news program or sound that was augmented with a barely noticeable noise can give hidden voice commands, e.g. to open the door, to the ASR system of a personal assistant [5,6]. From the perspective of ASR research, a network should be robust against such small perturbations that can change the transcription of an utterance; its speech recognition capability shall relate more closely to what humans understand.

In speech recognition domain, working Audio Adversarial Examples (AAEs) were already demonstrated for CTC-based [6], as well as for attention-based ASR systems [27]. The contribution of this work is a method for generation of untargeted adversarial examples in feature domain for the hybrid CTC/attention ASR system. For this, we propose two novel algorithms that can be used to generate AAE for attention-based encoder-decoder architectures. We then combine these with CTC-based AAEs to introduce an algorithm for joint CTC/attention AAE generation. To further evaluate our methods and exploit the information within AAEs, the ASR network training is then augmented with generated AAEs. Results indicate improved robustness of the model against adversarial examples, as well as a generally improved speech recognition performance by a moderate 10% relative to the baseline model.

2   Related Work

Automatic Speech Recognition (ASR) Architecture. Our work builds on the hybrid CTC/attention ASR architecture as proposed and described in [30,31], using the location-aware attention mechanism [9]. This framework combines the most two popular techniques in end-to-end ASR: Connectionist Temporal Classification (CTC), as proposed in [13], and attention-based encoder-decoder architectures. Attention-based sequence transcription was proposed in the field of machine language translation in [4] and later applied to speech recognition in Listen-Attend-Spell [7]. Sentence transcription is performed with the help of a RNN language model (RNNLM) integrated into decoding process using shallow fusion [14].

Audio Adversarial Examples (AAEs). Adversarial examples were originally porposed and developed in the image recognition field and since then, they have been amply investigated in [18,19,28]. The most known method for generation is the Fast Gradient Sign Method (FGSM) [12]. Adversarial examples can be prompt to *label leaking* [19], that is when the model does not have difficulties finding the original class of the disguised sample, as the transformation from the original is "simple and predictable". The implementation of AAEs in ASR systems has been proven to be more difficult than in image processing [10]. Some of them work irrespective of the architecture [1,22,29]. However, these examples are crafted and tested using simplified architectures, either RNN or CNN. They lack an attention mechanism, which is a relevant component of the framework used in our work. Other works focus on making AAEs remain undetected by human subjects, e.g., by psychoachustic hiding [23,25]. Psychoacoustic hiding limits adversarial noise within the spectral envelope of the human hearing threshold derived from MP3 compression; in converse, as a countermeasure, MP3 compression was shown to reduce adversarial noise [3]. Carlini et al. [5] demonstrated how to extract AAEs for the CTC-based DeepSpeech architecture [15] by applying the FGSM to CTC loss. Hu et al. gives a general overview over adversarial attacks on ASR systems and possible defense mechanisms in [16]. In it, they observe that by treating the features matrix of the audio input as the AAE seed, it is possible to generate AAE with algorithms developed in the image processing field. However, this leads to the incapacity of the AAE to be transformed back to audio format, as the feature extraction of log-mel f-bank features is lossy. Some have proposed ways to overcome this problem [2,3]. AAEs on the sequence-to-sequence attention-based LAS model [7] by extending FGSM to attention are presented in [27]. In the same work, Sun et al. also propose adversarial regulation to improve model robustness by feeding back AAEs into the training loop.

3   Audio Adversarial Example (AAE) Generation

The following paragraphs describe the proposed algorithms to generate AAEs (a) from two attention-based gradient methods, either using a static or a moving window adversarial loss; (b) from a CTC-based FGSM, and (c) combining both previous approaches in a joint CTC/attention approach. In general, those methods apply the single-step FGSM [12] on audio data and generate an additive adversarial noise $\delta(x_t)$ from a given audio feature sequence $X = x_{1:T}$, i.e.,

$$\hat{x}_t = x_t + \delta(x_t), \quad \forall t \in [1, T]. \tag{1}$$

We assume a *whitebox* model, i.e., model parameters are known, to perform backpropagation through the neural network. For any AAE algorithm, its reconstructed label sentence $y_{1:L}^*$ of length L is derived from the network by decoding the input features $x_{1:T}$ for T time steps. Instead of the ground truth sequence, the reconstructed label sentence is used to avoid label leaking, i.e., to prevent that the adversarial example carries information of the correct transcription.

3.1   Attention-Based Static Window AAEs

For attention-based AAEs, the cross-entropy loss $J(X, y_l; \boldsymbol{\theta})$ w.r.t. $x_{1:T}$ is extracted by iterating over sequential token posteriors $p(y_l^* | y_{1:(l-1)}^*)$ obtained from the attention decoder. Sequence-to-sequence FGSM, as proposed in [27], then calculates $\delta(\boldsymbol{x}_t)$ from the *total* sequence as

$$\delta_{\text{SW}}(\boldsymbol{x}_t) = \epsilon \cdot \text{sgn}(\nabla_{\boldsymbol{x}_t} \sum_{l=1}^{L} J(\boldsymbol{X}, y_l^*; \boldsymbol{\theta})), \quad l \in [1; L]. \tag{2}$$

As opposed to this algorithm, our approach does not focus on the total token sequence, but only a portion of certain sequential steps. This is motivated by the observation that attention-based decoding is auto-regressive; interruption of the attention mechanism targeted at one single step in the sequence can change the corresponding portion of the transcription as well as throw off further decoding up to a certain degree. A sum over all sequence parts as in Eq. 2 may dissipate localized adversarial noise. From this, the first attention-based method is derived that takes a single portion out of the output sequence. We term this algorithm in the following as *static window* method. Gradients in the sentence are summed up from the start token, which is given by a chosen parameter γ, on to the following l_w tokens as chosen window length, such that

$$\delta_{\text{SW}}(\boldsymbol{x}_t) = \epsilon \cdot \text{sgn}(\nabla_{\boldsymbol{x}_t} \sum_{l=\gamma}^{l_w} J(\boldsymbol{X}, y_l^*; \boldsymbol{\theta})), \quad l \in [1; L]. \tag{3}$$

3.2 Attention-Based Moving Window AAEs

As observed from experiments with the static window, the effectiveness of the static window method strongly varies depending on segment position. Adversarial loss from some segments has a higher impact than from others. Some perturbations only impact local parts of the transcription. Therefore, as an extension to the static window gradient derived from Eq. 3, multiple segments of the sequence can be selected to generate $\delta_{MW}(\boldsymbol{x}_t)$. We term this the *moving window* method. For this, gradients from a sliding window with a fixed length l_w and stride ν are accumulated to $\nabla_{\text{MW}}(\boldsymbol{x}_t)$. The optimal values of length and stride are specific to each sentence. Similar to the iterative FGSM based on momentum [11], gradient normalization is applied in order to accumulate gradient directions.

$$\nabla_{\text{MW}}(\boldsymbol{x}_t) = \sum_{i=0}^{\lceil L/\nu \rceil} \left(\frac{\nabla_{\boldsymbol{x}_t} \sum_{l=i \cdot \nu}^{l_w} J(\boldsymbol{X}, y_l^*; \boldsymbol{\theta})}{||\nabla_{\boldsymbol{x}_t} \sum_{l=i \cdot \nu}^{l_w} J(\boldsymbol{X}, y_l^*; \boldsymbol{\theta})||_1} \right), \quad l \in [1; L] \tag{4}$$

$$\delta_{MW}(\boldsymbol{x}_t) = \epsilon \cdot \text{sgn}(\nabla_{\text{MW}}(\boldsymbol{x}_t)) \tag{5}$$

3.3 AAEs from Connectionist Temporal Classification

From regular CTC loss $\mathcal{L}_{\text{CTC}}$ over the total reconstructed label sentence $\boldsymbol{y}^*$, the adversarial noise is derived as

$$\delta_{\text{CTC}}(\boldsymbol{x}_t) = \epsilon \cdot \text{sgn}(\nabla_{\boldsymbol{x}_t} \mathcal{L}_{\text{CTC}}(\boldsymbol{X}, \boldsymbol{y}^*; \boldsymbol{\theta})). \tag{6}$$

3.4 Hybrid CTC/Attention Adversarial Examples

A multi-objective optimization function [21] is then applied to combine CTC and attention adversarial noise δ_{att}, that was either generated from δ_{SW} or from δ_{MW}, by introducing the multi-objective factor $\xi \in [0; 1]$.

$$\delta_{\text{Hybrid}}(\boldsymbol{x}_t) = (1 - \xi) \cdot \delta_{\text{att}}(\boldsymbol{x}_t) + \xi \cdot \delta_{\text{CTC}}(\boldsymbol{x}_t), \quad \forall t \in [1, T] \qquad (7)$$

The full process to generate hybrid CTC/attention AAEs is shown in Fig. 1.

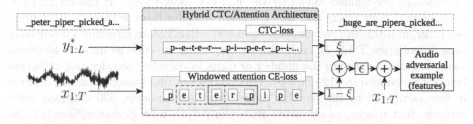

Fig. 1. Generation of AAEs. The unmodified speech sentence $\boldsymbol{x}_{1:T}$ and the label sequence $\boldsymbol{y}_{1:L}^*$ are given as input. Then, using the hybrid CTC/attention model, the adversarial loss by the CTC-layer as well as the windowed attention sequence parts are calculated. Those are then combined by the weighting parameter ξ and noise factor ϵ to obtain the adversarial example.

Extensions to This Algorithm. Whereas this formula calculates the adversarial noise based on FGSM, the extension to targeted and iterative methods can be done as proposed in [6], e.g., by accumulation of multiple steps. To extend this method to psychoacoustical hiding, the adversarial noise $\delta_{\text{Hybrid}}(\boldsymbol{x}_t)$ can be restricted to the psychoacoustic hearing threshold in the spectral or feature domain as in [25]. Furthermore, a method to convert the AAEs from the feature domain to audio is given in [3].

3.5 Adversarial Training

Similar as data augmentation, Adversarial Training (AT) augments samples of a minibatch with adversarial noise $\delta_{\text{AAE}}(x_t)$. The samples for which we create the AAEs are chosen randomly with a probability of p_a, as proposed by Sun et al. [27]. Because of its successive backpropagation in a single step, this method is also termed *adversarial regularization* and is applied not from the beginning of the neural network training but after the Nth epoch. Sun et al. additionally included a weighting factor α to distinct sequence components that we omit, i.e., set to 1; instead, the gradient is calculated for the minibatch as a whole. Furthermore, our AT algorithm also samples randomly the perturbation step size ϵ to avoid overfitting, as originally described in [19]. The expanded loss function for sequence-based training is then calculated by

$$\hat{J}(\boldsymbol{X}, \boldsymbol{y}; \boldsymbol{\theta}) = \sum_i (J(\boldsymbol{X}, y_i; \boldsymbol{\theta}) + J(\hat{\boldsymbol{X}}, y_i; \boldsymbol{\theta})). \qquad (8)$$

4   Evaluation

Throughout the experiments, the hybrid CTC/attention architecture is used with an LSTM encoder with projection neurons in the encoder and location-aware attention mechanism, classifying from log-mel f-bank features [31]. As we evaluate model performance, and not human perception on AAEs, we limit our investigation to the feature space. Evaluation is done on the TEDlium v2 [24] speech recognition task consisting of over 200h of speech. The baseline model is provided as pre-trained model by the ESPnet toolkit [30][1]. It has an LSTMP encoder with four layers and one attention decoder layer with each 1024 units per layer, in total 50m parameters. We use the BLSTM architecture for our later experiments (see Table 2), each with two layers in the encoder and the location-aware attention decoder; the number of units in encoder, decoder and attention layers was reduced to 512 units [8]. This model has 14m parameters, one quarter in size compared to the baseline model. For both models, 500 unigram units serve as text tokens, as extracted from the corpus with SentencePiece [17]. In all experiments, the token sequence y^* is previously decoded using the attention mechanism, as this is faster than hybrid decoding and also can be done without the RNNLM. We also set $\epsilon = 0.3$, as done in AAE generation for attention-based ASR networks [27]. Throughout the experiments, we use the decoded transcription y^* as label sequence to generate the adversarial examples. The dataset used in the evaluation, TEDlium v2, consists of recordings from presentations in front of an audience and therefore is already noisy and contains reverberations. To better evaluate the impact of adversarial noise generated by our algorithms, two noise-free sample sentences are used for evaluation. Both sample sentences are created artificially using Text-to-Speech (TTS) toolkits so that they remain noise-free.

4.1   Generation of AAEs: Two Case Studies

The first noise-free sentence *Peter* is generated from the TTS algorithm developed by Google named Tacotron 2 [26]. It was generated using the pre-trained model by Google[2] and reads *"Peter Piper picked a peck of pickled peppers. How many pickled peppers did Peter Piper pick?"* The second sentence *Anie* was generated from the ESPNet TTS[3] and reads *"Anie gave Christina a present and it was beautiful."* We first test the CTC-based algorithm. The algorithm outputs for *Peter* an AAE that has 41.3% CER w.r.t. the ground-truth, whereas an error rate of 36.4% for *Anie*. For our experiments with the static window algorithm, we observe that it intrinsically runs the risk of changing only local tokens. We take, for example, the sentence *Anie* and set the parameter of $l_w = 3$ and $\gamma = 4$. This gives us the following segment, as marked in bold font, out of the previously decoded sequence y^*

---

[1] Model *tedlium2.rnn.v1*, retrieved from https://github.com/espnet/espnet#asr-demo on ESPnet version 1.5.1.

[2] https://google.github.io/tacotron/publications/tacotron2/index.html.

[3] The `ljspeech.tacotron2.v2` model.

any gave christina present and it was beautiful.

After we compute the AAE, the ASR system transcribes

any game christian out priasant and it was beautiful

as the decoded sentence. We obtain a sequence that strongly resembles the original where most of the words remain intact, while some of them change slightly. Translated to CER and WER w.r.t the original sequence, we have 50.0 and 55.6 respectively. We also test its hybrid version given $\xi = 0.5$, which is analogue to the decoding configuration of the baseline model. It outputs a sequence with rates of 31.8% CER, lower than its non-hybrid version. The moving window method overcomes this problem, as it calculates a non-localized AAE. For example, a configuration with the parameters $\nu = 3$ and $l_w = 3$ applied to *Peter* generates the pattern

peter *piper* **pick**ed *a* p*eck* **of** *pickle* **pep***pers.*
how **many** **pick***le* pe**ppers** *did* pe**ter** **pi***per* pa**ck**

for which we obtain the decoded sentence

huge her piper okapk pickple take her techners
harmony pittle tipers stayed peter paper pick

This transcribed sentence then exhibits a CER of 80.4% w.r.t the ground-truth. The same parameter configuration applied in the hybrid version with $\xi = 0.5$ achieves error rates of 41.3% CER. Throughout the experiments, higher error rates were observed on the moving window than static window or CTC-based AAE generation.

Table 1 lists examples obtained using the proposed algorithms. We take $\xi = 0.5$ for all hybrid algorithms.

4.2   Evaluation of Adversarial Training

Throughout the experiments, we configured the moving window method with $\nu = 2$ and $l_w = 4$ as arbitrary constant parameters, motivated by the observation that those parameters performed well on both sentences *Peter* and *Ani*. By inspection, this configuration is also suitable for sentences of the TEDlium v2 dataset. Especially for its shorter utterances, a small window size and overlapping segments are effective. Each model is trained for 10 epochs, of which $N = 5$ epochs are done in a regular fashion from regular training data; then, the regular minibatch is augmented with its adversarial counterpart with a probability $p_a = 0.05$. Adversarial training examples are either attention-only, i.e. $\xi = 0$, or hybrid, i.e. $\xi = 0.5$. Finally, the noise factor ϵ is sampled uniformly from a range of $[0.0; 0.3]$ to cover a wide range of possible perturbations. We compare our modified BLSTM model, regularly trained with unperturbed data, to the baseline model as reported in [30]. Regarding the adversarially trained models, we use the moving window and its hybrid in the AT algorithm, because

Table 1. Examples with CER and WER w.r.t. the correct transcription in percent

Type	Sentence	CER	WER
Peter y^*	Peter piper picked a peck of pickle peppers. how many pickle peppers did peter piper pack	–	–
Anie y^*	Any gave christina a present and it was beautiful	–	–
SW, $l_w = 3$, $\gamma = 4$	Any gave christina a present and it was beautiful	–	–
SW-AAE	Any game christian out priasant and it was beautiful	50.0	55.6
Hybrid SW-AAE	Any game christian a present and it was beautiful	31.8	33.3
SW, $l_w = 4$, $\gamma = 26$	Peter piper picked a peck of pickle peppers. how many pickle peppers did peter piper pack	–	–
SW-AAE	Either piper cricker ticket tickless techners came a typical turkished plea piper pick	73.9	87.5
Hybrid SW-AAE	Peter piper picked a pick of tickle tappers how many tickle tapper stood plea piper pick	50.0	56.3
MW, $l_w = 2$, $\nu = 3$	**any** gave christina a **present** and it was **beautiful**	–	–
MW	Any canada's crystall out current since and it was	109.1	77.8
Hybrid MW	Any gaitians crystain out a present and it was beautiful	45.5	44.4
MW, $l_w = 3$, $\nu = 3$	**peter** piper **pick**ed a **peck of** pickle **peppers.**how **many** **pick**le **peppers** did **pet**er **pip**er **pack**	–	–
MW-AAE	Huge her piper okapk pickple take her techners harmony pittle tipers stayed peter paper pick	80.4	93.8
Hybrid MW-AAE	Feater piper a picked depic of tapled tapper how many pickles pepper state peter piper pick	41.3	62.5
CTC	Any dove christian no presented it was beautiful	36.4	66.7
CTC	Peter piper a picked a pack of tackled tappers how many piggle peppers didn't pay their piper pick	41.3	56.3

we hypothesize that both can benefit the training process of the hybrid model. Both models use the simplified structure. The RNNLM language model that we use is provided by the ESPnet toolkit [30]; it has 2 layers with each 650 units and its weight in decoding was set to a constant value of $\beta = 1.0$ in all experiments; i.e., we did not use language model rescoring. The proposed adversarial training process can be combined with additional data augmentation methods, such as SpecAugment; we omitted this step for comparable results. Results are collected by decoding four datasets: (1) the regular test set of the TEDlium v2 corpus; (2) AAEs from the test set, made with the attention-based moving window algorithm; (3) the test set augmented with regular white noise at 30 dB SNR; and (4) the test set with clearly noticeable 5 dB white noise.

General Trend. Some general trends during evaluation are manifested in Table 2. Comparing decoding performances between the regular test set and the AAE test set, all models perform worse. In other words, the moving window technique used for creating the AAEs performs well against different model configurations. Setting the CTC weight lowers error rates in general. The higher error rates in combination with a LM are explained by the relatively high language model weight $\beta = 1.0$. Rescoring leads to improved performance, however, listed results are more comparable to each other when set to a constant language model weight in all decoding runs.

Table 2. Decoding results for all models. The baseline model is a pre-trained *LSTMP* model obtained by the ESPnet toolkit [30]. The second row lists results obtained with a *BLSTMP* model that has the same model configuration than the adv. trained models. The parameter λ determines the weight of the CTC model during the decoding. Trained models with attention-only AAEs are marked with $\xi = 0$; with hybrid AAEs with $\xi = 0.5$. The first value in each cell corresponds to the CER and the second to the WER, both given as percentage

CER/WER	ξ	λ	LM	Dataset			
				Test	Test AAE	Noise 30dB	Noise 5dB
pre-trained model [30]	–	0.0	–	20.7/22.8	90.7/89.1	23.6/25.8	78.8/78.8
	–	0.5	–	15.7/18.6	86.1/89.9	18.1/21.3	66.1/68.3
	–	0.5	✓	16.3/18.3	**98.5/92.2**	19.2/20.8	73.2/72.7
regular same-parameter model	–	0.0	–	18.5/20.2	69.4/67.7	21.1/22.7	76.9/76.0
	–	0.5	–	14.4/16.8	58.0/60.8	16.7/19.2	63.9/66.1
	–	0.5	✓	15.4/17.0	65.9/62.7	17.8/19.2	71.2/69.6
adv. trained with att.-only AAE	0.0	0.0	–	17.7/19.6	63.6/63.3	21.0/22.8	74.7/74.4
	0.0	0.5	–	14.3/16.9	**53.5/56.8**	16.5/18.9	62.6/65.0
	0.0	0.5	✓	15.1/16.9	60.3/58.3	17.5/18.9	69.0/68.0
adv. trained with hybrid AAE	0.5	0.0	–	17.9/19.8	65.2/65.0	20.4/22.3	74.9/75.0
	0.5	0.5	–	**14.0/16.5**	54.8/58.6	**16.2/18.7**	**63.5/65.8**
	0.5	0.5	✓	14.8/16.6	61.8/59.9	17.0/18.5	70.0/69.2

Successful AAEs. Notably, the baseline model performed worst on this dataset with almost 100% error rates, even worse when decoding noisy data. This manifests in wrong transcriptions of around the same length as the ground truth, with on average 90% substitution errors but only 20% insertion or deletion errors. Word loops or dropped sentence parts were observed only rarely, two architectural behaviors when the attention decoder looses its alignment. We report CERs as well as WERs, as a relative mismatch between those indicates certain error patterns for CTC and attention decoding [20]; however, the ratio of CER to WER of transcribed sentences was observed to stay roughly at the same levels in the experiments with the AAE test set [20].

Adv. Trained Models Are More Robust. Even though we can see a significant improvement on the performance of the regularly trained model compared to the baseline model, both models obtained from AT perform better in general, especially in the presence of adversarial noise; the model trained with hybrid AAEs achieves a WER of 16.5% on the test set, a performance of absolute 2.1% over the baseline model. At the same time, the robustness on regular noise and specially on adversarial noise was improved. For the latter we have an improvement of 24–33% absolute WER compared to the baseline. The most notable difference is in decoding in combination with CTC and LM, where the regularly trained model had a WER of 92.2%, while the corresponding adv. trained model had roughly 60% WER. Both methods for generating AAE and injecting them into the training process seem to have similar results in terms of performance. This is unexpected, as we hypothesized that the information contained in an attention-only method was greater than in the hybrid, which would translate to a greater contribution to the training process.

5 Conclusion

In this work, we demonstrated audio adversarial examples against hybrid attention/CTC speech recognition networks. The first method we introduced was to select a *static window* over a selected segment of the attention-decoded output sequence to calculate the adversarial example. This method was then extended to a *moving window* that slides over the output sequence to better distribute perturbations over the transcription. In a third step, we applied the fast gradient sign method to CTC-network.

AAEs constructed with this method induced on a regular speech recognition model a word error rate of up to 90%. In a second step, we employed these for adversarial training a hybrid CTC/attention ASR network. This process improved its robustness against audio adversarial examples, with 55% WER, and also slightly against regular white noise. Most notably, the speech recognition performance on regular data was improved by absolute 1.8% from 18.3% compared to baseline results.

References

1. Abdoli, S., Hafemann, L.G., Rony, J., Ayed, I.B., Cardinal, P., Koerich, A.L.: Universal adversarial audio perturbations. ArXiv abs/1908.03173 (2019)
2. Andronic, I.: MP3 Compression as a means to improve robustness against adversarial noise targeting attention-based end-to-end speech recognition. Master's thesis, Technical University of Munich, Germany (2020)
3. Andronic, I., Kürzinger, L., Rosas, E.R.C., Rigoll, G., Seeber, B.U.: MP3 compression to diminish adversarial noise in end-to-end speech recognition. arXiv preprint arXiv:2007.12892, To be published at SPECOM 2020 (2020)
4. Bahdanau, D., Cho, K., Bengio, Y.: Neural machine translation by jointly learning to align and translate. arXiv preprint arXiv:1409.0473 (2014)
5. Carlini, N., et al.: Hidden voice commands. In: 25th {USENIX} Security Symposium ({USENIX} Security 16), pp. 513–530 (2016)
6. Carlini, N., Wagner, D.: Audio adversarial examples: Targeted attacks on speech-to-text. In: 2018 IEEE Security and Privacy Workshops (SPW), pp. 1–7. IEEE (2018)
7. Chan, W., Jaitly, N., Le, Q., Vinyals, O.: Listen, attend and spell: a neural network for large vocabulary conversational speech recognition. In: 2016 IEEE International Conference on Acoustics, Speech and Signal Processing (ICASSP), pp. 4960–4964. IEEE (2016)
8. Chavez Rosas, E.R.: Improving robustness of sequence-to-sequence automatic speech recognition by means of adversarial training. Master's thesis, Technical University of Munich, Germany (2020)

9. Chorowski, J.K., Bahdanau, D., Serdyuk, D., Cho, K., Bengio, Y.: Attention-based models for speech recognition. In: Neural Information Processing Systems, pp. 577–585 (2015)
10. Cisse, M., Adi, Y., Neverova, N., Keshet, J.: Houdini: fooling deep structured prediction models. ArXiv abs/1707.05373 (2017)
11. Dong, Y., et al.: Boosting adversarial attacks with momentum. In: Proceedings of the IEEE Conference on Computer Vision and Pattern Recognition, pp. 9185–9193 (2018)
12. Goodfellow, I.J., Shlens, J., Szegedy, C.: Explaining and harnessing adversarial examples. CoRR abs/1412.6572 (2014)
13. Graves, A., Fernández, S., Gomez, F., Schmidhuber, J.: Connectionist temporal classification: labelling unsegmented sequence data with recurrent neural networks. In: Proceedings of the 23rd International Conference on Machine Learning, pp. 369–376. ACM (2006)
14. Gulcehre, C., et al.: On using monolingual corpora in neural machine translation. arXiv preprint arXiv:1503.03535 (2015)
15. Hannun, A.Y., et al.: Deep speech: scaling up end-to-end speech recognition. arXiv abs/1412.5567 (2014)
16. Hu, S., Shang, X., Qin, Z., Li, M., Wang, Q., Wang, C.: Adversarial examples for automatic speech recognition: attacks and countermeasures. IEEE Commun. Mag. **57**(10), 120–126 (2019)
17. Kudo, T.: Subword regularization: improving neural network translation models with multiple subword candidates. ArXiv abs/1804.10959 (2018). https://doi.org/10.18653/v1/P18-1007
18. Kurakin, A., Goodfellow, I., Bengio, S.: Adversarial examples in the physical world. CoRR abs/1607.02533 (2016). http://arxiv.org/abs/1607.02533
19. Kurakin, A., Goodfellow, I., Bengio, S.: Adversarial machine learning at scale. CoRR abs/1611.01236 (2016). http://arxiv.org/abs/1611.01236
20. Kürzinger, L., Watzel, T., Li, L., Baumgartner, R., Rigoll, G.: Exploring hybrid CTC/Attention end-to-end speech recognition with Gaussian processes. In: Salah, A.A., Karpov, A., Potapova, R. (eds.) SPECOM 2019. LNCS (LNAI), vol. 11658, pp. 258–269. Springer, Cham (2019). https://doi.org/10.1007/978-3-030-26061-3_27
21. Lu, L., Kong, L., Dyer, C., Smith, N.A.: Multitask learning with CTC and segmental CRF for speech recognition (2017). 10/gf3hs6
22. Neekhara, P., Hussain, S., Pandey, P., Dubnov, S., McAuley, J., Koushanfar, F.: Universal adversarial perturbations for speech recognition systems. ArXiv abs/1905.03828 (2019). https://doi.org/10.21437/interspeech.2019-1353
23. Qin, Y., Carlini, N., Goodfellow, I., Cottrell, G., Raffel, C.: Imperceptible, robust, and targeted adversarial examples for automatic speech recognition. ArXiv abs/1903.10346 (2019)
24. Rousseau, A., Deléglise, P., Esteve, Y.: Enhancing the TED-LIUM corpus with selected data for language modeling and more ted talks. In: LREC, pp. 3935–3939 (2014)
25. Schönherr, L., Kohls, K., Zeiler, S., Holz, T., Kolossa, D.: Adversarial attacks against automatic speech recognition systems via psychoacoustic hiding. ArXiv abs/1808.05665 (2018). https://doi.org/10.14722/ndss.2019.23288
26. Shen, J., et al.: Natural TTS synthesis by conditioning WaveNet on Mel spectrogram predictions. In: 2018 IEEE International Conference on Acoustics, Speech and Signal Processing (ICASSP), pp. 4779–4783. IEEE (2018)

27. Sun, S., Guo, P., Xie, L., Hwang, M.Y.: Adversarial regularization for attention based end-to-end robust speech recognition. IEEE/ACM Trans. Audio Speech Lang. Process. **27**(11), 1826–1838 (2019)
28. Szegedy, C., et al.: Intriguing properties of neural networks. CoRR abs/1312.6199 (2013)
29. Vadillo, J., Santana, R.: Universal adversarial examples in speech command classification. ArXiv abs/1911.10182 (2019)
30. Watanabe, S., et al.: ESPnet: end-to-end speech processing toolkit. In: Interspeech, pp. 2207–2211 (2018). https://doi.org/10.21437/Interspeech.2018-1456
31. Watanabe, S., Hori, T., Kim, S., Hershey, J.R., Hayashi, T.: Hybrid CTC/Attention architecture for end-to-end speech recognition. IEEE J. Sel. Top. Sig. Process. **11**(8), 1240–1253 (2017)

CTC-Segmentation of Large Corpora for German End-to-End Speech Recognition

Ludwig Kürzinger[✉][iD], Dominik Winkelbauer, Lujun Li[iD], Tobias Watzel[iD], and Gerhard Rigoll[iD]

Institute for Human-Machine Communication, Technische Universität München, Munich, Germany
{ludwig.kuerzinger,dominik.winkelbauer}@tum.de

Abstract. Recent end-to-end Automatic Speech Recognition (ASR) systems demonstrated the ability to outperform conventional hybrid DNN/HMM ASR. Aside from architectural improvements in those systems, those models grew in terms of depth, parameters and model capacity. However, these models also require more training data to achieve comparable performance.

In this work, we combine freely available corpora for German speech recognition, including yet unlabeled speech data, to a big dataset of over 1700 h of speech data. For data preparation, we propose a two-stage approach that uses an ASR model pre-trained with Connectionist Temporal Classification (CTC) to boot-strap more training data from unsegmented or unlabeled training data. Utterances are then extracted from label probabilities obtained from the network trained with CTC to determine segment alignments. With this training data, we trained a hybrid CTC/attention Transformer model that achieves 12.8% WER on the Tuda-DE test set, surpassing the previous baseline of 14.4% of conventional hybrid DNN/HMM ASR.

Keywords: German speech dataset · End-to-end automatic speech recognition · Hybrid CTC/Attention · CTC-segmentation

1 Introduction

Conventional speech recognition systems combine Deep Neural Networks (DNN) with Hidden Markov Models (HMM). The DNN serves as an acoustic model that infers classes, or their posterior probabilities respectively, originating from hand-crafted HMMs and complex linguistic models. Hybrid DNN/HMM models also require multiple processing steps during training to refine frame-wise acoustic model labels. In comparison to hybrid DNN/HMM systems, end-to-end ASR simplifies training and decoding by directly inferring sequences of letters, or tokens, given a speech signal. For training, end-to-end systems only require the raw text corresponding to an utterance. Connectionist Temporal Classification

---

L. Kürzinger and D. Winkelbauer—Contributed equally to this work.

© Springer Nature Switzerland AG 2020
A. Karpov and R. Potapova (Eds.): SPECOM 2020, LNAI 12335, pp. 267–278, 2020.
https://doi.org/10.1007/978-3-030-60276-5_27

(CTC) is a popular loss function to train end-to-end ASR architectures [5]. In principle, its concept is similar to a HMM, the label sequence is modeled as sequence of states, and during training, a slightly modified forward-backward algorithm is used in the calculation of CTC loss. Another popular approach for end-to-end ASR is to directly infer letter sequences, as employed in attention-based encoder-decoder architectures [3]. Hybrid CTC/attention ASR architectures combine these two approaches [18].

End-to-end models also require more training data to learn acoustic representations. Many large corpora, such as Librispeech or TEDlium, are provided as large audio files partitioned into segments that contain speech with transcriptions. Although end-to-end systems do not need frame-wise temporal alignment or segmentation, an utterance-wise alignment between audio and text is necessary. To reduce training complexity, previous works used frameworks like sphinx [8] or MAUS [15] to partition speech data into sentence-length segments, each containing an utterance. Those frameworks determine the start and the end of a sentence from acoustic models (often HMMs) and the Viterbi algorithm. However, there are three disadvantages in using these for end-to-end ASR: (1) As only words in the lexicon can be detected, the segmentation tool needs a strategy for out-of-vocabulary words. (2) Scaling the Viterbi algorithm to generate alignments within larger audio files requires additional mitigations. (3) As these algorithms provide *forced* alignments, they assume that the audio contains only the text which should be aligned; but for most public domain audio this is not the case. So do for example all audio files from the Librivox dataset contain an additional prologue and epilogue where the speaker lists his name, the book title and the license. It might also be the case that the speaker skips some sentences or adds new ones due to different text versions. Therefore, aligning segments of large datasets, such as TEDlium [14], is done in multiple iterations that often include manual examination. Unfortunately, this process is tedious and error prone; for example, by inspection of the SWC corpus, some of those automatically generated transcriptions are missing words in the transcription.

We aim for a method to extract labeled utterances in the form of correctly aligned segments from large audio files. To achieve this, we propose CTC-segmentation, an algorithm to correctly align start and end of utterance segments, supported by a CTC-based end-to-end ASR network [1]. Furthermore, we demonstrate additional data cleanup steps for German language orthography.
Our contributions are:

- We propose CTC-segmentation, a *scalable* method to extract utterance segments from speech corpora. In comparison to other automated segmentation tools, alignments generated with CTC-segmentation were observed to more closely correspond to manually segmented utterances.
- We extended and refined the existing recipe from the ASR toolkit kaldi with a collection of open source German corpora by two additional corpora, namely

---

[1] The source code underlying this work is available at https://github.com/cornerfarmer/ctc_segmentation.

Librivox and *CommonVoice*, and ported it to the end-to-end ASR toolkit ESPnet.

2 Related Work

2.1 Speech Recognition for German

Milde et al. [10] proposed to combine freely available German language speech corpora into an *open source* German speech recognition system. A more detailed description of the German datasets can be found in [10], of which we give a short summary:

- The Tuda-DE dataset [13] combines recordings of multiple sentences concerning various topics spoken by 180 speakers using five microphones.
- The Spoken Wikipedia Corpus (SWC, [2]) is an open source summary of recordings of different Wikipedia articles made by volunteers. The transcription already includes alignment notations between audio and text, but as these alignments were often incorrect, Milde et al. re-aligned utterance segments using the Sphinx speech recognizer [8].
- The M-AILABS Speech Dataset [16] mostly consists of utterances extracted from political speeches and audio books from Librivox. Audio and text has been aligned by using synthetically generated audio (TTS) based on the text and by manually removing intro and outro.

In this work, we additionally combine the following German speech corpora:

- CommonVoice dataset [1] consists of utterances recorded and verified by volunteers; therefore, an utterance-wise alignment already exists.
- Librivox [9] is a platform for volunteers to publish their recordings of reading public domain books. All recordings are published under a Creative Common license. We use audio recordings of 579 books. The corresponding texts are retrieved from Project Gutenberg-DE [6] that hosts a database of books in the public domain.

Milde et al. [10] mainly used a conventional DNN/HMM model, as provided by the kaldi toolkit [12]. Denisov et al. [4] used a similar collection of German language corpora that additionally includes non-free pre-labeled speech corpora. Their ASR tool *IMS Speech* is based on a hybrid CTC/attention ASR architecture using the BLSTM model with location-aware attention as proposed by Watanabe et al. [18]. The architecture used in our work also is based on the hybrid CTC/attention ASR of the ESPnet toolkit [19], however, in combination with the Transformer architecture [17] that uses self-attention. As we only give a short description of its architecture, an in-detail description of the Transformer model is given by Karita et al. [7].

2.2   Alignment and Segmentation Methods

There are several tools to extract labeled utterance segments from speech corpora. The Munich Automatic Segmentation (MAUS) system [15] first transforms the given transcript into a graph representing different sequences of phones by applying predefined rules. Afterwards, the actual alignment is estimated by finding the most probable path using a set of HMMs and pretrained acoustic models. Gentle works in a similar way, but while MAUS uses HTK [20], Gentle is built on top of Kaldi [12]. Both methods yield phone-wise alignments. Aeneas [11] uses a different approach: It first converts the given transcript into audio by using text-to-speech (TTS) and then uses the Dynamic Time Warping (DTW) algorithm to align the synthetic and the actual audio by warping the time axis. In this way it is possible to estimate begin and end of given utterance within the audio file.

We propose to use a CTC-based network for segmentation. CTC was originally proposed as a loss function to train RNNs on unsegmented data. At the same time, using CTC as a segmentation algorithm was also proposed by Graves et al. [5]. However, to the best knowledge of the authors, while the CTC algorithm is widely used for end-to-end speech recognition, there is not yet a segmentation tool for speech audio based on CTC.

3   Methodology

3.1   CTC-Segmentation of Utterances

The following paragraphs describe CTC-segmentation, an algorithm to extract proper audio-text alignments in the presence of additional unknown speech sections at the beginning or end of the audio recording. It uses a CTC-based end-to-end network that was trained on already aligned data beforehand, e.g., as provided by a CTC/attention ASR system. For a given audio recording the CTC network generates frame-based character posteriors $p(c|t, x_{1:T})$. From these probabilities, we compute via dynamic programming all possible maximum joint probabilities $k_{t,j}$ for aligning the text until character index $j \in [1; M]$ to the audio up to frame $t \in [1; T]$. Probabilities are mapped into a trellis diagram by the following rules:

$$k_{t,j} = \begin{cases} \max(k_{t-1,j} \cdot p(blank|t),\ k_{t-1,j-1} \cdot p(c_j|t)) & \text{if } t > 0 \wedge j > 0 \\ 0 & \text{if } t = 0 \wedge j > 0 \quad (1) \\ 1 & \text{if } j = 0 \end{cases}$$

The maximum joint probability at a point is computed by taking the most probable of the two possible transitions: Either only a blank symbol or the next character is consumed. The transition cost for staying at the first character is set to zero, to align the transcription start to an arbitrary point of the audio file.

The character-wise alignment is then calculated by backtracking, starting off the most probable temporal position of the last character in the transcription, i.e, $t = \arg\max_{t'} k_{t',M}$. Transitions with the highest probability then determine the alignment a_t of the audio frame t to its corresponding character from the text, such that

$$a_t = \begin{cases} M - 1 & \text{if } t \geqslant \arg\max_{t'}(k_{t',M-1}) \\ a_{t+1} & \text{if } k_{t,a_{t+1}} \cdot p(blank|t+1) > k_{t,a_{t+1}-1} \cdot p(c_j|t+1) \\ a_{t+1} - 1 & \text{else} \end{cases} \quad (2)$$

As this algorithm yields a probability ρ_t for every audio frame being aligned in a given way, a *confidence score* s_{seg} for each segment is derived to sort out utterances with deviations between speech and corresponding text, that is calculated as

$$s_{seg} = \min_j m_j \quad \text{with} \quad m_j = \frac{1}{L} \sum_{t=jL}^{(j+1)L} \rho_t. \quad (3)$$

Here, audio frames that were segmented to correspond to a given utterance are first split into parts of length L. For each of these parts, a mean value m_j based on the frame-wise probabilities ρ_t is calculated. The total probability s_{seg} for a given utterance is defined as the minimum of these probabilities per part m_j. This method inflicts a penalty on the confidence score on mismatch, e.g., even if a single word is missing in the transcription of a long utterance.

The complexity of the alignment algorithm is reduced from $O(M \cdot N)$ to $O(M)$ by using the heuristic that the ratio between the aligned audio and text position is nearly constant. Instead of calculating all probabilities $k_{t,j}$, for every character position j one only considers the audio frames in the interval $[t - W/2, t + W/2]$ with $t = jN/M$ as the audio position proportional to a given character position and the window size W.

3.2 Data Cleaning and Text Preparation

The ground truth text from free corpora, such as Librivox or the SWC corpus, is often not directly usable for ASR and has therefore to be cleaned. To maximize generalization to the Tuda-DE test dataset, this is done in a way to match the style of the ground truth text used in Tuda-DE, which only consists of letters, i.e. a-z and umlauts (ä, ü, ö, ß). Punctuation characters are removed and all sentences with different letters are taken out of the dataset. All abbreviations and units are replaced with their full spoken equivalent. Furthermore, all numbers are replaced by their full spoken equivalent. Here it is also necessary to consider different cases, as this might influence the suffix of the resulting word. Say, "*18 00 Soldaten*" needs to be replaced by "*eintausendachthundert Soldaten*", whereas "*Es war 18 00*" is replaced according to its pronunciation by "*Es war achtzehnhundert*". The correct case can be determined from neighboring words with simple heuristics. For this, the NLP tagger provided by the spacy framework [7] is used.

Another issue arised due to old German orthography. Text obtained from Librivox is due to its expired copyright usually at least 70 years old and uses old German spelling rules. For an automated transition to the reformed German orthography, we implemented a self-updating lookup-table of letter replacements. This list was compiled based on a list of known German words from correctly spelled text.

4   Evaluation and Results

4.1   Alignment Evaluation

In this section, we evaluate how well the proposed CTC-segmentation algorithm aligns utterance-wise text and audio. Evaluation is done on the dev and test set of the TEDlium v2 dataset [14], that consist of recordings from 19 unique speakers that talk in front of an audience. This corpus contains labeled sentence-length utterances, each with the information of start and end of its segment in the audio recording. As these alignments have been done manually, we use them as reference for the evaluation of the forced alignment algorithms. The comparison is done based on three parameters: the mean deviation of the predicted start or end from ground truth, its standard deviation and the ratio of predictions which are at maximum 0.5 s apart from ground truth. To evaluate the impact of the ASR model on CTC-segmentation, we include both BLSTM as well as Transformer models in the comparison. The pre-trained models [2] were provided by the ESPnet toolkit [19]. We compare our approach with three existing forced alignment methods from literature: MAUS, Gentle and Aeneas. To get utterance-wise from phone-wise alignments, we determine the begin time of the first phone and the end time of the last phone of the given utterance. As can be seen in Table 1, segment alignments generated by CTC-segmentation correspond significantly closer to ground truth compared to the segments generated by all other tested alignment algorithms.

Figure 1 visualizes the density of segmentation timing deviations across all predictions. We thereby compare our approach using the LSTM-based model trained on TEDlium v2 with the Gentle alignment tool. It can be seen that both approaches have timing deviations smaller than one second for most predictions. Apart from that, our approach has a higher density in deviations between 0 and 0.5 s, while it is the other way around in the interval from 0.5 to 1 s. This indicates that our approach generates more accurately aligned segments when compared to Viterbi- or DTW-based algorithms.

As explained in Sect. 3.1, one of the main motivations for CTC-segmentation is to determine utterance segments in a robust manner, regardless of preambles or deviating transcriptions. To simulate such cases using the TEDlium v2 dev and

---

[2] Configuration of the pre-trained models: The Transformer model has a self-attention encoder with 12 layers of each 2048 units. The BLSTM model has a BLSTMP encoder containing 4 layers with each 1024 units, with sub-sampling in the second and third layer.

Table 1. Accuracy of different alignment methods on the dev and test set of TEDlium v2, compared via the mean deviation from ground truth, its standard deviation and the ratio of predictions which are at maximum 0.5 seconds apart from ground truth.

	Mean	Std	< 0.5 s
Conventional segmentation approaches			
MAUS (HMM-based using HTK)	1.38 s	11.62	74.1%
Aeneas (DTW-based)	9.01 s	38.47	64.7%
Gentle (HMM-based using kaldi)	0.41s	1.97	82.0%
CTC-segmentation (Ours)			
Hybrid CTC/att. BLSTM trained on TEDlium v2	0.34 s	1.16	90.1%
Hybrid CTC/att. Transformer trained on TEDlium v2	0.31 s	0.85	88.8%
Hybrid CTC/att. Transformer trained on Librispeech	0.35 s	0.68	85.1%

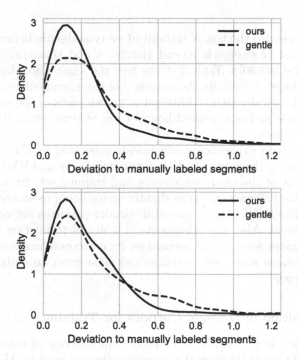

Fig. 1. Relative deviation, denoted in seconds, of segments generated by Gentle and our CTC-segmentation compared to manually labeled segments from TEDlium 2. CTC-segmentation exhibited a greater accuracy to the start of the segment (top) in comparison with Gentle; an also was observed to be slightly more accurate towards the end of the segments (bottom). The y axis denotes density in a histogram with 60 bins.

Table 2. Different alignment methods on the augmented dev and test set of TEDlium v2. Similar to the evaluation procedure as in Table 1, but the audio samples are augmented by adding random speech parts to their start and end. In this the robustness of the different approaches is evaluated.

	Mean	Std	< 0.5 s
Existing methods			
MAUS (HMM-based using HTK)	3.18 s	18.97	66.9%
Aeneas (DTW-based)	10.91 s	40.50	62.2%
Gentle (HMM-based using kaldi)	0.46 s	2.40	81.7%
CTC-segmentation (ours)			
BLSTM trained on TEDlium v2	0.40 s	1.63	89.3%
Transformer trained on TEDlium v2	0.35 s	1.38	89.2%
Transformer trained on Librispeech	0.40 s	1.21	84.2%

test set, we prepended the last N seconds of every audio file before its start and appended the first M seconds to its end. Hereby, N and M are randomly sampled from the interval $[10, 30]$ s. Table 2 shows how the same algorithms perform on this altered dataset. Especially the accuracy of the alignment tools MAUS and Aeneas drops drastically when additional unknown parts of the audio recording are added. Gentle and our method however are able to retain their alignment abilities in such cases.

To conclude both experiments, alignments generated by CTC-segmentation correspond closer to the ground truth compared to DTW and HMM based methods, independent of the used architecture and training set. By inspection, the quality of obtained alignments varies slightly across domains and conditions: The Transformer model with a more powerful encoder performs better compared to the BLSTM model. Also, the alignments of a model trained on the TEDlium v2 corpus are more accurate on average on its corresponding test and dev set; this corpus contains more reverberation and noise from an audience than the Librispeech corpus.

4.2   Composition of German Corpora for Training

Model evaluation is performed on multiple combinations of datasets, listed in Table 3. Thereby we build upon the corpora collection used by Milde et al. [10], namely, Tuda-DE, SWC and M-AILABS. As [10], we also neglect recordings made by the Realtek microphone due to bad quality. Additional to these three corpora, we train our model on Common Voice and Librivox. Data preparation of the Common Voice dataset only required to post-process the ground truth text by replacing all numbers by their full spoken equivalent. As the Viterbi-alignment provided by [10] for SWC is not perfect, with some utterances missing its first words in the transcription, we realign and clean the data using CTC-segmentation, as in Sect. 3.1. Utterance alignments with a confidence score s_{seg}

lower than 0.22, corresponding to −1.5 in log space, were discarded. To perform CTC-segmentation on the Librivox corpus, we combined the audio files with the corresponding ground truth text pieces from Project Gutenberg-DE [6]. Comparable evaluation results were obtained from decoding the Tuda-DE dev and test sets, as also used in [10].

In total, the cumulative size of these corpora spans up to 1772 h, of which we use three partially overlapping subsets for training: In the first configuration that includes 649 h of speech data, we use the selection as provided by Milde et al. that includes Tuda-DE, SWC and M-AILABS. The second subset is created by adding the CommonVoice corpus, resulting in 968 h of training data. The third selection conjoins the Tuda-DE corpus and CommonVoice with the two CTC-segmented corpora, SWC and Librivox, to 1460 h of speech data.

Table 3. Datasets used for training and evaluation.

Datasets		Length	Speakers	Utterances
Tuda-DE train [13]	TD	127 h	147	55497
Tuda-DE dev [13]	Dev	9 h	16	3678
Tuda-DE test [13]	Test	10 h	17	4100
SWC [2], aligned by [10]	SW	285 h	363	171380
M-ailabs [16]	MA	237 h	29	118521
Common voice [1]	CV	319 h	4852	279516
CTC-segmented SWC	SW*	210 h	363	78214
CTC-segmented Librivox [6,9]	LV*	804 h	251	368532

4.3 ASR Configuration

For all experiments, the hybrid CTC/attention architecture with the Transformer is used. It consists of a 12 layer encoder and a 6 layer decoder, both with 2048 units in each layer; attention blocks contain 4 heads to each 256 units[3]. All models were trained for 23 epochs using the noam optimizer. We did not use data augmentation, such as SpecAugment. At inference time, the decoding of the test and dev set is done using beam search with beam size of 16. To further improve the results on the test and dev set, a language model was used to guide the beam search. Language models with two sizes were used in decoding. The RNNLM language models were trained on the same text corpus as used in [10] for 20 epochs. The first RNNLM has two layers with 650 LSTM units per layer. It achieves a perplexity of 8.53. The second RNNLM consists of four layer of each 1024 units, with a perplexity of 6.46.

---

[3] The default configuration of the Transformer model at ESPnet v.0.5.3.

Table 4. A comparison of using different dataset combinations. Word error rates are in percent and evaluated on the Tuda-DE test and dev set.

Datasets							ASR model	LM	Tuda-DE	
TD	SW	MA	CV	SW*	LV*	h			dev	test
✓	✓	–	–	–	–	412	TDNN-HMM [10]	4-gram KN	15.3	16.5
✓	✓	–	–	–	–	412	TDNN-HMM [10]	LSTM (2 × 1024)	13.1	14.4
✓	✓	✓	–	–	–	649	TDNN-HMM [10]	4-gram KN	14.8	15.9
✓	✓	✓	–	–	–	649	Transformer	RNNLM (2 × 650)	16.4	17.2
✓	✓	✓	✓	–	–	986	Transformer	RNNLM (2 × 650)	16.0	17.1
✓	✓	✓	✓	–	–	986	Transformer	RNNLM (4 × 1024)	14.1	15.2
✓	–	–	✓	✓	✓	1460	Transformer	None	19.3	19.7
✓	–	–	✓	✓	✓	1460	Transformer	RNNLM (2 × 650)	14.3	14.9
✓	–	–	✓	✓	✓	1460	Transformer	RNNLM (4 × 1024)	**12.3**	**12.8**

4.4 Discussion of Results

The benchmark results are listed in Table 4. First, the effects of using different dataset combinations are inspected. By using the CommonVoice dataset in addition to Tuda-DE, SWC and M-AILABS, the test WER decreases to 15.2% WER. Further replacing SWC and M-AILABS by the custom aligned SWC and Librivox dataset decreased the test set WER down to 12.8%.

The second observation is that the language model size and also the achieved perplexity on the text corpus highly influences the WER. The significant improvement in WER of 2% can be explained by the better ability of the big RNNLM in detection and prediction of German words and grammar forms. For example, Milde et al. [10] described that compounding poses are a challenge for the ASR system; not recognized compounds resulted in at least two errors, a substitution and an insertion error. This was also observed in a decoding run without the RNNLM, e.g., *"Tunneleinfahrt"* was recognized as *"Tunnel_ein_fahrt"*. By inspection of recognized transcriptions, most of these cases were correctly determined when decoding with language model, even more so with the large RNNLM.

Table 4 gives us further clues how the benefits to end-to-end ASR scale with the amount of automatically aligned data. The benchmark results obtained with the small language model improved by absolute 0.1% WER on the Tuda-DE test set, after addition of the CommonVoice dataset, 319 h of speech data. The biggest performance improvement of 4.3% WER was obtained with the third selection of corpora with 1460 h of speech data. Whereas the composition of corpora is slightly different in this selection, two main factors contributed to this improvement: The increased amount of training data and better utterance alignments using CTC-segmentation.

5   Conclusion

End-to-end ASR models require more training data as conventional DNN/HMM ASR systems, as those models grow in terms of depth, parameters and model capacity. In order to compile a large dataset from yet unlabeled audio recordings, we proposed CTC-segmentation. This algorithm uses a CTC-based end-to-end neural network to extract utterance segments with exact time-wise alignments.

Evaluation of our method is two-fold: As evaluated on the hand-labeled dev and test datasets from TEDlium v2, alignments generated by CTC-segmentation were more accurate compared to those obtained from Viterbi- or DTW-based approaches. In terms of ASR performance, we build on a composition of German speech corpora [10] and trained an end-to-end ASR model with CTC-segmented training data; the best model achieved 12.8% WER on the Tuda-DE test set, an improvement of 1.6% WER absolute in comparison with the conventional hybrid DNN/HMM ASR system.

References

1. Ardila, R., et al.: Common voice: a massively-multilingual speech corpus. arXiv preprint arXiv:1912.06670 (2019)
2. Baumann, T., Köhn, A., Hennig, F.: The spoken Wikipedia corpus collection (2016)
3. Chan, W., Jaitly, N., Le, Q., Vinyals, O.: Listen, attend and spell: a neural network for large vocabulary conversational speech recognition. In: 2016 IEEE International Conference on Acoustics, Speech and Signal Processing (ICASSP), pp. 4960–4964. IEEE (2016)
4. Denisov, P., Vu, N.T.: IMS-speech: a speech to text tool. Studientexte zur Sprachkommunikation: Elektronische Sprachsignalverarbeitung **2019**, 170–177 (2019)
5. Graves, A., Fernández, S., Gomez, F., Schmidhuber, J.: Connectionist temporal classification: labelling unsegmented sequence data with recurrent neural networks. In: Proceedings of the 23rd International Conference on Machine Learning, pp. 369–376. ACM (2006)
6. Gutenberg, n.: Projekt gutenberg-de (2019). https://gutenberg.spiegel.de
7. Karita, S., et al.: A comparative study on transformer vs RNN in speech applications. arXiv preprint arXiv:1909.06317 (2019)
8. Lamere, P., et al.: The CMU sphinx-4 speech recognition system. In: IEEE International Conference on Acoustics, Speech and Signal Processing (ICASSP 2003), Hong Kong, vol. 1, pp. 2–5 (2003)
9. Librivox, N.: Librivox: free public domain audiobooks (2020). https://librivox.org/
10. Milde, B., Köhn, A.: Open source automatic speech recognition for German. In: Speech Communication; 13th ITG-Symposium, pp. 1–5. VDE (2018)
11. Pettarin, A.: Aeneas (2017). https://www.readbeyond.it/aeneas/
12. Povey, D., et al.: The kaldi speech recognition toolkit. In: IEEE 2011 Workshop on Automatic Speech Recognition and Understanding. IEEE Signal Processing Society, December 2011. IEEE Catalog No.: CFP11SRW-USB
13. Radeck-Arneth, S., et al.: Open source German distant speech recognition: corpus and acoustic model. In: Král, P., Matoušek, V. (eds.) TSD 2015. LNCS (LNAI), vol. 9302, pp. 480–488. Springer, Cham (2015). https://doi.org/10.1007/978-3-319-24033-6_54
14. Rousseau, A., Deléglise, P., Esteve, Y.: Enhancing the TED-LIUM corpus with selected data for language modeling and more ted talks. In: LREC, pp. 3935–3939 (2014)

15. Schiel, F.: Automatic phonetic transcription of non-prompted speech. In: Proceedings of the ICPhS, pp. 607–610. San Francisco, August 1999
16. Solak, I.: The M-AILABS speech dataset (2019). https://www.caito.de/2019/01/the-m-ailabs-speech-dataset/
17. Vaswani, A., et al.: Attention is all you need. In: Advances in Neural Information Processing Systems, pp. 5998–6008 (2017)
18. Watanabe, S., Hori, T., Kim, S., Hershey, J.R., Hayashi, T.: Hybrid CTC/Attention architecture for end-to-end speech recognition. IEEE J. Sel. Top. Sig. Process. **11**(8), 1240–1253, December 2017. https://doi.org/10.1109/JSTSP.2017.2763455
19. Watanabe, S., et al.: ESPnet: end-to-end speech processing toolkit. In: Interspeech, pp. 2207–2211 (2018). https://doi.org/10.21437/Interspeech.2018-1456. http://dx.doi.org/10.21437/Interspeech.2018-1456
20. Young, S.J., Young, S.: The HTK hidden Markov model toolkit: design and philosophy. University of Cambridge, Department of Engineering Cambridge, England (1993)

Stylometrics Features Under Domain Shift: Do They Really "Context-Independent"?

Tatiana Litvinova(✉)

RusProfiling Lab, Voronezh State Pedagogical University, Voronezh, Russia
centr_rus_yaz@mail.ru

Abstract. The problem of determining the author of a text, i.e. authorship attribution (AA), is based on the assumption on the stability and uniqueness of the authorial style. Since it is obvious that a lot of content words bear topical information, researchers widely use the so-called "context-independent" style markers which are believed to be resistant to the shift of topic, genre, mode, etc. ("domain"), e.g., character n-grams, part-of-speech n-grams, function words, readability indices, etc. However, there is a lack of systematic studies on the stability of such markers in idiolects under "context" (domain) change, mainly due to the lack of appropriate corpora. This paper aims to partially fill this gap by revealing the effect of domain shift (simultaneous change of genre and topic) on several groups of stylometric markers using a carefully constructed author and domain-controlled corpus. The effect of domain shift was studied separately for each group of markers (function words, lexical complexity indices, lexical and morphological diversity markers, readability indices, morphological markers, combined n-grams of part-of-speech and punctuation marks) using a combination of unsupervised and supervised methods: principal component analysis (PCA), ANOVA on PCA, joint PCA and cluster analysis, class (domain) prediction by nearest shrunken centroid. Only the markers of lexical complexity, lexical and morphological diversity were revealed to be untouched by domain shift, which stresses the necessity to choose the markers for cross-domain AA with special cautiousness due to the risk of domain, not authorship detection.

Keywords: Stylometry · Authorship attribution · Domain classification · Principal component analysis · Clustering

1 Introduction

The problem of determining the author of a text (i.e. authorship attribution, AA) has been studied for a long time. However, in the modern world with tons of texts produced by millions of people online on a daily basis, this issue became more urgent because of the amount of harmful content among produced texts related to trolling, bullying, etc. Meanwhile, the modern statement of problem of AA became even more complicated in comparison with the traditional (original) task of determining the author of a fiction text: as of now, the texts whose authorship are needed to be revealed are usually short and often come from different domains, i.e. they differ in topic, genre, register or even mode.

© Springer Nature Switzerland AG 2020
A. Karpov and R. Potapova (Eds.): SPECOM 2020, LNAI 12335, pp. 279–290, 2020.
https://doi.org/10.1007/978-3-030-60276-5_28

Despite its practical importance, researchers started to approach the problem of cross-domain AA only recently. For example, the organizers of PAN, a widely known series of hackathons in digital text forensics, did not include this task in their hackathons until 2018 [7]. It is obvious that in a such scenario we need stylometric features immune to domain shift since information about, for instance, the topic of texts can be misleading if each author in the dataset writes on a distinctive topic. However, the commonly used markers for this type of task are the same as used in all the other types of AA tasks: character and word n-grams, part-of-speech (POS) n-grams, function word statistics. The features such as readability indices (limited to word and sentence length, word length distribution, though) and lexical diversity markers (usually limited to type-token ratio and its variations) are also used, but much more rarely [7]. Another type of features used in cross-domain AA is related to the different types of text distortion. The general idea behind this type of features is to remove content words or letters leaving only function words or non-alphabetical symbols (e.g., punctuation), respectively [10, 18].

Nevertheless, several problems arise. First, research on the stability of style markers in idiolects under domain shift are very scarse. The existing research mostly focuses on the stability of style markers under topic change and shows that the features traditionally considered as "context-independent" (such as function word or different types of character or POS n-grams) are at least partially topic-dependent [13, 14, 16]. Research on the efficiency of AA predictive models under different topic shift scenarios [15] has shown that the accuracies of the models with the commonly used stylometric features (character, word and POS n-grams) decrease dramatically in the scenario where the test document belongs to a different topic than both training documents (cross-topic scenario) and even a larger drop in the accuracy is observed when "the 'correct' training document by the same author as the test document belongs to a different topic, whereas the 'wrong' document by a different author belongs to the same topic as the test document: thus the topic information interferes with authorship" [15, p. 307], which clearly shows the difficulty to infer true authorship (not topical) information.

Second, a closer inspection of the corpora used in cross-domain AA shows that texts under investigation usually did not differ in genre – they only differ in topics, although we assume that the genre + topic shift causes even more dramatical changes in style markers.

Third, the use of poorly interpretable features such as character n-grams adds little value to the general understanding of the concept of idiolect and its structure, which prevents practitioners from using the results of such research in forensic settings. A closer look at each linguistic level of idiolect under the change of factors related to communication situation (topic, genre, mode, etc.) is crucial to gaining insights into the real idiolect structure.

All abovementioned clearly states the necessity to inspect the effect of domain change on different groups of stylometric markers in the idiolect with a two-fold aim in mind: first, to reevaluate the efficiency of the already used features; second, to possibly broaden the list of such features with a special focus on the interpretability of the obtained results, so we deliberately exclude the character-based features from our list of markers. To the best of our knowledge, this is the first study to have aimed to systematically assess the linguistic behavior of the wide range of stylometric features under domain (topic+genre) shift, some of which have been introduced to a stylometric task for the first time.

2 Stylometric Features Under Domain Shift

2.1 Corpus

To arrive at the purpose we have set in this paper, it is crucial to construct a corpus of texts controlled for topic and genre produced by the same authors. Texts should be free from plagiarism of any kind, citations, etc. For a variety of methods to be applicable to the data, it is necessary to get substantial instances per class. Obviously, it is easier to control for all the above mentioned conditions if the data are gathered in the course of a specially designed experiment (writing tasks). A lot of resources are needed to gather such data, which hurdles research progress in this area. We used a dataset derived from the RusIdiolect database [15]. This is a freely available resource[1] constructed in RusProfiling Lab and specially designed to study the effect of different factors of idiolectal variation on style markers. The database contains both characteristics of texts and authors. We selected texts produced as a writing assignment by university students (N = 206, all native Russian speakers). Each author wrote 2 texts: the first text is a description of a series of pictures (one of three sets of pictures on choice, with most authors (80%) choosing the same picture to describe), the second one is an essay (10 topics to choose from, e.g. "What does friendship mean to you?", "What is your attitude to social media?", with no more than 15% of the authors choosing the same topic), so that we could study the effect of a joint topic and register (genre) change on idiolect markers. This combination is chosen since it resembles a real-world situation where train and test texts often come from different topics and genres simultaneously. All the texts were written in the presence of a researcher (40 min is a mean time authors needed to produce two texts) and subsequently checked for plagiarism. The mean text length is 331 words (Std = 136).

2.2 Style Markers

We analyzed each type of possibly context-independent style markers separately to reveal the effect of the topic + genre shift on idiolect markers. These types include:

1. **The most frequent tokens (word forms) of the corpus.** To extract these features, we used R package **quanteda** [1] and set document frequency threshold to 0.7 (scheme = "prop"), which leads to 8 features in this category, all of which are percentage of function words in each text: *что* ("that"), *как* ("how"), *на* ("on"), *с* ("with"), *и* ("and"), *к* ("to"), *в* ("in"), *не* ("no").
2. **Lexical complexity indices** based on the percentage in texts (both in the whole text length and in the first 100 word forms or lemmas of each text in the dataset) of the

---

[1] https://rusidiolect.rusprofilinglab.ru (accessed on June 10, 2020).

most frequent Russian words (with and without pronouns). Lists of such words and lemmas were constructed manually based on RuTenTen thesaurus IPM rank with subsequent removal of content words[2]. Indices calculated on the whole length of texts from non-lemmatized dataset are **8_MFW** (percentage of the sum of raw values of features from the first group), **100_Forms_Pron** (percentage of function words including pronouns from 100 most frequent Russian word forms list), **1000_Forms_Pron** (percentage of function words including pronouns from 1000 most frequent Russian word forms list), **100_Forms_strict** (same as **100_Forms_Pron** but excluding pronouns), **1000_Forms_strict** (same as **1000_Forms_Pron** but excluding pronouns), **8_MFW_100** (same as **8_MFW** but in the first 100 word forms), **100_Forms_Pron_100** (same as **100_Forms_Pron** but in the first 100 word forms), **1000_Forms_Pron_100** (same as **1000_Forms_Pron** but in the first 100 word forms), **100_Forms_strict_100** (same as **100_Forms_strict** but in the first 100 word forms), **1000_Forms_Strict_100** (same as **1000_Forms_strict** but in the first 100 word forms). We also extract the same features on lemmatized dataset (except for **8_MFW**) using the list of most frequent Russian lemmas with subsequent removal of content words: **100_Lemms_Pron, 1000_Lemms_Pron, 100_Lemms_Strict, 1000_Lemms_Strict, 100_Lemms_Pron_100, 1000_Lemms_Pron_100, 100_Lemms_Strict_100, 1000_Lemms_Strict_100.**

3. **Lexical diversity markers** calculated both on non-lemmatized and lemmatized dataset separately. These markers included all indices calculated with function **textstat_lexdiv** of R package **quanteda: TTR, CTTR, C, R, U, S, K, I, D, Vm, Maas** and two indices calculated with R package **koRpus** [12]: **lgV0, lgeV0**. Due to the restricted space, we refer the reader to the **quanteda** and **koRpus** manuals for detailed description of the indices. We also include indices calculated with QUITA software and least dependent on text length according to software developers [8]: **ATL, R1, RRmc, Lambda, WritersView, CurveLength.** All the above mentioned indices were calculated on the whole text length. In addition, all QUITA indices calculated on the first 100 word forms and lemmas were added to enrich the analysis: **TTR_100, h-Point_100, Entropy_100, ATL_100, R1_100, RR_100, RRmc_100, Lambda_100, Adj_Mod_100, G_100 R4_100, Hapax_100, WritersView_100, L_100, CurveLength_100.** For the lemmatized corpus, we also added **quanteda** features with the most frequent Russian function word lemmas with pronouns from top-1000 list removed.

4. **Readability indices** extracted with R package **koRpus** [12]: letters.all, syllables, punct, avg.sentc.length, avg.word.length, avg.syll.word, sntc.per.word, ARI, Coleman.Liau, Danielson.Bryan.DB1, Danielson.Bryan.DB2, Dickes.Steiwer, ELF, Farr.Jenkins.Paterson, Flesch, Flesch.Kincaid, FOG, FORCAST, Fucks, Linsear. Write, LIX, nWS1, nWS2, nWS3, nWS4, RIX, SMOG, Strain, TRI, Tuldava, Wheeler.Smith. Due to the lack of space, we cannot describe them in details and refer the reader to the **koRpus** manual where all these features are well-documented.

---

[2] https://www.sketchengine.eu/rutenten-russian-corpus/#toggle-id-1 (accessed on June 10, 2020).

5. **Morphological diversity indices** which have been, to the best of our knowledge, introduced for the first time in stylometry. These are indices of lexical diversity calculated with **quanteda**, but this time not on word forms or lemmas but on POS tags: **C, R, U, S, K, I, D, Vm, Maas** and two indices calculated with R package **koRpus**: **lgV0, lgeV0**.

6. **Morphological indices** which reflect different textual properties derived from POS tags: **c_punct** (ratio of conjunctions to punctuation marks (PM)), **prep_conj** (ratio of prepositions to conjunctions), **prep** (ratio of prepositions to punctuation marks), **verb_sent** (ratio of verbs to the number of sentences), **c_sent** (ratio of conjunctions to the number of sentences), **adv_sent** (ratio of adverbs to the number of sentences), **pron** (ratio of conjunctions to punctuation marks), **q** (ratio of particles to punctuation marks), **Lex_Dens** (ratio of function words to text length), **Lex_Dens_Mod** (ratio of function words to the sum of nouns and verbs), **Prep_Noun** (ratio of prepositions to nouns), **Quality** (ratio of adjectives to noun), **Quality_Mod** (ratio of adjectives to adverbs), **Details** (ratio of the sum of adjectives and adverbs to the sum of nouns and verbs), **Deicticity** (ratio of pronouns to the sum of adjectives, nouns and adverbs), **F_measure** (noun + adjective + preposition − pronoun − verb − adverb - interjection + 100)/2), **ID** (the ratio of the sum of verbs, adjectives, adverbs and prepositions to text length), **Connection** ((prepositions + conjunctions)/number of sentences multiplied by 3), **SUBJ** (ratio of nouns to the sum of adjective and verbs), **QUAL_Mod1** (ratio of adverbs to adjectives), **QUAL_Mod2** (ratio of adverbs to verbs), **DYNAMICS** (ratio of verbs to the sum of adjectives, nouns, adverbs and pronouns).

7. **Combined POS and punctuation marks n-grams.** When performing POS-tagging, we did not remove PM except for inverted commons which we believe are topic-dependent. Instead, we substituted all PMs occurring at the end of sentence with SENT, all other PMs as PUNCT. We set n to [2:3], since higher values of n lead to high sparsity of the data. With minimum document frequency = 190 we got 10.8% sparsity and 31 features in this group (normalized on text length).

Lemmatization and POS-tagging were performed with Treetagger [17]. We used a short list of morphological tags in our experiments (nouns, verbs, numerals, adjectives, particles, pronouns, prepositions, conjunctions, adverbs).

2.3 Methods

To get more solid results, we used a combination of methods. First, we built a correlation plot to reveal the level of intercorrelations (Pearson's correlation coefficient r, $p < 0.05$) of markers inside groups as well as their correlation with text length. The latter was used as supplementary quantitative variable in PCA.

Since we are interested in linguistics behavior of the whole groups of features, we apply principal component analysis (PCA) to each group of features with a supplementary qualitative variable (factor) "domain" with two levels (essay and picture) using R packages **ade4** [2] and **FactoMineR** [9] with default feature standardization. Supplementary variables do not participate in factorial axes construction but help in the interpretation of the results. We make a one-way analysis of variance for the coordinates of the individuals (i.e. texts) on the axis and qualitative variable (factor) as implemented in R package **FactoMineR** with p cut-off value of F-test equal to 0.01. Although it is well-known that "univariate methods offer the advantage of directly investigating differences on endpoints of interest whereas multivariate tests are applied on a linear combination of the original variables" [19], we did not perform individual ANOVAs for each variable since most of them correlate with each other (inside groups). "Correlated variables reflect overlapping variance and therefore univariate tests provide little information about the unique contribution of each dependent variable" [19]. Confidence ellipsis were constructed to assess the stability of division of individuals by groups (domains) with the help of **FactoMineR.**

PCA results were also visualized with R package **factoextra** [6].

We also used new method for joint dimension reduction and clustering as implemented in R package **clustrd** [11] to reveal hidden structure in the data since our preliminary experiments showed that the cluster quality measured with the average silhouette width is higher when low-dimensional rather than raw data are used (text length was excluded from this analysis). Since PCA does not take into account group membership and differences between groups do not necessarily become apparent in reduced space, we added classification tool for our analysis, namely nearest shrunken centroid (NSC) as implemented in R package 'Stylo' [3] for class (domain) prediction. This is a simple, yet efficient classifier which provide feature weights [5]. 10-fold cross-validation protocol was used. Such combination of methods enables one to analyze data from different viewpoints and to draw more solid conclusions.

2.4   Results and Discussion

Contrary to our expectations, most of the markers were not totally resistant to domain change. First, the coordinates of individuals on both Dimensions 1 and 2 of PCA performed on the most frequent words were linked to the factor "domain" ($p < 0.0001$). The classifier divided texts from different domains with a high mean accuracy (0.825), with prepositions being the features with the highest discriminative power. Biplot (Fig. 1) clearly shows patterns of using function words depending on the domain as well as tendency of separation of essays and pictures. Joint PCA and cluster analysis resulted in the two-cluster solution as the most stable with a clear tendency to domain-based clustering: 79.6% of all the texts in cluster 1 are essays and 77.7% in cluster 2 are picture descriptions.

Contrary to our expectations, lexical complexity indices did not correlate with text length. No one dimension was linked to the factor. A further analysis showed confidence ellipsis total overlap (Fig. 2). The classifier did not manage to divide the texts according to their domain (mean accuracy = 0.5821).

Among the lexical diversity indices, calculated with **quanteda** on non-lemmatized corpus, **TTR** и **CTTR** (Carroll's Corrected TTR) had a correlation with text length higher than 0.5 and were excluded from the analysis as well **Entropy, L, h.Point, RR и Adj_Mod** calculated with QUITA. PCA on non-lemmatized corpus showed no relation of coordinates of individuals on Dim 1 explaining large part (54.17%) of variance, unlike Dim 2 which was related to the factor ($p = 0.0004$). The confidence ellipsis partially overlapped.

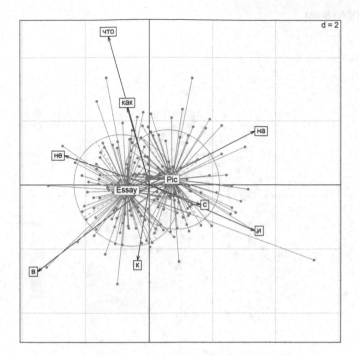

Fig. 1. PCA with most frequent tokens of the corpus.

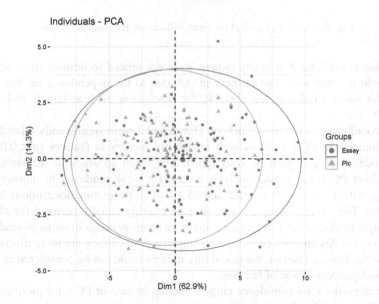

Fig. 2. PCA with lexical complexity indices.

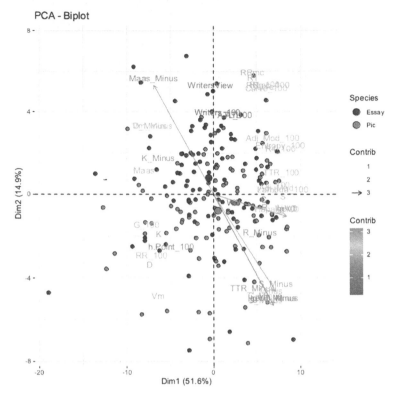

Fig. 3. PCA with lexical diversity indices on lemmatized corpus.

Therefore, the lexical diversity indices were not related to domain shift, although they should be used with cautiousness in AA due to the dependence on text length typical for some of them (see also [4] for discussion on text length and lexical diversity).

The coordinates of the individuals on Dim 1 and 2 were significantly related to the factor "domain" in PCA analysis with readability indices as features (p < 0.00001). The confidence ellipsis partially overlapped. The mean accuracy of the classifier is 0.6911. Joint PCA and cluster analysis revealed a weak tendency to domain-based clustering, with a two-cluster solution and 85.4% of all the picture descriptions fall into one cluster. The essays, however, were divided roughly equally between the clusters, which might be due to the fact that they are more heterogeneous in terms of readability than the picture descriptions (which is not surprising, since they are more diverse with respect to the topics). Overall, the readability indices could not be considered as totally domain-independent group of features.

We observed a total confidence ellipsis overlap in case of PCA for morphological diversity features (Fig. 4).

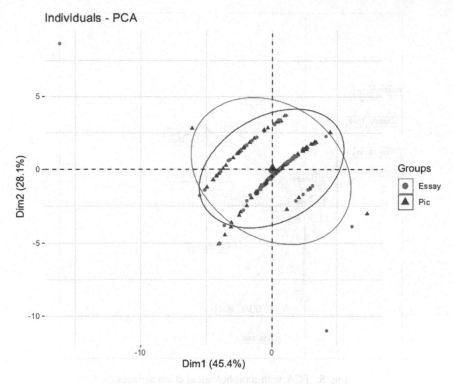

Fig. 4. PCA with morphological diversity features.

The classification accuracy is 0.5. However, although we used only the features which demonstrated a low or no correlation with text length in the experiment on word forms and lemmas, this time 4 features of this group (**R, C, Maas, U**) were highly correlated with text length (Pearson's r > 0.5, p < 0.001), which should be taken into account when using this type of features.

The morphological indices were weakly correlated with text length. The coordinates of individuals on both Dim 1 and Dim 2 were related to the factor 'domain' (p < 0.0001). The confidence ellipses showed a slight overlap, the classification accuracy was 0.7375. The morphological patterns are not unrelated to domain (Fig. 5).

N-grams of POS and punctuation marks showed a low correlation with text length. Dim 1 and 2 were related to the factor. The confidence ellipsis were slightly overlapped (Fig. 6). The classifier was able to divide the texts according to domain with a high mean accuracy 0.8125.

Overall, the morphological features (both POS ratios and POS n-grams) were not totally domain-independent.

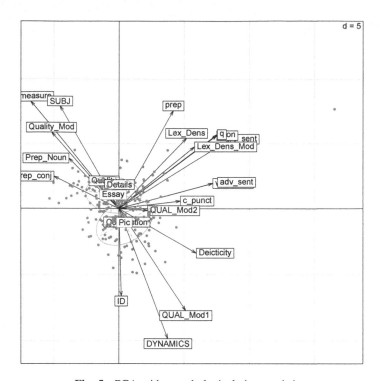

Fig. 5. PCA with morphological characteristics.

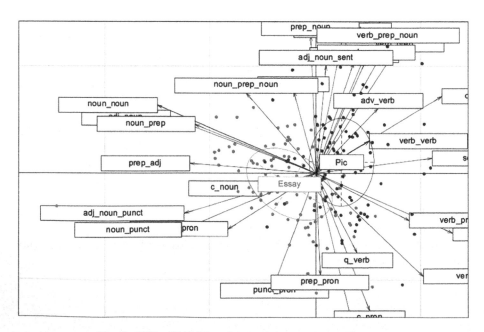

Fig. 6. PCA with POS and punctuation n-grams as features.

3 Conclusions and Future Work

Our research performed on the carefully constructed domain- and author-controlled corpus with the help of both supervised and unsupervised data mining techniques has clearly shown that most groups of stylometric features widely used in authorship attribution studies and considered to be "context independent" (function words, POS n-grams) are not really domain-independent. Only the markers related to the concept of lexical complexity and lexical diversity as well as morphological diversity (the latter introduced in this paper for the first time) were immune to domain shift. However, the features related to lexical and morphological diversity should be selected in future AA studies bearing in mind high correlations of some of them with text length. Only the features related to lexical complexity were domain and text-length independent, which could be possibly linked to the fact that they reflect some fairly stable characteristic of language ability.

Our future plans include further investigation of the stability of analyzed groups of features in different domain-shift scenarios as well as their relation to the factor "author", i.e. their ability to distinguish idiolects taken into account different factors of idiolectal variation. Along with the already discussed groups of features, we aim at performing sequence mining to investigate the patterns of punctuational behavior of the authors more thoroughly.

The main takeaway of this study is that it is crucial to carefully select stylometric markers for cross-domain authorship attribution as well as to take into account the factors of intraidiolectal variation (genre, topic, register, mode, etc.) in such experiments. Otherwise, there is a risk that the domain rather than authorship is determined.

Acknowledgement. The research has been performed in Voronezh State Pedagogical University under the support of Russian Science Foundation, grant number 18-78-10081, which is gratefully acknowledged.

References

1. Benoit, K., et al.: An R package for the quantitative analysis of textual data. J. Open Source Softw. **3**(30), 774 (2018)
2. Dray, S., Dufour, A.: The ade4 package: implementing the Duality Diagram for Ecologists. J. Stat. Softw. **22**(4), 1–20 (2007)
3. Eder, M., Rybicki, J., Kestemont, M.: Stylometry with R: a package for computational text analysis. R J. **8**(1), 107–121 (2016)
4. Fergadiotis, G., Wright, H.H., Green, S.: Psychometric evaluation of lexical diversity indices: assessing length effects. J. Speech Lang. Hear. Res. **58**, 52–840 (2015)
5. Jockers, M., Witten, D.: A comparative study of machine learning methods for authorship attribution. LLC **25**, 215–223 (2010)
6. Kassambara, A.: Practical Guide to Principal Component Methods in R (Multivariate Analysis Book 2). CreateSpace Independent Publishing Platform (2017)
7. Kestemont, M., et al.: Overview of the cross-domain authorship attribution task at PAN 2019. In: CLEF 2019 Labs and Workshops, Notebook Papers (2019). http://ceur-ws.org/Vol-2380/paper_264.pdf

8. Kubát, M., Matlach, V., Čech, R.: QUITA – Quantitative Index Text Analyzer. RAM, Lüdenscheid (2014)
9. Lê, S., Josse, J., Husson, F.: FactoMineR: an R package for multivariate analysis. J. Stat. Softw. **25**, 1–18 (2008)
10. Litvinova, T., Litvinova, O., Panicheva, P.: Authorship attribution of Russian forum posts with different types of n-gram features. In: Proceedings of the 2019 3rd International Conference on Natural Language Processing and Information Retrieval, pp. 9–14 (2019)
11. Markos, A., D'Enza, A.I., van de Velden, M.: Beyond tandem analysis: joint dimension reduction and clustering in R. J. Stat. Softw. **91** (2019)
12. Michalke, M.: An R package for text analysis (Version 0.11-5). https://reaktanz.de/?c=hacking&s=koRpus. Accessed 21 May 2018
13. Mikros, G.K.: Investigating topic influence in authorship attribution. In: SIGIR 2007 Amsterdam, Workshop on Plagiarism Analysis, Authorship Identification, and Near-Duplicate Detection (2007)
14. Panicheva, P., Litvinova, O., Litvinova, T.: Author clustering with and without topical features. In: Salah, A.A., Karpov, A., Potapova, R. (eds.) SPECOM 2019. LNCS (LNAI), vol. 11658, pp. 348–358. Springer, Cham (2019). https://doi.org/10.1007/978-3-030-26061-3_36
15. Panicheva, P., Litvinova, T.: Authorship attribution in Russian in real-world forensics scenario. In: Martín-Vide, C., Purver, M., Pollak, S. (eds.) SLSP 2019. LNCS (LNAI), vol. 11816, pp. 299–310. Springer, Cham (2019). https://doi.org/10.1007/978-3-030-31372-2_25
16. Sapkota, U., Bethard, S., Montes, M., Solorio, T.: Not all character n-grams are created equal: a study in authorship attribution. In: Proceedings of the 2015 Conference of the North American Chapter of the Association for Computational Linguistics: Human Language Technologies, pp. 93–102. ACL. Denver (2015)
17. Schmid, H.: Probabilistic part-of-speech tagging using decision trees. In: Proceedings of International Conference on New Methods in Language Processing, Manchester, UK (1994)
18. Stamatatos, E.: Masking topic-related information to enhance authorship attribution. J. Assoc. Inf. Sci. Technol. **69**(3), 461–473 (2018)
19. Todorov, H., Searle-White, E., Gerber, S.: Applying univariate vs. multivariate statistics to investigate therapeutic efficacy in (pre)clinical trials: a Monte Carlo simulation study on the example of a controlled preclinical neurotrauma trial. PLoS ONE **15**(3), e0230798 (2020)

Speech Features of 13–15 Year-Old Children with Autism Spectrum Disorders

Elena Lyakso$^{(\boxtimes)}$ [iD], Olga Frolova [iD], Aleksey Grigorev,
Viktor Gorodnyi, Aleksandr Nikolaev, and Anna Kurazhova

The Child Speech Research Group,
St. Petersburg State University, St. Petersburg, Russia
lyakso@gmail.com

Abstract. The goal of the study is to determine the effect of the child's age and the severity of autistic disorders on the speech features of children with autism spectrum disorders (ASD) aged 13–15 years and the impact of these factors in recognition by adults of the information contained in the child speech. Participants of the study were 10 children with ASD and 107 adults – listeners. The design of the study included: spectrographic, phonetic, linguistic analyses of children's speech; two series of perceptual experiments with determining the children's speech intelligibility and children's state – psychophysiological and emotional via speech. The data on the determining of the articulation clarity, normative intonation, recognition of the psychoneurological state and the emotional state of children via speech, and the acoustic features of the speech material are presented. The impact of child's age and severity of ASD on the speech features and the listeners' responses is shown. The obtained data indicate a complex trajectory of speech development in ASD informants. The future approach of this research will be automatic classification of ASD child state via speech.

Keywords: Autism spectrum disorders · Perceptual experiment · Acoustic analysis · Child age · Gender · Severity of the disease

1 Introduction

Speech impairment is one of the symptoms of autism spectrum disorders (ASD). This characteristic is included in the triad of autistic disorders based on which the diagnosis is made. Regarding the existence of specific speech disorders in individuals with ASD [1], most researchers have the same opinion; the differences concern particular features of speech [e.g., 2–5]. A wide range of speech disorders has been described in children with ASD: from a severe delay to a faster rate of speech development in children with high-functioning autism. The acoustic features of sounds and speech constructions of infants are significant as a possible predictor of the development of autism [6], and the acoustic features of child speech could be considered as possible biomarkers of autism [3]. Prosodic violations in children with ASD that may occur in the spontaneous speech [3, 7, 8] and in the repetition speech [8] were described. Specific prosody, which is caused by abnormal accentuation and emphasis of the stressed syllable in the word and

© Springer Nature Switzerland AG 2020
A. Karpov and R. Potapova (Eds.): SPECOM 2020, LNAI 12335, pp. 291–303, 2020.
https://doi.org/10.1007/978-3-030-60276-5_29

of the word in the phrase, is a specific peculiarity of children with ASD [9, 10] and makes verbal communication difficult. A few studies indicate patients' inability to control the height of voice, atypical word and phrasal stress [2, 8]. The question about the height of voice (pitch) in informants with ASD is widely discussed. The monotony of speech of Japanese children aged 4–9 years with ASD [4] and high values of the pitch and its range in 8–9 year-old Portuguese children [11], 4–10 year-old bilinguals (Hindi – English) [10], English schoolers with high-functioning autism [12], 4–6.5 year-old Israeli children [3] have been described. The child age is the most studied, so the symptoms are more expressed than in adulthood and in elderly age. It is also shown that the speech of 4–15 year-old Russian children is characterized by high values of the pitch and its range [13].

ASD are a combined term for a group of disorders that, along with the general features, have their own characteristics. According to the World Health Organization, ASD are accompanied by intellectual disabilities in about 50% of children. Therefore, the participants of the study were children with ASD of various severity.

The goal of the study is to determine the effect of child's age and severity of autistic disorders on the speech features of ASD children aged 13–15 years and the impact of these factors in recognition by adults of the information contained in the child speech. The study tested two hypotheses. The first hypothesis was that the age of children mainly affects the pitch values, and the severity of ASD affects the pitch values and the linguistic component of speech. The second hypothesis was aimed at testing the assumption about the influence of the severity of ASD on the listeners' determination of children's speech intelligibility and about the significance of the linguistic component of speech for determining the state of children with ASD via their speech.

2   Methods

2.1   Data Collection

Participants of the study were 10 children with ASD aged 13–15 years and 107 adults – listeners.

Speech material of children was taken from the speech database "AD-Child.Ru" [14]. Speech files are stored in the database in Windows PCM format WAV, 44.100 Hz, 16 bits per sample.

Children were selected based on the criteria of age and the presence of ASD. For this study, the ASD participants were divided into two groups according to developmental specificities: the presence of development reversals at the age of 1.5–3.0 years (the first group - ASD-1) – ASD are the leading symptom, and children with developmental risk diagnosed at the infant birth (the second group - ASD-2). For these children with intellectual disabilities, ASD are a symptom of neurological diseases associated with brain damage.

The sample of autistic children included 6 children (boys) (ASD-1) and 4 children (3 girls, 1 boy) with intellectual disabilities for whom the symptoms of ASD resulted from the underlying disease (ASD-2).

All children swam in the swimming pool individually with their instructor for 40 min and then took part in the study of their speech and cognitive development.

2.2 Data Analysis

The design of the study: 1. Spectrographic, phonetic, and linguistic analyses of children's speech. 2. Two series of perceptual experiments.

Spectrographic analysis of the children's speech material was carried out based on the algorithms implemented in the Cool Edit Pro sound editor. The acoustic features: the duration of phrases, words, stressed vowels, and pauses between words in the phrase, pitch values (F0) of phrases, words, stressed and unstressed vowels, formants (F1 – first formant, F2 – second formant) in the words were measured. Vowel articulation index [15] was calculated. The phonetic transcription of words was conducted on the basis of SAMPA for the Russian language. A comparison of speech features depending on the gender, age, group, emotional state was carried out.

The main goal of the perceptual study was to reveal the possibility of listeners to determine the children's speech intelligibility and correct intonation; the children's psychoneurological state – developmental disorders or typical development via speech; the emotional state via speech. The speech of ASD individuals is characterized by atypical prosody and articulation, so the ability of listeners to correctly determine speech features is important for specialists working with ASD informants and for society where ASD individual lives.

Two experts with professional experiences in the field of child speech and emotional states analysis selected speech samples, which were pronounced by children at the discomfort, neutral, and comfort state (by viewing video and using the recording protocol). The speech sample was considered as attributed to the corresponding category only when two experts gave the same answers.

Two tests contained the speech samples of ASD children (70 speech samples were annotated into three states "comfort – neutral – discomfort"). The division of speech samples into two tests was made for the convenience of listeners.

The listeners were 107 adults (Table 1):

20 adults with professional experiences in interacting with children (professional experience is more than 5 years) and household experiences (their own children, younger brothers and/or sisters in the family) in interacting with children – experts; 48 first-year pediatric Russian students; 29 foreign first-year pediatric students; 10 master students – IT (Information technology).

Table 1. Summary of listeners group demographics. Age is given in years.

	Experts	Russian pediatric students	Foreign pediatric students	Russian master students – IT
n	20	48	29	10
mean	38.4	19.5	20.5	25.1
SD	13	1.9	2.6	4.2

Every group of listeners had their specific task. Two perceptual studies were carried out. Study 1: Experts recognized the children's speech intelligibility and correct intonation.

Intelligibility is a critical component of effective communication. For children with significant speech disorders, intelligibility often has detrimental impact on functional communication and social participation. Intelligibility has been defined as the extent to which an acoustic signal, generated by a speaker, can be correctly recovered by a listener [16]. In our study, intelligibility was determined on the basis of the listener's response - whether he understood what the children said or not.

A rating scale with two poles (0 – no, 1 – yes) was used in three tasks – the determination of articulation accuracy, intelligibility (it is clear what the child said or not), intonation (corresponds to the normative for the Russian language or not). There was no preliminary training for the experts.

Study 2. Pediatric students and master students–IT recognized the psychoneurological state and emotional state of the children.

The selection of listeners – pediatric students – Russian and foreign, was carried out to test the significance of the linguistic component in determining the state of children.

The selection of Russian listeners in two groups – pediatric students and students with technical specialization will determine the influence of the listener's predisposition to recognizing the state of children. First-year pediatric students have minimal experience in interacting with children with atypical development, but their choice of medical specialization allows comparing their answers with those of students with technical specialization.

The test sequences were presented for groups of listeners.

All statistical tests were conducted using Statistica 10.0.

Ethical approval was obtained from the Ethics Committee (Health and Human Services, HHS, IRB 00003875, St. Petersburg State University).

3   Results

3.1   Groups of ASD Children

The discriminant analysis showed that the assignment of the child to the group is due to $F(5,154) = 237.49$ the child's gender (Wilk's Lambda = 0.42 $p < 0.1$) and the Child Autism Rating Scale (CARS [17]) scores (reflects the severity of autistic disorders) (Wilk's Lambda = 0.142 $p < 0.3$). The correlation between the CARS scores F $(2,41) = 15.944$ $R^2 = 0.437$, the group (ASD-2) ($\beta = -1.223$ $p < 0.003$), and the child's gender ($\beta = 0.614$ $p < 0.1$) was found. The correlation between the CARS-verbal scores $F(1,159) = 102.72$ $p < 0.00001$ ($R^2 = 0.392$ $\beta = -0.626$) and the child's gender was found.

3.2   Speech Features of ASD Children

The speech samples of children included words (24.6%), phrases (49.3%), and speech constructions, which meaning is difficult to identify without situation context (26.1%).

The correlation between the speech complexity and high values for CARS–verbal scores (r = 0.703 – Spearman p < 0.05) was found. Parts of speech were presented by nouns (34.8%), verbs (24.1%), adjectives (12.4%), adverbs (7%), pronouns (7.8%), numerals (3.1%), interjections (0.8%), and function words (10%). The correlation between the use of the function words in the child's speech: prepositions, conjunctions, particles, and the child's phonemic hearing (r = −0.523 – Spearman p < 0.05) was revealed.

Differences between the acoustic features of children's speech depending on the age, emotional state, gender, and group were found (Table 2).

Table 2. The characteristic of significant differences in the speech of children depending on the age, emotional state, gender, and group (Mann-Whitney test).

Features	Speech material	Factors			
		Age, y	Emotional state	Gender m/f	Group
Duration	Phrase		D > N*		
	Word				
	Stressed vowel (av)			m < f***	1 < 2*
	Stressed vowel (st.p.)		D > C*; D > N*		
Pitch (F0)	Phrase			m < f***	
	Word	13 > 14*; 13 > 15*		m < f***	1 < 2***
	Stressed vowel (av)	13 > 14*; 13 > 15*	D > N*	m < f***	1 < 2***
	Stressed vowel (st.p.)	13 > 14*; 13 > 15*	D > N*	m < f***	1 < 2***
	Posttonic vowel 1	13 > 15*	C > N*	m < f***	1 < 2***
	Posttonic vowel 2	13 > 14*			
F0max-F0min	Stressed vowel (av)		C > N*; D > N*	m < f***	
	Posttonic vowel 1		C > N*		
	Posttonic vowel 2		D > N*		

Notes: * – p < 0.05; *** p < 0.001 – Mann-Whitney test; av – average, st.p. – stationary part; C – comfort state, D - discomfort state, N - neutral state.

For all children, the correlation F(1,159) = 53.679 p < 0.0001 (R^2 = 0.161 β = −0.402) between the pitch values for the phrase and the CARS-verbal scores was shown.

Differences between boys and girls were revealed in pitch values (significantly lower in boys than in girls), therefore, the further determination of the influence of age and CARS scores on speech features was carried out separately for each group.

For ASD-1 children: The regression analysis revealed a correlation between the CARS scores and: $F(1,90) = 9.936$ $p < 0.002$ ($R^2 = 0.99$ $\beta = 0.315$) average pitch values of stressed vowels; $F(1,90) = 15.494$ $p < 0.00003$ ($R^2 = 0.147$ $\beta = 0.383$) maximum pitch values of stressed vowels; $F(1,90) = 13.770$ $p < 0.00003$ ($R^2 = 0.133$ $\beta = 0.364$) minimum pitch values of stressed vowels. The CARS-verbal scores are correlated: $F(1,90) = 17.046$ $p < 0.00001$ ($R^2 = 0.159$ $\beta = 0.399$) with average pitch values of stressed vowels and with $F(12,53) = 1.368$ $p < 0.008$ ($R^2 = 0.237$ $\beta = 0.373$) duration of phrases. The age factor is reflected in the correlation between the age $F(3,88) = 2.465$ $R^2 = 0.775$ and the CARS scores ($p < 0.03$ $\beta = 0.31$) and the CARS-verbal scores ($p < 0.02$ $\beta = -0.375$).

Group ASD-2 is mostly represented by girls; children from ASD-2 were 13 and 14 years old. Therefore, the revealed acoustic features (Table 2) were compared for boys and girls. At the age of 14 years (ASD-2 – 1 boy, 1 girl): pitch values for the phrase, word, average values of the stressed vowel, the stationary part of the stressed vowel, and unstressed vowel, F0max–F0min values, duration of the stressed vowel in girls' speech are significantly higher than in boys' speech ($p < 0.001$). That is, within the group, differences between girls and boys via speech features were revealed. At the age of 13 years (ASD-2–2 girls): pitch values of the phrase, word, stressed vowel (average), stressed vowel (the stationary part), and unstressed vowel are higher ($p < 0.001$) in girls' speech than in boys. Thus, these comparative data showed that the differences between the groups (ASD-1, ASD-2) are mainly due to the presence of three girls in group 2.

All the speech material was annotated into three states "comfort – neutral – discomfort". The correlation between the emotional state $F(2,78) = 5.536$ $R^2 = 0.124$ and the heart rate ($\beta = 0.734$ $p < 0.001$) and blood oxygen saturation ($\beta = 0.652$ $p < 0.004$) was found. These vegetative parameters indicate the excitement of the children in an emotional state. The VAI for stressed vowels from words uttered in the comfort state is higher (1.09) than the VAI for stressed vowels in the neutral (0.96) and discomfort states (0.94). The discriminant analysis showed that the classification of ASD-1 children's speech samples to emotional states is correlated $F(12,120) = 3.466$ Wilk's Lambda = 0.551 with the duration of the phrase (Wilks' = 0.788 $p = 0.00002$), pitch values of the phrase (Wilks' = 0.635 $p = 0.01$), pitch values of the word (Wilks' = 0.631 $p = 0.01$).

3.3   Perceptual Study

Study 1. Speech Intelligibility and Normative Intonation. The agreement of expert answers was the greatest in recognizing the articulation accuracy and intelligibility of the children's speech. According to the experts, the speech of 13–14 year-old children is mostly intelligible; articulation is the clearest at the age of 13 and 14 years. The speech of 15 year-old children is mainly unclear to the experts, intonation is abnormal, and articulation is fuzzy (Fig. 1).

The phonetic analysis showed that, in the speech material whose meaning is clear to the experts (the speech is intelligible), there are substitutions of /S':/for /S/, /t/for /d/, / p'/for /b'/, /d'/for /t'/. Reduced vowels /I, @/occur on average 1:5–6 (one reduced vowel of 5–6 vowels). In the speech material described as illegible, speech is unclear and characterized by substitutions of /tS'/for /k'/and /t'/, /r'/for /R'/, /r/for /G/. Reduced vowels /I, @/occur on average 1:8–9 (1 reduced vowel of 8–9 vowels). 30% of children's phrases cannot be described, 20% of phrases are partially described.

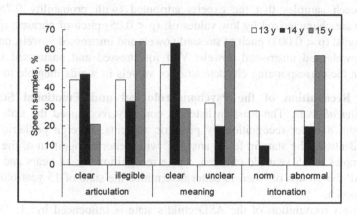

Fig. 1. The number of speech samples determined by the experts (with probability 0.75–1.0) depending on children's age (%).

The determination of articulation accuracy in children, meaning of children's speech, normal intonation correlates with some information of children (Table 3).

Table 3. Characteristics of children influencing correct recognition of the children's articulation skills, meaning of speech, and intonation - Regression analysis.

Indicators F (1,67)	Experts answers								
	Articulation			Meaning			Normal intonation		
	P	R^2	β	p	R^2	β	p	R^2	β
Gender	.0009	.154	−.392	.05	.164	−.392			
Group	.0005	.071	−.267	.05	.085	−.291	.05	.09	.30
Child age							.002	.129	−.359
Emotional state							.002	.139	−.373
CARS total							.005	.118	−.343
CARS-verbal							.0001	.535	−.732

Given the CARS scores (high total CARS scores and CARS-verbal scores), it was shown that the experts identified the speech samples of these children as characterized by unclear articulation (100%), illegible (100%), with abnormal intonation (100%).

The speech samples that the experts attributed to clear articulation (recognition probability 0.75–1.0) are characterized by lower values of: (p < 0.001) pitch of phrases, pitch of words, pitch of stressed vowels, pitch of unstressed vowels, duration of stressed vowels and unstressed vowels; (p < 0.01) pitch range of stressed and unstressed vowels. VAI for stressed vowels is higher than in words with unclear articulation; VAI for unstressed vowels is lower vs. stressed vowels. VAI for unstressed vowels in words with unclear articulation is higher than VAI for stressed vowels.

The speech samples that the experts attributed (with probability 0.75–1.0) as intelligible are characterized by low values of: (p < 0.05) pitch of phrases; (p < 0.01) pitch of words; (p < 0.001) pitch of stressed vowels and unstressed vowels, duration of stressed vowels and unstressed vowels; VAI for stressed and unstressed vowels is higher than the corresponding characteristic for vowels in words illegible to experts.

Study 2. Recognition of the Psychoneurological and Emotional State. *Psychoneurological State.* The Russian listeners correctly recognized the state of ASD children with a better recognition by pediatric students. Foreign pediatric students correctly identified the state in fewer answers, with better recognition of the state via speech samples of 15 year-old children, worse recognition for 14 years and 13 years (Fig. 2). All the groups of listeners better determined the state of 15 year-old children via their speech.

The correct recognition of the ASD child's state is influenced by: 1. The native language of the listeners – Russian-speaking pediatric students accurately recognized the psychoneurological state of children more successfully than foreign students $F(1,138) = 93.049$ p < 0.00001 ($R^2 = 0.403$ β = −0.635); 2. Specialization – Russian pediatric students better recognized the state of children than IT students $F(1,138) = 19.013$ p < 0.00003 ($R^2 = 0.115$ β = −0.348).

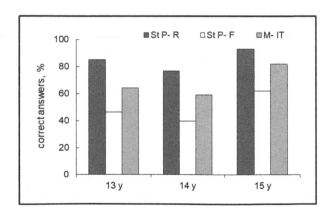

Fig. 2. The number of listeners' correct answers on the psychoneurological state recognition of ASD children "typical development – developmental disorders". St P-R – pediatric students, native Russian-speakers; St P-F – pediatric students, foreigners; M-IT – master students, IT.

The determination of the children's psychoneurological state is influenced by the indicators of the children's development (Table 4).

Table 4. Developmental indicators influencing correct recognition of the children's psychoneurological state (Regression analysis).

Indicators F (1,68)	Listeners' groups								
	St P-R			St P-F			M-IT		
	p	R^2	β	p	R^2	B	P	R^2	β
Gestation	.001	.14	−.375	.0008	.154	−.392	.0001	.271	−.521
Birth	.001	.137	−.37	.0001	.256	−.504	.0005	.166	−.408
Gender	.026	.71	.266				.005	.208	.329
CARS-verbal				.0006	.159	.398	.01	.089	.298
Group	.02	.068	.261				.008	.099	.314
Emotional state				.004	.107	.337			

Notes: St P-R – pediatric students, native Russian-speakers; St P-F – pediatric students, foreigners; M-IT – master students, IT.

Foreign students more correctly recognized the psychoneurological state by speech samples of discomfort (p < 0.05) than by neutral and comfort speech samples.

The speech samples correctly categorized by listeners are characterized by (in comparison with speech samples assigned to typical development): high pitch values (p < 0.001) of words, of stressed and unstressed vowels in words; pitch range values of stressed (p < 0.01) and unstressed vowels (p < 0.001); longer duration of stressed (p < 0.001) and unstressed vowels (p < 0.01); VAI for stressed vowels is higher than VAI for unstressed vowels. VAI for unstressed vowels in words is lower than VAI of unstressed vowels in words from utterances attributed to typical development.

Emotional State. Russian listeners – pediatric students and IT students – equally correctly recognized the emotional state of children (Table 5). The listener's native language affects the emotional state recognition of children – foreign pediatric students recognized the emotional state of children worse than Russian pediatric students F (1,138) = 6.33 p < 0.05 (R^2 = 0.044 Beta = −0.209).

Table 5. Confusion matrix for recognition of the emotional state of children with ASD by groups of Russian and Foreign listeners.

	Comfort			Neutral			Discomfort		
	St P-R	M-IT	St P-F	St P-R	M-IT	St P-F	St P-R	M-IT	St P-F
Comfort	54.6	56.7	38.4	28.3	28.8	31.5	17.1	14.6	30.1
Neutral	23.1	20.7	21.7	59.8	62.8	56.6	17.1	16.5	21.7
Discomfort	22.8	12.6	22.2	17.7	32.7	22.2	59.6	54.7	55.6

Notes: St P-R – pediatric students, native Russian-speakers; St P-F – pediatric students, foreigners; M-IT – master students, IT.

The speech samples attributed to the state of discomfort are characterized by higher pitch range values of stressed vowels than neutral speech samples ($p < 0.05$), higher pitch range values of unstressed vowels than neutral speech samples ($p < 0.001$) and samples related to the state of comfort ($p < 0.05$), longer duration of unstressed vowels in the words than speech samples in the neutral ($p < 0.01$) and comfort state ($p < 0.05$), lower values of VAI for stressed vowels and higher values of VAI for unstressed vowels compared to the neutral and comfort states.

The speech samples related to the comfort state are characterized by higher pitch range values of stressed and unstressed vowels than neutral speech samples ($p < 0.01$), higher values of VAI for stressed vowels and lower values of VAI for unstressed vowels than discomfort speech samples.

4 Discussion

The study is part of the project to investigate the speech features of children with atypical development and the factors influencing the process of speech formation and speech peculiarities. In previous studies on a sample of children with ASD aged 4-14 years, the main differences were revealed in the speech features of children with ASD and typically developing children [18], the features of spontaneous and repeated speech of children with ASD aged 7–12 years were described [13], the ability of listeners to determine the psychoneurological state of 11–12 year-old children with ASD and Down syndrome has been shown [19]. In this work, the emphasis is on the study of the speech of children with ASD at the age of 13–15 years.

The results of the study showed difficulty in describing the speech features of children with ASD diagnosis in anamnesis. It was shown that with this choice of sample, the speech material of children is determined not only by the age, the child's gender, and the severity of autistic disorders, but also by the accompanying diagnosis. In general, the obtained data are in line with the common trajectory of the specific speech development in children with ASD [3, 10]. The correlation between high CARS scores and high pitch values in children with autism (group 1) traced in our study was also described for the younger children [20]. The data on differences between girls and boys in pitch values of speech and speech samples duration are partially comparable with the data for typically developing children aged 13 years, for whom the pitch values of stressed vowels in words are significantly higher in girls than in boys [21]. In typically developing boys, by the age of 15 years, a "voice break" ends, the voice becomes rougher and lower than in girls [22]. In children with ASD, the severity of ASD significantly affects the voice pitch. Low intelligibility of a part of speech samples [23] is due to a large number of consonant phoneme substitutions and reduction of vowel phonemes. This is also noted in a few works describing the atypical or incorrect pronunciation of phonemes [24], the lack of formation of affricates, and the incorrect pronunciation of consonant clusters [25]. The attribution of intonation to mainly atypical emphasizes the specificity of speech prosody of children with ASD [2, 3]. The violation in articulation accuracy and abnormal intonation are the basis for determining the psychoneurological state by Russian listeners who correctly recognized developmental disorders in ASD children aged 13–15 years. This study is part of the research of ASD children's speech development trajectory in order to reveal acoustic markers of autism. The future approach of this research will be automatic classification of ASD child state via speech.

5   Conclusion

The results of this study showed the age impact in a nonlinear change in pitch values of the ASD child's speech. High total and verbal scores characterize children using speech constructions, which meaning is difficult to determine without situation context, and having high pitch values.

Recognition by listeners is influenced by the gender and group of children (articulation accuracy), age (characteristics of intonation) and CARS scores; developmental features, the listener's native language and specialization (psychoneurological state recognition); gender, group of children and the listener's native language (the emotional state recognition).

The acoustic features of children's speech that characterize speech samples correctly assigned by listeners to the corresponding category: for speech samples with clear articulation – low pitch and pitch range values, low values of duration of stressed and unstressed vowels in words, high values of VAI; for speech samples that correctly characterize impaired development: opposite values of pitch and duration, VAI for stressed vowels is higher than VAI for unstressed vowels. The speech samples classified to the discomfort state differ from comfort speech samples in higher pitch range values of unstressed vowels, longer duration, low values of VAI for stressed vowels, and high values of VAI for unstressed vowels.

The data obtained indicate a complex trajectory of speech development in informants with ASD aged 13–15 years and the need to consider the multiplicity of factors to assess the level of speech formation.

Acknowledgements. This study is financially supported by the Russian Science Foundation (project 18-18-00063).

References

1. Kanner, L.: Autistic disturbances of affective contact. Nervous Child **2**, 217–250 (1943)
2. Diehl, J., Paul, R.: Acoustic and perceptual measurements of prosody production on the profiling elements of prosodic systems by children with autism spectrum disorders. Appl. Psycholinguist. **34**(1), 135–161 (2013)
3. Bonneh, Y.S., Levanov, Y., Dean-Pardo, O., Lossos, L., Adini, Y.: Abnormal speech spectrum and increased pitch variability in young autistic children. Front. Human Neurosci. **4**(237), 1–7 (2011)
4. Nakai, Y., Takashima, R., Takiguchi, T., Takada, S.: Speech intonation in children with autism spectrum disorder. Brain Develop. **36**(6), 516–522 (2014)
5. Stefanatos, G., Baron, I.S.: The ontogenesis of language impairment in autism: a neuropsychological perspective. Neuropsychol. Rev. **21**(3), 252–270 (2011)

6. Hubbard, K., Trauner, D.A.: Intonation and emotion in autistic spectrum disorders. J. Psycholinguist. Res. **36**(2), 159–173 (2007)
7. Shriberg, L.D., Paul, R., Black, L.M., van Santen, J.P.: The hypothesis of apraxia of speech in children with autism spectrum disorder. J. Autism Dev. Disord. **41**(4), 405–426 (2011)
8. Diehl, J.J., Paul, R.: Acoustic differences in the imitation of prosodic patterns in children with autism spectrum disorders. Res. Autism Spectr. Disord. **6**(1), 123–134 (2012)
9. Olivati, A.G., Assumpção Jr., F.B., Misquiatti, A.R.: Acoustic analysis of speech intonation pattern of individuals with autism spectrum disorders. CoDAS **29**(2), e20160081 (2017)
10. Sharda, M., et al.: Sounds of melody—pitch patterns of speech in autism. Neurosci. Lett. **478** (1), 42–45 (2010)
11. Filipe, M.G., Frota, S., Castro, S.L., Vicente, S.G.: Atypical prosody in asperger syndrome: perceptual and acoustic measurements. J. Autism Dev. Disord. **44**(8), 1972–1981 (2014)
12. Diehl, J.J., Watson, D.G., Bennetto, L., McDonough, J., Gunlogson, C.: An acoustic analysis of prosody in high-functioning autism. Appl. Psycholinguist. **30**, 385–404 (2009)
13. Lyakso, E., Frolova, O., Grigorev, A.: Perception and acoustic features of speech of children with autism spectrum disorders. In: Karpov, A., Potapova, R., Mporas, I. (eds.) SPECOM 2017. LNCS (LNAI), vol. 10458, pp. 602–612. Springer, Cham (2017). https://doi.org/10.1007/978-3-319-66429-3_60
14. Lyakso, E., Frolova, O., Kaliyev, A., Gorodnyi, V., Grigorev, A., Matveev, Y.: AD-Child.Ru: Speech Corpus for Russian Children with Atypical Development. In: Salah, A.A., Karpov, A., Potapova, R. (eds.) SPECOM 2019. LNCS (LNAI), vol. 11658, pp. 299–308. Springer, Cham (2019). https://doi.org/10.1007/978-3-030-26061-3_31
15. Roy, N., Nissen, S.L., Dromey, C., Sapir, S.: Articulatory changes in muscle tension dysphonia: evidence of vowel space expansion following manual circumlaryngeal therapy. J. Commun. Disord. **42**(2), 124–135 (2009)
16. Kent, R., Weismer, G., Kent, J., Rosenbek, J.: Toward phonetic intelligibility testing in dysarthria. J. Speech Hear. Disord. **54**, 482–499 (1989)
17. Schopler, E., Reichler, R.J., DeVellis, R.F., Daly, K.: Toward objective classification of childhood autism: childhood autism rating scale (CARS). J. Autism Dev. Disord. **10**(1), 91–103 (1980)
18. Lyakso, E., Frolova, O., Grigorev, A.: A comparison of acoustic features of speech of typically developing children and children with autism spectrum disorders. In: Ronzhin, A., Potapova, R., Németh, G. (eds.) SPECOM 2016. LNCS (LNAI), vol. 9811, pp. 43–50. Springer, Cham (2016). https://doi.org/10.1007/978-3-319-43958-7_4
19. Frolova, O., Gorodnyi, V., Nikolaev, A., Grigorev, A., Grechanyi, S., Lyakso, E.: Developmental Disorders Manifestation in the Characteristics of the Child's Voice and Speech: Perceptual and Acoustic Study. In: Salah, A.A., Karpov, A., Potapova, R. (eds.) SPECOM 2019. LNCS (LNAI), vol. 11658, pp. 103–112. Springer, Cham (2019). https://doi.org/10.1007/978-3-030-26061-3_11
20. Lyakso, E., Frolova, O., Grigorev, A., Gorodnyi, V., Nikolaev, A.: Strategies of speech interaction between adults and preschool children with typical and atypical development. Behav. Sci. **9**(12), 159 (2019)
21. Grigorev, A., Frolova, O., Lyakso, E.: Acoustic features of speech of typically developing children aged 5–16 years. In: AINL 2018, Communications in Computer and Information Science, vol. 930, pp. 152–163 (2018)
22. Busch, A.S., et al.: Voice break in boys—temporal relations with other pubertal milestones and likely causal effects of BMI. Hum. Reprod. **34**(8), 1514–1522 (2019)
23. Wild, A., Vorperian, H.K., Kent, R.D., Bolt, D.M., Austin, D.: Single-word speech intelligibility in children and adults with Down syndrome. Am. J. Speech Lang. Pathol. **27** (1), 222–236 (2018)

24. Wolk, L., Brennan, C.: Phonological difficulties in children with autism: an overview. Speech Lang. Hear. **19**(2), 121–129 (2016)
25. Cleland, J., Gibbon, F.E., Peppé, S.J., O'Hare, A., Rutherford, M.: Phonetic and phonological errors in children with high functioning autism and asperger syndrome. Int. J. Speech Lang. Pathol. **12**(1), 69–76 (2010)

Multi-corpus Experiment on Continuous Speech Emotion Recognition: Convolution or Recurrence?

Manon Macary[1,2](✉), Martin Lebourdais[1](✉), Marie Tahon[1](✉),
Yannick Estève[3](✉), and Anthony Rousseau[2](✉)

[1] LIUM, Le Mans, France
{martin.lebourdais,marie.tahon}@univ-lemans.fr
[2] Allo-Média, Paris, France
{m.macary,a.rousseau}@allo-media.fr
[3] LIA, Avignon, France
yannick.esteve@univ-avignon.fr

Abstract. Extraction of semantic information from real-life speech, such as emotions, is a challenging task that has grown in popularity over the last few years. Recently, emotion processing in speech moved from discrete emotional categories to continuous affective dimensions. This trend helps in the design of systems that predict the dynamic evolution of affect in speech. However, no standard annotation guidelines exist for these dimensions thus making cross-corpus studies hard to achieve. Deep neural networks are nowadays predominant in the task of emotion recognition. Almost all systems use recurrent architectures, but convolutional networks were recently reassessed as they are faster to train and have less parameters than recurrent ones. This paper aims at investigating pros and cons of the aforementioned architectures using cross-corpus experiments to highlight the issue of corpus variability. We also explore the best suitable acoustic representation for continuous emotion, together with loss functions. We concluded that recurrent networks are robust to corpus variability and we confirm the power of cepstral features for continuous Speech Emotion Recognition (SER), especially for satisfaction prediction. A final post-treatment applied on prediction brings very nice result ($ccc = 0.719$) on AlloSat and achieves new state of the art.

Keywords: Continuous speech emotion recognition · Deep neural networks · Acoustic features

1 Introduction

Semantic information extraction from real-life human-human conversations, such as concepts, emotions or intents, is a major topic for speech processing researchers and companies, especially in call-center related activities. In this context, Speech Emotion Recognition (SER) remains unsolved while being a

© Springer Nature Switzerland AG 2020
A. Karpov and R. Potapova (Eds.): SPECOM 2020, LNAI 12335, pp. 304–314, 2020.
https://doi.org/10.1007/978-3-030-60276-5_30

research field of interest for many years. Predicting naturalistic emotions is a challenging task, especially if the scope is to capture subtle emotions characterizing real-life behaviors.

In the discrete theory [3], emotion categories are usually defined on a word, a segment or a conversation level. However, this approach does not permit to extract the evolution of affect along a conversation. In the continuous theory, the complex nature of emotion in speech is described with continuous dimensions, notably arousal and valence [15], but also dominance, intention, conductive/obstructive axis [16]. As of now, there are only a few available realistic SER corpora, as they need to respect strong ethical and legal issues and collecting emotional speech demands tremendous efforts. Among these, very few has been annotated continuously according to emotional dimensions because it is a difficult and expensive task. Among the most popular, we can cite SEMAINE [11] (English interactions with virtual agent), RECOLA [12] (on-line conversations). The cross-cultural emotion database (SEWA) [8] (human-human conversations) was presented for the 2018 Audio/Visual Emotion Challenge [13] which aimed to retrieve arousal, valence and liking dimensions across different cultures. Recently, AlloSat [10] (French call-centers telephone conversations) was annotated along the satisfaction dimension. The present study aims at continuously predicting affect in naturalistic speech conversations: this explains why we focus on the SEWA and AlloSat databases.

In SER, speech signals are traditionally represented with acoustic vectors which represent the entire emotional segment. Finding the better acoustic feature set is still an ongoing active sub-field of research in the domain [7]. Most of existing sets intend to describe prosody in the signal, with low level descriptors capturing intensity, intonation, rhythm or voice quality. These features have the advantage of being easily interpretable, however their extraction in degraded signals are error-prone. The HUMAINE association also took an inventory of acoustic features in the CEICES initiative [1] which conducts to a set of a hundred of descriptors selected over several corpora with various techniques [5]. Another option is to extract spectral features; mel frequency cepstral coefficients (MFCCs) are clearly the most often used as they are robust to noisy signals. Previous studies have shown that some implementations of these MFCCs perform better than others for SER [18].

To perform SER tasks, deep neural networks are more and more used, especially for continuous emotion recognition. Recurrent neural networks (RNNs), especially long short term memory (LSTM)-based architectures, are particularly convenient thanks to their memory properties, enabling to retrieve the evolution of long-term time series [6] such as emotions or intents. Convolutional neural networks (CNNs) are widely used in image recognition as they are designed to exploit spatially local correlations. This type of network has not been used in continuous SER to the best of the author's knowledge, until [17] in which the authors conclude to the advantage of CNNs over RNNs to continuously predict arousal and valence in SEWA. As CNNs are faster to train and have less parameters to optimize, they could be interesting to use for continuous SER tasks instead of RNNs.

The main difficulty encountered in SER is due to the task itself: emotions are complex, subtle and subjective thus making the reproduction of the results and standardization very hard. Cross-corpus approaches imply to compare automatic predictions given by a single system on different corpora. This task has already been investigated to retrieve emotion categories [2,4,18], but as far as we know, not to retrieve continuous affective dimensions. The standardization problem is addressed by AVEC challenges [13,14] which aim at defining standard models and evaluations metrics for continuous SER tasks.

A first goal is to estimate in what extent the experiments realised by [17] are consistent on other data. To do so, the two studied datasets SEWA and AlloSat are introduced in Sect. 2. Section 3 compares system architectures (CNN or RNN) while Sect. 4 explores the impact of input acoustic features on performances. In Sect. 5 we also discuss on results and on the introduction of post-treatment applied to the neural network output.

2   Emotional Data and Acoustic Features

2.1   AlloSat

AlloSat [10] is composed of 303 call-centers telephone real-life conversations. Speakers are French native adult callers (i.e. customer) asking for information such as contract information, complaints on multiple domain company (energy, insurance, etc.). Each signal contains only the caller's speech as the part of the receiver (i.e agent) has been discarded from the corpus for ethical and commercial reasons. A continuous annotation among the satisfaction dimension was realized by 3 annotators on all conversations with a time step of 250 ms. The unique gold reference is the mean of each annotator's values. The corpus is divided into train, development and test sets as shown in Table 1 totalling 37h17' of conversations. For each conversation the speaker is different, ensuring a speaker independent partition.

Table 1. Number of mono speaker conversations (and duration) in AlloSat and the two configurations of SEWA (German only or German and Hungarian), in train, development and test sets

Corpus	Language	Train	Dev	Test
AlloSat	French	201 (25h26')	42 (5h55')	60 (5h58')
SEWA	German	34 (1h33')	14 (37')	82 (3h)
SEWA	Ger + Hun	68 (2h41')	28 (1h05')	104 (4h40')

2.2   SEWA

The cross-cultural Emotion Database (SEWA) [8] consists of 48 audiovisual recordings of elicited reactions between unique pairs of subjects. Pairs are discussing for less than 3 min about an advert seen beforehand. The database is now a reference in the community, as it has been used in the two last Audio/Visual Emotion Challenges (AVEC). In this study, only a subset of the database, containing German and Hungarian records, is investigated according to the guidelines of AVEC 2018 and 2019 workshop [13,14]. A continuous annotation among three dimensions (arousal, valence and liking) was made by 6 annotators and a unique gold reference has been computed, for every 100 ms. The additional liking axis describes how much the subjects liked the commercial. The corpus is divided into train, development and test sets as shown in Table 1. Test gold references are not distributed. Predictions have to be sent to AVEC organizers to get the final performances.

2.3   Acoustic Features

This paper mainly tries to reproduce the experiments from AVEC challenges and Schmitt et al. [17] with different data, to analyze the robustness to corpus variability. To do so, the acoustic feature sets used in these challenges are used as well as an additional feature set. In the end, either hand crafted expert features or Mel Frequency Cepstral Coefficients (MFCC) are used in the following experiments. First, low-level descriptors (LLD) are extracted directly from the speech signal, each 10 ms. These LLDs are then summarized over a fixed time window of 100 ms for SEWA and 250 ms for AlloSat.

- **eGeMAPS-88** contains 88 features from the extended Geneva Minimalistic Acoustic Parameter Set [5]. This set consists of LLDs capturing spectral, cepstral, prosodic and voice quality information from the speech signal, which are then summarised over the time window with a set of statistical measures. This feature set has been extensively used for SER, especially thanks to AVEC challenges. This feature set is extracted with the toolkit OpenSmile[1].
- **eGeMAPS-47** is a subset of GeMAPS which includes 23 LLDs. Mean and standard deviation of these 23 LLDs are computed over the time window. This feature set is extracted with the toolkit OpenSmile. An additional binary feature denoting speaker presence extracted from speech turns, is also included.
- **Mfcc-Os** consists of MFCC1-13, and their first and second derivatives, also extracted with the toolkit OpenSmile. Mean and standard deviation are computed on these 39 features over the time window. In total, we use a 78-dimensional feature vector.
- **Mfcc-lib** is an alternative implementation of cepstral coefficients from librosa[2] that is used in many speech processing experiments. In this set, 24 MFCCs are extracted each 10 ms on a 30 ms window and summarized

---

[1] http://audeering.com/technology/opensmile/.
[2] https://librosa.github.io/librosa/.

with mean and standard deviation over the time window. In total, we use a 48-dimensional feature vector.

In SER, acoustic representation of emotion usually tends to capture prosody. That is the reason why expert features are more often used than single MFCC. However, in the context of telephone conversations, the audio signal is severely degraded and expert features can not avoid estimation errors. Therefore MFCC features can be more robust and reliable than expert features and thus gives us better performance in this context. Experiments are conducted in Sect. 4 to analyze the impact of the features set chosen.

3   Convolutional or Recurrent Models?

To estimate in what extent the experiments published in [13, 14, 17] are consistent on different corpus, we decided to reproduce them on SEWA and compare them to those obtained on AlloSat.

3.1   Network Architectures

In Schmitt et al. [17], CNN and RNN architectures are investigated. The CNN is composed of 4 convolutional layers with a ReLU activation. The RNN is composed of 4 bidirectional Long Short Term Memory (biLSTM-4) layers of respectively 200, 64, 32 and 32 units, and a tanh activation. A single output neuron is used to predict the regression samples each 250 ms for AlloSat, respectively 100 ms for SEWA. In addition, a second RNN with 2 bidirectional LSTM (biLSTM-2) layers of respectively 64 and 32 units, with a tanh activation, proposed in AVEC 2018 and 2019 challenge [13, 14], is experimented. In the end 3 different networks are tested as shown in Fig. 1.

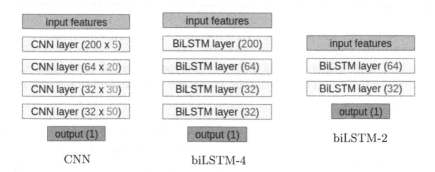

Fig. 1. Description of the models used: units are in red and filter span in blue. (Color figure online)

3.2   Protocol

The networks, summarized in Fig. 1, are implemented with the Keras framework[3] using the Tensorflow backend[4]. The learning rate has been empirically set to 0.001, the number of epochs is fixed to 500 and ADAgrad optimiser is used. The Concordance Correlation Coefficient (CCC) [9] was established as a standard metric in the two last AVEC challenges and is generally used to compute the network loss function. As the random initialization of the system can alter the prediction, each network is trained five times with different seeds and all following results represent the average of these 5 systems and the best score. Table 2 reports the reproduction results in terms of CCC, obtained with the following configurations:

- AVEC 2018: network: biLSTM-2; features: Mfcc-Os, eGeMAPS-88; train/dev: SEWA (Ger), AlloSat
- AVEC 2019: network: biLSTM-2; features: Mfcc-Os, eGeMAPS-88; train/dev: SEWA (Ger+Hun), AlloSat
- Schmitt et al.: networks: CNN, biLSTM-4; features: eGeMAPS-47; train/dev: SEWA (Ger), AlloSat

3.3   Cross-Corpus Experiments: CNN Vs. RNN

According to Table 2, our models perform in average slightly better than AVEC 2018 baseline on all SEWA dimensions. Although our performances are in average almost below AVEC 2019 baseline, our best models trained with Mfcc-Os features are in the range of the baseline results. This underlines the importance of initialization and seed choice. The comparison with Schmitt et al. shows that we did not managed to reproduce the published results on both arousal and valence dimensions except on the prediction of valence with CNN. We also report the results obtained on liking which are better than the ones obtained with AVEC systems.

Comparison of average and best performances allows us to find the systems which have low dependency with weight initialization, which are biLSTM-2 trained with Mfcc-Os on SEWA Ger+Hun and CNN trained with eGeMAPS-47 on SEWA Ger. Generally, the best performances seem to be more difficult to reproduce on liking than arousal or valence.

The performances on satisfaction prediction are comparable to those obtained on other dimensions when using biLSTM. However, the results with CNN completely differ. More precisely, with 5 different seeds, 3 models diverged, while all biLSTM converged to satisfactory results.

It shows that satisfaction prediction performs better with RNNs than CNNs. Therefore, continuous SER does require a biLSTM architecture to be competitive on multiple dimensions and resolve the corpus variability issue.

---

[3] https://keras.io.
[4] https://www.tensorflow.org/.

Table 2. Comparison of averaged CCC scores of AlloSat and SEWA development sets on the 4 dimensions: satisfaction, arousal, valence and liking with CCC as loss function. Reports of the results from [13, 14, 17] are also included.

Models	Features	AlloSat	SEWA		
		Satisfaction	Arousal	Valence	Liking
Our systems					
CNN	eGeMAPS-47	.178(.458)	**.528**(.541)	**.515**(.527)	**.304**(.321)
biLSTM-4	eGeMAPS-47	.437(.458)	.487(.527)	.428(.468)	.258(.346)
biLSTM-2	eGeMAPS-88	.480(.564)	.280(.357)	.174(.212)	.095(.171)
biLSTM-2	Mfcc-Os	.364(.439)	.395(.438)	.325(.373)	.158(.208)
biLSTM-2	eGeMAPS-88*	.480(.564)	.244(.273)	.118(.155)	.082(.132)
biLSTM-2	Mfcc-Os*	.364(.439)	.325(.326)	.186(.192)	.125(.126)
biLSTM-4	eGeMAPS-88	**.564**(.634)	.316(.429)	.237(.309)	.119(.188)
Schmitt et al. : Train and Dev on German conversations					
CNN	eGeMAPS-47		.571	.517	
biLSTM-4	eGeMAPS-47		.568	.561	
AVEC 2018 : Train and Dev on German conversations					
biLSTM-2	eGeMAPS-88		.124	.112	.001
biLSTM-2	Mfcc-Os		.253	.217	.136
AVEC 2019 : Train and Dev on German and Hungarian conversations					
biLSTM-2	eGeMAPS-88*		.371	.286	.159
biLSTM-2	Mfcc-Os*		.326	.187	.144

*Train and prediction has been made on the concatenation of German and Hungarian SEWA conversations

We noticed that satisfaction (resp. liking) varies very slowly (resp. slowly) in time in comparison to arousal and valence. This can be due to the annotation protocol (mouse vs. joystick) and the affective content. To investigate if the dynamics in annotation was responsible for poor results on satisfaction, we run additional experiments with smoothed references. The results (not reported here) show that smoothing the reference up to 1 s helps in increasing performances on AlloSat (average $ccc = 0.185$, best $ccc = 0.475$) with the systems which were already converging. To conclude, CNN can not be used on every kind of speech data as their convergence is not straightforward, probably because of their dependency on filter initialization.

Focusing on satisfaction prediction, eGeMAPS-88 ($ccc = 0.480$) appears to perform better than MFcc-Os ($ccc = 0.364$) with 2 biLSTM layers in the network. However, eGeMAPS-47 with 4 biLSTM layers also reaches good results ($ccc = 0.437$). To conclude on the best number of layers, we run a final experiment with 4 biLSTM layers and eGeMAPS-88 ($ccc = 0.564$) which achieves our best result.

This first experiment concludes that biLSTM-4 is the best architecture regarding variability robustness over different emotion corpora. However, the structure of the network is not the only component implicated in a SER module. The representation of the speech data in input and the loss function are also crucial. The following section explores acoustic features and loss functions, using biLSTM-4 models.

4   Impact of Input Features and Loss Function

In this section, we study the impact of the acoustic representation of speech in input together with loss functions used during the training phase.

4.1   Loss Functions

Traditionally in continuous emotion prediction, the loss function is computed with the CCC as this metric was established as a standard metric in the two last AVEC challenges. However, the pertinence of the CCC as loss function is reassessed in our experiments. CCC is given by Eq. 1, where x and y are two variables, μ_x and μ_y are their means and σ_x and σ_y are the corresponding standard deviations. ρ is the correlation coefficient between the two variables.

$$\rho_c = \frac{2\rho\sigma_x\sigma_y}{\sigma_x^2 + \sigma_y^2 + (\mu_x - \mu_y)^2} \tag{1}$$

Actually, when the reference is constant over time, $(\sigma_y = 0)$, CCC is zero. More generally, when the reference varies slowly, CCC will be almost zero. Consequently, the loss function penalizes conversations where the reference varies slowly $(\sigma_y \simeq 0)$, and the trained network will have difficulties to predict correctly such references.

We decided to use the root mean square error (RMSE) (see Eq. 2 where x_i is a prediction, y_i a reference and n the number of values), as loss function to neutralize this effect.

$$RMSE = \sqrt{\frac{1}{n}\Sigma_{i=1}^n \left(y_i - x_i\right)^2} \tag{2}$$

4.2   Cross-Corpus Experiments: Acoustic Features and Loss Functions

In this experiment, the 4 acoustic feature sets described in Sect. 2 are explored as inputs of a biLSTM-4 network. The results are consigned in Table 3. Clearly, and whatever the loss is, eGeMAPS-47 performs better on SEWA dimensions, following by Mfcc-Os. At the other end, Mfcc-lib performs better on AlloSat satisfaction, followed by eGeMAPS-88.

This is probably due to the fact that AlloSat contains telephone conversations with diverse background noises, which can alter the extraction of fine-tuned features present in eGeMAPS sets. Moreover, RMSE loss increases performance only when combined with Mfcc features, for satisfaction and liking. Interestingly, these two dimensions are the ones that vary the less according time, consequently have the lowest σ_y. Mfcc-lib with RMSE loss achieve new state-of-art performance on AlloSat.

Table 3. Average CCC scores on AlloSat and SEWA dev sets with 4 acoustic feature sets and 2 loss functions (*l-ccc* and *l-rmse*). The network is a biLSTM-4. Training and predictions on SEWA has been made on the German conversations.

| Features | AlloSat | | SEWA | | | | | |
| | Satisfaction | | Arousal | | Valence | | Liking | |
	l-ccc	*l-rmse*	*l-ccc*	*l-rmse*	*l-ccc*	*l-rmse*	*l-ccc*	*l-rmse*
eGeMAPS-47	.437	.381	**.487**	.438	**.428**	.404	**.258**	.252
eGeMAPS-88	.564	.514	.316	.201	.237	.211	.119	.077
Mfcc-lib	.675	**.698**	.258	.222	.192	.103	.180	.192
Mfcc-Os	.382	405	.394	.377	.373	.357	.221	.234

5   Analysis of the Results and Post-processing

Even if the score achieved by the best model on satisfaction is high ($ccc = 0.698$), we can observe that predictions vary rapidly with time (see Fig. 2).

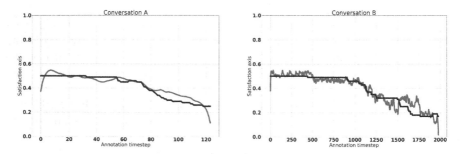

Fig. 2. Evolution of reference satisfaction (red) and its prediction (gray) in two conversations from AlloSat test subset. $ccc(A) = 0.564$, $ccc(B) = 0.903$. (Color figure online)

We propose to post treat the prediction with Savistky-Golay smoothing algorithm and polynomial degree of 0. Table 4 confirms that this post-treatment improves the results, achieving new state-of-art results on the satisfaction dimension.

Table 4. CCC scores without and with a smoothing function computed on development and test set of AlloSat on our best model: biLSTM-4, Mfcc-lib features and RMSE loss function

| Features | Dev | | Test | |
	Raw	Smoothed	Raw	Smoothed
Mfcc-lib (*l-rmse*)	.698	**.719**	.513	**.570**

6   Conclusion

In this paper, we estimate how much continuous SER is robust to variability issues with cross-corpus experiments.

CNN and RNN were evaluated in the continuous SER task and we conclude that RNNs are robust to cross-corpus conditions and achieve the best results on satisfaction. CNNs perform better on arousal, valence and liking but seem to be very sensitive to filter initialization. We also show that the best feature set (and its implementation) depends on the studied dimension and/or corpus: satisfaction is better represented by MFCCs ($ccc = 0.698$) while arousal, valence and liking are better represented by eGeMAPS-47 features. Indeed, the extraction of fine-tuned features as those in eGeMAPS is probably very sensitive to noisy signals such as telephone. All these results highlight the issue of variability robustness as performances are corpus-dependant. To go further, a post-treatment has been applied on satisfaction predictions showing a significant improvement. The very nice result ($ccc = 0.719$) obtained on AlloSat achieves new state of the art.

References

1. Batliner, A., Steidl, S., Schuller, B., Seppi, D,, Laskowski, K., Vogt, T., et al.: CEICES : combining efforts for improving automatic classification of emotional user states: a forced co-operation initiative. In: Language and Technologies Conference, pp. 240–245 (2006)
2. Devillers, L., Vaudable, C., Chasatgnol, C.: Real-life emotion-related states detection in call centers: a cross-corpora study. In: Proceedings of Interspeech, pp. 2350–2355 (2010)
3. Ekman, P.: Basic Emotions, pp. 301–320. Wiley, New-York (1999)
4. Eyben, F., Batliner, A., Schuller, B.W., Seppi, D., Steidl, S.: Cross-corpus classification of realistic emotions - some pilot experiments. In: LREC 2010 (2010)
5. Eyben, F., Scherer, K., Schuller, B., Sundberg, J., André, E., et al.: The Geneva minimalistic acoustic parameter set (GeMAPS) for voice research and affective computing. IEEE Trans. Affect. Comput. **7**(2), 190–202 (2016)
6. Hochreiter, S., Schmidhuber, J.: Long short-term memory. Neural Comput. **9**, 1735–80 (1997)
7. Jing, S., Mao, X., Chen, L.: Prominence features: effective emotional features for speech emotion recognition. Digital Sig. Process. **72**, 216–231 (2018)
8. Kossaifi, J., Walecki, R., Panagakis, Y., Shen, J., Schmitt, M., et al.: SEWA DB: a rich database for audio-visual emotion and sentiment research in the wild. IEEE Trans. Pattern Anal. Mach. Intell. 1–20 (2019)
9. Lin, L.I.K.: A concordance correlation coefficient to evaluate reproducibility. Biometrics **45**(1), 255–268 (1989)
10. Macary, M., Tahon, M., Estève, Y., Rousseau, A.: AlloSat: a new call center French corpus for satisfaction and frustration analysis. In: Language Resources and Evaluation Conference, LREC 2020 (2020)
11. McKeown, G., Valstar, M., Cowie, R., Pantic, M., Schröder, M.: The SEMAINE database: annotated multimodal records of emotionally colored conversations between a person and a limited agent. IEEE Trans. Affect. Comput. **3**(1), 5–17 (2012)

12. Ringeval, F., Sonderegger, A., Sauer, J., Lalanne, D.: Introducing the RECOLA multimodal corpus of remote collaborative and affective interactions. In: Proceedings of International Conference on Automatic Face and Gesture Recognition (FG), pp. 1–8 (2013)
13. Ringeval, F., Schuller, B., Valstar, M., Cowie, R., Kaya, H., Schmitt, M., et al.: AVEC 2018 Workshop and Challenge: Bipolar Disorder and Cross-Cultural Affect Recognition. In: Proceedings of the 2018 on Audio/Visual Emotion Challenge and Workshop, pp. 3–13. ACM (2018)
14. Ringeval, F., Schuller, B., Valstar, M., Cummins, N., Cowie, R., Tavabi, L., et al.: AVEC 2019 Workshop and Challenge: State-of-Mind, Detecting Depression with AI, and Cross-Cultural Affect Recognition. In: Proceedings of the 9th International on Audio/Visual Emotion Challenge and Workshop, pp. 3–12 (2019)
15. Russel, J.: Reading Emotions from and into Faces: Resurrecting a Dimensional-Contextual Perspective, pp. 295–360. Cambridge University Press, Cambridge (1997)
16. Scherer, K.R.: What are emotions ? and how can they be measured ?. In: Social Science Information, pp. 695–729 (2005)
17. Schmitt, M., Cummins, N., Schuller, B.W.: Continuous emotion recognition in speech - do we need recurrence? In: Proceedings of Interspeech, pp. 2808–2812 (2019)
18. Tahon, M., Devillers, L.: Towards a small set of robust acoustic features for emotion recognition: challenges. IEEE/ACM Trans. Audio Speech Lang. Process. **24**, 16–28 (2016)

Detection of Toxic Language in Short Text Messages

Olesia Makhnytkina[1]([⊠])(iD), Anton Matveev[1]([⊠])(iD), Darya Bogoradnikova[1]([⊠])(iD),
Inna Lizunova[1]([⊠])(iD), Anna Maltseva[2]([⊠])(iD), and Natalia Shilkina[2]([⊠])(iD)

[1] ITMO University, Saint Petersburg 197101, Russian Federation
{makhnytkina,aymatveev,dabogoradnikova,lizunova}@itmo.ru
[2] Saint Petersburg State University, Saint Petersburg 191124, Russia
{st801923,n.shilkina}@spbu.ru

Abstract. The ever-increasing online communication landscape provides circumstances for people with significant differences in their views to cross paths unlike it was ever possible before. This leads to the raise of toxicity in online comments and discussions and makes the development of means to detect instances of such phenomenon critically important. The toxic language detection problem is fairly researched and some solutions produce highly accurate predictions when significantly large datasets are available for training. However, such datasets are not always available for various languages. In this paper, we review different ways to approach the problem targeting transferring knowledge from one language to another: machine translation, multi-lingual models, and domain adaptation. We also focus on the analysis of methods for word embedding such as Word2Vec, FastText, GloVe, BERT, and methods for classification of toxic comment: Naïve Bayes, Random Forest, Logistic regression, Support Vector Machine, Majority vote, and Recurrent Neural Networks. We demonstrate that for small datasets in the Russian language, traditional machine-learning techniques produce highly competitive results on par with deep learning methods, and also that machine translation of the dataset to the English language produces more accurate results than multi-lingual models.

Keywords: Toxic language · Machine learning · Natural language processing · Classification methods · Multi-lingual models · Domain adaptation · Word embedding · Machine translation

1 Introduction

The communication landscape has changed drastically in recent years: education; discussion of news, movies, and goods; everyday communication–all handled by online media and social platforms. One adverse consequence of that is the increased volume of profane and offensive comments given by the freedom from direct contact with an opponent and the possibility to "hide behind a username" which blurs the boundaries of morality for some people and produces

© Springer Nature Switzerland AG 2020
A. Karpov and R. Potapova (Eds.): SPECOM 2020, LNAI 12335, pp. 315–325, 2020.
https://doi.org/10.1007/978-3-030-60276-5_31

"spammers" and "trolls"; various research initiatives estimate toxic messages to reach up to 10% of all communication instances.

Although currently there is no established definition for a "toxic message" [1], the common agreement is messages with profane language, threats, excessive negative opinions about people or situations, insults, and other hostilities that hurt the target of the message or a reader. The term "Hate speech" is often applied to such messages, but in this case the statement is racist, sexist or fanatical and can form unfavorable images of various social and cultural groups [2]. Toxic comments, in turn, are more abstract and do not have such common association with social groups or phenomena.

More generally, automated detection and recognition of toxic messages can be helpful in a variety of fields: from social networks to distance education platforms. For example, the virtual dialog assistant for conduction of online exams based on argumentation approach and machine-learning, developed by the authors on this paper [3], can detect toxic messages in comments and also in the answers, given by a student to the system, and can trigger abortion of the session.

Detection of toxic messages can be formulated as a binary classification problem for text messages, and this kind of problem has been shown with increasing success to be solvable by machine-learning techniques, often based on neural networks. These methods provide high accuracy and precision but have a significant limiting factor in that they require volumes of data for training, and availability of the data depends on the language. There are very high performing solutions for the problem for the English language [2, 4–6], German [7] and Bengali [8], but solutions for many other languages are missing, particularly for Russian. This leads us to believe that developing methods for adaptation of models in one language to another is a promising and relevant way to approach the problem.

In this paper, we present a comparative analysis of various methods for the detection of toxic messages in English and transferring the models to Russian.

2   Related Works

Presently, the classification of comments or text messages in social networks into negative, neutral, and positive categories is the key problem for sentiment analysis [9] since it allows to assess the attitude of a person towards an object or a situation. An example of an application of this approach is the evaluation of a person's opinion about the quality of goods or services. Less often, we can see the classification of text messages by emotional state. [10–12] present the classification by base emotions: happiness, sadness, anger, fear, disgust, and surprise. Detection of specifically toxic messages is a rather recent field of research.

One of the main obstructions for such research is the lack of acceptably large labeled datasets for toxic messages. In 2019, for the "Jigsaw unintended bias in toxicity classification"[1] competition, Google and Jigsaw presented a dataset with more than 1.8 million annotated commentaries in the English language. To

---

[1] https://www.kaggle.com/c/jigsaw-unintended-bias-in-toxicity-classification.

improve the quality of models, some researchers additionally employed augmentation techniques to further increase the volume of the dataset. For example, machine translation was used to translate the dataset to French or German and then back to English. That created new text messages with similar meanings, but different structures [13]. Another approach is to generate new messages by concatenating original messages with a similar context [14].

In general, the toxic messages classification problem can be separated in two, depending on the specific goal: a binary classification that predicts whether the message toxic or not, and a multi-class classification for detection of specific offenses, such as "threat", "profanity", "insult", "hate speech", etc. [15].

For solving the toxic messages detection problem, one of the key decisions is the choice of the method for obtaining vector representations of words (embeddings). Variety of research [2,16] investigate different methods for word embedding: bag-of-word, term frequency-inverse document frequency (TF-IDF), Word2vec, FastText, GloVe, Bert. Some authors suggest to simultaneously employ combinations of embedding models [17].

Some works demonstrate an increase in accuracy of detection by extracting additional features from a text. [18] and [19] examine the influence of syntactical dependencies (proper nouns and possessive pronouns) on the detection performance. Syntactic dependencies are relationships with proper nouns, personal pronouns, possessive pronouns, etc. Twenty syntactic features of sentences have been verified in the total. The authors suggest 3 additional specific features that significantly improve the quality of toxic comments detection: the number of dependencies with singular proper nouns, the number of dependencies that contain "bad" (toxic) words, and the number of dependencies between personal pronouns and "bad" words. By tracking the history of messages written by a selected author, it is possible to improve understanding of the language and the behavior of the author [15]. In [5], the authors investigate a user's predisposition to toxic language, i.e., the more often a user engages in toxic speech the higher the chance their next message will be toxic [20].

Presently, the most common approach to the problem is to employ traditional machine-learning techniques, such as logistic regression, random forest, support vector machine, decision trees, and their modifications that work in combination with neural networks [4,21] or by themselves [19,22]. Among the best deep neural network-based architectures are convolutional and recurrent neural networks LSTM, GRU, and their combinations and assemblies [23].

The lack of datasets of toxic messages for different languages, particularly Russian, is a significant limitation to the applicability of the listed pipelines and it necessitates a search for methods for transferring models from one language to another. Several methods are employed for specific natural language processing problems: machine translation [24–26], multi-language models [27–29], and domain adaptation [30–33]. In this paper, we present a comparative analysis of methods for transferring models from the English language to Russian with the consequent application of these models to toxic messages detection pipelines.

3   Datasets

3.1   The Training Dataset in English

For the training dataset, we chose one offered to the community for the "Jigsaw Unintended Bias in Toxicity Classification" challenge on Kaggle in 2019. All comments from the dataset were marked with 7 labels: neutral, severe toxicity, obscene, identity attack, insult, threat, sexually explicit. At the same time, there was a subjective assessment, in which a message without toxic words in its semantic content could be marked by most annotators as toxic, including if it contained irony. In this work, we consider the overall toxicity value, which is determined by whether or not a comment is marked by at least one of the non-neutral labels. The distribution between classes is not uniform and 92% (1660540) of the comments are only marked by a single label "neutral" while 8% (144334) is marked by at least one toxicity label. Here, we work with a sub-sample of 180500 comments split equally into 2 classes: toxic and neutral.

3.2   Target Russian Datasets

The Russian dataset contains comments from the social platform "Vkontakte" collected via Vkontakte API. The comments are related to one of the three topics: education, news, and entertainment. Each topic is represented by two groups. All comments were collected between 1st and 30th of April 2020. Representative sub-samples were selected from each group with a 5% error. The selected comments were manually annotated by three experts with Cohen's kappa coefficient for their agreement between 40% and 60% depending on the group.

The dataset of users' comments from the Stepik education platform and Coursera. This dataset contains 1709 comments and was augments with artificial toxic comments.

4   Methods

4.1   Preprocessing

For this work, the following procedures were executed:

1. Transformation into lowercase, elimination of punctuation.
2. Tokenization, i.e., splitting long lines into smaller sections. Particularly in our case, similar to how it is most often implemented, tokenization results in a separation of sentences into words.
3. Lemmatization: the process of transforming inflected forms into dictionary forms (lemmas). In Russian, for example, the dictionary forms are nominative singular nouns; masculine nominative singular adjectives; infinitive verbs, participles, and transgressives.
4. Elimination of stop words, which do not affect the meaning of a message and only increase its length. There is no commonly accepted definition for stop words, but, generally speaking, it is the most common, short function words, such as the, is, at, which, on, etc.

4.2   Text Embedding

At this step, text messages are converted into a series of vectors of real numbers. Embedding is the common term for various techniques for modeling languages in natural language processing pursuing the transformation of words or phrases into low-dimensional vectors. In this work, we demonstrate the application of Word2Vec, FastText, GloVe, and BERT models.

For the datasets in Russian, there are two options: 1) translation of the comments into English, a then using the pre-trained Word2Vec [34], FastText, and GloVe models [35], 2) using the Multilingual BERT [36] which works for all languages required in this research. The authors of [37, 38] analyze the model and conclude than Multilingual BERT can make cross-lingual generalizations and its vector space works sufficiently accurate for cross-lingual problems. BERT is a context-dependent language model based on the Transformer architecture and the idea of word masking: in training, some words are hidden behind the mask, and the model attempts to recover the hidden words from the context.

4.3   Domain Adaptation

The adversarial domain adaptation method is often employed when the target data which need to be classified and the source data are similar in structure and meaning but originates from different domains. For example, in biometry, data for a training dataset might be obtained by collecting voice samples with different microphones and in the varying environment: each configuration is a separate domain. Text messages with similar structure and meaning coming from different languages reflect the same situation, where each language constitutes a separate domain. To train a model with adversarial domain adaptation there need to be annotated source data from one domain and non-annotated target data from another. Essentially, the model is a combination of three machine-learning models. The first model is a feature extractor that generates input for the other two. The second model is the classifier target label, and the third is a domain-level classifier. The domain-level classifier, however, is not trained to separate domains, but instead to mix them via a special gradient reversal layer.

In our work, the source dataset was in English, and the target – in Russian. We then tested the model with data in Russian.

4.4   Classification Methods

We tested the following methods:

Naïve Bayes (NB): a simple probabilistic classifier based on applying Bayes' theorem with strong (naïve) independence assumptions between the features.

Random forest classifier (RFC): a machine-learning technique where an assembly of weak classifiers (decision trees) makes a collaborative decision by "voting".

Logistic regression (LR): a linear classifier that allows estimating posterior probabilities and is widely used in classification problems.

Support vector machine (SVM): one of the most popular tools for classification and regression. Based on a simple idea to find a hyperplane that optimally separates objects, it is very agile and can be modified to fit a particular task.

Majority vote (MV): one of the most popular assembly-based methods where an object is predicted to belong to a class when more than 50% of the participants of the assembly predict this class.

Long short-term memory (LSTM): a special kind of Recurrent Neural Networks, capable of learning long-term dependencies. The main distinctive feature of LSTM is the presence of feedback connections which allows it to process not just singular objects but series of data.

5   Results

Table 1 illustrates a comparison between the accuracies of Word2Vec, FastText, GloVe, and BERT for solving the toxic comments detection problem.

Table 1. Accuracies for the dataset in English

Classification method	Embedding method	Precision	Recall	F1-score
NB	word2vec	0.74	0.74	0.74
	FastText	0.72	0.72	0.72
	GloVe	0.73	0.73	0.73
	BERT	0.68	0.66	0.65
RFC	word2vec	0.79	0.79	0.79
	FastText	0.79	0.79	0.79
	GloVe	0.79	0.79	0.79
	BERT	0.73	0.73	0.73
SVM	word2vec	0.83	0.83	0.83
	FastText	0.81	0.81	0.81
	GloVe	0.81	0.80	0.80
	BERT	0.76	0.76	0.76
LR	word2vec	0.83	0.83	0.83
	FastText	0.81	0.81	0.81
	GloVe	0.80	0.80	0.80
	BERT	0.76	0.76	0.76
MV	word2vec	0.82	0.81	0.81
	FastText	0.81	0.80	0.80
	GloVe	0.80	0.80	0.80
	BERT	0.76	0.76	0.76
LSTM	word2vec	0.85	0.85	0.85
	FastText	0.83	0.83	0.83
	GloVe	0.83	0.83	0.83
	BERT	0.86	0.86	0.86

A sub-sample from the dataset in English was split into sections for training and validation. The most accurate results were obtained using pre-trained Word2Vec. Among the methods based on machine learning, LSTM gave the best results. The results obtained show the competitiveness of our models with other studies and even surpass some of them [39,40]

Table 2 demonstrates the accuracies for the dataset in Russian processed by BERT.

Table 2. Accuracies for the dataset in Russian processed by BERT

	Precision	Recall	F1-score
NB	0.99	0.89	0.95
RFC	0.98	0.91	0.95
SVM	0.98	0.96	0.97
LR	0.94	0.80	0.88
MV	0.98	0.91	0.95
LSTM	0,98	0,97	0,97
ADA	0,98	0,98	0.98

Transferring knowledge from one language to another via the multi-lingual model BERT appears to result in high accuracy rates for each machine-learning approach. It is interesting to point out that methods based on recurrent neural networks are significantly more complex and require much more computational resources but do not produce noticeably better results than the classic, more simple, methods. However, in this case, it might be attributed to the relatively small dataset. The following tests with these datasets were performed without them for simplicity.

The dataset containing comments from the social platform "Vkontakte" in Russian. Tables 3 illustrate testing results for the classification of comments in Russian via BERT and via machine translation combined with Word2Vec, Fast-Text, and GloVe.

Reviewing the statistics, the most accurate results are produced by methods that use vector representations obtained by translating comments in Russian into English and applying Word2Vec. Majority vote appears to produce the best classification outcome. The multi-lingual model BERT produces less accurate results with transferring knowledge from one language to another for the toxic comments detection problem.

Table 3. Classification accuracy for the target Russian dataset.

Classification method	Embedding method	Precision	Recall	F1-score
NB	word2vec (en)	0.75	0.72	**0.73**
	FastText (en)	0.76	0.69	0.72
	GloVe (en)	0.76	0.63	0.68
	BERT (ru)	0.82	0.50	0.57
RFC	word2vec (en)	0.75	0.74	**0.75**
	FastText (en)	0.76	0.73	0.74
	GloVe (en)	0.75	0.69	0.72
	BERT (ru)	0.78	0.64	0.69
SVM	word2vec (en)	0.75	0.73	**0.74**
	FastText (en)	0.75	0.55	0.62
	GloVe (en)	0.76	0.61	0.66
	BERT (ru)	0.81	0.60	0.66
LR	word2vec (en)	0.75	0.73	**0.74**
	FastText (en)	0.76	0.57	0.63
	GloVe (en)	0.76	0.60	0.66
	BERT (ru)	0.80	0.62	0.67
MV	word2vec (en)	0.75	0.76	**0.76**
	FastText (en)	0.76	0.71	0.73
	GloVe (en)	0.76	0.70	0.72
	BERT (ru)	0.80	0.67	0.72

6   Conclusions

In this paper, we review different ways to approach the problem of automatic toxic language detection focusing on transferring knowledge from one language to another. Based on our analysis of methods for word embedding, machine-learning techniques, and domain adaptation, we build an efficient pipeline for the detection of toxic comments in the Russian language which is trained on a dataset in English combined with techniques for transferring knowledge from one language model to another. One key feature of the pipeline is uniformity in the way of preprocessing datasets and word embedding for different languages. For transferring knowledge from one language to another, the best performance was shown by Word2Vec embeddings of machine-translated comments, and the Majority vote performed on par with Recurrent Neural Networks and domain adaptation. In general, the accuracy for data sets in Russian is lower than for data sets in English, which can be explained by differences in models and, possibly, syntactic and other features characteristic of each language, including cultural ones. For example, Russian-language resources do not discuss racism, ecology and various aspects of tolerance, as well as significantly less attention to gender identity issues. Interestingly, the automatic classifiers outperformed experts in the detection of toxic language which is quite impressive for such an abstract concept. For future work, we are interested in testing our pipeline on datasets in other languages.

Acknowledgements. This work was partially financially supported by the Government of the Russian Federation (Grant 08-08).

References

1. Risch, J., Krestel, R.: Toxic comment detection in online discussions. In: Deep Learning-Based Approaches for Sentiment Analysis, pp. 85–109 (2020)
2. Badjatiya, P., Gupta, S., Gupta, M., Varma, V.: Deep learning for hate speech detection in tweets. In: Proceedings of the International Conference on World Wide Web (WWW), pp. 759–760. International World Wide Web Conferences Steering Committee (2017)
3. Matveev, A., et al.: A virtual dialogue assistant for conducting remote exams. In: Proceedings of the 26th Conference of Open Innovations Association FRUCT, pp. 284–290 (2020)
4. Elnaggar, A., Waltl, B., Glaser, I., Landthaler, J., Scepankova, E., Matthes, F.: Stop illegal comments: a multitask deep learning approach. In: ACM International Conference Proceeding Series, pp. 41–47 (2018)
5. Pitsilis, G.K., Ramampiaro, H., Langseth, H.: Effective hate-speech detection in Twitter data using recurrent neural networks. Appl. Intell. **48**(12), 4730–4742 (2018). https://doi.org/10.1007/s10489-018-1242-y
6. Wang, C.: Interpreting neural network hate speech classifiers. In: Proceedings of the 2nd Workshop on Abusive Language Online, Brussels, Belgium, pp. 86–92. Association for Computational Linguistics (2018)
7. Risch, J., Krebs, E., Loser, A., Riese, A., Krestel, R.: Fine-grained classification of offensive language. In: Proceedings of GermEval (co-located with KONVENS), pp. 38–44 (2018)
8. Banik, N., Rahman, M.H.H.: Toxicity detection on Bengali social media comments using supervised models. In: International Conference on Innovation in Engineering and Technology (ICIET) (2019)
9. Kharlamov, A.A., Orekhov, A.V., Bodrunova, S.S., Lyudkevich, N.S.: Social network sentiment analysis and message clustering. In: El Yacoubi, S., Bagnoli, F., Pacini, G. (eds.) INSCI 2019. LNCS, vol. 11938, pp. 18–31. Springer, Cham (2019). https://doi.org/10.1007/978-3-030-34770-3_2
10. Zucco, C., Calabrese B., Agapito, G., Hiram Guzzi, P., Cannataro M.: Sentiment analysis for mining texts and social networks data: methods and tools. Wiley Interdiscip. Rev. Data Min. Knowl. Discov. **10**(1), 1–32 (2020)
11. Gupta, S., Singh, A., Ranjan, J.: Sentiment analysis: usage of text and emoji for expressing sentiments. In: Advances in Data and Information Sciences, pp. 477–486 (2020)
12. Sarkar, D.: Sentiment analysis. In: Text Analytics with Python, pp. 567–629 (2019)
13. Risch, J., Krestel, R.: Aggression identification using deep learning and data augmentation. In: Proceedings of the First Workshop on Trolling, Aggression and Cyberbullying (2018)

14. Morzhov, S.V.: Modern approaches to detect and classify comment toxicity using neural networks. Model. Anal. Inf. Syst. **27**(1), 48–61 (2020)
15. Qian, J., ElSherief, M., Belding, E.M., Yang Wang, W.: Leveraging intra-user and inter-user representation learning for automated hate speech detection. In: Proceedings of the 2018 Conference of the North American Chapter of the Association for Computational Linguistics: Human Language Technologies, vol. 2, pp. 118–123 (2018)
16. D'Sa, A., Illina, I., Fohr, D.: Towards non-toxic landscapes: automatic toxic comment detection using DNN (2019)
17. Saia, R., Corriga, A., Mulas, R., Recupero, D.R., Carta, S.: A supervised multiclass multi-label word embeddings approach for toxic comment classification. In: 11th International Joint Conference on Knowledge Discovery, Knowledge Engineering and Knowledge Management (KDIR-2019), Vienna, Austria (2019)
18. Shtovba, S., Petrychko, M., Shtovba, O.: Detection of social network toxic comments with usage of syntactic dependencies in the sentences. In: The Second International Workshop on Computer Modeling and Intelligent Systems, CEUR Workshop 2353 (2019)
19. Shtovba, S., Shtovba, O., Yahymovych, O., Petrychko, M.: Impact of the syntactic dependencies in the sentences on the quality of the identification of the toxic comments in the social networks. In: SWVNTU, no. 4 (2019)
20. Obadimu, A., Mead, E.L., Hussain, H., Agarwal, N.: Identifying toxicity within YouTube video comment text data (2019)
21. Saif, M.A., Medvedev, A.N., Medvedev, M.A., Atanasova, T.: Classification of online toxic comments using the logistic regression and neural networks models. In: AIP Conference Proceedings, vol. 2048, no. 1, p. 060011 (2018)
22. Hosam, O.: Toxic comments identification in Arabic social media. Int. J. Comput. Inf. Syst. Ind. Manage. Appl. 219–226 (2019)
23. Haralabopoulos, G., Anagnostopoulos, I., McAuley, D.: Ensemble deep learning for multilabel binary classification of user-generated content. Algorithms **13**(4), 83 (2020)
24. Banitz, B.: Machine translation: a critical look at the performance of rule-based and statistical machine translation. In: Cad. Tradução, val. 40, pp. 54–71 (2020)
25. López-Pereira, A.: Neural machine translation and statistical machine translation: perception and productivity. In: Tradumàtica Tecnol. la traducció (2019)
26. Wang, X., Lu, Z., Tu, Z., Li, H., Xiong, D., Zhang, M.: Neural machine translation advised by statistical machine translation (2016)
27. Liu C.L., Hsu T.Y., Chuang, Y.S., Lee, H.: A study of cross-lingual ability and language-specific information in multilingual BERT (2020)
28. Virtanen, A., et al.: Multilingual is not enough: BERT for Finnish (2019)
29. Vries, W., Cranenburgh, A., Bisazza, A., Caselli, T., Noord, G., Nissim, M.: BERTje: a Dutch BERT model (2019)
30. Ghosh, S., Singh, R., Vatsa, M., Ratha, N., Patel, V.M.: Domain adaptation for visual understanding. In: Domain Adaptation for Visual Understanding, pp. 1–15 (2020)
31. Kouw, W.M.: On domain-adaptive machine learning (2018)
32. Li, Z., Tang, X., Li, W., Wang, C., Liu, C., He, J.: A two-stage deep domain adaptation method for hyperspectral image classification. Remote Sens. **12**(7), 1054 (2020)
33. Xu, S., Mu, X., Zhang, X., Chai, D.: Unsupervised remote sensing domain adaptation method with adversarial network and auxiliary task. In: Cehui Xuebao/Acta Geod. Cartogr. Sin., pp. 1969–1977 (2017)

34. Mikolov, T., Corrado, G.S, Chen, K., Dean, J.: Efficient estimation of word representations in vector space. In: Proceedings of the International Conference on Learning Representations (ICLR 2013), pp 1–12 (2013)
35. Bojanowski, P., Grave, E., Joulin, A., Mikolov, T.: Enriching word vectors with subword information (2016)
36. Devlin, J., Chang, M.-W., Lee, K., Toutanova, K.: BERT: pretraining of deep bidirectional transformers for language understanding (2018)
37. Pires, T., Schlinger, E., Garrette, D.: How multilingual is multilingual BERT? (2019)
38. Wu, S., Dredze, M.: Beto, Bentz, Becas. The Surprising Cross-Lingual Effectiveness of BERT (2019)
39. Vaidya, A., Mai, F., Ning, Y.: Empirical analysis of multi-task learning for reducing model bias in toxic comment detection (2020)
40. Reichert, E., Qiu, H., Bayrooti, J.: Reading between the demographic lines: resolving sources of bias in toxicity classifiers (2020)

Transfer Learning in Speaker's Age and Gender Recognition

Maxim Markitantov[✉]

St. Petersburg Institute for Informatics and Automation of the Russian Academy
of Sciences (SPIIRAS), St. Petersburg, Russia
m.markitantov@yandex.ru

Abstract. In this paper, we study an application of transfer learning approach
to speaker's age and gender recognition task. Recently, speech analysis systems,
which take images of log Mel-spectrograms or MFCCs as input for classification, are gaining popularity. Therefore, we used pretrained models that showed
good performance on ImageNet task, such as AlexNet, VGG-16, ResNet18,
ResNet34, ResNet50, as well as state-of-the-art EfficientNet-B4 from Google.
Additionally, we trained 1D CNN and TDNN models for speaker's age and
gender recognition. We compared performance of these models in age (4
classes), gender (3 classes) and joint age and gender (7 classes) recognition.
Despite high performance of pretrained models in ImageNet task, our TDNN
models showed better UAR results in all tasks presented in this study: age
(UAR = 51.719%), gender (UAR = 81.746%) and joint age and gender
(UAR = 48.969%) recognition.

Keywords: Age and gender recognition · Computational paralinguistics ·
Transfer learning · Convolutional neural networks · Time delay neural networks

1 Introduction

Modern technologies of artificial intelligence are focused on the analysis of various
characteristics of a person, including their behavior, facial expressions and speech used
in biometric verification and identification systems. For instance, speech analysis is
useful as it does not need a direct contact with a client (user). Human speech contains a
wide range of non-verbal information (paralinguistic information) about speaker's
states and traits. Such speaker characteristics as psycho-emotional state [1, 2], age,
gender [3], and dialect [4] can be detected using this information. In addition, we can
recognize intoxication and sleepiness [5], detect 'unhealthy' speech in cold conditions [6], as well as tell apart whether a speaker wears a surgical mask or not [7].

An automatic system for speaker's age and gender recognition is useful for forensic
medical purposes, for example, to narrow down the list of suspects after committing a
crime when speech samples are available. In addition, such system can be used in
speaker's verification and identification tasks, to improve human-machine interaction,
to increase the effectiveness of targeted advertising, as well as in the work of telephone
contact centers and healthcare institutions. This year, the last two applications are
especially relevant due to wide spreading and prevention efforts of coronavirus

A. Karpov and R. Potapova (Eds.): SPECOM 2020, LNAI 12335, pp. 326–335, 2020.
https://doi.org/10.1007/978-3-030-60276-5_32

pandemic (COVID 19). By June 2020, the number of people infected by this virus has exceeded 8 million. Therefore, speaker's age and gender recognition system allows identification of seniors at risk.

In this study, we research Deep Neural Networks (DNNs) that receive a lot of attention nowadays as powerful machine learning algorithms that can be used for multiple purposes. Transfer learning can significantly increase DNNs performance [8]. They are effectively used for feature extraction [9–11], automatic speech recognition [12], and speech synthesis [13]. In addition, Time-Delay Neural Networks (TDNN) that can extract x-vectors [14], showed great performance in speaker's verification and identification [15], speaker's diarization [16], language recognition [17] as well as speaker's age recognition [18] tasks.

The paper is organized as following: Sect. 2 surveys transfer learning and TDNN application; Sect. 3 gives a brief description of the a Gender corpus used. Section 4 describes the proposed approach for speaker's age and gender recognition. Section 5 shows the results of the experiments conducted. Section 6 contains the discussion, conclusion and future work directions.

2 Related Works

The 2010 Computational Paralinguistics Challenge (ComParE), which has been regularly held in the framework of the INTERSPEECH conference, included two sub-challenges: speaker's age (4 classes) and gender (3 classes) recognition [3]. The participants of this challenge used the aGender corpus [19] to tackle speaker's age and gender recognition problem. Unweighted Average Recall (UAR) was the challenge performance measure. The baseline system used a set of 450 features, extracted with the open-source platform openSMILE [20] and was based on Support Vector Machine (SVM) reaching a test set performance of 49.91% and 81.21% for speaker's age and gender recognition, respectively [3]. On the validation set the baseline system showed UAR = 47.11% and UAR = 77.28% in speaker's age and gender recognition tasks, respectively.

The majority of systems presented in 2010 and after that used both spectral and prosodic features [21, 22]. The classifiers included Gaussian Mixture Models – Universal Background Model (GMM-UBM), Multi-Layer Perceptrons (MLP) and SVM. All presented systems used score-level fusion [23]. Yücesoy and Nabiyev reached best results in both sub-challenges [24]. They proposed a system, which consists of 11 sub-systems trained on various feature sets. In speaker's gender recognition task, the best UAR of the fusion system was 90.39%. The recall of the single system was 87.88%. In speaker's age recognition sub-challenge, the performance using score-level fusion was 54.10% and a single system accuracy was 49.72%.

Sarma et al. tried to recognize children's age and gender [25]. The proposed approach was trained and tested on the OGI Kids' corpus [26]. TDNN layers along with LSTM layers were applied to raw speech waveform. Both single task and multitask models were proposed, the last one predicts age and gender at the same time. In age detection task, mean absolute error (MAE) was 0.91 and 1.39 in the single-task and in the multi-task modes, respectively. In gender recognition task UAR was applied and showed performance of 82.30% and 70.55% in the single-task and in the multi-task modes, respectively.

In [18], authors proposed end-to-end approach based on Mel-Frequency Cepstral Coefficients (MFCCs) features and TDNN, that allowed to extract x-vectors. The system was trained on the NIST SRE08 [27] dataset and was evaluated on SRE10 [28]. The final system used i-vectors and x-vectors as features for 2-layers DNN. MAE showed a value of 4.92 in age evaluation task. The x-vector system significantly outperformed i-vector by 14%. Both i-vector and x-vector fusion improved by 9% with regard to the i-vector system.

The pretrained models can significantly increase performance of the recognition system. There are a lot of pretrained models. Simpler models are AlexNet [29] and VGG [30] that showed great performance in ImageNet [31] problem. To overcome the vanishing gradient problem (VGP), using a lot of convolutional layers and making DNN more complex, He et al. introduced skip connections and ResNet models [32]. To go even further, Tan and Le used neural architecture search to design a new baseline network and scaled it up to obtain a family of models, called EfficientNets [33], which achieve much better accuracy and efficiency than previous convolutional neural networks (CNNs).

These pretrained models are widely used in sound and speech analysis. Images of log Mel-spectrograms (log-Mels) and MFCCs are used for training these models. For example, ImageNet pretrained models are applied in acoustic scene classification [34], in speech emotion recognition [35, 36], as well as in ASR systems [37]. These models are used even in the bird sound recognition [38].

3   Speech Corpus

For the proposed system training and testing we used the aGender Corpus [19] introduced in the ComParE-2010 Challenge [3]. It consists of 49 h of telephone speech, which was recorded from 945 speakers. Every utterance is annotated with the speaker's age and gender. This corpus has seven groups (classes) of speakers: children, youth (male and female), adults (male and female), and seniors (male and female) [39]. The developers of the aGender Corpus do not divide children group into female and male gender due to similarity of voices.

4   Proposed System

The pipeline of the proposed system for age and gender recognition is presented in Fig. 1.

First, we converted raw files to wav format and removed silence from each audio file. After that, we cut all wav files into N chunks with duration of 1 s for pretrained models. As well as in previous work [39], we extracted such features as 128 log-Mels and 30 MFCCs with their Δ and $\Delta\Delta$ from the speech signal using Python library librosa [40]. These features showed good performance in state-of-the-art paralinguistics systems [14–18]. The window width was 22 ms and an overlap was 5 ms. All features were normalized with Z-score normalization.

Spectrogram with X log-Mel descriptors

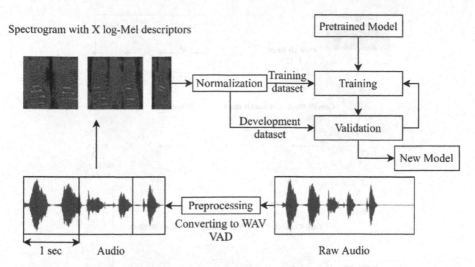

Fig. 1. Pipeline of the proposed system for speaker's age and gender recognition.

For these features, we used the pretrained models, which showed good performance in the ImageNet classification task. We used AlexNet, VGG-16, ResNet18, ResNet34, ResNet50, as well as state-of-the-art EfficientNet-B4 from Google. The classification layer was replaced by a new one with the number of outputs that corresponded to our tasks. For instance, in the age recognition task it was 4, in speaker's age and gender recognition task it was 7. After that, we fine-tuned all the layers.

We also developed our own 1D CNN and TDNN models. The first one consists of 5 convolutional blocks, each of them has convolutional layer, batch normalization layer and ReLU activation function. After the third convolutional block, we applied the dropout layer with probability of 0.15. Then, the statistical pooling layer aggregates information after last convolutional block. This layer combines mean and standard deviation of outputs of the previous layer. The outputs from the statistical pooling layer are passed through the last fully connected layer used as a classifier. The architecture of this network is shown in Fig. 2.

TDNN receives a sequence of feature frames, which are processed by time-delay neural layers. The first five layers operate on frame-level with a small temporal context centered at the current frame t. The last time-delay layer observes a total context of 15 frames. The results of these layers are combined by the statistical pooling layer, as in the previous model. After that, we consistently applied 3 fully-connected layers with different number of neurons. The parameters of this network are summarized in Table 1.

All models were implemented with the PyTorch programming platform [41]. All neural networks were trained to minimize the cross-entropy objective. The mini-batch size was chosen from 16 to 256 depending on the neural network memory size. We set an initial learning rate of 0.0001 for the Adam optimizer. We also decreased the learning rate, when the validation loss does not improve for two successive epochs. The training process was stopped after 15 epochs. The models with the smallest validation loss were chosen.

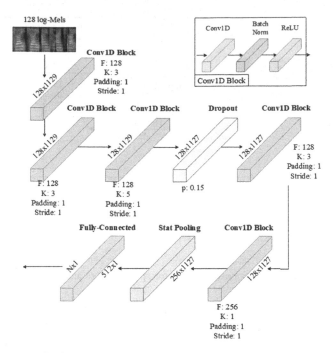

Fig. 2. 1D CNN architecture with 5 Conv1D Blocks.

Table 1. Parameters of created TDNN. **T**: Length of sequences. **N**: Number of classes.

Layer	Layer context	Total context	Input × Output
Frame1	$[-2, 2]$	5	196 × 512
Frame2	$\{-2, 0, 2\}$	9	512 × 512
Frame3	$\{-3, 0, 3\}$	15	512 × 512
Frame4	$\{0\}$	15	512 × 512
Frame5	$\{0\}$	15	512 × 1500
Statistical Pooling	$[0, T)$	T	1500 × 3000
Dense1	$\{0\}$	T	3000 × 400
Dense2	$\{0\}$	T	400 × 400
Dense3	$\{0\}$	T	400 × N

5   Experimental Results

The proposed systems were tested on the validation dataset of the aGender corpus (the same as at ComParE-2010). The results of speaker's age (4 classes), gender (3 classes) and joint recognition of age and gender (7 classes) using methods described above are presented in Table 2. The pretrained models showed better performance on 128 log-Mel than on MFCC features. The UAR of the pretrained models has increased with the increasing complexity of the models.

Table 2. Results of speaker's age and gender recognition, UAR (%). **A**: Age. **G**: Gender.

Model Features		AlexNet	VGG16	ResNet18	ResNet34	ResNet50	Efficient-Net B4	1D CNN	TDNN
128 log-Mel	A	47.348	49.486	49.990	51.169	49.949	49.756	49.500	45.039
	G	75.694	77.711	78.743	78.951	78.356	78.263	80.756	56.856
	A & G	44.994	48.102	47.350	48.876	48.837	47.347	47.165	41.549
30 MFCCs	A	47.139	50.003	49.285	49.216	49.602	48.792	49.842	50.516
	G	76.156	77.456	77.709	77.953	77.854	77.684	80.591	81.129
	A & G	44.023	47.664	46.742	47.202	48.168	46.411	46.688	47.237
30 MFCCs only Δ	A	46.599	49.301	48.505	48.167	48.639	46.385	50.603	**51.719**
	G	74.105	75.892	76.723	76.112	76.469	74.713	80.606	81.155
	A & G	45.040	47.317	47.262	47.806	47.798	45.381	47.996	**48.969**
30 MFCCs + Δ + ΔΔ	A	47.229	49.042	49.003	48.867	49.252	48.693	50.197	51.024
	G	75.737	76.696	77.334	77.515	77.650	76.895	81.203	**81.746**
	A & G	44.710	47.650	47.103	46.933	48.181	47.005	47.295	46.235

The 1D models demonstrated better UAR accuracy on 30 Δ MFCCs, as these features describe MFCC change over time, which is important in age and gender recognition. For instance, children and seniors speak more slowly than youth or adults do. The best performance was achieved by TDNN on 30 MFCCs with Δ and ΔΔ features in all tasks: age (UAR = 51.719%), gender (UAR = 81.746%) and in joint age and gender (UAR = 48.969%) recognition.

In Figs. 3 and 4, we report the confusion matrices for all tasks described above.

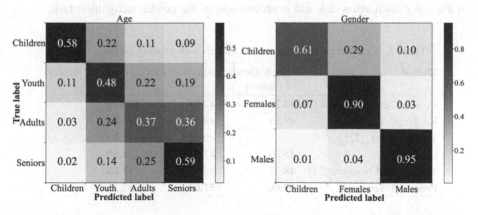

Fig. 3. Confusion matrices for speaker's age and gender recognition tasks

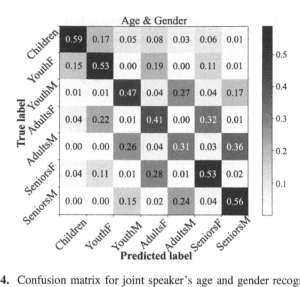

Fig. 4. Confusion matrix for joint speaker's age and gender recognition

Confusion matrices demonstrate that it is difficult to distinguish children's voices from female voices in the ternary gender recognition task due to similarity of voices. In speaker's age recognition task, most mistakes were made when distinguishing between youth, adults and seniors, due to blurry boundaries between these three age groups. The same holds true for joint recognition of age and gender.

Table 3 demonstrates comparison of the proposed system with existing systems using single model on development set of the aGender corpus. Note, we do not use development set for training our models. Proposed system is the first among all others in the age classification task and is second one in the gender recognition task.

Table 3. Comparison of the proposed system with existing systems, UA (%)

Method	Age & Gender (7 classes)	Age (4 classes)	Gender (3 classes)
ComParE-2010 Baseline [3]	44.24	47.11	77.28
Kockmann et al. [21]	47.55	49.16	79.88
Meinedo et al. [22]	–	49.20	82.20
Li et al. [23]	45.80	47.20	78.00
Yücesoy and Nabiyev [24]	48.37	49.52	**87.88**
Proposed Approach	**49.66**	**51.02**	82.26

6 Conclusion and Future Work

In this study, we modified the pretrained models such as AlexNet, VGG-16, ResNet18, ResNet34, ResNet50, as well as state-of-the-art EfficientNet-B4 from Google, and created 1D CNN and TDNN models for speaker's age and gender recognition. We compared performance of these models. Despite high performance of the pretrained models in ImageNet task, TDNN models showed better performance in all tasks presented in this study: age (UAR = 51.719%), gender (UAR = 81.746%) and in joint age and gender (UAR = 48.969%) recognition.

In the follow-up work, we aim to investigate recurrent neural networks, in particular, long short-term memory. In addition, we plan to explore multimodal approaches in speaker's age and gender recognition tasks.

Acknowledgements. This research is supported by the Russian Science Foundation (project No. 18-11-00145).

References

1. Schuller, B., Steidl, S., Batliner, A.: The INTERSPEECH 2009 emotion challenge. In: INTERSPEECH 2009, pp. 312–315 (2009)
2. Kaya, H., et al.: Predicting depression and emotions in the cross-roads of cultures, paralinguistics, and non-linguistics. In: Proceedings of International on Audio/Visual Emotion Challenge and Workshop, pp. 27–35 (2019)
3. Schuller, B. et al.: The INTERSPEECH 2010 paralinguistic challenge. In: INTERSPEECH 2010, pp. 2794–2797 (2010)
4. Schuller, B. et al.: The INTERSPEECH 2019 computational paralinguistics challenge: Styrian dialects, continuous sleepiness, baby sounds & orca activity. In: INTERSPEECH 2019, pp. 2378–2382 (2019)
5. Schuller, B., Steidl, S., Batliner, A., Schiel, F., Krajewski, J.: The INTERSPEECH 2011 speaker state challenge. In: INTERSPEECH 2011, pp. 3201–3204 (2011)
6. Schuller, B. et al.: The INTERSPEECH 2017 computational paralinguistics challenge: Addressee, cold & snoring. In: INTERSPEECH 2017, pp. 3442–3446 (2017)
7. Schuller, B. et al.: The INTERSPEECH 2020 computational paralinguistics challenge: elderly emotion, breathing & masks. In: INTERSPEECH 2020, p. 5 (2020)
8. Pan, S., Yang, Q.: A survey on transfer learning. IEEE Trans. Knowl. Data Eng. **22**(10), 1345–1359 (2009)
9. Yu, D., Seltzer, M.: Improved bottleneck features using pretrained deep neural networks. In: INTERSPEECH 2011, pp. 237–240 (2011)
10. Dalmia, S., Li, X., Metze, F., Black, A.: Domain robust feature extraction for rapid low resource ASR development. In: Proceedings of IEEE Spoken Language Technology Workshop (SLT), pp. 258–265 (2018)
11. Sainath, T., Mohamed, A., Kingsbury, B., Ramabhadran, B.: Deep convolutional neural networks for LVCSR. In: Proceedings of IEEE International Conference on Acoustics, Speech and Signal Processing (ICASSP), pp. 8614–8618 (2013)
12. Dahl, G., Yu, D., Deng, L., Acero, A.: Context-dependent pre-trained deep neural networks for large-vocabulary speech recognition. IEEE Trans. Audio Speech Lang. Process. **20**(1), 30–42 (2011)

13. Wang, Y. et al.: Tacotron: towards end-to-end speech synthesis. arXiv preprint arXiv:1703. 10135, p. 10 (2017)
14. Snyder, D., Garcia-Romero, D., Povey, D., Khudanpur, S.: Deep neural network embeddings for text-independent speaker verification. In: INTERSPEECH 2017, pp. 999–1003 (2017)
15. Snyder, D., Garcia-Romero, D., Sell, G., Povey, D., Khudanpur, S.: X-vectors: robust DNN embeddings for speaker recognition. In: Proceedings of IEEE International Conference on Acoustics, Speech and Signal Processing (ICASSP), pp. 5329–5333 (2018)
16. Garcia-Romero, D., Snyder, D., Sell, G., Povey, D., Mccree, A.: Speaker diarization using deep neural network embeddings. In: Proceedings of IEEE International Conference on Acoustics, Speech and Signal Processing (ICASSP), pp. 4930–4934 (2017)
17. Snyder, D., Garcia-Romero, D., Mccree, A., Sell, G., Povey, D., Khudanpur, S.: Spoken language recognition using x-vectors. In: Proceedings of Odyssey, pp. 105–111 (2018)
18. Ghahremani, P., et al.: End-to-end deep neural network age estimation. In: INTERSPEECH 2018, pp. 277–281 (2018)
19. Burkhardt, F., Eckert, M., Johannsen, W., Stegmann, J.: A database of age and gender annotated telephone speech. In: Proceedings of 7th International Conference on Language Resources and Evaluation (LREC), pp. 1562–1565 (2010)
20. Eyben, F., Wöllmer, M., Schuller, B.: OpenSMILE – the Munich versatile and fast open-source audio feature extractor. In: Proceedings of ACM Multimedia International Conference, pp. 1459–1462 (2010)
21. Kockmann, M., Burget, L., Cernocký, J.: Brno university of technology system for INTERSPEECH 2010 Paralinguistic Challenge. In: INTERSPEECH 2010, pp. 2822–2825 (2010)
22. Meinedo, H., Trancoso, I.: Age and gender classification using fusion of acoustic and prosodic features. In: INTERSPEECH 2010, pp. 2818–2821 (2010)
23. Li, M., Han, K., Narayanan, S.: Automatic speaker age and gender recognition using acoustic and prosodic level information fusion. Comput. Speech Lang. **27**(1), 151–167 (2013)
24. Yücesoy, E., Nabiyev, V.: A new approach with score-level fusion for the classification of a speaker age and gender. Comput. Electr. Eng. **53**, 29–39 (2016)
25. Sarma, M., Sarma, K., Goel, K.: Children's age and gender recognition from raw speech waveform using DNN. In: Proceedings of Advances in Intelligent Computing and Communication (ICAS), pp. 1–9 (2020)
26. Shobaki, K., Hosom, J.: The OGI kids speech corpus and recognizers. In: INTERSPEECH 2000, p. 4 (2000)
27. Martin, A., Greenberg, C.: NIST 2008 speaker recognition evaluation: performance across telephone and room microphone channels. In: INTERSPEECH 2009, pp. 2579–2582 (2009)
28. Martin, A., Greenberg, C.: The NIST 2010 speaker recognition evaluation. In: INTERSPEECH 2010, pp. 2726–2729 (2010)
29. Krizhevsky, A., Sutskever, I., Hinton, G.: ImageNet classification with deep convolutional neural networks. In: Proceedings of Advances in Neural Information Processing Systems, pp. 1097–1105 (2012)
30. Simonyan, K., Zisserman, A.: Very deep convolutional networks for large-scale image recognition. arXiv preprint arXiv:1409.1556, p. 14 (2014)
31. Russakovsky, O., Deng, J., Su, H., et al.: Imagenet large scale visual recognition challenge. Int. J. Comput. Vis. **115**, 211–252 (2015)
32. He, K., Zhang, X., Ren, S., Sun, J.: Deep residual learning for image recognition. In: Proceedings of IEEE Conference on Computer Vision and Pattern Recognition, pp. 770–778 (2016)

33. Tan, M., Le, Q.: EfficientNet: rethinking model scaling for convolutional neural networks. arXiv preprint arXiv: 1905.11946, p. 10 (2019)
34. Zhou, H., Bai, X., Du, J.: An investigation of transfer learning mechanism for acoustic scene classification. In: Proceedings of International Symposium on Chinese Spoken Language Processing (ISCSLP), pp. 404–408 (2018)
35. Tang, D., Zeng, J., Li, M.: An end-to-end deep learning framework for speech emotion recognition of a typical individuals. In: INTERSPEECH 2018, pp. 162–166 (2018)
36. Tripathi, S., Kumar, A., Ramesh, A., Singh, C., Yenigalla, P.: Focal loss based residual convolutional neural network for speech emotion recognition. arXiv preprint arXiv:1906. 05682 (2019)
37. Hartmann, W., et al.: Improved single system conversational telephone speech recognition with VGG bottleneck features. In: INTERSPEECH 2017, pp. 112–116 (2017)
38. Sankupellay, M., Konovalov, D.: Bird call recognition using deep convolutional neural network, ResNet-50. Acoustics Australia 7(9), 8 (2018)
39. Markitantov, M., Verkholyak, O.: Automatic recognition of speaker age and gender based on deep neural networks. In: Proceedings of International Conference on Speech and Computer (SPECOM), pp. 327–336 (2019)
40. McFee, B. et al.: Librosa: audio and music signal analysis in Python. In: Proceedings of 14th Python in Science Conference, pp. 18–24 (2015)
41. Paszke, A. et al.: Automatic differentiation in PyTorch. In: Proceedings of Autodiff Workshop of Neural Information Processing Systems (NIPS), p. 4 (2017)

Interactivity-Based Quality Prediction of Conversations with Transmission Delay

Thilo Michael[1(✉)] and Sebastian Möller[1,2]

[1] Quality and Usability Lab, Technische Universität Berlin, Berlin, Germany
thilo.michael@tu-berlin.de, sebastian.moller@dfki.de
[2] German Research Center for Artificial Intelligence (DFKI), Kaiserslautern, Germany

Abstract. While the experienced quality of a conversation transmitted over a telephone network is dependent on the parameters of the network and possible degradations, it has been shown that the sensitivity to these degradations is influenced by the type of the conversation and its contents. Especially for delayed speech transmission the conversational scenario and the adaption of turn-taking behavior of the interlocutors factor into the conversational quality rating of a specific conversation. Parametric Conversation Analysis has been proven to be a good method to extract parameters from a recorded conversation that are representable of the interactivity. While the narrowband version of a popular and standardized quality prediction model, the E-model, uses the interactivity to predict a conversational quality MOS for the given conditions, there are no attempts to predict the conversational quality of individual conversations so far. In this paper, we propose a model to predict the quality of a conversation under the influence of transmission delay based on the interactivity parameters extracted from that conversation. We evaluate which parameters are most suited for such a prediction, and compare our results to the parameter-based predictions of the E-model.

Keywords: Speech quality · Conversation analysis · Model · Delay

1 Introduction

Today's communication networks mostly rely on packet-switched networks to transmit speech over the Voice-over-IP (VoIP) protocol. In this procedure, the coded speech is split into packets that are transmitted over a routed network. Often these networks are heterogeneous, and the provider of the speech service has no control over the reliance and speed of the transmission. With current speech codecs using the full audible frequency band (0–20,000 Hz), the most prominent degradations that influence the conversational quality are packet-loss and transmission delay. Because delay has only an effect on the structure of the conversation and is not audible in listening-only tests, the conversational quality is measured most realistically with conversation tests, where two participants

© Springer Nature Switzerland AG 2020
A. Karpov and R. Potapova (Eds.): SPECOM 2020, LNAI 12335, pp. 336–345, 2020.
https://doi.org/10.1007/978-3-030-60276-5_33

interact with each other over an (oftentimes simulated) telephone network. While packet-loss is audible and thus can be measured in listening-only tests, it too can have an effect of the structure of the conversation, for example, if information is lost during packet-loss bursts.

The influence of echo-free transmission delay on the conversational quality is mainly determined by the amount of delay. However, it has been shown that the sensitivity to it varies depending on the type and interactivity of the conversation [1]. Specifically, transmission delay has a higher impact on the structure of conversations with high interactivity and more frequent speaker alternations than on low-interactivity conversations.

Parametric Conversation Analysis (P-CA) is a framework to instrumentally extract parameters and metrics from full-reference conversation recordings and quantify the interactivity of the conversation [3]. These parameters are based on a 4-state conversation model (speaker A, speaker B, mutual silence, and double talk) [15] and range from low-level information like the sojourn times [7] of these states to high-level metrics like the conversational temperature, that tries to model how *heated* a conversation is [14].

Different algorithms are available to predict conversational quality under degradations. The E-model is the most popular of these models standardized by the International Telecommunication Union (ITU-T) as ITU-T Recommendation G.107 in 1996, and extended several times since then. The most recent standard, the fullband E-model [6], predicts conversational quality in the full audio band, but does not yet consider the interactivity of a conversation as a parameter for its predictions. However, the narrowband E-model includes parameters that distinguish between different levels of conversational interactivity [4]. In [13], Raake et al. proposed a model to estimate those E-model parameters based on interactivity parameters of a conversation.

In this paper, we focus the prediction of conversational quality of delay-affected conversations based on interactivity parameters obtained from P-CA. In contrast to previous work, we predict the absolute category rating of the overall conversational quality given a single conversation, not the mean opinion score over multiple conversations. While single quality ratings are affected by strong noise due to individual conversational behavior and the high subjectivity of the ratings, we believe that this approach is able to capture the differences that lead to the quality rating of the specific conversation. We average over the predictions of our model and compare these results with the predictions of the E-model. In Sect. 2 describes the related work and briefly outlines the fundamentals of the delay impairment factors of the narrowband and fullband E-model. Section 3 describes the databases we used to train and validate our model and the conversational parameters we extracted from those conversations. In Sect. 4, we correlate these parameters to the individual quality ratings, show our model, and discuss the results.

2   Related Work

The ITU-T describes subjective measures for the evaluation of conversational quality in [8]. This is usually done with conversation tests stimulated by standardized scenarios, such as the Short Conversation Test (SCT) and the Random Number Verification test (RNV) [8,9]. It has been shown that while delay is not perceivable in the speech signal itself, it affects the flow of a conversation [3]. However, the degree to which the conversation is affected by delay and thus the sensitivity to delay in quality ratings is influenced by the interactivity of the conversation itself [2,13].

The interactivity of conversations can be assessed using the parametric conversation analysis (P-CA) framework [3]. This framework is based on the 4-state model of Sacks et al. [15], where a conversation between two speaker A and B is split up into four states: mutual silence (ms) where neither A nor B talks, double talk (dt) where A and B talk at the same time, speakers A (sa) and speaker (sb) where only the respective speaker is active [7,11]. This can be done instrumentally by using the voice activity of the speakers. These four states yield state probabilities, that determine what percentage the conversation was in that state, and also sojourn times, that express how long the speakers sojourned in the respective states. Based on the four states, parameters like the speaker alternation rate (sar, speaker alternation per minute), and interruption rate (ir, interruptions per minute) can be calculated [3]. Higher-level metrics like the conversational temperature [14] combine different parameters to assess the interactivity of the conversation specifically. Also, parameters for delayed conversations have been developed to measure the changes in the conversation: the intended and unintended interruption rate (iir, uir) measure how often an interruption was intended by the speaker and how often the delay caused an unintended interruption [1]. The corrected speaker alternation rate (SARc) corrects the SAR for longer pauses caused by delay [16].

The E-model is the most popular parametric model for the prediction of conversational quality. While initially only applicable to narrowband voice services in ITU-T Rec. G.107 [4], it has been extended for wideband communication in ITU-T Rec. G.107.1 [5] and recently fullband communication in ITU-T Rec. G.107.2 [6]. The E-model assumes that impairments are independent of each other and can be quantified as impairment factors on a so-called "psychological scale", that is the scale of the transmission rating R. The transmission rating R itself can be transformed into an average judgment of a test participant on a 5-point absolute category rating scale, the Mean Opinion Score (MOS) using an S-shaped monotonous relationship. While in the narrowband version, the maximum transmission rating is 100, it increases to 148 in the fullband version.

The fullband E-model [6] calculates the impairment factor Idd for effects of pure delay (without echo) as a function of the delay time Ta:
for $Ta \leq 100\,\mathrm{ms}$:

$$Idd_{FB} = 0 \tag{1a}$$

for $Ta > 100$ ms:

$$Idd_{FB} = 1.48 \cdot 25\{(1 + X^6)^{\frac{1}{6}} - 3(1 + [\frac{X}{3}]^6)^{\frac{1}{6}} + 2\} \tag{1b}$$

and:

$$X = \frac{\log(\frac{Ta}{100})}{\log(2)} \tag{2}$$

The narrowband E-model calculates the impairment factor for delay in a similar fashion, but includes two parameters that depend on the interactivity of the conversation: sT denotes the delay sensitivity of the participants, and mT reflects the minimal perceivable delay in milliseconds. Both parameters are dependent on the type of conversation that the prediction should reflect:
for $Ta \leq mT$:

$$Idd_{NB} = 0 \tag{3a}$$

for $Ta > mT$:

$$Idd_{NB} = 25\{(1 + X^{6 \cdot sT})^{\frac{1}{6 \cdot sT}} - 3(1 + [\frac{X}{3}]^{6 \cdot sT})^{\frac{1}{6 \cdot sT}} + 2\} \tag{3b}$$

and:

$$X = \frac{\log(\frac{Ta}{mT})}{\log(2)} \tag{4}$$

The narrowband E-model also includes delay sensitivity classes to be used. In the default case sT is set to 1 and mT is set to 100 ms, while in the case of very low sensitivity $sT := 0.4$ and $mT := 150$.

In [13], where Raake et al. propose the extended Idd formula for the narrowband E-model, the following equations for calculating mT and sT from the corrected speaker alternation rate (SARc) are given:

$$mT = 436.02 - 71.56 \cdot \log(16.76 + SARc) \tag{5}$$
$$sT = 0.246 + 0.02 \cdot \exp(0.053 \cdot SARc) \tag{6}$$

3   Data and Setup

Two conversation tests have been conducted, one at the Quality and Usability Labs, Technische Universität Berlin, Germany (TUB) [12], the other one at the University of Western Sydney, Australia (UWS) [17]. As shown in Table 1, both experiments used a fullband PCM simulated telephone network and introduced delay levels of 0, 800, and 1600 ms. While the UWS experiment was conducted only using SCT scenarios, the TUB experiment features SCT as well as RNV scenarios. In both experiments, the participants were located in separate sound-proofed rooms and communicated through stereo headsets, as to minimize the potential for an acoustic echo. The conversational quality was assessed with the extended continuous scale (ECS) and later transformed into MOS based on [10].

Table 1. Settings for the TUB and UWS experiment.

	TUB	UWS
Subjects	58	20
Conversations	534	126
Age	18–71 (mean 32)	23–45 (mean 27)
Language	German	English
Codec	16-bit PCM at 44.1 kHz	16-bit PCM at 44.1 kHz
Scenario	SCT, RNV	SCT
Delay [ms]	0, 800, 1600	0, 800, 1600

We use the data from the TUB experiment to fit our model and the data from the UWS experiment for validation only.

From the recorded conversations obtained from the experiments we extracted 16 parameters that describe a range of different interactivity characteristics: the state probabilities (ms, dt, sa, sb) and sojourn times (st_ms, st_dt, st_sa, st_sb) of mutual silence, double talk, speaker a, and speaker b, the speaker alternation rate (sar), interruption rate (ir), double talk rate (dtr), pause rate (pr), intended interruption rate (iir), unintended interruption rate (uir), and corrected speaker alternation rate (sarc), and finally the conversational temperature (temp).

We calculated the MOS_{CQEF} as predicted by the fullband version of the E-model with the Idd_{FB} Eqs. 1a, 1b, and 2. To make use of the conversational interactivity-specific parameters sT and mT of the narrowband E-model, we adapted the Idd formula to work in the extended fullband context:
for $Ta \leq mT$:

$$Idd_{FB} = 0 \tag{7a}$$

for $Ta > mT$:

$$Idd_{FB} = 1.48 \cdot 25\{(1 + X^{6 \cdot sT})^{\frac{1}{6 \cdot sT}} - 3(1 + [\frac{X}{3}]^{6 \cdot sT})^{\frac{1}{6 \cdot sT}} + 2\} \tag{7b}$$

and:

$$X = \frac{\log(\frac{Ta}{mT})}{\log(2)} \tag{8}$$

With these updated equations, we calculated one MOS per condition (delay level and conversation scenario) based on the standardized values of mT and sT parameters of these conditions provided by the narrowband E-model in ITU-T Rec. G.107 [4]. Additionally, we calculated a quality prediction for each conversation based on the mT and sT values obtained from the Eqs. 5 and 6.

4   Model and Results

To select the features for our model, we first looked at the correlation between the features and the ACR rating of their conversations (Fig. 2), as well as the correlation between the features themselves (Fig. 1). The correlation map shows positive correlations between the state probabilities (ms, dt, sa, sb) and their sojourn times (st_ms, st_dt, st_sa, st_sb), which is to be expected. Also, a positive correlation between the state probability of double talk (dt) and the double talk rate (dtr) exists. The speaker alternation rate strongly correlates positively with the pause rate, but also with the conversational temperature and the corrected speaker alternation rate. Strong negative correlations can be seen between the state probability mutual silence and all other state probabilities and similar patterns can be found with the sojourn times.

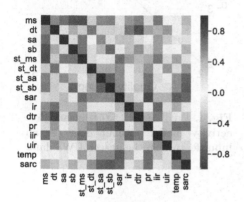

Fig. 1. Correlation matrix of the 16 interactivity parameters.

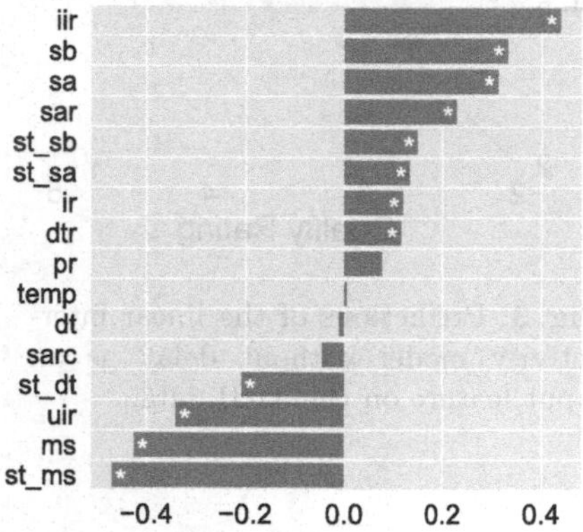

Fig. 2. Spearman correlation of each interactivity parameter with the respective quality rating of the conversation.

Figure 2 shows the correlation between the interactivity features and the conversational quality of the conversation. While the intended interruption rate has a positive correlation, the unintended interruption rate has a lower, negative correlation with the conversational quality. The strength of the correlation is rather low at 0.44 for the intended interruption rate and −0.47 for the sojourn times on mutual silence. However, this is to be expected due to the strong variations in the conversational quality ratings of the participants. The pause rate, the state probability for double talk, the corrected speaker alternation rate, and the conversational temperature do not correlate significantly with the conversational quality ratings.

We created two linear models to predict the conversational quality based on the interactivity parameters with high correlations: one model solely on the interactivity parameters and one model that includes the end-to-end transmission delay in seconds as a parameter.

$$\widehat{CQ}_1 = 5.328 - 1.6877 \cdot ms - .9699 \cdot st\_dt - 0.526 \cdot st\_ms - 0.0374 \cdot uir \quad (9)$$

$$\widehat{CQ}_2 = 4.4824 - 1.4055 \cdot ms - 0.5665 \cdot delay + 0.113 \cdot (sar \cdot delay) \quad (10)$$

The model in Eq. 9 resulted in an adjusted R^2 of 0.401 and the model in Eq. 10 resulted in an adjusted R^2 of 0.44 on the training data.

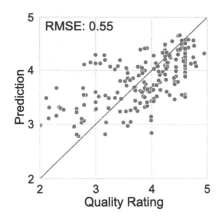

Fig. 3. Predictions of the linear interactivity model without delay as an input feature on the TUB data.

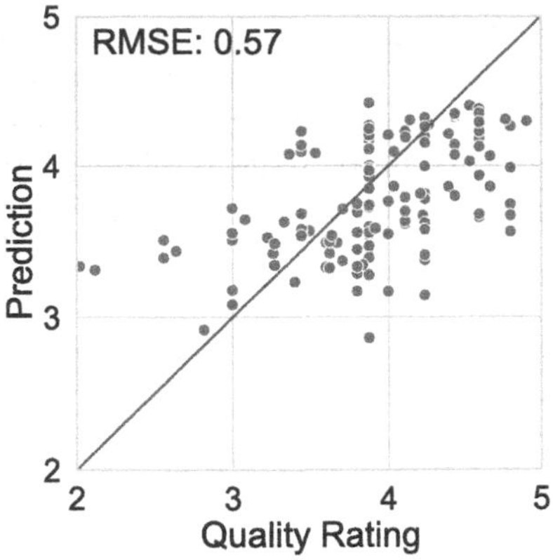

Fig. 4. Predictions of the linear interactivity model without delay as an input feature on the unseen UWS data.

Figures 3 and 4 show the performance of the model without delay as an input parameter (Eq. 9) on the training and test data. The RMSE is rather high at 0.55 for the training data and 0.57 for the test data. The model with delay as an input parameter (Eq. 10) has a slightly lower RMSE for the training data at 0.53 and also a slightly higher RMSE on the test data with 0.58. This high RMSE is to be expected, because the labels the models were trained on are noisy due to the high interpersonal variance in the subjective ratings.

To further evaluate the model, we averaged the predictions over delay and conversation scenarios to produce a MOS and compared the results of the two linear models with different E-model versions (Table 2). Overall, the two interactivity-based models perform better than the E-model predictions. While the interactivity model that includes delay as an input parameter performs better on the training data, the other interactivity model performed better on the unseen data (UWS). The RMSE as seen in Table 2 is comparable to the standard deviation of the data, with values ranging from 0.36 to 0.96. As expected, the RMSE for the unseen data from the UWS experiment is slightly higher than for the training data from the TUB experiment.

Figure 5 compares the different prediction approaches on the MOS scale to the subjective test results. The fullband E-model (light orange) in its current state does not account for the different interactivity (SCT and RNV) of the conversations and is in all data sets and conditions too pessimistic. The E-model extension (dark orange), where the sT and mT parameters of the narrowband version are included, performs better in all cases and accounts for the difference

Table 2. RMSE for the two interactivity-based models ("interactivity model" and "interactivity model (+delay)", the fullband E-model ("E-model"), the fullband E-model extended by the Idd-formula of the narrowband E-model ("E-model extension") and the extended E-model prediction where the sT and mT parameters are predicted from the conversations ("E-model est. sT/mT") on the training (TUB) and test dataset (UWS).

| Dataset | Delay (s) | Scenario | SD | RMSE | | | | |
			Data	Interactivity model	Interactivity model (+delay)	E-Model est. sT/mT	E-Model	E-Model extension
TUB	0	sct	0.4730	0.4752	**0.4667**	0.6080	0.6910	0.6080
		rnv	0.5032	0.5337	**0.4869**	0.5546	0.5546	0.5546
	0.8	sct	0.5623	0.5325	**0.5214**	0.6976	0.9000	0.5897
		rnv	0.5351	0.5390	**0.5368**	0.6179	0.8246	0.5462
	1.6	sct	0.6466	**0.6325**	0.6448	0.8846	0.8684	0.7143
		rnv	0.6635	0.6359	**0.5986**	0.8043	0.6910	0.7144
UWS	0	sct	0.3886	**0.3636**	0.3652	0.4931	0.4931	0.4931
	0.8	sct	0.5697	0.6203	0.6545	**0.6172**	1.0820	0.5711
	1.6	sct	0.6835	**0.7060**	0.7121	0.9561	0.9026	0.7438

in interactivity. The E-model prediction where the sT and mT parameters are calculated for every conversation based on Eq. 5 and 6 (green) is often too optimistic. The interactivity model of Eq. 9 (without delay) fits the data the closest, while being slightly too pessimistic for $0.8s$ delay and too optimistic for $1.6\,s$ delay.

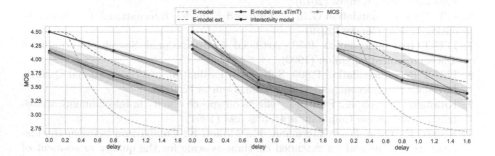

Fig. 5. MOS predictions of the E-model variants and the interactivity model without delay on the TUB SCT conversations (left), the TUB RNV conversations (center) and the UWS SCT conversations (right). Shaded areas are the standard deviations.

5   Conclusion

In this paper, we analyzed interactivity parameters of fullband conversations regarding their suitability to predict the conversational quality of single conversations. Based on these interactivity parameters, we built two models (one with the transmission delay as an input parameter and one without) and compared them to the parametric fullband E-model and modifications of it. As part of this, we proposed and evaluated an extension to the fullband E-model to account for differences in interactivity based on the narrowband version of the E-model.

In future work, we plan to extend the dataset and validate the model for more delay levels and additional conversation scenarios (e.g., interactive SCT and timed RNV). We also plan to extend the model to account for changes in conversations due to bursty packet-loss as described in [12].

References

1. Egger, S., Schatz, R., Scherer, S.: It takes two to tango-assessing the impact of delay on conversational interactivity on perceived speech quality. In: Eleventh Annual Conference of the International Speech Communication Association, pp. 1321–1324. ISCA (2010)
2. Egger, S., Schatz, R., Schoenenberg, K., Raake, A., Kubin, G.: Same but different? - Using speech signal features for comparing conversational VoIP quality studies. In: 2012 IEEE International Conference on Communications (ICC), pp. 1320–1324. IEEE (2012)
3. Hammer, F.: Quality aspects of packet-based interactive speech communication. Forschungszentrum Telekommunikation Wien (2006)
4. ITU-T Recommandation G.107: The E-model: a computational model for use in transmission planning. International Telecommunication Union, Geneva (2011). http://handle.itu.int/11.1002/1000/12505
5. ITU-T Recommandation G.107.1: Wideband E-model. International Telecommunication Union, Geneva (2015)
6. ITU-T Recommandation G.107.2: Fullband E-model. International Telecommunication Union, Geneva (2019)
7. ITU-T Recommendation P.59: Artificial Conversational Speech. International Telecommunication Union (1993)
8. ITU-T Recommendation P.805: Subjective Evaluation of Conversational Quality. International Telecommunication Union, Geneva (2007)
9. Kitawaki, N., Itoh, K.: Pure delay effects on speech quality in telecommunications. IEEE J. Sel. Areas Commun. 9(4), 586–593 (1991)
10. Köster, F., Guse, D., Wältermann, M., Möller, S.: Comparison between the discrete ACR scale and an extended continuous scale for the quality assessment of transmitted speech. Fortschritte der Akustik-DAGA (2015)
11. Lee, H., Un, C.: A study of on-off characteristics of conversational speech. IEEE Trans. Commun. 34(6), 630–637 (1986)
12. Michael, T., Möller, S.: Effects of delay and packet-loss on the conversational quality. Fortschritte der Akustik-DAGA (2020)
13. Raake, A., Schoenenberg, K., Skowronek, J., Egger, S.: Predicting speech quality based on interactivity and delay. In: Proceedings of INTERSPEECH, pp. 1384–1388 (2013)
14. Reichl, P., Hammer, F.: Hot discussion or frosty dialogue? Towards a temperature metric for conversational interactivity. In: Eighth International Conference on Spoken Language Processing (2004)

15. Sacks, H., Schegloff, E., Jefferson, G.: A simplest systematics for the organization of turn-taking for conversation. Language **50**(4), 696–735 (1974). https://doi.org/ 10.2307/412243. http://www.jstor.org/stable/412243?origin=crossref
16. Schoenenberg, K.: The Quality of Mediated-Conversations under Transmission Delay. Ph.D. thesis, TU Berlin (2015). http://dx.doi.org/10.14279/depositonce-4990
17. Uhrig, S., Michael, T., Möller, S., Keller, P.E., Voigt-Antons, J.N.: Effects of delay on perceived quality, behavior and oscillatory brain activity in dyadic telephone conversations. In: 2018 Tenth International Conference on Quality of Multimedia Experience (QoMEX), pp. 1–6. IEEE (2018)

Graphic Markers of Irony and Sarcasm in Written Texts

Polina Mikhailova[✉]

National Research University Higher School of Economics,
Saint-Petersburg, Russia
pmmikhaylova@edu.hse.ru

Abstract. Even in real-life communication detecting irony in someone's speech is a challenging task sometimes. The main reason is that ironists often say something straight opposite to what they actually mean. Verbal irony can be detected through prosodic and paralinguistic features of speech. However, normative writing does not have such, more or less, obvious markers. That is why most papers on this topic are based on syntax, words frequency, context, semantics and other linguistic features of texts. Simultaneously, automatic detection of irony and sarcasm still remains one of the most difficult and only partially resolved questions. This paper gives an overview of the main irony markers that are usually used to infer ironic intent and suggests an approach to different graphic signs used in everyday computer-mediated communication as potential irony markers. In addition an experimental study is held to compare these markers and examine their relevance and accuracy.

Keywords: Irony · Sentiment analysis · Irony recognition · Verbal irony

1 Introduction

The automatic detection of irony and sarcasm is one of the most difficult and interesting topics in the field of sentiment analysis, in which at the moment there are still quite a lot of unresolved issues and tasks. One of the most serious difficulties is the very nature of irony, which does not allow the use of standard methods applied to sentiment analysis for its identification, such as, for example, a bag-of-words model. The fact is that irony and sarcasm imply the presence of an additional meaning diametrically opposed to the literal. Moreover, a statement that will be perceived as neutral in one situation may be sarcastic in another. Therefore, the task of identifying irony requires the use of many other properties of sentences, in addition to lexical, including contextual, syntactic, punctuation and semantic ones.

This work is devoted to the study of these properties and consists of several sections: first of all, existing studies will be discussed, then oral and written irony will be compared, and finally, an experimental study of some markers of irony and sarcasm will be described.

A. Karpov and R. Potapova (Eds.): SPECOM 2020, LNAI 12335, pp. 346–356, 2020.
https://doi.org/10.1007/978-3-030-60276-5_34

2 Previous Work

There are several articles on the topic of automatic irony detection demonstrating different approaches to data selection and annotation, usage of different markers, features analysis and selection and methods of classification task resolution. Here the methodology of the four of the existing research papers and their main ideas and principles will be described.

Wallace and colleagues give a complete overview on the problem of context analysis in irony detection tasks and bring forward strong arguments in favor of contextualization [5]. There are few research papers on automatic irony detection because this task requires methods that are different from sentiment analysis on the whole. The main difference is that irony and sarcasm detection is nearly impossible without context analysis while one of the basic methods for sentiment analysis is bag-of-words-based text classification models. In a bag-of-words model, the text is represented as a set of words it contains, regardless to grammar or word order. Wallace and colleagues provide empirical evidence that context is often essential to recognize an ironic intent [5]. The authors created a corpus in which every sentence is annotated manually, and the process of annotation itself gives some answers to the question whether context is important or not. Annotators have a chance to request extra context to decide if the sentence contains an ironic intent or not. The main observation is that "annotators request context significantly more frequently for those comments that (are ultimately deemed to) contain an ironic sentence" [5].

The work by Reyes and colleagues is an experimental research on automatical irony detection based on four data sets, "that were retrieved from Twitter using content words (e.g. "Toyota") and user-generated tags (e.g. "#irony")" [2]. The corpus consists of tweets containing hashtags #irony, which makes the fact of ironic intent obvious, #humour, #education and #politics, so that there is a great variety of non-ironic examples. The classification model relies on four types of conceptual features: signatures, unexpectedness, style, and emotional scenarios, and then its representativeness and relevance are assessed.

The next article by Van Hee et al. is a complex research of irony, its conceptualizations and computational approaches to its detection [4]. To operationalize the task the authors create a dataset of 3000 English tweets from Twitter containing the hashtags #irony, #sarcasm and #not. All tweets are annotated manually and separated into three main categories: ironic by means of a clash, "the text expresses an evaluation whose literal polarity is opposite to the intended polarity", other type of irony, "there is no contrast between the literal and the intended evaluation, but the text is still ironic" and not ironic [4]. Next in every tweet annotators mark evaluative expressions, modifiers and targets. Finally, a series of binary classification experiments are conducted "using a support vector machine (SVM) that exploits a varied feature set and compare this method to a deep learning approach that is based on an LSTM network and (pre-trained) word embeddings" [4]. Evaluation of these two models shows that SVM performs much better according to the combination of lexical, semantic and syntactic information used in its work.

The article by Sulis et al. is an experimental study on irony, sarcasm and different methods of their detection in written texts [3]. The research is based on a data set consisting of 12,532 tweets "annotated in a fine-grained sentiment polarity from −5 to +5" [3]. There are 5114 hashtags in the corpus the most used ones are #not, #sarcasm and #irony, which occur in 7244 tweets. Next the analysis of several linguistic features and some of the hypothesis tests are held and classification models are built. The list of classification algorithms used in this study includes Naïve Bayes, Decision Tree, Random Forest, Logistic Regression and Support Vector Machine, from which the best result is performed by Random Forest model [3].

As it can be seen, these works have several things in common:

- the use of hashtags (excepting the first paper) for data collection and annotation of the examples as ironic, sarcastic or neutral;
- the use of semantic or syntactic properties of the evaluated sentence, which are further used in a complex manner, in various combinations or separately;
- the use of statistics and machine learning models to assess the significance of the selected variables and the effectiveness of the proposed solution to the problem;
- according to the results of all the mentioned studies, those models that use not a separate indicators, but a whole complex of parameters as the basis for the classification models demonstrate the greatest efficiency and accuracy.

Thus, there are many methods and features that can vary and be used in various combinations. However, one of the problems is that many of them have not been studied in sufficient detail. For example, some factors with only two levels could give a more accurate result if they were transformed or combined with others into longer scale values. Therefore, later in this work several of such factors in particular will be considered in a more precise way.

3   Verbal Irony Markers and Their Written Alternatives

The main feature of irony and sarcasm is that a statement in which in one situation their presence is obvious in another can have a different sentiment, or it can be completely neutral. That is why most researchers consider the identification of irony without regard to the context to be almost impossible. However, such situations, when the use of irony by one of the speakers in no way depends on the context, and is motivated mostly by the will of the speaker, are also frequent. If, in such a situation, another participant in the dialogue is not able to recognize the ironic intention, the probability of which is rather high in the absence of contextual motivation for such an event, then a communicative failure can occur, characterized by a lack of understanding between the speakers [1]. Such a development of events is not preferable for any of the interlocutors, therefore, in oral speech, to indicate the presence of irony in the utterance, intonation, facial expressions, gestures and other prosodic and paralinguistic means are used.

There are no such tools in normative written language, however, in informal written communication adaptations of those markers of irony that are used in oral speech can be found:

- punctuation and lengthening of vowels as a means of transmitting intonation;
- brackets and text emojis as a means of transmitting facial expressions;
- emoji as a means of transmitting facial expressions and gestures.

Despite the utter importance of the context for identifying ironic intentions, the most difficult task is to identify those factors that may allow it to be identified outside the context. Even in the same context, the same statement can be neutral and ironic. Therefore, the purpose of this study is to determine precisely those factors that will allow, "other things being equal," to distinguish ironic from neutral ones. As we are talking about written communication, their search will be carried out among the graphic markers mentioned above.

Another important observation is that most of the studies reviewed in Sect. 2 make a distinction between irony and sarcasm. Irony and sarcasm, indeed, differ from each other in intent: unlike irony, sarcasm usually implies a rather rude mockery, taunting, and sometimes a desire to offend the other person. Ironic statements do not have such a message, most often representing an example of a language game. However, sarcasm is often called "the highest degree of irony", thus putting these linguistic phenomena on a par. This is explained by the fact that, despite the difference in intention, both irony and sarcasm are based on a double meaning and the opposition of subtext to the literal meaning of the phrase. Thus, the same statement in different contexts and situations may have different meanings. It is important to note that this is not only about the linguistic context: for example it may also depend on the relationship between the speakers, whether the statement is more likely to be regarded as irony, sarcasm, or even roughness.

Since in the current study it is graphic markers that are considered, and the distinction between irony and sarcasm is at the level of intention and is encoded to a greater extent at the contextual and lexical level, therefore, in this study, this border will not be drawn.

4   Experimental Study

Since graphic markers were chosen for further research, it is necessary to highlight the oppositions, with the help of which the degree of their presence in the sentences can be characterized. The three main oppositions that form the basis for what is described in this section:

- normal spelling vs. vowel repetition;
- normal punctuation vs. excessive punctuation;
- a small number of brackets (1–2) vs. a large number of them (3 or more).

Speaking about these oppositions, several features should be noted. Firstly, "normal" punctuation in this case does not imply an ideally correct punctuation marking that conforms to all the rules, but the absence, for example, of repetition of exclamation marks at the end of a sentence. This is due to the fact that in everyday written communication people rarely seek to comply with the literary norms of the language. The same applies to "normal" spelling, but to a lesser extent, since some spelling distortions besides the mentioned lengthening of vowels (which, however, are not considered in this study and are not found among experimental examples) can be interpreted as marking irony and/or other emotions and subtexts. Secondly, the border between a small and excessive number of brackets in the text is nominal, since this graphic phenomenon has no described usage standards and at the same time strongly depends on the individual style. Therefore, in order to avoid the influence of the latter, in the experimental examples the use of, for example, 3–4 brackets was avoided, since this number for different carriers may be of a different nature or may be borderline and cause doubt.

The experiment was conducted on the basis of the Russian language, so all the examples in this section will be provided with a translation and, if necessary, other explanations.

4.1   Experimental Design and Hypotheses

For the experiment, a factorial design with two variables was chosen: markers of irony (independent; factor with eight levels) and the level of irony of the statement (dependent; rating on a scale). The levels of the independent variable include not only the previously mentioned markers individually, but also their pairwise combinations (Table 1). The scale for assessing the irony of experimental sentences includes the following answer options:

- the sentence is neutral;
- the sentence is neutral rather than ironic;
- the sentence is ironic rather than neutral;
- the sentence is ironic/sarcastic;
- other.

Subsequently, the answers obtained on this scale were translated into a four-point system with values from 1 to 4 for the first four options, respectively, and a value of 0 for the "other" option.

The main hypotheses (in generalized form, since forming H1, H2, etc. for each pair of relevant factors will give a list that is too long):

1. The average rate of all factors will be higher than the average rate of factor A;
2. The average rate of factor C will be higher than that of factor B;
3. Average rates of factors combining several markers (F, G, H) will be higher than that of factors containing single markers (B, C, D, E).

H0: there will be no significant difference in the average rates of different factors.

Table 1. The set of factors with examples.

Factor	Example
A – normal punctuation and spelling, no brackets	Konechno! *Of course!*
B – normal punctuation and spelling, a small number of brackets	Konechno)) *Of course))*
C – normal punctuation and spelling, a large number of brackets	Konechno))))))) *Of course)))))))*
D – excessive punctuation, normal spelling, no brackets	Konechno!!!! *Of course!!!!*
E – vowel repetition, normal punctuation, no brackets	Koneeeeeechno *Of cooooourse*
F – excessive punctuation, repetition of vowels, no brackets	Koneeeeeechno!!!!!!! *Of cooooourse!!!!!!!*
G – repetition of vowels, a large number of brackets, normal punctuation	Koneeeeeechno)))))))))) *Of cooooourse))))))))))*
H – excessive punctuation, a large number of brackets, normal spelling	Konechno!!!!))))))) *Of course!!!!)))))))*

As experimental examples, it was decided to use short dialogues on neutral everyday topics (Table 2), consisting of two replicas, the first of which always remained unchanged, and the second was joined by various combinations of the supposed markers of ironic context.

Table 2. Stimuli examples.

Example	Translation
A: Zaplatish' za menya? B: Da, bez problem	A: Will you pay for me? B: Yes, no problem
A: Chto dumaesh' ob etom? B: Po-moemu, prelestno	A: What do you think about this? B: I think it's lovely
A: Kak tak vyshlo? B: Sovershenno sluchajno	A: How did it happen? B: Quite by accident
A: Kak tvoi dela? B: Kak vsegda - luchshe vsekh!	A: How are you? B: As always - the best!

To create questionnaires, 24 such dialogue-bases were invented, to which markers and their combinations were subsequently added. Thus, 192 experimental examples were obtained, which were further distributed among eight questionnaires so that in each of them the respondent saw each dialogue-base only once and each of the factors three times, that is, each questionnaire contained 24 examples. The experimental examples within each questionnaire were randomly mixed.

4.2 Participants

The experiment was attended by 143 respondents, whose average age was about 18–23 years. Informants were randomly assigned to eight groups, each of which was assigned to one of eight questionnaires.

When processing the received answers, the mean rates of each respondent were checked, during which no abnormal results (the so-called "bad respondents" that marked absolutely all experimental examples, for example, as neutral) were received, so all 143 answers were held for further analysis.

The mean rates of each dialogue-base were also checked to prevent distortion of the results due to the fact that any of the examples has abnormally high or low ratings, regardless of which markers were used with it. However, despite the fact that some examples had higher or lower grades on average, no dialogues were found for which fluctuations in grades were not observed.

4.3 Data Analysis and Results

When analyzing the data, first of all, descriptive statistics were applied and measures of the central tendency, including medians, and mean rates for each of the factors were calculated. The visualization of these calculations is demonstrated on the boxplot (Fig. 1). As it can be seen, the lowest median value is observed for factor A, and two groups of factors with approximately equal medians, as well as similar values of the upper and lower quartiles and the maximum and minimum values, are distinguished. Fluctuations in mean rates are slightly more significant compared to the median, but a similar trend is also observed for them.

Next, it is necessary to carry out a statistical test of hypotheses. The first step was to use the ANOVA test, which allows to determine the presence or absence of at least one significant difference between several factors. According to the results of this test, the P-value was less than 2e-16, which means that the difference in the rates of at least one of the factors is statistically significant and we can proceed to pairwise compare them.

For this, the Tukey test (Tukey HSD: Tukey Honest Significant Differences) was used, which uses the results of the ANOVA test and makes a pairwise comparison of all factors. Table 3 contains the results of this test: in the first column there are the pairs of factors, in the second the difference between their mean rates, then the lower and upper boundaries of the confidence intervals and in the fifth column the p-value for each pair of factors is indicated.

The grey color indicates the lines containing the factors, the difference between which according to the P-value can be considered statistically significant and corresponding to one of the main hypotheses.

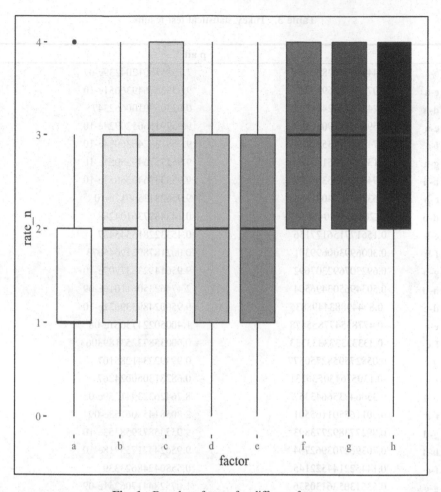

Fig. 1. Boxplot of rates for different factors.

Thus, according to the results of this test, the first and second hypotheses that the rates of factor A are significantly lower than the rest, and that the average rating of factor C are higher than that of factor B, are fully confirmed. The third hypothesis about the difference between factors F, G, H and B, C, D, E, however, for some pairs this difference was not convincing enough, and for some it was negative instead of positive.

Below is the summary of statistics on the characteristics of each of the factors (Table 4). Factors are ordered by increasing mean rate, and the percentage of "other" choices is shown for related examples. Thus, the factors under consideration form a kind of "scale of irony" with a fairly stable step. At the same time, it is important to note that precisely the factors following each other (except for the first two) are located, the difference between which was not recognized as statistically significant, that is why the gradation of the irony of the examples with these markers is quite smooth.

Table 3. Tukey statistical test results.

	diff	p adj
b-a	0.442890442890443	2.13357801204239e-07
c-a	1.07692307692308	9.95535764936051e-10
d-a	0.242424242424242	0.0330670100662347
e-a	0.599067599067599	9.95791560320924e-10
f-a	0.743589743589744	9.95613924636984e-10
g-a	1.13519813519814	9.95535764936051e-10
h-a	0.946386946386947	9.95535764936051e-10
c-b	0.634032634032634	9.956281354917e-10
d-b	-0.200466200466201	0.148487736104288
e-b	0.156177156177156	0.453322815545487
f-b	0.300699300699301	0.00218789319646073
g-b	0.692307692307692	9.95614923837707e-10
h-b	0.503496503496504	2.47462350611016e-09
d-c	-0.834498834498835	9.95602489339831e-10
e-c	-0.477855477855478	1.4005022475061e-08
f-c	-0.333333333333333	0.000358722531893063
g-c	0.0582750582750577	0.994922341285107
h-c	-0.130536130536131	0.682813060024267
e-d	0.356643356643357	8.76426323941359e-05
f-d	0.501165501165501	2.8058145806753e-09
g-d	0.892773892773893	9.95535875958353e-10
h-d	0.703962703962704	9.95624471755718e-10
f-e	0.144522144522145	0.55804248623339
g-e	0.536130536130536	1.07528441706251e-09
h-e	0.347319347319348	0.00015578725414811
g-f	0.391608391608391	8.86877913819362e-06
h-f	0.202797202797203	0.138094983378878
h-g	-0.188811188811188	0.20916309373349

Table 4. Summary statistics

	Factor	Mean rate	"other" %
1	A - normal punctuation and spelling, no brackets	1.792541	1.86%
2	B - normal punctuation and spelling, a small number of brackets	2.034965	2.1%
3	C - normal punctuation and spelling, a large number of brackets	2.235431	0.7%
4	D - excessive punctuation, normal spelling, no brackets	2.391608	1.63%
5	E - vowel repetition, normal punctuation, no brackets	2.536131	1.86%
6	F - excessive punctuation, repetition of vowels, no brackets	2.738928	2.56%
7	G - repetition of vowels, a large number of brackets, normal punctuation	2.869464	0.93%
8	H - excessive punctuation, a large number of brackets, normal spelling	2.927739	1.4%

5   Conclusion and Future Work

As a result, we can conclude that the most reliable markers for the task of detecting irony are a large number of brackets and a combination of a large number of brackets with repetition of vowels or excessive punctuation; a combination of vowel lengthening with excessive punctuation also shows a good result.

It is important to note that only some of the most obvious markers of irony and sarcasm have been examined in the current study. Further, other markers such as different use of capital letters, excessive punctuation not only at the end of a phrase, but also in the middle of a sentence (for example, the use of periods after each word), intentional spelling distortion and many others should also be considered, and other sentiment variations represented by the same means.

Nevertheless, the described results can be used, for example, to include the described markers as additional arguments in the classification models, and the resulting scale will allow us to assign weights to these arguments corresponding to the position of the markers on it. Moreover, the results of this study can not only be used in further studies of the automatic recognition of irony and sarcasm, but can also be integrated into other studies related to automatic text processing, text analysis and tonality analysis.

References

1. Bosco, F.M., Bucciarelli, M., Bara, B.G.: Recognition and repair of communicative failures: a developmental perspective (2005)
2. Reyes, A., Rosso, P., Veale, T.: A multidimensional approach for detecting irony in Twitter. Lang. Resour. Eval. **47**(1), 239–268 (2013)

3. Sulis, E., et al.: Figurative messages and affect in Twitter: differences between #irony, #sarcasm and #not. Knowl. Based Syst. **108**, 132–143 (2016)
4. Van Hee, C., Lefever, E., Hoste, V.: Exploring the fine-grained analysis and automatic detection of irony on Twitter. Lang. Resour. Eval. **52**(3), 707–731 (2018). https://doi.org/10.1007/s10579-018-9414-2
5. Wallace, B.C., et al. : Humans require context to infer ironic intent (so computers probably do, too). In: Proceedings of the 52nd Annual Meeting of the Association for Computational Linguistics (Volume 2: Short Papers), pp. 512–516 (2014)

Digital Rhetoric 2.0: How to Train Charismatic Speaking with Speech-Melody Visualization Software

Oliver Niebuhr[(✉)] and Jana Neitsch

MCI/CIE, University of Southern Denmark, Alsion 2,
6400 Sønderborg, Denmark
olni@sdu.dk

Abstract. The present study deals with a key factor of speaker charisma: prosody or, to put it less technically, speech melody. In a contrastive analysis, we investigate the extent to which computer-aided real-time visualizations of acoustic-melodic parameters in a self-guided training task help speakers use these parameters more charismatically. Fifty-two speakers took part in the experiment, subdivided into four equally large, gender-balanced groups. Each group received the same introduction and instruction, but then trained their melodic charisma with a different software tool. One tool – the "Pitcher" – was specifically developed for visualizing melodic parameters and providing feedback on their proper use in charismatic speech. The other software tools serve language- or singer-training purposes. Results show that three out of the four tools in fact support speakers in adopting a more charismatic speech melody after training, whereby the "Pitcher" outperforms all other tools by more than 25%. We discuss the implications of our findings for refining speech-melody visualization strategies (also for L2 training) and for digitizing rhetorical speaker coaching.

Keywords: Charisma · Pitcher · Voice · Prosody · Public speaking · Rhetoric

1 Introduction

1.1 What Is Charisma?

Everyone has already intuitively experienced charisma in everyday situations. At the same time, however, charisma is such a complex and versatile phenomenon that those who experience it can hardly explain it or assign it to a specific perceptual feature or personality trait. This is the reason why charisma has been described as "easy to sense but hard to define" [1: 305] in previous studies.

Nevertheless, at least since the Aristotle era, researchers have tried to itemize, systematize and, ultimately, to objectify charisma. Aristotle himself already pointed out that a speaker's persuasive impact on others critically depends on emotions, values and symbolic language. However, the term 'charisma' was not introduced by him, but by the German sociologist Max Weber in the early 20th century [2]. Moving away from Aristotle's idea, Weber defined charisma as an extraordinary and almost magical gift

© Springer Nature Switzerland AG 2020
A. Karpov and R. Potapova (Eds.): SPECOM 2020, LNAI 12335, pp. 357–368, 2020.
https://doi.org/10.1007/978-3-030-60276-5_35

that is innate in a speaker. By this, he turned charisma into a categorical concept. That is, Weber thought of charisma as a feature that a speaker either does or does not have. In contrast, modern definitions of charisma, derived from decades of empirical and experimental research in politics, psychology, and management, take a few steps back to Aristotle's original idea. Charisma is nowadays considered neither innate nor magical anymore. Rather, it is defined as a skill that relies on concrete, measurable signals and that, like any other gradually pronounced skill, can be learned and improved by its users and assessed by its trainers.

In terms of the influential charisma definition by Antonakis [3] and a recent terminological refinement by Michalsky and Niebuhr [4], charisma represents a particular communication style. It gives a speaker leader qualities through symbolic, emotional, and value-based signals. Three classes of charisma effects are to be distinguished in this context: (i) conveying emotional involvement and passion inspires listeners and stimulates their creativity; (ii) conveying self-confidence triggers and strengthens the listeners' intrinsic motivation; (iii) conveying competence creates confidence in the speakers' abilities and hence in the achievement of (shared) goals or visions. Inspiration, motivation, and trust together have a strongly persuasive impact by which charismatic speakers are able to influence their listeners' attitudes, opinions, and actions.

1.2   A Charismatic Speech Melody and Its Impact on Listeners

Since charisma has been defined as a way of communicating, studies investigated the relative contributions of verbal and nonverbal communication features on a speaker's charismatic impact. As it turned out, speech melody – i.e. the complex interplay of intonation, stress, rhythm, loudness and voice quality in the acoustic signal – is anything but a negligible factor for perceived speaker charisma [5, 6]. As Amon puts it: "there is a clear superiority of the audible impression over the visible" [7: 19–20] (our translation).

A number of recent studies support Amon's statement and demonstrate the strong effects of a charismatic speech melody on listeners. For instance, [8] investigated the acoustic characteristics of the US presidential candidates between 1960 and 2000. To that end, they analyzed the famous American TV debates that always take place between the two remaining opponents of the republican and democratic parties just before the actual election. From the extracted signals that were low-pass filtered to cover the range from f0 to roughly the first formant, they calculated a measure of the spectral energy distribution. This measure was able to give a 100% accurate prediction of the outcomes of all eight US presidential elections over the forty-year period.

Similarly, [9] analyzed the speech-melody profiles of 175 founders who presented their business ideas to investors at major startup events. Results showed that about 80% of the investors' funding/no-funding decisions could be explained by speech-melody patterns alone. In an earlier study, [10] and [11] used such speech-melody patterns to evoke striking effects on people's thoughts and behavior. For instance, in experiments with talking robots, [10] and [11] superimposed the exceptionally charismatic speech melody of Steve Jobs or the moderately charismatic speech melody of Mark Zuckerberg on a robot's synthesized speech, all else equal, including the spoken words.

Results showed that participants stuck to what the robot with Jobs' speech melody said and largely disregarded suggestions or commands made by the robot with Zuckerberg's speech melody. For example, Jobs' speech melody made participants fill out longer questionnaires or book different sightseeing trips; and implemented in a car-navigation system's voice, it made the participants follow a suggested route for longer, even if this route meant a considerable detour for them.

1.3   The Pitcher: A Software for Melodic Charisma Training

Regarding 1.2, we have now reached a point in science at which charismatic effects in terms of changes in listeners' thoughts and actions have become a learnable, improvable, and measurable skill. Moreover, we have arrived at a solid understanding of how speech melody, as a key element of charisma, contributes to these changes in thoughts and actions. This includes that we know where the "sweet spots" are located, i.e. narrow charisma-enhancing target-value ranges that speakers have to hit along each parameter of their speech melody [12]. In other words, we can start asking how we can train speakers such that they adopt a more charismatic speech melody.

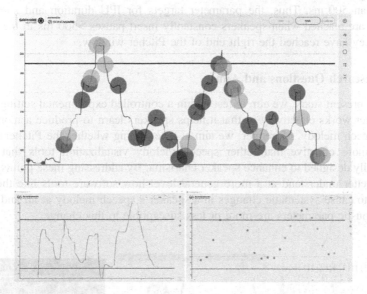

Fig. 1. Three screenshots of the Pitcher software with dots embedded in the pitch contour (top) as well as with pitch contour and dots displayed separately for focused learning (bottom).

Based on these findings, we developed the "Pitcher" – a tool for the computer-aided learning of a more charismatic speech melody (see Fig. 1). The Pitcher measures major charisma-related acoustic parameters of a speaker's speech melody, for example, while giving a presentation. Based on these measurements, the Pitcher provides simple, color-coded real-time feedback for the individual parameters, with "green" and "red" indicating whether the respective parameter is within or outside its charisma-enhancing "sweet-spot" target range. Target ranges are defined gender-specifically. The parameters taken into account by the Pitcher are: f0 range, mean f0 level, lowness of phrase-final falling f0, duration of prosodic phrases or interpausal units (IPUs), silent pause duration, speaking rate, and acoustic energy.

Furthermore, dots can be shown in the Pitcher display. Each dot marks a syllable. Depending on the display mode chosen by the speaker, the dots provide feedback on speaking rate or acoustic energy. In the speaking-rate mode, they change their color from green through yellow to red if the speaker becomes too fast or slow in his/her speech. In the acoustic-energy mode, the dots additionally change their size, with larger dots indicating a higher acoustic-energy level. If speakers get too soft in their voice, the smaller dots turn red. Note that speakers are able to hide all dots in order to focus on the f0-related feedback only. Inversely, they can hide the stylized f0 contour in order to focus on the dot-related tempo or loudness feedback only.

Finally, feedback on IPU and silent-pause durations is provided by means of the Pitcher window itself that represents the time axis. Both the pitch contour and the dots emerge over time from left to right along this time axis. The time axis is chosen such that the right end of the Pitcher window coincides with the end of the charisma-enhancing IPU duration. The time axis is reset when the speaker makes a silent pause larger than 500 ms. Thus, the parameter targets for IPU duration and silent pause duration are reached when speakers constantly insert pauses >500 ms in their speech before they have reached the right end of the Pitcher window.

1.4   Research Questions and Aims

With the present study, we aim to test within a controlled experimental setting whether the Pitcher works effectively in that it helps speakers learn to produce a more charismatic speech melody. Moreover, we aim at determining whether the Pitcher is in this respect more effective than other speech melody visualization tools that are not specifically designed to enhance speaker charisma. By addressing these points, we also aim to better understand at a more general level how software tools like the Pitcher manage to cause systematic changes in a speaker's speech melody at all and whether some acoustic parameters are more or less susceptible to this change.

Fig. 2. Screenshots of the three software tools to which the Pitcher is compared.

The other melody-visualization tools to which the Pitcher is compared are (1) "See & Sing Professional 1.5.8", (2) "AmPitch 1.2", and (3) "RTPitch 1.3a", see Fig. 2. Tool (1) displays f0 as a continuous non-stylized line that is, like for all others tools, drawn from left to right across the screen. It is automatically reset after 5 s and stops during silent pauses. F0 contours are associated with sequences of musical notes and, in addition, a separate vertical acoustic-energy meter is shown on the right edge of the screen. Tool (2) displays f0 by a discontinuous line (interrupted by silent pauses and voiceless consonants). The representation of acoustic energy is integrated in the thickness of that line. Unlike for tool (1), this line is not automatically reset for 60 s (and horizontal scrolling was switched off as well). Thus, tool (2) creates an overview of the frequency distribution of a speaker's f0. For tool (3), the horizontal-scrolling function is mandatory, but, like for tool (2), the f0 contour is drawn in a discontinuous, non-stylized fashion. Acoustic-energy information is provided indirectly by the waveform amplitudes that are displayed in a separate window at the top.

Based on the comparison of these four different software tools (i.e., the Pitcher + tools (1)–(3)), the following questions are addressed in the present paper:

- Can speakers learn to adopt a different speech melody through software-supported real-time visualizations of acoustic-melodic features?
- If so, does the Pitcher outperform the other visualization tools?
- Does the comparison between the four tools provide additional indications about
  - ... how the difference between a continuously and a discontinuously displayed pitch contour affects the learning of melodic parameter settings?
  - ... how the difference between pitch-contour reset and horizontal scrolling affects the learning of melodic parameter settings?
  - ... how the different ways of displaying acoustic energy affect the learning of melodic parameter settings, a strong and loud voice in particular?

2   Method

2.1   Participants

Overall, 52 participants took part in the study (19 female, 33 male, all between 19 and 22 years old). All of them were Bsc students of electrical engineering in their first semester at the Mads-Clausen Institute of the University of Southern Denmark (SDU) in Sønderborg. All participants were proficient L2 English speakers, i.e. at level B2 or higher on the European CEFR scale [16] according to SDU-internal entry tests. Furthermore, no participant had former experience in public speaking or in using any of the tested/compared software tools. There were also no (semi)professional singers among the students and none of the students reported any speech or hearing disorders.

2.2   Procedure

The experiment began in a lecture hall at SDU (Fig. 3). All 52 students received a 90-min plenary lecture about speaker charisma and the specific key characteristics of a charismatic speech melody (see Sect. 2.3). The lecture, given by the first author, has been successfully tried and tested in numerous professional charisma coachings with managers,   entrepreneurs,   and   sales   agents   by   the   consulting   company

AllGoodSpeakers ApS of the first author. After this introduction, participants presented, in a random order, a short elevator pitch in front of their fellow students. An elevator pitch is a short outline of a business idea meant to attract customers and/or investors. The given elevator pitches were about 2–3 min long and recorded with a head-mounted mic (Sennheiser HSP2-3) and a Zoom-H4 recorder (48 kHz, 24-bit). The 52 recorded pitches constituted the before-training condition of the experiment.

In the next step, the participants were randomly split up into 4 groups of 13 speakers each. Speaker gender was balanced across the groups such that there were 4–5 female speakers per group. Each group was instructed to practice the elevator pitch according to the melodic key characteristics that they had learned in the lecture, using – as an additional resource – a software tool that would visualize their speech melody in real-time. Based on this instruction, each group was assigned to *one* of the four software tools (Speech & Sing, AmPitch, RTPitch, Pitcher) and *not* allowed to use any other or additional tool for their training. Then, the group members were sent out in order to distribute themselves across the campus for individual self-guided training, having the respective software tool on their student laptop.

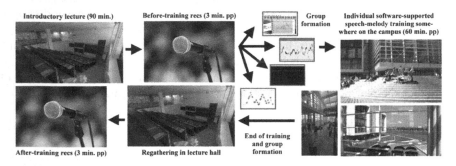

Fig. 3. Steps of the experimental procedure (photos by SDU, Oliver Niebuhr, pxfuel.com).

After one hour of this individual software-supported presentation training, all students met again in the lecture hall. In a differently randomized order, they held the same elevator pitches as before and were again recorded with the same equipment as before. These further 52 recordings constituted the experiment's after-training condition.

Note that in the chosen experiment design, the before/after-training effect is to a certain extent confounded with a familiarization effect that concerns the task and the presented text. The reason why we nonetheless deliberately chose this setup is that, firstly, this familiarization effect inevitably also occurs in real training scenarios. Our design is hence ecologically valid. Secondly, familiarization effects are not simply positive for speaker charisma. For one thing, familiarization means less stress and, thus, a lower f0 level [17], whereas a higher f0 level is required for charismatic speech [4]. For another thing, [18] have shown that intensive presentation training quickly causes routine/boredom, which also has unfavorable effects on the melodic aspects of speaker charisma. Therefore, it is reasonable to ask whether and which training software can improve melodic speaker charisma, also and in particular in the chosen experiment design. Another point is that the determination of the relative performance of the "Pitcher" software compared to the other three software tools is a between-subjects matter in the chosen experiment design. Thus, any potential within-subjects familiarization effects have no influence on this important question.

2.3   Data Analysis

The 2×52 elevator pitches were acoustically analyzed by means of PRAAT with respect to the following parameters that reflect the state-of-the-art knowledge about melodic charisma (see [4, 12] for an overview of the corresponding findings):

- (1) f0 range (semitones, st): a larger f0 range is more charismatic;
- (2) Lowness of phrase-final f0 (st relative to a speaker's individual baseline f0): lower final falls are more charismatic;
- (3) Net speaking rate (syllables/s, disregarding silent pauses): a faster tempo is more charismatic;
- (4) IPU duration (s): a shorter IPU duration is more charismatic;
- (5) Acoustic-energy level (RMS, decibel, dB): a louder voice is more charismatic.

Measurements were taken automatically based on PRAAT (praat.org) scripts but checked for implausible values, which were then replaced by manual measurements. All measurements were made in increments of 10 ms, and mean values were calculated per IPU. Each elevator-pitch performance and, thus, each speaker is represented by between 33 and 68 IPUs or mean parameter values per condition (before/after training). Note that disfluency phenomena like false starts or fillers were not excluded from the measurements; firstly, because their status for perceived speaker charisma is unclear [19] and, secondly, because they were rare enough throughout all recorded presentations (<10 on average per speaker) to not bias the analysis.

In addition, all mean parameter values of an elevator pitch were integrated and translated into a total acoustic-melodic charisma (PASCAL) score per speaker. The assessment procedure underlying the PASCAL score was developed on an experimental-phonetic basis. Numerous stimulus series were generated, 50% each with male and female speakers. In one type of series, individual melodic parameters such as speaking rate, f0 level, or f0 range were raised and lowered, for example, with PSOLA resynthesis (http://cnx.org/content/m12474/latest/) in at least 20 equidistant steps. Subsequently, in another type of stimulus series, two melodic parameters were raised or lowered in parallel or in opposite directions. With all stimulus series, online perception experiments were carried out, in which listeners rated the speakers' charisma with values between 0–100. A total of over 500,000 such listener ratings were analyzed over a period of four years. In this way, the gender-specific charisma "sweet spots" per parameter were determined, together with the exact charisma-lowering effect of parameter values outside these "sweet spots". Moreover, a set of numbers (multipliers) was worked out that expresses the relative importance of each parameter in the interplay of all charisma-relevant parameters in a speaker's speech melody.

The PASCAL score quantifies the overall charismatic performance of a speaker as the sum of the parameter-specific performance values, weighted with the individual multipliers per parameter. The score itself is then specified in relation to the performance of the approximately 1,000 speakers whose melodic charisma has so far been assessed by the PASCAL system.

PASCAL scores have repeatedly proven to be reliable and precise in predicting the listener ratings of perceived speaker charisma [20, 21]. On this basis, we computed PASCAL scores for all before/after-training presentations in our study and used them in place of formal listener ratings to quantify the 52 speakers' perceived melodic charisma levels and their magnitude of change from before to after training.

3 Results and Discussion

For the inferential statistics, we used two-way repeated-measures analyses of variance (in SPSS v26) with Training (before vs. after) and Software as the two fixed factors and Speaker as a random factor. Two analyses were conducted, one with the five acoustic measurements (MANOVA) and one with the speaker's PASCAL scores (ANOVA) as dependent variable(s). Analyses of variance could be used as previous explorative data analyses, including Kolmogorov-Smirnov and Henze-Zirkler tests, showed that the individual datasets underlying the Training and Software factor levels do not deviate significantly from a normal distribution. Moreover, Greenhouse-Geisser corrections were applied in the analyses of variance, if required.

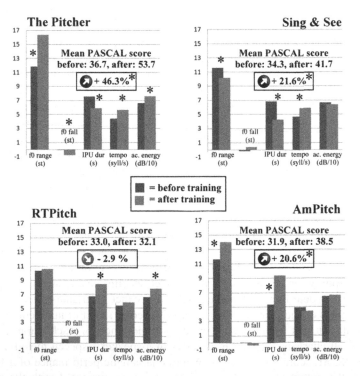

Fig. 4. Mean values for PASCAL scores and the 5 acoustic measures in the 4 Software conditions before/after training. Each value = 13 speakers and about 650 IPUs; <*> = p < 0.05.

First of all, we found a significant interaction between Training and Software for both the acoustics-based MANOVA (F[3,408] = 53.451, p < 0.001, η_p^2 = .717) and the PASCAL-based ANOVA (F[3,408] = 38.992, p < 0.001, η_p^2 = .663). The interaction reflects that, while the 4 groups or software-conditions did not differ *before* training, they did differ significantly *after* training. This interaction is a key result as it enables us to safely interpret group differences after training as being caused by the group-specific software tools used for melodic charisma training. Note that the absence of group differences before training does not mean that all speakers performed equally in their initial elevator pitch. It merely means that the students in one group were not generally better or worse than in another group. In fact, the effect of Speaker was significant in both domains acoustics and perception (MANOVA: F[3,408] = 12.018, p < 0.001, η_p^2 = .195; ANOVA: F[3,408] = 16.833, p < 0.001, η_p^2 = .246).

Given the significant Training*Software interaction, we split the total sample into the 4 software groups and then separately analyzed the effect of Training for each group. Multiple-comparisons t-tests with Sidak corrections of alpha-error levels were conducted to examine the effects of Training in detail. Only significant differences at p < 0.05 are reported below with reference to the results summary in Fig. 4.

Regarding the acoustic results, Fig. 4 shows that training with the Pitcher had significant charisma-supporting effects along all speech-melody parameters. After training, the f0 range was larger, the phrase-final f0 movements ended lower, the tempo and acoustic-energy levels were higher, and the IPU durations were shorter.

AmPitch was the only tool with a similarly positive effect on f0 as the Pitcher. That the f0 range became significantly larger and the phrase-final fall slightly but significantly lower with AmPitch is probably due to the f0 distribution that gradually emerged on y-axis over the 60 s time window. The downside of being able to see for a longer time period how (much) f0 spreads along the y-axis is the lack of f0-contour resets at short periodic time intervals. This is probably the reason why AmPitch did not motivate speakers to talk faster, and why the IPU duration significantly increased with AmPitch, both of which is very unfavorable for a speaker's charismatic impact.

An unfavorable increase in IPU duration occurred for RTPitch as well whose horizontal-scrolling function also lacked a periodic f0-contour reset. However, while the approach of AmPitch to integrate the acoustic-energy visualization in the thickness of the f0 contour failed to make speakers talk louder after training, the separate waveform-amplitude display of RTPitch was successful in that respect. Like with the dots of the Pitcher, RTPitch users were significantly louder after training. But, besides that RTPitch had no favorable effects on any melodic charisma parameter, including f0.

The charisma-supporting performance of Sing&See was in between those of the other tools. Like for the Pitcher, the Sing&See visualization was sensitive to silent pauses and featured a f0-contour reset at short periodic time intervals; and, like the Pitcher, Sing&See had a charisma-supporting effect on speech timing. That is, it significantly increased the speakers' tempo and reduced their IPU durations after training. In contrast, Sing&See was the only tool that significantly narrowed speakers' f0 range after training and hence reduced their charisma in this respect. Note in this context that Sing&See was also the only tool whose visualized f0 contour was both continuous *and* non-stylized. Also note that, unlike for RTPitch and the Pitcher, the separate acoustic-energy meter of Sing&See had no effect on speakers' loudness level after training, perhaps because the meter is placed too far from the focused f0 contour.

As regards the PASCAL scores that are used here to quantify the speakers' charismatic impact on listeners, Fig. 4 shows a three-part gradation of the software-tools' performances. RTPitch yielded the lowest after-training scores. The speakers even became on average slightly (but not significantly) less charismatic when they trained with this tool. The other three tools all performed better than RTPitch and led to a significant increase in PASCAL scores – and, thus, in the speakers' charismatic impact – from before to after training. Additional t-test comparisons between the after-training samples of the 4 groups showed moreover that Sing&See and AmPitch performed equally well at improving speaker charisma (by 21.6% or 20.6%, respectively), while the Pitcher was still significantly better ($p < 0.01$) and outperformed both Sing&See and AmPitch by another 25%, yielding a total increase in speakers' PASCAL scores of on average 46.3% after only a single hour of self-guided training.

4   Conclusion and Outlook

Based on our results, the research questions put forward in 1.2 can be answered as follows: Yes, speakers can learn to use a different, more charismatic speech melody by means of software-supported real-time visualizations of acoustic-melodic features. Yes, the Pitcher does in fact outperform the other visualization tools. This applies to the acoustic measurements (the Pitcher improved all measured parameters and, moreover, proved outstandingly effective for f0-range improvement) as well as to the PASCAL score that quantifies a speaker's charismatic impact on his/her audience and that was used here in place of real listener ratings.

The significantly better charisma-training performance of the Pitcher compared to the other software tools is remarkable, but also not surprising given that the Pitcher was the only one among the 4 compared tools that was specially developed for training a charismatic speech melody. Furthermore, it was the only tool that provided speaker feedback not just in terms of a visualization, but also in terms a color-coded evaluation.

From this point of view, it is striking that the other tools achieved improvements in the speakers' after-training presentation charisma at all. Given how big and software-specific these improvements are, it is unlikely that they stem from a general familiarization effect whose effect on charisma is not consistently positive anyways. Rather, what comes into play here is probably an effect that the authors already know from their coaching experience: The breakdown of a phenomenon as intangible as the speech melody into individual elements (i.e., parameters) such as range, tempo, final fall, etc., in combination with the mere visualization of these elements, gives speakers a meta-understanding of the cause-and-effect relationships for these elements and, thus, a more conscious control over their speech melody. Combined with targets such as "faster", "lower", "louder" etc., this more conscious control can already lead to improvements in a self-guided training task. In this context, the results of our study suggest that the effect size or average improvement potential of this meta-understanding alone is around 20%. For additional evaluative feedback (like the red and green lines and dots in the Pitcher), it is larger and approximately at 30%. An interesting question for future research is to what extent these numbers change relatively and absolutely when the self-guided training session is longer than one hour.

Furthermore, the present results have the following implications for the visualization and learning of speech melody, also with respect to applications beyond charisma training, such as second-language learning: First, all elements of speech melody can be learned and changed through software-based real-time visualizations. Second, in accord with [13] and [14], a continuous but stylized f0 contour is most effective in guiding speakers to specific f0 changes. In contrast, a contour that is both continuous and non-stylized is least effective and even worse than discontinuous, non-stylized contours. Third, an overview of the f0 distribution along the y-axis (as provided by AmPitch) is instructive to learn changes in the f0 range. Fourth, also in accord with [13] and [14], integrated visualizations like that of acoustic-energy and f0 in AmPitch are not effective. Acoustic-energy visualizations (like those of stress and rhythm) must be kept separate from f0 visualizations. Fifth, waveform visualizations are an effective way to provide feedback on acoustic energy or loudness, as are changes in the size and color of syllable-related dots. Sixth, f0-contour resets (both their presence or absence) are effective ways to guide speakers into desired speaking-rate and/or pausing behaviors.

It is up to future studies to examine and substantiate these implications empirically. The authors' next step will be to look at the feedback potential of waveform visualizations, especially for an element of speech melody that plays a role in perceived speaker charisma [22] but is difficult to measure, visualize and learn: voice quality.

References

1. Heide, F.J.: Easy to sense but hard to define: charismatic nonverbal communication and the psychotherapist. J. Psychother. Integr. **23**, 305–319 (2013)
2. Weber, M.: On Charisma and Institution Building. University Press, Chicago (1968)
3. Antonakis, J., Bastardoz, N., Jacquart, P., Shamir, B.: Charisma: an ill-defined and ill-measured gift. Ann. Rev. Organ. Psychol. Organ. Behav. **3**, 293–319 (2016)
4. Michalsky, J., Niebuhr, O.: Myth busted? Challenging what we think we know about charismatic speech. Acta Universitatis Carolinae Philologica **2**, 27–56 (2019)
5. Chen, L,. et al.: Towards automated assessment of public speaking skills using multimodal cues. In: Proceedings of the 16th International Conference on Multimodal Interaction, Istanbul, Turkey (2014)
6. Wörtwein, T., et al.: Multimodal public speaking performance assessment. In: Proceedings of the ACM on International Conference on Multimodal Interaction, Seattle, USA, pp. 43–50 (2015)
7. Amon, I.: Die Macht der Stimme. Redline, Munich (2016)
8. Gregory Jr., S.W., Gallagher, T.J.: Spectral analysis of candidates' nonverbal vocal communication: predicting U.S. presidential election outcomes. Soc. Psychol. Q. **65**, 298–308 (2002)
9. Tegtmeier, S., Schweisfurth, T., Niebuhr, O.: Gatekeepers' biases and the role of voice in start-up pitches. In: Proceedings of the 1st SEFOS, Sonderborg, Denmark (2019)
10. Fischer, K., Niebuhr, O., Jensen, L.C., Bodenhagen, L.: Speech melody matters – how robots profit from using charismatic speech. ACM THRI **9**, 1–21 (2019)

11. Niebuhr, O., Michalsky, J.: Computer-generated speaker charisma and its effects on human actions in a car-navigation system experiment - or how Steve Jobs' tone of voice can take you anywhere. In: Misra, S., et al. (eds.) ICCSA 2019. LNCS, vol. 11620, pp. 375–390. Springer, Cham (2019). https://doi.org/10.1007/978-3-030-24296-1_31

12. Niebuhr, O., Tegtmeier, S., Schweisfurth, T.: Female speakers benefit more than male speakers from prosodic charisma training – a before-after analysis of 12-weeks and 4-h courses. Front. Commun. 4, 12 (2019)

13. Niebuhr, O., Alm, M., Schümchen, N., Fischer, K.: Comparing visualization techniques for learning second language prosody: first results. Int. J. Learn. Corpus Res. 3, 250–277 (2017)

14. Fischer, K., Niebuhr, O., Schümchen, N.: Visualizing intonation for foreign language teaching. J. Lang. Teach. Res. (submitted)

15. Wagner, P., et al.: Different parts of the same elephant: a roadmap to disentangle and connect different perspectives on prosodic prominence. In: Proceedings of the 18th International Congress of Phonetic Sciences, pp. 1–5 (2015)

16. Alderson, J.C.: The CEFR and the need for more research. Mod. Lang. J. 91(4), 659–663 (2017)

17. Sondhi, S., Khan, M., Vijay, R., Salhan, A.K.: Vocal indicators of emotional stress. Int. J. Comput. Appl. 122(15) (2015)

18. Niebuhr, O., Tegtmeier, S.: Virtual reality as a digital learning tool in entrepreneurship: how virtual environments help entrepreneurs give more charismatic investor pitches. In: Baierl, R., Behrens, J., Brem, A. (eds.) Digital Entrepreneurship. FSSBE, pp. 123–158. Springer, Cham (2019). https://doi.org/10.1007/978-3-030-20138-8_6

19. Niebuhr, O., Fischer, K.: Do not hesitate! - Unless you do it shortly or nasally: how the phonetics of filled pauses determine their subjective frequency and perceived speaker performance. In: Proceedings of the 20th International Interspeech Conference, Graz, Austria, pp. 544–548 (2019)

20. Niebuhr, O., Michalsky, J.: PASCAL and DPA: a pilot study on using prosodic competence scores to predict communicative skills for team working and public speaking. In: Proceedings of the 20th Interspeech Conference, pp. 306–310 (2019)

21. Schmidt, N.: Die charismatische Stimme - Der Ton macht die Musik. https://www.daserste.de/information/wissen-kultur/w-wie-wissen/charismatische-Stimme-100.html. Accessed 22 July 2020

22. Niebuhr, O., Skarnitzl, R., Tylečková, L.: The acoustic fingerprint of a charismatic voice – initial evidence from correlations between long-term spectral features and listener ratings. In: Proceedings of the 9th Speech Prosody, pp. 359–363 (2018)

Generating a Concept Relation Network for Turkish Based on ConceptNet Using Translational Methods

Arif Sırrı Özçelik and Tunga Güngör[✉]

Computer Engineering, Boğaziçi University, Istanbul, Turkey
arifsozcelik@gmail.com, gungort@boun.edu.tr

Abstract. ConceptNet is a large-scale network of concepts and relationships, based on various common sense knowledge bases. Turkish is a language that lacks similar resources for processing texts and extracting meaning. This study discusses various methods to create a Turkish ConceptNet using translational techniques based on English ConceptNet and explains the results herewith obtained. Multiple models were tested, using different knowledge sources and tools including WordNet, Wikipedia, and Google Translate. Results obtained from each model and the approaches to improve these results are discussed, while also explaining details, assumptions, and drawbacks relevant to each relation.

Keywords: Common sense knowledge base · ConceptNet · Turkish · Word sense disambiguation

1 Introduction

Common sense databases are resources used for extracting deeper semantic knowledge in texts. They express common sense knowledge in simple sentence forms. Examples are "birds have wings", "the sun is very hot", "candy tastes sweet", and so on. When it comes to text processing, humans naturally and extensively use this source of knowledge in understanding and drawing conclusions. So, to actually capture the semantics of any given text, an apriori existence of this knowledge would help immensely.

ConceptNet [5] is a large-scale network of concepts and relations built initially on common sense databases. These assertions include examples like "pottery is made of clay" or "a cactus is capable of surviving with little water". ConceptNet spans 12.5 million edges which represent 8.7 million assertions connecting 3.9 million multilingual nodes [10]. English is the most represented language by 11.5 million edges including at least one English concept.

ConceptNet defines a "Common Language" to be a language included in the network with a vocabulary size of at least 10 thousand terms. There are 68 languages considered to be a "Common Language" and Turkish is one of them. Turkish has a vocabulary size of around 66 thousand terms [11]. There are around 10.4 thousand assertions where both concepts are in Turkish.

© Springer Nature Switzerland AG 2020
A. Karpov and R. Potapova (Eds.): SPECOM 2020, LNAI 12335, pp. 369–378, 2020.
https://doi.org/10.1007/978-3-030-60276-5_36

Languages such as English, French, Portuguese have a good amount of ontological and common sense resources, but when it comes to Turkish there is an apparent need for similar sources. Turkish as a language lacks studies that either create common sense knowledge or somehow translate existing resources in other languages. In this work, we develop a concept relation network for Turkish by employing a method that translates from the ConceptNet resource. Our goal is to build an initial concept relation network that will benefit text processing systems which currently lack knowledge resources of this kind for the Turkish language.

2   Related Work

Studies with a similar aim of creating language resources for Turkish have been published in the past. Balkanet [2, 12] is a collective attempt to gradually create multilingual WordNet lexicons similar to WordNet [3] that spans Greek, Turkish, Romanian, Bulgarian, Czech, and Serbian. The work makes use of local monolingual WordNets if available, otherwise sources like dictionaries, corpora or language specific lexicons. The process then links each monolingual WordNet to an Inter-Lingual-Index that serves as a centralized index relating synsets among all languages.

Turkish WordNet [7] is a project lead by the Turkish team in the Balkanet project. The team started by translating base concepts into Turkish. Later on, a monolingual dictionary was used to extract synonyms, antonyms and hyponyms for these base concepts. In the second phase, the team gathered a "defining vocabulary" of most frequent words in the English language and compared these words to Turkish WordNet synsets. Missing terms were then used to extend the Turkish synset collection through hyperonym-hyponym relations.

In their study titled SentiTurkNet, Oflazer, Dehkharghani, Saygın and Yanıkoğlu [8] aimed to create a lexicon of polarity for words in Turkish similar to SentiWordNet for English that could be used by sentiment analysis methods. They used the Turkish version of WordNet [7] by semi-automatically assigning polarity values to create SentiTurkNet.

In their attempt to create a similar common sense list of assertions like what ConceptNet was built on, Özcan and Amasyalı [9] used an online game approach that would ask users to play a game and as a result generate common sense knowledge for Turkish. In this study, they look into a number of games previously implemented for English. They proposed using a game site called CSOYUN which they kept online for 4 years with 5 different games and reported that 57 thousand reliable concept relations were generated through these online games.

In another study aiming to create a Turkish WordNet named KeNet,[1] Yıldız, Solak and Ehsani [13] start by extracting synonym candidates from an online dictionary for Turkish. Then they verify synonyms by manually annotating them and create a graph where nodes represent senses connected by synonymy relations. Finally, by looking at

---

[1] http://haydut.isikun.edu.tr/kenet.html.

clusters they create synsets. They also mined Turkish Wikipedia for hypernym relations that increased the set of such relations obtained using only a dictionary.

3   Methods

ConceptNet relations are assertions describing relations between two different concepts. They represent everyday common sense information. Some example assertions are:

- a bowl – *MadeOf* – steel;
- an organism – *MadeOf* – cells;
- chip – *PartOf* – computer;
- edinburgh – *PartOf* – scotland;
- brain – *UsedFor* – think;
- breathing – *UsedFor* – meditating.

We develop a method that translates similar assertions into Turkish. Before starting translating a relation, the following preparations and assumptions were made:

- Only English to English relations in ConceptNet were considered.
- Nodes on each side of the relation were preprocessed to remove initial stop words like "a", "an", "the", etc.
- Any translation of a concept to Turkish that fails is assumed to be a technical or domain specific term, so can be accepted as it is in Turkish.
- Except a few specific relations, all concepts were translated in their singular forms.
- Depending on relations certain Part of Speech (POS) categories (noun, adjective, verb, etc.) were used to filter senses while translating concepts.
- English terms were lemmatized using the Stanford Core NLP tool [6].
- Turkish terms were lemmatized using Zemberek [1].
- Crawlers were used to extract data from sites like Tureng,[2] Wiktionary,[3] Wordreference,[4] Wikipedia,[5] and Google Translate.[6]

ConceptNet includes 58 relations. Some of the relations were not included in this work because either there were no English to English examples in the relation (e.g. *TranslationOf*) or there were too few examples (e.g. *ParticipleOf* or *LocatedNear*).

Various models for translating English concepts into Turkish were tested, initially starting with using online bilingual dictionaries and then gradually introducing sources like WordNet, Wikipedia, Google Translate, and Google Search API.

---

[2] https://tureng.com/tr/turkce-ingilizce.

[3] http://en.wiktionary.org.

[4] https://www.wordreference.com/.

[5] https://wikipedia.org/.

[6] https://translate.google.com/.

Using only online dictionaries does not incorporate context and it is challenging to disambiguate between various translation candidates given that each example in a ConceptNet relation is short and lacks sufficient context.

WordNet was used in an attempt to extend context with best matching synsets. Synsets for one concept were filtered using the other concept by applying the Lesk [4] algorithm. An augmented version of the Lesk algorithm [4] was tested to enrich WordNet based contexts by adding hypernyms, hypernym ancestries, hyponyms, and part meronyms to the glosses of synsets for one concept before comparing them to synsets generated for the other concept.

As an example, given the concept "Laptop - *MadeOf* - Chip", "Chip" is translated into Turkish in the sense of "French Fries". The correct sense in WordNet includes terms like "Microchip", "Silicon Chip", and "Microprocessor Chip". But "Laptop" reveals the terms "Laptop", "Computer" and "Portable". However, hypernym hierarchy for "Laptop" includes the terms "Microprocessor" and "Microcomputer". By using these hypernym terms it is possible to translate "Chip" in its correct sense in Turkish.

The final model to translate ConceptNet into Turkish used Google Translate, Google Search API, and Wikipedia. Figure 1 shows the pseudocode of the algorithm.

FOR a relation type RELTYPE
FOR an instance of RELTYPE named RELATION
FOR each of the two concepts in RELATION named CONCEPT

1. IF CONCEPT consists of more than 2 words, QUERY Google Translate for "CONCEPT" and RETURN the result as the correct translation. Otherwise proceed to next step.
2. QUERY Google Translate as described in Figure 2.
3. LEMMATIZE all English terms in CONCEPT. LEMMATIZE all Turkish terms in TRANSLATION.
4. QUERY Google Custom Search API for RELATION in Wikipedia specifically. COLLECT the first 10 ARTICLES.
5. TRANSLATE and ALIGN sentences for Wikipedia ARTICLES as described in Figure 3.
6. ASSIGN scores to ARTICLE - TRANSLATION pairs as described in Figure 4. CHOOSE the highest scoring ARTICLE – TRANSLATION pair among all and RETURN TRANSLATION as the correct translation.
7. REPEAT steps (1) through (7) for both concepts.

Fig. 1. Translation model.

The proposed model uses Google Translate to generate a list of translations for each concept, searches for Wikipedia articles related to the example in the form "concept1 relatesTo concept2", translates Wikipedia article extracts using Google Translate, and finally scores each translation candidate on the basis of matching terms within both English and Turkish extracts.

Google Search API was used to search for Wikipedia articles. Google Translate would be capable of translating articles into Turkish with a certain degree of success as article extracts will contain many terms, hence a large context.

USING Google Translate:
1. QUERY CONCEPT.
2. COLLECT definitions, examples for source terms and the top 6 ranking TRANSLATIONS.
3. IF there are no translations returned, default to Tureng.
   (a) IF Tureng does not return a TRANSLATION, RETURN CONCEPT.
   (b) Otherwise RETURN the first Tureng entry.

Fig. 2. Translation model – step 2.

For each Wikipedia ARTICLE:
1. TRANSLATE an extract of the ARTICLE using Google Translate including the abstract.
2. ALIGN English and translated Turkish sentences for both versions of the ARTICLE.
3. For each aligned sentence, EN and TR of ARTICLE:
   (a) LEMMATIZE all English terms in EN.
   (b) LEMMATIZE all Turkish terms in TR.

Fig. 3. Translation model – step 5.

For each ARTICLE – TRANSLATION pair:
1. For each lemmatized aligned sentence, EN and TR of ARTICLE:
   (a) COUNT how many times EN includes CONCEPT and TR includes TRANSLATION.
   (b) Assign the number of times both sentences had matches as the ALIGNED SENTENCE SCORE for TRANSLATION.
2. SUM all ALIGNED SENTENCE SCORES to assign the ARTICLE – TRANSLATION pair a score.

Fig. 4. Translation model – step 6.

Figure 5 lays out the algorithm of translating a concept in a flow chart.

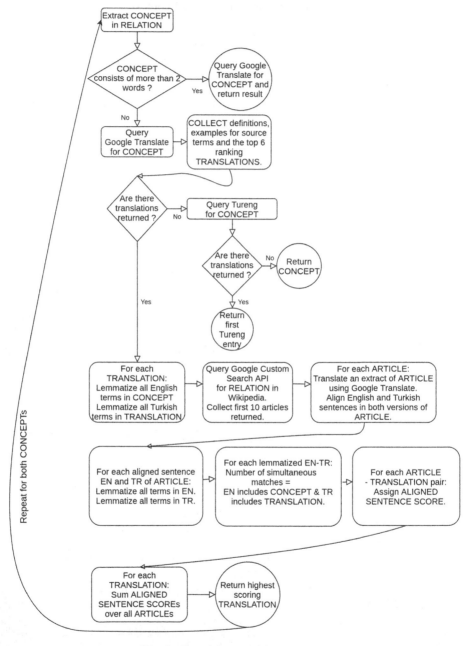

Fig. 5. Translation model – flowchart.

4 Experiments and Discussion

Out of the 58 relations, 37 relations were translated. 10 relations were not included because they were small in size and 9 were not translated because they were huge and the resources of this study were limited. 2 of the relations, namely *influenced* and *influencedBy* were not translated because the majority of nodes for these relations were entity names, specifically famous people in various domains.

The following results assume a translation to be correct only if it makes sense in Turkish. Some translations are grammatically slightly incorrect or are seemingly correct but semantically not sensible either in the source example or translation. Some of the grammatical errors are caused by the tools used and others are results of source examples with poor quality.

Nearly all relations have examples that do not make much sense in English. There are also many examples which are hard to translate. Some examples are actually asymmetrically divided long sentences. They do not conform to the assumption that there are two concepts on both sides of a relation that are isolated units of meaning. These are considered to be incorrect.

Some of these unexpected examples are:

- difference between an entranceway and a patio door: patio door – *MadeOf* – glass;
- pizza usually – *MadeOf* - tomato sauce, cheese and crust;
- stabbing to death may – *SymbolOf* – dead domination to some person;
- graph – *MadeOf* – set of vertices and a set of edge.

Examples above do not actually satisfy the assumption made in this study that each side of a relation consists of simple isolated concepts.

Relations like *NotHasProperty*, *adjectivePertainsTo*, *adverbPertainsTo* and *NotCapableOf* did not seem to perform well under the model proposed. This is because either they contain many hard to translate or noisy examples or are very domain specific and it is hard to translate without a domain specific resource.

Some hard to translate examples for these relations are:

- accidents can happen to someone who – *NotHasProperty* – careful;
- fenestral – *adjectivePertainsTo* – fenestra;
- ravishingly – *adverbPertainsTo* – ravish;
- television – *NotCapableOf* – need to be watered.

For relations like *mainInterest*, *MemberOf*, and *notableIdea*, the concept to be translated can be very domain specific. Some examples are:

- martin heidegger – *notableIdea* – desein;
- bomarea – *MemberOf* – amaryllidaceae.

Examples where either concept consists of only stop words were discarded (not translated). Some examples of this type are:

- almost – *NotCapableOf* – count except in horseshoes;
- they – *HasA* – nice smell;
- it – *HasProperty* – dirty or clean.

Random samples of size 150 were selected and evaluated by the authors. Estimated accuracies were also based on annotations done by the authors.

Table 1 lists results after applying the proposed model to 37 relations. Size is the number of examples in a relation, Coverage is the number of these examples that were

Table 1. Results for all relations.

Relation	Size	Coverage	Accuracy
SymbolOf	166	165	84%
DesireOf	280	275	83%
Entails	408	404	79%
NotHasA	409	390	61%
NotIsA	478	402	73%
CreatedBy	503	499	74%
Attribute	639	624	85%
notableIdea	908	908	72%
NotHasProperty	1144	1085	72%
MadeOf	2198	2177	82%
mainInterest	2764	2764	85%
adverbPertainsTo	2880	2841	61%
NotCapableOf	2915	2440	60%
HasLastSubevent	3065	3063	66%
adjectivePertainsTo	3313	3297	56%
HasFirstSubevent	4208	4202	62%
NotDesires	4280	4239	71%
Desires	5062	4870	74%
CausesDesire	5176	5158	69%
DefinedAs	6406	6179	61%
HasContext	8851	8615	71%
HasA	9762	9283	62%
ReceivesAction	10429	10090	61%
SimilarTo	11061	10679	74%
MemberOf	12190	12052	55%
PartOf	14151	13791	65%
spokenIn	15590	15427	53%
MotivatedByGoal	15960	15605	68%
Causes	18355	18143	55%
languageFamily	19713	19504	60%
HasProperty	19823	18615	67%
HasPrerequisite	24545	24155	69%
Antonym	26551	24478	71%
Field	26732	26450	83%
HasSubevent	26911	26602	62%
knownFor	27519	27224	75%
UsedFor	46522	45381	64%

processed, and Accuracy is the relative frequency of successful translations obtained in randomly selected samples of size 150.

Higher scoring relations seem to have a tendency to contain shorter concepts while mid or lower scoring relations are more spread out. This is consistent with the assumption made throughout this study assuming ConceptNet consisted of short concepts on both sides of a relation. Concepts that are longer in size were not considered to be disambiguated and Google Translate or Tureng results were accepted instead.

Table 2 shows how successful translations are distributed when examples are grouped by combined concept lengths, for the 10 top scoring relations. A combined concept length is the sum of the number of words in each concept. As an example, 65% of *MadeOf* relations having a combined concept length of 4, were accurately translated. In most of the relations there seems to be a decrease of performance as the concepts become larger. For sizes larger than 4, at least one of the concepts are translated directly through Google Translate, which explains the increase in performance. The relation *Field* is the only exception to this observation and this is mainly caused by technical or domain specific terms in concepts.

Table 2. Estimated accuracies vs. concept lengths.

Relation	Combined concept length				
	2	3	4	5	6
mainInterest	100%	82%	83%	96%	75%
SymbolOf	89%	83%	72%	100%	40%
DesireOf	88%	77%	93%	82%	
Field	81%	83%	90%	100%	
MadeOf	88%	88%	65%	83%	25%
Entails	82%	81%	100%		
knownFor	77%	80%	66%	95%	
CreatedBy	89%	71%	83%	44%	40%
Desires	86%	67%	60%	80%	78%
SimilarTo	79%	56%			

5 Conclusion

Building resources for Turkish like WordNet, ConceptNet or other common sense knowledge bases manually is time and resource consuming. Instead, attempting to translate resources from other languages is more feasible.

The work described throughout this study attempts to translate as many examples of ConceptNet relations as possible from English into Turkish by making use of different existing tools. The goal was to create a network of everyday knowledge for Turkish, a language that lacks a proper common sense knowledge base. Looking at the results, it could be said that the method used to translate performed slightly better with relatively small examples, consisting of simple nodes.

Future work could integrate KeNet [13] and possibly cross lingual WordNet links into the algorithm. It is also possible to improve incorrect translations through feedback implementations or manual corrections.

References

1. Akın, A.A.: Zemberek-NLP (2019). https://github.com/ahmetaa/zemberek-nlp
2. Cristeau, D., Tufis, D.I., Stamou, S.: BalkaNet: aims, methods, results and perspectives. A general overview. Rom. J. Inf. Sci. Technol. **7**, 9–43 (2004)
3. Fellbaum, C.: WordNet and wordnets. In: Encyclopedia of Language and Linguistics, vol. 3, pp. 665–670. Elsevier (2006)
4. Lesk, M.: Automatic sense disambiguation using machine readable dictionaries: how to tell a pine cone from a ice cream cone. In: Proceedings of SIGDOC, pp. 24–26 (1986)
5. Liu, H., Singh, P.: ConceptNet: a practical commonsense reasoning tool-kit. BT Technol. J. **22**, 211–226 (2004). https://doi.org/10.1023/B:BTTJ.0000047600.45421.6d
6. Manning, C., et al.: The Stanford coreNLP natural language processing toolkit. In: Proceedings of 52nd Annual Meeting of the Association for Computational Linguistics: System Demonstrations, pp. 55–60 (2014)
7. Oflazer, K., Bilgin, O., Çetinoğlu, Ö.: Building a wordnet for Turkish. Rom. J. Inf. Sci. Technol. **7**, 163–172 (2004)
8. Dehkharghani, R., Saygin, Y., Yanikoglu, B., Oflazer, K.: SentiTurkNet: a Turkish polarity lexicon for sentiment analysis. Lang. Resour. Eval. **50**(3), 667–685 (2015). https://doi.org/10.1007/s10579-015-9307-6
9. Özcan, S., Amasyalı, M.F.: Turkish commonsense database (CSDB) and csoyun (a game with a purpose). Sigma J. Eng. Nat. Sci. **32**, 116–127 (2014)
10. Speer, R., Havasi, C.: Representing general relational knowledge in ConceptNet 5. In: Language Resources and Evaluation (2012)
11. Speer, R.: ConceptNet5 languages (2016). https://github.com/commonsense/conceptnet5/wiki
12. Stamou, S., Azer, O., Pala, K., Christoudoulakis, D.: BalkaNet: a multilingual semantic network for Balkan languages. In: 1st International Wordnet Conference, India, pp. 12–14 (2002)
13. Yıldız, O.T., Solak, E., Ehsani, R.: Constructing a WordNet for Turkish using manual and automatic annotation. ACM Trans. Asian Low-Resour. Lang. Inf. Process. **17**(3), 1–15 (2018)

Bulgarian Associative Dictionaries in the LABLASS Web-Based System

Dimitar Popov(✉) [ID], Velka Popova[ID], Krasimir Kordov[ID],
and Stanimir Zhelezov[ID]

Konstantin Preslavsky University of Shumen,
Universitetska str. 115, 9700 Shumen, Bulgaria
popovstyle59@gmail.com, labling@shu.bg
http://labling.fhn-shu.com/home.htm

Abstract. The current article focuses on the miniscule Bulgarian trace in worldwide associative lexicography. The three published Bulgarian associative dictionaries are represented in the terms of LABLASS, the first Bulgarian web-based system, whose development is ensured within the working program of the ClaDa-BG project. At present, users have access to the pilot version of the system, which is periodically updated. Its functionalities are not limited to the inclusion of existing lexicographic sources but are quite broader, and include the creation of new dictionaries, the visualization and juxtaposition if data from different sources.

Keywords: Associative lexicography · Bulgarian associative dictionaries · Web-based system

1 Introduction

The present article deals with the Bulgarian trace in associative lexicography. It must be noted right away that there are very few dictionaries and an immense amount of unsystematized associative data gathered as a result of separate specific studies in various fields, which does not allow the modern researcher to gain a realistic overview of the existing empirical material. At the same time, considering the time- and energy- consuming nature of the process of gathering associative data as well as the ill-fitting in terms of analysis method of representation of data in associative dictionary articles, the conclusion can be drawn concerning the necessity of creating a web-based system which would integrate to the largest degree the existing Bulgarian material by representing it in a systematic form appropriate for statistic analysis.

The present article presents the Bulgarian associative dictionaries in the first Bulgarian web-based system LABLASS, which is being developed within the working program of the ClaDa-BG project. At this stage, users have access to the pilot version of the system which is being updated periodically. It is not limited to including only already existing lexicographic sources but has a much wider range of possibilities, including the creation of new dictionaries, visualization and comparison of data from various sources.

© Springer Nature Switzerland AG 2020
A. Karpov and R. Potapova (Eds.): SPECOM 2020, LNAI 12335, pp. 379–388, 2020.
https://doi.org/10.1007/978-3-030-60276-5_37

2 Associative Lexicography – A General Overview

One of the priority fields of research in contemporary psycholinguistics deals with the gathering of associative data and the developing on their basis of associative dictionaries and thesauri. They are extremely important for modern inter-disciplinary research which is developed within the terms of the anthropocentric paradigm as they provide sufficient "raw" material for studying the mental lexicon of human beings nowadays.

The present article focuses on some specific problems and applications of associative lexicography which can be considered a relatively young field of research, at the boundary between psycholinguistics and lexicography. Th article includes a short description of the existing Bulgarian associative dictionaries featured in the LABLASS web-based system developed within the working program of the ClaDa-BG project. Their share in Bulgarian lexicography is rather small, which can be explained by the fact that this is a very young tradition worldwide. The beginning of associative lexicography lies not more than a hundred years back. Its history begins only at the end of the 19th and the beginning of the 20th century, which necessitates the experimental approach to the study of these phenomena, although word associations have been the focus of the human spirit in search of knowledge since Aristotelian times.

A great number of associative experiments have been conducted during the last hundred years and numerous dictionaries and thesauri have been published on their basis, which points to the great interest toward word associations. Especially valuable and important are the experiments for isolating associative norms. The first dictionary determining the standard in associative lexicography is known as the Kent-Rosanoff dictionary [7]. Published in 1910, it includes the first associative norms for the English language. They were established as a result of an associative experiment including 1000 participants, all native speakers of English, including women and men of various professional and educational backgrounds. The stimulus-words used were 100 nouns and adjectives. This list is still in use in order to ensure comparability of the materials. Shortly after the establishment of these norms it became clear that they are a valuable source of data not only for psychiatry but also for a number of other fields which study language and thought, semantics, the influence of sociocultural factors on personality formation, intellectual development etc. To this points the ensuing boom in mass-conducted associative experiments on whose basis within applied psycholinguistics are created a number of mono- and multilingual *dictionaries of associative norms* (such as [2–4,6–10,14]) containing the typical, most frequent associative reactions of the members of a single or multiple linguistic communities of different languages respectively.

If we attempted to draw a general overview of the current state of affairs in associative lexicography what would become immediately obvious are the multitude of variations, the large number and precision of the development of the existing dictionaries and thesauri. To this needs to be added the considerable number of different languages they represent. Along with the list of advantages, how-ever, there need to be mentioned some disadvantages which result in scepticism and disappointment in researchers. They are related to the time- and effort-consuming nature of associative experiments, as well as the lexicographic representation of the material which does not facilitate easy description and analysis. Concurrently, the large quantity of gathered associative data is not generalized and systematized in dictionaries but are instead scattered in separate specific studies. This way the valuable material ends up excluded from the general associative palette. In this context it is understandable and entirely natural that there should be a turn to modern information technologies which offer great possibilities for gathering and processing data, as well as for the representation of the results from the associative experiments and effective means of creating lexicographic materials in the form of various electronic dictionaries and databases such as the Polish Word Association Network (https://www.klk. uj.edu.pl/sssjp/en), WEB service for visualization of associative and semantic networks (http://it-claim.ru/Products/VisualNet/index.html) etc.

3   The Bulgarian Trace in Associative Lexicography

Bulgarian associative lexicography is represented by three published dictionaries, namely: *Bulgarian Norms of Word Associations* [6], *Bulgarian Associative Dictionary* [1], *Dictionary of Child Word Associations* [13]. In addition, it needs to be noted that Bulgarian is represented on a comparative basis with Russian, Serbian, Ukrainian, Belarusian in the Slavic Associative Dictionary [15]. In addition to that there is an immense amount of gathered Bulgarian associative data outside these lexicographic sources, which is scattered among separate specific studies of researchers from various fields such as psychology, pedagogy, psycho-linguistics, FLT, special pedagogy etc. It is exactly within this context that the idea originates of creating the LABLASS web-based system, coming as a response to the need for reliable and dynamic systematizing of existing data and their representation in a format which is suitable for the work of the researcher. The following paragraph describes the associative data entered in LABLASS with the data being split into two groups depending on their status in the lexicographic source. Thus in Table 1 can be found information regarding the published dictionaries, whereas Table 2 features the dictionaries which are being developed and are at different stages of completion.

The three Bulgarian dictionaries whose online version is represented in LABLASS are also distinctly different from one another with regards to the age and number of surveyed persons (see Table 1), as well as regarding the method of conducting the associative experiment (oral-written in BNWA, written-written in BAD and oral-oral in DCHWA), as well as regarding the stimulus words.

The remaining 4 dictionaries are being developed. They are planned and executed only in an online version (see Table 2).

The dictionaries described in Table 2 are developed on the basis of associations elicited by means of a test realized in a written-written format (i.e., written stimulus - written response). The only exception is the NDCHWA where the surveyed children give oral answers to orally presented stimuli. In this case the list

Table 1. Online version of Bulgarian associative dictionaries in the LABLASS web-system

Dictionary	Surveyed persons	Status
Gerganov E. (ed.). Bulgarian norms of word associations. Sofia (1984) **BNWA** [6]	1000 adults of different ages, professions and education (an equal number of men and women)	Printed version; Online version in the LABLASS web-system (data entry ongoing)
Baltova, P., Lipovska, Petrova, K., Eftimova, A. Bulgarian associative dictionary Bidirectional dictionary. Sofia (2003) **BAD** [1]	500 university students from St. Kliment of Ohrid University in Sofia aged between 18 and 25 (an equal number of men and women)	Printed version; Online version in the LABLASS web-system (data entry ongoing)
Popova, V. Dictionary of associations. Shumen (2020) **DCHWA** [13]	100 children aged between 5 and 6 (an equal number of girls and boys)	Printed version; Online version in the LABLASS web-system (data entry ongoing)

of stimulus words is used which was used for the 1990s experiment underlying the DCHWA [13], which would form a solid foundation for research on the dynamics of the child mental lexicon.

The DWAMB has the option of differentiating between the associations based on age (children-adults), as well as with the status of the surveyed persons as monolinguals (Bulgarian language) and bilinguals (Bulgarian-Turkish). The last lexicographic source in Table 2 concerns a special study of the process of aging of words in the mental lexicon of the inhabitants of a Bulgarian village, due to which the stimulus wordlist is more specific (it includes words whose status is obsolete and archaic respectively).

4   Description of the LABLASS – First Bulgarian Web-Based System for Word Association Research

The existing Bulgarian dictionaries of word associations are few and published over a broad period of time - see [1,6,13]. Meanwhile, there is a great amount of collected Bulgarian associative data but those are dispersed in separate specific studies of researchers from a variety of fields such as psychology, pedagogy, psycholinguistics, FLT, special pedagogy, etc. - for example, [5,6,11,12,16]. In this sense there is a necessity for a reliable and dynamic systematizing of the existing data and their representation in a format which is convenient for research work.

Table 2. Online version of associative dictionaries in development.

Dictionary	Surveyed persons	Status
Dictionary of Bulgarian word associations from The early 2020s of the **DBWA 2020**	500 adult native speakers of Bulgarian (monolinguals)	Ongoing associative Odata gathering; Data entry ongoing in the LABLASS web-system
New dictionary of child word associations **NDCHWA (2020–2021)**	100 children aged 5 to 6 (monolinguals)	Ongoing associative data gathering
Dictionary of word associations of monolingual (Bulgarian) and bilingual (Bulgarian and Turkish) persons (2010–2020) **DWAMB**	431 persons total Monolinguals (231): Adults - 121 Children - 110; Bilinguals (200): Adults - 100 Children - 100	Online version in the LABLASS web-system
Nedyalkova, A., Popova, V. Dictionary of Associations of Bulgarian Native Speakers from the area of the village of Osmar, Shumen Region 2019–2020 **DABO**	100 adults (monolinguals)	Ongoing associative data gathering

In 2020 is given the start of the newly developed web-based system entitled **LABLASS**, whose name is an abbreviation of **LABL**ING (**Lab**oratory of Applied **Ling**uistics) and **ASS**OCIATIONS.

Figure 1 illustrates schematically the structure of the newly developed web-based system LABLASS, specialized in studying word associations.

The system is divided into 4 main modules - dictionary management module, word-stimulus management module, association management module, visualization module.

The **dictionary management module** allows for new dictionaries to be created, existing ones to be modified or unnecessary ones to be deleted. When a dictionary is created, along with its name is entered information about the number of surveyed persons. This is done for the purposes of automatic computation of particular statistical indicators about the data entered in the dictionary.

The **word-stimulus management module** allows to add new stimulus words, to modify existing ones or delete unnecessary ones. Initially, an already existing dictionary is selected, to which the newly added words are associated or already existing ones are modified and information regarding the number of

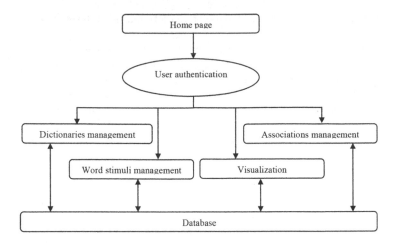

Fig. 1. Schematic representation of the web-based system structure.

persons surveyed is entered. This is once again done in order to compute automatically specific statistical indicators for the data included in the dictionary.

Dictionary name	Number surveyed	Position #	Edit	Delete
BNWA	100		Edit	Delete
BAD	500		Edit	Delete
DCHWA	100		Edit	Delete
DBWA 2020	175 (out of 500)		Edit	Delete
NDCHWA	10 (out of 100)		Edit	Delete
DWAMB: Adult monolinguals 2020	121		Edit	Delete
DWAMB: Adult bilinguals 2020	25 (out of 100)		Edit	Delete
DWAMB: Bilingual children 2020	16 (out of 100)		Edit	Delete
DWAMB: Monolingual children 2020	110		Edit	Delete
DABO	40 (out of 100)		Edit	Delete
Add new dictionary				
				Add

Fig. 2. The dictionary management module of LABLASS web-based system.

The **associations management module** allows the addition of new associations, the modification of existing ones or deletion of unnecessary ones. Initially, as with the previous module, an existing dictionary is selected. Then a word-stimulus already entered in this dictionary is selected from the drop-down menu, to which the new associations are connected or the already entered ones are modified, as well as information is entered about the number of persons who responded to this association. This is once again done in order to compute automatically specific statistical indicators for the data included in the dictionary.

The **visualization module** illustrates visually the available dictionaries in the database with the system being developed in a way that allows both for a tree-view representation of each separate dictionary and the summarized table-view of the information from several dictionaries simultaneously. In addition, statistical information is displayed for each association.

As an illustration, Fig. 2 presents the dictionaries whose online version is featured in the LABLASS web-based system.

МРАВОЯД (ANTEATER)
96 (49/47); P = 25; A = 96 (49/47)

мравка 33 (18/15); мравки 14 (6/8); яде 11 (5/6); яде мравки 11 (6/5); мравуняк 6 (3/3); къщата на мравките 2 (1/1); блъска 1/0; жито 1/0; звяр 0/1; изхвърля люспи 0/1; има мравки по него 0/1; калинка 1/0; къде 0/1; мармалад 0/1; много мравки 1/0; мухояд 1/0; пеперуда 1/0; птица 0/1; птичка 1/0; слон 0/1; стол 0/1; страх 0/1; уши 0/1; ютия 1/0; яде трохички 1/0.

Fig. 3. Dictionary article for the *anteater* stimulus [13].

Figure 2 clearly shows how 3 of the dictionaries (DWAMB, DBWA and DABO) are still being populated with associative data in batches which in the end will be automatically integrated, yielding the online version of each of the dictionaries.

It is important to point out the advantages of the standards for representation and visualization of associations in LABLASS, with fragments of the printed and online formats of the DCHWA [13] serving as an illustration for comparison. Compare the dictionary article of the Fig. 3 stimulus and the virtual representation of the respective data in Fig. 4 and Fig. 5.

Even the most basic comparison between the three figures (Fig. 3, Fig. 4, Fig. 5) reveals that the improved visibility of the quantitative data and relations most important for the analysis is achieved exactly in the two modules of the online version of DCHWA.

LABLASS offers different formats for structuring associative dictionaries. One of them is the modular one, which is applied to the DWAMB data. This allows each of its building modules (associative data about *adult monolinguals, child monolinguals, adult bilinguals, child bilinguals* respectively) to be compared to another dictionary (or module of a different dictionary). Thus, for example, if the reactions to the stimulus *grandmother/baba* from DCHWA and the "Adult mono-linguals" from DWAMB are selected in the "Visualization" section, the respective data is shown in table-view clearly and unambiguously on a comparative basis with a view of the age criterion (Fig. 6).

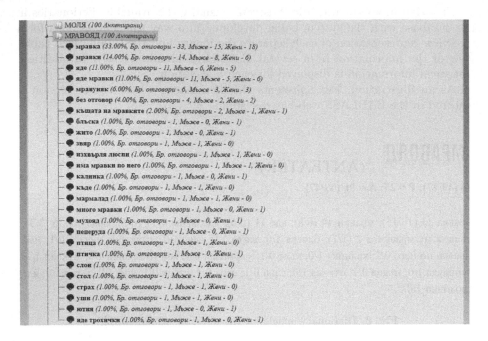

Fig. 4. Tree-view of the associations of the word *anteater* in the LABLASS "Associations visualization" module.

Дума-стимул МРАВОЯД ⌄ Асоциация	Брой отговори	Брой момчета/ мъже	Брой момичета/ жени	Поредна позиция	Редактиране	Изтриване
мравка	33	15	18	0	Редактирай	Изтрий
мравки	14	8	6	0	Редактирай	Изтрий
яде	11	6	5	0	Редактирай	Изтрий
яде мравки	11	5	6	0	Редактирай	Изтрий
мравуняк	6	3	3	0	Редактирай	Изтрий
без отговор	4	2	2	0	Редактирай	Изтрий
къщата на мравките	2	1	1	0	Редактирай	Изтрий
блъска	1	0	1	0	Редактирай	Изтрий
жито	1	0	1	0	Редактирай	Изтрий
звяр	1	1	0	0	Редактирай	Изтрий
изхвърля люспи	1	1	0	0	Редактирай	Изтрий
има мравки по него	1	1	0	0	Редактирай	Изтрий
калинка	1	0	1	0	Редактирай	Изтрий
къде	1	1	0	0	Редактирай	Изтрий
мармалад	1	1	0	0	Редактирай	Изтрий
много мравки	1	0	1	0	Редактирай	Изтрий
мухояд	1	0	1	0	Редактирай	Изтрий
пеперуда	1	0	1	0	Редактирай	Изтрий

Fig. 5. Table-view of the responses to the word *anteater* in the LABLASS "Associations" module.

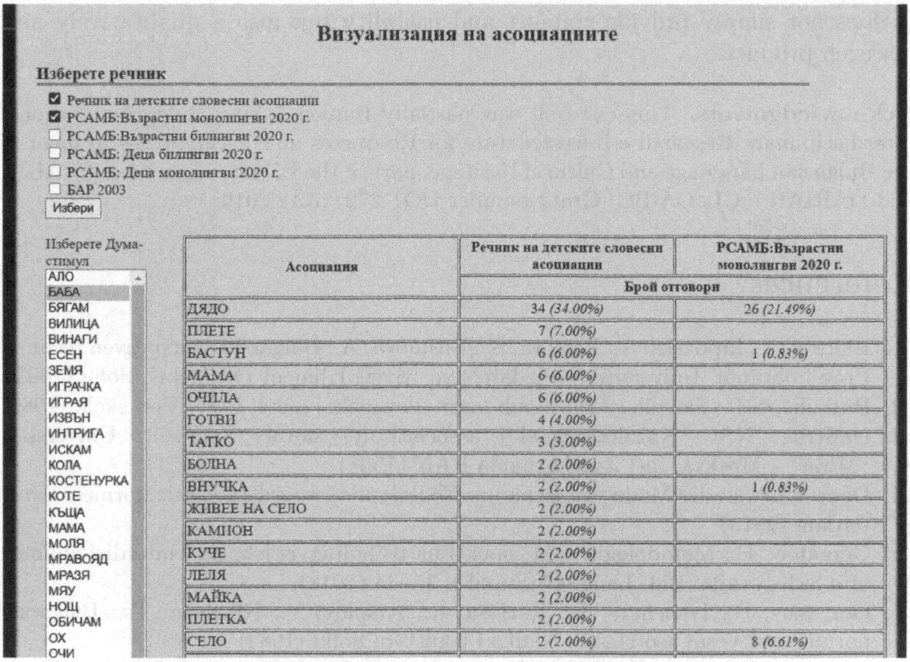

Визуализация на асоциациите

Изберете речник

- ☑ Речник на детските словесни асоциации
- ☑ РСАМБ:Възрастни монолингви 2020 г.
- ☐ РСАМБ:Възрастни билингви 2020 г.
- ☐ РСАМБ: Деца билингви 2020 г.
- ☐ РСАМБ: Деца монолингви 2020 г.
- ☐ БАР 2003

[Избери]

Изберете Дума-стимул

		Речник на детските словесни асоциации	РСАМБ:Възрастни монолингви 2020 г.
АЛО БАБА БЯГАМ ВИЛИЦА ВИНАГИ ЕСЕН ЗЕМЯ ИГРАЧКА ИГРАЯ ИЗВЪН ИНТРИГА ИСКАМ КОЛА КОСТЕНУРКА КОТЕ КЪЩА МАМА МОЛЯ МРАВОЯД МРАЗЯ МЯУ НОЩ ОБИЧАМ ОХ ОЧИ	**Асоциация**		
		Брой отговори	
	ДЯДО	34 (34.00%)	26 (21.49%)
	ПЛЕТЕ	7 (7.00%)	
	БАСТУН	6 (6.00%)	1 (0.83%)
	МАМА	6 (6.00%)	
	ОЧИЛА	6 (6.00%)	
	ГОТВИ	4 (4.00%)	
	ТАТКО	3 (3.00%)	
	БОЛНА	2 (2.00%)	
	ВНУЧКА	2 (2.00%)	1 (0.83%)
	ЖИВЕЕ НА СЕЛО	2 (2.00%)	
	КАМИОН	2 (2.00%)	
	КУЧЕ	2 (2.00%)	
	ЛЕЛЯ	2 (2.00%)	
	МАЙКА	2 (2.00%)	
	ПЛЕТКА	2 (2.00%)	
	СЕЛО	2 (2.00%)	8 (6.61%)

Fig. 6. Comparative representation of the associations of the *grandmother/baba* stimulus according to DCHWA and DWAMB: Adult monolinguals.

5   Conclusion

The present article is a brief introduction to the associative dictionaries in the first Bulgarian web-system LABLASS, which outlines a new opportunity for valuable work in which the researcher would never lose sight of the idea of the en-tire picture of associative lexicography. In addition, the further development of LABLASS would help overcome the fragmentary and scattered nature of a multitude of data accumulated in separate specific studies from various fields of knowledge., which would establish it as a wide platform for solid research in linguistics, psycholinguistics, psychology, pedagogy, psychiatry, cognitive linguistics, sociology etc. In conclusion, it is advisable to attempt to find an answer to question if the online version of the Bulgarian associative dictionaries in the LABLASS web-system is necessary and useful or if it is just a fashionable replica of the respective originals. First of all, there needs to be mentioned the advantages of the system itself, whose open, dynamic nature and excellent abilities for visualization and statistical comparison of data creates optimal conditions for the researcher's work in the process of creating a new dictionary, as well as in using the existing dictionaries available in the system. In view of the above, the conclusion can be drawn that the development of the web-based system LABLASS marks a new stage in the development of Bulgarian associative lexicography as it does not simply provide comfort and usability but also a qualitatively new research product.

Acknowledgments. This research was partially funded by the Bulgarian National Interdisciplinary Research e-Infrastructure for Resources and Technologies in favor of the Bulgarian Language and Cultural Heritage, part of the EU infrastructures CLARIN and DARIAH – CLaDA-BG, Grant number DO1-272/16.12.2019.

References

1. Baltova, P., Lipovska, A., Petrova, K., Eftimova, A.: Bălgarski asotsiativen rechnik. Prav i obraten. Universitetsko izdatelstvo "Sveti Kliment Ohridski", Sofia (2003)
2. Butenko, N.P.: Slovnik asociativnih norm ukrainskoj movi, Lviv, Visa skola (1989)
3. Dmitruk, N.V.: Kazahsko-russkij associativnyj slovar. Symkent: Universitet "Miras" - Moskva: In-t Jazykoznania RAN (1998)
4. Door, I.J.: van der Made-van Bekkum. Nederlandse woordassociatie normen, Amsterdam (1973)
5. Gerganov, E.: Metodologicheski aspekti na psiholingvistichnite i interdistsiplinarnite izsledvanija. Bul. Lang. **65**(Suppl.), 15–34 (2018)
6. Gerganov, E., Ivancheva, L., Karlova, R., Nikolov, V., Nikolova, Ts.: Balgarski normi na slovesni asotsiatsii. Nauka i izkustvo, Sofia (1984)
7. Kent, H., Rosanoff, J.: Study of association in Insanity. Am. J. Insanity **67**(1), 37–96 (1910)
8. Kurcz, I.: Polskie normy powszechności skojarzeń swobodnych na 100 słów z listy Kent-Rosanoffa. In: Tomaszewski, T. (ed.) Studia Psychologiczne, vol. VIII, Wrocław-Warszawa-Kraków, pp. 122–255 (1967)
9. Leontév, A.A.: (red.) Slovarássociativnyh norm russkogo jazyka. MGU, Moskva (1977). Cerkasova, G.A.: Elektronnoe izdanie (2006)
10. Palermo, D.S., Jenkins, J.J.: Word Associations Norms: Grade School through College, Minneapolis (1964)
11. Pehlivanova, P.: Kak se klasificirat poananijata za dumite - svoboden asotsiativen experi-ment. Pedagogicheski almanah **11**(1), 156–178 (2003)
12. Popova, V.: Slovesnite asotsiatsii i tjahnata (ne)dostatachnost pri izsledvaneto na detskija lexicon. In: Teorija i praktika yazykovoj komunikatsii, pp. 241–252. RIK UGATU, Ufa (2018)
13. Popova, V.: Rechnik na detski asotsiatsii. Universitetsko izdatelstvo "Episkop Konstantin Preslavski", Shumen (2020)
14. Postman, L. (ed.): Norms of Word Association. Academic Press, New York (1970)
15. Ufimceva, N.V., Cherkasova, G.A., Karaulov, Ju.N., Tarasov, E.F.: Slavjanskij associa-tivnyj slovar, Moskva (2004)
16. Vasileva, N.: Neuropsychological analysis of some aspects of the language functioning in the pre-school age period. In: Verbal Communication Quality. Interdisciplinary Research II Belgrade: LAAC & IEPSP, pp. 396–412 (2013)

Preliminary Investigation of Potential Steganographic Container Localization

Rodmonga Potapova and Andrey Dzhunkovskiy

Institute of Applied and Mathematical Linguistics, Moscow State Linguistic
University, Ostozenka Street 38, 119034 Moscow, Russia
RKPotapova@yandex.ru, Vetinari01@gmail.com

Abstract. In this paper we present the findings discovered when conducting a
perceptual experiment with the goal of analyzing Russian native speaker text
juncture partition. This experiment is one of the first forays into applying
experimental methods to discovering optimal methods of steganography in
general and the first to do so on the basis of Russian written texts. The participants were to read through a semantically cohesive phonetically neutral and
balanced Russian text that was modified to exclude any spaces, capital letters,
paragraph indicators and punctuation symbols. The participants were further
asked to divide the resulting string of symbols into paragraphs, phrases, syntagmas and words. The experiment goal was to discover statistically notable
junctures with the prospect of analyzing and comparing their discovery rates in
all four mentioned juncture paradigms. The results allow to determine
prospective optimal minimal text modifications for the use of steganographic
trigger-containers.

Keywords: Experimental linguistics · Perceptual experiment · Experimental
steganography · Trigger-containers · Russian text steganography

1 Introduction

In this paper we describe the continuation of our line of experimental inquiry into
optimizing linguistic steganographic text containers for Russian language. In our
previous works we have conceptualized trigger-containers as minimal, yet distinct text
alterations that refer to a previously agreed-upon message that is to be received upon
discovering the agreed upon alteration. Such an approach requires experimental
research into discovering corresponding alterations that are simultaneously nigh-
undetectable to human steganalyst and highly durable against automated and computer-
assisted methods of steganalysis.

We have additionally created a tri-stage steganalysis method for the purpose of
having a reference framework for testing the durability of possible linguistic
steganography containers [4]. The initial forays described a preliminary experiment in
which we analyzed the possibility of using spelling, grammar, style and punctuation
errors as such containers. In the present study a perceptual experiment with the goal of
analyzing natural text juncture manipulation is conducted.

A. Karpov and R. Potapova (Eds.): SPECOM 2020, LNAI 12335, pp. 389–398, 2020.
https://doi.org/10.1007/978-3-030-60276-5_38

2 Background and Method

The scientific approach to steganography is in its inception stage. Existing research mostly focuses on large-scale, semi-automated steganographic methods, namely digital watermarks and is genetically very similar to classic cryptography (e.g. [7, 8]). At the same time, there exists a number of investigations regarding steganography and steganalysis, most notably [1–3, 9].

There are two important caveats. The first is that the aforementioned research deals with automated and semi-automated mass-scale steganography algorithms. The second is that there exists a severely limited number of research concerning text steganography and steganalysis for Russian language. There exists a preliminary conceptual base and research on steganography in Russian speech, but text has been untouched by the scientific community.

We define the concept of a trigger container as a steganographic container in the form of a linguistic alteration deliberately interwoven into a text that is in all other instances identical to a natural language text. The message in the container is known to the sender and the receiver beforehand and the appearance of the container triggers its actualization. As such, it allows the users of this approach to covertly deliver messages of potentially unlimited length with minimal alterations to the text.

The subtlety and the minimal appearance of such alterations allows to create durability not only against analysis proper, but also against attempts at destroying the container instead of unlocking its contents. At the same time, the implementation of such containers requires research into what minimal alterations would be optimal for such a purpose.

Effectively, our research approach is unique and allows to procure valuable data due to using experimental perceptual methods and gleaming insight into how, on average, native speakers of a language perceive what constitutes a normal text and an altered one.

As part of this experiment, we have taken a natural language Russian text created by R.K. Potapova [6]. The text is phonetically balanced and semantically sensical, containing a coherent story about a sports event. The text has been modified for research purposes by way of removing all spaces between words, paragraph indicators, capital letters and punctuation symbols. The resulting visual stimulus consisted of a string of letters and was as follows (transcribed from Russian):

«nikogdaeshcheleningradcynebylisvidetelyamitakihinteresnyhproisshestvijnavodes hirochajshienaberezhnyeurechnyhvokzalovvetoiyul'skoeutrozapolnenylyud'midet'miy unymiipozhilymibolel'shchikamidvadcat'luchshihskuteristovizshestistranangliigvineifr anciigermaniigollandiirossiivoshedshievelitusportsmenovvodnikovs"ezzhayutsyakbere gamnevytrassaprobegasverhmoshchnyhskuterovrastyagivaetsyanenaodinkilometrvdlin uslegkarasshiryayas'kust'yukakvihr'pronessyapervyjkorabl'nerasschitavusilijszhutkim skrezhetomstolknulis'anglijskijifrancuzskijlideryvodnomgnoveniezayadlyegonshchiki bylisbroshenyvglub'mutnyhvodnuzhnaekstrennayapomoshch'podumalizriitelipomoch' byimnobezchuvstvastrahaonibezzaderzhkivzbirayutsyanaugolobbitojkormytshchetnop ytayas'obmanut'bezzhalostnuyusud'bunadzemlejugrozhayushchesgushchalas'mglaneb oobvoloklotuchamipodtverdilsyaprognozobservatoriiidetciklonsdozhdemsoshpicberge

navtret'emprobegeuchastvovalilish'vpyateromchtoihzhdetktozhevyigraetsmogbyhudos
hchavyjshutnikgeologorazvedchikspoeticheskimimenemolegajvazovskijnopodderzhiva
yabeshenyjtempchtobypartneroveffektnaob"ekhat'sbokukomandirzhongliruyarulemnes
orientirovalsyaiekipazhochutilsyavpuchinebushuyushchihvolnvykarabkavshis'izpodzat
onuvshegosudnaprobezhchikivplav'napravilis'kshlyupkepodzhidavshejihufinishnogop
unktanakonecsud'yaocenivkatastroficheskuyusituaciyussozhaleniemosushchestvilobsh
cheepozhelaniezakonchit'etizhestochajshieispytaniya»

The participants of our experiment were then asked to carefully acquaint themselves with the stimulus material and complete four independent tasks. These consisted
of separating the text into paragraphs, phrases, syntagmas and words, effectively
inserting juncture points for four different paradigm levels. This approach was partially
inspired by our previous work on multimodal polycode texts [5].

The participants were presented with a questionnaire in which they were asked to
relay personal information (but not their names), information about their preferences in
literature, and then the four aforementioned juncture tasks. The participants were not
informed of the goal of the experiment.

The resulting sample consisted of 102 participants. All of the participants were
Russian native speakers. Their mean age was 20,10 years. The age range was between
17 and 25 years old. All participants were students in the field of linguistics. Participant
sex ratio was skewed: there were four times as many women as men.

After the data was processed for each of the paradigms, we built graphs and
conducted comparative cross-paradigm analysis for pairs "paragraphs-phrases",
"paragraphs-syntagmas", "paragraphs-words", "phrases-syntagmas", "phrases-words"
and "syntagmas-words", thus encompassing every possible paradigm pair.

Thereafter we built graphs displaying the junctures present in all four paradigms at
their minimal and maximum significance points. This allowed us to draw conclusions
on what juncture types in the text are most promising as possible trigger-containers.

3   Results and Discussion

We will begin by providing the graphs for the four analyzed paradigms. On their own,
their data is insufficient but proved useful for observing the trends for cross-paradigm
same-juncture discovery rates. It additionally allowed us to come to the conclusion that
the results for the paradigm "words" required an additional step in the analysis which
will be described below. X-axis contains lexical units corresponding to juncture
location transliterated and translated (in parenthesis) from Russian

In and of itself, the single-paradigm analysis was an important stepping stone for
conducting comparative cross-paradigm analysis and eventually synthesizing a comprehensive view of the minimal and maximal values for cross-paradigm junctures
present in every single paradigm.

Our first step has been the analysis of relative value of juncture detection for the
paradigm "paragraphs" (Fig. 1).

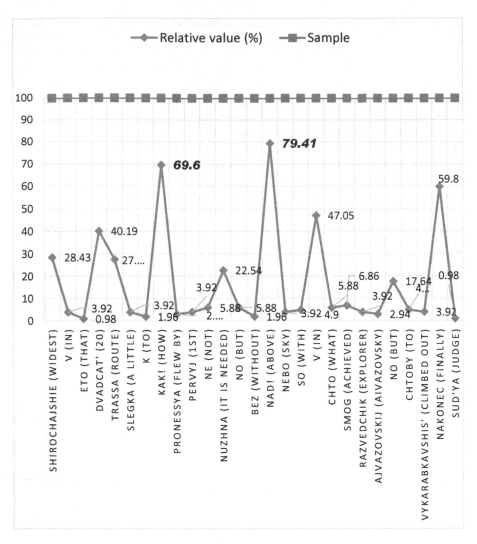

Fig. 1. Paragraph juncture detection rate.

The notable data in Fig. 1 is that there are five junctures that share relatively high indication rates, two of which (in bold) are identical with the ones put forward by the original author of the text. Proceeding to juncture detection for phrases (Fig. 2).

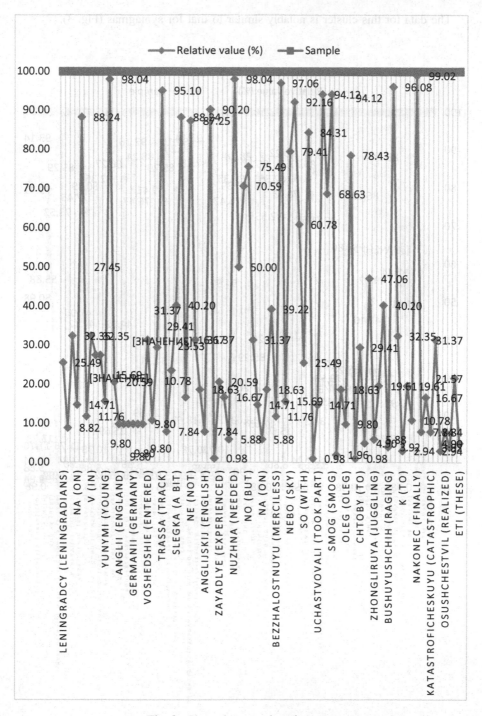

Fig. 2. Phrase juncture detection rate.

The data for this cluster is notably similar to that for syntagmas (Fig. 3).

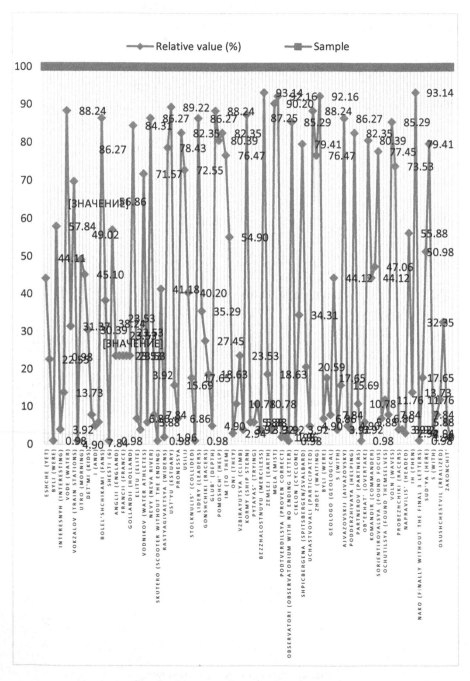

Fig. 3. Syntagmas juncture detection rate.

The data in Fig. 2 and Fig. 3 is remarkably similar, however not identical. The overall pattern of the graphs bears high resemblance in these paradigms. The number of juncture points in the "words" paradigm required us to create a subset of the paradigm in which we included only the junctures with anomalous values (non-100%) (Fig. 4).

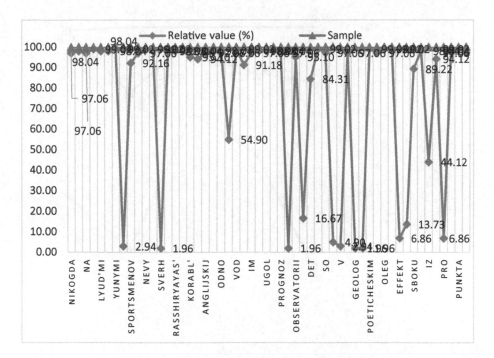

Fig. 4. Word juncture anomaly detection rate.

The presented data is much more concise. Having created graphs for all four paradigms (and one additional graph), we conducted comparative pair-by-pair comparative analysis. The junctures are nikogda (never), na (on), lyud'mi (by people), sportsmenov (athletes), nevy (Neva river), sverh (over), rasshiryayas' (widening), korabl' (ship), anglijskiy (English), odno (one), vod (waters), im (to them), ugol (angle), prognoz (forecast), observatorii (observatory), det (last three letters of the word idet (going); det is the stem of the word deit (children)), so (with), v (in), geology (geologist), poeticheskim (poetic), oleg (Oleg), effekt (effect), sboku (from the side), iz (from), pro (about) and punkta (point). The results allowed to create a table with statistical information about the parameters (Table 1).

Table 1 affords insight into how the paradigms fluctuate. In bold we have highlighted the notable parameters that allow us to conclude that the size of the elements analyzed by our participant is inversely proportional to the sample variance for the corresponding paradigm. The only exception is found in the "words" paradigm, but the pattern is restored once we look at the mean, count and confidence level (95,0%) for this paradigm and analyze only the anomalous junctures.

Finally, we present the cross-paradigm juncture maximum analysis (Fig. 5).

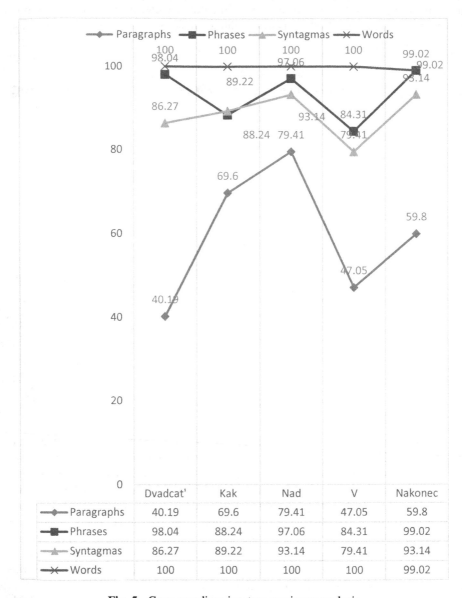

Fig. 5. Cross-paradigm juncture maximum analysis.

Figure 5 contains the most notable analysis results. The junctures are Dvadcat' (20), Kak (how), Nad (above), V (in) and Nakonec (finally). There are a number of notable peculiarities to highlight. Except for the paradigm "words" and one anomalous instance in the "phrases" paradigm in the second juncture point, the curves resemble each other and follow the same trends of increasing or decreasing in value for any particular juncture.

The "words" paradigm displays very high detection rates, with the minimal rate being 99,02% and others possessing 100% discovery rates. The "phrases" and "syntagmas" paradigms are nearly identical both in trends, visual representation of the curve and juncture detection values with the exception of the second juncture point. It is noteworthy that the final results included two juncture points for the «paragraphs» paradigm that were used by the original author of the text.

Ultimately, it is important to note that adapting phrase and syntagma junctions for the purposes of steganography is extremely challenging in Russian, but the data may prove useful for other languages. In Russian, there are no special symbols to consistently denote the junction between neighboring phrases and syntagmas, an issue further complicated by free word order.

On the other end of the spectrum, words and paragraphs have distinct juncture indicators in Russian and therefore offer a constant to be manipulated. The detection rate for the paradigm "words" has been highly inconsequential. Most of the anomalies, with a few interesting exceptions, were most likely caused by the participant fatigue. One of the most interesting cases was a juncture in the middle of a complex (multiroot) Russian word, where more participants reported a paragraph juncture than a word juncture. To clarify, Russian language does not contain means to start a new paragraph in the middle of a word.

Table 1. Comparison of paradigm statistical data.

Parameters	Paragraphs	Phrases	Syntagmas	Words	Word Anomalies
Mean	17,10	33,58	34,75	**94,07**	76,51
Standard error	4,36	3,46	3,04	1,45	5,09
Median	4,90	19,61	21,57	100,00	97,06
Mode	3,92	9,80	3,92	100,00	98,04
Standard deviation	22,67	31,92	33,05	21,28	37,41
Sample variance	**513,82**	**1019,11**	1092,57	**452,93**	1399,82
Kurtosis	1,71	−0,48	−1,28	13,23	−0,02
Skewness	1,65	1,01	0,60	−3,84	−1,36
Range	78,43	98,04	92,16	98,04	97,06
Minimum	0,98	0,98	0,98	1,96	1,96
Maximum	79,41	99,02	93,14	100,00	99,02
Sum	461,69	2853,92	4099,99	20131,37	4131,37
Count	**27,00**	**85,00**	**118,00**	**214,00**	**54,00**
Confidence level (95,0%)	8,97	6,89	6,03	2,87	10,21

It is difficult to conceptualize a trigger-container for a variable that enjoys nigh-universal agreement. Therefore, paragraph junctions become the foremost prospect for future research.

4 Conclusion

In summation, the described perceptual pilot experiment allows us to hypothesize that manipulating paragraph junctures in Russian texts may possess potential for being an optimal location for trigger-containers. This result requires further confirmation on a larger set of texts with explicit focus on determining paragraph junctions. Furthermore, the results require additional verification to confirm their robustness. We further hypothesize that the findings of further experiments may yield a pattern connecting the location of a paragraph juncture to the length of the text as a determining factor for its detection levels which needs additional study. The pilot experiment described lays the groundwork for creating an extensive framework for using perceptual experiments as a means of determining the durability and robustness of steganographic containers.

Acknowledgments. The reported study was funded by RFBR, project number 19-312-90066.

References

1. Bennet, K.: Linguistic steganography: survey, analysis, and robustness concerns for hiding information in text. CERIAS Tech Report 4, pp. 1–30 (2013)
2. Chandramouli, R.A.: Mathematical framework for active steganalysis. ACM Multimedia Syst. J. **9**, 303–311 (2003). Special Issue on Multimedia Watermarking
3. Chen, Z., Huang, L., Meng, P., Yang, W., Miao, H.: Blind linguistic steganalysis against translation based steganography. In: Kim, H.-J., Shi, Y.Q., Barni, M. (eds.) IWDW 2010. LNCS, vol. 6526, pp. 251–265. Springer, Heidelberg (2011). https://doi.org/10.1007/978-3-642-18405-5_21
4. Dzhunkovskiy, A.V.: Steganography: three-stage analysis methodology applied to Russian written texts. Vestnik Moscow State Linguist. Univ. Humanit. Sci. **6**(797), 117–123 (2018)
5. Potapova, R., Potapov, V., Komalova, L., Dzhunkovskiy, A.: Some peculiarities of internet multimodal polycode corpora annotation. In: Salah, A.A., Karpov, A., Potapova, R. (eds.) SPECOM 2019. LNCS (LNAI), vol. 11658, pp. 392–400. Springer, Cham (2019). https://doi.org/10.1007/978-3-030-26061-3_40
6. Potapova, R.K.: Speech: Communication, Information, Cybernetics, 5th edn. URSS, Moscow (2015). (in Russ.)
7. Si, H., Li, C.-T.: Fragile watermarking scheme based on the block-wise dependency in the wavelet domain. In: Proceedings of ACM Multimedia and Security Workshop, pp. 214–219 (2004)
8. Su, J.K., Eggers, J.J., Girod, B.: Capacity of digital watermarks subjected to an optimal collusion attack. In: Proceedings of European Signal Processing Conference, pp. 1–4. IEEE, Tampere (2000)
9. Xiang, L., Sun, X., Luo, G.: Linguistic steganalysis using the features derived from synonym frequency. Multimedia Tools Appl. **3**, 1893–1911 (2014). https://doi.org/10.1007/s11042-012-1313-8

Some Comparative Cognitive and Neurophysiological Reactions to Code-Modified Internet Information

Rodmonga Potapova[1](✉) ⓘ and Vsevolod Potapov[2] ⓘ

[1] Institute of Applied and Mathematical Linguistics, Moscow State Linguistic University, 38 Ostozhenka Street, Moscow 119034, Russia
RKPotapova@yandex.ru
[2] Centre of New Technologies for Humanities, Lomonosov Moscow State University, Leninskije Gory 1, Moscow 119991, Russia
volikpotapov@gmail.com

Abstract. The overall decoding of the daily Internet information mainstream by human may be affected by such a factor – and first of all – as cognitive background of the linguistically and non-linguistically structured phenomenon. Other ways of the communication codes help redress the cognitive, neuro- and physiological balance. This paper considers preliminary results of two pilot experiments: a) cognitive aspects of human monocode and polycode perception of the semantic focus (rheme); and b) neurophysiological aspects of the same task. The background information about the two experiments presented by means of questionnaire testing and polygraph analysis provides an opportunity to use these analytic methods for further research approaches to solving the task, which is related to the influence of the monocode and polycode stimuli on the cognitive theme-rheme information on the neurophysiological behavior of the perceivers.

Keywords: Monocode and polycode perception · Theme-rheme focus · Cognitive and neurophysiological reactions to Code-Modified stimuli

1 Introduction

For the modern network information space, it is particularly important to find a solution to the problem of coordinating the actions of a large number of Internet-communicants [1]. In this regard, interdisciplinary research aimed at the perception and understanding of semantic Internet-information begins to play a leading role. The perception and semantic interpretation of multimodal polycode Internet-information are directly related to the problems of sociocognitive and discursive analysis [11–15]. The multimodal polycode virtual reality of the Internet imitates physical reality and affects the Internet user who perceives this entire complex set of stimuli. The logical conclusion is the point of view of some Internet researchers, for example [5], according to which the intersection of old and new means of communication leads to a convergence effect in human activity. Research aimed at analyzing the hidden (latent) meanings of Internet-messages [12, 13] is becoming increasingly important. The polycode nature of various

© Springer Nature Switzerland AG 2020
A. Karpov and R. Potapova (Eds.): SPECOM 2020, LNAI 12335, pp. 399–411, 2020.
https://doi.org/10.1007/978-3-030-60276-5_39

media was also highlighted by F. Kittler [6–8]. At the same time, it is worth mentioning F. Kittler's assumption, according to which the audience's perception of messages and the medium itself are at the same time dependent variables of the technical repro-ducibility of the real image. Based on the aforesaid, the task of studying the Internet user's cognitive and neurophysiological perceptual reactions to multimodal polycode Internet-information is of particular interest and of interdisciplinary nature.

2 Experiments in Rheme Location

In this investigation three types of multimodal perception – that of auditory, visual and combination of auditory and visual stimuli – were analysed with regard to multimodal search process of the semantic focus. This term is used by some linguists in the context of semantic analysis of sentences where it comes to the distinction between the information assumed by the communicants and that which is at the centre (or «focus») of their communicative interest [3]. On the basis of special test corpus where analysed: auditory Internet stimuli, visual Internet stimuli of the same Internet fragment and combinations of auditory (auditive) and visual Internet stimuli. The task for recipients was to define the localisation of semantic focus with regard to the localization positions in the Internet content fragment: initial, medial or final.

A series of pilot experiments was started to evaluate the parameters of the actual semantic division (theme-rheme segmentation) of perceived polycode texts.

2.1 Experiment 1

Method. The first experiment involved 24 test subjects aged 17 to 24 years, males (n = 6) and females (n = 18); their average age was 20 years (MSLU students). The average period of use of the Internet was 9.25 years for female subjects and 12 years for male subjects. The average number of hours spent on the Internet per day was 4 h for female subjects (min = 1.5, max = 8) and 4.5 h for male subjects (min = 1.5, max = 6). It is planned that in the future a series of similar experiments will be performed in order to verify the stability of the results, which correlates with the methodology for conducting longitudinal studies. For the experiment, a special test questionnaire was developed, which included specific questions relating to three different Internet content fragments (see Table 1).

This questionnaire was a cognitive test for evaluation of the mental processing (taking into account the definition of a logical predicate, semantic focus, theme-rheme segmentation) of perceived stimuli – multimodal polycode texts taking into account element-by-element and complex perception. Each experimental fragment was pre-sented to the test subjects three times in each of three modes: only an audio stimulus, only a video stimulus and an audio + video stimulus together (see Fig. 1). At the end of the experiment, it was required to determine the degree of difficulty in determining the rheme (semantic focus) of this fragment.

Table 1. The questionnaire used in Experiment 1.

Background information:

Age:
Gender:
Internet use experience (years):
Average time online per day:
Experiment date:

Audio
---
1. Listen 3 times to the fragment
2. Locate the semantic focus of the fragment. Mark with a +:
    Beginning
    Middle
    End
3. With what phonetic means the semantic focus is highlighted?
    Stress
    Melodic changes:
        expansion of pitch range
        contraction of pitch range
        switching to another pitch register (high, middle, low)
    Tempo increase or decrease
    Loudness increase or decrease
4. Is it possible to distinguish between the "known, presupposed" and "unknown, new" to the communicants by just listening?
    Yes (why?)
    No (why?)
    Cannot give an answer

5. Were 3 times enough for you to decide on all the questions posed?
    Yes
    No
    Write your suggestions here:

Video
---
1. View the video-only fragment 3 times
2. Locate the semantic focus of the fragment. Mark with a +:
    Beginning
    Middle
    End
3. What non-verbal means help you to locate the semantic focus
    Mimic
    Gestures
    Distance between characters
4. Is it possible to distinguish between the "known, presupposed" and "unknown, new" to the communicants by just non-verbal features?
    Yes (why?)
    No (why?)
    Cannot give an answer
5. Were 3 times enough for you to decide on all the questions posed?
    Yes
    No
    Write your suggestions here:

Audio+Video
---
1. View the video-only fragment 3 times
2. Locate the semantic focus of the fragment. Mark with a +:
    Beginning
    Middle
    End
3. What was already known to you (theme) and what was totally new (rheme)?
4. When is it easier to determing the known (given) and unknown (new):
    When just listening to the fragment, without the video
    When just watching the fragment, without the audio
    When watching the fragment with the audio
5. What required the most efforts from you? In what mode of fragment presentation? Explain why.

Results. Some examples of test results are presented in Fig. 2, 3, 4 and 5 (note: colors in the figure do not carry any special meaning and only serve to make the figures more readable).

Of particular difficulty in determining the localization of the rheme (semantic focus) were audio stimuli (presentation of the Internet fragment in the "audio only" mode). The trajectory of the rheme localization process when the stimulus was presented three times (only audio, only video, audio + video), as a rule, is in the direction from the beginning or middle of the fragment to its end (position of the semantic focus, rheme). The correlation of information on the decision regarding the localization of the semantic focus (rheme) in the context of the element-by-element (audio or video) and complex (audio + video) presentation of three Internet fragments obtained during the experiment makes it possible to conclude that the final decision is in positive correlation with the information processing channel: mono-audio, mono-video, poly-audio-video. The maximum number of errors when making a decision is related to the perception of a mono-audio signal. The semantic information on the localization of the rheme would be probably defined more accurately and consistently at the stage of mono-video and poly-audio-video stimuli, which is accompanied by a transition from entropy and information noise to the information center.

The preliminary results of the above pilot experiments correlating with two types of tasks (cognitive and neurophysiological) make it possible to obtain preliminary conclusions, according to which the theme-rheme identification of a multimodal polycode Internet-message can vary depending on the type of Internet-stimuli (audio, video, audio + video). The percipients' greatest difficulties arose in the perception of monochannel and monocode auditory information. As a rule, in most cases, the perception of rheme (the semantic focus) reveals a tendency to shift towards the end of the presented experimental Internet-fragment, which may be hypothetically associated with the implementation of the "ending law" known in communicative psychology.

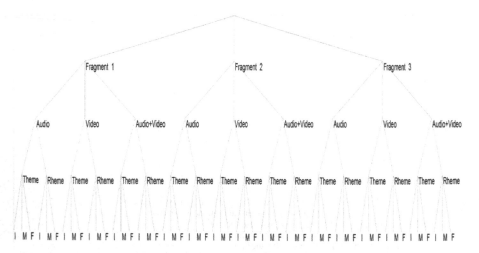

Fig. 1. General scheme of the experiment on determining the localization of the rheme with the element-by-element and complex presentation of polycode stimuli.

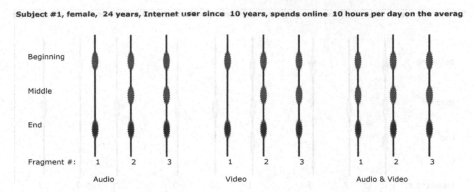

Fig. 2. Localization of the rheme according to the results of the experiment with the element-by-element and complex presentation of polycode stimuli.

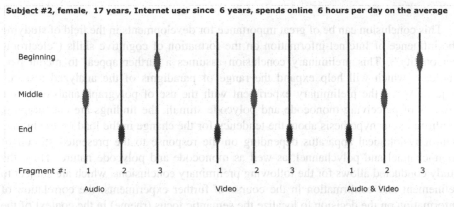

Fig. 3. Localization of the rheme according to the results of the experiment with the element-by-element and complex presentation of polycode stimuli.

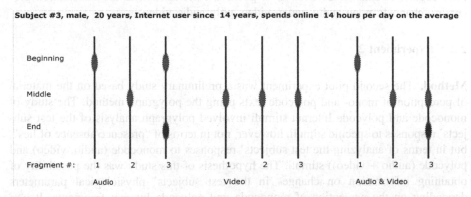

Fig. 4. Localization of the rheme according to the results of the experiment with the element-by-element and complex presentation of polycode stimuli.

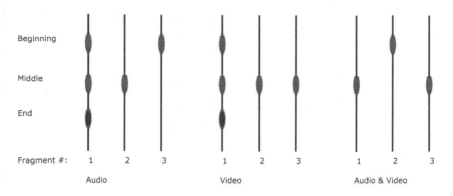

Fig. 5. Localization of the rheme according to the results of the experiment with the element-by-element and complex presentation of polycode stimuli.

This conclusion can be of great importance for developments in the field of studying the influence of Internet-information on the formation of cognitive skills ("electronic personality"). This preliminary conclusion assumes a further appeal to media psychology, which will help expand the range of paradigms of the analyzed research object. As for the preliminary experiment with the use of polygraph analysis in the process of perceiving monocode and polycode stimuli, the findings are encouraging, confirming our hypothesis about the tendency for the change in the load on the human neurophysiological apparatus depending on the response to the presented stimuli of monochannel and polychannel, as well as monocode and polycode nature. Thus, the study conducted allows for the following preliminary conclusions, which are subject to refinement and confirmation in the course of further experiments: the correlation of information on the decision to localize the semantic focus (rheme) in the context of the element-by-element (audio or video) and complex (audio + video) presentation of Internet fragments obtained during the experiment allows for the conclusion that the final cognitive-semantic decision is in positive correlation with the information processing channel: mono-audio, mono-video, poly-audio-video.

2.2   Experiment 2

Method. The second pilot experiment was a preliminary study based on the material of perception of mono- and polycode texts using the polygraph method. The study of monocode and polycode Internet stimuli involved polygraph analysis of the test subjects' responses to specific stimuli, however, not in terms of "presence/absence of lies", but in terms of analyzing the test subjects' responses to monocode (audio, video) and polycode (audio + video)) stimuli. The hypothesis of this study was the possibility of obtaining information on changes in the test subjects' physiological parameters depending on the perception of monocode and polycode Internet fragments. It was interesting and made sense to make an attempt to identify how physiological parameters only can be indicators of the perception of monocode and polycode stimuli along with neurophysiological and mental-cognitive parameters. Is it possible in the future to trace some types of connections between cognitive, neurophysiological and purely physiological parameters? In this regard, a searching experiment was performed, the results of which are somewhat encouraging in an effort to clarify and deepen our knowledge about the reaction of the human body to mono- and polycode information. It is known that vegetative and physiological reactions in emotions include the following:

- a change (in the absence of physical activity) of the heart rate, as well as a change in the rhythm of pulse contractions;
- a change in blood pressure (hypertonic or hypotonic type);
- changes in the rhythm of breathing (especially a sharp reduction in the expiratory phase and distortion of the usual respiratory cycle);
- marked changes in the galvanic skin reaction (GSR), appearance of hypersweating without temperature and physical stress;
- changes in the electrocardiogram (ECG);
- changes in the electroencephalogram (EEG);
- increased intestinal motility; increased diuresis;
- various changes in the elements of blood, urine, saliva [10].

Correct interpretation of polygrams is of great importance for establishing the degree of reliability of the results obtained during the study. Preliminary assessment is performed by a specialist visually, during the study. Following the changes in the psycho-physiological state of the test subjects, the polygraph examiner can draw certain conclusions. However, such an assessment is not sufficient to reach a final verdict, since it is not free from subjectivity. Qualitative and quantitative processing is performed in several stages: a) general inspection and assessment of the quality of polygrams; b) identification and measurement of psychophysiological reactions; c) reaction ranking and final mathematical data processing. During the preliminary assessment of the polygram, the specialist visually analyzes the results obtained, without calculations and the use of algorithms. In the case of pronounced reactions, such a method can be quite efficient.

The polygram reflects respiratory reactions, plethysmogram, galvanic skin reaction, cardiogram, micromotor muscle tremor. The test subject's response is a pronounced change in a parameter in response to a stimulus. The reaction can be expressed in several ways: increase in amplitude, decrease or stabilization over a period of time. The manifestation of the reaction is purely individual; it depends on the subject's body idiosyncrasies [2]. There is also such a thing as a "relief reaction", which is expressed in a clearly readable change in physiological data over a period of time after stress. In this research, however, the increase in amplitude of a parameter was regarded as a marker of prominent reaction to the stimulus showing that there was a connection between cognitive and neurophysiological perception and decision-making mechanisms.

An artefact is a noticeable change in a person's data against the background of a "calm" polygram that is not related to stimuli. The nature of the artefact can be internal (endogenous) or external (exogenous) exposure to various destabilizing factors.

Internal factors include any movement of the subject (conscious or unconscious), cough, respiratory rhythm disturbance, pain, yawning, etc. External factors include noise, light, etc. Of great difficulty is the inability to accurately determine the nature of a particular reaction. The only way to determine that the polygraph examiner's stimulus became the cause of the reaction is to accumulate statistics on the dynamics of the reaction to a particular stimulus. In polygraphology, the term "test" is usually understood as "a particular implementation of the basic logical principle of a particular technique" [9]. It can also be said that "in the most general form, the tests used for interrogation with a polygraph are a system of specially formulated and sequentially arranged questions, individual words (concepts), drawings, plans, photographs, things or objects" [4]. In our study, an attempt was made to use mono-audio stimuli, mono-video stimuli and polycode audio-video stimuli focused on the test subjects' answers.

The preliminary experiment involved 4 persons (MSLU students). Three pieces of Internet information were used as experimental material. For the study, the polygraph of the "Archont" Center (Moscow) was used. The experiment included two stages: a) "Zero" (inception phase) for each test subject (the "Background" polygram) with answers to two questions: "Is it an academic school day today?"; "Is it a holiday today?", and b) "Working" – with an appropriate response to three Internet stimuli: mono-audio stimulus, mono-video stimulus, poly-audio-video stimulus.

Results. The results of processing the obtained data are presented in Table 2. Also, below (Table 3) are summarized data on the dynamics of changes in the recorded parameters (characteristics). Printouts of polygrams are presented in Fig. 6, 7, 8 and 9.

Table 2. Representation of various patterns for dynamics of changes in polygraph parameters.

Pattern	Count
lack of dynamics (=)	18
increase (<)	4
decrease (>)	27
decrease-increase > <	2
increase-decrease < >	9

Analysis of the results revealed the following general trends: in most cases, the general dynamic trend was the decline or preservation of values of recorded parameters, which correlates with the effect of psychological adaptation of the test subjects to the situation of the experiment. At the same time, in a number of cases, complex dynamics were observed, indicating a possible correlation of changes in these parameters with a change in a stimulus. Thus, as a result of the searching study of the test subjects' reaction to monocode and polycode Internet signals, it was previously found out that: a) there are certain changes in polygraphic parameters associated with a change in the type of stimulus (mono-audio -> mono-video -> poly-audio-video); b) the most relatively stable physiological indicators are such parameters as respiration (breathing), GSR and ECG; c) least correlation is observed between the change in the type of stimulus and GSR. These new experiments on the psycho- and neurophysiological reactions of test subjects to polycode Internet stimuli using a polygraph showed that polygraphology tools can be used to study the effects of polycode stimuli on test subjects.

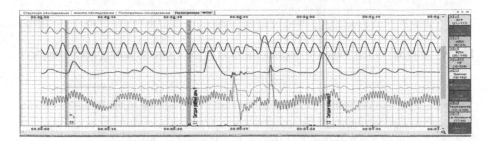

Fig. 6. Polygram of test subject 1 (background).

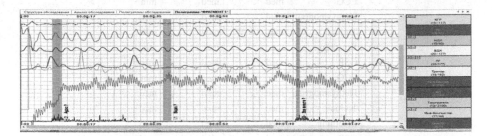

Fig. 7. Polygram of test subject 1 (Internet-fragment 1, three stimuli: audio, video, complex).

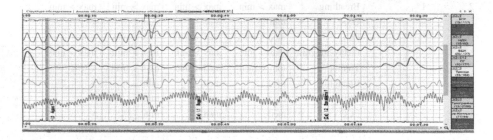

Fig. 8. Polygram of test subject 1 (Internet-fragment 2, three stimuli: audio, video, complex).

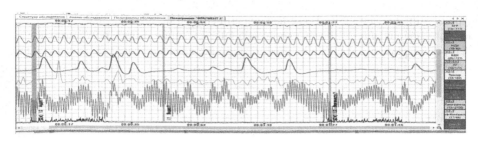

Fig. 9. Polygram of test subject 1 (Internet-fragment 3, three stimuli: audio, video, complex).

Table 3. Dynamics of changes in polygraph parameters as response to the presented stimuli.

Subject #	Fragment No.	Dynamics of changes in polygraph parameters				Pattern type
		Parameters	*Mono-audio*	*Mono-video*	*Polystimulus*	
1	1	Breathing	med	med	med	=
		GSR	med	med	med	=
		ECG	med	med	min	>
		Plethysmogram	med	med	med	=
		Tremor	min	med	med	>
	2	Breathing	med	med	med	=
		GSR	med	med	med	=
		ECG	med	med	min	>
		Plethysmogram	med	med	med	>
		Tremor	med	med	med	=
	3	Breathing	med	med	med	=
		GSR	med	med	med	=
		ECG	med	med	min	>
		Plethysmogram	med	max	min	< >
		Tremor	min	med	med	=
2	1	Breathing	max > min	max > min	max > min	max > min
		GSR	min	min	min	=
		ECG	med < max	max < maxx	max > min	< < >
		Plethysmogram	max > min	max > min	max > min	max > min
		Tremor	min < med	med < max	max < maxx	< < <
	2	Breathing	med	med	min	>
		GSR	med	med	min	>
		ECG	med	med	min	>
		Plethysmogram	max > min	max > min	max > min	max > min
		Tremor	med < max	max < maxx	max > min	< < >
	3	Breathing	max > min	med	max > min	>
		GSR	med	med	max > min	>
		ECG	maxx[1] > min	max > min	max > min	> >
		Plethysmogram	med	max	min	< >
		Tremor	min	med < max	max < maxx	< < <

(continued)

Table 3. (continued)

Subject #	Fragment No.	Dynamics of changes in polygraph parameters				Pattern type
		Parameters	Mono-audio	Mono-video	Polystimulus	
3	1	Breathing	med	med	med	=
		GSR	min	med	max	< <
		ECG	max	med	min	> >
		Plethysmogram	med	min	med	> <
		Tremor	min	med	maxx	< < <
	2	Breathing	med	med	min	>
		GSR	med	med	med	=
		ECG	med < max	max > med	min	< < > >
		Plethysmogram	med < max	max > med	min	< < > >
		Tremor	max	med	min	>[2]
	3	Breathing	med	med	med	=
		GSR	med	med	med	=
		ECG	med < maxx	med > min	min	≪ >
		Plethysmogram	max	med	min	>
		Tremor	med	med	med	=
4	1	Breathing	med	maxx	med	< < < >
		GSR	max	max > med	med < max	= > <
		ECG	maxx	max	min	> > >
		Plethysmogram	med	med	min	>
		Tremor	max	med	min	>
	2	Breathing	med	med < maxx	med	< >
		GSR	med	med > min	med	> =
		ECG	maxx	med	min	>
		Plethysmogram	med	med	med	=
		Tremor	max	med	min	>
	3	Breathing	med	med	med	=
		GSR	med	med > min	min	> >
		ECG	med	med > min	min	>
		Plethysmogram	max	med	min	>
		Tremor	med	med	med	=

[1]Note: maxx stands for a very prominent maximum.

[2]But for the complex stimulus in this case, the attenuation of the amplitude is accompanied by an increase in the frequency.

3 Conclusion

The results lead to the hypothesis that our method gives us an opportunity to analyze two arts of human reactions on the stimuli: cognitive and psychophysiological. The results indicate that all of the perceptual arts of analysis carry a certain degree of specific information. A recipient`s attribution of the semantic focus is based on the general impression of the multimodal polycode stimuli, but not on speech single acoustic parameters. This paper proposes a general approach towards the objective measurement of different semantic focus localisation types on Internet.

Semantic focus recognition may be defined as the ability to recognize the semantic focus of the multimodal polycode Internet fragment. Multichannel neurophysiological systems (e.g., EEG, ECG and others) could be useful for semantic focus recognition by subjects, but not only on the basis of spoken dialogues, texts and so on. The preliminary results of our pilot experiment showed there are differences between some meanings usually expressed by the same or similar Internet stimuli. First of all the are differences between the auditory semantic focus localization and visual and combined auditory and visual semantic focus localization. It would be very useful to investigate the independence between the rhematic (semantic focus) function and emphatic function. The emphatic markedness can change the subjects evaluation of the semantic focus. This paper proposes a general approach towards the subjective and objective measurements of Internet different multimodal stimuli types. The acoustic speech signal analyses is not the only one method for the definition of the semantic focus localization. In natural communication situations speech features are combinations of different types of human behavior. Emotional arousal can influence the speaker`s voice quality or type of phonation and listeners use this knowledge to attribute the semantic focus of the Internet stimulus. Hence the analysis of acoustic speech signals cannot serve as a reliable method of recognition of semantic focus regarding multimodal polycode Internet information.

Acknowledgments. The research is supported by the Russian Science Foundation, grant #18-18-00477, scientific supervisor Prof. Rodmonga Potapova.

References

1. Benkler, Y.: The Wealth of Network: How Social Production Transforms Markets and Freedom. Yale University Press, London (2006)
2. Brown, R.: Auditory Speaker Recognition. Helmut Buske Verlag, Hamburg (1987)
3. Crystal, D.: A Dictionary of Linguistics and Phonetics, 6th edn., pp. 192–193. Blackwell, Hoboken (2019)
4. Gauzhaeva, V.A.: Methodological foundations of conducting research with a polygraph. Theory and Practice of Social Development, no. 1, pp. 57–60 (2015)
5. Jenkins, H.: Convergence Culture: Where Old and New Media Collide. New York University Press, New York (2006)
6. Kittler, F.: Gramophone, Film, Typewriter. Brinkmann & Bose, Berlin (1986)
7. Kittler, F.: Hardware, das unbekannte Wesen. Medien - Computer - Reality, Frankfurt-am-Main, pp. 119–123 (2000)
8. Kittler, F.: Optical Media. Logos/Gnosis, Moscow (2009)
9. Ogloblin, S.I., Molchanov, A.Yu.: Instrumental "lie detection". Academic course. "Nuance", Yaroslavl (2004)
10. Potapova, R.K., Potapov, V.V.: Semantic field of "drugs": discourse as an object of applied linguistics, Moscow (2004)
11. van Dijk, T.A.: Discourse and power. Representation of dominance in language and communication. "Librokom", Moscow (2013)
12. van Dijk, T.A.: Discourse as Social Interaction. Discourse Studies: A Multidisciplinary Introduction, vol. 2. Sage Publication, London (1997)

13. van Dijk, T.A.: Discourse as Structure and Process. Discourse Studies: A Multidisciplinary Introduction, vol. 1. Sage Publication, London (1997)
14. van Dijk, T.A.: News of Discourse. Erlbaum, Hillsdale (1988)
15. van Dijk, T.A.: Text and Context. Explorations in the Semantic and Pragmatics of Discourse. Longman, London and New York (1977)

The Influence of Multimodal Polycode Internet Content on Human Brain Activity

Rodmonga Potapova[1] (iD), Vsevolod Potapov[2] (iD), Nataliya Lebedeva[3],
Ekaterina Karimova[3] (iD), and Nikolay Bobrov[1]([⊠]) (iD)

[1] Institute of Applied and Mathematical Linguistics, Moscow State Linguistic
University, 38 Ostozhenka Street, Moscow 119034, Russia
RKPotapova@yandex.ru, arctangent@yandex.ru,
n.bobrov@linguanet.ru
[2] Centre of New Technologies for Humanities, Lomonosov Moscow State
University, Leninskije Gory 1, 119991 Moscow, Russia
volikpotapov@gmail.com
[3] Institute of Higher Nervous Activity and Neurophysiology of RAS,
(IHNA&NPh RAS), 5A Butlerova Street, Moscow 117485, Russia
lebedeva@ihna.ru

Abstract. The present paper seeks to answer the question of the multimodal polycode Internet influence on the brain functional activity of adolescent users generation of three types of Internet stimuli: auditory, visual and combined ones. The results obtained on the basis of neurophysiological and psychophysiological investigation of the influence of the stimuli on the human brain give the opportunity to define which neurophysiological changes occur as the result of the Internet content variability perception and to define the valeological impact of the multimodal polycode Internet use on adolescent Internet users.

Keywords: Multimodal polycode Internet · Brain functional activity · Auditive, visual and combined stimuli · Adolescent Internet users

1 Introduction

Today many interdisciplinary investigations questioned the concept of the "internet addiction". This problem is dominating in the field of Internet psychology [9]. In some psychological researches there are three types of Internet extreme conceptions: at one extreme, are Internet enthusiasts who view Internet use as the panacea for all the plagues of society; at the other extreme, there are the Internet alarmists who view Internet use as undermining the very fabric of society including the healthy development of its children, most people fall somewhere between these extremes [18]. A few studies have examined the relationship between adolescents' Internet use and psychological outcomes [1, 13, 19]. Modern society is embracing online interaction more and more. Young people spend most of their free time with gadgets, listening to or browsing blogs, news, programs, photos, chatting and communicating in social networks. The continuous information load on the brain affects the psychological and physical condition of people [11, 12, 14–17]. The purpose of this study was to identify

A. Karpov and R. Potapova (Eds.): SPECOM 2020, LNAI 12335, pp. 412–423, 2020.
https://doi.org/10.1007/978-3-030-60276-5_40

which neurophysiological and psychophysiological changes occur when perceiving stimuli of various modality. The problem of multimodal polycode social network communication and its influence upon psychophysiological and cognitive characteristics of Internet users is of both academic and practical interest. Polycode communication in the Internet is a major trend of the recent years, and the necessity of researching it is determined by the fact that its nature renders the traditional linguistic methods of analysis are only partially applicable. This explains the importance of the problem for theoretical linguistics. Results obtained on the basis of cross-disciplinary approach will help to establish a perspective for development of an ecological model of health, in particular, for student-age population.

2   Method

The study involved 19 people, 17 women and 2 men, aged 19-21, all students of Moscow State Linguistic University. When selecting stimulus material, we focused on the interests of the studied social group. In this regard, one subgroup was offered political debate (6 people), another subgroup – a popular youth program about celebrities and musicians (13 people). Each participant took place in 3 experiments on three different days. In each experiment, the subject was asked to listen to or to view the stimulus material of a certain modality: only an audio track (sound), only a video series (video), or a video with sound (video + sound). In order for the stimulus material to be perceived under the same conditions of new information, in each session new material was proposed, but in context and format similar to two others (three different presentations of the same TV program). The order of the stimulus material of each modality was determined randomly. The duration of each stimulus was 45 min. Before and after the perception of the stimulus material, psychophysiological testing and recording of the background EEG were performed with each subject in each session. To assess the functional state of the subjects, the following tests were proposed: a) simple visual-motor reaction (SVMR); b) complex visual-motor reaction (CVMR); c) ECG recording with analysis of heart rate variability (HRV).

SVMR and CVMR. The SVMR test makes it possible to check simple visual-motor (hand-eye) reaction. When the light turned on, one had to press the button as soon as possible. The CVMR test reveals the reaction rate under the conditions of choice. When the green light turned on, one had to press the right button as soon as possible, and when the red light turned on, one had to press the left button. A simple sensori-motor reaction is a simple response to signals in the same specific way (for example, by pressing a specific button). A complex sensorimotor reaction includes signal distinguishing and, accordingly, the choice of different methods of behavioral response. It is believed that the time of a simple sensorimotor reaction reflects the functional state of the central nervous system (CNS), as well as some features of the human nervous system (for example, the mobility of processes involving nerves) [10]. The time required to complete complex sensorimotor tests is always longer than the time taken to complete simple sensorimotor reactions, which is due to a complication of the central link of mental activity. In addition, it also depends on such factors as the attention span and switching, short-term memory, thinking, personality characteristics of the subjects [6, 8]. Therefore, the analysis of complex sensorimotor reactions is very informative in assessing cognitive processes. It is assumed that the intelligence level largely reflects the level of differentiation of representative cognitive structures and their potential for further differentiation. In turn, the time of a complex sensorimotor reaction is an important indicator of the discriminatory capacity of the brain [5].

Variational Cardiometry. Both heart rate, cardiac contractions, cardiac minute output, and blood pressure, regional blood flow change naturally in different conditions. The hemodynamics is regulated using the sympathetic and parasympathetic parts of the nervous system. The sympathetic nervous system is intended to provide mobilization of the body to activity; therefore, the mobilization and action will take place in the presence of vegetative changes in the sympathetic system. With a decrease in the level of tension and increasing calming, the tone of the sympathetic nervous system will decrease, and the tone of the parasympathetic one will increase, while all changes in the body systems will have corresponding dynamics. However, in a state of chronic stress, when a person is in a state of permanent stress, the normal balance of vegetative regulation may shift towards the predominance of a sympathetic drive [4]. Heart rate variability (HRV) reflects the work of the cardiovascular system and the functioning of the regulation mechanisms of the whole organism. HRV is the most convenient method that enables evaluation of the effectiveness of the interaction between the cardiovascular and other body systems. This method is becoming popular due to its simplicity, since it is non-invasive.

EEG Analysis. The use of spectral analysis makes it possible to quantitatively evaluate the various frequency components of heart rate fluctuations that reflect the activity of certain parts in the regulatory mechanism [3]. The dynamics of bioelectric activity of the brain was evaluated on the basis of electroencephalographic data. EEGs were recorded in a calm state with open and closed eyes before, after, and during the perception of the stimulus material using the Neurovisor EEG analyzer with 11 electrodes (F3, F4, Fz, C3, C4, Cz, P3, P4, Pz, O1, O2) located according to the "10–20" system, unipolarly with respect to the combined ear electrodes A1 and A2. For all leads, the sampling frequency was set to 256 Hz, the filtering bands were 0.5–70 Hz, and the impedance was less than 30 kΩ. At the first stage, the EEG was recorded in the "Typology" program, and then the frequency spectra from epochs equal to 1216 Hz or 4.86 s were calculated in the same program for further processing. As a result of spectral analysis, the values of the amplitudes of the spectral power and peak values of the frequencies in the frequency bands of theta-, alpha- and beta-1-, and beta-2 rhythms were obtained for each of 13 leads. Further, the rhythm values of the various leads were averaged over the frontal, central and parietal-occipital areas. Thus, the average power values were obtained for four rhythms (theta, alpha, beta-1, beta-2) for three areas of the brain (frontal, central, parietal-occipital) in each trial (eyes closed (EC), eyes open (EO) and during the perception of the stimulus material).

To analyze the dynamics of rhythm power during the perception of the stimulus material, the entire session (45 min) was divided into 3 periods (15 min each), that is, the beginning of the session, the middle of the session and the end of the session. Further, the spectral power values recorded in the middle and at the end of the session were normalized to the values obtained at the beginning. Relative changes in the rhythm power were also obtained after the session in trials with closed and open eyes. Statistical analysis was performed using Statistica 64 software (StatSoft Inc.). The following analysis methods were involved: Wilcoxon test for pairwise comparisons, ANOVA analysis (analysis of variance) with repeated measurements, and ANOVA factor analysis [2, 7]. All methods used in the experiments were approved by the ethics committee of the Institute of Higher Nervous Activity and Neurophysiology of RAS (IHNA&NPh RAS).

3 Results

Statistical analysis was performed according to the "top-down" principle, i.e. at first, no factors (types of stimulus and content) were taken into account; only the values of the psychophysiological indicators "before" and "after" each session were compared using the paired Wilcoxon test. Then, using the ANOVA analysis of variance with repeated measurements, the "stimulus" factor was added, and the influence of the stimulus material type on the dynamics of psychophysiological indicators was identified. In addition to the "stimulus" factor, the effect of a particular day of participation in the experiments on the dynamics of indicators was also analyzed, since it can be assumed that the experiment held for the first, second, and third time will differ for the subject in terms of emotional reactions. The influence of the "content" factor was also checked: did the content of the programs (about politics and celebrities) affect the psychophysiological state of the subjects.

"Simple Visual-Motor Reaction" Test. Statistical analysis using the paired Wilcoxon test showed that after the experiment the number of errors in the subjects increases for sure ($p = 0.04$), and the average time of the simple visual-motor reaction decreases ($p = 0.03$). The use of ANOVA analysis of variance with repeated measurements (before-after) and the "stimulus" factor revealed the influence of the stimulus material on the average reaction time (Fig. 1a) and the minimum reaction time of the simple visual-motor reaction (Fig. 1b). Moreover, the ANOVA analysis taking into account the "day" factor (that is, the effect of the day of the experiment on the results) showed no significant differences (Fig. 1c). In this case, one could observe that after experiments with the simultaneous perception of both the video and the audio channels, the simple visual-motor reaction improved.

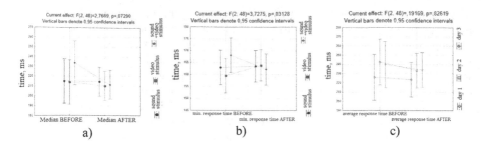

Fig. 1. a) Time median for the simple visual-motor reaction before and after the session in experiments with various stimulus materials; b) Minimum time for the simple visual-motor reaction before and after the session in experiments with various stimulus materials; c) Average time for the simple visual-motor reaction before and after the session on various experimental days.

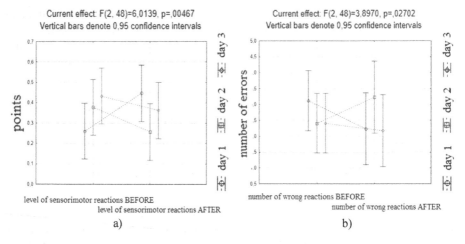

Fig. 2. a) level of the complex visual-motor reaction before and after the session on various experimental days; b) number of incorrect responses for the complex visual-motor reaction before and after the session on various experimental days.

"Complex Visual-Motor Reaction" Test. A paired comparison of the results "before" and "after" the experiment did not reveal any significant differences. ANOVA analysis of variance with the "stimulus" factor also did not show a significant effect of the factor on the dynamics of indicators of the complex visual-motor reaction. However, with the "day" factor taken into account, the influence of the day of the experiment on the level of sensorimotor reactions (Fig. 2a) and the number of incorrect responses (Fig. 2b) were revealed. These results show that only on the first day of the experiment the subjects had an increased level of sensorimotor reactions after the session, which suggests the learning process. It should come as no surprise that during the very first test on the first day of the experiment, the subjects showed the worst result, while the second test on the same day, but after the session, demonstrated significant improvements. Moreover, on the second and third day there is even a slight deterioration in the sensorimotor reactions after the experiment.

Variational Cardiometry. Analysis of variation cardiometry indicators before and after the experiment using the paired Wilcoxon test revealed significant ($p < 0.01$) and certain differences in all indicators: after the experiment, the duration of the R-R intervals in the subjects increased significantly; the dispersion of average also increased; the tension index (TN) decreased; and the total spectral power increased. All this speaks of the relaxation and strengthening of parasympathetic regulation. These results are quite predictable, since after an hour of sitting in a chair in a relaxed state, indicators of cardiovascular activity will demonstrate an increase in parasympathetic regulation. However, ANOVA analysis of variance did not reveal the effect of the stimulus material type on the dynamics of variational cardiometry (Fig. 3a). At the same time, a significant effect of the "day" factor on the total spectral power of the cardiogram was identified: no changes were observed on the first day of the experiment, while on the second and third days the total power increased after the session (Fig. 3b), which corresponds to a greater state of relaxation of the body.

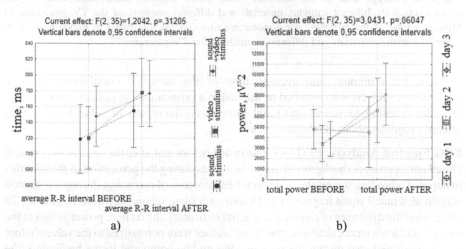

Fig. 3. a) Average duration of R-R intervals before and after the session in experiments with various stimulus materials; b) Total power of the cardiogram before and after the session on various experimental days.

Effect of the Content Type on Psychophysiological Reactions. To check whether the interaction of factors such as the stimulus type (sound, video or video plus sound) and the content of the program (about politics or celebrities) affected the change in indicators, the indicators were presented in relative terms, while the values of the indicators "after" the experiment were normalized to the values "before" the experiment. After this, ANOVA analysis was also performed with repeated measurements, but now the values of experiments with different types of stimulus materials and the "content" factor were used as repeated measurements. The analysis revealed the influence of the "content" factor only on the average and minimum time of the complex visual-motor reaction (Fig. 4a and 4b).

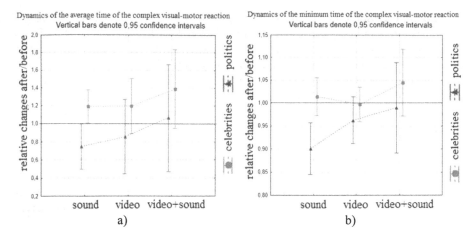

Fig. 4. a) Average time of the complex visual-motor reaction (after/before ratio) in the experiments with different stimulus materials and different contents of the TV programs; b) Minimum time of the complex visual-motor reaction (after/before ratio) in experiments with different stimulus materials and different contents of the TV programs.

Thus, the minimum and average time of the complex visual-motor reaction decreased in subjects who watched or listened to a program about politics (and, most of all, after listening to sound only) and more likely increased in subjects who were offered programs about celebrities.

EEG Spectral Analysis. EEGs were recorded before and after the session at rest with closed and open eyes (background trials), as well as during the perception of the stimulus material. To process the results, a 45-min EEG process of recording during the session was divided into 3 equal fragments of 15 min each (first, second and third fragment). To understand the dynamics of changes in spectral indicators, the relative power values in the background trials were calculated (the "after" values were normalized to the values before the session), and also during the session—the middle compared to the beginning (the second fragment to the first), and the end compared to the beginning (the third fragment to the first). Changes in the power of rhythms in various parts of the brain (frontal, central, temporal, parietal, occipital) were considered as a "location" factor of ANOVA analysis in each comparison. There were no significant effect of "location" factor in a large number of comparisons, except in some cases which we will mention below. The background trial with closed eyes was the main trial during the EEG registration. At this moment, at rest, the alpha-rhythm had to increase and the activity of fast rhythms had to decrease. However, after the session with listening to the audio channel, a decrease in the power of the alpha-rhythm and an increase in the power of faster beta rhythms were observed (Fig. 5a). The greatest changes in the faster beta rhythms after listening to audio content occurred in the parietal-occipital regions: the relative power of the rhythm in the parietal-occipital regions reached 1.36 rel. units, while in the central and frontal regions it increased only to 1.12 rel. units. Such a distribution of activity indicates an increase in the tension in the subjects, which was a result of the perception of the sound channel only. After the experiment with watching a video without sound, an increase in the power of the alpha-rhythm was observed, while the other rhythms practically did not change (their relative values are close to one), which indicates a state of relaxation. Obviously, watching a video without sound did not cause psycho-emotional stress. After the session involving watching a video with sound, an increase in the power of the alpha-rhythm is also observed, but theta- and beta rhythms increase. An increase in the power of beta- and theta-rhythms indicates increased attention and interest; in this case, that there is an emotional reaction, while there is no state of tension, as is the case with the audio channel. It can be assumed that in the experiment with sound only, the subjects did not have a pronounced interest in what was going on, since there was no informative sound channel, but only a picture. The dynamics of the alpha-rhythm with closed eyes and after perceiving different stimuli and on different days is shown in Fig. 5b and 5c.

While experiments with different stimuli were randomly ranked in different subjects, the distribution of power changes for the first, second and third days showed that on the first day a decrease in the power of the alpha-rhythm is observed. As already mentioned, the first day is always the most uncomfortable for the subjects, since the new situation and familiarization with the experimental conditions, of course, cause an orientating response and stress, which can be seen even after the session.

The dynamics of the power of EEG rhythms during the perception of stimuli is presented in Fig. 6a. Analysis of variance showed a significant effect of the "stimulus" and "day" factors on the relative changes in the middle of the session compared to the beginning (the ratio of the indicators of the second fragment to the first). Firstly, it is worth noting that the power values of all rhythms increase in the middle of the session and these changes mostly occurred in the parietal-occipital leads. This is due to the perception of information, emotional reactions and increased tension. A breakdown of power changes by days shows that the rhythms increased most significantly on the first and second days of the experiment; and these days theta- and beta-rhythms increased to a greater extent, which indicates an increase in the concentration of attention and activity of the visual and sensory areas of the brain.

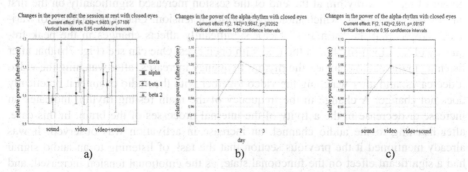

Fig. 5. Changes in the spectral power of EEG rhythms after the session in a trial with closed eyes: a) with various stimuli (all rhythms); b) on different days (alpha-rhythm only); c) with different stimuli (alpha-rhythm only).

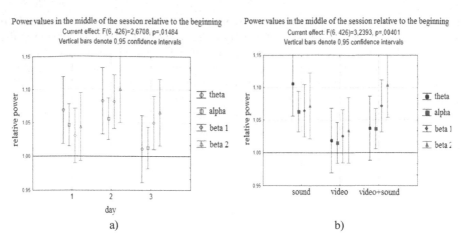

Fig. 6. a) Changes in the spectral power of EEG rhythms in the middle of the session relative to the beginning on different days; b) Changes in the spectral power of EEG rhythms in the middle of the session relative to the beginning with the perception of various stimuli.

A breakdown of the EEG power indicators by various stimuli (Fig. 6b) demonstrates that the theta-rhythm power increased significantly in the experiment with listening to the audio channel only. At the same time, the powers of alpha- and beta-rhythms on this day also increased sufficiently. The least increase in the power of rhythms is observed when watching videos only. These data also testify to the fact that the greatest load and attention was required when perceiving only an audio signal. At the same time, the perception of videos, most likely, did not cause such an increase in attention, since the main information channel in this context was the audio channel, and the video picture was perceived without information analysis. When perceiving a video along with audio, an increase in the power of fast beta-rhythms was observed to a greater extent, i.e. in this case, the perception and analysis of information and context take place, however, without increased stress, as in the case with the audio channel. An analysis of changes in the final fragment of the session compared to the beginning showed significant changes only in the power of the theta-rhythm (Fig. 7a), while the power of the theta-rhythm at the end of the session increased significantly on the first day of the experiment, which indicates emotional tension. One-way analysis of variance showed that the "stimulus" factor significantly affects changes in the peak frequency of the alpha-rhythm in the trial with open eyes. One can see (Fig. 7b) that after listening to the audio channel, the rhythm frequency increases, after watching the video it decreases, and after watching the video together with the audio channel it practically does not change. A change in the frequency of the main resting rhythm indicates an increase or decrease in the activity of the internal processes of the brain. In this case, after listening to the audio channel, an increase in activation was observed. It was already mentioned it the previous section that the task of listening to an audio signal had a significant effect on the functional state, as the emotional tension increased, and the lack of a video channel required increased attention to analyze the context. This situation led to increased activation and an increase in the frequency of the alpha-rhythm. A decrease in the frequency when watching a video indicates a decrease in the activity of brain structures; most likely, boredom and a reduction in tone are possible. And only with simultaneous watching and listening processes, the activity did not change.

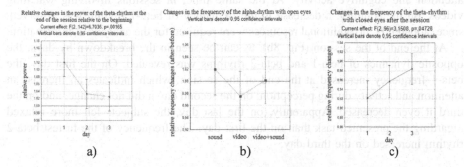

a)   b)   c)

Fig. 7. a) Changes in the spectral power of the theta-rhythm at the end of the session relative to the beginning on different days; b) Changes in the peak frequency of the alpha-rhythm after the session with different stimuli in the trial with open eyes; c) Changes in the peak frequency of the theta-rhythm after the session in the trial with closed eyes on different days.

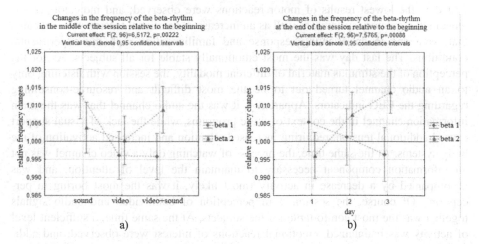

a)   b)

Fig. 8. Changes in the peak frequency of beta-rhythms in the middle of the session with different stimuli (a) and at the end of the session on different days (b).

The one-way analysis of variance also revealed that the frequency of the theta-rhythm varies reliably taking into account the "day" factor (Fig. 7c). It can be seen that on the first day of the experiment after the session, the frequency of the theta-rhythm decreased significantly, and the smallest changes were observed on the last (third) day.

This may indicate that, on the first day, emotional tension manifested itself most remarkably, which is consistent with previous data. Interestingly, in the background trials, changes were observed only in the slower theta- and alpha-rhythms, and fast beta-rhythms showed significant changes during the session.

In the middle of the session (Fig. 8a), the frequency of beta-1 rhythm significantly increased with the perception of only an audio signal, which indicates an increase in attention and cognitive activity. At the same time, in sessions involving watching videos, on the contrary, a decrease in the frequency of beta-1 rhythm was observed, since, apparently, no additional resources were required for the analysis of information.

At the end of the session (Fig. 8b), as can be seen in the breakdown by days, the opposite dynamics of beta-1 and beta-2 rhythms was revealed. On the first day, the beta-1 frequency increased at the end of the session, which indicates an increase in attention and activity during perception; on the second day it did not change, and on the third it even decreased. Apparently, on the last day, the subjects felt more relaxed regarding the presented task than on the first day. The frequency of the fastest beta-2 rhythm increased on the third day.

4   Conclusion

In experiments on the perception of stimuli of different modality in the subjects, all psychophysiological and EEG indicators revealed several key points.

First of all, of course, the results were affected by the day of the experiment. On the first day, the lowest results of motor reactions were observed; and indicators of the tension of regulatory systems, as well as an increase in emotional tension were revealed that were due to an orienting response and familiarization with the experimental conditions. The last day was the most emotionally stable for all subjects. As for the perception of the stimulus material of different modality, the session with listening only to an audio channel turned out to be the most difficult and resource-consuming regarding the EEG indicators. Apparently, it was the audio channel that was the main information channel in the context of these programs, while the lack of visual support caused additional tension, requiring increased attention and increased activation of the body systems. At the same time, the process of watching only a video channel was not an information component necessary to maintain the level of attention, and was accompanied by a decrease in activity (most likely, it was the most boring in perception). Of course, the session with perception of both video and audio signals together was the most comfortable for the subjects. At the same time, a sufficient level of activity was maintained, emotional reactions of interest were observed, and additional resources were not required to analyze the situation and the content.

Acknowledgments.   The research is supported by the Russian Science Foundation, grant #18-18-00477.

References

1. Anderson, M., Jiang, J.: Teens, social media and technology 2018. Pew Research Center (2018). http://pewinternet.org/2018/05/31/teens-social-media-technology-2018/
2. Anscombe, F.J.: The validity of comparative experiments. J. Roy. Stat. Soc. Ser. A (General) **111**(3), 181–211 (1948)
3. Bayevskiy, et al.: Analysis of heart rhythm variability using different electrocardiographic systems (metodic recommendations). Vestnik aritmologii **24**, 65–87 (2001). (in Russian)
4. Dmitrieva, N.: Stress: information/wave mechanisms and diagnostics. Librokom, Moscow (2012). (in Russian)
5. Endrikhovsky, S., Shamshinova, A., Sokolov, E., Nesteryuk, L.: Time of sensomotoric reaction of the human in modern psychophysiological research. Sensornyje sistemy **10**(2), 13–29 (1996). (in Russian)
6. Gribanov, et al.: 2006 Etudes on sensomotoric activity of a child with ADHD. Arkhangelsk (2006). (in Russian)
7. Gelman, A.: Analysis of variance? Why it is more important than ever. Ann. Stat. **33**, 1–53 (2005)
8. Ilyin, E.: Psychophysiology of human states. Piter, Saint-Petersburg (2005). (in Russian)
9. Jackson, L.A., et al.: Personality, cognitive style, demographic characteristics and Internet use – findings from the HomeNetToo project. Swiss J. Psychol. **62**(2), 79–90 (2003)
10. Korobeynikova, I.: Parameters of senso-motoric reactions, psychophysiological characteristics, academic aptitude and EEG of the human. Psihologicheskij zhurnal **21**(3), 132–136 (2000)
11. Krämer, N.C.: Hostile media effect. In: Medienpsychologie, Schlüsselbegriffe und Konzepte, pp. 151–157 (2016)
12. Kraut, R., Scherlis, W., Mukhopadhyay, T., Manning, J., Kiesler, S.: The HomeNet field trial of residential internet services. Commun. ACM **39**, 55–65 (1996)
13. Kraut, R., Shields, M.K., Behrman, R.E.: Children and computer technology: analysis and recommendations (2001)
14. Maria, C.: National Science Foundation (NSF) report 2001. Division of science resources studies, the applications and implications of information technologies in the home. Papadakis: Virginia Polytechnic Institute and State University, Arlington (2001)
15. Potapova, R., Potapov, V., Lebedeva, N., Karimova, E., Bobrov, N.: EEG investigation of brain bioelectrical activity (regarding perception of multimodal polycode internet discourse). In: Salah, A.A., Karpov, A., Potapova, R. (eds.) SPECOM 2019. LNCS (LNAI), vol. 11658, pp. 381–391. Springer, Cham (2019). https://doi.org/10.1007/978-3-030-26061-3_39
16. Potapova, Potapov, Lebedeva, Agibalova: Interdisciplinarity in the research of speech polyinformativity. Languages of the Slavic Cultures, Moscow (2015). (in Russian)
17. Potapova, Potapov, Lebedeva, Agibalova: Polycode Internet medium: valeology problems. Languages of the Slavic Cultures, Moscow (2020). (in Russian)
18. Subrahmanyam, K., Kraut, R.E., Greenfield, P.M., Gross, E.F.: The impact of computer use on children's and adolescents' development. Appl. Dev. Psychol. **22**, 7–30 (2001)
19. Subrahmanyam, K., Kraut, R.E., Greenfield, P.M., Gross, E.F.: The impact of home computer use on children's activity and development. Future Child. **10**, 123–144 (2001)

Synthetic Speech Evaluation by Differential Maps in Pleasure-Arousal Space

Jiří Přibil[1,2(✉)], Anna Přibilová[1], and Jindřich Matoušek[2]

[1] Institute of Measurement Science, SAS, Dúbravská cesta 9,
841 04 Bratislava, Slovakia
{Jiri.Pribil,Anna.Pribilova}@savba.sk
[2] Department of Cybernetics, Faculty of Applied Sciences,
University of West Bohemia, Univerzitní 8, 306 14 Plzeň, Czech Republic
jmatouse@kky.zcu.cz

Abstract. The paper deals with automatic evaluation of the quality of synthetic speech using Gaussian mixture models (GMM) for classification in the Pleasure-Arousal (P-A) scale and subsequently calculated 2D and 3D P-A differentials maps. The speech synthesized from the voice of a speaker is compared with the original voice of the same speaker. Three methods of speech synthesis are ordered by descending 3D perceptual distances from the original speech material. Basic experiments confirm the principal functionality of the developed system. The detailed analysis shows a great influence of the number of mixture components, the size of the processed speech material, and the type of the database for GMM creation on partial results of the continual P-A detection and the final results. The objective evaluation results are finally compared with the subjective ratings by human evaluators.

Keywords: GMM classification · Objective and subjective evaluation · Quality of synthetic speech

1 Introduction

Subjective or objective approaches can be used to assess synthetic speech quality. In the subjective ones a physical property is sensed by an assessor as a perceptual feature with a constrained regular perception range on the psycho-physical level within the physiological perception range on the basic physical level [1]. Five perceptual quality dimensions can be identified as naturalness of voice, prosodic quality, fluency and intelligibility, absence of disturbances, and calmness [2]. The objective methods for evaluation of speech synthesis produced by text-to-speech (TTS) systems may use automatic speech recognition and/or hidden Markov models [3, 4]. A more recent TTS-related metric based on support vector regression as an application of the support vector machine principle for regression to solve different quality prediction models was successfully used for TTS synthetic speech evaluation [1, 5].

At present, we are aimed at automatic evaluation of the quality of the Czech synthetic speech generated by different methods. Instead of the quality dimensions defined in [2], in our research we use only naturalness and a term "acoustic purity"

© Springer Nature Switzerland AG 2020
A. Karpov and R. Potapova (Eds.): SPECOM 2020, LNAI 12335, pp. 424–434, 2020.
https://doi.org/10.1007/978-3-030-60276-5_41

specifying absence of disturbances in the synthetic speech. The current work is focused on the comparison between sentences produced by the TTS system based on the unit selection (USEL) synthesis [6], and two methods based on neural networks: the synthesis using a deep neural network (DNN) paradigm – a simpler long short-term memory (LSTM) neural network [7] with a conventional WORLD vocoder and the LSTM with a more advanced recurrent neural network (WaveRNN) [8] vocoder. The quality of the speech synthesis produced by the neural networks is typologically different from the USEL one. While the USEL errors manifest themselves more or less on the border chained units [9], in the neural network synthesis a problem can occur in fidelity and purity of an acoustic representation of the produced speech signal. Therefore, in the development of an automatic evaluation system, we must accept the requirements for comparison of fundamentally different approaches and different acoustic realizations of the synthetic speech.

The motivation of this work was to verify whether the proposed method of continual perceptual detection in the Pleasure-Arousal (P-A) scale based on the Gaussian mixture models (GMM) [10] classifier and derived 2D as well as 3D maps of differences is feasible for automatic evaluation of the synthetic speech quality. The original speech material of a speaker used for synthesis is compared with the synthesized one to find similarities/differences between them. The final evaluation order is done by 3D distances between the tested synthesis and the original speech material. In this way, we try to eliminate subjectivity and great time-consuming difficulty of the standard evaluation method based on the listening test. The realized experiments confirm the suitability of the used method for this type of task as well as the principal functionality of the developed system. The additional detailed analysis shows a great impact of the used number of mixture components, the used type of speech features, and the type of databases for GMM creation on partial results of continual P-A detection and on overall evaluation results. Finally, the obtained objective evaluation results are compared with the subjective ratings by human evaluators participating in the MUltiple Stimuli with Hidden Reference and Anchor (MUSHRA) listening test [11].

2   Description of Proposed Automatic Evaluation System

Various sounds (noise, speech, music, etc.) can invoke various emotions in listeners. In the dimensional theory, each emotion can be uniquely identified by a number of dimensions. The choice of these dimensions depends on their statistical capability to characterize perceptional ratings by a number of dimensions that is as least as possible [12]. Usually, three dimensions are used to capture Pleasure (Valence), Arousal (Activation), and strength (Dominance). These emotional attributes may be illustrated by self-assessment manikins (SAMs) for perceptual evaluation of the valence from negative through neutral to positive, the activation from calm to excited, and the dominance from weak to strong [13]. These SAMs are often used in the listening test-based subjective evaluation of speech quality, naturalness, etc. Beside this three-dimensional basic categorization of emotions, psychology knows a discrete categorization to six primary emotions (joy, sadness, fear, anger, disgust, surprise) [14]. These discrete emotions can be mapped in the two-dimensional emotional space of

Valence (ranging from pleasant to unpleasant) and Arousal (ranging from calm to excited) when the dominance factor is excluded. Here, the Valence (Pleasure) is the lowest for anger or sadness, and the highest for surprise or joy. The Arousal is the lowest for passive emotions for boredom and the highest for frantic excitement [15].

The proposed evaluating algorithm carries out determination and statistical analysis of distances between the original speaker's speech and the tested synthesis in the 2D P-A space. For continual P-A detection, the standard GMM classifier is used to obtain the scores quantized into discrete levels corresponding to the output P-A classes. In the preparation phase, the GMMs are created and trained for each of pleasure and arousal classes. The training data consist of the sound/speech material from the databases (DB$_1$/DB$_2$) labelled in the P-A scale without any relation to the used original speech or the evaluated synthetic speech sentences processed in the next phase of classification. The input sound/speech signal is processed in frames to obtain three basic types of speech features: time duration (TDUR), prosodic (PROS), and spectral (SPEC) – see the left part of the block diagram in Fig. 1. The TDUR features contain time durations of voiced/unvoiced parts and their ratios. The PROS parameters are represented by microintonation, jitter, shimmer, etc. determined from the fundamental frequency F_0 together with energetic levels. The SPEC properties – first two formants and their ratio, spectral decrease, harmonics-to-noise ratio (HNR), spectral entropy (SE), etc. – are determined from the spectral envelope and the power spectral density. These features are used for subsequent calculation of their statistical parameters (mean, median, rel. max/min, std, flatness, skewness, kurtosis, etc.) that constitute the feature input vectors (with the length of N_{FEAT}) for training of the GMMs – see the right part of the block diagram in Fig. 1.

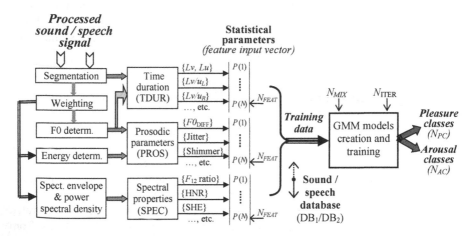

Fig. 1. The principal block diagram of the preparation phase – creation and training of GMMs for P-A classes.

The GMMs approximate the input data statistically as a linear combination of a number of functions with Gaussian probability distributions. The covariance matrix, the vector of means, and the weights must be calculated from the input training data. The expectation-maximization iteration algorithm is used to determine the maximum likelihood function of the GMM [10]. This process depends on the number of mixtures N_{MIX} and the number of iterations (N_{ITER}).

The main evaluation phase begins with the analysis of the input sentences yielding various speech properties. The GMM classifier works with models for each of P-A classes (with the total number of N_{PC} / N_{AC}) created with the help of the training data from the DB_1/DB_2 database. During the classification process, the feature vector obtained from the analysed sentence is processed by the GMM classifier giving the scores quantized into N_{PC} / N_{AC} discrete levels corresponding to the finally detected output P-A classes. In this way, each of M frames is classified individually to obtain the output vector of P-A classes and the winner class corresponds to the maximum probability within the whole analysed sentence. 2D maps are calculated from histograms of perceptual P-A classes for the whole analysed speech material – see the block diagram in Fig. 2. Next, from the built 3D differential maps (differences between 2D maps of the original and the synthesis), Euclidean distances between minima and maxima in the Cartesian coordinate system (ED_{2D} and ED_{3D} parameters) are determined for the final evaluation process – see the demonstration example in Fig. 3. Subsequently, the sorted relative differential values ($SRD_{2/3D}$) expressed in [%] after mean subtraction from the ascending sorted $ED_{2/3D}$ values are used for making decision about the quality of the tested synthesis. Differences between 1st-2nd and 2nd-3rd positions ($\Delta_{1-2,\ 2-3}$2D/3D) of $SRD_{2/3D}$ values are used for judging similarity between evaluated synthesis types. They have always positive values because they originate from the sorted $SRD_{2/3D}$ parameters. The final order (FO) parameter represents proximity of the evaluated synthesis to the original (*Orig*). For example the FO output combination of symbols "*1, 2, 3*" means that the synthesis *Synt 1* has the maximum resemblance to *Orig* (the "best" quality), the synthesis *Synt 2* is less similar to *Orig* than *Synt 1* but more than *Synt 3* ("mean" quality), and the synthesis *Synt 3* differs from *Orig* the most (the "worst" quality). The notation "1.5" means similarity between the "best" and "mean" quality and the notation "2.5" represents similarity between "mean" and the "worst" quality. These similarities are detected in the case of smaller differential parameters $\Delta_{1-2,\ 2-3}$ (lying below the threshold value D_{thresh}). As documented by the visualization example in Fig. 4, this situation occurs when the ED_{2D} parameter is used (see the right bar-graph and graph comparison of the $SRD_{2/3D}$ parameters); for the chosen similarity threshold the result using FO-2D was "1.5, 1.5, 3".

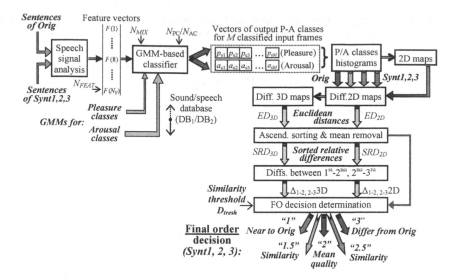

Fig. 2. Block diagram of the proposed automatic evaluation system.

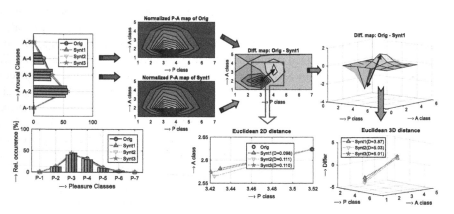

Fig. 3. Graphical example of statistical processing of detected P-A classes: histograms, 2D maps in P-A space, differential 2D and 3D maps, determination of Euclidean (2D/3D) distances between sentences of the original and the synthesis (*Synt 1, 2, 3*).

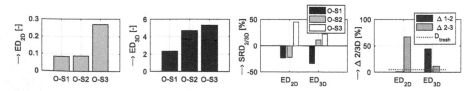

Fig. 4. Visualization of: Euclidean 2D/3D distances between originals and three synthesis types (O-S1, O-S2, O-S3), relative differences $SRD_{2D/3D}$ from ascending sorted $ED_{2D/3D}$ values after mean removal, differences between 1st-2nd and 2nd-3rd positions ($\Delta_{1-2, 2-3}$2D/3D); FO-2D(Synt 1, 2, 3) = "1.5, 1.5, 3", and FO-3D(Synt 1, 2, 3) = "1, 2, 3"; Dthresh = 5%.

3   Material, Experiments, and Results

Two databases were used for training of the GMM models of Pleasure/Arousal classes. The first database DB_1 is the International Affective Digitized Sounds (IADS-2) [16] consisting of 167 sound and noise records produced by humans, animals, simple instruments, industrial environment, weather, music, etc. (no real speech records). This database contains also mean and standard deviation (std) values of perceptual evaluation by 100 male/female listeners using the Pleasure, Arousal, and Dominance parameters, all in the range from 1 to 9. All the used sound records with the duration of 6 s were resampled to 16 kHz. The second used database DB_2 is the MSP-IMPROV consisting of audio and video records in the English language [17]. This corpus includes also mean, and std values for the emotional attributes of the Valence (Pleasure), Activation (Arousal), and Dominance. For purpose of this work, only declarative sentences from the audio part of the database labeled in the P-A scale were used. In total, 2×240 sentences uttered by 3 male and 3 female speakers with the duration from 0.5 to 6.5 s were used for GMM creation and training. For compatibility with DB_1, these speech signals were also resampled to 16 kHz and the mean P-A values were recalculated to the range from 1 to 9.

For continual P-A classification, a separate speech corpus was collected consisting of four parts: the natural speech uttered by original speakers and three variations of speech synthesis: the USEL based TTS system (assigned to *Synt 1*) [7] and two LSTM based ones with different vocoders: conventional WORLD (further called as *Synt 2*), WaveRNN (marked as *Synt 3*) [8]. Both original and synthetic speech originates from four professional speakers (male M1, M2, and female F1, F2). 200 original sentences and 600 synthetic ones (200 for each of synthesis types) were used for this work. The processed speech signals with the duration from 2 to 12 s were resampled to 16 kHz.

To obtain approximately evenly distributed sentences for both tested databases and each of P-A classes, the number was practically reduced to $N_{PC} = 7$ and $N_{AC} = 5$. The similarity threshold D_{thresh} for FO determination was set to 5%. In correspondence with [18] we use the input feature vector with $N_{FEAT} = 16$ representative statistical parameters from the mentioned three types of speech features (TDUR, PROS, and SPEC) – see the detailed description in Table 1. A diagonal covariance matrix was used in GMMs due to its lower computational complexity and $N_{ITER} = 1500$ was finally applied. The described analysis and speech signal processing were currently realized in the Matlab 2016b environment and the Ian T. Nabney "Netlab" pattern analysis toolbox [19] was used for the implementation of the GMM classifier block.

Table 1. Used speech features and their representative statistical parameters.

Type	Feature name	Statistical parameters	No.
TDUR	{T_{DUR} of voiced/unvoiced parts and their ratios}	{range, rel. max, std}	3
PROS	{differential F_0, F_0 zero crossing, jitter, shimmer, signal energy}	{max, rel. max, mean, median, std}	5
SPEC	{F_1, F_2, F_1/F_2 ratio, HNR, spectral decrease, centroid, flatness, entropy}	{skewness, kurtosis, max, min, rel. max, mean, std, dispersion}	8

The detailed preliminary analysis was motivated by seeking an appropriate setting of control parameters for the whole GMM-based evaluation process. The following investigations were performed:

1. Influence of the number of tested synthetic sentences on partial $SRD_{2D/3D}$ and $\Delta_{1-2,\ 2-3}$2D/3D parameters, and final evaluation results – compare values in Table 2 for $N_{TS} = \{15, 25, 40, 50\}$ applied on *Synt 3* of the male speaker M1.
2. Influence of the number of applied Gaussian mixtures on partial results as well as the FO decision about the evaluated synthetic speech quality using $N_{MIX} = \{8, 16, 32, 48, 64, 96,$ and $128\}$ – see the visualization of partial output class trajectories for the voices M1 and F1 in Fig. 5.
3. Influence of different databases for creation/training of the GMMs corresponding to P-A classes on the evaluated quality of two speech synthesis types for the voices M1 and F1 – see differential parameters calculated from the sorted ED_{3D} values and FO results in Table 3.

The objective evaluation results were finally matched with the subjective ones using the MUSHRA listening test published in [8] – see comparison of the bar-graphs in Fig. 6. For every speaker, each of 10 sets of utterances consisted of four sentences: one *Orig* and three synthesized ones by *Synt 1, Synt 2, Synt 3* methods. The speech quality was assessed on a scale between 0 (the worst) and 100 (the best).

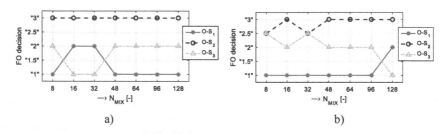

a)   b)

Fig. 5. Comparison of the influence of N_{MIX} parameter on FO decision determined from the 3D Euclidean distances between Orig-Synt$_{1,\ 2,\ 3}$: for male M1 (a), and female F1 (b) voices; DB_1 (IADS-2); FO decisions: "1" = the best, "1.5"/"2.5" = similarity, "2" = mean, "3" = the worst.

Table 2. Comparison of the impact of the number of tested sentences N_{TS} in the *Synt3* group on partial results and FO decisions for M1 voice.

N_{TS} [A]	Determined from 2D maps			Determined from 3D maps		
	ED (S1,2,3)	$\Delta_{1-2,\ 2-3}$ [%]	FO (S1,2,3)[B]	ED (S1,2,3)	$\Delta_{1-2,\ 2-3}$ [%]	FO (S1,2,3)[B]
10	0.118, 0.339, 0.124	1.7, 63.4	**1.5, 3, 1.5**	5.24, 10.92, 5.08	1.6, 51.0	**1.5, 3, 1.5**
25	0.118, 0.339, 0.125	1.9, 63.2	**1.5, 3, 1.5**	5.24, 10.92, 6.22	8.9, 43.1	**1, 3, 2**
40	0.118, 0.339, 0.195	22.6, 42.5	**1, 3, 2**	5.24, 10.92, 6.46	11.1, 40.9	**1, 3, 2**
50	0.118, 0.339, 0.207	26.3, 38.8	**1, 3, 2**	5.24, 10.92, 6.83	14.8, 37.2	**1, 3, 2**

[A] for the *Synt 3* type, sentences randomly taken from the whole set of 50; $N_{MIX} = 48$ and DB_2 (MSP-IMPROV) database were used.

[B] FO decisions: "1" = the best, "1.5"/"2.5" = similarity, "2" = mean, "3" = the worst.

Table 3. Comparison of partial results and FO decisions for M1 and F1 voices using different speech databases in GMM creation/training phases.

Speech database [A]	Male voice from speaker M1			Female voice from speaker F1		
	ED_{3D} (S1,2,3)	$\Delta_{1\text{-}2,\ 2\text{-}3}$ [%]	FO (S1,2,3)[B]	ED_{3D} (S1,2,3)	$\Delta_{1\text{-}2,\ 2\text{-}3}$ [%]	FO (S1,2,3)[B]
IADS-2	5.5, 12.4, 9.6	33.5, 22	**1, 3, 2**	4.6, 6.8, 4.9	1.6, 27.5	**1.5, 3, 1.5**
MSP-IMPROV	5.2, 10.9, 6.8	14.8, 37.2	**1, 3, 2**	6.6, 10.4, 9.3	25.5, 11	**1, 3, 2**

[A] used N_{MIX} = 48 in all cases.

[B] FO decisions: "1" = the best, "1.5"/"2.5" = similarity, "2" = mean, "3" = the worst.

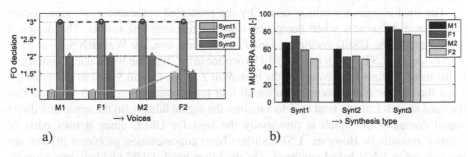

Fig. 6. Summary comparison: (a) FO decisions by GMM evaluation from ED_{3D} parameter, using DB_2 (MSP-IMPROV) and N_{MIX} = 48, (b) results by MUSHRA listening test introduced in [8] for all four evaluated voices; FO decisions: "1" = the best, "1.5"/"2.5" = similarity, "2" = mean, "3" = the worst.

4   Discussion and Conclusion

The performed experiments show that the proposed system based on Euclidean distances of 2D/3D differentials maps obtained after GMM classification in the P-A scale can be applicable in our current task – the automatic objective evaluation of the synthetic speech quality. The main advantage of this approach consists in the fact, that there is no necessity of relation between the sound/speech material used for creation and training of GMMs of each of P-A classes and the speech corpus to be subsequently evaluated. In addition, the baseline sentences from the original speaker can have diametrically different content, time duration, structure, etc. than the evaluated sentences produced by the TTS system.

The first auxiliary analysis has shown that at least 25 sentences (one half of the full set) must be processed to achieve correct evaluation parameters as demonstrated by the results in Table 2. For lower number of sentences, the FO parameter would not be correct or stable, giving the wrong evaluation order or no useful information by the similarity of the categories "1.5" and "2.5". As the ED_{2D} parameter achieves principally low absolute value (in comparison with the ED_{3D} ones), in further experiments the ED_{3D}, the derived SRD_{3D}, and the $\Delta_{1\text{-}2,\ 2\text{-}3}3D$ parameters were used first of all. The detailed analysis of the partial results shows principal importance of the correct choice of the number of mixture components as well as the used speech feature set for the stability of the final decision. The results differ slightly for male and female voices but for both of them the lowest values of $N_{MIX} = 8 \div 32$ give a similarity decision in the majority of cases and an increase of N_{MIX} up to 128 mixtures (more computationally complex) has also a negative impact on the results, as documented by the partial output class trajectories in Fig. 5. In further analysis, the number of GMM components was set to $N_{MIX} = 48$ as a stable solution. A comparison of different sound/speech corpora for GMM training has shown better results for DB_2 (from the MSP-IMPROV speech material) as can be seen in Table 3.

The final objective results obtained by the presented automatic evaluation system show some differences when compared with the listening test results. The designed system marks the USEL synthesis (*Synt 1*) as the best one, the WaveRNN (*Synt 3*) as worse, and the LSTM with the WORLD vocoder (*Synt 2*) as the worst one. Contrary to it, the listening tests rated *Synt 3* as the best, *Synt 1* as worse, and *Synt 2* as the worst [8] – see the final graphical comparison for all four tested speakers in Fig. 6. It seems that the used speech features tend to detect mainly the signal fidelity (in the sense of a direct signal comparison), which is principally the best for USEL since it uses parts of original recordings. However, USEL suffers from concatenation problems [9] that are not present in DNN based synthesis. On the other hand, DNN yields worse acoustic signal properties because it simply generates speech from a model. Despite this fact, WaveRNN yields quite good acoustics as it acts as a DNN-based vocoder and thus generates speech by learning of direct mapping between parametric representation and speech samples through a complex network. For this reason, we suppose that the simpler LSTM with the WORLD vocoder is more smoothed and suppresses fast changes (presented for example in explosives) which are necessary for good naturalness of the generated speech signal. Even the result that *Synt 3* is better than *Synt 2* is good – according to our expectations. Evaluation of WaveRNN versus USEL methods can be more subjective in general.

Our future work will be focused on a detailed analysis of influence of different types of speech features in the input vectors for GMM classification on Euclidean 2/3D distances between originals and three types of synthesis. In addition, we will try to collect speech databases in the Czech (Slovak) languages with sentences labelled in the P-A scale, for next creation of GMM models used in continuous P-A classification. The Slovak language similar to Czech is also planned for use in the tested TTS systems described in [20] and in the proposed GMM-based automatic evaluation.

Acknowledgements. This work was been supported by the Czech Science Foundation GA CR, project No GA19-19324S (J. Matoušek, and J. Přibil), by the Slovak Scientific Grant Agency project VEGA 2/0003/20 (J. Přibil), and the COST Action CA16116 (A. Přibilová).

References

1. Norrenbrock, C.R., Hinterleitner, F., Heute, U., Möller, S.: Quality prediction of synthesized speech based on perceptual quality dimensions. Speech Commun. **66**, 17–35 (2015)
2. Hinterleitner, F., Moller, S., Norrenbrock, C., Heute, U.: Perceptual quality dimensions of text-to-speech systems. Proc. Interspeech **2011**, 2177–2180 (2011)
3. Vích, R., Nouza, J., Vondra, M.: Automatic speech recognition used for intelligibility assessment of text-to-speech systems. In: Esposito, A., Bourbakis, N.G., Avouris, N., Hatzilygeroudis, I. (eds.) Verbal and Nonverbal Features of Human-Human and Human-Machine Interaction. LNCS (LNAI), vol. 5042, pp. 136–148. Springer, Heidelberg (2008). https://doi.org/10.1007/978-3-540-70872-8_10
4. Cerňak, M., Rusko, M., Trnka, M.: Diagnostic evaluation of synthetic speech using speech recognition. In: Proceedings of the 16th International Congress on Sound and Vibration (ICSV 16) (2009)
5. Hinterleitner, F., Norrenbrock, C., Möller, S., Heute, U.: Predicting the quality of text-to-speech systems from a large-scale feature set. In: Proceedings of Interspeech 2013, ISCA, pp. 383–387 (2013)
6. Hunt, A.J., Black, A.W.: Unit selection in a concatenative speech synthesis system using a large speech database. In: Proceedings of the IEEE International Conference on Acoustics, Speech and Signal Processing (ICASSP), Atlanta (Georgia, USA), pp. 373–376 (1996)
7. Tihelka, D., et al.: Current state of text-to-speech system ARTIC: A decade of research on the field of speech technologies. In: Sojka, P., Horák, A., Kopeček, I., Pala, K. (eds.) TSD 2018. LNAI, vol. 11107, pp. 369–378. Springer, Cham (2018). https://doi.org/10.1007/978-3-030-00794-2_40
8. Vít, J., Hanzlíček, Z., Matoušek, J.: Czech speech synthesis with generative neural vocoder. In: Ekštein, K. (ed.) TSD 2019. LNCS (LNAI), vol. 11697, pp. 307–315. Springer, Cham (2019). https://doi.org/10.1007/978-3-030-27947-9_26
9. Vít, J., Matoušek, J.: Concatenation artifact detection trained from listeners evaluations. In: Habernal, I., Matoušek, V. (eds.) TSD 2013. LNCS (LNAI), vol. 8082, pp. 169–176. Springer, Heidelberg (2013). https://doi.org/10.1007/978-3-642-40585-3_22
10. Reynolds, D.A., Rose, R.C.: Robust text-independent speaker identification using gaussian mixture speaker models. IEEE Trans. Speech Audio Process. **3**, 72–83 (1995)
11. International Telecommunications Union: Method for the Subjective Assessment of Intermediate Quality Level of Coding Systems. ITU Recommendation ITU-R BS.1534-2 (2014)
12. Bradley, M.M., Lang, P.J.: Measuring emotion: The self-assessment manikin and the semantic differential. J. Behav. Ther. Exp. Psychiatry **25**, 49–59 (1994)
13. Grimm, M., Kroschel, K.: Evaluation of natural emotions using self assessment manikins. In: Proceedings of IEEE Automatic Speech Recognition and Understanding Workshop, San Juan, Puerto Rico, pp. 381–385 (2005)
14. Teodorescu, H.-N., Feraru, S.M.: A study on speech with manifest emotions. In: Matoušek, V., Mautner, P. (eds.) TSD 2007. LNCS (LNAI), vol. 4629, pp. 254–261. Springer, Heidelberg (2007). https://doi.org/10.1007/978-3-540-74628-7_34
15. Nicolau, M.A., Gunes, H., Pantic, M.: Continuous prediction of spontaneous affect from multiple cues and modalities in valence-arousal space. IEEE Trans. Affect. Comput. **2**(2), 92–105 (2011)
16. Bradley, M.M., Lang, P.J.: The international affective digitized sounds (2nd Edition; IADS-2): Affective ratings of sounds and instruction manual. Technical report B-3. University of Florida, Gainesville, Fl (2007)

17. Busso, C., et al.: MSP-IMPROV: An acted corpus of dyadic interactions to study emotion perception. IEEE Trans. Affect. Comput. **8**(1), 67–80 (2017)
18. Přibil, J., Přibilová, A., Matoušek, J.: Automatic evaluation of synthetic speech quality by a system based on statistical analysis. In: Sojka, P., Horák, A., Kopeček, I., Pala, K. (eds.) TSD 2018. LNCS (LNAI), vol. 11107, pp. 315–323. Springer, Cham (2018). https://doi.org/10.1007/978-3-030-00794-2_34
19. Nabney, I.T.: Netlab Pattern Analysis Toolbox, Release 3.3. Accessed 2 Oct (2015)
20. Matoušek, J., Tihelka D., Psutka J.: New Slovak unit-selection speech synthesis in ARTIC TTS system. In: Proceedings of the World Congress on Engineering and Computer Science, pp. 485–490 (2011)

Investigating the Effect of Emoji in Opinion Classification of Uzbek Movie Review Comments

Ilyos Rabbimov[1]([✉]), Iosif Mporas[2], Vasiliki Simaki[3], and Sami Kobilov[1]

[1] Applied Mathematics and Informatics Department,
Samarkand State University, Samarkand, Uzbekistan
ilyos.rabbimov91@gmail.com, kobsam@yandex.ru
[2] School of Engineering and Computer Science,
University of Hertfordshire, Hatfield, UK
i.mporas@herts.ac.uk
[3] Centre for Languages and Literature, Lund University, Lund, Sweden
vasiliki.simaki@englund.lu.se

Abstract. Opinion mining on social media posts has become more and more popular. Users often express their opinion on a topic not only with words but they also use image symbols such as emoticons and emoji. In this paper, we investigate the effect of emoji-based features in opinion classification of Uzbek texts, and more specifically movie review comments from YouTube. Several classification algorithms are tested, and feature ranking is performed to evaluate the discriminative ability of the emoji-based features.

Keywords: Opinion classification · Sentiment analysis · Emoji

1 Introduction

Over the last decades the use of Internet has dramatically increased with online activities like e-commerce, social media and blogs becoming extremely popular. Extraction of information from structured or unstructured online text is performed using text mining methodologies with applications among others in sentiment analysis, opinion mining [1–3], emotions [4] and stance classification [5]. In opinion classification, the positive or the negative opinion of users is automatically identified in data usually extracted from social media platforms such as Twitter, YouTube, Reddit and Facebook. The topics of discussion vary as well, and there are various studies that analyze movie reviews [6], political debates [7], the presence of offensive language in online texts [8] etc. Most of the methodologies for automatic opinion classification that have been proposed in the literature are based on machine learning algorithms for classification, such as support vectors machines (SVMs) [1, 9, 10], Bayesian classifiers [1, 11], decision trees [1, 6] and neural networks [1, 12, 13]. In those approaches, the users' posts are represented by vectors of text features like language model based [1, 14, 15], word level [1, 6] and part-of-speech-based [1] statistical parameters. Other

© Springer Nature Switzerland AG 2020
A. Karpov and R. Potapova (Eds.): SPECOM 2020, LNAI 12335, pp. 435–445, 2020.
https://doi.org/10.1007/978-3-030-60276-5_42

approaches are lexicon-based that rely on the presence in the data of words characterized as positive or negative [6, 12, 16, 17]. More recent approaches use word embeddings-based solutions to address the task [13, 14, 17].

In online text, and apart from the sentences consisting purely of word sequences, users frequently use emoticons or emoji to express themselves, emphasize and reinforce or mitigate the illocutionary force of their text. Emoticons are combinations of letters, numbers and symbols available on the keyboard, while emoji are pictures rather than typographic approximations of facial expressions, both expressing user's mood. The role of emoticons and emoji in sentiment analysis has been examined in previous studies. In [18], emoji sentiment ranking and sentiment map of the 751 most frequently used emoji was developed for automated sentiment analysis based on human annotators and tweets in 13 European languages. The authors found that most emoji are positive, especially the most popular ones, and the emotional perception of tweets changes significantly depending the presence or not of emoji in the text. In [19], a sociolinguistic study was presented exploiting emoji information for sentiment analysis. Similarly, in [20], the effect of emoji in sentiment analysis was studied, observing that taking into account emoji in sentiment analysis improves the recognition accuracy of the sentiment. The role of emoticons in the overall meaning of the text and their use in lexicon-based sentiment analysis was investigated in [21], where Dutch tweets and forum messages contained at least one emoticon were used. The results showed that the overall document polarity classification accuracy significantly improved when emoticons are considered. In [22], the problem of sarcasm detection in tweets using emoji was studied. In [23], the emoji2vec pre-trained embeddings of 1,661 unicode emoji was presented, and sentiment analysis on twitter posts was performed using word2vec and emoji2vec. In [24, 25], the effect of emoji in twitter posts was studied and in [26], the effect of using combined emoji-based features with textual features of Arabic tweets on sentiment classification task was presented.

While several corpora and language processing tools have been developed for the major languages of the world (i.e. languages spoken in many countries and/or by significant proportion of global population), not many language resources – that is the key component for any language-based technology – exist for other languages [27]. One of the less-resourced languages is Uzbek, which is the second most widely spoken Turkic language after the Turkish language. In Uzbek language until the first decades of the 20th century an Arabic-based script was used (the Yaña imlâ alphabet), then from 1928 to 1940 Latin-based Yañalif was used officially, and from 1940 the Cyrillic alphabet became the official script of the Uzbek language. In 1991, Uzbekistan's official script became the Yañalif-based Latin alphabet again. Despite the official status of the Latin script, the Cyrillic alphabet is still widespread and used in various occasions, especially by people who received education in the Soviet Union before the 1990's.

In the existing literature, there are natural language processing studies in the Uzbek language [28–35], but very few can be found in Uzbek opinion mining. More specifically, in [36], the authors present an opinion mining dataset with 4,300 review comments about the top 100 applications from Google Play App Store used in Uzbekistan. The data is annotated according to the comment's positive/negative polarity, and baseline binary opinion classification results using support vector machines, logistic regression, recurrent neural networks and convolutional neural networks are presented. The same corpus and algorithms are used in [14], where the authors evaluate deep learning models for binary (positive vs. negative) opinion classification. In [37], a multilingual collection of sentiment lexicons including Uzbek is presented. In addition, there are few papers for sentiment/opinion mining in Turkish [38–40] and Kazakh [41, 42], which are languages belonging to the same language family. No emoji-related study in Uzbek opinion mining has been reported in the literature.

In this paper, we investigate the effect of emoji-based features in opinion classification of Uzbek movie review comments posted on YouTube. The reminder of the paper is organized as follows. In Sect. 2, the evaluated architecture for opinion classification is presented. In Sect. 3, the experimental setup is described. In Sect. 4, the experimental results are presented. Finally, Sect. 5 concludes our paper.

2   Uzbek Movie Reviews Opinion Classification

The architecture we used for the opinion classification of YouTube movie review comments and the investigation of the importance of emoji in the identification of user's opinion follows a standard approach. This approach is adopted in most opinion mining studies found in the bibliography and consists of the preprocessing of users' posts, the extraction of features and the classification experiments. The block diagram of the architecture is presented in Fig. 1.

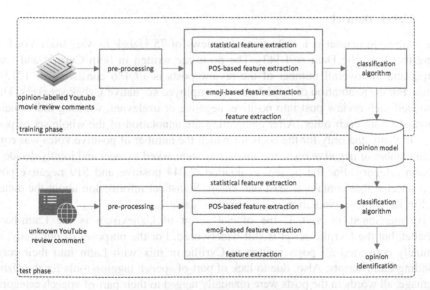

Fig. 1. Block diagram of the architecture for opinion classification of YouTube movie reviews, using text features based on word-based, part-of-speech-based and emoji-based statistics.

During the training phase, a set of movie review comments in Uzbek, extracted from YouTube, with known opinion labels are used to train an opinion classification model. The opinion labels are attributed manually by annotators that are native Uzbek speakers. More specifically, each review post is initially preprocessed and then features are extracted thus representing it by a feature vector. The text features are divided into statistical features, part-of-speech-based (henceforth, POS-based) features and emoji-based features, with the first ones consisting of statistical parameters on word level, the POS-based features consisting of statistical parameters on part-of-speech level and the last one consisting of statistical and opinion characteristic parameters of emoji. In the test phase, new movie reviews are preprocessed, and features are extracted similarly as in the training phase. The pretrained opinion classification model is used to classify the unknown movie review as positive or negative.

In the present evaluation, we have considered two opinion types, namely positive and negative, and not a scaled opinion ranking thus the classification models are binary. The architecture is generalized and thus can be used in multiclass opinion classification as well as in opinion mining of text from other Internet sources.

3   Experimental Setup

The architecture presented in Sect. 2 for opinion classification of Uzbek YouTube movie reviews was evaluated using the dataset, the text features and the classification algorithms described below.

3.1   Uzbek Dataset

The evaluation dataset is a collection of reviews of 75 Uzbek movies from YouTube using the YouTube Data API [43]. The posts are written in both Cyrillic and Latin scripts and the overall number of the review posts is 17,486 consisting of 121,941 words. For the annotation of the dataset, we employed six native Uzbek speakers. They annotated each review post into positive, negative or irrelevant, working individually, separately from each other. After completing the annotation of the whole set of posts given to them, and only for the posts for which the number of positive votes was equal to the number of negative ones, the annotators were asked to meet and jointly decide an opinion category. For the present evaluation, 2,044 positive and 519 negative posts were selected that contain at least one emoji. Statistical information about the dataset used in the present evaluation is presented in Table 1.

As mentioned in Sect. 1, the official script of Uzbekistan is the Latin-based alphabet, but the Cyrillic script is still widely used. For the purposes of our study, we manually converted all posts written in Cyrillic or mix with Latin into their corresponding Latin scripts. Also, due to lack of part-of-speech tagging tools for the Uzbek language, all words in the posts were manually tagged to their part-of-speech categories by two expert Uzbek linguists.

Table 1. Statistical information of the evaluation dataset.

Characteristic	Statistical information
# posts in Cyrillic	1,453
# posts in Latin	1,100
# posts in mixed Cyrillic – Latin	10
total # posts	2,563
# positive posts	2,044
# negative posts	519
minimum # emoji per post	1
maximum # emoji per post	213
average # emoji per post	5.26

3.2 Feature Extraction

For each preprocessed movie review post, the statistical, POS-based and emoji-based features were calculated. More specifically, the calculated features are:

Statistical Features. Total number of characters; total number of characters without spaces; number of special characters ('(', ')', '[', ']', '{', '}', '-', '/', '&', '|', '–', '—', '#', '%', '+', '*', '@', '$', '~', '=', '_', '«', '»', '<', '>', '^'); number of lower case characters; number of upper case characters; number of digits characters; number of all words; number of unique words; mean length of all unique words; maximum length of all words; minimum length of all words; mean length of all words; standard deviation of the length of all words; variance of the length of all words; kurtosis of the length of all words; skewness of the length of all words; percentile 25% of the length of all words; percentile 50% (median) of the length of all words; percentile 75% of the length of all words; number of punctuation characters ('.', ',', '!', '?', ':', ';'); number of words with length less than 4 characters; number of the hapax-legomena; number of the hapax-dislegomena.

POS-Based Features. Number of nouns; number of proper nouns; number of verbs; number of adjectives; number of numerals; number of pronouns; number of adverbs; number of helping words; number of coordinating conjunctions; number of subordinating conjunctions of review; number of modal words; number of imitative words; number of interjections; number of auxiliaries; number of other words (x) like undefined or incomprehensible/meaningless cases.

Emoji-Based Features. Number of emoji; average of sentiment score of all emoji per post [18]; number of positive emoji; number of negative emoji.

The number of statistical features is 23, the number of POS-based features is 15 and the number of emoji-based features is 4. The total dimensionality of the feature vector is equal to 42.

3.3   Classification Algorithms

For the classification stage, we used different well-known machine learning algorithms that are extensively used in several text classification tasks. In particular, we used:

Instance Based Classifier (IBk). A k-nearest neighbor classifier with linear search of the nearest neighbor and without weighting of the distance [44].

Neural Networks (NN). A multilayer perceptron neural network [45] with two hidden layers architecture (30 sigmoid nodes per hidden layer) trained with 5,000 iterations, using the back-propagation algorithm.

Support Vector Machines (SVM). The support vector machines using the sequential minimal optimization algorithm [46], which was tested using two different kernels, namely the radial basis kernel (rbf) and polynomial kernel (poly).

Decision Trees. Three tree algorithms were tested, namely the pruned C4.5 decision tree (J48) [47], the random forest (RandForest) [48] constructing a multitude of decision trees and the fast decision tree learner (RepTree) [49] that builds a decision tree using information gain or variance and prunes it using reduced-error pruning with back-fitting.

Bayesian Classifier (BayesNet). We used the Bayes network learning [49], with simple estimator (alpha = 0.5) and the K2 search algorithm (maximum number of parents = 1), which is a probabilistic graphical model that represents a set of random variables and their conditional dependencies via a directed acyclic graph and the naive Bayes multinomial updateable, in which feature vectors represent the frequencies with which certain events have been generated by a multinomial.

All classifiers were implemented using the WEKA software [45]. For all algorithms, the free parameters were empirically selected, while parameter values not reported here were kept in their default values. For all classification algorithms two versions of opinion classification models were trained, one including emoji-based features (dimensionality equal to 42) and one without emoji-based features (dimensionality equal to 38).

4   Experimental Results

The architecture presented in Sect. 2 for opinion classification from YouTube movie review posts was evaluated according to the experimental setup described in Sect. 3. The performance of the evaluated algorithms was measured in terms of classification accuracy, i.e.

$$accuracy = \frac{TP + TN}{TP + FP + TN + FN},$$ (1)

where *TP* are the true positives, *TN* are the true negatives, *FP* are the false positives and *FN* are the false negatives. The opinion classification accuracy for all evaluated classification algorithms and for two setups, namely with and without emoji-based text features, are presented in Table 2. In both setups, 10-fold cross validation was followed to avoid the overlap between the training and the test data. The best performance is indicated in bold.

Table 2. Opinion classification accuracy for different classification algorithms, with and without emoji-based text features.

Classification algorithm	Accuracy (%) without emoji-based features	Accuracy (%) with emoji-based features
IBk	75.89	80.26
NN	79.32	82.72
SVM-poly	79.63	84.55
SVM-rbf	79.75	84.39
J48	77.10	83.46
RandForest	80.69	**85.25**
REPTree	78.62	84.12
BayesNet	64.30	75.34

As can be seen in Table 2, the best performing classification algorithm is the Random Forest (85.25%) when using emoji-based features, followed by REPTree and SVMs performing approximately 1% worse. The use of emoji-based features improved significantly the opinion classification accuracy across all evaluated algorithms. For the best performing classification algorithm (RandForest decision tree), the opinion classification accuracy improvement was approximately 5% when using the emoji-based features.

In a further step, we evaluated the discriminative ability of the 42 calculated features using the ReliefF [50] feature ranking algorithm. The feature ranking results for the top-10 ranked features are shown in Table 3.

Table 3. Opinion classification feature ranking of the top 10 ranked features using the ReliefF algorithm.

Ranking	ReliefF score	Feature name
1	0.0671251	Average sentiment score of all emoji per post
2	0.0192251	Skewness of the length of all words
3	0.0185649	Number of adjectives
4	0.017307	Minimum length of all words
5	0.0162653	Maximum length of all words
6	0.0157003	Mean length of all words
7	0.0156075	Mean length of all unique words
8	0.0148784	Percentile 25% of the length of all words
9	0.014131	Number of words with length less than 4 characters
10	0.0139635	Percentile 50% (median) of the length of all words

As can be seen in Table 3, the most discriminative feature is the 'average sentiment score of all emoji per post', indicating the importance of emoji-based information in opinion classification. The remaining three emoji-based features where ranked in positions 24 ('number of negative emoji', ReliefF score: 0.0034666), 30 ('number of positive emoji', ReliefF score: 0.0019034) and 31 ('number of emoji', ReliefF score: 0.0017307), i.e. not in the first ten ranked features, however having positive ReliefF score which indicated that they also carry information with respect to users' opinion in the YouTube posts. The results from Tables 2 and 3 show that the positive effect of emoticons and emoji in opinion classification of other languages reported in previous studies [18–26] is also valid in the case of Uzbek opinion mining.

5   Conclusion

The identification of user's opinion in social media posts is of increasing interest in the research community as well as in commercial applications and services. Users often use emoticons and emoji apart from words in order to express their opinions. We investigated the effect of emoji-based features in opinion classification of Uzbek text from movie review comments from YouTube, and we evaluated their discriminative ability using the ReliefF feature ranking algorithm. The experimental results showed that Random Forest decision tree outperformed with accuracy equal to 85.25%, and the emoji-based features improved the opinion classification accuracy by approximately 5%. Also, the feature ranking evaluation showed that the most discriminative feature is the 'average sentiment score of all emoji per post'. The results in Uzbek opinion mining are in agreement with previous similar studies in other languages.

Acknowledgement.   The authors are thankful to the six annotators and two linguists who contributed to the creation and processing of the dataset used in this work. This work was partially supported by the "El-Yurt Umidi" Foundation under the Cabinet of Ministers of the Republic of Uzbekistan (Internship Programme no. ST-2019-0080).

References

1. Koumpouri, A., Mporas, I., Megalooikonomou, V.: Opinion recognition on movie reviews by combining classifiers. In: Ronzhin, A., Potapova, R., Fakotakis, N. (eds.) SPECOM 2015. LNCS (LNAI), vol. 9319, pp. 309–316. Springer, Cham (2015). https://doi.org/10.1007/978-3-319-23132-7_38
2. Poria, S., Cambria, E., Gelbukh, A.: Aspect extraction for opinion mining with a deep convolutional neural network. Knowl. Based Syst. **108**, 42–49 (2016)
3. Sun, S., Luo, C., Chen, J.: A review of natural language processing techniques for opinion mining systems. Inf. Fusion **36**, 10–25 (2017)
4. Dvoynikova, A., Verkholyak, O., Karpov, A.: Analytical review of methods for identifying emotions in text data. In: Proceedings of the III International Conference on Language Engineering and Applied Linguistics (PRLEAL-2019), pp. 8–21. CEUR-WS (2020)

5. Simaki, V., Paradis, C., Kerren, A.: Stance classification in texts from blogs on the 2016 British referendum. In: Karpov, A., Potapova, R., Mporas, I. (eds.) SPECOM 2017. LNCS (LNAI), vol. 10458, pp. 700–709. Springer, Cham (2017). https://doi.org/10.1007/978-3-319-66429-3_70

6. Koumpouri, A., Mporas, I., Megalooikonomou, V.: Evaluation of four approaches for "sentiment analysis on movie reviews: the Kaggle competition. In: Proceedings of the 16th International Conference on Engineering Application of Neural Networks (INNS), pp. 1–5. ACM (2015)

7. Simaki, V., et al.: Annotating speaker stance in discourse: the Brexit Blog corpus. In: Corpus Linguistics and Linguistic Theory, vol. 1, ahead-of-print (2017)

8. Çöltekin, Ç.: A corpus of Turkish offensive language on social media. In: Proceedings of the 12th Language Resources and Evaluation Conference, pp. 6176–6186. ELRA (2020)

9. Sunitha, P.B., Joseph, S., Akhil, P.V.: A study on the performance of supervised algorithms for classification in sentiment analysis. In: TENCON 2019, pp. 1351–1356. IEEE (2019)

10. Rinaldi, E., Musdholifah, A.: FVEC-SVM for opinion mining on Indonesian comments of youtube video. In: Proceedings of the 2017 ICoDSE, pp. 1–5. IEEE, Indonesia (2017)

11. Saif, H., Fernández, M., He, Y., Alani, H.: On stopwords, filtering and data sparsity for sentiment analysis of twitter. In: Proceedings of the Ninth International Conference on Language Resources and Evaluation (LREC 2014), pp. 810–817. ELRA, Iceland (2014)

12. Cunha, A.A.L., Costa, M.C., Pacheco, M.A.C.: Sentiment analysis of Youtube video comments using deep neural networks. In: Rutkowski, L., Scherer, R., Korytkowski, M., Pedrycz, W., Tadeusiewicz, R., Zurada, J.M. (eds.) ICAISC 2019. LNCS (LNAI), vol. 11508, pp. 561–570. Springer, Cham (2019). https://doi.org/10.1007/978-3-030-20912-4_51

13. Sido, J., Konopík, M.: Curriculum learning in sentiment analysis. In: Salah, A.A., Karpov, A., Potapova, R. (eds.) SPECOM 2019. LNCS (LNAI), vol. 11658, pp. 444–450. Springer, Cham (2019). https://doi.org/10.1007/978-3-030-26061-3_45

14. Kuriyozov, E., Matlatipov, S., Alonso, M., Gómez-Rodríguez, C.: Deep learning vs. classic models on a new Uzbek sentiment analysis dataset. In: Proceedings of the Human Language Technologies as a Challenge for Computer Science and Linguistics, pp. 258–262 (2019)

15. Jagdale, R.S., Shirsat, V.S., Deshmukh, S.N.: Sentiment analysis on product reviews using machine learning techniques. In: Mallick, P.K., Balas, V.E., Bhoi, A.K., Zobaa, A.F. (eds.) Cognitive Informatics and Soft Computing. AISC, vol. 768, pp. 639–647. Springer, Singapore (2019). https://doi.org/10.1007/978-981-13-0617-4_61

16. Esuli, A., Sebastiani, F.: Sentiwordnet: a publicly available lexical resource for opinion mining. In: Proceedings of the 5th Conference on Language Resources and Evaluation (LREC06), pp. 417–422. ELRA (2006)

17. Rezaeinia, S., Rahmani, R., Ghodsi, A., Veisi, H.: Sentiment analysis based on improved pre-trained word embeddings. Expert Syst. Appl. **117**, 139–147 (2019)

18. Novak, P.K., Smailović, J., Sluban, B., Mozetič, I.: Sentiment of emojis. PloS One **10**(12), e0144296 (2015)

19. Guibon, G., Ochs, M., Bellot, P.: From emojis to sentiment analysis (2016)

20. Shiha, M., Ayvaz, S.: The effects of emoji in sentiment analysis. IJCEE **9**(1) (2017)

21. Hogenboom, A. et al.: Exploiting emoticons in sentiment analysis. In: Proceedings of the 28th Annual ACM Symposium on Applied Computing, pp. 703–710. ACM, New York (2013)

22. Karthik, V., Nair, D., Anuradha, J.: Opinion mining on emojis using deep learning techniques. Procedia Comput. Sci. **132**, 167–173 (2018)

23. Eisner, B., Rocktäschel, T., Augenstein, I., Bošnjak, M., Riedel, S.: Emoji2vec: learning emoji representations from their description. arXiv preprint arXiv:1609.08359 (2016)

24. Dandannavar, P.S., Mangalwede, S.R., Deshpande, S.B.: Emoticons and their effects on sentiment analysis of twitter data. In: Haldorai, A., Ramu, A., Mohanram, S., Onn, C.C. (eds.) EAI International Conference on Big Data Innovation for Sustainable Cognitive Computing. EICC, pp. 191–201. Springer, Cham (2020). https://doi.org/10.1007/978-3-030-19562-5_19

25. Wegrzyn-Wolska, K., Bougueroua, L., Yu, H., Zhong, J.: Explore the effects of emoticons on Twitter sentiment analysis. Comput. Sci. Inf. Technol. **2**, 65 (2016)

26. Al-Azani, S., El-Alfy, E.S.M.: Combining emojis with Arabic textual features for sentiment classification. In: Proceedings of the 9th International Conference on Information and Communication Systems (ICICS), pp. 139–144. IEEE (2018)

27. Besacier, L., Barnard, E., Karpov, A., Schultz, T.: Automatic speech recognition for under-resourced languages: a survey. Speech Commun. **56**, 85–100 (2014)

28. Li, X., Tracey, J., Grimes, S., Strassel, S.: Uzbek-English and Turkish-English morpheme alignment corpora. In: Proceedings of the 10th LREC 2016, pp. 2925–2930. ELRA, Portorož (2016)

29. Baisa, V., Suchomel, V.: Large corpora for Turkic languages and unsupervised morphological analysis. In: Proceedings of the 8th International Conference on Language Resources and Evaluation (LREC 2012), pp. 28–32. ELRA, Turkey (2012)

30. Ismailov, A. Jalil, M.M.A., Abdullah Z., Rahim N.H.A.: A comparative study of stemming algorithms for use with the Uzbek language. In: Proceedings of the 3rd International Conference on Computer and Information Sciences (ICCOINS), pp. 7–12. IEEE (2016)

31. Xu, R., Yang, Y., Liu, H., Hsi, A.: Cross-lingual text classification via model translation with limited dictionaries. In: Proceedings of the 25th ACM International on Conference on Information and Knowledge Management (CIKM 2016), pp. 95–104. ACM (2016)

32. Abdurakhmonova, N.: Dependency parsing based on Uzbek Corpus. In: Proceedings of the International Conference on Language Technologies for All (LT4All) (2019)

33. Chew, Y.C., Mikami, Y., Marasinghe, C.A., Nandasara, S.T.: Optimizing n-gram order of an N-gram based language identification algorithm for 63 written languages. Int. J. Adv. ICT Emerg. Reg. (ICTer) **2**(2), 21–28 (2009)

34. Uzbek text corpora page of Sketch Engine. https://www.sketchengine.eu/corpora-and-languages/uzbek-text-corpora. Last Accessed 10 Jun 2020

35. Kuriyozov, E., Doval, Y., Gómez-Rodríguez, C.: Cross-lingual word embeddings for Turkic languages. In: Proceedings of the 12th LREC 2020, pp. 4047–4055. ELRA (2020)

36. Kuriyozov, E., Matlatipov, S.: Building a new sentiment analysis dataset for Uzbek language and creating baseline models. Multi. Digit. Publishing Inst. Proc. **21**(1), 37 (2019)

37. Chen, Y., Skiena, S.: Building sentiment lexicons for all major languages. In: Proceedings of the 52nd Annual Meeting of the Association for Computational Linguistics (vol. 1: Long Papers), pp. 383–389. ACL, Baltimore (2014)

38. Kaya, M., Guven, F., Toroslu, I.H.: Sentiment analysis of Turkish political news. In: Proceedings of the 2012 IEEE/WIC/ACM International Joint Conferences on Web Intelligence and Intelligent Agent Technology, vol. 1, pp. 174–180. IEEE (2012)

39. Dehkharghani, R., Yanikoglu, B., Saygin, Y., Oflazer, K.: Sentiment analysis in Turkish at different granularity levels. Nat. Lang. Eng. **23**(4), 535–559 (2017)

40. Vural, A.G., Cambazoglu, B.B., Senkul, P., Tokgoz, Z.O.: A framework for sentiment analysis in Turkish: application to polarity detection of movie reviews in Turkish. In: Gelenbe, E., Lent, R. (eds.) Computer and Information Sciences III, pp. 437–445. Springer, Cham (2012). https://doi.org/10.1007/978-1-4471-4594-3_45

41. Yergesh, B., Bekmanova, G., Sharipbay, A., Yergesh, M.: Ontology-based sentiment analysis of kazakh sentences. In: Gervasi, O., et al. (eds.) ICCSA 2017. LNCS, vol. 10406, pp. 669–677. Springer, Cham (2017). https://doi.org/10.1007/978-3-319-62398-6_47

42. Sakenovich, N.S., Zharmagambetov, A.S.: On one approach of solving sentiment analysis task for Kazakh and Russian languages using deep learning. In: Nguyen, N.-T., Manolopoulos, Y., Iliadis, L., Trawiński, B. (eds.) ICCCI 2016. LNCS (LNAI), vol. 9876, pp. 537–545. Springer, Cham (2016). https://doi.org/10.1007/978-3-319-45246-3_51
43. YouTube Data API documentation page. https://developers.google.com/youtube/v3/docs/commentThreads. Last Accessed 10 Jun 2020
44. Aha, D., Kibler, D.: Instance-based learning algorithms. Mach. Learn. 6, 37–66 (1991)
45. Frank, E., Hall, M.A., Ian, H.: The WEKA Workbench. Online Appendix for "Data Mining: Practical Machine Learning Tools and Techniques", Morgan Kaufmann, (2016)
46. Keerthi, S.S., Shevade, S.K., Bhattacharyya, C., Murthy, K.R.K.: Improvements to Platt's SMO algorithm for SVM classifier design. Neural Comput. 13(3), 637–649 (2001)
47. Quinlan, R.: C4.5: Programs for Machine Learning. Morgan Kaufmann Publishers, San Francisco (1993)
48. Breiman, L.: Random forests. Mach. Learn. 45, 5–32 (2001)
49. Bouckaert, R.R.: Bayesian networks in Weka. Technical report 14/2004. Computer Science Department. University of Waikato (2004)
50. Robnik-Šikonja, M., Kononenko, I.: An adaptation of Relief for attribute estimation in regression. In: Machine Learning: Proceedings of ICML 1997, vol. 5, pp. 296–304 (1997)

Evaluation of Voice Mimicking
Using I-Vector Framework

Rajeev Rajan[✉], Abhijith Girish, Adharsh Sabu,
and Akshay Prasannan Latha

College of Engineering, Trivandrum, Thiruvananthapuram, Kerala, India
rajeev@cet.ac.in

Abstract. Intentional speech modifications by mimicking cause degradation in speaker verification systems. Improving the robustness of speaker verification systems against most competent mimicked speech is extremely important. The fusion of mel-frequency cepstral coefficient (MFCC)-based i-vectors and phonation-based features is utilized to identify the most competent imitator who imitates a target voice in the proposed work. Two scoring mechanisms namely, deep neural network (DNN) and dynamic time warping (DTW) has been employed. The phonation features are computed from the characteristics related to perturbation measures of fundamental frequency and amplitude of speech The proposed system evaluates the competence of artists in mimicking voices target voices and ranks them according to the metric obtained from the scoring mechanism. The results of the scoring frameworks are evaluated through a perception test conducted on twenty listeners. If the artist with the highest mean opinion score is identified as rank-1 by the proposed system, a hit occurs. The performance is evaluated using top X-hit criteria on a mimicry dataset. On the top-2 rate, the DNN-based experiment reports an accuracy of 72.72% with i-vector-phonation feature fusion.

Keywords: Mimicry · MFCC · Phonation · I-vector · Perception

1 Introduction

Mimicry comprises imitation of public figures, actors, actresses, singers, birds, animals, and sound of music instruments[1]. The entertainment industry witnessed an impressive history of people who have made mimicry as their profession with their skills of impression. Synchronised imitation, covering the voice as well as gestures has the highest entertainment value for the audience. Voice disguise by mimicking is often employed in voice-based crimes by perpetrators who try to evade identification while sounding genuine. From the experimental results [9], it was seen that the impersonator mimicking the target speakers caused significant performance degradation as compared to that of speaker verification systems

---

[1] http://www.voiceartistes.com/.

A. Karpov and R. Potapova (Eds.): SPECOM 2020, LNAI 12335, pp. 446–456, 2020.
https://doi.org/10.1007/978-3-030-60276-5_43

without mimicked voices. Since there are a few actual cases of fraud by mimicry in the courts, it is reasonable to say that impersonation attack is a challenging issue in a speaker verification system [5].

In mimicking speech, the imitator modifies the position of articulators like lips, tongue, speech duration, and pitch to reproduce another person's voice [14]. Since the primary objective is perceptual, the mimicry itself tends to focuson or to even overemphasize, perceptually distinctive aspects of the target's speech rather than the fine details. Mimics are also reportedly able to imitate key formant patterns for several phonemes [18]. For example, the variation of formant locations for a speech segment during voice mimicking is illustrated in Fig. 1. The variation of the formant locations for the best imitator from the target is minimal in most of the time as pointed out in [5]. The formant frequencies of the mimicked vowels are closer to the target than those of the artist's voice. Since professional artists are trying to imitate a target voice during performances, an automatic method to identify the best artist who mimics the target voice intact is worth assessing the aesthetic values. The proposed system mainly evaluates the ability of imitators in mimicking voices by analyzing the mimicked speech using the feature-scoring framework.

Analysis of imitation using auditory, acoustic and prosody features can be found in [20]. The study in [20] shows that a professional voice imitator can approximate the system parameters of a well-known target speaker. Intonation duration and energy features are effectively used from segmented phrases in [14]. In [19], relative phase information computed from a Fourier spectrum is used to detect human and spoofed speech. Various features at segmental, suprasegmental and subsegmental levels of natural and mimicked speech are investigated in [4]. Evidence and literature suggest that large scale characteristics like prosody and style are also well-imitated in impersonations [12,18].

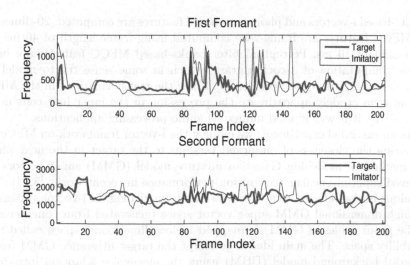

Fig. 1. First and second formant of target and imitator for a speech segment.

The i-vector modelling has been effectively used in speaker recognition [8], emotion recognition [13] and speech-music classification [7]. The state-of-the-art i-vector subspace modelling provides the benefit of modelling both the intra-domain and inter-domain variabilities into the same low dimensional space [10,21]. The effect of i-vector computed from the mel-frequency cepstral coefficient and the role of phonation features have bee experimented in the proposed experiment for evaluating mimicked speech.

The rest of the paper is organized as follows: Sect. 2 describes the proposed system. The performance evaluation is described in Sect. 3. The analysis of results is given in Sect. 4. Finally, the paper is concluded in Sect. 5.

2   Proposed System

MFCC, i-vectors (i_{MFCC}) and phonation-based features are frame-wise computed and inputted to the scoring framework in the feature extraction phase. For the scoring mechanism standard DTW and DNN methodologies have been adopted. DTW finds the closest mimicked version to the reference by computing the distance between the target and test descriptors. For the DNN experiment, each node in the output of trained-DNN (trained with target voice feature vectors) corresponds to one celebrity. During the testing phase, the probability value (score) at the corresponding target node in the output layer is examined by inputting all the mimicked versions (artist's voices) of that target and rank based on the scores. The artist, who is getting the highest probability score at the target node, is assumed to be the best artist among five. The performance is also compared with a baseline support vector machine (SVM)-framework.

2.1   Feature Extraction

MFCC-based i-vectors and phonation-based features are computed. 20-dimensional MFCC features are frame-wise computed using frame-length of 40 ms and frame-shift of 10 ms. Perceptual filter banks-based MFCC features are based on the computation of a cochleagram, which in some sense try to model the frequency selectivity of the cochlea [6]. Since the transformations in the MFCC computation crudely approximate the processing in the inner hair cells in the cochlea [16], it is widely used numerous audio processing applications.

As an extended experiment, we employ the i-vector framework on MFCC for measuring the closeness of mimicked versions to the target in the next phase. The method of modelling Gaussian mixture model (GMM) super vectors has achieved superior speaker recognition performance in recent works [3,21]. This technique is also referred to as total variability modelling. In i-vector system [2], the high dimensional GMM super vector space (generated from concatenating all the mean values of GMM) is mapped to low dimensional space called total variability space. The main idea is to adapt the target utterance GMM from a universal background model (UBM) using the eigenvoice adaption introduced in [11]. The target GMM super vector can be viewed as shifted from the UBM. Formally, a target GMM super vector M can be written as:

$$M = m + Tw \tag{1}$$

where m represents the UBM super vector, T is a low dimensional rectangular total variability matrix, and w is termed as i-vector. Using training data, the UBM and total variability matrix will be modeled expectation maximization. 10-dimensional i-vectors (i_{MFCC}) are computed for each utterance in the proposed experiment.

Prosody, phonation, articulation, and intelligibility are considered as the more important speech dimensions. Prosody shows the variation of loudness, pitch, and timing in speech production. Phonation is defined as the capability to make the vocal folds vibrate to produce sound; articulation comprises changes in position, stress, and shape of the organs, tissues, and limbs involved in speech production. Prosodic features computed from Legendre polynomial expansions had been utilized for the valuation of mimicking speech evaluation in [14]. As a different approach, phonation features are introduced in the proposed work. Phonation modes have three dimensions, namely, pitch, loudness and laryngeal adjustments. In the proposed work, phonation features have been computed as complementary information. Seven descriptors [15], namely, first derivative and second derivative of the fundamental frequency, jitter (temporal perturbation of the fundamental frequency), shimmer (temporal perturbation of the amplitude of the signal), amplitude perturbation quotient (APQ), and pitch perturbation quotient (PPQ), and logarithmic energy are computed with framesize of 40 ms and hop size of 10 ms. Further to the perturbation measures, the degree of unvoiced is also included. Feature matrix with 29 features formed with four metrics mean, standard deviation, skewness and kurtosis are extracted from the seven descriptors.

Jitter and shimmer represent temporal perturbations in the frequency and amplitude of speech, respectively. APQ measures the long–term variability of the peak-to-peak amplitude of the speech signal. Its computation includes a smoothing factor of consecutive voice periods, and it is calculated as the absolute average difference between the amplitude of a frame, and the amplitudes averaged over its neighbours, divided by the average amplitude. PPQ measures the long–term variability of the fundamental frequency, with a smoothing factor of five periods [15]. The degree of unvoiced is calculated as the ratio between the duration of unvoiced frames and the total duration of the utterance. It is calculated upon sustained phonations, thus it gives information about the amount of aperiodicity in the phonation [15].

2.2 Scoring Framework

Two types of scoring mechanisms have been employed, namely, dynamic time warping and DNN. Earlier, the kernel of automatic speech recognition (ASR) was based on feature-based template matching called dynamic time warping (DTW).

It belongs to the category technique of dynamic programming (DP). DTW algorithm compares the parameters of an unknown word with the parameters of one reference template. In the proposed experiment, the distance between the target and mimicked descriptors are computed using DTW and ranked based on the distance metric. The typical diagram of computing the DTW with i-vectors is shown in Fig. 2.

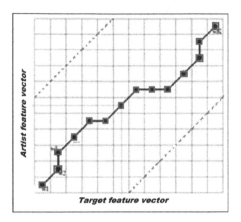

Fig. 2. Distance computation between target feature vectors and artist's feature vectors using dynamic time warping.

In the second phase, the deep learning strategy is incorporated. The recent success of deep neural networks (DNNs) for automatic speech recognition (ASR) motivated the application of DNNs to the proposed task. Our DNN is based on four hidden layers with the number of nodes 64, 64, 64, and 32 per layer. Stochastic gradient descent (SGD) algorithm is used for optimization. Relu and softmax have been chosen as the activation function for the hidden layer and output layer, respectively. Support vector machine (SVM)-based framework is used as a baseline system.

3   Performance Evaluation

3.1   Dataset

The proposed experiment is conducted with audio samples recorded in studio environments. In addition to the samples obtained from the corpus used in the study of Mary *et al.* [14], we recorded more test cases to improve the efficacy of the experimental setup. In our experiment, 22 celebrities were tested against professional artists for text-independent cases. Artists were given the flexibility to speak any text of their choice while mimicking the celebrities. Data were collected at a sampling frequency of 44.1 kHz with a resolution of 16 bits per sample. The utterances were from the Malayalam language spoken in the southern part of India.

3.2   Perceptual Evaluation

A subjective quality evaluation is conducted with twenty listeners using mean opinion score test. Listeners are presented with a group of speech stimuli consisting of the target stimuli and the mimicked stimuli. Listeners are asked to rate the mimicked stimuli based on voice similarity. In the perception test, listeners are presented with the target speech stimuli and asked to rate the mimicked stimuli on a 5 point numerical scale, with one indicating highly dissimilar and a five indicating highly similar. Mean opinion score for each mimicked utterance is computed by taking the average of the scores given by all the twenty listeners. The artist who gets the maximum score is identified as the best one in mimicking the celebrity well. We have also collected the feed-backs from the listeners. It is observed that the similarity of voice characteristics, rather than the sentences uttered, influenced them in rating the quality of the artist's voice.

3.3   Experimental Framework

i_{MFCC} are extracted using 128 mixture GMM built using MFCC (delta and delta-delta features) through Alize open source speaker recognition tool-kit [1]. First, a UBM-GMM model is trained using MFCCs computed from the auxiliary database comprising audio samples other than the celebrities in the corpus. Total variability matrix, T is estimated in the successive phase using the training data. A python framework DisVoice [15] is employed to compute the phonation feature set. DTW-based scoring mechanism followed the typical distance metric based ranking, and DNN-based system uses probability-score metric for rating the artists. During DNN based experiment, training is carried out using target data for 6000 epochs with a batch size of 512 and learning rate of 0.01. 5% of the data is used for validation. The performance of i_{MFCC} and phonation-fusion experiments are carried out in subsequent phases. SVM and DNN classifiers are implemented using LibSVM and Keras-TensorFlow, respectively.

   Based on the score, each system ranks the artists from the best mimic to worst (Rank 1 to Rank 5). If the artist who is ranked first by the proposed method, matches the outcome of MOS test, a hit occurs and assumes that the system correctly identified the best mimicking speaker. The performance is evaluated using the top-X hit rate criteria on five artists who imitate 22 celebrities. The top-X hit rate reports the proportion of queries for which $r_i \leq X$, where r_i denotes the rank of an imitator given by the proposed system [17]. One point is scored for a hit in the top X outcome and zero is scored otherwise.

4   Results and Analysis

The impersonator controls vocal tract acoustic characteristics, as well as those of the glottal source and pitch frequency to imitate the target voice [12]. In addition to the correlation between the formant frequencies of target and imitator, it is apparent that there is a strong tendency of the imitator to adjust pitch

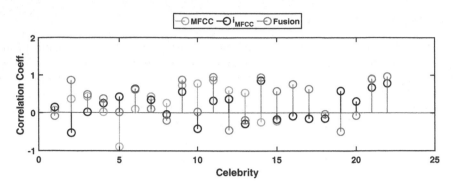

Fig. 3. Correlation coefficient computed between MOS with scores from three phases of experiments using DNN.

trajectory during imitation. The aim of the proposed experiment is to identify the best imitator who imitates each target with maximum similarity. It is worth noting that most of the previous studies focus on prosodic features than formant characteristics in addressing voice disguise research.

The correlation with MOS for the three phases of experiments using DNN is given in Fig. 3. It is seen that scoring pattern of fusion system for 11 out of 22 cases show a positive correlation with the MOS score. It means that scoring pattern of DNN matches with the trend of MOS. It is also important to note that even if the first rank matches the MOS and others vary, the correlation coefficient may result in a low value. The significance of Fig. 2 is to show the similarity in relative scoring both in perception test and machine scoring. The variation of scores for the rank-1 position artist (obtained in the MOS) from the rank-1 position artist obtained in the fusion system is shown in Fig. 4. It is observed that up to 30% of score variation is noticed for top rate-2 candidates.

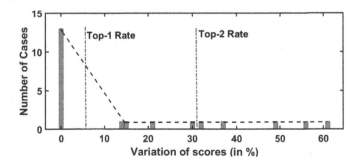

Fig. 4. Variation of scores in %. zero % variation means top-1 position artist of fusion matches with the highest MOS artist. 15% means 15% variation of scores noticed for fusion scheme-top-1 candidate from the highest MOS artist.

The scores obtained during the performance evaluation for four cases are shown in Table 1. The probability scores for all the artists for all feature set are listed in the 4th, 5th and 6th columns of Table 1. The MOS is listed in the 3rd column. Let us examine one case in Table 1. As per the table, for celebrity-4, the MOS suggests that artist-2 is the best imitator among five performers. MFCC based system rates the performance of artist-2 in the third position by giving the first rank to artist-1. The i_{MFCC} and its fusion framework correctly identify the best performer by giving a high score to artist-2. In the case of celebrity-2, the i_{MFCC} system fails to identify the best performer, but the i_{MFCC}-phonation fusion result matches with that of MOS result. Scores obtained for the fusion system (i_{MFCC} + phonation) for entire targets is shown in Fig. 5. It is noticed that out of 22 cases, 11 are identified correctly for top-1 rate.

Table 1. The results of DNN-based experiments with mean opinion score. The red coloured entry is the rank one position. Fusion represents early integration of i_{MFCC} and phonation features.

Target	Imitator	MOS	MFCC	i_{MFCC}	i_{MFCC} +phonation
Celebrity 1	Art-1	**4.6**	0.63	0.28	0.47
	Art-2	2.1	0.54	0.61	**0.75**
	Art-3	3.9	0.24	**0.68**	0.43
	Art-4	2.0	0.23	0.22	0.33
	Art-5	2.4	**0.73**	0.24	0.34
Celebrity 2	Art-1	2.80	0.29	0.33	0.35
	Art-2	**4.1**	0.43	0.31	**0.76**
	Art-3	3.5	**0.65**	0.21	0.44
	Art-4	3.7	0.38	0.29	0.66
	Art-5	2.9	0.40	**0.59**	0.50
Celebrity 3	Art-1	3.3	0.42	0.02	0.42
	Art-2	2.6	0.48	0.43	0.34
	Art-3	2.9	0.67	0.21	0.56
	Art-4	1.7	0.56	0.51	0.52
	Art-5	**4.2**	**0.75**	**0.63**	**0.69**
Celebrity 4	Art-1	4.0	**0.87**	0.21	0.54
	Art-2	**4.5**	0.61	**0.37**	**0.63**
	Art-3	3.9	0.56	0.26	0.21
	Art-4	2.5	0.78	0.26	0.33
	Art-5	3.1	0.39	0.32	0.55

The results are tabulated in Table 2. The identification accuracy with the Rank-2 margin for the MFCC and MFCC-phonation feature fusion is 40.90%

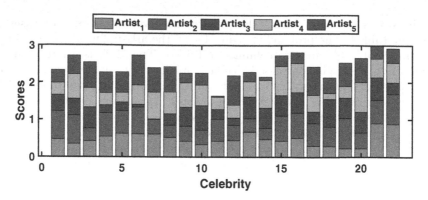

Fig. 5. Scores obtained for i$_{\text{MFCC}}$+phonation features-DNN for all artists in mimicking 22 celebrities.

Table 2. Accuracy in top X-hit rate (%).

Top-X rate			
Features	Scoring	X = 1	X = 2
MFCC	DTW	31.81	40.90
MFCC	DNN	31.81	54.54
MFCC+Phonation	DTW	36.36	40.90
MFCC+Phonation	DNN	31.81	63.63
i$_{\text{MFCC}}$	SVM	54.54	68.18
i$_{\text{MFCC}}$	DNN	40.90	68.18
i$_{\text{MFCC}}$+ Phonation	DTW	45.45	59.09
i$_{\text{MFCC}}$+ Phonation	**DNN**	**50.00**	**72.72**

with DTW. Phonation feature fusion shows significant improvement as compared to MFCC alone in the DNN framework. But in the i$_{\text{MFCC}}$ framework, both the baseline SVM and DNN framework report accuracy of 68.18% with improvement over MFCC-phonation fusion. Slight degradation in the result is observed in i$_{\text{MFCC}}$+phonation fusion with DTW. In the case of the top-2 hit rate, the i$_{\text{MFCC}}$+phonation fusion DNN scheme identifies 16 out of 22 cases correctly with an accuracy of 72.72%. For top-2 hit rate, 4.54% improvement is observed for i$_{\text{MFCC}}$ phonetic fusion-DNN framework over baseline SVM-based system. It is worth noting that, DNN framework outperforms DTW framework in most of the phases for rank-2 margin. Due to the unavailability of test-corpus with sufficient performers, the corpus preparation with is in progress for further study.

As part of the future work, we would like to experiment with other deep learning architecture such as a convolutional neural network (CNN) and long short-term memory (LSTM) by exploring a larger corpus. The experimental results demonstrate that i_{MFCC} and its fusion with phonation features have merit in evaluating the performance of voice imitators and related applications.

5 Conclusion

The proposed system uses i-vector framework and its fusion with phonation features in evaluating the competency of mimicking voice. Two scoring frameworks, DTW and DNN are employed in the experiment. DTW ranked the artist based on the distance metric. DNN classifier is trained using a feature set computed from the target voices of 22 celebrities and tested with mimicry voices of professional mimicry artists. The artist with the maximum probability score is declared as the best mimicking speaker by the system. It is observed that DNN outperforms DTW in most of the experiments. Besides i-vector fusion with phonation features shows promise in the evaluation of mimicked speech. DNN system gives a top-2 hit rate of 72.72% for the fusion approach. The study shows the potential of i-vectors and phonetic features in identifying the most competent mimicking artist during voice mimicking.

References

1. Bonastre, J.F., Wils, F., Meignier, S.: Aliźe, a free toolkit for speaker recognition. Proc. Interspeech **1**, I-737 (2005)
2. Dehak, N., Kenny, P., Dehak, R., Dumouchel, P., Ouellet, P.: Front-end factor analysis for speaker verification. IEEE Trans. Audio Speech Lang. Process. **19**, 788–798 (2011)
3. Dehak, N., Torres-Carrasquillo, P.A., Reynolds, D., Dehak, R.: Language recognition via i-vectors and dimensionality reduction. Proc. Interspeech, 857–860 (2011)
4. Gomati, D., Thati, S.A., Sridaran, K.V., Yegnanarayana, B.: Analysis of mimicry speech. In: Proceedings of Interspeech (2012)
5. Eriksson, A., Wretling, P.: How flexible is the human voice?- a case study of mimicry. In: Proceedings of European Conference on Speech Technology, pp. 1043–1046 (1997)
6. Francesc, A., Joan, C.S., Sevillano, X.: A review of physical and perceptual feature extraction techniques for speech, music and environmental sounds. Appl. Sci. **65**, 1–44 (2016)
7. Hao, Z., Yang, X.K., Zhang, W.Q., Zhang, W.L., Liu, J.: Application of i-vector in speech and music classification, pp. 1–5. https://doi.org/10.1109/ISSPIT.2016.7885999 (2016)
8. Ibrahim, N.S., Ramlia, D.A.: I-vector extraction for speaker recognition based on dimensionality reduction. Procedia Comput. Sci. **126**, 1534–1540 (2018)
9. George, K.K., Kumar, C., Ramachandran, K.I., Panda, A.: Improving robustness of speaker verification against mimicked speech. In: Proceedings of Odessey, pp. 245–251 (2016). https://doi.org/10.21437/Odyssey.2016-35
10. Kanagasundaram, A., Vogt, R., Dean, D., Sridharan, S., Mason, M.: I-vector based speaker recognition on short utterances. In: Proceedings of Interspeech, pp. 2341–2344 (2011)

11. Kenny, P., Boulianne, G., Dumouchel, P.: Eigenvoice modeling with sparse training data. IEEE Trans. Speech Audio Process. **13**, 345–354 (2005)
12. Kitamura, T.: Acoustic analysis of imitated voice produced by a professional impersonator. In: Proceedings of Interspeech, pp. 813–816 (2008)
13. Lopez-Otero, P., Docio-Fernandez, L., Garcia-Mateo, C.: I-vectors for continuous emotion recognition. In: Proceedings of Interspeech (2014)
14. Mary, L., Babu, A., Joseph, A., George, G.M.: Evaluation of mimicked speech using prosodic features. In: Proceedings of IEEE International Conference on Acoustics, Speech, and Signal Processing, pp. 7189–7193 (2013)
15. Orozco-Arroyave, J.R., Vásquez-Correa, J.C., et al.: Neurospeech: an open-source software for parkinson's speech analysis. Digit. Signal Process. **77**, 207–221 (2017)
16. Richard, G., Sundaram, S., Narayanan, S.: An overview on perceptually motivated audio indexing and classification. Proc. IEEE **101**, 1939–1954 (2013)
17. Ryynanen, M., Klapuri, A.: Query by humming of midi and audio using locality sensitive hashing. In: Proceedings of IEEE International Conference on Acoustics, Speech, and Signal Processing, pp. 2249–2252 (2008)
18. Singh, R., Jiménez, A., Øland, A.: Voice disguise by mimicry: deriving statistical articulometric evidence to evaluate claimed impersonation. IET Biometrics **6**(4), 282–289 (2017)
19. Wang, L., Yoshida, Y., Kawakami, Y., Nakagawa, S.: Relative phase information for detecting speech and spoofed speech. In: Proceedings of Interspeech, pp. 2092–2096 (2015)
20. Zetterholm, E.: Same speaker-different voices, a study of one impersonator and some of his different imitations. In: Proceedings of International Conference on Speech Science and Technology, pp. 70–75 (2006)
21. Zhong, J., Hu, W., Soong, F., Meng, H.: DNN i-vector speaker verification with short, text-constrained test utterances. In: Proceedings of Interspeech, pp. 1507–1511 (2017). https://doi.org/10.21437/Interspeech.2017-1036

Score Normalization of X-Vector Speaker Verification System for Short-Duration Speaker Verification Challenge

Ivan Rakhmanenko[1]([⊠]) (iD), Evgeny Kostyuchenko[1] (iD),
Evgeny Choynzonov[1] (iD), Lidiya Balatskaya[1,2] (iD),
and Alexander Shelupanov[1] (iD)

[1] Tomsk State University of Control Systems and Radioelectronics,
Lenina str. 40, 634050 Tomsk, Russia
ria@keva.tusur.ru, nii@oncology.tomsk.ru
[2] Tomsk Cancer Research Institute, Kooperativniy av. 5, 634050 Tomsk, Russia

Abstract. In this paper we present our contribution to the task 2 of the short-duration speaker verification (SdSV) challenge. The main task for this challenge is to find new technologies for text-dependent and text-independent speaker verification in short duration scenario. Some of the approaches used by the authors during participation in the challenge are presented. Described speaker verification systems include baseline x-vector system with PLDA backend and score normalization, x-vector system with neural PLDA backend and fusion of both systems.

The main goal of this paper is to analyze influence of different score normalization methods on x-vector based speaker verification systems performance. We found that system with PLDA backend and ZT-normalization method (single system) gives superior performance in Farsi trials, but gives lower performance improvement in English trials. Overall, in terms of minDCF single system performs 46.3% better than baseline x-vector system. We found that enroll data augmentation is useless for Neural PLDA backend, as performance of the system does not improve after adding augmented enroll data. Single system with ZT-score normalization and additional enroll audio augmentation performs 14.8% better than Neural PLDA backend system.

Keywords: Speaker recognition · Speaker verification · X-vectors · Speech processing · Deep learning · Score normalization · Neural networks

1 Introduction

Speaker verification is one of the most challenging tasks in speech processing researches. Despite the great success in this field, nevertheless, this problem is not solved and requires the development of new methods and approaches. In recent years, deep learning is used more and more for developing speaker verification systems. Speaker verification systems described in this paper were submitted for task 2 of a short-duration speaker verification (SdSV) challenge 2020.

© Springer Nature Switzerland AG 2020
A. Karpov and R. Potapova (Eds.): SPECOM 2020, LNAI 12335, pp. 457–466, 2020.
https://doi.org/10.1007/978-3-030-60276-5_44

The main goal of the challenge is to evaluate new technologies for text-dependent (TD) and text-independent (TI) speaker verification (SV) in a short duration scenario [1]. The proposed challenge evaluates SdSV with varying degree of phonetic overlap between the enrollment and test utterances (cross-lingual). It is the first challenge with a broad focus on systematic benchmark and analysis on varying degrees of phonetic variability on short-duration speaker recognition. Task 2 of the SdSV Challenge is speaker verification in text-independent mode: given a test segment of speech and the target speaker enrollment data, automatically determine whether the test segment was spoken by the target speaker.

In this work we wanted to explore how far to the limits could be pushed the performance of the baseline x-vector speaker verification system. To achieve this we used several techniques such as train and test data augmentation, score normalization, different backend scoring, gender recognition, language recognition.

End-to-end architectures of neural networks for speaker verification are becoming more popular [2–6], but they have a large number of parameters and require a lot of computational resources for network training and inference. Such architectures include end-to-end DNNs with raw waveform input [2, 3], architectures that are using parts of a image recognition DNNs [4, 5], approaches that are inspired by i-vector [6] speaker verification systems [7], etc. Considering speaker verification systems based on x-vector networks have less parameters, they could be trained faster and faster verification speed could be achieved.

X-vectors network is one of the most well-known modern neural networks that are used in speaker verification systems [8]. It is based on the E-TDNN architecture that is typically used as a frontend for deep embedding extraction. This embeddings are called x-vectors. They are used as a features for different backends, such as PLDA [9], cosine distance scoring [10]. In this paper we use two different types of backends: PLDA and Neural PLDA [11].

The organization of the paper is as follows. In Sect. 2, we present a short description of the proposed methods including features and systems we used. Data used in different parts of our speaker verification pipeline are described in Sect. 3. Results of the experimental evaluation and performance analysis are described in Sect. 4. Conclusions are presented in Sect. 5.

2   Proposed Methods

In this section, we present components of speaker verification systems used in SdSV 2020 challenge. It includes complete description of the submitted systems components, including front-end and back-end modules along with their configurations.

2.1   X-Vector PLDA (Single) System

First system is baseline x-vector [8] with several modifications. It consists of three main components: x-vector frontend, gender classifier and gender-dependent PLDA backend with score normalization (Fig. 1). It is considered as single system, because it has simple linear structure, it does not have score fusion and only one gender-dependent modeling is used.

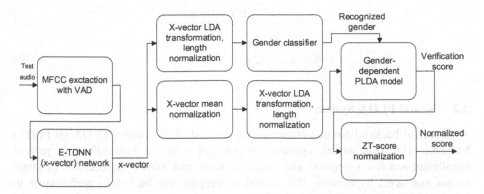

Fig. 1. X-vector PLDA (single) system diagram.

X-vector frontend for this system is based on provided baseline Kaldi [12] recipe without any significant modifications. As a features 30 mel-frequency cepstral coefficients (MFCC) are used, with 25 ms frame length, 20–7600 Hz frequency range. After MFCC extraction we apply sliding window cepstral mean and variance normalization (CMVN) and remove silence frames. Silence frames are removed using Kaldi energy-based voice activity detection script.

X-vector extraction network follows E-TDNN structure [13], but first TDNN-ReLU layer has bigger context, 9-th Dense-ReLU layer is deleted and last dense layer before pooling has size = 1536. Size of the last layer is 7323, equals to number of speakers in the training dataset. X-vector network was trained for 6 epochs with batch size = 128 at first epoch and batch size = 164 from second to sixth epochs. First dense layer after pooling layer was used for x-vector extraction, resulting in x-vector size = 512.

Next component of the single system is gender classifier. It consists of logistic regression classifier that was trained on x-vectors without x-vector mean normalization, with LDA transformation and Kaldi x-vector length normalization. LDA dimension is 200. Output of this classifier was used to select proper gender-dependent PLDA model for every evaluation x-vector.

Last component of the system is PLDA model. It gets mean normalized, LDA-transformed and length-normalized x-vectors as input features. X-vector mean and LDA parameters calculated using train dataset. There were two PLDA models trained, one for male and one for female speakers. Scores for this models were normalized using Z- and T-normalization methods [14]. Z-normalization includes normalizing PLDA scores using target speaker averaged enroll x-vector against a set of impostor speakers statistics (1). T-normalization includes normalizing PLDA scores using test segments x-vectors against a set of impostor speakers statistics (2). Both of this methods use mean μ and standard deviation σ as normalization statistics. We selected 10% of the best scores for normalization statistics calculation. Single system's final score is sum of Z- and T-normalized PLDA scores (3).

$$S_{Z-norm} = \frac{S(X, \lambda) - \mu_\lambda}{\sigma_\lambda}, \tag{1}$$

$$S_{T-norm} = \frac{S(X, \lambda) - \mu_X}{\sigma_X}, \tag{2}$$

$$S_{ZT-norm} = S_{Z-norm} + S_{T-norm}. \tag{3}$$

2.2 Neural PLDA System

As a second backend for this challenge we decided to use Neural PLDA (N-PLDA) backend [11]. This model, operates on pairs of x-vector embeddings (a pair of enrollment and test x-vectors) and outputs a score that allows the decision of target versus non-target hypotheses. We wanted to compare this backend's performance to standard PLDA backend.

For N-PLDA network initialization we used gender-dependent PLDA models trained for X-vector PLDA system. We used the same network parameters as in [4], except batch size = 4096 and beta = 9.9. N-PLDA model was trained for 20 epochs with learning rate = 0.0001, Adam optimizer and learning rate halving after error on the validation set increased for 2 epochs.

2.3 Fusion (Primary) System

Primary system consists of two components described earlier: X-vector PLDA and Neural PLDA systems. Primary system score (4) was computed as sum of two subsystems normalized scores. We thought that this fusion could give better performance stability on full test set and could reduce overfitting on the training data.

$$S_{primary} = S_{PLDA} + S_{N-PLDA}. \tag{4}$$

3 Data Usage

For x-vector network training we used Kaldi scripts of the x-vector baseline system. There was only two datasets used for network training: full VoxCeleb1 and VoxCeleb2 datasets [15]. Audio augmentation methods provided with the Kaldi were used. This includes adding reverberation, music, noise and babble from RIRS noises and MUSAN corpuses to the training audio. Two millions of augmented audio was selected in addition to the original VoxCeleb datasets.

For gender classifier training we used part of the VoxCeleb dataset (650k audio segments) and whole SdSVC task 2 train dataset [16].

For PLDA and Neural PLDA models training we used x-vectors extracted from VoxCeleb1 and VoxCeleb2 datasets with 2 million augmented audio, SdSVC train dataset and LibriSpeech [17] (clean-100, clean-360, other-500) datasets adding audio augmentations (reverberation, music, noise and babble).

Every audio in enroll dataset was augmented using the same audio augmentation techniques as for the x-vector DNN training. Thus, augmentation gave us five times more data for every speaker in enroll dataset. Enroll augmentation gives 12.9% minDCF decrease comparing to PLDA system without enroll data augmentation.

For score normalization whole SdSVC train set was used as gender-dependent impostor speakers data.

4 Experimental Evaluation

Performance of the three presented systems on the progress and evaluation subsets is shown in Table 1. Despite our suggestion that fusion of the PLDA and N-PLDA x-vector systems could improve verification results, primary system showed worse performance than our single system. Single system had better results on both the progress and the evaluation subset. Overall, results for primary and single system are similar (Fig. 2).

Table 1. Submitted systems evaluation results.

System	Progress		Evaluation	
	m-DCF	EER	m-DCF	EER
Baseline	0.4319	10.67	0.4324	10.67
PLDA (single)	**0.2313**	**5.47**	**0.2324**	**5.46**
N-PLDA	0.2706	6.19	0.2727	6.21
Fusion (primary)	0.2352	5.55	0.2361	5.55
Competition winner	0.0654	1.45	0.0651	1.45

We could see significant improvement in minDCF and EER comparing SdSV challenge winner primary system [18] and our systems. We could assume that different improvements in x-vector frontend including Squeeze-Excitation SE blocks [19], multiscale Res2Net [20] features, multi-layer feature aggregation [21], channel-dependent attentive statistics poolings [22] and Angular Additive Margin softmax loss [23] gave superior results to their system.

In Table 2, we could see that there is no overfitting on the progress subset, as performance on progress and evaluation subsets are similar. In addition, our single system performs better on male and Farsi speech (Fig. 3). It seems that our system recognized cross-lingual speakers much worse than speakers with native Farsi language. One of the main reasons is that there was no English speech in SdSVC Train and Enroll datasets and our system overfitted to Farsi speech.

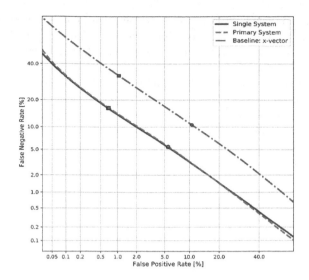

Fig. 2. Detection error trade-off (DET) curves for presented systems on evaluation dataset.

Table 2. Single and primary systems evaluation results.

System	Single progress		Primary progress		Single evaluation		Primary evaluation	
	m-DCF	EER	m-DCF	EER	m-DCF	EER	m-DCF	EER
Male	0.1825	4.16	**0.1799**	**4.12**	0.1845	4.17	**0.1812**	**4.12**
Female	**0.2606**	**6.19**	0.2611	6.28	**0.2610**	**6.17**	0.2669	6.28
English	**0.2468**	**5.46**	0.2601	5.75	**0.2442**	**5.46**	0.2587	5.76
Farsi	0.1423	2.98	**0.1417**	**2.93**	0.1435	2.96	**0.1427**	**2.93**
EN-male	**0.1908**	**4.21**	0.2032	4.44	**0.1898**	**4.23**	0.2022	4.49
EN-female	**0.2797**	**6.19**	0.2899	6.40	**0.2764**	**6.17**	0.2875	6.41
FA-male	0.1083	1.99	**0.1005**	**1.88**	0.1087	2.00	**0.1019**	**1.89**
FA-female	**0.1627**	3.55	0.1637	**3.50**	**0.1640**	3.52	0.1645	**3.47**
Farsi-TC-vs-IC-subset	0.1334	3.21	**0.1241**	**3.04**	0.1362	3.20	**0.1264**	**3.06**

We tried to apply audio augmentations to test data, but it gave us much worse results than after applying enroll data augmentation (Table 3). For N-PLDA backend enroll data augmentation gave worse results than without this augmentation.

After scores submission deadline we tried to vary fraction of data used for score normalization and found that ZT-score normalization with selecting top 40% of the highest impostor scores gave us a little boost to our single system's performance (Table 3). In addition, it could be seen that applying Z- and T-normalization separately gives worse performance than applying ZT-score normalization simultaneously.

We tried to train language classifier for selecting language-dependent PLDA models, but this classifier was seriously overfitted to training data. The main reason for

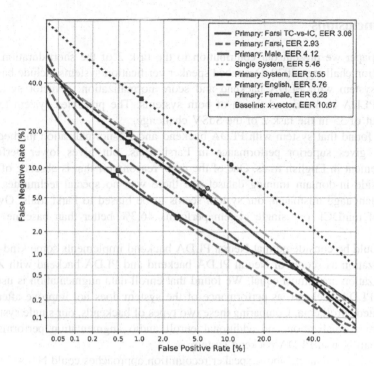

Fig. 3. Detailed DET-curves for primary system.

Table 3. Additional augmentation and normalization experiments results.

System	Aug.	Norm.	Evaluation	
			m-DCF	EER
PLDA (single)	Enroll	ZT	0.2324	5.46
N-PLDA	–	–	0.2727	6.21
N-PLDA (enr aug)	Enroll	–	0.2802	6.40
PLDA (top 40% norm)	**Enroll**	**ZT**	**0.2306**	**5.45**
PLDA (Z-norm)	Enroll	Z	0.2371	5.32
PLDA (T-norm)	Enroll	T	0.2439	5.90
PLDA (without norm)	–	–	0.2962	7.38
PLDA (test aug)	Test	ZT	0.3174	7.51
PLDA (no aug)	–	ZT	0.2669	6.49

this is that there was no English speech in the SdSVC train set. We tried to use English speech from VoxCeleb datasets, but it resulted in high performance degradation.

We could conclude, that ZT-score normalization and enroll data augmentation could significantly improve x-vector PLDA speaker verification system's performance.

5   Conclusions

In this paper we present our contribution to the task 2 of the short-duration speaker verification challenge 2020. Described speaker verification systems include baseline x-vector system with PLDA backend and score normalization, x-vector system with neural PLDA backend and fusion of both systems. The presented system took 22nd place out of 35 in the task 2 of the SdSV challenge.

We found that system with PLDA backend and ZT-normalization method (single system) gives superior performance in Farsi trials, but gives lower performance improvement in English trials. One of the main reasons for that is absence of English data inside in-domain training dataset. As there was no special techniques used to reduce language mismatch, our system seems to be biased to Farsi trials. Overall, in terms of minDCF our single system performs 46.3% better than baseline x-vector system.

It could be considered that Neural PLDA backend implements some kind of score normalization as scores of Neural PLDA backend and PLDA backend with ZT-score normalization are almost equal. We found that enroll data augmentation is useless for Neural PLDA backend, as performance of the system does not improve after adding augmented enroll data. Comparing these two types of backends, our single system with ZT-score normalization and additional enroll audio augmentation performs 14.8% better than Neural PLDA backend system.

We suspect that the above speaker recognition approaches could be used to assess the quality of pronunciation. In order to check this an experiment is planned to determine the degree of correspondence between the patient's speech before the surgery, after the surgery and after undergoing rehabilitation. If we come up with the use of records before and after the surgery (for one speaker) as different speakers, we could verify that presented approach could be used for solving the problem of assessing speech quality.

Acknowledgements. The study was performed by a grant from the Russian Science Foundation (project 16-15-00038).

References

1. Zeinali, H., Lee, K.A., Alam, J., Burget L.: Short-duration Speaker Verification (SdSV) Challenge 2020: The Challenge Evaluation Plan. arXiv preprint https://arxiv.org/abs/1912.06311 (2019)
2. Jung, J.W., Heo, H.S., Kim, J.H., Shim, H.J., Yu, H.J.: RawNet: advanced end-to-end deep neural network using raw waveforms for text-independent speaker verification. In: Proceedings Interspeech 2019, pp. 1268–1272 (2019)
3. Yun, S., Cho, J., Eum, J., Chang, W., Hwang, K.: An end-to-end text-independent speaker verification framework with a keyword adversarial network. In: Proceedings Interspeech 2019, pp. 2923–2927 (2019)
4. Li, C., et al.: Deep speaker: an end-to-end neural speaker embedding system. arXiv preprint https://arxiv.org/abs/1705.02304 (2017)

5. Xie, W., Nagrani, A., Chung, J.S., Zisserman, A.: Utterance-level aggregation for speaker recognition in the wild. In: 2019 IEEE International Conference on Acoustics, Speech and Signal Processing (ICASSP), pp. 5791–5795. IEEE (2019)
6. Dehak, N., Kenny, P.J., Dehak, R., Dumouchel, P., Ouellet, P.: Front-end factor analysis for speaker verification. IEEE Trans. Audio Speech Lang. Process. **19**(4), 788–798 (2010)
7. Rohdin, J., et al.: End-to-end DNN based speaker recognition inspired by i-vector and PLDA. In: 2018 IEEE International Conference on Acoustics, Speech and Signal Processing (ICASSP), pp. 4874–4878 (2018)
8. Snyder, D., Garcia-Romero, D., Sell, G., Povey, D., Khudanpur, S.: X-vectors: rRobust DNN embeddings for speaker recognition. In: 2018 IEEE International Conference on Acoustics, Speech and Signal Processing (ICASSP), pp. 5329–5333 (2018)
9. Prince, S.J., Elder, J.H.: Probabilistic linear discriminant analysis for inferences about identity. In: 2007 IEEE 11th International Conference on Computer Vision, pp. 1–8 (2007)
10. Garcia-Romero, D., et al.: X-vector DNN refinement with full-length recordings for speaker recognition. In: Proceedings Interspeech 2019, pp. 1493–1496 (2019)
11. Ramoji, S., Krishnan, P., Ganapathy, S.: NPLDA: a deep neural PLDA model for speaker verification. In: Proceedings Odyssey 2020 The Speaker and Language Recognition Workshop, pp. 202–209 (2020)
12. Povey, D., Ghoshal, A., Boulianne, G., et al.: The Kaldi speech recognition toolkit. In: IEEE 2011 Workshop on Automatic Speech Recognition and Understanding. IEEE Signal Processing Society (2011)
13. Snyder, D., et al.: Speaker recognition for multi-speaker conversations using x-vectors. In: 2019 IEEE International Conference on Acoustics, Speech and Signal Processing (ICASSP), pp. 5796–5800. IEEE (2019)
14. Barras, C., Gauvain, J.L.: Feature and score normalization for speaker verification of cellular data. In: 2003 IEEE International Conference on Acoustics, Speech, and Signal Processing, ICASSP 2003, vol. 2, pp. 49–52. IEEE (2003)
15. Nagrani, A., Chung, J.S., Zisserman, A.: VoxCeleb: a large-scale speaker identification dataset. arXiv preprint https://arxiv.org/abs/1706.08612 (2017)
16. Zeinali, H., Burget, L., Černocký, J.: A multi purpose and large scale speech corpus in Persian and English for speaker and speech recognition: the DeepMine database. arXiv preprint https://arxiv.org/abs/1912.03627 (2019)
17. Panayotov, V., Chen, G., Povey, D., Khudanpur, S.: LibriSpeech: an ASR corpus based on public domain audio books. In: 2015 IEEE International Conference on Acoustics, Speech and Signal Processing (ICASSP), pp. 5206–5210. IEEE (2015)
18. Thienpondt, J., Desplanques, B., Demuynck, K.: Cross-lingual speaker verification with domain-balanced hard prototype mining and language-dependent score normalization. arXiv preprint https://arxiv.org/abs/2007.07689 (2020)
19. Hu, J., Shen, L., Sun, G.: Squeeze-and-excitation networks. In: Proceedings of the IEEE Conference on Computer Vision and Pattern Recognition, pp. 7132–7141. IEEE (2018)
20. Gao, S.: Res2Net: a new multi-scale backbone architecture. IEEE Trans. Pattern Anal. Mach. Intell. (2019)
21. Gao, Z., et al.: Improving aggregation and loss function for better embedding learning in end-to-end speaker verification system. In: Proceedings Interspeech 2019, pp. 361–365 (2019)

22. Thienpondt, J., Desplanques, B., Demuynck, K.: ECAPA-TDNN: Emphasized Channel Attention, Propagation and Aggregation in TDNN Based Speaker Verification. arXiv preprint https://arxiv.org/abs/2005.07143 (2020)
23. Deng, J., Guo, J., Xue, N., Zafeiriou, S.: ArcFace: additive angular margin loss for deep face recognition. In: Proceedings of the IEEE Conference on Computer Vision and Pattern Recognition, pp. 4690–4699. IEEE (2019)

Genuine Spontaneous vs Fake Spontaneous Speech: In Search of Distinction

Ekaterina Razubaeva[1] and Anton Stepikhov[2(✉)]

[1] The Russian Language Department, St. Petersburg State University,
7/9 Universitetskaya emb., St. Petersburg 199034, Russia
katy1806@mail.ru

[2] Research Institute for Applied Russian Studies, Herzen University,
48 Moika emb., St. Petersburg 191186, Russia
astepihov@herzen.spb.ru

Abstract. The paper examines the forensic problem of identifying genuine spontaneous speech as opposed to its imitation (quasi-spontaneous speech) during the interrogation of a suspect or in the process of testifying. We sought to establish criteria for distinguishing between genuine and fake spontaneous speech. In the course of the linguistic experiment three different types of speech (prepared, quasi-spontaneous, spontaneous) were recorded from six participants belonging to three professional groups. Recordings were examined in terms of disfluencies, constructions of conversational syntax and speech rate. The paper reports the initial results based on the performed analysis.

Keywords: Spontaneous speech · Quasi-spontaneous speech · Imitation · Disfluencies · Hesitation · Speech rate · Conversational syntax · Forensic phonetics · Russian

1 Introduction

At present, properties of spontaneous speech have been studied in various fields – linguistics, natural language processing, speech technologies etc. A better understanding of its features, especially as compared to other types of oral speech, proves to be crucial for forensic phonetics as well [1,4,5,9].

Since an investigation is, among other things, intended to verify testimony and establish truth, investigators need tools to detect signs of falsification. Such signs may be related to speech intelligibility, the reproduction of a preexisting text, as well as a speaker's deliberate intentions at feigning spontaneity. In this regard, the practises employed by linguists in judicial proceedings – forensic linguistics – should be used to help distinguish between spontaneous speech and its imitation during interview, interrogation or testimony.

However, neither the number of forensically significant linguistic features, nor the limits of the inter-individual variability of each particular feature, are clearly defined [5]. In our research we aim to establish criteria for distinguishing

© Springer Nature Switzerland AG 2020
A. Karpov and R. Potapova (Eds.): SPECOM 2020, LNAI 12335, pp. 467–478, 2020.
https://doi.org/10.1007/978-3-030-60276-5_45

between genuine and fake spontaneous speech. This paper reports initial results based on the linguistic analysis of different speech types.

2   Method Description

2.1   Experimental Design

To elicit different types of speech for our research, we used two extracts from a comparative study of read and spontaneous Russian speech [2]. Within that project, spontaneous speech of ten Russian standard native speakers of different genders and ages (18–55 years old) was recorded. Some months later the read versions of speech transcripts were produced by the same speaker. Hesitations, repetitions, incomplete words, and filled pauses were removed from the text before recording to ensure fluent reading. The grammar of the texts was not corrected [11].

For our study we selected two extracts from the read versions of spontaneous speech transcripts. The text fragments were thematically limited; i.e. we used only those fragments of speech in which they describe a certain event or situation and the emotions associated with it. Each excerpt amounted to about 200 words. The speech was produced by one male and one female speaker under 20 years of age.

Subsequently, those excerpts were given to six participants in the experiment so they could learn them by heart. Participants were matched to discourses by sex.

During the experiment all participants were given three tasks:

1. Reproduce the memorised text by heart with no instructions for style of delivery.
2. Reproduce the text trying to imitate spontaneity using any means available. They did not receive any instruction for this task and were permitted to change the text of the discourse at their discretion.
3. Describe a film they had watched recently. The opportunity to begin retelling the description was not given to the participants to provide maximum spontaneity.

In this way, we obtained three types of speech with different levels of preparedness from each participant (18 texts in total): 1) prepared speech (completely prepared beforehand); 2) quasi-spontaneous speech (imitation of spontaneous speech at the moment of recording); 3) spontaneous speech (unprepared monologue).

2.2   Participants

The participants of the experiment were six Russian native speakers ranging between 23 and 25 years old. They were well acquainted with the person making the recording, which made their speech natural to a maximum extent. The participants were split into three groups (two persons in each) based on their field:

- users (students of non-linguistic specialisations);
- linguists (students specialising in analysis of spontaneous speech);
- actors.

We decided to enlist actors since feigned naturalness is an aspect of their professional training [14]. For the second group, it was assumed that, when trying to imitate spontaneous speech, linguists would sooner draw upon the features of spontaneous speech with which they are familiar. Users' speech should have shown how the properties of spontaneous speech would appear in imitation by a "naïve" speaker.

Thus, we hypothesised that actors and linguists would complete the task better than users without special knowledge.

The groups of actors and users consisted of one male and one female speaker. The group of linguists consisted of two female speakers since we could not find a male speaker who would agree to participate at the experiment.

3 Data Analysis

3.1 Summary Statistics. Approach to the Linguistic Analysis of the Data

The duration of recorded speech was 30 min, with the length of each text from one up to two minutes. Summary statistics for the recorded texts' size is given in Table 1. It shows that participants did not reproduce the initial texts verbatim.

Table 1. Summary statistics for the recorded texts' size in words for each participant (P).

Type of speech	Users		Linguists		Actors	
	P1	P2	P3	P4	P5	P6
Prepared speech	153	142	143	140	160	170
Quasi-spontaneous speech	142	149	148	134	178	166
Spontaneous speech	213	208	149	173	252	124

After transcription, an auditory and linguistic analysis was performed. We focused on two types of linguistic features. The first one indicates unpreparedness directly, and is therefore used in automatic speech recognition systems for detecting spontaneous speech. Direct indicators include disfluencies of different kinds [3,6,10,15,16] and conversational syntactical constructions [8,12,17]. The second type – namely speech rate – may serve as an indirect indicator of speech spontaneity [3,6,13].

3.2 Disfluencies

We analysed the following types of disfluencies:

1. Unfilled hesitation pauses: *считаю это / очень / очень перспективной вещью для себя ('I consider it /a very / very promising thing for me').*
2. Vowel lengthening: *<...> уже поехать в качестве-е / наверно сопро-вождающего лица ('then to go as a-a / perhaps accompanying person').*
3. Repetitions: *Господи / гениальный фильм / э-э фильм / молодого режиссера ('Lord / a genius film / uh film / of a young director').*
4. Discourse markers: *<...> чего я пока что / очень боюсь / ну / хотя надеюсь на это ('which I'm very afraid of so far / well / though I hope for it').*
5. Self-repairs: *его сложности с общением пациентов / с пациентами ('his difficulties in communicating of patients / with patients').*

Unfilled Hesitation Pauses. Unfilled hesitation pauses as detected during auditory analysis were observed in each type of speech and, moreover, in all texts. To better understand the frequency of hesitation phenomena across speakers and types of speech, we computed the coefficient of hesitation (CH) – the number of disfluencies of a certain type per 10 words. CH for unfilled pauses across participants is shown in Table 2.

Table 2. Coefficient of hesitation for unfilled pauses (number of pauses per 10 words) across participants (P) in different types of speech.

Type of speech	Users		Linguists		Actors		Mean value
	P1	P2	P3	P4	P5	P6	
Prepared speech	0.4	0.3	0.1	0.1	0.7	0.3	0.32
Quasi-spontaneous speech	0.5	0.7	0.4	0.1	0.4	0.5	0.43
Spontaneous speech	1.2	0.5	0.5	0.9	1	1.5	0.93

Unfilled hesitation pauses turned out to be the most frequent type of hesitation pauses in our material. Their amount in quasi-spontaneous speech increases in comparison with prepared speech for 67% of participants. The same trend is observed for spontaneous speech as compared to quasi-spontaneous speech for 83% of cases.

The mean values of CH for unfilled pauses in different types of speech show that the amount of unfilled pauses in quasi-spontaneous speech is 34% higher than in prepared speech. At the same time, the number of unfilled pauses in spontaneous speech increases dramatically. It is more than twice (116%) higher than in quasi-spontaneous speech and almost three times higher as compared to prepared speech. Thus, the amount of unfilled hesitation pauses correlates with the extent of spontaneity.

It should be mentioned, however, that in the spontaneous speech of one of the participants (P2), the number of unfilled hesitation pauses decreased as compared to quasi-spontaneous speech. This made it impossible to distinguish these types of speech in this respect.

Analysis also revealed that only one linguist (P3) drew upon his professional competence while trying to fake spontaneous speech. The number of unfilled hesitation pauses in her speech is four times higher in comparison with prepared speech. The amount of unfilled pauses in spontaneous speech, however, is only 25% higher, which does not allow us to differentiate between these two types of speech (Table 2).

Interestingly, the hesitation pauses in one of the actors' (P5) quasi-spontaneous speech were almost half that of the prepared speech. Most likely, the speaker, as a professional actor, took the 'prepared' speech as an authentic performance. Presumably the pauses in the quasi-spontaneous speech represent the actors' impression of 'bad acting'. Another reason may be that the speaker did not know about this linguistic feature of speech and, hence, did not pay attention to this characteristic. Whatever the case, the maximum amount of hesitation pauses in P5's speech is observed in her spontaneous speech.

To sum up, we may assume that, in general, the professional characteristics of the participants are quite insignificant for successful imitation of spontaneous speech when considering unfilled hesitation pauses (Table 3).

Table 3. Coefficient of hesitation for unfilled pauses (mean value) in different types of speech across groups.

Type of speech	Users	Linguists	Actors
Prepared speech	0.35	0.1	0.5
Quasi-spontaneous speech	0.6	0.25	0.45
Spontaneous speech	0.85	0.7	1.25

Filled Hesitation Pauses. To evaluate the frequency of filled hesitation pauses we similarly computed the coefficient of hesitation for filled pauses (CHF). The results are shown in Table 4.

Table 4. Coefficient of hesitation for filled pauses (number of pauses per 10 words) across participants (P) in different types of speech.

Type of speech	Users		Linguists		Actors		Mean value
	P1	P2	P3	P4	P5	P6	
Prepared speech	0.1	0.1	0.5	0	0.1	0.2	0.17
Quasi-spontaneous speech	0	0.2	0.5	0.2	0.06	0.4	0.22
Spontaneous speech	0.4	0.05	0.9	0.7	0.2	1.2	0.57

The analysis showed that in 83% of cases the amount of filled hesitation pauses in spontaneous speech significantly increases in comparison with prepared and quasi-spontaneous speech. At the same time, in half of all cases the number of filled pauses is higher in quasi-spontaneous speech as compared to prepared one. Thus, three participants might realise this feature of spontaneous speech and try to fake it in their imitations.

The mean values for CHF reveal that while the amount of filled pauses in quasi-spontaneous speech is 29% higher than in prepared speech, their frequency in spontaneous speech is more than 2.5 times higher than in quasi-spontaneous speech.

The analysis of inter-speaker variability in the use of fillers shows that while the number of filled pauses increases in quasi-spontaneous speech of P2, in her spontaneous speech there was only one filled pause, with CHF close to zero (see Table 4). Since P2 uses fillers mostly in quasi-spontaneous speech we may assume that she does it deliberately though she is not a linguist.

As for linguists, there are no fillers in prepared speech of P4. They appear in her quasi-spontaneous speech and then their frequency dramatically increases in spontaneous speech, which allows for considering this feature a marker of spontaneity for P4's speech. The use of fillers in a linguist's fake spontaneous speech may indicate her intention to successfully imitate spontaneity.

One of actors (P5) does not use fillers to imitate spontaneous speech. While they are regular in her spontaneous speech, their amount in quasi-spontaneous speech is even lower than in prepared one. In P6's spontaneous speech the frequency of filled pauses is six times higher than in his prepared speech. At the same time, the amount of fillers in his quasi-spontaneous speech is only twice higher compared to prepared speech, which marks the difference between genuine spontaneous and fake spontaneous speech.

The inter-group comparison (Table 5) shows that, even though fillers are used to imitate spontaneous speech, their quantity in quasi-spontaneous speech is not enough to fake spontaneity.

Table 5. Coefficient of hesitation for filled pauses (mean value) in different types of speech across groups.

Type of speech	Users	Linguists	Actors
Prepared speech	0.1	0.25	0.15
Quasi-spontaneous speech	0.1	0.35	0.23
Spontaneous speech	0.23	0.8	0.7

Vowel Lengthening. In our recordings vowel stretching (i. e. non-phonological duration of certain sounds (cf. [6,7]) was observed in all participants' speech, both in single-syllable and multi-syllable words. Their statistics (in absolute numbers) is shown in Table 6.

Table 6. Frequency of vowel lengthening across participants (P) in different types of speech.

Type of speech	Users		Linguists		Actors		Mean value
	P1	P2	P3	P4	P5	P6	
Prepared speech	2	1	2	0	2	1	1.3
Quasi-spontaneous speech	4	1	5	3	4	3	3.3
Spontaneous speech	5	1	4	3	3	2	3

This reveals that vowel lengthening in spontaneous speech is the same or even lower than quasi-spontaneous speech by 83%. At the same time the frequency of vowel lengthening in quasi-spontaneous speech is 83% higher than in prepared speech. This allows us to assume that vowel lengthening is used by speakers to fake spontaneous speech. It should be also mentioned that for one speaker (P2) the amount of vowel lengthening remains stable between texts. Taking into account a similar amount of vowel lengthening in spontaneous and quasi-spontaneous speech this feature cannot be considered a distinctive factor between these two types of speech.

Table 7 describes the frequency of vowel lengthening in different types of speech across control groups and shows the similar trend in each group regardless of presence or absence professional knowledge.

Table 7. Frequency of vowel lengthening (mean) in different types of speech across groups.

Type of speech	Users	Linguists	Actors
Prepared speech	1.5	1	1.5
Quasi-spontaneous speech	2.5	4	3.5
Spontaneous speech	3	3.5	2.5

Repetitions. They may indicate hesitations in the choice of a following word or a construction [13]. Table 8 shows that while repetition is mostly a feature of spontaneous speech, it was not used to imitate spontaneity.

Table 8. Frequency of repetitions across participants (P) in different types of speech.

Type of speech	Users		Linguists		Actors	
	P1	P2	P3	P4	P5	P6
Prepared speech	0	0	1	0	0	1
Quasi-spontaneous speech	0	0	0	0	0	0
Spontaneous speech	0	1	1	2	0	1

Discourse Markers. While all participants used discourse markers in spontaneous speech, only two thirds inserted filler words into its imitation (Table 9).
The 67% difference between the two discourses indicates that this feature may
be used to distinguish faked speech.

Table 9. Frequency of discourse markers across participants (P) in different types of
speech.

Type of speech	Users		Linguists		Actors		Mean value
	P1	P2	P3	P4	P5	P6	
Prepared speech	0	0	0	0	0	0	0
Quasi-spontaneous speech	1	1	1	0	2	0	0.8
Spontaneous speech	6	1	1	2	7	4	3.5

Self-repairs (Restarts and False Starts). In our recordings self-repairs
proved to be a distinctive feature between spontaneous speech and its imitation since in quasi-spontaneous speech not a single self-repair was detected. At
the same time, three restarts were observed in the prepared speech of two participants (Table 10).

Table 10. Self-repairs across participants (P) in different types of speech.

Type of speech	Users		Linguists		Actors		Mean value
	P1	P2	P3	P4	P5	P6	
Prepared speech	0	0	1	0	2	0	0.5
Quasi-spontaneous speech	0	0	0	0	0	0	0
Spontaneous speech	5	2	3	3	2	1	2.7

3.3   Conversational Syntax

In this section we consider syntactical constructions typical for unprepared and
informal oral discourse in the process of direct communication. The substantial
part of these constructions is observed only in speech and, hence, cannot be used
in written texts.

Since part of the tasks involved the reproduction of the earlier memorised
texts, most speakers tried not to deviate much from the original and did not bring
any changes to it, both in prepared speech and in the imitation of spontaneous
speaking (Table 11).

Table 11. Frequency of consructions of conversational syntax across participants (P) in different types of speech.

Type of speech	Users		Linguists		Actors	
	P1	P2	P3	P4	P5	P6
Prepared speech	0	0	0	0	3	0
Quasi-spontaneous speech	0	1	0	0	3	0
Spontaneous speech	5	5	4	4	5	5

Five out of six speakers left the initial text completely unchanged in prepared speech and four out of six did so in quasi-spontaneous speech. While the only non-standard syntactical construction in quasi-spontaneous speech of P2 may be considered a mistake (*если бы / она у меня сломается по дороге / то я бы так и осталась 'if it had / it breaks on my way / I would have stayed there'*), inversions[1] and left-dislocations revealed in P5's speech are obviously conversational.

In contrast, the spontaneous speech of each speaker contained constructions typical for unprepared discourse. The most frequent ones are as follows:

1. Parenthetical constructions (8): *как они э-э // создавались в // **по-моему это в 70-е–80-е годы был сделан фильм** / вот в то время* ('how they er // were created in // **I think it was in 70-s–80-s when the film was made** / just at that time').
2. Repetitions (6): *<...> в фильме говорится о-о // **мальчике** / маленьком мо... **мальчике** а-а Джорджио* ('the movie tells about // a **boy** / a little **boy** er Georgio').
3. Multi-clausal asyndetic constructions (5): *если даже спросить моя любимая / условно / драма <...>* ('even if to ask my favourite / say / drama <...>').

Thus, we argue that constructions of conversational syntax may serve as reliable markers of genuine spontaneity.

3.4 Speech Rate

In our study we focused on speech rate (words per minute) rather than articulation rate. As speech rate can be estimated much more quickly and without special training, it is a more useful metric for judicial practise and expertise. The information about speech rate in our recordings is given in Table 12.

---

[1] All inversions in our recordings were observed in quasi-spontaneous speech of P5.

Table 12. Speech rate (words per minute) across participants (P) in different types of speech.

Type of speech	Users		Linguists		Actors		Mean value
	P1	P2	P3	P4	P5	P6	
Prepared speech	176	149	153	175	129	128	152
Quasi-spontaneous speech	136	117	161	143	152	117	138
Spontaneous speech	119	107	114	112	128	102	114

The analysis showed that in 67% of cases, participants trying to imitate spontaneous speech began to speak slower and their speech rate decreased by 10% on average. The slowest speech rate was observed in all speakers' spontaneous speech where it decreased by 21% on average as compared to quasi-spontaneous speech.

4   Conclusions and Discussion

In this paper we sought to establish criteria which can help distinguish between genuine spontaneous and fake spontaneous speech. The linguistic analysis of these types of oral discourse revealed the number of features which may be considered markers of spontaneity. Among disfluencies these are, foremost, unfilled and filled hesitation pauses, discourse markers, and self-repairs (restarts and false starts). Other types of hesitations (vowel lengthening and repetitions) cannot reliably distinguish between spontaneous speech and its imitation.

Another distinctive feature between genuine and fake spontaneity is speech rate. On the one hand, speech rate in quasi-spontaneous speech is slower compared to prepared speech, which is effected by disfluencies, deliberately used by speakers to imitate spontaneity. On the other hand, speech rate in spontaneous speech is the slowest within the examined types of speech. It may be explained by a necessity for a speaker to plan and produce an utterance simultaneously, which requires more time and leads to higher amount of disfluencies having an effect on speaking rate.

Syntactic analysis of the data revealed that prepared and quasi-spontaneous speech scarcely employed any constructions implying the speaker's intention to improvise or change the original text. At the same time, spontaneous speech expectedly showed quite a large number of constructions typical for conversational syntax. Thus, we consider them a strong marker of spontaneity.

The participants of our experiment belonged to different professional groups as we wanted to check whether actors and linguists researching spontaneous discourse would fake spontaneous speech better than ordinary language users. However, we did not find any sustainable correlation between speech properties of a speaker and their professional skills. At the same time, this conclusion should be taken with caution because of the limited data and requires further verification.

The results of our study help to specify a question to a forensic linguist regarding a speaker's spontaneity or preparedness: "Which features indicate and to what extent may we allow for considering N's speech spontaneous or prepared?"

In future, we plan to compare prosodic properties of genuine and fake spontaneous speech and to conduct a perceptual experiment to verify the obtained results and to likely discover new distinctive features between spontaneous speech and its imitation. Another prospective task would examine the issue of disfluencies associated with speakers' emotional state in genuine and fake spontaneous speech.

Acknowledgments. The paper has benefited greatly from the valuable comments and suggestions of Svetlana Balyberdina, Anna Shapovalova, Walker Trimble and three anonymous reviewers. We also thank all the participants of the experiment.

References

1. Azarchenkova, E.I., Zhenilo, V.R., Lozhkevich, A.A.: Expert speaker identification by the means of phonograms of voice and speech [Ekspertnaya identifikatsija cheloveka po fonogrammam ego golosa i rechi]. Moscow (1987). (in Russian)
2. Bondarko, L.V., Volskaya, N.B., Tananaiko, S.O., Vasilieva, L.A.: Phonetic properties of Russian spontaneous speech. In: ICPhS-15, pp. 2973–2976 (2003)
3. Dufour, R., Estève, Y., Deléglise, P.: Characterizing and detecting spontaneous speech: application to speaker role recognition. Speech Commun. **56**, 1–18 (2014)
4. Galyashina, E.I.: The basics of forensic speech science [Osnovy sudebnogo rechevedenija]. Moscow (2003). (in Russian)
5. Galyashina, E.I.: Linguistic support for forensic research into oral and written discourse [Lingvisticheskoe obespechenie kriminalisticheskogo issledovanija ustnogo i pismennogo teksta]. Moscow (2001). (in Russian)
6. Kibrik, A.A., Podlesskaya, V.I. (eds.): Night dream stories: A corpus study of spoken Russian discourse [Rasskazy o snovidenijakh: Korpusnoe issledovanie ustnogo russkogo diskursa]. Moscow (2009). (in Russian)
7. Kodzasov, S.V.: The phonetic symbolism of space (the semantics of duration) [Foneticheskaya simvolika prostranstva (semantika dolgoty i kratkosti)]. In: Arutyunova, N.D. (ed.) Logical Analysis of Language. Languages of Spaces [Logicheskij analiz jazyka. Jazyki prostranstv]. Moscow (2000). (in Russian)
8. Lapteva, O.A.: Russian conversational syntax [Russkij razgovornyj sintaksis]. Moscow (1976). (in Russian)
9. Potapova, R., Potapov, V.: On individual polyinformativity of speech and voice regarding speakers auditive attribution (forensic phonetic aspect). In: Ronzhin, A., Potapova, R., Németh, G. (eds.) SPECOM 2016. LNCS (LNAI), vol. 9811, pp. 507–514. Springer, Cham (2016). https://doi.org/10.1007/978-3-319-43958-7_61
10. Shriberg, E.: Phonetic consequences of speech disfluency. In: ICPhS-14, pp. 619–622 (1999)
11. de Silva, V., Iivonen, A., Bondarko, L.V., Pols, L.C.W.: Common and language dependent phonetic differences between read and spontaneous speech in Russian, Finnish and Dutch. In: ICPhS-15, pp. 2977–2980 (2003)
12. Sirotinina, O.B.: Contemporary conversational speech and its properties [Sovremennaya razgovornaya rech i ee osobennosti]. Moscow (1974). (in Russian)

13. Svetozarova, N.D. (ed.): Phonetics of spontaneous speech [Fonetika spontannoj rechi]. Leningrad (1988). (in Russian)
14. Vasilyev, Y.A.: Stage speech: the acting voice [Stsenicheskaya rech: golos dejstvujushchij]. Moscow (2010). (in Russian)
15. Verkhodanova, V., Shapranov, V.: Automatic detection of speech disfluencies in the spontaneous Russian speech. In: Železný, M., Habernal, I., Ronzhin, A. (eds.) SPECOM 2013. LNCS (LNAI), vol. 8113, pp. 70–77. Springer, Cham (2013). https://doi.org/10.1007/978-3-319-01931-4_10
16. Verkhodanova, V.O., Shapranov, V.V., Kipyatkova, I.S., Karpov, A.A.: Automatic detection of vocalized hesitations in Russian speech. Voprosy Jazykoznanija **6**, 104–118 (2018). (in Russian)
17. Zemskaya, E.A., Kitajgorodskaya, M.V., Shiryaev, E.N.: Russian conversational speech. General issues. Derivation. Syntax [Russkaya razgovornaya rech. Obshchie voprosy. Slovoobrazovanie. Sintaksis]. Moscow (1981). (in Russian)

Mixing Synthetic and Recorded Signals for Audio-Book Generation

Meysam Shamsi[✉], Nelly Barbot[✉], Damien Lolive[✉], and Jonathan Chevelu[✉]

Univ Rennes, CNRS, IRISA, Lannion, France
{meysam.shamsi,nelly.barbot,damien.lolive,jonathan.chevelu}@irisa.fr

Abstract. Using TTS systems helps to reduce the cost of audio-book generation. This paper investigates the idea of mixing synthetic and recorded natural speech signals to control the trade-off between the overall quality of audio book and its production cost. Firstly, fully synthetic signals and mixed synthetic and natural signals are compared perceptually using different levels of synthetic quality. The listeners' perception shows that mixed signals are preferred. Next, the order and configuration of mixed signals are studied. The perceptual test does not show any significant difference between the different configurations. Finally, the synthetic quality and the bias of a starting and ending part of mixed signals in perceptual test are investigated.

Keywords: Audio-book generation · Text-to-Speech · Quality evaluation

1 Introduction

The conventional way to produce an audio book is to record a speaker who reads the entire book. This process is costly and time consuming, and the use of a Text-to-Speech (TTS) system could be a lead. However, in case of a specific speaker, a recording phase remains necessary and the problem of audio-book generation can be defined, like in [11], as the optimization of the vocalization quality when a mixture of synthetic voice and recorded natural one is used. Selection of the script to record (extracted from the book) which has to be used as the TTS voice corpus to synthetically vocalize the rest of the book has been addressed in [10,11]. Due to the parsimony constraint, the required speech naturalness and expressiveness, a hybrid TTS has been employed in this study.

The idea of mixing synthetic and human voice signals is not new in the literature. Previous studies have agreed that fully synthetic signals are preferred rather than hybrid ones (mixed TTS signal and human voice) by listeners. In [4], users' liking and clarity of fully synthetic signals are higher than mixed TTS-human signals. As for [7], the preference of listeners is asked about mixed synthetic and recorded voice when dynamic part of a message is synthesized by TTS. Participants have indicated that they prefer the fully synthetic signals.

© Springer Nature Switzerland AG 2020
A. Karpov and R. Potapova (Eds.): SPECOM 2020, LNAI 12335, pp. 479–489, 2020.
https://doi.org/10.1007/978-3-030-60276-5_46

Recently, authors of [3] have investigated the naturalness of synthetic sentences in three different contexts: isolated sentence, full paragraph, context-stimulus pairs. It has turned out that two successive synthetic signals get higher scores than a sequence composed of a natural signal and a synthetic one. Furthermore, considering some aspects of synthetic speech quality such as intelligibility, [13] has found out that synthetic speech produced by recent TTS systems could be as good as human voice.

Audio-book generation is different and more challenging than tasks done in [3,4,7] that do not require expressiveness (news or message reading). Even advanced TTS systems are not yet as good as professional speakers for generating expressive books in terms of overall quality.

The sub-set selection problem has been investigated by taking into account the richness of the voice corpus and synthetic quality. It could be expected that synthetic signals would have less overall quality than recorded signals of professional speakers. Regardless of the signals' quality achieved by a TTS and its voice corpus, the length of synthetic portion in final audio book has been the only constraint in audio-book generation problem which is considered in [10,11]. Although the order or configuration of synthetic and recorded speech signals could be important too.

In this study, we investigate the configuration of mixed signals in expressive audio-book generation using a hybrid TTS system. There are two main motivations for the experiments conducted in this paper. First, the initial idea of audio-book generation as mixed signals will be examined by comparing fully synthetic signals to mixed synthetic and natural signals based on their perceptual quality. All synthetic signals are produced using a TTS voice corpus recorded by same speaker as natural voice. In other words, the main aim of this study is to answer these questions: *In terms of overall quality, is it helpful to generate an audio book with mixed synthetic and natural signals? Or do listeners prefer a fully synthetic audio book?* Second, in the case of preference for mixed signals, the impact of the order of synthetic and natural signals on the perceived quality will be investigated. These experiments will be done considering different levels of synthetic quality.

2   Perceptual Evaluation

In [5], a protocol for subjective evaluation of TTS in audio book reading tasks has been presented. The authors have suggested asking listeners to evaluate the quality of an audio book using 11 criteria such as listening pleasure, listening effort, intonation, emotion, etc. We believe that these terms are not always clear and do not have common definitions among listeners. Since the target of audio-book generation task is ordinary people, not expert voice quality annotators, we suggest asking for overall quality or overall preference of listeners.

The goal of the first experiment is to compare the overall quality of fully natural speech, fully synthetic speech, and a mix of natural and synthetic signals for expressive audio-book generation using a hybrid TTS system. The experimental framework of studies previously cited was completely different from the one considered here. For example, the naturalness of synthetic signals generated by a vocoder-based TTS in [3] or synthetic quality of an HMM-based TTS in a non-expressive context in [4] have been evaluated. Considering our objective, it seems to be necessary to evaluate again the hypothesis of listeners' preference for fully synthetic signals and mixed signals. Moreover, mixed audio-book generation needs to take into account the impact of signal type order (synthetic or recorded first) in overall quality of the final audio book.

The quality of synthetic signals using different TTS settings may vary, especially in expressive tasks. In order to take into account this variability, different quality levels of synthetic signals have been considered in this study. Since the TTS voice corpus size has a direct impact on synthetic quality [9], we propose to conduct all perceptual tests by using synthetic speech built from 3 voice corpus sizes (30 min, 1 h, 5 h).

Regardless of the quality degradation of synthetic signals in comparison with recorded speech, the change of signal type in a mixed signal sequence could be disturbing for audio book listeners. We call this change a transition. Transitions can happen from synthetic speech to recorded natural speech or the contrary. In order to examine their impact on overall perceptual quality, 8 different configurations are evaluated (see Table 1). In total, each sample is prepared with these 3 voice corpus sizes and these 8 transition configurations.

Table 1. Different transition configurations for 4 parts (each part is a breath group). Natural parts are indicated by N and synthetic parts by S.

Without transition	1 transition	2 transitions	3 transitions
NNNN	NNSS	NSSN	NSNS
SSSS	SSNN	SNNS	SNSN

Some studies such as [2,6] have emphasized on the importance of context in voice perception. For instance, in [6], it has been found that, without context, listeners do not always prefer the signals produced by humans. It leads to evaluate the speech perception using long-form speech. A possible drawback of this approach is the exhausting nature for listeners to assess an entire chapter or even several paragraphs of an audio book and thus a reduction of the evaluation reliability. We propose then four consecutive breath groups to construct perceptual test samples: signal for each breath group can be synthetic or natural one as mentioned in Table 1). The long (more than 70 phones) and short (less than 45 phones) breath groups are filtered out from the candidate list for listening test samples. The average duration of breath groups is 3.49 ± 0.40 s. It helps listening test samples to have a reasonable duration (around 14 s) and have almost same synthetic and recorded lengths. In order to provide some context, the transcription of the signal plus the script of the utterance just before and after the test sample are provided to testers.

3   Experiments and Results

Although, in TTS, vocoder-based approaches – like end-to-end DNN systems – are more and more prevalent, hybrid systems are still well-adapted to take into account the data parsimony constraint. The TTS system for the experiments is a hybrid TTS based on the acoustic model described in [12]. The target cost in this TTS system is computed based on an euclidean distance in an embedding space. This embedding is learned from an encoder-decoder trained on the voice. In the following experiments, the *hybrid* TTS uses only one DNN per speaker, which is trained with the full voice corpus (10 h) once and for all, to compute the target cost.

A French expressive audio book (*Albertine disparue* by Marcel Proust) spoken by a male speaker is used for this experiment. More information on the annotation process can be found in [1]. From this voice corpus, we extract randomly three voice corpora (30 min, 1 h and 5 h). The smaller corpora are included in the larger ones.

The listening test transcriptions are extracted from the rest of the book which is not selected for voice corpora according the following methodology. The sequences of four consecutive breath groups with a duration between 3 and 6 s are listed. They are not limited to only one utterance and can belong to several consecutive utterances. Out of this list, 20 transcriptions (80 breath groups) have been selected randomly for the listening test samples. These breath groups are synthesized using the three voice corpora. Each configuration of mixed signals in Table 1 is prepared by using synthetic and natural signals.

3.1   Overall Quality Evaluation

A MOS (Mean Opinion Score) test [8] is designed for evaluation. Listeners are asked to rate the overall quality of each sample on a scale from 1 to 5 with a step of 0.5. The cognitive load of a long perceptual test causes unreliability of evaluation. Consequently, in order to keep the quality of evaluation, only 25 samples are provided to each listener which takes around 12 min to be evaluated.

In total, 29 non-expert listeners participated to the evaluation which gives 725 scores. Table 2 details the main results of this perceptual test and the confidence intervals of score average calculated using the bootstrap method with $\alpha = 0.05$.

This MOS scores are not comparable between different languages and test settings. The average score for human voice (4.32) is lower than in previous studies [3] (around 4.6). Moreover, the question evaluated in our experiment is the overall quality while the MOS score in [3] corresponds to naturalness.

Based on these results, the mixed signals have significantly higher scores than fully synthetic signals in all voice corpus sizes. This observation is contrary to the previous studies [3,4,7] which showed superiority of fully synthetic signals. On the other hand, Fig. 1 does not show any significant difference when the number of transitions changes. However, mixed signals with 3 transitions have slightly higher score in comparison with others in 30 min and 5 h voice corpora. In case

Table 2. MOS test results for evaluating mixed synthetic/natural signals.

Num. of transitions	Config.	5 h	1 h	30 min
Fully synthetic	SSSS	2.70 ± 0.41	1.64 ± 0.36	1.31 ± 0.39
1 transition	NNSS	3.67 ± 0.28	3.08 ± 0.38	2.86 ± 0.30
	SSNN	2.90 ± 0.39	2.30 ± 0.38	1.81 ± 0.41
2 transitions	NSSN	3.30 ± 0.33	3.03 ± 0.26	2.20 ± 0.38
	SNNS	3.30 ± 0.34	2.87 ± 0.35	2.58 ± 0.41
3 transitions	NSNS	3.48 ± 0.31	2.74 ± 0.29	2.67 ± 0.42
	SNSN	3.68 ± 0.31	2.62 ± 0.34	2.42 ± 0.41
Human voice	NNNN	4.32 ± 0.17		

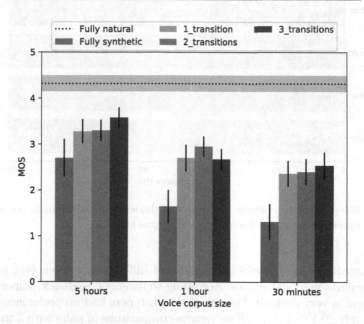

Fig. 1. MOS test results for evaluating mixed synthetic/natural signals (aggregated based on voice corpus size).

of long synthetic part (*SSNN, NNSS, SSNN*), the following ranking between MOS scores can also be observed: *NNSS ≥ NSSN ≥ SSNN*.

3.2   Preference Test

In order to investigate more the impact of transitions, another perceptual test is proposed. The configurations given in Table 1 are categorized into two groups: the first one corresponds to configurations with a *long* synthetic part and the second one to the *short* synthetic part configurations (*SNNS, SNSN, NSNS*).

We propose to use a simple protocol, an AB test, to directly compare the *long* and *short* categories. This test is designed with 3 levels of preference (no preference, slightly, strongly). For a same transcription and a same TTS voice corpus size, a sample with *long* synthetic part is compared with a sample stemming from the *short* category. The 3 voice corpus sizes are considered. Listeners are asked to compare 25 pairs of samples which takes around 25 min time. 26 listeners have done the test, which resulted in 595 comparisons. Results are shown in Fig. 2.

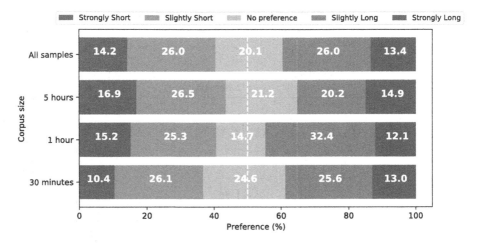

Fig. 2. AB test result to evaluate the impact of the length of continuous synthetic part in mixed signals (aggregated based on voice corpus sizes).

The result does not reveal any significant difference between *long* and *short* synthetic parts in mixed signals. According to listeners' feedback, sometimes the comparison is very difficult. Despite of this, listeners had no preference between samples only 20.1% of times. If we remove comparisons of pairs with 2 transitions (*NSSN* and *SNNS*), the AB test changes to a direct comparison between one transition and three transitions. In this case, the preference results do not show any difference between those two configurations.

4   Results Analysis

Two considerations about perceptual test results will be followed. First the result of the perceptual test will be considered based on signal quality instead of the size of voice corpus. Second, we will investigate the impact of starting and ending part types.

4.1   Investigate of Synthetic Quality

All combinations composed of two synthetic parts and two natural ones are evaluated in the listening test. The average length of synthetic parts is same as the average length of natural parts in mixed signals. But the length of synthetic parts has variations among listening test samples. It could be claimed that samples with longer synthetic parts would be evaluated with small MOS scores. To examine this hypothesis we calculate the correlation coefficient of MOS scores and synthetic part lengths after removing *NNNN* and *SSSS*. A low correlation (Pearson: −0.20, Spearman ranking correlation: −0.21) rejects the relation between MOS score and synthetic part length. On the other hand the Pearson correlation coefficient of MOS scores with TTS global cost is −0.47 (Spearman: −0.50).

The main question of this experiment was to compare the quality of mixed signals with fully recorded speech using different levels of synthetic quality. We used voice corpus sizes to simulate the synthetic quality levels. Now we have evaluated the samples, we can sort the results based on the perceived synthetic quality which can be obtained according to MOS scores of fully synthetic signal (*SSSS* configuration). The *SSSS* samples with different voice corpus sizes are grouped to three levels of high, medium, and low quality (20 samples for each group) based on their MOS scores. In this way, mixed signals are categorized to new quality levels based on the label of their synthetic parts. For example if *SSSS* configuration of a script with 1 h voice corpus has been labeled as high quality, all mixed configurations of this script with 1 h voice corpus should be labeled as high quality. Indeed this new aggregation causes the script of samples in each category to be potentially different. Nevertheless, it helps to consider the samples in different perceptual quality levels.

Figure 3 displays the MOS test result (3a) and the AB test result (3b) based on quality levels. Figure 3a shows that by improving the quality level from low to high the difference between mixed signal and fully synthetic signals decreased. The MOS score of fully synthetic signals in high synthetic quality level is comparable with mixed signals. Figure 3b confirms that *long* and *short* synthetic parts with different quality levels in mixed signals do not have any overall preference. Although a narrow band for *no preference* (about 20%) shows that sometimes listeners prefer *long* synthetic parts and sometimes *short* synthetic parts. It means that it was not a difficult task for listeners to tell their preference about a single sample. Anyway, their preference was not caused by the length of continuous synthetic part.

4.2   Impact of Starting and Ending Parts

In these perceptual tests, mixed signals comprising 4 parts with total duration of 12–18 s are evaluated. While each configuration in Table 1 accompanies its inverse configuration, there is a hypothesis that starting part or ending part could bias the listeners assessment. In order to investigate this hypothesis, the MOS scores of mixed signals are aggregated into four groups: the configurations

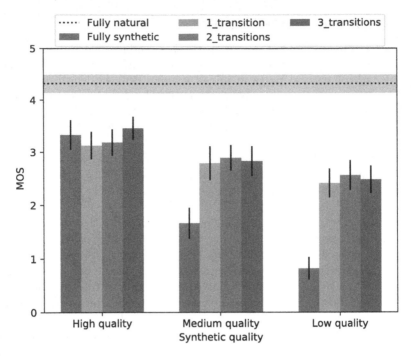

(a) MOS test result for evaluating mixed signals based on perceptual quality level.

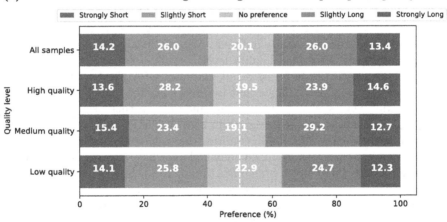

(b) AB test result for evaluation the impact of the length of continuous synthetic part in mixed signals based on perceptual quality level

Fig. 3. Aggregating previous perceptual tests result based on perceptual quality level of fully synthetic signals instead of voice corpus size.

that start with natural/synthetic part and the configurations that end with natural/synthetic part.

The MOS score related to these groups for different voice corpus sizes are shown in Table 3.

Table 3. Aggregating MOS test results based on starting end ending parts for mixed signals. Signals which start with natural part and end with synthetic part are evaluated with higher quality.

	5 h	1 h	30 min
Total average	3.38 ± 0.10	2.79 ± 0.10	2.43 ± 0.12
Start with synthetic (SNSN,SNNS,SSNN)	3.30 ± 0.21	2.60 ± 0.22	2.28 ± 0.25
Start with natural (NNSS,NSSN,NSNS)	3.48 ± 0.18	2.96 ± 0.18	2.58 ± 0.21
End with synthetic (NNSS,SNNS,NSNS)	3.48 ± 0.18	2.90 ± 0.20	2.70 ± 0.23
End with natural (SNSN,NSSN,SSNN)	3.30 ± 0.21	2.66 ± 0.20	2.15 ± 0.23
Start with natural and end with synthetic (NSNS,NNSS)	3.58 ± 0.21	2.92 ± 0.25	2.77 ± 0.26

This result shows a trend that mixed signals which start with a natural part and end with a synthetic part have been evaluated with a higher score. This bias indicates the weakness of our protocol for evaluating the mixed synthetic and natural speech of audio books. In the final audio book all parts are in middle (except the first and the last) and starting part and ending part will have less impact on listeners' perception. In any case, it is not possible to evaluate a complete audio book with listeners.

5   Conclusion

In this study, the idea of the mixing synthetic and recorded human voice for expressive audio-book generation has been investigated. A perceptual test has shown that mixed signals are preferred by listeners in comparison with fully synthetic signals. This has been observed with different levels of synthetic speech quality which was controlled by TTS voice corpus sizes.

Regardless of synthetic length in mixed audio book, the change of signal type may impact the listeners perception. Consequently, the impact of transition times in mixed signals, half synthetic and half natural, has been investigated. The MOS scores and a direct comparison in an AB test do not show that the number of transitions could change the listeners' perception and preference. The AB test has been originally designed to study the effect of the length of continuous synthetic part in mixed signal on listeners perception. Listeners do not have any preference between a long synthetic signal and those contain two short synthetic signals.

Investigation of the perceptual quality differences, between mixed signals and fully synthetic signals, also reveals that by improving the quality of synthetic signal, these two kinds of signals become comparable. As a future work, this comparison could be done with different kinds of TTS systems and bigger voice corpora which could result in higher synthetic quality.

Our analyses on results show that listeners have a bias on starting and ending parts of 4 breath groups mixed signals. It reveals that listeners have preferred mixed signals which start with a natural part and end with a synthetic part. This result emphasizes that evaluating a longer part of mixed signals is needed. Due to perceptual test duration, longer signals limit the number of evaluations per listener and need more listeners.

References

1. Boeffard, O., Charonnat, L., Le Maguer, S., Lolive, D., Vidal, G.: Towards fully automatic annotation of audio books for TTS. In: Eighth International Conference on Language Resources and Evaluation (LREC), Istanbul, Turkey, pp. 975–980 (2012)
2. Chiaráin, N.N., Chasaide, A.N.: Effects of educational context on learners' ratings of a synthetic voice. In: Seventh ISCA Workshop on Speech and Language Technology in Education, Stockholm, Sweden, pp. 47–52 (2017)
3. Clark, R., Silen, H., Kenter, T., Leith, R.: Evaluating long-form text-to-speech: Comparing the ratings of sentences and paragraphs. In: Tenth ISCA Workshop on Speech Synthesis (SSW10), Vienna, Austria (2019)
4. Gong, L., Lai, J.: To mix or not to mix synthetic speech and human speech? Contrasting impact on judge-rated task performance versus self-rated performance and attitudinal responses. Int. J. Speech Technol. **6**(2), 123–131 (2003)
5. Hinterleitner, F., Neitzel, G., Möller, S., Norrenbrock, C.: An evaluation protocol for the subjective assessment of text-to-speech in audiobook reading tasks. In: Blizzard Challenge Workshop. International Speech Communication Association (ISCA), Florence, Italy (2011)
6. Latorre, J., Yanagisawa, K., Wan, V., Kolluru, B., Gales, M.J.: Speech intonation for tts: Study on evaluation methodology. In: Fifteenth Annual Conference of the International Speech Communication Association (INTERSPEECH), Singapore, pp. 2957–2961 (2014)
7. Lewis, J.R., Commarford, P.M., Kotan, C.: Web-based comparison of two styles of auditory presentation: All TTS versus rapidly mixed TTS and recordings. In: Human Factors and Ergonomics Society Annual Meeting, San Francisco, vol. 50, pp. 723–727 (2006)
8. P.800.2, I.T.R.: Mean opinion score interpretation and reporting. International Telecommunication Union (2013)
9. Shamsi, M.: TTS voice corpus reduction for audio-book generation. Rencontre des Étudiants Chercheurs en Informatique pour le Traitement Automatique des Langues [Young Researchers Meeting in Computer Science for Automatic Language Processing], RÉCITAL, Nancy, France, 22 edn., vol. 3, pp. 193–204 (2020)
10. Shamsi, M., Chevelu, J., Lolive, D., Barbot, N.: Corpus design for expressive speech: Impact of the utterance length. In: Tenth International Conference of Speech Prosody, pp. 955–959 (2020)
11. Shamsi, M., Lolive, D., Barbot, N., Chevelu, J.: Corpus design using convolutional auto-encoder embeddings for audio-book synthesis. In: Twentieth Annual Conference of the International Speech Communication Association (INTERSPEECH), Graz, Austria, pp. 1531–1535 (2019)

12. Shamsi, M., Lolive, D., Barbot, N., Chevelu, J.: Investigating the relation between voice corpus design and hybrid synthesis under reduction constraint. In: Martín-Vide, C., Purver, M., Pollak, S. (eds.) SLSP 2019. LNCS (LNAI), vol. 11816, pp. 162–173. Springer, Cham (2019). https://doi.org/10.1007/978-3-030-31372-2_14

13. Wester, M., Watts, O., Henter, G.E.: Evaluating comprehension of natural and synthetic conversational speech. In: Eighth International Conference of Speech Prosody, Boston, USA, pp. 736–740 (2016)

Temporal Concord in Speech Interaction: Overlaps and Interruptions in Spoken American English

Tatiana Shevchenko[1] and Anastasia Gorbyleva[2(✉)]

[1] Moscow State Linguistic University, 38 Ostozhenka St., Moscow 119034,
Russian Federation
tatashevchenko@mail.ru
[2] Diplomatic Academy of the Ministry of Foreign Affairs of the Russian
Federation, 53/2, Build. 1 Ostozhenka St., Moscow 119992, Russian Federation
nastyagorbyleva@gmail.com

Abstract. Dynamicity of natural human interaction, such as the incidence of overlaps, cause difficulties for creating man-machine dialogue systems [4]. Overlaps may lead either to continued talk (positive effect) or to interruptions (negative effect). The study is concerned with temporal organization of speech due to participants' interaction in natural discourse, based on Santa Barbara corpus of spoken American English [5]. The goal is to examine the ways speakers in a two-party conversation intuitively coordinate their production time through the lengths of turns, pauses, overlaps and interruptions. Cultural norms are based on acoustic data in homogeneous (both men; both women) and heterogeneous (man and woman) gender-related compositions of dyads. The basic findings: irrespective of the dyad's composition there is one dominant speaker who takes up twice as much speaking time, while the other party initiates most of the overlaps; in the mixed pairs women take the lead; the time limits for the overlaps before they turn into interruptions vary across the groups but the values are close for the two partners. Correlation analysis provides evidence of negative correlation between the total turn length and the number of overlaps and interruptions; between turn length and the length of overlaps and interruptions. Gender differences are significant with regards to the number of overlaps and interruptions, as well as turn length and the length of overlaps and interruptions. Women speak longer, whereas men overlap and interrupt more often.

Keywords: English · Interaction · Gender · Temporal concord · Overlaps · Interruptions

1 Introduction

Research aimed at improving the efficiency of spoken dialogue systems meets with difficulties in processing prosodic accommodation of speakers in conversation and its dynamics in social interaction [4]. Some of the problems are being solved, while others, as we see them today, need a more fine-grained method of natural conversation analysis

© Springer Nature Switzerland AG 2020
A. Karpov and R. Potapova (Eds.): SPECOM 2020, LNAI 12335, pp. 490–499, 2020.
https://doi.org/10.1007/978-3-030-60276-5_47

before technical decisions are made for automatic speech recognition. In particular, cases of overlaps and interruptions that accompany natural conversation have to be objectively measured, specified for gender roles and socially evaluated.

Overlaps, for instance, are treated indiscriminately as cases of synchrony by automatic measurements, while in a natural conversation they may actually mean two different things: either backchannel signs of empathy, support, encouragement, attention and comprehension, or an attempt to claim the floor, which, in case they exceed a certain time limit, may cause interruption. That time limit remains to be found: we have to explore on what condition overlaps may be crucial for the conversation flow. We expect the time limits to be culturally specific, just like there is a complex relationship between length of pause, structure of pause and turn-taking which manifest themselves in interaction [16, 21–23]. We will consider these events in a socio-cultural perspective, correlated with the role [1] and gender of the speakers.

Previous research based on Gravano Games corpus [11] showed that interruptions were detrimental to the success of the game, on account of lack of 'entrainment' (accommodation) of the participants. We hypothesize that overlaps actually predict interruptions, and if the speaker fails to notice them, or ignores them consciously, the more insistent partner will eventually break the flow of the conversation. We also believe that the rules of overlaps and interruptions are culturally-based: in Japan, for instance, backchannel cues ('aizuchi') are expected to be permanent, and they are, therefore, positively evaluated by the interlocutor. In other cultures the effect of simultaneous talk could be quite different: American women gave way to French women in case the latter started overlapping; so did Californians to New-Yorkers ('high-considerateness' style is opposed to 'high-involvement' style) [19].

Another culturally determined side of verbal interaction rests with the expectations based on gender stereotypes [12, 13, 17]. American cultural gender stereotypes used to attribute leadership, assertiveness and aggression to men's talk [18], while women were expected to demonstrate cooperation, support and compliance. Women also were accused of talkativeness [10] and their communication strategy was interpreted as manipulative [19]. However, when the issue was discussed in the 80-s and 90-s, a lot of research data was found to demonstrate men's dominance in discussion, including their tendency to interrupt the partner, whereas women's stereotypes were rejected as 'myths' and prejudice [8, 9, 23]. It turned out that women's verbal behavior was evaluated differently from that of men. Our goal is to obtain facts by objective methods.

We are aware of multimodality in verbal, para-verbal and non-verbal communication of face-to-face interaction [15]. Multimodal means of accommodation include lexical and grammatical [14, 21], as well as visual and auditory signs of convergence between the speakers [2]. In our previous work we found prosodic features of speakers' convergence on pitch and intensity levels locally, at turn boundaries, and dynamics of prosody globally, in the three consecutive parts of the conversation [7]. In the current study we will focus on global temporal features of turn-taking determined by the speakers' interaction in conversation. We will measure overlaps and pauses correlated with the total turn length to establish the role of the leader. Overlaps are to be correlated with interruptions to establish cultural constraints on their length.

Thus, we will look for the online timing concord means in the temporal organization of a two-party conversation, as manifested in the lengths of turns, pauses, overlaps and the incidence of interruptions in friendly (unison) conversations between acquaintances, educated speakers of American English [5]. There are a number of questions we posed for ourselves:

- With regards to convergence: do speakers opt to balance their speaking time?
- With regards to possible dominance: who dominates the talk?
- Who initiates overlaps and interruptions? How often?
- How long is it before overlaps turn into interruptions?
- What are gender-related features in conversational timing?

2 Methodology

2.1 Data

The material under analysis is based on the Santa Barbara Corpus of Spoken American English [5]. The Corpus presents a large body of recorded conversations of naturally occurring spoken interaction from all over the United States and as a whole consists of 60 recordings approximately 20 min each. The audio data as published by the Linguistic Data Consortium consist of 16-bit, stereo, 22.05 kHz audio files in WAV format. The predominant form of language use represented is face-to-face conversation. The corpus is sampled according to gender, age, state of origin, education levels, occupation, social and ethnic backgrounds (white, black, Chicano, Crow Indian, Hispanic, Latino). Personal names of speakers on the recordings, as well as other identifying information such as telephone numbers, have been altered to preserve the anonymity of the speakers.

Table 1. Speakers' data.

#Speaker	Gender	Age	State	Education	Occupation	Race/Ethnicity
S1	m	45	CA	college	Sound editor	white
S2	m	34	CA	B.A.	Musician/computers	white
S3	f	30	CA	B.A.	student	white
S4	f	40	CA	B.A.	housewife	white
S5	m	33	CA	B.A.	computers	white
S6	f	38	CA	B.A.	actress	white
S7	f	28	MT	college	student	Crow Indian
S8	f	27	MT	college	cook fire	Crow Indian
S9	m	32	CA	high school	Car salesman	Chicano
S10	m	–	CA	–	Factory worker	Chicano
S11	f	22	MA	B.A.	Bookstore employee	white
S12	m	24	MA	B.A.	Library public service	white

For our pilot study we selected six dialogues that total to around 2 h of recorded speech in order to balance sexes and age groups. The pairs we chose include all three gender combinations (male–male, female–female, female–male) with similar education level, annotated as 'B.A.' or 'college' (with one exception only). The subjects are native speakers of American English from the West Coast (California), Northwest (Montana) and the East Coast (Massachusetts). The age ranges from 22 to 45 (mean 31.4) (Table 1). The subset of the corpus includes dialogues in the form of free conversation and speaker pairs who are acquainted with one another and who were both interested in the topic of their friendly chat.

2.2 Measurements

All the conversations were analyzed by means of computer program PRAAT [3] with pauses and phonation durations calculated manually. The measures comprise the following variables:

- Turns duration: the length of speaking time between turn shifts annotated for every speaker;
- Total turns duration: the overall length of speaking time within a conversation annotated for every speaker;
- Pause duration: the time attributed to silent inter-speaker pauses. Mid (inner-speaker) pauses were not taken into account. The inter-speaker pauses were classified into two groups: inaudible pauses (<200 ms) and audible pauses (>200 ms) (classified according to [6]);
- Overlaps duration: the duration of simultaneous talk, including backchannels, where S2 starts to talk while S1 is still talking, but S1 completes the turn. Overlap duration was counted in each speaker 's' turn;
- Interruptions duration: the duration of simultaneous talk, where S2 starts to talk while S1 is still talking, and as a result S1 does not complete the turn;
- Simultaneous talk duration: the duration of simultaneous talk including both overlaps and interruptions. Simultaneous speech was counted as part of both speakers' time;
- Number of overlaps and interruptions: the number of times the speaker talked simultaneously with his partner, with either positive or negative effect.

The total time of the narrow corpus is 6918.8 s (approximately 2 h). Over the time of six conversations we registered 939 turn-shifts, 313 pauses, 270 overlaps and 139 interruptions.

Statistical processing included one-way analysis of variance (ANOVA) that was conducted to test the difference in the speech parameters between six dyads and two gender groups. Furthermore, the correlation analysis (Pearson correlation coefficient) was run to test the relationship between the extracted measures. Statistical procedures were performed in Jamovi program [20].

3   Results

3.1   Temporal Concord Features: How Two People Regulate Their Timing

In this section we will look at the data to find out how two people regulate their timing in the conversation. By comparing the data computed from the six conversations we can observe that in five cases out of six there is a dominant speaker in the conversation who takes twice as much time as the other party (Table 2).

There is a leader in each group, irrespective of the gender composition in a dyad, i.e. both genders may be favored. The most unexpected fact is that in the two mixed dyads women took the lead.

Correlation analysis gave evidence of the overlaps initiated by the speakers with the least share of the speaking time.

According to ANOVA results the values of mean interruption duration significantly vary in six conversations ($p < .001$) (Fig. 1). The time spent in overlaps before they turn into interruptions varies considerably between the pairs, and it is hard to find a common acceptable norm for overlaps to stop before turn-taking. However, within a single conversation the two speakers appear to be in full agreement about that limit, which suggests their online coordination in timing. In a friendly chat the values of mean interruption durations are very close with average 10.7% of deviation. A certain length of the overlap is taken for a signal to give way to the interlocutor, a regulating technique of the listener.

Table 2. Temporal characteristics in six conversations (12 speakers).

Conver-sation	Speaker	Turns mean, s	Turns median, s	Total turns-duration, s	Overlap N	Overlap mean, s	Inter-ruption N	Interruption mean, s	Overlaps + interruptions N	Total overlaps + Interruptions duration, s
1	S1 m	22.38	8.89	1151.70	14	1.39	11	1.25	25	33.26
	S2 m	10.88	3.71	257.59	49	0.87	21	1.02	70	63.76
2	S3 f	2.60	1.40	148.40	31	0.87	14	0.81	45	38.18
	S4 f	46.04	23.69	1371.83	8	0.54	11	0.81	19	13.19
3	S5 m	3.62	1.86	311.62	12	1.13	15	0.83	27	25.98
	S6 f	8.72	4.33	715.39	9	0.55	6	0.75	15	9.46
4	S7 f	8.15	3.22	521.72	9	1.30	5	0.60	14	12.58
	S8 f	8.16	2.78	636.73	24	0.75	3	0.69	27	20.06
5	S9 m	3.53	2.22	338.62	50	0.84	16	0.65	66	52.17
	S10 m	8.65	5.31	822.00	30	0.74	22	0.48	52	32.96
6	S11 f	10.13	4.61	942.19	2	1.27	4	0.48	6	4.44
	S12 m	2.82	1.62	259.52	32	1.09	11	0.47	43	40.09

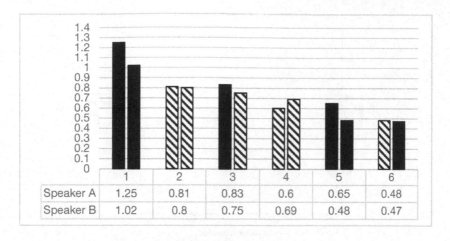

	1	2	3	4	5	6
Speaker A	1.25	0.81	0.83	0.6	0.65	0.48
Speaker B	1.02	0.8	0.75	0.69	0.48	0.47

■Male ⊠Female

Fig. 1. Mean interruption durations for six two-party interactions.

Alongside the analyses of variance Pearson correlation coefficient was tested for total turns duration viewed as the leader's characteristic and other temporal parameters revealing the following relationships:

- significant negative correlation between total turns duration and overlap duration ($r = -.640$, $p = .012$);
- significant negative correlation between total turns duration and number of overlaps ($r = -.608$, $p = .018$);
- significant negative correlation between total turns duration and overlaps and interruptions duration ($r = -.564$, $p = .028$);
- significant negative correlation between total turns duration and number of overlaps and interruptions ($r = -.556$, $p = .030$).

The speakers who appeared to be passive on account of their modest share of the speaking time, were, in fact, active in claiming the floor and gaining it by overlaps and interruptions. The findings indicate that the more interlocutor speaks, the less one resorts to overlaps and interruptions and vice versa (Fig. 2). That is, less active participants make significantly more attempts to interfere with their partner's speech. The result is valid for both male and female speakers.

Another particularly important factor for speakers' coordination is turn-taking. The more speakers understand each other, the smoother the conversation flow and the shorter their inter-speaker pauses are. We realize that in face-to-face interaction coordination is aided by visual signs, while in auditory data we find three types of turn-taking: short or inaudible pauses (up to 0.2 s), long or audible pauses (0.2 s and more), and interruptions which follow overlaps. In the analyzed dialogues out of 392 turn-shifts 96 pauses (24.5%) were short or inaudible (0–0.2 s) and 218 (55.6%) were audible pauses (more than 0.2 s). The remaining 78 turns (19.9%) were interrupted due to overlapping.

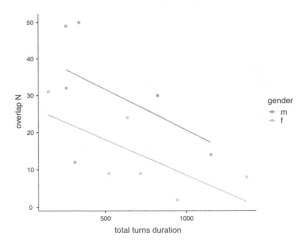

Fig. 2. Correlation of total turns duration and number of overlaps.

3.2 Gender Differences: Speaking Time Shares and Interruptions Number

As is illustrated in Fig. 3, the mean duration of interruptions in gender-related data shows the difference in time limits between the sexes: men are prepared to put up with their partner's interference for a longer time before overlaps turn into interruptions, i.e. they are more reluctant to give way.

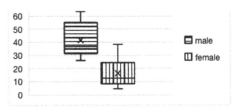

Fig. 3. Mean duration of overlaps and interruptions for male and female speech groups.

ANOVA results also reveal gender-related difference in the following values:

- overlaps duration (p = .025)
- interruptions duration (p = .025)
- number of overlaps made during conversation (p = .037)
- number of interruptions made during conversation (p = .016)
- total duration of overlaps and interruptions (p = .016)
- total number of overlaps and interruptions made during conversation (p = .03)

The results are manifested in more frequent overlaps and interruptions occurring in male speech and longer time spent on simultaneous talk.

4   Conclusions and Discussion

Starting from previous research data on the dynamicity, accommodation and gender-related features in verbal interaction, we focused on overlaps which reflect all the above-mentioned tendencies in interaction. In the smoothly flowing friendly chats on topics of common interest we could still find:

- unequal shares of the speaking time, with one party taking up two thirds of the time;
- in homogeneous two-party conversations (two men, two women) one party may be called dominant as far as the amount of speaking time is concerned; in case of a heterogeneous composition of the conversation women are evidenced to speak twice as much as men;
- the party who speaks less initiates most of the overlaps;
- one third of the overlaps result in interruptions;
- the time limit for an overlap to have a positive effect of continued talk, as well as the length of simultaneous talk which results in interruption varies across pairs of speakers but has much similarity within the pair;
- overall comparison of turn lengths, number of overlaps and interruptions, total lengths of overlaps and interruptions proves that women speak more, whereas men initiate most of the overlaps and interruptions;
- men spend more time in simultaneous talk before they give way to the partner.

Thus, the temporal regulation of speaking activity in a two-party conversation shows that the speaking shares are not actually equal but at least through the technique of overlapping the party who has a lesser amount of speaking time can demonstrate interest and involvement, and claim the turn. As regards the time limit for simultaneous talk and the evident similarity between the two parties in setting that limit, it appears to be coordinated online.

Summing up the data collected in our pilot study of spoken American corpus we can evaluate the results:

- from the social norms perspective
- as a means of social accommodation
- from gender roles stereotypes
- from man-machine interaction perspective.

First, culturally relevant for American spoken dialogue are the tendencies for overlaps and interruptions to occur in American friendly talks. Contrary to D. Tannen's descriptions of American women being interrupted by French women who started overlapping, or the story about Californian speakers shocked by New Yorkers on similar occasions [19], we found that all American English speakers, including Californians and ethnic minorities' representatives, tend to overlap, though to a varying extent. The results seem to be dependent on their place of origin associated with local cultural conventions, gradually diminishing from the West to the East of the country. Californians tend to be most patient and are ready to put up with their partners' interference for approximately 2.32 s. (in the range of 1.58–3.1 s.) before they give the floor to their interlocutor, while Montanans (the Northwest) can tolerate up to 1.65 s. of simultaneous talk (0.75–2.55 s.) before interruption. Massachusettsans (the East Coast) have patience only for 0.95 s. of overlapping.

Second. J. Hirshberg and her co-authors [11] when analyzing verbal interaction in games, gave a negative assessment of interruptions as a sign of unsuccessful 'entrainment' which may lead to the game lost by the partners. Our data, however, may suggest another interpretation of interruptions as a regulating mechanism for establishing temporal parity of the two interlocutors.

Three. There is plenty of evidence reported in American research about men dominating discussion with women, and to claim the opposite would be treated as gender prejudice [8–10, 18, 19, 21, 23]. Nevertheless, our random choice of two mixed dyads from the spoken corpus, as well as the overall scores from twelve participants, provided the data on women talking more than men. The only plausible explanation may be the one suggested by J. Holmes [8] about men generally talking more in formal, public contexts which have a potential for increasing their social status, whereas women are more likely to contribute in private, informal interactions to maintain relationships in situations where they feel more socially confident.

Finally, from man-machine interaction perspective we could predict that spoken dialogue system has to reckon with dynamicity of human talks, including the incidence of overlaps, interruptions, their frequencies and durations governed by social norms appropriate for certain cultures, communities, styles and gender roles if speaker/listener-friendly effect is desirable. On the other hand, knowledge of the social norms and conventions as regards the incidence and duration of overlaps and interruptions associated with certain cultures and subcultures may serve as a means of automatic speaker identification.

References

1. Babel, M.: Evidence for phonetic and social selectivity in spontaneous phonetic imitation. J. Phon. **40**, 177–189 (2012)
2. Bernieri, F.J., Rosenthal, R.: Interpersonal coordination: behavior matching and interactional synchrony. In: Feldman, R.S., Rime, B. (eds.) Fundamentals of Nonverbal Behavior, pp. 401–432. Cambridge University Press, Cambridge (1991)
3. Boersma, P., Weenink, D.: Praat: doing phonetics by computer (Version 6.0.14). http://www.praat.org/. Accessed 22 Jan 2018
4. De Looze, C., Scherer, S., Vaughan, B., Campbell, N.: Investigating automatic measurements of prosodic accommodation and its dynamics in social interaction. Speech Commun. **58**, 11–34 (2014)
5. Du Bois, J.W., et al.: Santa Barbara Corpus of Spoken American English, Parts 1–4. Linguistic Data Consortium, Philadelphia (2005)
6. Fletcher, J.: The Prosody of Speech: Timing and Rhythm. The Handbook of Phonetic Sciences, 2nd edn. pp. 523–602. Wiley-Blackwell, Chichester (2013)
7. Gorbyleva, A.: Prosodic convergence as a result of speakers' interaction in spontaneous dialogue. Teoreticheskayaiprikladnayalingvistika **5**(3), 41–52 (2019). (in Russia)
8. Holmes, J.: Women talk too much. In: Bauer, L., Trudgill, P. (eds.) Language Myths, pp. 41–49. Penguin Books, London (1998)

9. James, D., Drakich, J.: Understanding gender differences in amount of talk: a critical review of research. In: Tannen, D. (ed.) Gender and Conversational Interaction, pp. 281–312. Oxford University Press, New York (1993)

10. Karpf, A.: The Human Voice: How this Extraordinary Instrument Reveals Essential Clues About Who We Are. Bloomsbury, New York (2006)

11. Levitan, R., Hirschberg, J.: Measuring acoustic-prosodic entrainment with respect to multiple levels and dimensions. In: Interspeech, pp. 3081–3084 (2011)

12. Namy, L., Nygaard, L., Sauerteig, D.: Gender differences in vocal accommodation: the role of perception. J. Lang. Soc. Psychol. **2**, 422–432 (2002)

13. Pardo, J.S.: On phonetic convergence during conversational interaction. J. Acoust. Soc. Am. **119** (4), 2382–2393 (2006)

14. Pickering, M.J., Garrod, S.: Toward a mechanistic psychology of dialogue. Behav. Brain Sci. **27**, 69–226 (2004)

15. Potapova, R., Potapov, V., Komalova, L., Dzhunkovskiy, A.: Some Peculiarities of internet multimodal polycode corpora annotation. In: Salah, A.A., Karpov, A., Potapova, R. (eds.) SPECOM 2019. LNCS (LNAI), vol. 11658, pp. 392–400. Springer, Cham (2019). https://doi.org/10.1007/978-3-030-26061-3_40

16. Sacks, H., Schegloff, E.A., Jefferson, G.: A simplest systematics for the organization of turn-taking for conversation. Language **50**(4), 696–735 (1974)

17. Shevchenko, T., Sokoreva, T.: First minute timing in American telephone talks: a cognitive approach. In: Salah, A.A., Karpov, A., Potapova, R. (eds.) SPECOM 2019. LNCS (LNAI), vol. 11658, pp. 451–458. Springer, Cham (2019). https://doi.org/10.1007/978-3-030-26061-3_46

18. Swann, J.: Talk control: an illustration from the classroom of problems in analyzing male dominance of conversation. In: Coates, J., Cameron, D. (eds.) Women in Their Speech Communities, pp. 122–140. Longman, London (1988)

19. Tannen, D.: Language and Culture. An Introduction to Language and Linguistics, pp. 343–372. Cambridge University Press, Cambridge (2008)

20. The jamovi project. jamovi (Version 1.2) (2020). https://www.jamovi.org. Accessed 22 Jan 2018

21. Wennerstrom, A., Siegel, A.F.: Keeping the floor in multiparty conversations: Intonation, syntax, and pause. Discourse Process. **36**, 77–107 (2003)

22. Wilson, T., Zimmerman, D.: The structure of silence between turns in two-party conversation. Discourse Process. **9**, 375–903 (1986)

23. Zimmerman, D.H., West, C.: Sex roles, interruptions and silences in conversations. Gend. Soc. 105–129 (1983)

Cognitively Challenging: Language Shift and Speech Rate of Academic Bilinguals

Tatiana Shevchenko[ID] and Tatiana Sokoreva[(✉)][ID]

Moscow State Linguistic University, 38 Ostozhenka Street, Moscow 119034,
Russian Federation
tatashevchenko@mail.ru, jey-t@yandex.ru

Abstract. Speech rate characteristics of academic bilinguals are critically assessed from the linguistic and socio-cultural perspectives. We argue that language shift in typologically different languages is especially challenging as it involves shifting the focus to linguistic units of a different order, such as the syllable in Mandarin and the sentence structure in English. Failure to cope with major semantic and syntactic structures results in randomized pause locations, while focus on the syllable increases its duration. Slow down or misplaced pauses, as purely linguistically determined temporal features, transpire in speaking and reading in a second or a third language. We assume that they may be the reason for a drop in social attractiveness ratings in the situation of intercultural communication. As evidenced by a number of temporal features, Mandarin students speak English much slower than their American English peers. In reading Chinese wisdom phrases Mandarin speakers are faster than in English. Russian learners significantly slow down in reading Chinese phrases but are more fluent in English.

Keywords: Bilingualism · Speech rate · Mandarin · English

1 Introduction

Studies in neural physiology of syllable, word and sentence prosody processing in typologically different languages suggest that Mandarin prosody processing calls for a more complex mechanism involving both hemispheres simultaneously, together with attention and working memory centers, which accounts for a longer reaction time in sentence processing, compared to English listeners (See review in [5]). That was interpreted as the interaction of lexical tone and intonation information based on the experiments with one-syllable-morpheme/one-sentence linguistic unit.

However, still more complex processes affecting temporal organization of speech, are bound to operate in connected speech data, and the relevant factors for Mandarin are reported as syllable duration coordination in a two-syllable word, the position of the word in a sentence, focus and final syllable lengthening [4, 13, 20], metric organization and *sandhi* [16], rhythm cadences [19], emotional intonation impact [10, 18], mode and style of speech [9, 13]. While most of the context factors are known to be operating in a

© Springer Nature Switzerland AG 2020
A. Karpov and R. Potapova (Eds.): SPECOM 2020, LNAI 12335, pp. 500–508, 2020.
https://doi.org/10.1007/978-3-030-60276-5_48

number of languages, though to a varying extent [4, 14, 15], there are such specifically Mandarin properties as, for instance, the durational 'etymological' hierarchy of the four Mandarin tones (3 > 2 > 1 > 4) [13] and their associations with certain emotional states [18].

The overall effect of either slow or fast tempo of a learner's speech is of socio-cultural relevance, and together with segmental errors in the target language [3, 11] may become either positively or negatively assessed in intercultural communication [2, 7, 20]. In case of the Mandarin learners of English, the general impression of slow tempo is most common. Now that the Russian undergraduates started to learn both Mandarin and English, we set ourselves the task of finding out whether it is the language structure of a particular language, Mandarin or English, or their proficiency in either of them, that is responsible for the slow down. We hypothesize that in cases of academic multilingualism, language shift is always a cognitively challenging task [17], and finding a more powerful factor for each learners group is our main goal. The subjects are Russian learners of Mandarin who spent a year in China (n = 10), and a group of Mandarin speakers, also university undergraduates, who spent a year in Moscow learning Russian (n = 10). In both groups English was their major foreign language in secondary school, and a means of international communication before they could achieve fluency in their second target language. So, at the time of the recording they were all trilingual. Other Mandarin speakers were learners of English who sent their recordings from China (n = 7). We also compared our data on American English (n = 10) with that of Mandarin speakers (n = 10) reported in [2].

The overall impression of speaking rate (tempo) depends, as is well known, on the articulation rate, pause time and their proportions [8, 12]. To this list a comparison of the temporal components in the speech of 10 British and 10 Mandarin English speakers in an interview added a few other items which are determined by the syntactic com-position of discourse and the mental processes involved in speech production: the length of between-pauses chunks, the amount and the length of vocalized pauses [2]. Summing up the impressionistic views of native speakers of English associated with Mandarin English temporal features the authors made very important sociocultural conclusions: slow tempo has negative impact on social attractiveness of speakers. This is certainly not the effect we are aiming at in introducing Mandarin together with English in the academic curriculum. We only hoped that American English which is known to be slower than British English [14], with its slurred sound and less pointed rhythm might be closer to Mandarin English in temporal features. We also meant to avoid the complexity of spontaneous speech production by focusing on reading as a more prepared mode of speech and less cognitively challenging.

Thus, the *goal* of the current research consists in exploring the *speech rate* of Mandarin and Russian learners in Mandarin and English with the purpose of estimating their potential in sociocultural evaluation after language shift, and finding the cause of the (un)desired effect.

2 Methodology

2.1 Procedure

The research comprises two steps. The preliminary research – Step 1 – presents a range of temporal factors which distinguish the speech of young Mandarin speakers compared with their American peers. The articulation rate, however, which is the mean/median syllable duration was not considered yet. Most of the researchers hold that articulation rate is basically an individual characteristic which is prone to situational and stylistic/pragmatic variability [12]. In Step 2 we intend to find evidence of nationally and language-specific features of articulation rate which transpire in the speech of bilinguals due to language shift from the intonation language to the syllable-based one. Here the observations will be focused on the syllable and its duration.

2.2 Material

The material of the first experiment represents 10 speech samples of American dialogue speech derived from the Switchboard Corpus of American English telephone conversations [6]. The duration of analyzed recordings totals to over 10 min (about 1 min per speaker). The speakers are young people whose age (20–39) and education level (graduate) are annotated; the group is also gender-balanced. The received results are compared to those of 10 Mandarin English speakers obtained by H.C. Chen and Q. Wang in 2018 [2].

The material of the second experiment is the corpus of 20 speech samples of 10 Russian learners reading the Chinese wisdom phrases (n = 4) in Mandarin and in English. The recordings were made at MSLU with the help of professional voice recording equipment. All speakers are female, aged 21–22, students learning Mandarin Chinese and English as foreign languages, spent a year in Beijing learning Mandarin.

The data also comprise the corpus of 20 speech samples of 7 Chinese learners reading the Chinese wisdom phrases (n = 4) in Mandarin and in English. The speakers include 5 women and 2 men, aged 21–22, university undergraduates, who spent a year in Moscow learning Russian (n = 10).

In both groups English was their major foreign language in secondary school, and a means of international communication before they could achieve fluency in their second target language.

We admit that the data set would benefit if the samples included Russian and Chinese speakers reading in Russian, thus, this will be amended in our forthcoming research.

The material of the third experiment includes speech samples of 10 Mandarin speakers reading the standard text (the English fable "Arthur the Rat", totaling to 385 syllables) in English which is contrasted to Russian and native British speakers' results in reading rates. The speakers are on exchange program learning Russian in Moscow.

2.3   Measurements

The temporal characteristics of American English unprepared speech are studied through the following parameters measured by PRAAT [1]:

- averaged pause duration (APD) – the duration of physical pause that equals to 100 ms and above;
- number of vocalized pauses per minute (VP) – the total number of vocalized pauses divided by total duration of utterances;
- phonation ratio (%P) – the ratio of phonation duration (including the vocalized pauses duration) to total duration of the utterances;
- articulation rate (AR) – mean number of syllables per minute of speech (excluding physical pauses and including vocalized pauses and partial words containing at least one vowel and one consonant sounds);
- mean length of run (MLR) – mean number of syllables in IP (Intonation Phrase – a pause-to-pause chunk, which corresponds to an English clause or a sentence, surrounded by pauses of 100 ms and above).

The reading rate of Chinese wisdom phrases (pronounced both by Russian and Chinese learners) as well as the English fable (read by Chinese learners) were investigated by measuring the following parameters:

- averaged syllable duration (ASD) – the mean/median duration of syllable in English/morpheme in Chinese;
- averaged pause duration (APD) – the mean/median duration of pauses of 200 ms and above;
- number of silent pauses (SP).

Statistical processing included one-way analysis of variance (ANOVA) that was conducted to test the difference in the above-mentioned speech and reading rate parameters between two nationality groups of speakers and two languages of the read phrases. Furthermore, the Mann-Whitney test was run to verify the statistical difference between each pair of variables. Statistical procedures were performed in JAMOVI program.

3   Results

3.1   Temporal Features in Mandarin English Versus American English

Our first step consisted in investigating the components of Mandarin English temporal organization of speech which distinguish it from American English native speakers' performances.

The first experiment results covering the temporal characteristics of young Americans' speech together with pairwise data on Mandarin English (taken from [2]) are represented in Table 1.

As is seen, four out of five parameters differ significantly in two groups of speakers. The average pause duration in Mandarin English is 1.3 longer than that of American English (0.59 and 0.44 s respectively). The average number of vocalized pauses per

Table 1. Temporal characteristics of American English and Mandarin English speech.

Measure	American English				Mandarin English (from Chen, Wang 2018)			
	Mean	Max	Min	SD	Mean	Max	Min	SD
APD (sec)	0.44	1.225	0.113	0.24	0.59	0.92	0.44	0.14
VP	4.1	9.85	0.97	5.45	9.68	19.69	2.55	4.3
%P	0.88	1	0.63	0.1	0.87	0.95	0.7	0.06
AR	303	473	167	65	191.14	217.95	153.23	22.97
MLR	9.5	29	2	5.09	14.01	34.31	5.72	7.14

minute of speaking in Mandarin English more than twice exceeds the same parameter in American English (9.68 and 4.1 respectively). American English articulation rate is 1.6 times higher than the one of Mandarin English speakers (303 vs. 191 syllables per minute respectively). And the number of syllables in IP in Mandarin English 1.5 times larger than the value of the same parameter in American English (14.01 and 9.5 syllables respectively).

The only temporal characteristic with similar values, that appeared to have no significant difference in two groups of subjects, is the ratio of phonation duration to the total duration of utterance.

Thus, the speech of young Chinese talking English proved to have slower tempo due to the lower articulation rate values; it sounds lengthy and lingering due to a larger number of syllables in IP, longer duration of physical pauses as well as bigger amount of vocalized pauses. All that, together with slower syllable-to-syllable articulation, adds to the impression that the English speech of the Chinese has too long and monotonous utterances which prosodic organization is violated as compared to that of native English speech. Hence, the phrase prosody in Mandarin Chinese does not reflect the speech rhythm characteristic of accent-based English language, which rests on the contrast between accented and unaccented syllables duration where the latter is much shorter than the former, thus, accounting for a larger number of syllables which can be pronounced per minute by AmE speakers. Also, it is hard to identify a communicative center of the utterance in extra-long IPs in the English speech of Chinese.

3.2   Russian and Chinese learners' Reading Rates of Chinese Wisdom Phrases in Mandarin and in English

The second experiment was aimed at defining the reading rate of Chinese wisdom phrases pronounced in Mandarin and in English by Russian and Chinese learners.

As for the average syllable duration, the results revealed a significant difference ($p < .001$) in the values of Russian speakers in their Mandarin and English readings where the former considerably exceeds the latter (see Fig. 1) at 360 ms and 245 ms respectively.

The average pause duration in the two variants of reading by Russian learners did not prove to have significant distinctions.

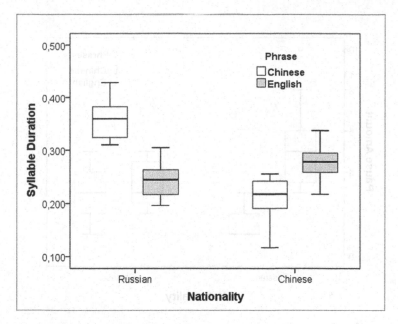

Fig. 1. The boxplots of syllable duration values in Mandarin Chinese and English phrases pronounced by Russian (n = 10) and Chinese (n = 7) learners.

However, the number of pauses in the Mandarin and English phrases, read by Russian subjects, differs greatly (see Fig. 2) and amounts to 5 vs 1.5 respectively.

Regarding the three parameters in Chinese learners' reading, the significant difference was discovered only in the average syllable duration values where ASD in reading the Mandarin phrase is much shorter than ASD in reading the English phrase, 217 ms and 278 ms respectively (see Fig. 1).

Comparing the results of the Russian and Chinese learners the statistics revealed the significant difference in the average syllable duration (p < .05) and the average pause duration (p < .05) values between the two national groups of speakers (Russians and Chinese), disregarding the language used for reading.

The general comparison of the languages employed for reading the phrases (disregarding the nationality of the speakers) showed that there exists a significant difference in the average syllable duration (p < .05) and the number of pauses (p < .001) used in Mandarin and English readings.

Thus, the Russian learners appeared to have slower tempo, longer duration and larger amount of pauses when they read in Mandarin as compared to reading in English which can be accounted for by a more complicated, cognitively challenging task for Russian learners, to acquire the prosodic system of the Mandarin Chinese language.

At the same time, while reading in English, which is foreign both for Russian and Chinese learners, the Russian subjects employed faster tempo and fewer pauses in contrast to the Chinese respondents reading in English.

In their turn, the Chinese learners proved to have lower values of average syllable duration when they read in their native language, though the pause duration values as

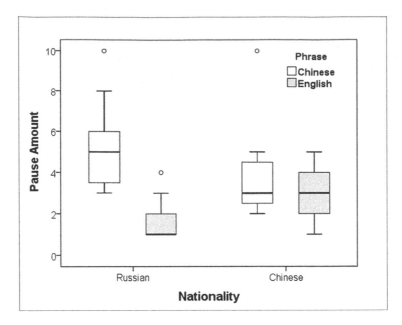

Fig. 2. The boxplots of pauses amount in Mandarin Chinese and English phrases pronounced by Russian (n = 10) and Chinese (n = 7) learners.

well as their amount do not differ significantly and have almost the same values both in reading in native Mandarin and in reading in English.

3.3   Reading Rate of the Standard Text by Mandarin and Russian Learners Compared to Native British Speakers

The third experiment implied the comparison of articulation rate in reading a standard English text by Mandarin and Russian speakers compared to the British model (RP) reported in [14] (Table 2).

Table 2. English reading articulation rate of Mandarin, Russian and British speakers.

	Mandarin	Russian	British (RP)
Articulation rate (ms)	279	245	213

In reading the English text the Chinese readers are the slowest as regards the average syllable duration which could be accounted for by a largely commented lack of reduction in unstressed syllables [3, 11]; the Russian learners do follow the stress-based rhythm but not to the extent the British model requires. As the accent-based English language rests on the contrast between accented and unaccented syllables duration, the

AR measure (the mean duration of syllables) in British English is expected to be lower due to a much shorter unaccented syllables duration.

4 Conclusions and Discussion

Psychological research in Laboratory Phonology revealed the nature of cognitive challenge in language shift by manipulating the prosody of one-syllable word which in a tone language like Mandarin might convey both the lexical tone and the intonation of either a statement or a question. Code-switching in natural running speech is a much more complex phenomenon which we attempted to analyze by following the techniques of Chen and Wang [2]. American English and Mandarin English data proved to be distinctive in pause and IP duration. That accounts, we assume, for Mandarin speakers' difficulties in coping with the syntactical composition and the pragmatic meaning of English.

We also assume that the overall effect of slow Mandarin English is due to articulation rate and loss of accent-based rhythm which rest on the specific segmental properties of Mandarin English found by D. Deterding [3]: frequent introduction of an epenthetic e-sound inside the English consonant clusters and lack of vowel reduction in unaccented syllables. The presence of long open syllables in the unstressed positions, especially in form words, destroys the contrast between accented and unaccented syllables which is the basis of English rhythm.

For the Russian learners the shift from Russian to Mandarin entails the introduction of a second focus of attention, i.e. the syllable and its prosodic shape, apart from the new foreign words and their meaning. Hence, they slow down by increasing syllable duration and pause lengths. Russian learners of English appear to be more fluent in English than Mandarin speakers, as evidenced by average syllable duration, pause amount and pause durations in reading both phrases and the text. The Mandarin Chinese speakers show stability in average syllable duration in reading both English phrases and the text. Syllable duration data, therefore, may be informative with regards to the language structure and the challenging task of language shift by academic bilinguals.

References

1. Boersma, P., Weenink, D.: Praat: doing phonetics by computer. Version 5.3.80. http://www. praat.org/. Accessed 20 May 2020
2. Chen, H.C., Wang, Q.: The effects of chinese learners' english acoustic-prosodic patterns on listeners' attitudinal judgements. SE Asian J. Engl. Lang. Stud. 22(2), 91–108 (2018)
3. Deterding, D.: The pronunciation of english by speakers from China. Engl. World-Wide 27(2), 175–198 (2006)
4. Fletcher, J.: The Prosody of Speech: Timing and Rhythm. The Handbook of Phonetic Sciences. 2nd edn. pp. 523–602 (2013)
5. Gandour, J.T.: Neural substrates underlying the perception of linguistic prosody. In: Tones and Tunes, Volume 2: Experimental Studies in Word and Sentence Prosody, pp. 3–25 (2007)

6. Godfrey, J., Holliman, E.: Switchboard-1 Release 2 LDC97S62. Linguistic Data Consortium, Web Download. Philadelphia (1993)
7. Huo, S.Y., Luo, Q.: Misuses of english intonation for Chinese students in cross-cultural communication. Cross-Cult. Commun. **13**(1), 47–52 (2017)
8. Kachkovskaia, T., Skrelin, P.: Intonational phrases and pauses in read and spontaneous speech: evidence from large speech corpora. Anal. Russ. Colloq. Speech **2019**, 46–54 (2019). (In Russian)
9. Kratochvil, P.: Intonation in Beijing Chinese. Intonation Systems: A Survey of Twenty Languages, pp. 417–431 (1998)
10. Li, A., Fang, Q., Dang, J.: Emotional intonation in a tone language: experimental evidence from Chinese. In: ICPhS XVII, pp. 1198–1201 (2011)
11. Postnikova, L.V., Buraya, E.A., Galochkina, Ye, I., Shevchenko, T.I.: Typology of Varieties in English Phonology: Monography. Gos. ped. Un-ta im. L.N. Tolstogo (2012). (In Russian)
12. Potapova, R.K., Potapov, V.V.: Language, Speech, Personality. Yazyki Slavyanskoy Kultury (2006). (In Russian)
13. Rumyantsev, M.K.: Phonetics and Phonology of Modern Chinese. Vostok-Zapad, AST (2007). (In Russian)
14. Shevchenko, T.I.: Sociophonetics: national and social identity in English pronunciation. URSS (2016). (In Russian)
15. Shevchenko, T., Sokoreva, T.: Corpus data on adult life-long trajectory of prosody development in American english, with special reference to middle age. In: Karpov, A., Jokisch, O., Potapova, R. (eds.) SPECOM 2018. LNCS (LNAI), vol. 11096, pp. 606–614. Springer, Cham (2018). https://doi.org/10.1007/978-3-319-99579-3_62
16. Shu-hui, P., et al.: Towards a Pan-Mandarin system for prosodic transcription. In: Jun, S.-A. (ed.) Prosodic Typology: The Phonology of Intonation and Phrasing, pp. 230–270 (2005)
17. Valian, V.: Bilingualism and cognition. Bilingualism: Lang. Cogn. **18**, 3–24 (2015)
18. Vykhrova, A.Y.: Emotionally colored intonations of happiness, fear and regret in modern Chinese (on universal and specific features in intonation). Vestnik MSU,13, pp. 97–107 (2011). (In Russian)
19. Zadoenko, T.P., Shuin H.: Chinese: an introductory course). M: Nauka (1993). (In Russian)
20. Zavyalova, V.L.: Sound system of English in Eastern Asia: conception of regional phonetic variation. Dokt. Diss. filol. nauk (2018). (In Russian)

Toward Explainable Automatic Classification of Children's Speech Disorders

Dima Shulga[1], Vered Silber-Varod[2](✉) ⓘ,
Diamanta Benson-Karai[1] ⓘ, Ofer Levi[1] ⓘ, Elad Vashdi[3],
and Anat Lerner[1] ⓘ

[1] Mathematics and Computer Science Department,
The Open University of Israel, Ra'anana, Israel
shudima@gmail.com, {diamanta,oferle}@openu.ac.il,
anat@cs.openu.ac.il
[2] Open Media and Information Lab (OMILab), The Open University of Israel,
Ra'anana, Israel
vereds@openu.ac.il
[3] Yael Center, Alonei Abba, Israel
yaelcenter@gmail.com

Abstract. Early and adequate diagnosis of speech disorders can contribute to the quality of the treatment and thus to treatment success rates. Using acoustic analysis of the speech of children with speech disorders may aid therapists in the diagnostic process by identifying the acoustic characteristics that are unique to a specific disorder and that distinguish it from normal speech development. The purpose of this work is to investigate the feasibility of the automatic detection of speech disorders based on children's voices. In this preliminary study, using a dataset of utterance recordings of 24 children whose mother tongue is Hebrew, we propose an automatic system that may facilitate accurate speech assessment by therapists by providing a preliminary diagnosis and explainable insights about the model's predictions. We built a serial, two-step network that is both powerful and possibly interpretable. The first step can model the complex relations between acoustic features and the speech disorder while the second can shed light on the utterances that make the greatest contribution to the final classification. Our preliminary results focus on the broad spectrum of speech disorders. In future work, we plan to design a system that will be able to detect childhood apraxia of speech (CAS) specifically and shed light on the differences in the speech of individuals with CAS and those with other speech disorders.

Keywords: Speech disorder · GeMAPS · Deep spectrum · Childhood Apraxia of Speech (CAS)

1 Introduction

The popularity of machine learning has increased markedly in recent years due to the wider availability of data, more computational power, and recent breakthroughs in learning algorithms (especially deep neural networks). These favorable conditions drove the massive adoption of machine learning based solutions for a variety of

© Springer Nature Switzerland AG 2020
A. Karpov and R. Potapova (Eds.): SPECOM 2020, LNAI 12335, pp. 509–519, 2020.
https://doi.org/10.1007/978-3-030-60276-5_49

problems. Machine learning applications have thus become ubiquitous across a range of fields such as image processing, natural language processing and speech analysis. Classifications based on speech signals are used for various applications, including gender classification, emotion recognition, and others. Methods include both classic machine-learning approaches with relevant feature extraction and deep neural network representation and classification. Among the fields that can benefit markedly by utilizing machine learning is that of speech disorder assessment and diagnosis. Many speech disorders are treated using repeated interventions with a speech therapist, and these treatments sometimes continue for several years [1]. Early and adequate diagnosis can contribute to the quality of the treatment and thus to its level of success. Exploiting machine learning early in the diagnostic process may improve both diagnostic accuracy and treatment success level.

In this work, we created a dataset that comprises children diagnosed with a speech disorder (suspected childhood apraxia of speech (CAS)) and children with typical speech. Our goal was to build a model that paves the way for future research on CAS assessment and that is potentially explainable to a human therapist. Explainable machine learning is an active field of research that is considered non-trivial. The challenge in explaining the model's predictions is to provide meaningful insights regarding the model's inputs while keeping them abstract and at higher levels of representation. In voice analysis, for example, using features like pitch, frequency, or jitter to explain a model will provide little to no help to a human evaluator. Instead, we aim to generate insights based on higher-level phonological features. Our protocol was designed to help us decompose the model's predictions and to provide information about the contribution to the analysis of each utterance/syllable. To that end, the 2-stage model we introduce here generates powerful and interpretable predictions.

1.1 Childhood Apraxia of Speech

Apraxia is a type of speech disorder defined as a difficulty in motor planning [2], and when it occurs in the context of children's speech, it is called childhood apraxia of speech (CAS). According to the professional literature [2, 3], CAS is a speech disorder with a pediatric neurological background characterized by impaired precision and consistency in speech-related motor movements in the absence of neuromuscular failures. This may occur as a result of neurological failure and in the context of neuro-behavioral syndromes with a known or unknown background. The main problem associated with CAS is the difficulty sufferers experience in planning the spatial coordination of the speech organs that are responsible for the articulatory process, and this ultimately impairs relevant motor learning processes during speech acquisition [2].

The prevalence of this phenomenon is unknown. Although a few studies have attempted to assess its frequency [3, 4], and [5], the lack of an accurate scientific diagnosis based on rigorous physiologic standards that can be used to distinguish CAS from similar impediments precludes accurate prevalence studies. Previous estimates from almost 20 years ago, which reported a prevalence of one per five hundred births, may indeed be too low [3] given the increase found in the prevalence of autism [6] and the observed relationship between autism and CAS [7].

Though many studies have been carried out in an attempt to identify CAS [8, 9], all concluded that no accurate cue can be used as a reliable discriminator with high success rates. These studies failed to isolate a physiological, genetic or behavioral dimension that would be common to all children with apraxia and thus enable the phenomenon to be accurately identified (for a Summary of errors associated with CAS see [1]). The CAS definition process has considered mostly verbal children who have some verbal skills, while excluding the non-verbal population almost entirely. The behavioral diagnosis relies on behavioral symptoms rather than on the child having a physiological marker or on clinician confirmation of the existence of a motor planning deficit as the primary indicator that the child has CAS. Asked to define the phenomenon clinically, speech therapists described a variety of different symptoms as characteristic of CAS. One symptom, however, pronunciation inconsistencies, was mentioned by 50% of respondents [10].

Acoustic analyses of the speech of children with suspected CAS may shed light on this phenomenon. Indeed, such analyses may enable us to discover the acoustic characteristics that are unique to this group and that distinguish its members from children with typical speech development.

1.2 Related Work

Several recent works have applied automatic techniques to facilitate speech disorder diagnostics and treatment. In Shahin et al. [1], the authors present a system called "Tabby Talks", a multi-tier system for the remote administration of speech therapy. The authors describe the speech processing pipeline they developed to automatically detect common errors associated with childhood apraxia of speech (CAS). The pipeline contains modules for voice activity detection (VAD), pronunciation verification, and lexical stress verification. Their work shows a system tailored to identify disorder-specific errors enables the use of automatic speech recognition techniques in speech pathology research [11, 12].

Another approach entails a system that produces an automatic classification of speech disorders. In a first step toward developing an intelligent system capable of providing feedback to patients with aphasia, the authors of [13] develop classifiers to automatically estimate speech quality based on human perceptual judgment. The results showed that their automatic prediction yields accuracies comparable to those of the average human evaluator.

Another study by Baird et al. [14] evaluates a range of speech-based classification approaches for the automatic detection of autism severity according to what is known as the Social Responsiveness Scale. Based on analyses of a novel dataset of 803 utterances recorded from 14 autistic children 4 to 10 years of age, their results demonstrate the suitability of support vector machines (SVMs), which use acoustic feature sets from multiple INTERSPEECH ComParE challenges [15] and deep spectrum features extracted via an image classification convolutional neural network (CNN) from the spectrogram of speech instances of these children.

An important part of automatic speech analysis is the feature extraction step. In this work, we experiment with two common feature extraction methods from raw audio and evaluate them. The first method, called GeMAPS [16], proposes a basic standard acoustic parameter set for various areas of automatic voice analysis, such as paralinguistic or clinical speech analysis. These features were selected based on three considerations: their potential to index affective physiological changes in voice production, their proven value in previous studies and their automatic extractability, and their theoretical significance. In the second method, [13] explore the use of CNN to represent voice recordings. This method is evaluated by classifying emotions from speech. In [17], the authors used deep spectrum features of the raw spectrogram as input to a CNN for the recognition of emotional speech. They exploited the fact that the outputs of the higher layers of a deep pre-trained CNN have consistently been shown to provide a rich representation of an image for use in recognition tasks. To that end, they treat the spectrograms as images and use a pre-trained AlexNet model [18] to "extract" the deep spectrum features. AlexNet is a Deep CNN that ran on the ImageNet dataset [19] to classify images.

2 Data

For this study, we recorded 24 Israeli children whose mother tongue is Hebrew (6 girls and 18 boys, mean age 8 years, range 4 to 16 years). We examined the acoustic features that distinguish children and adolescents with apraxia from children with typical speech development, regardless of age, gender or level of literacy. Sixteen children with speech disorders (all with suspected CAS) and 8 children with typical speech were recorded by the same examiner (one of the authors). This study received prior approval from the Open University ethics committee. Written informed consent was obtained from all of the participants' parents. All recordings were carried out using the same recording device – ZOOM H6 handy recorder with interchangeable microphone system [20]. Recordings were performed with two-channels (stereo) with a sampling frequency of 44100 Hz.

2.1 Recording Protocol

For the recordings, 15 syllable strings were chosen. Since several constructions are real words in Hebrew, we will refer to these words and pseudo-words as *utterances* (on the advantages of using pseudo-words, see [21]; on the use of real words in a similar setup for Hebrew speaking children, see [22]. The logic behind the construction of these utterances was to represent monosyllabic and disyllabic constructions consisting of both a consonant (C) and a vowel (V): a CV structure; a CVC structure that is common in the literature in the field [23]; and a CVCV structure, by repeating the same CV twice (for example, *tata*). It was also decided that each structure will have representations for both vocal and non-vocal consonants and that all five vowels in the Modern Hebrew phonological system – *i, e, a, o,* and *u* – will be part of the protocol.

The 15 target utterances were as follows:

CV structure set: *ba, be, bi, po, pu*;
CVC structure set: *dif, def, daf, dof, duf*; and
CVCV structure set: *kuku, gaga, shosho, zeze, titi.*

It was important to keep the protocol simple (compatible with the young ages of the study participants) and the utterances free of linguistic factors. We wanted the participants to simply repeat the therapist's target utterances without activating other cognitive abilities (i.e., to think of what to answer) (similar to [21] and [22]). Therefore, during the session, the therapist used a single sentence template as a prompt to elicit the child's speech production: "Say X please" (In Hebrew: *tagid _____ bevakasha*). From this priming, we expected the participant to mimic the target utterance. Each session comprised four sub-sessions with five-minute breaks between them. Utterances were divided by set, such that a CVC set was followed by a CVCV set, and so on, in a different order in each session. In addition, for each sub-session, the utterances were in different order. After completion of the first session, the therapist could play the recording from that session to motivate the child in subsequent sessions. Figure 1 illustrates the difference between typical and atypical speech production via a comparison of the spectrograms of productions of the same utterance, [dif]. Although the typical utterance (top panel) is only a bit shorter than the atypical (710 ms versus 740 ms, respectively), the durations of phonemes are significantly different in the two speech productions. Likewise, these speech samples also differ in terms of the waveforms and the F_0, the latter of which is almost absent from the atypical speech.

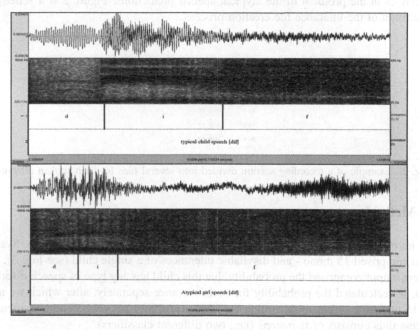

Fig. 1. Comparison of typical [dif] (top panel) and atypical [dif] (bottom panel) speech productions. Durations were 710 ms and 740 ms, respectively.

2.2 Data Preprocessing

All the recordings were manually segmented and labeled using the PRAAT textgrid tool [24] to produce $4 \times 24 = 96$ sound files with additional metadata, such as speaker annotation (tags), target utterance, child's ID, age, gender, and the therapist's diagnosis. Four children were randomly selected for the test set, which was not included in any of the analyses but was included in the reporting of the final results.

2.3 Data Parsing

As mentioned above, only the children's utterances were of interest in this work (and not, for example, silence durations or reaction time). To provide those, the complete recordings and their annotations were parsed using Python software to create new files. The parsing generated a separate file for each child utterance for a total of $96 \times 15 = 1,440$ files. Because we only used segments that were annotated with "good for modeling", we ended up with a total of 1,224 audio files, each of which contains a single utterance produced by one child. These files were used in the modeling, in which each child utterance was scored separately for atypical speech. This approach provided more data for the model, making it possible to analyze the contribution of individual utterances in the protocol to the atypical speech predictions. Figure 2 is a schematic illustration of the utterance file creation process.

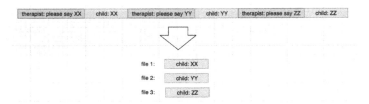

Fig. 2. Example of a recording session divided into several files for each spoken utterance.

3 Method

We introduce a machine learning based system with two classifiers (Fig. 3). System input comprised 15 mono - and disyllabic utterances of a single child (see Fig. 3), and system output comprised the probability that this child has any type of speech disorder. First, we calculated the probability for each utterance separately, after which we used the 15 utterance scores to determine the final probability per child. The classification process thus consists of two steps (i.e., two different classifiers).

3.1   First Classifier – Utterance Model

The first classifier – for "utterance-probability" – takes a single utterance as input and, based on this utterance alone, generates the probability that the speaker has a speech disorder. We used the labeled utterance recordings as training data. Each utterance recording was labeled typical/atypical in line with the label assigned to the child who produced it.

Feature Extraction. We use (and evaluate) two common feature extraction methods: GeMAPS [16]; and deep learning based on a pre-trained image classifier to represent the raw speech spectrograms [17].

a. GeMAPS features: for each recording of a child's utterance, we extracted GeMAPS features using the OpenSMILE software version 2.3 [25]. We used both the extended GeMAPS with 88 features and the minimalistic version with 62 features.
b. Deep spectrum representation: Similar to [17], we used a large pre-trained CNN model to obtain representations of our audio files. We extracted raw spectrograms and used them as input to a pre-trained ResNet [17] model. We used the final layer of 2,048 activations as the feature vector representing the audio segment.

We added age, gender, and the utterance characteristics (as a one-hot vector) to each feature set and then used a gradient boosting classifier to obtain the probabilities for every utterance.

3.2   Second Classifier – Decision Tree

The second classifier is a decision tree classifier. It receives as input the probabilities per 15 utterances calculated by the first classifier, and outputs the final speech disorder probability per child. We use the "gini" criterion with a maximum tree depth of 2 and minimum leaf size of 1. Using the decision tree classifier, we were able to create an interpretable model that may help us evaluate the importance of each utterance to the system generated decision as to whether the child has a speech disorder.

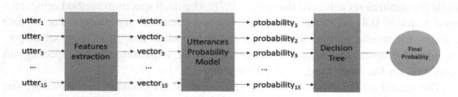

Fig. 3. Classification system flow chart for two classifiers: Utterance probability model and a decision tree.

3.3   Training

The training phase consisted of two parts following the design of the two classifiers. We randomly divided the 24 children into two groups such that 20 children were assigned to the classifier training group (median age 10.5) and 4 were in the classifier

testing group (median age 8). To prevent information leakage into the model evaluation, we divided the study participants at the children level (and not at the utterance
level). First, we trained the utterance model on a training set of utterances from 20
children using the single utterance files that were parsed from the recorded sessions of
the children, as described (Sect. 2.3). Each file refers to a single utterance of one child.
We labeled each utterance input with the label corresponding to the respective child,
i.e., typical or atypical. That is, the first step modeled the probability that a child
belonged to the atypical speech group given the production of a single utterance. We
used a gradient boosting classifier with 500 empirically chosen trees for this part of the
training.

In the second part of the training phase, we trained the second classifier. The inputs
to this classifier were vectors of size 15. The elements of a vector comprised the
probabilities of the utterances as described above. That is, the second classifier modeled
the probability that a child belonged to the atypical speech group given the probabilities
of all pronounced utterances. For this part of the training, we created 20 training
vectors, one for each child in the training set. The probabilities per utterance were
calculated as the average probability of that utterance over the four sessions. To create
the training vectors, we iterated over all the 20 children in the training set. For each
child, we trained the utterance model without the recordings of that child. In so doing,
we ensured that in calculating the probabilities for one child, the model does not rely on
the child's other utterances. This process can be described as leave-one-out cross
validation at the children's level. For the training process, we used a decision tree
classifier with the empirically chosen Gini criterion, a maximum depth of 2 and
minimum leaf size of 1.

4   Results

In the first part of our model, the utterance scores model, the F1 score obtained for each
feature set using child level cross-validation on the training set was used as the performance metric. The extended GeMAPS features set achieved the best results, 0.83,
while the minimal set achieved the lowest, 0.78. The deep spectrum method achieved a
good result of 0.81. Although we could not test statistical significance due to data
sparsity, the results suggest that, on the one hand, while using the large set of pre-
engineered features can promote improved modeling, exploiting the raw spectrogram
properties, on the other hand, is also beneficial.

The second and final part of our model comprised the decision tree classifier. The
output of this model was the system-generated decision, and hence, model performance
during this step represented the performance of the whole system. To understand this
performance, we used the children in the test set. First, we used the utterances model to
obtain four probability vectors that we then used as input to the decision tree classifier.
The results for the four children in the test set show that the model predicted the correct
values for each child (Table 1).

Table 1. Results of the deep spectrum test.

Child ID	Target	Predicted	Probability
1	True	True	0.93
2	True	True	0.934
3	False	False	0.0
4	False	False	0.0

As mentioned above, our use of a simple decision tree classifier enabled us to clearly identify which utterances made the greatest contributions to the final decision generated by the system. In the current setting, the most important utterances were *dif* and *zeze*. the final decision tree for which is shown in Fig. 4. It is interesting to note that those two utterances have dissimilar phonological properties, including different numbers of syllables, different vowels, different syllable structures, and more.

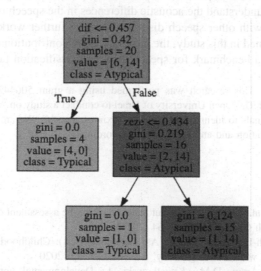

Fig. 4. The final decision tree. The first node checks for the speech disorder probability of *dif*; in the second node the probability of *zeze* is evaluated.

5 Discussion and Future Research

In this study, we presented the initial steps in how to exploit machine learning methods to recognize atypical speech. Although the results are not significant (due to dataset size), they are promising, as they suggest that the use of acoustic features of mono- and disyllabic pseudo-words to automatically recognize speech disorders is feasible. The GeMAPS features set has been proven to be effective in many voice-based tasks, including those in this work, but other methods used in atypical speech analysis are also valuable. By introducing a novel data set and a recording protocol, we built a serial model that is both powerful and possibly interpretable. Its first stage models the complex relations between acoustic features and speech disorder probability while its second stage can shed light on which utterances contribute the most to the final decision.

In such research, the main challenge—the collection of data from CAS children—necessitates cooperation and collaboration not only with clinical experts, but also with the parents and with the children themselves. Our findings constitute proof of concept with preliminary results and they must be confirmed on a larger dataset to demonstrate generalizability.

Furthermore, we are aware that while pronunciation disabilities are a salient feature of speech produced by children with CAS, Not all of the features in the acoustic feature sets provide direct information about segmental pronunciation; to produce an interpretable model, additional features that use linguistic knowledge to capture pronunciation disabilities should be explored, as demonstrated in Fig. 1.

In future studies, we also plan to expand our model from binary classification to a multi-class model and to explore its explainability power. Among the goals of such research will be to understand the acoustic differences in the speech of individuals with CAS and of those with other speech disorders. Indeed, further work may improve on the results we obtained in this study, the most beneficial contribution of which seems to be its provision of a benchmark for speech disorder classification tasks.

Acknowledgements. This research was performed using a grant 506442 (37183) from the Research Authority of The Open University of Israel to conduct a study on "Analysis of acoustic and physiological signals to identify childhood apraxia of speech". We are grateful to Daphna Amit for the segmentation and annotation of the recordings.

References

1. Shahin, M.A., et al.: Tabby talks: an automated tool for the assessment of childhood apraxia of speech. Speech Commun. **70**, 49–64 (2015)
2. American Speech-Language-Hearing Association (ASHA): Childhood apraxia of speech. Technical report. www.asha.org/policy. Accessed 21 Apr 2020
3. Shriberg, L.D., Aram, D.M., Kwiatkowski, J.: Developmental apraxia of speech: I. Descriptive and theoretical perspectives. J. Speech Lang. Hear. Res. **40**(2), 273–285 (1997)
4. Deal, J.L., Darley, F.L.: The influence of linguistic and situational variables on phonemic accuracy in apraxia of speech. J. Speech Lang. Hear. Res. **15**(3), 639 (1972)
5. Yoss, K.A.: Developmental apraxia of speech in children: familial patterns and behavioral characteristics. In: ASHA North Central Regional Conference, Minneapolis, MN (1975)
6. Hansen, S.N., Schendel, D.E., Parner, E.T.: Explaining the increase in the prevalence of autism spectrum disorders: the proportion attributable to changes in reporting practices. JAMA Pediatr. **169**(1), 56–62 (2015)
7. Tierney, C., et al.: How valid is the checklist for autism spectrum disorder when a child has apraxia of speech? J. Dev. Behav. Pediatr. **36**(8), 569–574 (2015)
8. Shriberg, L.D., et al.: A diagnostic marker for childhood apraxia of speech: the lexical stress ratio. Clin. Linguist. Phon. **17**(7), 549–574 (2003)

9. Strand, E.A., Duffy, J.R., Clark, H.M., Josephs, K.: The apraxia of speech rating scale: a tool for diagnosis and description of apraxia of speech. J. Commun. Disord. **51**, 43–50 (2014)
10. Malmenholt, A., Lohmander, A., McAllister, A.: Childhood Apraxia of Speech (CAS): a survey of knowledge and experience of Swedish Speech-Language Pathologists. In: ICPLA 2012 14th Meeting of the International Clinical Phonetics and Linguistics Association, p. 143 (2012)
11. Hosom, J.P., Shriberg, L., Green, J.R.: Diagnostic assessment of childhood apraxia of speech using automatic speech recognition (ASR) methods. J. Med. Speech-Lang. Pathol. **12**(4), 167–171 (2004)
12. Keshet, J.: Automatic speech recognition: a primer for speech language pathology researchers. Int. J. Speech-Lang. Pathol. **20**(6), 599–609 (2018)
13. Le, D., Licata, K., Persad, C., Provost, E.M.: Automatic assessment of speech intelligibility for individuals with aphasia. IEEE/ACM Trans. Audio Speech Lang. Process. **24**(11), 2187–2199 (2016)
14. Baird, A., et al.: Automatic classification of autistic child vocalisations: a novel database and results. In: Proceedings of INTERSPEECH 2017. International Speech Communication Association, Stockholm, Sweden (2017)
15. Schuller, B., et al.: The INTERSPEECH 2013 computational paralinguistics challenge: social signals, conflict, emotion, autism. In: Proceedings INTERSPEECH 2013, 14th Annual Conference of the International Speech Communication Association, Lyon, France (2013)
16. Eyben, F., et al.: The Geneva minimalistic acoustic parameter set (GeMAPS) for voice research and affective computing. IEEE Trans. Affect. Comput. **7**(2), 190–202 (2016)
17. Cummins, N., et al.: An image-based deep spectrum feature representation for the recognition of emotional speech. In: Proceedings of the 2017 ACM on Multimedia Conference. ACM (2017)
18. Krizhevsky, A., Sutskever, I., Hinton, G.E.: ImageNet classification with deep convolutional neural networks. Advances in Neural Information Processing Systems, pp. 1097–1105 (2012)
19. Deng, J., et al.: ImageNet: a large-scale hierarchical image database. In: 2009 IEEE Conference on Computer Vision and Pattern Recognition, pp. 248–255 (2009)
20. ZOOM H6. https://zoom-na.com/products/field-video-recording/field-recording/zoom-h6-handy-recorder-1. Accessed 30 May 2010
21. Liberman, M.Y., Streeter, L.A.: Use of nonsense-syllable mimicry in the study of prosodic phenomena. J. Acoust. Soc. Am. **63**(1), 231–233 (1978)
22. Icht, M., Ben-David, B.M.: Oral-diadochokinetic rates for Hebrew-speaking school-age children: real words vs. non-words repetition. Clin. Linguist. Phon. **29**(2), 102–114 (2015)
23. Gadesmann, M., Miller, N.: Reliability of speech diadochokinetic test measurement. Int. J. Lang. Commun. Disord. **43**(1), 41–54 (2008)
24. Boersma, P.: PRAAT, a system for doing phonetics by computer. Glot Int. **5**(9/10), 341–345 (2001)
25. Eyben, F., Weninger, F., Gross, F., Schuller, B.: Recent developments in openSMILE, the Munich open-source multimedia feature extractor. In: Proceedings of the 21st ACM Multimedia, pp. 835–838 (2013)

Recognition Performance of Selected Speech Recognition APIs – A Longitudinal Study

Ingo Siegert[1(✉)], Yamini Sinha[1], Oliver Jokisch[2], and Andreas Wendemuth[3]

[1] Mobile Dialog Systems, Otto von Guericke University Magdeburg,
Magdeburg, Germany
ingo.siegert@ovgu.de
[2] Institute of Communications Engineering,
Leipzig University of Telecommunications (HfTL), Leipzig, Germany
[3] Cognitive Systems Group, Otto von Guericke University Magdeburg,
Magdeburg, Germany

Abstract. Within the last five years, the availability and usability of interactive voice assistants have grown. Thereby, the development benefits mostly from the rapidly increased cloud-based speech recognition systems. Furthermore many cloud-based services, such as Google Speech API, IBM Watson, and Wit.ai, can be used for personal applications and transcription tasks. As these tasks vary in their domain, their complexity as well as in their interlocutor, it is challenging to select a suitable cloud-based speech recognition service. As the update-process of online-services can be completely handled in the back-end, client applications do not need to be updated and thus improved accuracies can be expected within certain periods. This paper contributes to the field of automatic speech recognition, by comparing the performance of speech recognition between the above-mentioned cloud-based systems on German samples of high-qualitative spontaneous human-directed and device-directed speech as well as noisy device-directed speech over a period of eight months.

Keywords: Evaluating speech recognition systems · Google Speech API · IBM Watson · Wit.ai · German conversational speech · Noisy environment · Unconstrained speech input

1 Introduction

In recent years, there was a continuous evolution of Automatic Speech Recognition (ASR) systems. Besides traditional applications of ASR, e.g. in dictation systems, ASR systems are commonly employed in everyday applications such as smart speakers [10,11]. This success was possible due to the use of sophisticated deep network architectures with efficient training methods for both acoustic and language modeling. Nowadays, mostly deep convolutional neural network architectures are used for acoustic modeling while variants of long-short-term memory

A. Karpov and R. Potapova (Eds.): SPECOM 2020, LNAI 12335, pp. 520–529, 2020.
https://doi.org/10.1007/978-3-030-60276-5_50

networks are employed for both acoustic and language modeling [16]. Furthermore, the evolution of ASR systems is carried on by the growth of cloud-based services and the Internet of Things (IoT) trend, where the local simple and 'unintelligent' sensor is capable of using a more powerful remote service. Among the various providers of speech recognition APIs (Google Web/Cloud Speech API, IBM Watson, Microsoft Azure, Amazon Alexa API, Nuance recognizer, Wit.ai, Houndify), some provide the possibility to use their speech recognition API also within personal products or with own speech files.

An area that can benefit from high-quality ASR services is the automatic transcription of speech data. Not only that this task is needed in many application domains: call-center, political debate, and conversational analysis to name just a few, it is also a time-consuming process that profits from automatic processing [2]. It is, therefore, reasonable to use an automatic ASR system for the transcription of voice data. But, due to the increasing number of available ASR services, it becomes challenging to select the one with the best performance in the given application context. Hereby the most important criterion is the recognition performance, mainly dependent on the type of conversation, e.g. human-directed speech (HDS) or device-directed speech (DDS), and the quality of data, e.g. clean versus noisy, is of interest. Another important aspect that has rarely been investigated so far is the performance improvement over time. As most cloud-based services are continuously developed in both underlying model architectures and data sources, one can assume that the recognition performance will increase over time.

To investigate both aspects, we conducted a longitudinal study over a period of eight months to analyze the ASR performance, using different speech material (test sets). Thereby, this paper has the following main contributions: a) a longitudinal investigation of speech recognition performance for three selected systems over a period of maximal eight months, b) the comparison of ASR performance for clean human-human as well as clean or noisy, and distant human-machine speech, and c) investigating the recognition performance for German.

2 Related Work

Many studies investigated the performance of different cloud-based ASR systems. For English, different subsets of the Switchboard telephone speech dataset [3] are often used to benchmark the ASR performance. Hereby, IBM Research reported a Word error rate (WER) of 5.1% on the Switchboard Hub5 2000 evaluation test with their ASRU'17 system introducing unsupervised Language Model (LM) adaptation [7]. Microsoft Research achieved the same WER performance on the NIST 2000 Switchboard task introducing ngram rescoring [16]. For Google's cloud-based ASR system no results on the Switchboard dataset are reported. Instead, an own traffic application dataset was used achieving 6.7% WER on voice search and 4.1% WER for a dictation task, see [1]. In comparison, human recognition rates are reported as 5.9% to 5.1% WER [12,16].

Recently, several studies compared the performance of different cloud-based ASR systems with specific test sets. In [17], the WER of Google Cloud Services, IBM Watson, and Microsoft Bing Cloud Systems is compared regarding the WER on a self-generated English test set acquired via a smartphone: Google was rated best and Microsoft Bing was rated as worst. Especially for noisy speech samples (generated by mixing with restaurant sound), Microsoft Bing was not able to recognize any noisy sample. The authors of [5] compared the WER of Google, Microsoft Azure, IBM Watson, Trint, and Youtube for the task of transcription service of simulated medical doctor/patient consultations. According to their analyzes, Youtube was rated best followed by Google, Microsoft Azure, and Trint. IBM Watson had the highest WER. Another study compared the WER of Microsoft Speech and Google Cloud using 15 samples from the TIMIT Corpus of read speech [8]. They reported that Google (WER 9%) surpassed the Microsoft system (WER 18%). But, it is not clear whether Microsoft Bing or the successor Azure has been used.

While the previously cited studies were conducted on English samples, just a few studies analyzed other language data. In [4], the authors analyzed the WER of Google Speech for Romanian. Based on a corpus of 20 YouTube videos with manual annotations, an average WER of 31.22% was achieved. A comparison to other cloud-based ASR services was not possible due to the missing support of the Romanian Language. In [6], the WER of Google Speech was analyzed for Japanese on various types of speech. The best performance was achieved for the reading task (WER: 8.34%) followed by emotionally reading (WER: 13.01%). For spontaneous conversations, a WER of 18.67% was achieved with similar results under noisy environment (crowded noise). Spontaneous HDS was poorly processed by the system, resulting in a WERs above 32%.

Additionally, some authors analyzed acoustic influences on the WER. In [5], it is stated that the recording quality as well as environmental noise and audio feedback are the main reasons for a high WER. In [17], the mixed restaurant noise leads to a much higher WERs.

Interestingly, we did not find investigations for German, although this is one of the major languages supported by all cloud-based systems.

3   Experimental Setup

All experiments were conducted using Python and the SpeechRecognition library (version 3.8.1)[1]. From the variety of speech recognition APIs, we selected those offering free usage or trials and supporting German. As test samples, we used three types of data from very recent German conversation studies: spontaneous clean HDS and DDS as well as spontaneous noisy DDS (see Sect. 3.2).

3.1   Selected Speech APIs

For our experiments, we selected four different speech APIs, accessible through the SpeechRecognition library.

---

[1] https://pypi.org/project/SpeechRecognition/.

Google Cloud Speech to Text API (GCS)[2] is capable to process real-time as well as prerecorded audio. It can handle more than 120 languages including variants. Google Speech API handles noisy audios from many environments without the requirement of additional noise cancellation. Furthermore, by using classes provided by the API, the user can convert spoken numbers into years, currencies, or addresses according to the context. With the use of Speech-to-Text, the user can identify the language which is spoken in the utterance (up to four languages). As input file, it accepts mono-channel FLAC files.

Google Web Speech API (GWS)[3] is for testing or personal purposes and it may be revoked by Google in the future. It has a default key or an API key can be obtained from Google Developers Console (limited to 50 requests per day). This API also allows recognition in more than 120 languages but does not have access to the latest cloud features. As input file, it accepts mono-channel FLAC files.

IBM Watson Speech to Text API (IBM)[4] is a cloud-based service using deep learning to understand grammar language structure and voice signal composition. It supports real-time transcription but can also be used with prerecorded audio files. The main drawback of IBM is the support of a very small amount of languages with custom models available for an even fewer number of languages. The API can be deployed on any cloud platform and thus, complies with high security and privacy standards. IBM will not collect, store, or use the data without explicit agreement.

Wit.ai Speech to Text API (WIT)[5] is an open-source chatbot framework with advanced natural language processing, or NLP, capabilities. Owned by Facebook, Wit.ai is a popular choice for Facebook Messenger bots powered by NLP. Wit.ai can be used to build intelligent chatbots for social channels, mobile apps, websites, and IoT devices. Facebook acquired the company in 2015, but Wit.ai is remaining an open-source project. Wit.ai is mostly intended to be used as a text-based chatbot but can also be used to build conversational applications that respond to voice commands.

3.2 Utilized Datasets

As we wanted to investigate different speaking styles (HDS and DDS), we used available datasets from earlier conversational experiments: one dataset comprising HDS and DDS, and one dataset comprising unconstrained DDS in a noisy environment. In comparison to previous investigations, we are using speech samples from spontaneous interactions [3,7,8,16], including scruffy grammatical phrases, self-corrections, and hesitations.

Voice Assistant Conversation Corpus (VACC) provides high-quality spontaneous HDS and DDS samples. The database consists of conversations between a

---

[2] https://cloud.google.com/speech-to-text/docs.
[3] https://developers.google.com/web/updates/2013/01/Voice-Driven-Web-Apps-Introduction-to-the-Web-Speech-API.
[4] https://www.ibm.com/cloud/watson-speech-to-text.
[5] https://medium.com/wit-ai.

participant with another human partner, and with a commercial voice assistant system (Amazon's ALEXA). The recordings were conducted in 2017 with high-quality neck-band microphones (Sennheiser HSP 2-EW-3), in a living-room-like surrounding, reducing possible background noise. They are stored in WAV format at 44.1 kHz sample rate with 16 bit resolution. Turn times and utterances were manually annotated. The conversations are about possible scheduling of events and the request for possible dates to both the human interlocutor and the technical device as well as questions and possible solution strategies. Further details are given in [15]. VACC comprises approximately 1,800 HDS (participant towards interlocutor) and 3,800 DDS utterances (participant towards Alexa) for the 27 subjects with an average length of 2.4 s (min: 0.06 s, max: 12.40 s) and 2.3 s (min: 0.02 s, max: 6.09 s), respectively.

Voice Assistant Conversations in the wild (VACW) consists of public unconstrained conversations between different visitors and a voice assistant at a 2019 science fair in Germany. The recordings were conducted with a far-field microphone array. An additional peculiarity is that the speakers are not aware of being recorded and thus do not fear pronunciation mistakes. This results in recordings of the speakers' voices heavily affected with background voices as well as environmental noise. Recording times and utterances were automatically provided. The topics range from questions of an employee quiz, over music and weather information as well as playing around with the features of a voice assistant, for more details see [14]. In total, this dataset consists of around 30,000 speech samples with a mean length of 5.5 s (Min: 0.2 s Max: 30 s), selected test samples are denoted as noisy Human-Computer interaction (HCI).

3.3   Experimental Procedure

The selection of test files followed a pseudo-random two-step approach. In the first step, all samples only having any no-speech content (laughter, background noise, etc.) were manually discarded. Afterwards from each dataset, 100 files were randomly selected from the remaining set of cleaned samples. Hereby, this selection is repeated three times without re-selecting already chosen files. Thus, in the end, we had three times 100 different files for each of the three conditions. To generate a "gold-standard" transcription, all of these selected files were manually transcribed followed by a post transcription processing: removing all punctuations, writing out numbers, and all words were set to lowercase.

As performance measure we have chosen Word error rate (WER), as this metric is most commonly used to evaluate the accuracy of such systems, see e.g. [13] and Sect. 2. WER is defined as the edit distance between the reference transcript and the hypothesis transcript. The edit distance is defined as the minimum number of insert, substitute, and delete operations to align both transcripts [9].

For pre-/post-processing all audio files were converted to the accepted input formats before sending it to the cloud-based ASR systems. To get the results for our longitudinal study, we conducted the test at the beginning of each prospective month. To exclude a possible bias due to the selected sub samples, we repeated this procedure with three independent trial rounds, see Table 1.

Table 1. Overview of conducted ASR tests with three trial rounds.

	Set 1	Set 2	Set 3
Nov/19	X	-	-
Dec/19	X	X	-
Jan/20	X	X	X
Feb/20	X	X	X
Mar/20	X	X	X
Apr/20	-	X	X
May/20	-	-	X
Jun/20	X	X	X

The first round with the first set of 100 files started in November'19, the second round started in December'19 with the second set of 100 files along with the first set, and the third round started in January'20 with the third set of 100 files with first two sets. For all sets, we made an additional run in June'20. Thus, we covered six runs over a span of at least half a year each. Each round comprises 100 selected files from each of the speaking styles, see Sect. 3.2.

The resulting transcriptions were post-processed by spelling out numbers, abbreviations, etc.

4   Results

The WERs for clean HDS are depicted in Fig. 1. As previously reported results on other languages and topics indicate, IBM achieves a relatively high WER, while GCS achieves a quite low WER. Surprisingly, WIT also achieves comparable results similar (Set 1 and Set 3) or just slightly above (Set 2) to GCS. Furthermore, although GWS is marked as deprecated, the results are still comparable to GCS with an average difference of only 3.3% absolute WER. But it has

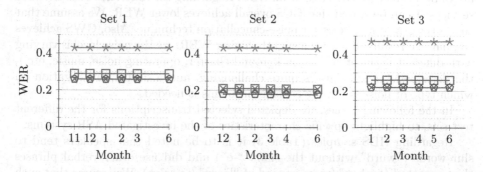

Fig. 1. Comparison of the development of WER values on clean HDS for the different test rounds utilizing GCS —⊖—, GWS —⊟—, IBM —✳— and WIT —✲— as ASR systems.

to be mentioned that this task is mostly out of scope for the various cloud-based ASR systems, as they are mostly intended to be used for voice-based search or command-control. The transcription of general human-to-human conversation is not part of the language modeling, as it will be further discussed in Table 3.

As we did not observe a change in the WER during our experiments, we will further only depict the latest result of June 2020 for the clean and noisy DDS tests. The correct recognition of clean DDS causes fewer problems for systems than noisy DDS, which is also reflected in the error rates, see Table 2. In line with previous results, IBM has the highest WER over all three sets. GCS and WIT achieve the lowest WER with 7% to 8% absolute.

Furthermore, the similarity between GCS and GWS is noticeable, as both systems have similar error rates. If GCS is good, GWS (with slight trade-offs) also delivers good results. On data that cause problems for GCS, GWS also faces problems and again achieves slightly worse (1% absolute WER) results.

Table 2. Comparison of the development of WER values on clean and noisy DDS for the different test rounds.

	Clean DDS			Noisy DDS		
	Set 1	Set 2	Set 3	Set 1	Set 2	Set 3
GCS	0.11	0.07	0.08	0.17	0.23	0.21
GWS	0.12	0.08	0.09	0.17	0.25	0.21
IBM	0.12	0.14	0.13	0.74	0.72	0.74
Wit	0.07	0.08	0.11	0.23	0.23	0.24

Regarding noisy data, it is apparent that both Google systems and Wit.ai use pre-processing to apply additional noise cancellation techniques, as both achieve a quite low WER, while IBM 's ASR system has massive problems with this kind of data, see Table 2. In general, the results on noisy DDS for Google and Wit.ai are in between clean DDS and clean HDS. Regarding the difference between both services, it can be stated that GCS overall achieves lower WER. We assume that this is the case due to a better noise-cancellation technique. Also, GWS achieves comparable results, but with a slightly higher WER. As the noise level is varying with different types of noises (background chatter, operating noise, music, etc.), the described type of data is quite challenging, and a deeper investigation on which type of noise is mainly causing problems is needed.

In the following tables, we depicted selected transcriptions for the different test sets, to highlight specific characteristics of the investigated ASR systems.

Regarding HDS samples (Table 3) it is to be noted that speakers tend to slur words ("würd" without the suffix "-e") and did use other verbal phrases than expected ("gehen"/to go instead of "lösen"/to solve). We assume that such misrecognitions are heavily influenced by the language modeling optimized for the use case of DDS.

Table 3. Example of recognized text for a clean HDS sample.

Ref	würd ich auch wieder über meine schöne rechenaufgabe gehen
GCS	werde ich auch wieder über meine schöne rechenaufgabe geben
GWS	werde ich auch wieder über meine schöne rechenaufgabe geben
IBM	werde ich auch wieder über meine schöne rechen auf vergehen
WIT	würde ich auch wieder über meine schöne rechenaufgabe gehen

Table 4. Example of recognized text for a clean DDS sample.

Ref	[ähm] alexa nenn mir meine termine am mittwoch den dreizehnten dezember
GCS	alexa nenn mir meine termine am mittwoch den dreizehnten dezember
GWS	alexa nenn mir meine termine am mittwoch den dreizehnten dezember
IBM	ähm alexander in mir meine termine am mittwoch den dreizehnten dezember
WIT	alexa nennen mir meine termine am mittwoch den dreizehnten dezember

For the DDS example (Table 4), which represents the use case optimized for different applications for HCI analyzes, only IBM has difficulties. Again, it can be assumed that this is due to the missing or less likely word "Alexa" in their language model. All other peculiarities (slurring, missing suffixes) do not pose any problems.

Noisy DDS samples had the additional difficulty that a lot of environmental noise and additional background speakers were present and thus, the cloud-based ASR systems have to rely on proper noise-cancellation techniques for pre-processing. As shown in Table 5, IBM Watson is not able to recognize the sample except the wake word "alexa". Especially the Google systems (GWS and GCS) demonstrated their capabilities and, apart from the human pronunciation error ("diplomatiter"), can recognize the sample correctly. This indicates that the Google system must have a well-established solution for out-of-vocabulary words. The Wit.ai results are slightly worse, as indicated by Table 2, since the system has a proper noise cancellation but its technique for out-of-vocabulary words falls behind the one implemented in the Google services.

Table 5. Example of recognized text for a noisy DDS sample.

Ref	alexa welcher schriftsteller erbaute als diplomatiter architekt ein freibad
GCS	alexa welcher schriftsteller erbaute als diplomierter architekt ein freibad
GWS	alexa welcher schriftsteller erbaute als diplomierter architekt ein freibad
IBM	alexa
WIT	alexa welcher schriftsteller erbaute halt die blume da ein freibad

5   Conclusion

We investigated the longitudinal development of the WER of selected cloud-based ASR systems. We focused on the German language for three different types of applications, clean HDS, clean DDS, and noisy DDS. While the first one appoints a representative for the transcription needed for conversational analyses, the two DDS test sets indicate different applications for HCI analyses.

Overall, GCS und WIT are recommendable for both investigated interaction types (HDS/DDS) and additionally can handle noisy DDS. Regarding both Google services, we cannot derive a clear recommendation, as both achieve almost identical performance, and therefore it has to be decided on an application-specific basis. Nonetheless, as long as there is the free basic service with GWS (without the obligation to register), another alternative, at least for smaller projects, is available.

It is apparent that all ASR services are optimized for clean DDS. Hereby, GCS and GWS are outstanding as they can correct mispronunciations, and thus afterwards, applied speech understanding units can concentrate to process the meaning of the speech input. Unfortunately, e.g. in conversational analyses, where a correct identification of the spoken input including mispronunciations is intended, this (non-deactivatable) feature can be rather disturbing. Beyond, we were surprised by WIT's competitive performance in comparison to both Google services, allowing users to have choices. Although IBM has the highest WER, it is worth to mention that this service was able to recognize nearly all German interjections (e.g. "äh", "ähm", "hm"). If these phrases are of interest, IBM can be applied additionally.

Regarding the longitudinal character of our study, we did not find any improvement over the period of investigation (eight months) for any of the analyzed ASR service. It can be assumed that improvements are only provided through larger roll-outs rather than in smaller improvement steps. Furthermore, for our experiments we could not observe an improved WER by testing the same material repeatedly, i.e. the same recognition errors still occurred after eight months. However, we can only speculate, whether our test period was too short for this kind of analysis, or whether the errors detected were not serious enough (frequency of occurrence, lack of manual correction, too little repetitive input) to require an intervention. Thus, we will continue with this study and also analyze the provided confidence values to analyze, whether we can detect changes here more sensitively.

References

1. Chiu, C., et al.: State-of-the-art speech recognition with sequence-to-sequence models. In: Proceedings IEEE ICASSP-2018, Calgary, Kanada, April 2018, pp. 4774–4778 (2018)
2. Egorow, O., Lotz, A., Siegert, I., Böck, R., Krüger, J., Wendemuth, A.: Accelerating manual annotation of filled pauses by automatic pre-selection. In: 2017 International Conference on Companion Technology (ICCT), September 2017, pp. 1–6 (2017). https://doi.org/10.1109/COMPANION.2017.8287079

3. Godfrey, J.J., Holliman, E.C., McDaniel, J.: SWITCHBOARD: telephone speech corpus for research and development. In: Proceedings of the IEEE ICASSP-1992, San Francisco, CA, USA, March 1992, vol. 1, pp. 517–520 (1992)

4. Iancu, B.: Evaluating Google speech-to-text API's performance for Romanian e-Learning resources. Informatica Economica **23**, 17–25 (2019). https://doi.org/10.12948/issn14531305/23.1.2019.02

5. Kim, J.Y., et al.: A comparison of online automatic speech recognition systems and the nonverbal responses to unintelligible speech. CoRR abs/1904.12403 (2019). http://arxiv.org/abs/1904.12403

6. Kimura, T., Nose, T., Hirooka, S., Chiba, Y., Ito, A.: Comparison of speech recognition performance between Kaldi and Google cloud speech API. In: Pan, J.-S., Ito, A., Tsai, P.-W., Jain, L.C. (eds.) IIH-MSP 2018. SIST, vol. 110, pp. 109–115. Springer, Cham (2019). https://doi.org/10.1007/978-3-030-03748-2_13

7. Kurata, G., Ramabhadran, B., Saon, G., Sethy, A.: Language modeling with highway LSTM. In: Proceedings of the IEEE ASRU, Okinana, Japan, pp. 244–251 (2017). https://doi.org/10.1109/ASRU.2017.8268942

8. Këpuska, V., Bohouta, G.: Comparing speech recognition systems (Microsoft API, Google API and CMU Sphinx). Int. J. Eng. Res. Appl. 20–24 (2017). https://doi.org/10.9790/9622-0703022024

9. Och, F.J.: Minimum error rate training in statistical machine translation. In: Proceedings of the 41st Annual Meeting on Association for Computational Linguistics, pp. 160–167. Association for Computational Linguistics, USA (2003). https://doi.org/10.3115/1075096.1075117

10. Petrock, V.: US Voice Assistant Users 2019 - Who, What, Where and Why. eMarketer (2019). Accessed 15 July 2019

11. Roberts, M.: OK Google, Siri, Alexa, Cortana; can you tell me some stats on voice search? The edit blog (2018). Accessed 8 Jan 2018

12. Saon, G., et al.: English conversational telephone speech recognition by humans and machines. In: Proceedings of the INTERSPEECH-2017, Stockholm, Sweden, pp. 132–136 (2017). https://doi.org/10.21437/Interspeech.2017-405

13. Sawakare, P.A., Deshmukh, R.R., Shrishrimal, P.P.: Speech recognition techniques: a review. Int. J. Sci. Eng. Res. **6**, 1693–1698 (2015)

14. Siegert, I.: "Alexa in the wild" - collecting unconstrained conversations with a modern voice assistant in a public environment. In: Proceedings of the 12th LREC, pp. 608–612. ELRA, Marseille, France (2020). https://www.aclweb.org/anthology/2020.lrec-1.77

15. Siegert, I., Krüger, J., Egorow, O., Nietzold, J., Heinemann, R., Lotz, A.: Voice Assistant Conversation Corpus (VACC): a multi-scenario dataset for addressee detection in human-computer-interaction using Amazon's ALEXA. In: Proceedings of the 11th LREC. ELRA, Paris, France (2018)

16. Xiong, W., Wu, L., Droppo, J., Huang, X., Stolcke, A.: The Microsoft 2017 conversational speech recognition system. In: Proceedings of the IEEE ICASSP-2018, Calgary, Kanada, April 2018, pp. 5934–5938 (2018)

17. Yurtcan, Y.: Performance evaluation of real-time noisy speech recognition for mobile devices. Master's thesis. Middle East Technical University, Turkey (2019)

Does *A Priori* Phonological Knowledge Improve Cross-Lingual Robustness of Phonemic Contrasts?

Lucy Skidmore[1] and Alexander Gutkin[2]([⊠])

[1] Speech and Hearing Research Group, University of Sheffield, Sheffield, UK
lskidmore1@sheffield.ac.uk
[2] Google Research, London, UK
agutkin@google.com

Abstract. For speech models that depend on sharing between phonological representations an often overlooked issue is that phonological contrasts that are succinctly described language-internally by the phonemes and their respective featurizations are not necessarily robust across languages. This paper extends a recently proposed method for assessing the cross-linguistic consistency of phonological features in phoneme inventories. The original method employs binary neural classifiers for individual phonological contrasts trained solely on audio. This method cannot resolve some important phonological contrasts, such as retroflex consonants, cross-linguistically. We extend this approach by leveraging prior phonological knowledge during classifier training. We observe that since phonemic descriptions are articulatory rather than acoustic the model input space needs to be grounded in phonology to better capture phonemic correlations between the training samples. The cross-linguistic consistency of the proposed method is evaluated in a multilingual setting on held-out low-resource languages and classification quality is reported. We observe modest gains over the baseline for difficult cases, such as cross-lingual detection of aspiration, and discuss multiple confounding factors that explain the dimensions of the difficulty for this task.

Keywords: Phonology · Cross-lingual models · Low-resource languages

1 Introduction

As the smallest constituents of phonological structure, distinctive features (DFs) can be used to provide a linguistically rich and language-independent representation schema for speech [7]. In contrast to the abstract and language-dependent phonemic representations commonly adopted in speech applications, such as automatic speech recognition (ASR) and text-to-speech (TTS), a unit of speech is instead represented by a set of phonologically derived characteristics. In a monolingual setting, one would typically use the phonemes of a language as the

© Springer Nature Switzerland AG 2020
A. Karpov and R. Potapova (Eds.): SPECOM 2020, LNAI 12335, pp. 530–543, 2020.
https://doi.org/10.1007/978-3-030-60276-5_51

basic sound units and derive their feature encodings from the corresponding DFs. Encoding speech in this way not only allows comparison of the structure of various phonemes, but also provides a flexible framework for modeling inter- and intra-speaker pronunciation variability.

Various DF representations have been integrated successfully into the ASR and TTS pipelines over the years. In ASR, adding DF detection to the recognition pipelines has been shown to increase recognition performance in monolingual [18,22,23,31,33,34,41], multilingual [35,38,39] and low-resource settings [5,40]. However, detection accuracy of individual DFs can vary widely, with the difference between lowest and highest detector accuracy reported to be as high as 60% [35]. In recent years, more accurate DF detection has been achieved using state-of-the-art deep learning methods [12,17,21,29]. In TTS, where one typically starts from monolingual phonemic pronunciation dictionaries and phoneme inventories, DFs were also shown to be beneficial, especially in multilingual scenario [3,30,42].

As noted in [16], it is not clear a priori whether all DFs will be useful or valid in a multilingual setting. If feature descriptions were *phonetic* rather than phonemic, and *acoustic* rather than articulatory, one would expect a close correspondence between phonetic features and the acoustic signal. Similar observations motivated other recent research aiming for more phonetic realism [25]. The reality, however, is different. In practice, one often starts with phoneme inventories, the pronunciation dictionaries based on these inventories and the DF representations based on the word-level dictionary-based phonemic transcriptions. This procedure potentially introduces multiple sources of problems such as suboptimal design of the original phoneme inventories and under-specified phonemic pronunciations. Additional complications arise due to the choice of DF system for featurization as many competing feature systems are in use today. These issues are further exacerbated in multilingual scenarios due to linguistic diversity among languages.

One possible way of framing the question of practical utility and empirical validity of the chosen features in a multilingual resource sharing setting was proposed in [16], where the consistency of DF descriptions was evaluated on a cross-lingual task in terms of classification quality on phoneme-size spans of connected speech. One of the main empirical findings of that work is that the postulated contrasts that generally hold within a language are not necessarily robust across languages. This method is useful in several application scenarios: design of multilingual phoneme inventories with optimal DF sharing, derivation of phoneme inventories from speech in low- and zero-resource language documentation and evaluation of DF detector errors in ASR.

In this paper we continue the line of research in [16] by examining some of the cases where phonemic contrasts do not hold cross-linguistically. We investigate whether extending the original method by integrating prior linguistic knowledge into the model can improve its performance across the board. Although there are several open-source linguistic ontologies providing useful types of typological information, such as aerial and phylogenetic features of Glottolog [10] and World Atlas of Language Structures (WALS) [11], in this work we limit the scope of prior knowledge sources to two phonological typologies: PHOIBLE [24] and PANPHON [26]. We hypothesize that the use of phonological grounding alone is sufficient for cross-lingual phonemic contrast resolution.

2   Distinctive Features and Their Typologies

Distinctive features were first established in phonological analysis in the 1950s [15], after which various approaches to their representation have been proposed. For this investigation, distinctive feature values are considered as binary — each speech sound is represented by a set of DFs that are either present or absent (see [7,9] for an overview of alternative feature systems). A sample representation for the phoneme /n/, taken from [9], which follows the binary representation scheme introduced in Chomsky and Halle's *The Sound Pattern of English* (SPE) [2] is [+CONSONANTAL, +SONORANT, −CONTINUANT, +NASAL, +CORONAL].

PHOIBLE is a free database of cross-linguistic phonological data compiled from many linguistic sources. The online 2014 edition [24] includes 2155 phoneme inventories with 2160 segment types found in 1672 distinct languages. The feature system in PHOIBLE aims to be descriptively adequate cross-linguistically and is likely to change as new languages are added. Overall the feature system consists of 37 "binary" features (such as LABIODENTAL and SPREADGLOTTIS that for the simple phonemic segments take the ternary values: present $(+)$, absent $(-)$ and not applicable $(\varnothing)$. For complex segments, such as diphthongs, tuples of the above values are used. For example, the value of a vowel feature SYLLABIC for diphthong /Ew/ is a pair $(+, -)$.

PANPHON is a resource consisting of a database that relates over 5,000 IPA segments (simple and complex) to their definitions in terms of about 23 articulatory features and a Python package to manipulate the segments and their feature representations [26]. Unlike PHOIBLE, which documents the actual snapshot of contemporary phonological knowledge of the world's languages from the standpoint of linguistic theory, PANPHON's mission is to develop a methodologically solid resource to facilitate research in NLP. One of the nice features of PANPHON is its great flexibility, which is achieved as follows: The resource contains a core set of approximately 146 segments represented in IPA and their corresponding features. The non-trivial segments are derived from this set using formal rules that describe the application of diacritics and modifiers, the feature specifications that provide the necessary context for the modification and articulatory feature changes required if the diacritic or modifier is applied. Similar to PHOIBLE, a ternary system is used to represent each of the articulatory features loosely based on well-established phonological classes.

3   Method and Corpora

We follow and extend the methodology proposed in [16]: to consider a phonemic contrast to be consistent or robust across languages, it needs to be easily predicted

Table 1. The six languages used in the experiments.

Name	Code	Family	Documentation	URL
Bengali	bn	Indo-Aryan	Kjartansson et al. [19]	http://www.openslr.org/37/
Gujarati	gu	Indo-Aryan	He et al. [13]	http://www.openslr.org/78/
Marathi	mr	Indo-Aryan	He et al. [13]	http://www.openslr.org/64/
Kannada	kn	Dravidian	He et al. [13]	http://www.openslr.org/79/
Telugu	te	Dravidian	He et al. [13]	http://www.openslr.org/66/
Sundanese	su	Malayo-Polynesian	Kjartansson et al. [19]	http://www.openslr.org/44/

on heldout languages. This is operationalized as follows: a particular phonemic contrast is presented as a binary classification problem. An instance of this problem consists of a span of a speech signal (e.g., a vowel in surrounding context) and a positive or negative label (e.g., front vowel vs. back vowel). A classifier is trained on a multi-speaker, multi-language dataset withholding one or more languages. We then evaluate the trained classifier on the held-out data and report its quality in terms of Area Under (resp. Over) the receiver operating characteristic Curve (AUC, resp. AOC). If the binary contrast in question is cross-linguistically consistent, we expect it to be readily predictable on held-out languages.

For cases where cross-linguistic consistency does not hold, we propose to extend this method by grounding the task on the contextual phonological knowledge provided by PHOIBLE and PANPHON. This is realized by augmenting the acoustic input features with dense categorical DF encodings. We hypothesize that a certain contrast that cannot be resolved cross-lingually from the speech signal alone may correlate with other contrasts that are robust. Such correlations may in theory be captured by including the full DF context in classifier training. At evaluation time, since the phonological context is unavailable, these categorical input features are set to 'not applicable' ($\varnothing$).

Languages and Phoneme Inventories. We use a smaller subset of the languages previously used for the experiments in [16]. Six languages from South and Southeast Asia were chosen for the experiments: three languages from the Indo-Aryan family (Bengali, Gujarati and Marathi), two languages from the Dravidian family (Kannada and Telugu) and Sundanese, a Malayo-Polynesian language. Open-source speech corpora for these languages are available, as shown in Table 1, which for each language shows its BCP-47 language code [27], corpus documentation reference and the corresponding location in the Open Speech and Language Resources (OpenSLR) repository [28]. All datasets consist of multi-speaker 48 kHz audio and the corresponding transcriptions. In this work we restrict the experiments to female speakers only to constrain the spectral variability due to gender-specific pitch differences. The Indo-Aryan and Dravidian languages are interesting to investigate because, on the one hand, they exhibit considerable phonological variation within each group, and on the other, share several cross-group similarities [4]. The inclusion of Malayo-Polynesian language is justified on the grounds of close historic contacts between the languages from this family, such as Javanese and Sundanese, with the Dravidian languages [14].

Table 2. Phoneme inventories grouped by language families

		Phonemes (in IPA notation)
Shared		a b dʒ e f g h i k l m n o p r s t ʃ u
	bn	bʰ dʒʰ ɖ ɖʰ ʝ kʰ n t͡ʃʰ ʈ ʈʰ æ ɔ ɖ ɖʰ gʰ ʃ t tʰ ʋ e ɵ
Indo-Aryan	gu	bʰ dʒʰ ɖ ɖʰ ʝ kʰ ŋ t͡ʃʰ ʈ ʈʰ æ ɔ ɖ ɖʰ gʰ ʃ t tʰ ʋ l ɳ ə
	mr	bʰ dʒʰ ɖ ɖʰ ʝ kʰ ŋ t͡ʃʰ ʈ ʈʰ æ ɔ ɖ ɖʰ gʰ ʃ t tʰ ʋ l ɳ ə dz dzʰ lʰ mʰ ɳʰ ts ʋʰ
Dravidian	te	ɖ ʝ ɳ ʈ ɖ l ɳ ɳ ʂ t ʋ bʰ ɖʰ kʰ t͡ʃʰ ʈʰ ɖʰ gʰ t͡ʃʰ ʃ æ
	kn	ɖ ʝ ɳ ʈ ɖ l ɳ ɳ ʂ t ʋ bʰ ɖʰ kʰ t͡ʃʰ ʈʰ ɖʰ gʰ t͡ʃʰ ʃ dʒʰ
Malayo-Polynesian	su	ɖ ʝ n t w x z ɑɪ ɑʊ ə ɲ ʃ ʔ ɔɪ ɤ

Table 3. Details of the corpora used in the experiments

Code	Speakers	Utterances	Words total	Words unique	Segments	Duration (seconds)
bn	23	7,499	42,177	7,684	51,945	45353.60
gu	18	2,853	23,065	8,172	27,481	15464.40
mr	10	1,719	17,103	2,889	20,131	10864.30
kn	24	2,897	14,780	8,050	18,882	15533.70
te	24	3,351	11,220	4,186	15,784	9819.65
su	20	2,401	21,848	3,169	26,742	11541.40
Total	119	20,720	130,193	–	160,965	108577.05

We reuse the phoneme inventories from prior work [16], which borrowed the South Asian phoneme inventories from [3] and Malayo-Polynesian phoneme inventories from [43]. These phoneme inventories were designed with multilingual speech applications in mind, where languages use a unified underlying phonological representation, which is leveraged to make the most of the available data and eliminate phonemic scarcity by conflating similar phonemes into a single representative phoneme. The phoneme inventories for all languages use International Phonetic Alphabet (IPA) and are shown in Table 2 grouped by their language families. A subset of phonemes that is common to all the inventories is shown as "Shared" in the first row of the table. While the inventories do not map one-to-one to the inventories provided by existing typological resources, such as PHOIBLE, there is nevertheless a significant correlation between them.

Basic overview of the corpora is provided in Table 3. There are 119 female speakers in the combined dataset of 20,720 utterances corresponding to just over 30 hours of speech and 130,193 words. Word-level phonemic transcriptions containing 160,965 segments in total were provided by proprietary lexicons using phoneme inventories from Table 2. In order to determine segment boundaries, transcriptions where force-aligned with the acoustic parametrization of the audio using standard Hidden Markov Model (HMM)-based recipe [44]. The acoustic parametrization was obtained by downsampling the audio to 16 kHz and parametrizing it into HTK-style Mel Frequency Cepstral Coefficients (MFCC) [6] using 10 ms frame shift. The dimension of the MFCC parameters is 39 (13 static $+ \Delta + \Delta\Delta$ coefficients).

Table 4. Distinctive features and corresponding phonemes

Feature		Corresponding Phonemes
FRONT	(+)	e eː æ i iː
	(−)	a aː o oː ɔ ʏ u uː ə
HIGH	(+)	i iː u uː
	(−)	e eː æ o oː ɔ ʏ ə a aː
SG	(+)	bʰ dzʰ ɖʰ h kʰ lʰ mʰ ɳʰ tʃʰ ʈʰ dʰ gʰ ʈʰ ʊʰ
	(−)	a aː b d dz dʒ ɖ e eː f i iː j k l m n ɳ o o p r s t ts tʃ ʈ u uː w x z æ ɳ ɔ ɖ ə g ʏ ɭ ɳ ɳ ʂ ʃ ʊ ʔ
CONT	(+)	a aː e eː f h i iː j l lʰ o oː r s u uː w x z æ ɔ ə ʏ ɭ ʂ ʃ ʊ ʊʰ
	(−)	b bʰ d dz dzʰ dʒ ɖ ɖʰ k kʰ m mʰ n ɳ ɳʰ p t ts tʃ ʈʃʰ ʈʰ ɳ ɖ ɖʰ g gʰ ɳ ɳ t ʈʰ ʔ

Phonemic Contrasts. Each of the DF contrasts can be represented by two sets of phonemes, one for which the feature is present, and one where it is absent. Table 4 shows a list of phoneme groups, together with the corresponding phonemes selected from our corpora, to study such contrasts. For the binary classification task, the former set of phonemes, provides the positive examples, while the later one provides the negative examples. PHOIBLE and PANPHON assign compatible feature values for the phonemes and contrasts shown.

We investigate four contrasts. The 'front–back' contrast, denoted FRONT in the table, is defined as a combination of features: front vowel (+) is taken to mean [+FRONT, −BACK] in both PHOIBLE and PANPHON, and back vowel (−) is based on [−FRONT, +BACK]. We reproduce this experiment from [16] as a sanity check because we use different training data. In the original work this contrast was found to be consistent cross-linguistically. We extend the vowel experiments to 'high-low' contrast (HIGH), which for high vowels (+) is defined as [+HIGH, −LOW] and for low vowels (−) as [−HIGH, +LOW] in both typologies. The class of low vowels contains the close-mid back unrounded vowel /ɤ/ , which is unique to Sundanese in our language set. The next two contrasts are particularly interesting to predict. In both cases the positive class (+) is formed by the set of spectrally diverse phonemes. The SPREADGLOTTIS laryngeal feature (SG) includes all the aspirated consonants in its positive class. The CONTINU-ANT manner of articulation feature (denoted CONT) specify the openness (+) or complete closure (−) of the vocal tract during the phonation. We don't restrict the +CONT class to consonants (fricatives and liquids) by also including all the vowels.

4   Experiments, Results and Discussion

Experiment Setup. We use MFCCs prepared during the phoneme alignment stage (described in Sect. 3) as acoustic parameters. Admittedly, the use of MFCCs may be too restrictive: other representations, such as F0 or auditory-derived features, may be better suited to model the acoustic cues that signal the contrasts in each scenario [32]. Although we previously demonstrated moderate gains of other acoustic features types over the MFCCs on a similar task [8], in

Table 5. AOC for HIGH, FRONT, CONT and SG DFs on held-out languages

L.	HIGH			FRONT		
	Baseline	PANPHON	PHOIBLE	Baseline	PANPHON	PHOIBLE
bn	6.68 (±0.41)	6.68 (±2.95)	5.64 (±0.27)	1.97 (±0.11)	2.14 (±1.86)	2.78 (±0.21)
gu	7.06 (±0.12)	7.96 (±1.41)	7.64 (±0.30)	0.69 (±0.12)	0.82 (±0.21)	0.72 (±0.05)
mr	7.03 (±0.66)	7.67 (±0.75)	7.93 (±0.52)	2.11 (±0.11)	2.58 (±0.79)	2.03 (±0.16)
kn	7.65 (±0.61)	9.30 (±1.78)	8.63 (±0.25)	0.60 (±0.01)	0.56 (±0.19)	**0.50** (±0.06)
te	6.00 (±0.09)	8.36 (±2.45)	7.50 (±0.44)	1.61 (±0.05)	1.44 (±0.20)	**1.32** (±0.07)
su	3.09 (±0.16)	5.32 (±1.02)	4.83 (±1.09)	1.90 (±0.13)	3.69 (±0.36)	3.57 (±0.23)

L.	CONT			SG		
	Baseline	PANPHON	PHOIBLE	Baseline	PANPHON	PHOIBLE
bn	2.00 (±0.16)	3.18 (±0.05)	3.00 (±0.88)	8.07 (±1.89)	6.61 (±5.00)	5.69 (±5.45)
gu	3.17 (±0.69)	3.16 (±0.09)	3.09 (±0.37)	10.79 (±3.62)	10.58 (±0.62)	10.54 (±0.21)
mr	2.94 (±0.33)	3.45 (±0.18)	3.57 (±0.57)	14.77 (±3.13)	14.36 (±0.71)	14.65 (±1.57)
kn	2.16 (±0.50)	2.29 (±0.24)	2.13 (±0.74)	13.45 (±3.32)	12.14 (±0.38)	12.34 (±0.25)
te	1.82 (±0.05)	**1.53** (±0.20)	**1.46** (±0.30)	13.39 (±3.20)	13.30 (±1.83)	13.42 (±0.64)
su	1.16 (±0.50)	1.83 (±0.20)	1.66 (±0.42)	18.68 (±2.86)	16.63 (±2.03)	18.15 (±0.59)

this work we limit the scope of investigation to MFCCs to keep the number of experiments manageable.

For each phonemic contrast three experiment configurations are constructed. For the baseline configuration, a single training example consists of 40 acoustic frames. It is constructed by stacking the frames corresponding to the particular phoneme plus its right and left context frames, possibly padding with zeros if the context is too short. Phonemes longer than 40 frames are ignored. The PHOIBLE and PANPHON configurations are constructed by extending the baseline input features with 37 and 23 categorical features describing the phonemic segment, respectively. For each DF, the input features corresponding to the classification labels are masked out (set to 'unspecified' value ∅) in the training data. At evaluation time, since no phonological information is available to PHOIBLE and PANPHON configurations, all the input categorical features are set to ∅. The training sets for PHOIBLE and PANPHON are doubled by the simple data augmentation technique: each training example is cloned once and the categorical portion of its input features is masked out, so that the model can also learn to generalize in the absence of phonological context.

The training and evaluation sets in our experiments always consist of disjoint sets of languages and speakers. For each dataset we also limit the number of training examples to 50,000 and evaluation examples to 10,000. In order to keep the overall set of training labels balanced, with equal number of positive and negative examples, we employ a simple under-sampling approach [20]. If enough examples are available, we sample an equal number of them from every language in the training set. Conversely, an imbalance in a language is preferred over the lack of training examples. It is important to note that we do not guarantee that the number of training examples is the same across speakers of a language.

We use mean and standard deviation computed over the training set input features to scale the training as well as evaluation sets. We employ vanilla feed-forward Deep Neural Network (DNN) binary classifier from TensorFlow [1]. A

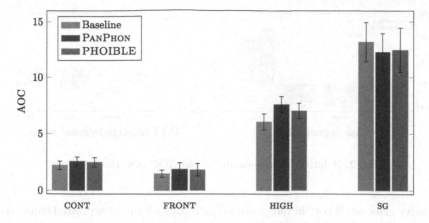

Fig. 1. Average AOC per DF classification across all held-out languages

simple two-layer architecture with 200 Softplus [46] units in each layer, dropout probability of 0.2 [37], Adadelta optimizer [45] and the learning rate of 0.6 with a large batch size of 6000 [36] were determined by tuning on the development set. A single classifier is trained on five languages and evaluated on the sixth held-out language. Overall we construct 72 classifiers: one for each of the six held-out languages, three input feature configurations (baseline, PHOIBLE and PANPHON) and four phonemic contrasts (HIGH, FRONT, CONT and SG). Each training/evaluation experiment is repeated three times resulting in 216 experiments overall and statistics for the Area Over the Curve (AOC) metric are accumulated. We use AOC for better readability, since the Area Under the ROC Curve (AUC) values are generally high.

Results and Discussion. Table 5 shows the AOC values for the detection HIGH, FRONT, CONT and SG DFs across multiple training configurations. Each row in the table represents the held-out language on which the classifier trained on five languages is evaluated. Each AOC value is the mean over three runs. Confidence interval (95%) range computed using t-test over sample size $n = 3$ is shown alongside each AOC mean. As can be seen from the table, the front vs. back vowel contrast FRONT is very robust across languages having the lowest AOC values among all the contrasts begin tested. This result confirms the result for FRONT reported in [16, Table 6] on different data. The second best contrast which is very consistent cross-lingually is the CONT manner of articulation contrast. This result is somewhat contrary to our expectations as the positive class +CONT is very heterogeneous including sounds like fricatives and vowels. The high vs. low vowel contrast HIGH is not as robust across languages as the FRONT contrast, but is also reasonably consistent cross-lingually, with the best predictions among the held-out configurations obtained for Sundanese. The worst performing configurations are found for the contrast SG that separates the aspirated sounds from the rest. With the exception of Bengali, this contrast is not robust across languages. We hypothesize that this contrast is hard to detect cross-lingually because the

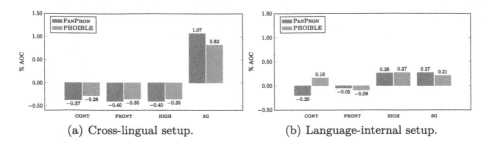

(a) Cross-lingual setup.   (b) Language-internal setup.

Fig. 2. Relative improvements (%) in AOC over the baselines

negative class -SG is very heterogeneous (including all the unaspirated consonants and vowels) and aspiration is acoustically more ambiguous compared to other contrasts we considered.

As can be seen from Table 5, the inclusion of phonological context in the classifier's input feature space leads only to the minor occasional statistically significant improvements over the baseline (shown in bold). In most of the other cases when the mean AOC values for the PHOIBLE and PANPHON configurations are lower than the baseline, the improvements are not statistically significant because of the overlap in confidence intervals. In order to summarize the performance of the classifiers across all the held-out languages for each phonemic contrast, we recomputed the statistics per each contrast, with Fig. 1 showing the AOC means and confidence intervals computed for the sample size $n = 18$ (three runs for six languages). As can be seen from the figure, the phonologically grounded classifiers do not improve (on average) over the corresponding baselines for CONT, FRONT and HIGH contrasts. This is likely due to the task being already unambiguous enough for the baseline classifier, where the introduction of additional context actually increases the confusion. Note that the inclusion of phonological context from both the PHOIBLE and PANPHON sources improves the detection of aspiration contrast SG. This improvement, however, is not statistically significant because of the confidence intervals overlap.

In order to further evaluate the influence of phonemic grounding on the detection of each phonemic contrast, we compared the relative improvements in AOC over the baselines for our current cross-lingual setup, shown in Fig. 2(a), with the language-internal setup shown in Fig. 2(b). The language-internal configurations are constructed similarly to the cross-lingual ones (also 216 configurations overall), with the only difference that the training and evaluation data is confined to one language. As can be seen from the figures, phonemic grounding has very little influence on the classifier performance in language-internal case. We hypothesize that this is due to the fact that the phonological context that DFs provide already exists implicitly in the structure of the acoustic training data (i.e. the phonemes in the positive and negative classes) and therefore representing it explicitly does not show an effect. In the cross-lingual case, however,

Table 6. Average AOC for phonological input features alone

Source	HIGH ($k = 4$)			FRONT ($k = 5$)			CONT ($k = 25$)			SG ($k = 14$)		
	NB	LR	SVM	NB	LR	SVM	NB	LR	SVM	NB	LR	SVM
PHOIBLE	22.92	25.00	31.25	30.00	30.00	25.00	10.00	6.00	16.00	33.93	50.29	57.44
PANPHON	29.17	37.50	37.50	25.00	40.00	20.00	12.00	14.00	16.00	46.13	54.46	54.46

phonological context helps detection of SG, the most "difficult" contrast under investigation.

In order to assess the discriminatory power of the phonological typologies alone for our task, we also constructed classifiers without relying on acoustics. For each phonemic contrast we constructed three types of classifiers for PHOIBLE and PANPHON typologies: Naive Bayes (NB), linear regression (LR) and support vector machine with linear kernel function (SVM). Stratified k-fold cross-validation with contrast task-dependent value of k constrained by the minimal negative or positive class size (as shown in Table 4) was employed. Crucially, the input features that directly correlate with the labels were masked out during training and evaluation. For example, when constructing and evaluating the classifier for contrast HIGH, both HIGH and LOW input features were set to $\varnothing$. The average AOC values over k runs are shown in Table 6 for each contrast, classifier and typology type. Comparing these classifiers' performance with the classifiers trained on the full acoustic and phonological data (Table 5) it is evident that the classifiers trained on phonology alone are significantly less accurate. Apart from CONT, the best performing contrast in Table 6, which also correlates reasonably with the results for the full training in Table 5, the rest of the classifiers struggle to detect the contrasts in question based on the phonological context alone. The classifier for SG performs the worst. It is interesting to note that, while these classifiers are generally useful on their own, apart from the very unreliable detector for SG, their accuracy increases significantly once we combine the phonological (articulatory) input space with the acoustics. One confounding factor that may explain detection inaccuracies in this scenario are the typological features themselves – in the case of PHOIBLE and PANPHON the remaining phonological features in the context of the contrast itself may not be enough (e.g., due to their potential ambiguity or wrong definitions) to signal the contrast. This merits further research into the design of feature inventories that are highly consistent in multilingual settings.

5   Conclusion

The results from this investigation provide a starting point for further research on the impact of a priori phonological knowledge on cross-lingual DF classification. The modest gains on the baseline recorded for SG classification require further contextualisation through experimentation on a wider group of DFs. In addition, it would be of value to explore the impact of phonological processes such as assimilation and co-articulation on DF detection accuracy. Exploring an alternative network configuration may also be beneficial, such as training a sequence model over feature detectors or embeddings. In summary, it is clear that it is not only the relationship between acoustic representations of speech and phonological feature inventories that is complex — the internal relationship between individual DFs within feature inventories is also impactful and should be taken into consideration when designing feature inventories for use in multilingual settings.

Acknowledgments.. The authors would like to thank Cibu Johny for his help with the experiments, and Işın Demirşahin and Rob Clark for fruitful discussions.

References

1. Abadi, M., et al.: TensorFlow: a system for large-scale machine learning. In: Proceedings of 12th Symposium on Operating Systems Design and Implementation (OSDI), pp. 265–283. USENIX Association (2016)
2. Chomsky, N., Halle, M.: The Sound Pattern of English. Harper & Row, New York (1968)
3. Demirsahin, I., Jansche, M., Gutkin, A.: A unified phonological representation of South Asian languages for multilingual text-to-speech. In: Proceedings of 6th Workshop on Spoken Language Technologies for Under-Resourced Languages (SLTU), pp. 80–84. ISCA, Gurugram, India (2018). https://doi.org/10.21437/SLTU.2018-17
4. Emeneau, M.: India as a linguistic area. Language **32**(1), 3–16 (1956). https://doi.org/10.2307/410649
5. Fu, T., Gao, S., Wu, X.: Improving minority language speech recognition based on distinctive features. In: Peng, Y., Yu, K., Lu, J., Jiang, X. (eds.) IScIDE 2018. LNCS, vol. 11266, pp. 411–420. Springer, Cham (2018). https://doi.org/10.1007/978-3-030-02698-1_36
6. Ganchev, T., Fakotakis, N., Kokkinakis, G.: Comparative evaluation of various MFCC implementations on the speaker verification task. In: Proceedings of 10th International Conference on Speech and Computer (SPECOM), vol. 1, pp. 191–194, Patras, Greece (2005)
7. Gussenhoven, C.: Understanding Phonology, 4th edn. Routledge, London (2017). https://doi.org/10.4324/9781315267982
8. Gutkin, A.: Eidos: an open-source auditory periphery modeling toolkit and evaluation of cross-lingual phonemic contrasts. In: Proceedings of 1st Joint Spoken Language Technologies for Under-Resourced Languages (SLTU) and Collaboration and Computing for Under-Resourced Languages (CCURL) Workshop (SLTU-CCURL 2020), pp. 9–20. European Language Resources Association (ELRA), Marseille (2020)
9. Hall, T.A.: Distinctive Feature Theory. Mouton de Grutyer, Berlin (2001). https://doi.org/10.1515/9783110886672
10. Hammarström, H., Forkel, R., Haspelmath, M., Bank, S.: Glottolog 4.2.1. Max Planck Institute for the Science of Human History, Jena, Germany (2020). https://doi.org/10.5281/zenodo.3754591
11. Haspelmath, M., Dryer, M.S., Gil, D., Comrie, B.: The World Atlas of Language Structures. Oxford University Press, Oxford (2005). https://doi.org/10.5281/zenodo.3731125

12. He, D., Yang, X., Lim, B.P., Liang, Y., Hasegawa-Johnson, M., Chen, D.: When CTC training meets acoustic landmarks. In: Proceedings of International Conference on Acoustics, Speech and Signal Processing (ICASSP), pp. 5996–6000. IEEE, Brighton (2019). https://doi.org/10.1109/ICASSP.2019.8683607

13. He, F., et al.: Open-source multi-speaker speech corpora for building Gujarati, Kannada, Malayalam, Marathi, Tamil and Telugu speech synthesis systems. In: Proceedings of 12th Language Resources and Evaluation Conference (LREC), pp. 6494–6503. European Language Resources Association (ELRA), Marseille (2020)

14. Hoogervorst, T.: Detecting pre-modern lexical influence from South India in Maritime Southeast Asia. Archipel: Études interdisciplinaires sur le monde insulindien (89), 63–93 (2015). https://doi.org/10.4000/archipel.490

15. Jakobson, R., Fant, G., Halle, M.: Preliminaries to Speech Analysis: The Distinctive Features and Their Correlates. MIT Press, Cambridge (1952)

16. Johny, C., Gutkin, A., Jansche, M.: Cross-lingual consistency of phonological features: an empirical study. In: Proceedings of Interspeech 2019, pp. 1741–1745. ISCA, Graz (2019). https://doi.org/10.21437/Interspeech.2019-2184

17. Karaulov, I., Tkanov, D.: Attention model for articulatory features detection. In: Proceedings of Interspeech 2019, pp. 1571–1575. ISCA, Graz (2019). https://doi.org/10.21437/Interspeech.2019-3020

18. Kirchhoff, K., Fink, G.A., Sagerer, G.: Conversational speech recognition using acoustic and articulatory input. In: Proceedings of International Conference on Acoustics, Speech, and Signal Processing (ICASSP), vol. 3, pp. 1435–1438. IEEE, Istanbul (2000). https://doi.org/10.1109/ICASSP.2000.861883

19. Kjartansson, O., Sarin, S., Pipatsrisawat, K., Jansche, M., Ha, L.: Crowd-sourced speech corpora for Javanese, Sundanese, Sinhala, Nepali, and Bangladeshi Bengali. In: Proceedings of 6th Workshop on Spoken Language Technologies for Under-Resourced Languages (SLTU), pp. 52–55. ISCA, Gurugram (2018). https://doi.org/10.21437/SLTU.2018-11

20. Krawczyk, B.: Learning from imbalanced data: open challenges and future directions. Prog. Artif. Intell. 5(4), 221–232 (2016). https://doi.org/10.1007/s13748-016-0094-0

21. Merkx, D., Scharenborg, O.: Articulatory feature classification using convolutional neural networks. In: Proceedings of Interspeech, Hyderabad, India, pp. 2142–2146 (2018). https://doi.org/10.21437/Interspeech.2018-2275

22. Metze, F., Waibel, A.: A flexible stream architecture for ASR using articulatory features. In: Proceedings of 7th International Conference on Spoken Language Processing (ICSLP), pp. 2133–2136. ISCA, Denver (2002)

23. Momayyez, P., Waterhouse, J., Rose, R.: Exploiting complementary aspects of phonological features in automatic speech recognition. In: Proceedings of IEEE Workshop on Automatic Speech Recognition and Understanding (ASRU), pp. 47–52. IEEE, Kyoto (2007). https://doi.org/10.1109/ASRU.2007.4430082

24. Moran, S., McCloy, D.: PHOIBLE 2.0. Max Planck Institute for Evolutionary Anthropology, Jena, Germany (2019). http://phoible.org/

25. Mortensen, D.R., et al.: AlloVera: a multilingual allophone database. arXiv preprint arXiv:2004.08031 (2020)

26. Mortensen, D.R., Littell, P., Bharadwaj, A., Goyal, K., Dyer, C., Levin, L.: PanPhon: a resource for mapping IPA segments to articulatory feature vectors. In: Proceedings of COLING, Osaka, Japan, pp. 3475–3484 (2016)

27. Phillips, A., Davis, M.: BCP 47 - Tags for Identifying Languages. IETF Trust (2009)

28. Povey, D.: Open SLR. John Hopkins University, Baltimore (2020). http://www. openslr.org/resources.php

29. Qu, L., Weber, C., Lakomkin, E., Twiefel, J., Wermter, S.: Combining articulatory features with end-to-end learning in speech recognition. In: Kůrková, V., Manolopoulos, Y., Hammer, B., Iliadis, L., Maglogiannis, I. (eds.) ICANN 2018. LNCS, vol. 11141, pp. 500–510. Springer, Cham (2018). https://doi.org/10.1007/978-3-030-01424-7_49

30. Rallabandi, S., Black, A.: Variational attention using articulatory priors for generating code mixed speech using monolingual corpora. In: Proceedings of Interspeech, pp. 3735–3739 (2019). https://doi.org/10.21437/Interspeech.2019-1103

31. Rasipurama, R., Magimai-Doss, M.: Articulatory feature based continuous speech recognition using probabilistic lexical modeling. Comput. Speech Lang. **36**, 233–259 (2016). https://doi.org/10.1016/j.csl.2015.04.003

32. Repp, B.H.: Categorical perception: issues, methods, findings. In: Speech and Language: Advances in Basic Research and Practice, vol. 10, pp. 243–335. Elsevier (1984)

33. Rose, R., Momayyez, P.: Integration of multiple feature sets for reducing ambiguity in ASR. In: Proceedings of International Conference on Acoustics, Speech and Signal Processing (ICASSP), pp. IV-325–IV-328. IEEE, Honolulu (2007). https://doi.org/10.1109/ICASSP.2007.366915

34. Siniscalchi, S.M., Lee, C.H.: A study on integrating acoustic-phonetic information into lattice rescoring for automatic speech recognition. Speech Commun. **51**(11), 1139–1153 (2009). https://doi.org/10.1016/j.specom.2009.05.004

35. Siniscalchi, S.M., Svendsen, T., Lee, C.H.: Toward a detector-based universal phone recognizer. In: Proceedings of International Conference on Acoustics, Speech and Signal Processing (ICASSP), pp. 4261–4264. IEEE, Las Vegas (2008). https://doi.org/10.1109/ICASSP.2008.4518596

36. Smith, S.L., Kindermans, P.J., Ying, C., Le, Q.V.: Don't decay the learning rate, increase the batch size. arXiv preprint arXiv:1711.00489 (2017)

37. Srivastava, N., Hinton, G.E., Krizhevsky, A., Sutskever, I., Salakhutdinov, R.: Dropout: a simple way to prevent neural networks from overfitting. J. Mach. Learn. Res. (JMLR) **15**(56), 1929–1958 (2014)

38. Stüker, S., Schultz, T., Metze, F., Waibel, A.: Integrating multilingual articulatory features into speech recognition. In: Proceedings of EuroSpeech, pp. 1033–1036. ISCA, Geneva (2003)

39. Stüker, S., Schultz, T., Metze, F., Waibel, A.: Multilingual articulatory features. In: Proceedings of International Conference on Acoustics, Speech and Signal Processing (ICASSP), pp. I144–I147. IEEE, Hong Kong (2003). https://doi.org/10.1109/ICASSP.2003.1198737

40. Stüker, S., Waibel, A.: Porting speech recognition systems to new languages supported by articulatory feature models. In: Proceedings of 13th International Conference on Speech and Computer (SPECOM). St. Petersburg, Russia (2009)

41. Tolba, H., Selouani, S., O'Shaughnessy, D.: Auditory-based acoustic distinctive features and spectral cues for automatic speech recognition using a multi-stream paradigm. In: Proceedings of International Conference on Acoustics, Speech, and Signal Processing (ICASSP), pp. I-837–I-840. IEEE, Orlando (2002). https://doi.org/10.1109/ICASSP.2002.5743869

42. Tsvetkov, Y., et al.: Polyglot neural language models: a case study in cross-lingual phonetic representation learning. In: Proceedings of 2016 Conference of the North American Chapter of the Association for Computational Linguistics, pp. 1357–1366. ACL, San Diego (2016). https://doi.org/10.18653/v1/N16-1161

43. Wibawa, J.A.E., et al.: Building open Javanese and Sundanese corpora for multi-lingual text-to-speech. In: Proceedings of 11th Conference on Language Resources and Evaluation (LREC), pp. 1610–1614. European Language Resources Association (ELRA), Miyazaki (2018)
44. Young, S., et al.: The HTK Book. Cambridge University Engineering Department, Cambridge (2006)
45. Zeiler, M.D.: ADADELTA: an adaptive learning rate method. arXiv preprint arXiv:1212.5701 (2012)
46. Zheng, H., Yang, Z., Liu, W., Liang, J., Li, Y.: Improving deep neural networks using softplus units. In: Proceedings of International Joint Conference on Neural Networks (IJCNN), pp. 1–4. IEEE (2015). https://doi.org/10.1109/IJCNN.2015.7280459

Can We Detect Irony in Speech Using Phonetic Characteristics Only? – Looking for a Methodology of Analysis

Pavel Skrelin, Uliana Kochetkova(✉), Vera Evdokimova,
and Daria Novoselova

Saint-Petersburg University, Universitetskaya emb. 11,
198034 Saint Petersburg, Russia
{skrelin, ukochetkova}@phonetics.pu.ru,
{v.evdokimova, st065112}@spbu.ru

Abstract. The current paper aims to investigate the perception of Russian ironic laboratory speech and to develop a methodological basis for the further research on perceptually relevant acoustic cues of irony. On the first stage of analysis we created short monologues and dialogues including ironic and homonymous non-ironic sentences. Two Russian native male speakers read this material. The recording was accomplished in a sound booth. Then, randomized fragments of the recordings were suggested to native listeners in an auditory perception experiment. The mean intensity, duration of the stressed vowel and F0 range in the reliably identified target fragments were analyzed, as well as functional intonation models in terms of the system suggested by E. Bryzgunova. The results showed that the listeners are able to detect irony without any contextual or lexical marker of it. The acoustic analysis of the reliably identified stimuli demonstrated the difference in intensity level and stressed vowel duration between ironic and non-ironic utterances. The obtained results allow creating a methodology of data collection and validation for the construction of ironic speech corpus and its analysis. They may also have applications in teaching Russian as a second language, human-machine communication and improving natural language processing systems.

Keywords: Phonetics · Irony · Auditory perception experiment · Acoustic analysis · Intonation models · Pitch · Intensity

1 Introduction

Nowadays, the developers of the human-machine dialogue systems show an increasing interest to the study of intonation patterns. However, there is still a lack of information on the expression of various kinds of modality in speech. Detecting irony remains an extremely challenging task. At the same time its correct understanding is crucial for the final semantic interpretation of an utterance. Not taking into account this type of modality may lead to an erroneous decision made by a machine and distort the whole process of communication. In order to bridge this gap and find perceptually relevant acoustic cues of irony, a research project has been launched. The final goal of this

© Springer Nature Switzerland AG 2020
A. Karpov and R. Potapova (Eds.): SPECOM 2020, LNAI 12335, pp. 544–553, 2020.
https://doi.org/10.1007/978-3-030-60276-5_52

project is to create a special corpus of ironic speech, provide a detailed analysis of its phonetic characteristics and establish a set of acoustic cues of irony using resynthesis and perception validation. We started with working out an appropriate methodology of the speech data collection. Thus, the present paper shows the results we obtained at this stage of our research. Its main objective is to collect the laboratory ironic speech issues and to validate them through an auditory perception experiment.

In the current study we set the following tasks:

- identifying lexical markers of irony;
- creating ironic utterances with or without markers based on the examples from fiction and media;
- creating homonymous non-ironic declarative, interrogative and exclamatory utterances in different contexts;
- recording this experimental material;
- auditory and acoustic analysis of the recordings and selecting fragments for a perception experiment;
- conducting the perception experiment in order to define how well listeners can identify irony in speech without any markers or context;
- statistical analysis.

2   Approaches to the Irony Analysis

2.1   Irony Definition

Despite the fact that irony as a multifaceted phenomenon has been studied for centuries, there is still no universally accepted definition. As was rightly pointed out by Harverkate [7, p. 106]: "<...> it is difficult, if not impossible, to register, describe and explain in an exhaustive manner all aspects of that fascinating phenomenon that we call 'irony'".

The most traditional approach is to reduce irony to antiphrasis, i.e. saying the opposite of what is really meant in a way that makes it obvious what the true intention is. In the XXth century several theories of verbal irony appeared as a result of an increasing scientific interest in this concept. Several scholars including Griece [6], Wilson [15], Sperber [13, 15] and Attardo [2], as well as Zhang [16], Simpson [12], Kreuz and Roberts [9] made valuable contributions to the development of the irony model and irony classification.

Taking into account the diversity of irony types and a large number of classifications, we proceed from defining verbal irony as antiphrasis when the intonation design negates or questions the lexical content of an utterance.

2.2   Irony Markers in Speech

Verbal irony can be expressed in multiple ways, which often leads to interactions of various means expressing irony on different language levels.

On the one hand, there often occurs a fusion of a certain intonation pattern and a concrete lexical content or a grammatical construction. Some lexical and grammatical markers can unambiguously define the interpretation of the preceding or the subsequent part of the utterance as ironic and lead to the negation of the content or doubt towards it. For example, the phrase "тоже мне" (some... you are) indicates that the rest of the sentence should be understood ironically: "тоже мне специалист" ("some expert you are"). However such markers are rarely found in speech. A much more common type is polysemantic markers. For instance, the word "какой" (what a...) can be used to refer to a range of different meanings from a neutral question or exclamation to an ironic exclamation or ironic echo question depending on the intonation design. For instance, the utterance "Какой он певец" can mean:

- What kind of singer is he? (non-ironic special question);
- What a great singer he is! (non-ironic exclamation);
- He can't be a singer! (ironic declarative utterance or exclamation);
- *He*'s a singer, are you kidding? (ironic echo question expressing disbelief).

On the other hand, the application of a certain intonation pattern to a sentence with a neutral meaning and a neutral syntactic construction (i.e. without any lexical or grammatical marker of irony) can lead to negation of the lexical meaning. For example, "I've done all the dishes and also washed the floor in the kitchen. – Great, thanks" and "I've left you some dishes that need to be done. – Great, thanks".

The present study considers both types of ironic utterances.

2.3 Studies of Ironic Speech Acoustic Characteristics in Different Languages

Phonetic expression of irony in the Russian language hasn't been a subject of particular interest amongst scholars. Some examples of intonation design used for expressing sarcasm can be found in the work 'Issledovaniia v oblasti russkoi prosodii' by Kodzasov [8], as well as in several Ph.D. theses dedicated to exploring the means of manifesting irony [1, 3, 14], the latter mainly focusing on lexical rather than phonetical ways of expressing irony. An integrated study of acoustic characteristics of irony in the Russian language is yet to be conducted.

Conversely, foreign studies on irony thoroughly examine a wide range of its phonetical characteristics from voice quality to intonation patterns. Research conducted on German material [11] showed that verbal irony is characterized by a decreased mean fundamental frequency (F0), raised energy levels and increased vowel duration, while there is no significant difference in F0-contours between ironic and literal target utterances. Another distinctive feature of ironic speech is vowel hyperarticulation – increasing articulatory effort in order to raise perceptual word clarity.

Not all studies, however, prove that hyperarticulation correlates with ironic tone of voice. Niebuhr [10] states that irony in German speech is characterized by segmental and prosodic reduction.

At the same time, there is a large number of papers showing that there is no particular ironic tone of voice, such as "Is there an ironic tone of voice?" by Bryant and Fox Tree [5]. One possible explanation is that seemingly contradictory results were obtained on fundamentally different data. It can also be due to a difference between two kinds of verbal irony ("genuine" sarcastic irony and the so-called "kind irony") which have been proven to have different acoustic properties [4].

3   Method and Material

3.1   Text Reading Task

In the course of scientific data analysis and literary texts study we selected 30 lexical markers of irony for our research. These utterance components expressing implicit negation were then used to make sentences (short monologues) and short dialogues with an ironic meaning. Subsequently, these utterances were transformed by cutting out the markers. Some of the markers turned out to be an indispensible part of the phrase, their omission affecting the syntactical integrity of the sentence. Due to this fact those markers were excluded. The transformed phrases were later completed with a remark indicating that the utterance should have an ironic intonation. Thus ironic meaning was suggested to the reader either by a lexical marker or by a remark.

On basis of the original ironic utterances a series of homonymous non-ironic utterances was created, as well as short monologues and dialogues, in which they were incorporated. These sentences were of different communicative types. The main part of ironic sentences was opposed to a declarative, an interrogative (general question) and an exclamatory non-ironic sentences. However, for the utterances beginning by an interrogative word, such as "какой", "какая" (what a...), "что" (what...) etc., no identical non-ironic declarative structure could be found, but only an interrogative (special question) and exclamatory non-ironic structures, as interrogative words can begin an exclamation in Russian.

The experimental material consisted of the following types of homonymous ironic and non-ironic utterances.

I. Ironic:
   a. sentences with lexical markers of irony;
   b. sentences without irony markers, but with remarks such as "he said ironically", "she asked with sarcasm" etc.;
II. Non-ironic:
   a. declarative sentences;
   b. interrogative sentences (general questions were opposed to the ironic sentences without interrogative word, special questions were opposed to the ironic sentences with an interrogative word);
   c. exclamatory sentences (with the same marker as in the ironic sentence, if it could be interpreted as a part of an exclamation, or without the marker elsewhere).

Thus, we obtained 86 short monologues and dialogues and randomized them in order to avoid repetition. This material was read by two Russian native speakers: young and middle-aged men. The recording was conducted in the sound booth at the Department of phonetics of Saint Petersburg State University, resulting in a total of

50 min of recording for each of the speakers. The recordings were later subject to auditory and acoustic analysis. Measurements were carried out using the Wave Assistant processing software, intonation labeling was done according to the model of intonation patterns by E. Bryzgunova.

3.2 Auditory Perception Experiment

Fragments from the recordings described above were presented to listeners in order to test whether they were able to detect irony from short speech stimuli. Snippets from the ironic contexts were manually extracted so that the remaining part would be devoid of any indication of the ironic context (irony markers and ironic remarks). In the course of the auditory expert analysis the brightest examples of irony expression were selected for the perception experiment. To each of the ironic utterances we added snippets from homonymous declarative, exclamatory or interrogative non-ironic sentences. As Speaker 1 provided more evident examples of irony than Speaker 2, the number of phrases pronounced by each of the two speakers in the experiment differed: 29 fragments by Speaker 1 and 16 by Speaker 2. This resulted in a total of 45 fragments that were then randomized.

As it was mentioned in Sect. 3.1, the communicative type of non-ironic utterances was implicated by the syntax and grammar of the corresponding ironic sentence. It led to the fact that the non-ironic fragments from the recordings of Speaker 1 included all communicative types, while in the non-ironic examples read by Speaker 2, there were only declarative and exclamatory non-ironic sentences. Nevertheless, our main goal in this experiment was to oppose ironic and non-ironic examples, the balance between the speakers and between the communicative types of the non-ironic sentences being of minor importance at this stage of the research. On the next step of our research project we are going to carry out a detailed analysis of the perceptive and acoustic difference between concrete communicative types of non-ironic utterances and their ironic pairs.

Only Russian native informants took part in the auditory perception experiment. By the time the results were collected, 91 people accomplished the entire test and filled in a questionnaire: 18 men and 73 women. The other participants who gave answers only to a part of this test or did not provide their personal data, were not considered in the analysis. There were 5 age groups of listeners: 11 people under the age of 20 years, 38 people from 21 to 30 years, 16 people from 31 to 40 years, 20 people from 41 to 50 years and 6 people above 50 years.

The listeners were supposed to chose the context, from which the given snippet was extracted. The three response options were created in a way that did not allow any direct indication of the ironic meaning of the utterance: the given contexts did not contain remarks such as "she said sarcastically". The context only implicitly indicated the meaning of the target fragment. This was done to ensure that the listeners were not aware of the aim of the experiment. The link to the survey was distributed online using the SoSci Survey platform (https://www.soscisurvey.de). Each participant could pass the test only once. This excluded reconsidering the decisions by the participants, as well as repeating answers by the same person.

4 Results

The statistical analysis showed that listeners can successfully detect irony in speech relying only on acoustic cues with no additional lexical or grammatical information.

52% of the ironic contexts (13 sentences out of 25) were reliably identified by more than 80% of the listeners. Another 40% of such contexts were adequately recognized as ironic by 60 to 80% of the listeners, which makes it 92% of the ironic utterances (23 out of 25) in total that were satisfactorily detected by the listeners.

The most reliably identified ironic fragments were then compared to homonymous non-ironic snippets that were also correctly recognized. The difference in intensity, duration of the stressed vowel and F0 range was analyzed.

The Wave Assistant processing software was used to identify the intonation pattern of an utterance as well as to measure its F0 range. The stressed vowel duration and intensity of the fragment were measured using the Praat software package. The obtained results led to the following generalized conclusions.

Nearly all the ironic contexts were characterized by a higher intensity than their neutral variants (87.5% of the stimuli). Table 1 shows the mean intensity of the reliably identified ironic utterances in comparison with reliably identified non-ironic contexts.

Table 1. Mean intensity level of fragments from reliably identified ironic and non-ironic utterances (Db).

Text of the fragment	Speaker	Ironic	Non-ironic
Он объяснит (*He will explain*)	S2	70	67
Раньше нельзя было это сделать (*It was impossible to do it before*)	S1	75	69
Друг (*<He's> a friend*)	S2	80	62
Будет она нам помогать (*She will help us*)	S2	75	70
Способный (*<He's> talented*)	S1	76	68
Он понимает (*He understands*)	S2	64	72
Советчик (*<He's> an adviser*)	S2	69	67
Сейчас (*In a moment*/sarcastically: *You wish*)	S2	74	65

The results of the study indicate that ironic fragments are often marked by a significant lengthening of the stressed vowels compared to the corresponding non-ironic declarative, exclamatory and interrogative stimuli (see Table 2). All the reliably identified ironic fragments had a longer stressed vowel.

The acoustic analysis showed that the most frequent intonation contours in ironic utterances are IP1 (the falling tone), IP2 (the falling tone with a certain prosodic emphasis) and IP4 (the falling-rising tone) in terms of the system of E. Bryzgunova.

Interestingly, special questions, unlike general questions, were recognized by the majority of listeners as ironic.

Table 2. Duration of the stressed vowels from reliably identified ironic and non-ironic utterances (ms).

Text of the fragment	Speaker	Ironic	Non-ironic
Он объяснит (*He will explain*)	S2	520	75
Раньше нельзя было это сделать (*It was impossible to do it before*)	S1	90	75
Друг (*<He's> a friend*)	S2	160	70
Будет она нам помогать (*She will help us*)	S2	105	55
Способный (*<He's> talented*)	S1	120	75
Он понимает (*He understands*)	S2	455	104
Советчик (*<He's> an adviser*)	S2	113	90
Сейчас (*In a moment*/sarcastically: *You wish*)	S2	165	115

Ironic stimuli were regularly distinguished from the non-ironic ones by a different melodic contour. Some of the ironic utterances differed from non-ironic ones by a wider F0 range or a shift of the intonation pattern. However, more research is required to establish a statistically significant difference in F0 range and other melodic characteristics between two types of utterances.

The following examples of the fragments from the reliably identified ironic and non-ironic stimuli (Figs. 1, 2, 3 and 4) illustrate the above-mentioned observations.

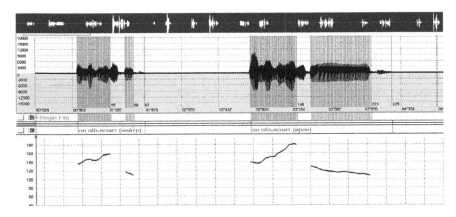

Fig. 1. Fragment "он объяснит" (he will explain) from non-ironic (at the left) and ironic (at the right) utterances; ironic fragment has a higher level of intensity, a wider F0 range and a longer stressed vowel.

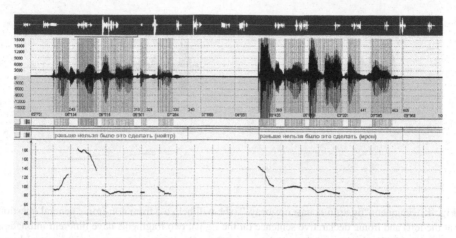

Fig. 2. Fragment "раньше нельзя было это сделать" (it was impossible to do it before) from non-ironic (at the left) and ironic (at the right) utterances; ironic fragment has a higher level of intensity and a wider F0 range; the phrasal accent is shifted from "нельзя" (impossible) in the non-ironic utterance to "раньше" (before) in the ironic one.

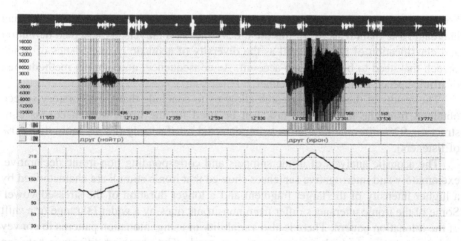

Fig. 3. Fragment "друг" (<he's> a friend) from non-ironic (at the left) and ironic (at the right) utterances; ironic fragment has a higher level of intensity, lengthened stressed vowel; the shift of the F0 level from middle to high is also apparent.

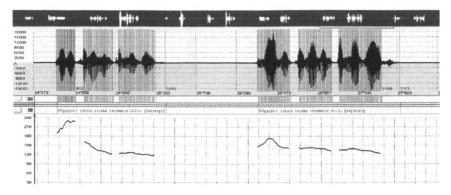

Fig. 4. Fragment "будет она нам помогать" (she will help us) from non-ironic general question (at the left) and ironic declarative sentence (at the right); two fragments have different intonation patterns – IP3 for the non-ironic question and IP2 for the ironic sentence; ironic sentence has a higher intensity level.

5 Conclusion and General Discussion

So far, there is no comprehensive description of ironic intonation in the Russian language. The main reasons for this are the absence of a universally accepted definition and classification of irony as well as the diversity of acoustic characteristics of ironic speech that need to be taken into account. As consequence, no common methodology of ironic speech collection exits either.

The results of the current research allow us to conclude that the laboratory speech obtained from the text-reading task can be used for the ironic speech corpus construction, 92% of ironic utterances being detected satisfactorily (i.e. by more than 60% of listeners).

The acoustic analysis of ironic snippets and corresponding non-ironic declarative, exclamatory and interrogative sentences showed that ironic stimuli are characterized by a higher intensity of the target fragment and a longer duration of the stressed vowel. Some of the ironic utterances differ from non-ironic ones by a wider F0 range or a shift of the intonation pattern. There is also variability in using intonation patterns to convey an ironic meaning, which requires further research. The observed difference between acoustic characteristics of ironic and non-ironic utterances extracted from the laboratory speech shows that the current method of analysis may be applied to a bigger corpus of linguistic data.

Although the suggested method may be developed and improved in future, its validation by the described perception experiment proves its reliability for the analysis of ironic speech. The next step will include conducting more recordings as well as listening experiments resulting in the extension of the existing corpus that could be used to address both research and application-oriented objectives.

Acknowledgements. This research was supported by the RFBR grant № 20-012-00552.

References

1. Arkhipetskaia, M.V.: Intonatsionnye frazeologizmy so znacheniem emotsionalnogo otritsaniia. Ph.D. thesis, Saint Petersburg (2012)
2. Attardo, S.: Irony as relevant inappropriateness. J. Pragmatics **32**(6), 793–826 (2000)
3. Baturskaia, L.A.: Intonatsionnoe otritsanie v dialogicheskoi rechi. Ph.D. thesis, Kiev (1975)
4. Braun, A., Schmiedel, A.: The phonetics of ambiguity: a study on verbal irony. In: Cultures and Traditions of Wordplay and Wordplay Research, pp. 111–136. De Gruyter, Berlin (2018)
5. Bryant, G., Fox Tree, J.: Is there an ironic tone of voice? Lang. Speech **48**, 257–277 (2005)
6. Grice, H.P.: Studies in the Way of Words. Harvard University Press, Cambridge (1989)
7. Haverkate, H.: A speech act analysis of irony. J. Pragmatics **14**, 77–109 (1990)
8. Kodzasov, S.V.: Issledovaniia v oblasti russkoi prosodii. Iazyki slavianskikh kultur, Moscow (2009)
9. Kreuz, J.R., Roberts, R.: Two cues for verbal irony: hyperbole and the ironic tone of voice. Metaphor Symb. Act. **10**, 21–31 (1995)
10. Niebuhr, O.: Rich Reduction: sound-segment residuals and the encoding of communicative functions along the hypo-hyper scale. In: 7th Tutorial and Research Workshop on Experimental Linguistics, pp. 11–24, St. Petersburg, Russia (2016)
11. Sharrer, L., Christman, U.: Voice modulations in German ironic speech. Lang. Speech **54**(4), 435–465 (2011)
12. Simpson, P.: 'That's not ironic, that's just stupid!': towards an eclectic account of the discourse of irony. In: The Pragmatics of Humour Across Discourse Domains, pp. 33–50. John Benjamins Publishing, Amsterdam (2011)
13. Sperber, D.: Verbal irony: pretense or echoic mention? J. Exp. Psychol. Gen. **113**, 130–136 (1984)
14. Shutova, T.A.: Semantika otritsaniia i sposoby ee implitsitnogo vyrazheniia v russkom iazyke. Ph.D. thesis, Saint Petersburg (1996)
15. Wilson, D., Sperber, D.: On verbal irony. Lingua **87**(1-2), 53–76 (1992)
16. Zhang, X.: English Rhetoric. Tsinghua University Press, Beijing (2005)

Automated Compilation of a Corpus-Based Dictionary and Computing Concreteness Ratings of Russian

Valery Solovyev[1] and Vladimir Ivanov[2(✉)]

[1] Kazan Federal University, 2, Tatarstan Street, Room 467, Kazan
, The Republic of Tatarstan 420021, Russian Federation
maki.solovyev@mail.ru
[2] Innopolis University, st. Universitetskaya, 1, Innopolis
, Republic of Tatarstan 420500, Russian Federation
nomemm@gmail.com

Abstract. The article presents new method implemented by the authors to generate dictionaries of concrete/abstract words for Russian. The method based on pretrained word embeddings computes concreteness ranking defined as a function of similarity between word vectors and the distance between a word in question and the 'seed' of concrete/abstract words. Implementation of the method resulted in generating the Russian Dictionary of Concreteness/Abstractness with concreteness ratings. Context information encoded by word embeddings useful to rank words by their concreteness score. The resulting C/A rankings strongly correlate with human experts' assessments.

Keywords: Word embedding · Concrete words · Abstract words · Experiment · Corpora

1 Introduction

In the modern paradigm, the concrete/abstract discrimination is based on the assumption that concrete words denote referents which can be experienced through sense, i.e. available to the senses, whereas referents nominated with abstract words lack the attribute and refer to ideas or concepts [6]. The notion of concreteness/abstractness is nowadays a focus of numerous studies [5] and for a few decades the problem of discriminating concrete and abstract words has been viewed as relevant by researchers in a number of areas: linguistics, psychology, pedagogy, medicine, neurophysiology, philosophy etc. At present, ratings of concrete/abstract words (hereinafter C/A ratings) are used in studies of statistical models of word distribution [3], Text Leveling Systems for ranging texts in difficulty thus profiling them for different categories of readers as well as in literacy education where ratings are implemented to help students with learning disabilities [19]. The spectrum of modern research in the area varies from mental performance and processing [11] to psychological disorders [4] and global aphasia [9].

© Springer Nature Switzerland AG 2020
A. Karpov and R. Potapova (Eds.): SPECOM 2020, LNAI 12335, pp. 554–561, 2020.
https://doi.org/10.1007/978-3-030-60276-5_53

2   Related Work

The modern scientific paradigm accumulated a number of methods and techniques on grading perception of concrete and abstract words. The data obtained through the study of lexical judgments on concrete and abstract words by Kroll [15] suggests that abstract words take more time to judge or decide on by a person and, thus, they are processed by human brain significantly longer as compared to concrete words. The works by Noppeney [18] and Kiehl [13] introduced a notion of 'the concreteness effect' which implies that concrete words are processed more efficiently and faster than abstract ones. Moffat et al. [16] determined that within conducted verbal semantic categorization, abstract words are processed faster through emotional experience, whereas context contributed to both, concrete and abstract words. A serious of associative experiments demonstrated that processing abstract words people experience emotions [14]. More affective associations with abstract words obtained from informants lead scholars to introduction of the so-called 'imageability variable'. It has also been proved that words with a higher degree (rating) of abstractness bear a higher dependence on associations [9,20]. Speakers tend to develop more associations with words bearing a higher degree of abstractness than concrete words [9].

The analysis conducted by Snefjella et al. [21] on the Corpus of Historical American English contributed to the understanding of increasing tendency to use concrete words more frequently than abstract words in English [21]. Based on the historical data analysis, B. Snefjella also argues that the degree of concreteness in words tends to grow over years. The metrics of concrete/abstract words as one of the means of artificial intelligence assessment are used to understand human language and develop training systems [8].

In education, C/A ratings are used to assess text readability/complexity [8,10]. E.g., in works of D. McNamara lists of abstract words and online ratings of abstract words are used as resources for the automated tools, such as Coh-Metrix, TAACO, SiNLP, developed to profile texts and teach reading and effective comprehension [8].

Modern online dictionaries of concrete/abstract words comprise lists words with C/A ratings in Chinese, German, Spanish, Italian, English [24]. One of the first English Dictionaries of 4,000 concrete/abstract words compiled in 1981 [7] still serves as a reference list in various research. The latest edition of Concrete/Abstract Words Dictionary comprises C/A ratings of 37,058 English words and 2,896 two-word expressions, obtained from over 4,000 participants by means of a forming study using Internet crowd sourcing for data collection [6].

Automated compilation of dictionaries is not a new notion. For instance, the authors of [21] suggested a method of designing a COHA (Corpus of Historical American English, https://www.english-corpora.org/coha) based dictionary. They started with eliciting a core list of obviously concrete and abstract words from COHA. For each word in the core list, with the help of word embeddings[1] method, they constructed a vector characterizing the word's joint occurrences

---

[1] Word embeddings are numeric vectors that represent words in a vector space.

with other words in the corpus. Further, they measured the distance between each word and vectors of other words in the core list. This allows assessing the degree of closeness of a given word to a concrete or an abstract extreme. The dictionary constructed in this way is compared with the manually created dictionary [6]. Spearman's correlation coefficient was estimated as 0.70.

A different method was proposed in [23]. It is based on the idea that abstract words occur together with abstract ones, and concrete words are used with concrete ones [22]. In [23], this method was implemented for the Russian language and the Dictionary compiled by the researchers registers 88.000 word forms with C/A ratings. This also allows, starting from a certain core list of words, to automatically generate a large corpus-based dictionary. The comparison with a manually constructed dictionary yielded the Spearman correlation coefficient of 0.71, i.e. the manually constructed dictionary almost failed to improve the quality of the automatically constructed dictionary.

3 Materials and Methods

The current study was performed using the results of a project aimed at computing C/A ratings of Russian words (visit the site of the project at [2]). The Research Question of the present study are as follows: *are concreteness/abstractness ratings computed using pretrained word embeddings and the ratings of words derived from human rankings surveys similar or different?* The materials and results were uploaded on a website thus providing availability of all the data used in the research (available from [1]).

3.1 Computing Concreteness Rankings of Abstract and Concrete Words

In this section we describe the method of pretrained word embeddings. To compute concreteness rankings of words, we use the following hypothesis. Given a concrete word z (adjective or noun), it will occur together with greater number of different concrete words, than with different abstract words. For abstract words we can formulate hypothesis symmetrically: Given an abstract word z, it will occur together with greater number of different abstract words, than with different concrete words. It is clear that testing both forms of the hypothesis imply existence of a large corpus where one can find occurrences of words. The hypothesis we stated does not contradict with common sense, but it is not trivial. The actual words occurrences are less informative. It is more productive to count different words, rather than a number of occurrences in a corpus. The strength of the hypothesis was measured using the real corpus. We base our results on the work presented in [12], where Solovyev et al. propose a new method to build C/A rankings (for English) and to extract abstract and concrete words from a large dataset of bigrams (Google Books NGram dataset). The resulting dictionary strongly correlates with human rankings dictionary [17]. Thus, their method could be applied to Russian part of the Google Books NGram. Using the method described above, we produced a list of 88.000 Russian tokens tagged with concreteness ratings. The whole list can be found online (using the link from the previous section).

It contains different word forms which have different concreteness scores, because in Google Books NGram different forms of the same word are treated as different tokens. In the context of the current study, it seems counter intuitive to have different ranks for forms of the same word. Therefore, we use a subset of the ranking that includes only initial forms of nouns; it contains 13500 words. Further we refer to this list as "NGRAMS list". In this study we present an alternative method for building C/A rankings. This method is based on pretrained word embeddings. Well-known and frequently used word embeddings for the Russian language, such as word2vec, GloVe, fastText, are available online. The main idea of the method is to measure similarity between word vectors. First, we define two sets: 10 abstract and 10 concrete nouns (further, we will call the sets seeds[2]). Then for any given word w in the vector space, the method compares cosine similarity between the word w and words from both seeds. Finally, the closer the word w is to the concrete seed (and farther from the abstract seed), the higher the concreteness rank it has. The corresponding concreteness score of word w can be calculated with the following formula:

$$concreteness_{score}(w) = \frac{sim(w, concrete\_seed)}{sim(w, abstract\_seed)} \qquad (1)$$

The score of the word w depends on the selection of the seeds. We call the corresponding ranked list derived from seeds an extension of seeds. The method was proposed in [5] and can be applied to the Russian language word embeddings. The key aspect of the method is selection of the abstract and concrete seeds. In [17] the authors proposed the following seeds: concrete (railroad, tree, bed, bird, neck, ball, baby, water, sand, cotton, knife, horse, clay, moon, student) and abstract (belief, existence, principle, responsibility, extent, justice, theory, purpose, courage, wisdom, chance, imagination, fate, glory, curiosity).

3.2 Expert Evaluation

For the purposes of the current study we designed an online survey and asked the respondents to assess concreteness/abstractness of 1000 most frequent Russian nouns on a 5-point scale. The scale was designed in such a way that the left side was used to mark concrete nouns and the right was used to mark abstract nouns. The 1st position (on the left) corresponds to 'the highest degree of concreteness', 2nd corresponds to 'a high degree of concreteness', the 3rd position corresponds to 'bearing equal degrees of concreteness and abstractness', 4th is 'a high degree of abstractness and the 5th position corresponds to 'the highest degree of abstractness' (see Fig. 1). Each respondent was asked to choose where, i.e. on the scale between two poles, concrete or abstract, his/her position about

---

[2] Size of the seeds is a parameter of the method. Further in the end of the Sect. 4 we show examples with different seed size.

the words under study lies. The research was held at Kazan Federal University and at Belarusian State Pedagogical University. The experts (N = 600) participating in the study are full-time University students, native speakers of Russian. All the respondents signed their consent to the experimental procedure. Their age ranges from 17 to 25 years old.

Fig. 1. Template of concreteness rating.

We held a total of 40 surveys in which each respondent rated 50 words. The C/A ratings of each Russian word were computed as an average of all the assessments received in the range from 1 to 5 (see Fig. 2). As a result of the study we computed the list of 1000 words ranked by concreteness. The survey methodology applied in our work follows the one from [21]. As a result of the study we computed the KFU-BSPU-1000 list (available from [1]).

4   Results

In our experiments, we use the pretrained word embeddings provided by fastText collection [17]. The fastText embeddings captures subword information which is significant for the Russian language with its rich morphology and, especially, for building C/A ranks, because many Russian suffixes are related to abstractness of nouns. To evaluate the performance of the method for the Russian language, we compare relative position of words in a computed ranked list to the word's position in the manually obtained C/A ranking. Spearman's rank provides a relevant measure for such comparison. To assess the variability of Spearman's rank as a function of seed words, we run a series of experiments with different seed size and randomly selected seed words for a fixed seed size. It is obvious that selection of seed size as well as the words in seeds affect final ranking. In our experiments, candidate seeds were randomly sampled from the NGRAMS-based ranking. Candidates for concrete words are sampled from top 1000 nouns; while candidates for abstract words are sampled from the bottom 1000 nouns of the NGRAMS list. So, the evaluation consists of two steps: (1) to sample seeds from the NGRAMS list; (2) using the seeds compute the C/A scores and calculate Spearman's rank. In the experiments p-values were less than 10^{-3}. Another type of seeds was sampled from the human assessments: candidates for concrete words are sampled from 100 most concrete nouns; while candidates for abstract words are sampled from 100 least concrete nouns of the 1000 words from the KFU-BSPU-1000 list. The resulting Spearman's correlation coefficient for different seed types and sizes presented in Fig. 2. As shown in Fig. 2, human assessments of concreteness in general provide better seeds. However, the correlation coefficient does not exceed 0.8 even for seeds with 100 words. Increasing the seed size over 20–30 words gives very small gain in the resulting correlation measure.

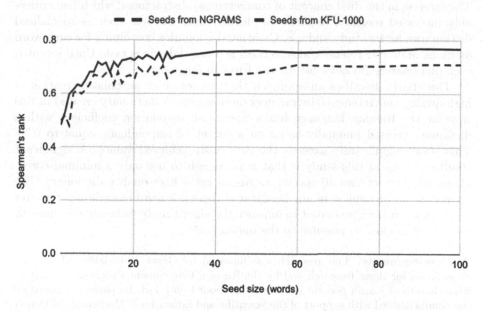

Fig. 2. Spearman's correlation calculated for different seed sizes (x-axis) and two sources of seeds: NGRAMS list (dashed); and 1000 words list (solid).

We provide an example of two seeds with Spearman's rank of 0.78:

Abstract seed = ['trenie' (friction) 'bozhestvo' (divinity) 'samodey-atel'nost'' (amateur work) 'uvelichenie' (increase) 'donor' (donor) 'vred' (harm) 'revolyutsionnost'' (revolutionism) 'velichie' (grandeur) 'chest'' (honour) 'mestopolozhenie' (location) 'vospitanie' (upbringing) 'selektsiya' (selection) 'shalost'' (trick) 'regulirovanie' (regulation) 'nanesenie' (application) 'nesta-bil'nost'' (instability) 'poetika' (poetics) 'sochinenie' (composition) 'rapsodiya' (rhapsody) 'svoistvo' (property) 'hrabrost'' (courage) 'podryv' (disruption) 'poema' (poem) 'ideologiya' (ideology) 'yazykoznanie' (linguistics) 'reaktsiya' (reaction) 'realizatsiya' (realization) 'diaspora' (diaspore) 'uklonenie' (deviation) 'integratsiya' (integration)];

Concrete seed = ['baton' (a loaf of white bread) 'telets' (taurean) 'ogurchik' (cucumber) 'stupka' (pounder) 'magnezit' (magnesite) 'smesitel'' (mixer tap) 'kishlak' (kishlak, type of a village) 'plitka' (tile) 'bortik' (border) 'shkafchik' (cabinet) 'kurok' (trigger) 'prihozhaya' (lobby) 'lesenka' (stairs) 'kord' (cord) 'kapor' (bonnet) 'kreml" (Kremlin) 'tanker' (tanker) 'kashitsa' (gruel) 'naberezh-naya' (quay) 'obrazok' (icon) 'nakonechnik' (tip) 'shipovnik' (hedge rose) 'kino-fabrika' (film studio) 'provoloka' (wire) 'nalichnye' (cash) 'stilet' (stiletto) 'kol-pachok' (cap) 'papaha' 'ublyudok' (bastard) 'gradonachal'nik' (city governor)].

5   Conclusion

The interest in the dual concept of concreteness/abstractness, which has notice-ably increased recently, requires the development of tools such as specialized dictionaries for its study and use. Creating dictionaries containing for each word an index of its concreteness/abstractness is a very laborious task. Until recently, such dictionaries did not exist for the Russian language.

The article describes an approach that allows quick automatic creation of high-quality concreteness/abstractness dictionaries. A dictionary created in this way for the Russian language has a Spearman correlation coefficient with a dictionary created manually based on a survey of respondents, equal to 0.78. This result significantly exceeds the previously achieved values. A significant result obtained in this study is that it is enough to use only a minimal core – about 30 abstract and 30 specific words to get a high-quality dictionary. The approach can be applied in any language if there is a word embeddings resource for it. In the future, we expect to improve the algorithm by bringing the machine dictionary as close as possible to the manual one.

Acknowledgments. This research was financed by Grant 19-07-00807 of Russian Foundation for Basic Research and by the Russian Government Program of Compet-itive Growth of Kazan Federal University. Sections 1 and 3 of the paper are based on the results derived with support of the Scientific and Educational Mathematical Center of KFU.

References

1. Technology for creating semantic electronic dictionaries. https://kpfu.ru/tehnologiya-sozdaniya-semanticheskih-elektronnyh.html. Accessed 29 July 2020 (in Russian)
2. Tehnologii sozdaniya elektronnykh slovarei [technologies of compiling electronic dictionaries]. https://kpfu.ru/tehnologiya-sozdaniya-semanticheskih-elektronnyh.html. Accessed 29 July 2020 (in Russian)
3. Balyan, R., McCarthy, K., McNamara, D.: Comparing machine learning classi-fication approaches for predicting expository text difficulty. In: The Thirty-First International Florida Artificial Intelligence Research Society Conference (FLAIRS-31), pp. 421–426 (2018)
4. Binney, R., Zuckerman, B., Reilly, J.: A neuropsychological perspective on abstract word representation: from theory to treatment of acquired language disorders. Curr. Neurol. Neurosci. Rep. **16**, 1–26 (2018)
5. Borghi, A., Binkofski, F., Castelfranchi, C., Cimatti, F., Scorolli, C., Tummolini, L.: The challenge of abstract concepts. Psychol. Bull. **143**(3), 263–292 (2017)

6. Brysbaert, M., Warriner, A., Kuperman, V.: Concreteness ratings for 40 thousand generally known English word lemmas. Behav. Res. Methods **46**(3), 904–911 (2014)
7. Coltheart, M.: The MRC psycholinguistic database. Q. J. Exp. Psychol. **33**(4), 497–505 (1981)
8. Crossley, S., Skalicky, S., Dascalu, M., McNamara, D., Kyle, K.: Predicting text comprehension, processing, and familiarity in adult readers: new approaches to readability formulas. Discourse Process. **54**(5–6), 340–359 (2017)
9. Crutch, S., Warrington, E.: The differential dependence of abstract and concrete words upon associative and similarity-based information: complementary semantic interference and facilitation effects. Cogn. Neuropsychol. **27**(1), 46–71 (2010)
10. Dashtestani, R.: EFL teachers' and students' perspectives on the use of electronic dictionaries for learning English. CALL-EJ **14**(2), 51–65 (2013)
11. Hines, D.: Recognition of verbs, abstract nouns and concrete nouns from the left and right visual half-fields. Neuropsychologia **14**(2), 211–216 (1976)
12. Ivanov, V., Solovyev, V.: Ranking concrete and abstract words using Google Books Ngram data. J. Intell. Fuzzy Syst., 1–9 (2020)
13. Kiehl, K., Liddle, P., Smith, A., Mendrek, A., Forster, B., Hare, R.: Neural pathways involved in the processing of concrete and abstract words. Hum. Brain Mapp. **7**(4), 225–255 (1999)
14. Kousta, S.T., Vigliocco, G., Vinson, D., Andrews, M., del Campo, E.: The representation of abstract words: why emotion matters. J. Exp. Psychol. Gen. **140**(1), 14–34 (2011)
15. Kroll, J., Merves, J.: Lexical access for concrete and abstract words. J. Exp. Psychol. Learn. Mem. Cogn. **12**(1), 92–107 (1986)
16. Moffat, M., Siakaluk, P.D., Sidhu, D.M., Pexman, P.M.: Situated conceptualization and semantic processing: effects of emotional experience and context availability in semantic categorization and naming tasks. Psychon. Bull. Rev. **22**(2), 408–419 (2014). https://doi.org/10.3758/s13423-014-0696-0
17. Naumann, D., Frassinelli, D., im Walde, S.S.: Quantitative semantic variation in the contexts of concrete and abstract words. In: Proceedings of the 7th Joint Conference on Lexical and Computational Semantics, pp. 76–85 (2018)
18. Noppeney, U., Price, C.: Retrieval of abstract semantics. Neuroimage **22**(1), 164–170 (2004)
19. Paivio, A., Clark, J.: Dual coding theory and education. Pathways to literacy achievement for high poverty children, pp. 1–20 (2006)
20. Ponari, M., Norbury, C., Vigliocco, G.: Acquisition of abstract concepts is influenced by emotional valence. Dev. Sci. **21**(2), e12549 (2018)
21. Snefjella, B., Généreux, M., Kuperman, V.: Historical evolution of concrete and abstract language revisited. Behav. Res. Methods **51**(4), 1693–1705 (2018). https://doi.org/10.3758/s13428-018-1071-2
22. Solovyev, V., Andreeva, M., Solnyshkina, M., Zamaletdinov, R., Danilov, A., Gaynutdinova, D.: Computing concreteness ratings of Russian and English most frequent words: contrastive approach. In: 12th International Conference on Developments in eSystems Engineering (DeSE), pp. 403–408 (2019)
23. Solovyev, V., Ivanov, V., Akhtyamov, R.: Dictionary of abstract and concrete words of the Russian language: a methodology for creation and application. J. Res. Appl. Linguist. **10**, 215–227 (2019)
24. Wang, J., Conder, J., Blitzer, D., Shinkareva, S.: Neural representation of abstract and concrete concepts: a meta-analysis of neuroimaging studies. Hum. Brain Mapp. **31**(10), 1459–1468 (2010)

Increasing the Accuracy of the ASR System by Prolonging Voiceless Phonemes in the Speech of Patients Using the Electrolarynx

Petr Stanislav, Josef V. Psutka[✉], and Josef Psutka

Department of Cybernetics, University of West Bohemia,
Technická 8, 306 14 Pilsen, Czech Republic
{pstanisl,psutka_j,psutka}@kky.zcu.cz
https://www.kky.zcu.cz/en

Abstract. Patients who have undergone total laryngectomy and use electrolarynx for voice production suffer from poor intelligibility. It may lead in many cases to fear of speaking to strangers, even over the phone. Automatic Speech Recognition (ASR) systems could help patients overcome this problem in many ways. Unfortunately, even state-of-the-art ASR systems cannot provide results comparable to those of conventional speakers. The problem is mainly caused by the similarity between voiced and unvoiced phoneme pairs. In many cases, a language model can help to solve the issue, but only if the word context is sufficiently long. Therefore adjustment of acoustic data and/or acoustic model is necessary to increase recognition accuracy. In this paper, we propose voiceless phonemes elongation to improve recognition accuracy and enrich the ASR system with a model that takes this elongation into account. The idea of elongation is verified on a set of ASR experiments with artificially elongated voiceless phonemes. To enriching the ASR system, the DNN model for rescoring lattices based on phoneme duration is proposed. The new system is compared with a standard ASR. It is also verified that the ASR system created using elongated synthetic data can successfully recognize the actual elongated data pronounced by the real speaker.

Keywords: Automatic speech recognition · Total laryngectomy · Phoneme duration · Electrolarynx

1 Introduction

Speech is one of the primary means of communication used by humans. Loss of it causes by total laryngectomy (TL) has several complications and has a significant impact on the quality of life. After such surgery, people are unable to produce a voiced sound. Their speech is called alaryngeal. The most used approaches for voice restoration include surgical-prosthetic and phoniatric methods.

© Springer Nature Switzerland AG 2020
A. Karpov and R. Potapova (Eds.): SPECOM 2020, LNAI 12335, pp. 562–571, 2020.
https://doi.org/10.1007/978-3-030-60276-5_54

The electrolarynx (EL) is commonly used nowadays. It uses an external device - the electrolarynx to produce the excitation necessary for proper voicing [3].

An EL is a battery-powered device that mechanically generates sound source signals, which are conducted into the oral cavity from either the soft parts of the neck or from the lower jaw [5]. In most cases, a monotone pitch is generated. Thus the produced speech sounds very mechanical. The significant disadvantage of using EL is that all pronounced phonemes tend to be voiced. Moreover, to make the resulting speech sufficiently audible, the EL needs to create rather strong vibrations that can be perceived as noise by other people. That's why the recognition of the EL speech is usually a tough task even for people.

Nevertheless, we can achieve respectable results, but only in case when we use a specific acoustic model for EL speech together with a robust language model (LM) during recognition. Unfortunately, the LM can fully unfold its strength only when recognizing longer sentence segments. But the problem is that speakers after TL almost do not communicate, because they are often ashamed. And when they talk, they only do it with short phrases or instructions. In these situations, LM cannot use the word context. Another reason for shorter sentences is that they are less exhaustive for EL speakers. Thus the benefit of LM to such short phrases seems negligible (it is difficult to determine the identity of words without the necessary context). However, such a task has excellent practical potential.

In this paper, we artificially elongate voiceless phonemes and apply phone duration probabilities via lattice restoring to improve ASR results. Such an approach can be used for training the EL speaker to properly prolong part of the words and helps obtain data for the creation of a real specific EL duration model. The phone duration model is a neural network to predict the phoneme duration. A cross-entropy loss function is used for training the model. This work is inspired by [2].

In the following section, we describe our approach of prolongation voiceless phonemes[1] in the Czech language. In Sect. 3, the experimental setup and results on artificial data are shown. In Sect. 4, the model's performance on the real data is presented. In Sect. 5 the idea of the simulator is briefly described and we conclude in Sect. 6.

2   Artificial Data Elongation

In our previous work [6], we presented poor results of ASR in recognition of isolated words compared to humans on EL speech. The main problem is that due to the loss of vocal cords, part of the information from the produced speech is lost. Recently, several approaches have been presented to recover lost data. A summary of the most promising is in [1]. Unfortunately, none of the presented methods got into production. Therefore, if it is not entirely realistic to retrieve lost information by completely changing the paradigm of how speech recognition

---

[1] It is essential to mention that all speech data are in the Czech language.

systems work, then all that remains is to work with the available information and adapt state-of-the-art methods. A particular possibility is to replace the lost data with a specifically targeted change in the speech produced, which is then taken into account by the acoustic model (AM). Of course, such an approach does not deprive the speaker of an electrolarynx, but it can help him in stressful situations and ultimately make his life more comfortable. The simplest solution is to prolongate specific phonemes.

2.1   Database Description

For our research, we recorded speech corpus, which contains 5589 sentences and 320 isolated words from one speaker. It consists of about 14 h of annotated speech. The speaker is an older woman who underwent TL more than ten years ago and uses electrolarynx (EL) daily. All the sentences and words are in the Czech language. The database of text prompts from, which the sentences were selected, was obtained in an electronic form from the web pages of Czech newspaper publishers [4].

The corpus consists of two parts. The major part contains 40 phonetically rich and 5036 phonetically balanced sentences. The other section contains 419 sentences and 320 isolated words. Separation of the corpus into two parts was necessary due to the long period between the recordings. During that period, the speaker changed the type of the electrolarynx. It plays an essential role in the quality of the speech, and also the recordings technology was upgraded.

Part of the corpus with isolated words contains 160 pairs of specially selected words that have different meanings, but they acoustically differ only in voicing. An example is a pair of Czech words "kosa[2]" - "koza[3]". At least one sentence containing each word was recorded.

All the speech utterances were recorded at a sampling frequency of 16 kHz with 16-bit resolution.

2.2   Phoneme Elongation

From the results presented in [5] and [6], it follows that it is the most difficult for ASR to recognize words that contain voiceless phonemes for which there is a voiced pair. In the Czech language, these are mainly the phonemes $/f/$, $/k/$, $/p/$, $/s/$ and $/t/$. Elongation of these phonemes can lead to better distinguish between the phoneme pairs and therefore improving the recognition accuracy. The recorded corpus does not contain words or sentences with elongated phonemes; for that reason, we used the TD-PSOLA method for the artificial elongating of phonemes.

For automatic phoneme elongation, it is necessary to have an alignment of the phoneme as precisely as possible. We trained by the Kaldi framework, a TDNN

---

[2] scythe.
[3] goat.

acoustic model (on 5040 sentences from the corpus), to obtain the required alignment. The AM is tested on 419 sentences and 320 isolated words. As already mentioned, if the ASR system has sufficient context, it can provide good performance. Still, it is not comparable to the state-of-the-art (speaker-independent) ASR system, because an acoustic model causes most errors. For that reason, we primarily evaluate the performance of the acoustic model with a phoneme zero-gram language model with $P(w_n) = \frac{1}{40}$. The TDNN model achieved phoneme recognition accuracy $Acc_p = 85.41$ %, and it is our baseline. Based on the alignment, voiceless phonemes were elongated with the TD-PSOLA algorithm. Elongation is considered $1.25x$, $1.50x$, $1.75x$, $2.00x$, $2.25x$, $2.50x$, $2.75x$, and $3.00x$, where $2.00x$ means that a voiceless phoneme is two times longer.

3 Data Modeling on Artificial Data

3.1 Model with Elongated Data

For each elongation, a new TDNN model was trained with the same sentences as the baseline model. The same phoneme zerogram language model was used for testing. The recognition accuracies for each elongation model are depicted in the Table 1. The best performance was achieved for $2.50x$ elongation, specifically, $Acc_p = 87.90\%$, which is an increase of 2.49% in absolute terms and 17.07% relative in the sense of error. Besides, models working with elongations from $1.75x$ to $2.75x$ achieve very close values of recognition accuracy as the best model.

Table 1. Influence of the degree of phoneme elongation on the accuracy of the TDNN model

Elongation									
	1.00x	1.25x	1.50x	1.75x	2.00x	2.25x	2.50x	2.75x	3.00x
Acc_p [%]	85.41	86.42	87.05	87.58	87.71	87.69	**87.90**	87.39	87.11

An essential question is the robustness of the model. To answer this question, we tested different elongation of test data to the best AM (2.5x elongation on training data). It can be said that the model is robust over a relatively wide range of elongations (from $2.00x$ to $3.00x$). Experiments with artificially elongated data proved the potential of the considered hypothesis. Elongation of one of the paired phonemes most probably sufficiently differentiate very similar sound representations and leads to different phoneme models in HMM.

3.2 Duration Model with Elongated Data

The second step in the improvement of the performance of the ASR system for EL speech uses the principle of lattice rescoring based on taking into account phone duration (number of frames) noted as d. The approach is in more detail described in [2]. The main idea is based on model $p(d)$ for each value of d, and to include $\log p(d)$ as one of the components of the score in the final lattice. To prediction $p(d)$ a neural network is used.

The function of the duration model is, therefore, to predict the sequence of durations based on the sequence of phonemes. This implies using both the left context (L) and the right context (R) of the current phoneme. However, only the duration of the phonemes of L context or R context can come into the input vector of the network, because of the definition of a circle. The model uses a softmax layer, therefore it is necessary to define the maximum length of the phonemes D. For all with durations $d \geq D$, $p(d) = p(D)$. The Fig. 1 shows an example of input and output of the network.

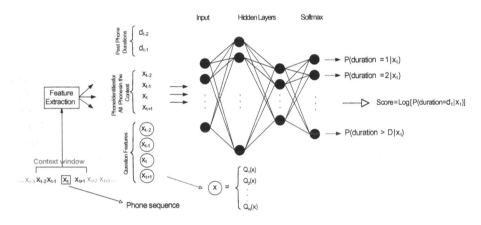

Fig. 1. The overview of the neural network duration model, along with inputs and outputs. A context size of $L = 2$ and $R = 1$ is assumed. The picture originated from [2].

The duration model is a standard type of feedforward network. The number of hidden layers depends on the domain, but usually 2 or 3 are used, see [2]. The size of these hidden layers is chosen as a multiple of the dimension of the input vector I. In this case, the value of $3I$ was determined, where $I = (L + R + 1) * (N_p + N_q) + L$, N_p is number of phonemes and N_q questions in the phonetic-context decision tree. The activation function is of the RELU type. The size of the output layer corresponds to the maximum number of microsegments D. In [2], the best results were obtained with $D = 50$. The Kaldi framework was used to create a duration model.

Verification of the functionality of the duration model was performed on $2.50x$ artificially elongated data. The context window has the value $(L, R) = (3, 3)$, $N_p = 42$ a $N_q = 6$. The size of the input vector $I = 339$. The model has 2 hidden layers with the 1017 neurons. The softmax output layer has dimension $D = 50$. The model is trained and tested using the same training and test set as the TDNN model. The language model is a phoneme zerogram. This model achieved $Acc_p = 88.79\%$, which represents an improvement of 0.89% in absolute and 7.35% relative in sense of error to the TDNN 2.50x model ($Acc_p = 87.90\%$).

The hyperparameters of the model include the size of L and R context, the number of network layers and the maximum duration D. In particular, the hyperparameter D theoretically provides the most significant opportunity to improve the model results, because $D = 50$ was chosen based on experiments in [2], however, the standard speech was used. Table 2 shows the effect of maximum duration on model accuracy. A special value is $D = 189$, which is determined automatically based on the alignment before the training. This model achieved the highest accuracy $Acc_p = 88.84\%$, which is a small improvement over the original model with $D = 50$.

Table 2. Influence of maximum duration D on model accuracy.

D	50	100	150	189	200
Acc_p [%]	88.79	88.80	88.74	**88.84**	88.82

Another hyperparameter is the number of neural network layers. In Table 3 the results of individual models are listed. Variant *1H* represents a model with 1 hidden layer, the *2H* model with 2 hidden layers, and *3H* with 3 hidden layers. A special case is models containing a bottleneck layer (*2H (bottleneck)* and *3H (bottleneck)*). Instead of the last hidden layer of size 3I, they contain a layer with only 10 neurons. This layer should help in generalization [2]. From the achieved results, it is clear that the size of the network is not a completely crucial parameter. The difference between the accuracy of a networks with 2 and 3 layers is minimal. The benefit of the bottleneck layer, compared to the results presented in [2], is also rather minimal. However, in general, this layer has a positive effect on accuracy.

Table 3. Influence of the number of hidden layers on the accuracy of the model ($D = 189$).

Model	1H	2H	3H	2H (bottleneck)	3H (bottleneck)
Acc_p [%]	88.79	88.81	88.84	**88.87**	88.85

The last hyperparameter that can affect the accuracy is the size of the L and R context. From Table 4 and Table 5 is evident that the best results were obtained by models that have the length $L + R = 6$. The very best result was

Table 4. Comparison of the effect of the size of the symmetric context ($D = 189$).

Context (L, R)					
	$(0,0)$	$(1,1)$	$(2,2)$	$(3,3)$	$(4,4)$
Acc_p [%]	87.52	88.36	88.75	**88.87**	88.84

Table 5. Influence of left and right context in case the total length $L+R = 6$ ($D = 189$).

Context (L, R)					
	$(5,1)$	$(4,2)$	$(3,3)$	$(2,4)$	$(1,5)$
Acc_p [%]	88.54	88.85	**88.87**	88.84	88.85

achieved by the model having a symmetrical context, but compared to models with an asymmetric context, the difference is rather negligible.

The best result was achieved by a model having $D = 189$, 2 hidden layers, where the last hidden layer has only 10 neurons and $L + R = 6$. However, the results show that the duration model is not significantly sensitive to parameter changes.

4   Phone Duration Model with Real Data

The obtained results with artificially elongated data confirmed the hypothesis that a model working with this data could achieve better results. The next step was to collect real data. The recording of the corpus is a relatively lengthy process, so it was necessary to choose the right way to best prolong the speech. If the goal is to get the whole word elongated, the speaker can be instructed to speak more slowly. But that is not the goal. The resulting speech should have only specific phonemes elongated, approximately twice as much.

The simplest option, and in a way elegant, turned out to be notation with double letters, which are supposed to represent an elongated phoneme, such as "kossa". The speaker was familiar with the fact that if a word contains this double notation, he should try to elongate the word accordingly. Besides, this notation subconsciously "forces" the speaker to pronounce the word differently than in the case of standard notation.

By this approach, a total of 731 sentences and 267 words (standard and elongated) were recorded. Figure 2 shows the comparison of amplitude and spectrogram of the word "kosa" elongated by the speaker and artificially ($2.00x$). The analysis of the recorded utterances showed that the process of artificial elongation produces very similar recordings to the real ones.

To verify the hypothesis, we used the duration model ($D = 189$, $2H$ (*bottleneck*), symetric context $(L, R) = (3, 3)$) and trained on artificially elongated data but tested with the real elongated data. Text for elongated utterances by the speaker was selected from 419 sentences and 320 words in the test set. This allows us to replace the original utterances in the test set and compare the results with previous experiments. Based on the analysis of real elongated data, we decided to use the model trained on 2.00x elongated data because this artificial data are the closest match to the real data. This model achieved $Acc_p = 86.23\%$ with the zerogram phoneme LM[4], which represents a deterioration of 2.31% in absolute. This may be seen as a poor result. But, if we consider the fact that we used only the artificial data for training the model and that there is at least one year between the recording of original data for elongation and real elongated data, it is quite a positive result for further work.

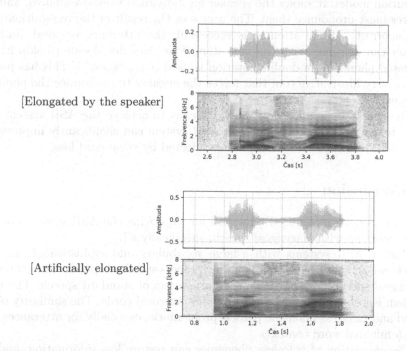

[Elongated by the speaker]

[Artificially elongated]

Fig. 2. Amplitude and spectrogram of the word "kosa" elongated by the speaker (a) and artificially (b)

5   Simulator

The results show that the presented approach significantly improves the accuracy of the ASR system, especially for utterances with a minimal context. The main problem with this approach is the elongation of the voiceless phonemes by the speaker himself. The presented method does not take into account the elongation of the word, but only the necessary part, the phoneme. However, with

---

[4] The duration model trained and tested only on artificial data achieved $Acc_p = 88.54\%$.

a little help, the speaker can elongate the desired portion of the word. Since we verified that the model trained on artificially elongated data can recognize well even prolonged data by the speaker, it is possible to create a simulator. Its primary function is to help the speaker learn to automatically elongate incriminated phonemes and incrementally increase the accuracy of recognition. At the same time, it is possible to adapt the acoustic model with gathered real data gradually. Over time, all errors in the data caused by artificial elongation should be eliminated.

The simulator itself is a graphical program containing an ASR system with a duration model. It shows the speaker an individual words/sentences, and the speaker must pronounce them. The user sees the result of the recognition (correct/incorrect). If the attempt is successful, the utterance is saved, and the speaker can continue with the next item if he does not decide to skip it. For elongated phonemes, a double notation is used (e.g., "kossa"). This has proven to be a very illustrative tool that forces the speaker to pronounce the phoneme "differently".

The simulator is, of course, only a means to achieve the ASR system that works best with EL speech. Such an ASR system can significantly improve the communication capabilities of speakers affected by vocal cord loss.

6 Conclusion

In this paper, we introduced a method for improving the ASR system's performance working with alaryngeal speech, specifically EL.

Modern ASR systems with a large vocabulary and sophisticated language models can achieve decent results if they process standard speech. Unfortunately, alaryngeal speech does not have the parameters of standard speech. The main problem is lost information caused by missing vocal cords. The similarity of the voiced and voiceless phoneme pairs is problematic, especially for utterances that have a minimal word context.

The elongation of voiceless phonemes can restore lost information, and the cooperation of the duration model can significantly improve the ASR system's performance in some situations. Problematic acquisition of really elongated data can be solved by artificial elongation and the creation of a simulator, which will facilitate the adaptation of the speaker.

The most fundamental problem so far is the duration model itself, which works offline. Lattice rescoring is done after the utterance ends. In the case of a simulator, it is not an issue, but it is in real situations. Our further work will focus on solving this issue and producing a system capable of working in real-life situations. Also, we will provide the first version of the simulator to speakers to gather more real data.

Acknowledgments. This research was supported by the Technology Agency of the Czech Republic, project No. TN01000024 and by the grant of the University of West Bohemia, project No. SGS-2019-027.

References

1. Denby, B., Schultz, T., Honda, K., Hueber, T., Gilbert, J.M., Brumberg, J.: Silent speech interfaces. Speech Commun. **52**(4), 270–287 (2010). https://doi.org/10.1016/j.specom.2009.08.002, http://linkinghub.elsevier.com/retrieve/pii/S0167639309001307
2. Hadian, H., Povey, D., Sameti, H., Khudanpur, S.: Phone duration modeling for LVCSR using neural networks. In: INTERSPEECH, pp. 518–522 (2017)
3. Liu, H., Ng, M.L.: Electrolarynx in voice rehabilitation. Auris Nasus Larynx **34**(3), 327–332 (2007). https://doi.org/10.1016/j.anl.2006.11.010
4. Radová, V., Psutka, J.: UWB-S01 corpus: a czech read-speech corpus. In: Proceedings of the 6th International Conference on Spoken Language Processing, pp. 732–735. ICSLP2000 (2000)
5. Stanislav, P., Psutka, J.V.: Influence of different phoneme mappings on the recognition accuracy of electrolaryngeal speech. In: Proceedings of the International Conference on Signal Processing and Multimedia Applications and Wireless Information Networks and Systems - Volume 1: SIGMAP, (ICETE 2012), pp. 204–207 (2012). https://doi.org/10.5220/0004129502040207
6. Stanislav, P., Psutka, J.V., Psutka, J.: Recognition of the electrolaryngeal speech: comparison between human and machine. In: Ekštein, K., Matoušek, V. (eds.) TSD 2017. LNCS (LNAI), vol. 10415, pp. 509–517. Springer, Cham (2017). https://doi.org/10.1007/978-3-319-64206-2_57

Leverage Unlabeled Data for Abstractive Speech Summarization with Self-supervised Learning and Back-Summarization

Paul Tardy[1,2(✉)], Louis de Seynes[1], François Hernandez[1], Vincent Nguyen[1], David Janiszek[2,3], and Yannick Estève[4]

[1] Ubiqus Labs, Paris, France
{ldeseynes,fhernandez,vnguyen}@ubiqus.com
[2] LIUM – Le Mans Université, Le Mans, France
pltrdy@gmail.com
[3] Université de Paris, Paris, France
david.janiszek@parisdescartes.fr
[4] LIA – Avignon Université, Avignon, France
yannick.esteve@univ-avignon.fr

Abstract. Supervised approaches for Neural Abstractive Summarization require large annotated corpora that are costly to build. We present a French meeting summarization task where reports are predicted based on the automatic transcription of the meeting audio recordings. In order to build a corpus for this task, it is necessary to obtain the (automatic or manual) transcription of each meeting, and then to segment and align it with the corresponding manual report to produce training examples suitable for training. On the other hand, we have access to a very large amount of unaligned data, in particular reports without corresponding transcription. Reports are professionally written and well formatted making pre-processing straightforward. In this context, we study how to take advantage of this massive amount of unaligned data using two approaches (i) self-supervised pre-training using a target-side denoising encoder-decoder model; (ii) back-summarization i.e. reversing the summarization process by learning to predict the transcription given the report, in order to align single reports with generated transcription, and use this synthetic dataset for further training. We report large improvements compared to the previous baseline (trained on aligned data only) for both approaches on two evaluation sets. Moreover, combining the two gives even better results, outperforming the baseline by a large margin of +6 ROUGE-1 and ROUGE-L and +5 ROUGE-2 on two evaluation sets.

Keywords: Abstractive Summarization · Semi-supervised learning · Self-supervised learning · Back-summarization · French

© Springer Nature Switzerland AG 2020
A. Karpov and R. Potapova (Eds.): SPECOM 2020, LNAI 12335, pp. 572–580, 2020.
https://doi.org/10.1007/978-3-030-60276-5_55

1   Introduction

Automatic Meeting Summarization is the task of writing the report corresponding to a meeting. We focus on so-called *exhaustive reports*, which capture, in a written form, all the information of a meeting, keeping chronological order and speakers' interventions. Such reports are typically written by professionals based on their notes and the recording of the meeting.

Learning such a task with a supervised approach requires building a corpus, more specifically (i) gathering *(transcription, report)* pairs; (ii) cutting them down into smaller segments; (iii) aligning the segments; and, finally, (iv) training a model to predict the report from a given transcription.

We built such a corpus using Automatic Speech Recognition (ASR) to generate transcription, then aligning segments manually and automatically [23]. This process is costly to scale, therefore, a large majority of our data remains unaligned. In particular, we have access to datasets with very different orders of magnitude; we have vastly more reports (10^6) than aligned segments (10^4). Reports are supposedly error-free and well formatted, making it easy to process on larger scale.

In this work, we focus on how to take advantage of those unaligned data. We explore two approaches, first a self-supervised pre-training step, which learns internal representation based on reports only, following BART [7]; then we introduce meeting *back-summarization* which is to predict the transcription given the report, in order to generate a synthetic corpus based on unaligned reports; finally, we combine both approaches.

2   Related Work

Abstractive Summarization has seen great improvements over the last few years following the rise of neural sequence-to-sequence models [1,22]. Initially, the field mostly revolved around headline generation [17], then multi-sentence summarization on news datasets (e.g. CNN/DailyMail corpus [11]). Further improvements include pointer generator [11,19] which learns whether to generate words or to copy them from the source; attention over time [11,12,19]; and hybrid learning objectives [12]. Also, *Transformer* architecture [24] has been used for summarization [4,25]. In particular, our baseline architecture, hyper-parameters and optimization scheme are similar to [25]. All these works are supervised, thus, useful for our baseline models on aligned data.

In order to improve performances without using aligned data, one approach is self-supervision. Recent years have seen a sudden increase of interest about self-supervised large Language Models (LM) [2,16] since it showed impressive transfer capabilities. This work aims at learning general representations such that they can be fine-tuned into more specific tasks like Question Answering, Translation and Summarization. Similarly, BART [7] proposes a pre-training based on input denoising i.e. reconstructing the correct input based on a noisy version. BART is no more than a standard *Transformer* model with noise functions applied to its input, making it straightforward to implement, train and fine-tune.

Another way of using unaligned data is to synthetically align it. While this is not a common practice for summarization (at the best of our knowledge) this has been studied for Neural Machine Translation as Back-Translation [3,13,20]. We propose to apply a similar approach to summarization to generate synthetic transcriptions for unlabeled reports.

3   Meeting Summarization

3.1   Corpus Creation

Using our internal Automatic Speech Recognition (ASR) pipeline, we generate transcriptions from meeting audio recordings.

Our ASR model has been trained using the Kaldi toolkit [15] and a data set of around 700 h of meeting audio. Training data was made of *(utterance text, audio sample)* pairs that were aligned with a Kaldi functionality to split long audio files into short text utterances with their corresponding audio parts (Sect. 2.1 of [5]). We end up with a train set of approximately 476 k pairs. We used a chain model with a TDNN-F [14] architecture, 13 layers of size 1024, an overall context of $(-28, 28)$ and around 1.7 million parameters. We have trained our model for 20 epochs with batch-normalization and L2-regularization. On the top of that, we added an RNNLM [18] step to rescore lattices from the acoustic model. This model consisted of three TDNN layers, two interspersed LTSMP layers resulting in a total of around 10 million parameters.

Our raw data, consisting in pairs of audio recordings (up to several hours long) and exhaustive reports (dozens of pages) requires segmentation without loss of alignment, i.e. producing smaller segments of transcription and report while ensuring that they match. This annotation phase was conducted by human annotators using automatic alignment as described in [23]. We refer to these aligned corpuses as, respectively, *manual* and *automatic*.

While this process allowed us to rapidly increase the size of the dataset, it is still hard to reach higher orders of magnitude because of time and the resource requirements of finding and mapping audio recordings with reports, running the Automatic Speech Recognition pipeline (ASR) and finally the automatic alignment process.

3.2   Baseline and Setup

Since automatic alignment has been showed to be beneficial [23], we train baselines on both manually and automatically aligned data. Models are standard *Transformer* [24] of 6 layers for both encoder and decoder, 8 attention heads and 512 dimensions for word-embeddings. Hyper-parameters and optimization are similar to the baseline model of [25] (just switching-off copy-mechanism and weight sharing as discussed in 6.1). During pre-processing, text is tokenized using *Stanford Tokenizer* [9], *Byte Pair Encoding* (BPE) [21]. Training and inference uses *OpenNMT-py*[1] [6].

---

[1] https://github.com/OpenNMT/OpenNMT-py.

In this paper, we study how simple models – all using the same architecture – can take advantage of the large amount of single reports (i.e. not aligned with transcription).

4 Self-supervised Learning

Self-supervised learning aims to learn only using unlabeled data. It has recently been applied – with great success – to Natural Language Processing (NLP) as a pre-training step to learn word embeddings using *word2vec* [10], encoder representations with *BERT* [2] or an auto-regressive decoder with *GPT-2* [16].

More recently, BART [7] proposed to pre-train a full *Transformer* based encoder-decoder model with a denoising objective. They experimented with various noise functions and showed very good transfer performances across text generation tasks like question answering or several summarization datasets. When applied to summarization, the best results were obtained with *text-infilling* (replacements of continuous spans of texts with special mask tokens) and *sentence permutation*.

The model is pre-trained on *reports* only, thus not requiring alignment. It allows us to use our unlabeled reports at this stage. We use the same setup as BART, i.e. two noise functions: (i) *text-infilling* with $p = 0.3$ and $\lambda = 3$; and (ii) *sentence permutation* with $p = 1$.

Since the whole encoder-decoder model is pre-trained, the fine-tuning process is straightforward: we just reset the optimizer state and train the model on the *transcription* to *report* summarization task using aligned data (*manual+auto*).

5 Back-Summarization

Instead of considering single documents as unlabeled data, we propose to reverse the summarization process in order to generate synthetic transcriptions for each report segment. We call this process *back-summarization* in reference to the now well known *back-translation* data augmentation technique for Neural Machine Translation [20].

Back-summarization follows three steps:

1. **backward training**: using manually and automatically aligned datasets (*man + auto*), we train a back-summarization model that predicts the source (=*transcription*) given the target (=*report*).
2. **transcription synthesis**: synthetic sources (=*transcriptions*) are generated for each single target with the backward model, making it a new aligned dataset, denoted *back*.
3. **forward training**: using all three datasets (*man + auto + back*), we train a summarization model.

Backward and forward models are similar to the baseline in terms of architecture and hyper-parameters. Synthetic transcriptions are generated by running inference on the best backward model with respect to the validation set in terms of ROUGE score while avoiding models that copy too much from the source (more details in 6.1). Inference uses beam search size 5 and trigram repetition blocking [12].

During forward training, datasets have different weights. These weights set how the training process iterates over each dataset. For example, if weights are $(1, 10, 100)$ respectively for $(man, auto, back)$, it means that for each *man* example, the model will also train over 10 *auto* and 100 *back* examples. This is motivated by [3] that suggests (in Sect. 5.5) *upsampling* – i.e. using more frequently – aligned data in comparison to synthetic data. To be comparable to other models, we set *man* and *auto* weights close to their actual size ratio i.e. $(2, 7)$. For synthetic data we experiment with 100, giving an approximate *upsample* rate of 6.

6   Results

Experiments are conducted against two reference sets, the `public_meetings` test set and the valid set. Summarization models are evaluated with ROUGE [8]. We also measure what proportion of the predictions is copied – denoted by copy% – from the source based on the ROUGE measure between the source and the prediction. Datasets for both training and evaluation are presented in Table 1.

6.1   Reducing Extractivity Bias

While training and validating models for back-summarization we faced the extractivity bias problem. It is a common observation that even so-called abstractive neural summarization models tend to be too extractive by nature, i.e. they copy too much from the source. This is true for first encoder-decoder models [19] but also for more recent – and more abstractive – models like BART [7].

Back-summarization amplifies this bias since it relies on two summarization steps: backward, and forward, both facing the same problem. At first, we trained typical *Transformer* models for summarization that involve copy-mechanism [19] and weight sharing between (i) encoder and decoder embeddings, (ii) decoder embedding and generator, similar to the baseline of [25]. When evaluating against the validation set – reference copy%: 55.38 –, the predictions had too much copy: 62.64%, (+7.26).

We found that turning off both copy mechanism and weight sharing reduces the copy% to 54.65, (−0.73). Following these interesting results, we also apply this to other models.

6.2   Datasets

Datasets used in this paper are presented in Table 1. We refer to the human annotated dataset as *manual* and the automatically aligned data as *automatic*. *Backsum* data consists of single reports aligned with synthetic transcription using *back-summarization* (see Sect. 5). The validation set – *valid*

– is made of manually aligned pairs excluded from training sets. Finally, we use public_meetings [23] as a test set. We observe differences between public_meetings in term of lengths of source/target and extractivity. We hypothesize that constraints specific to this corpus – i.e. the use of meetings with publicly shareable data only – introduced bias. For this reason, we conduct evaluation on both public_meetings and the validation set.

Table 1. About Ubiqus datasets: number of examples, lengths of source and target (words in average, first decile d_1 and last d_9), and the extractivity measured with ROUGE between target and source. (*) Extractivity of the back-summarization dataset is measured on a subset of 10,000 randomly sampled examples.

Dataset	#Pairs	src	tgt	Copy %
		$(avg, [d_1, d_9])$	$(avg, [d_1, d_9])$	R1
Manual	21 k	172, [42, 381]	129, [25, 297]	55.45
Automatic	68 k	188, [45, 421]	130, [25, 302]	54.72
Valid	1 k	178, [40, 409]	144, [22, 294]	55.38
Back	6.3 m	232, [53, 509]	90, [43, 153]	63.09*
public_meetings	1060	261, [52, 599]	198, [33, 470]	75.84

6.3 Summarization

Summarization results are presented in Table 2 on the test set and Table 3 on the valid set.

Self-supervised pre-training outperforms the baseline by a large margin: +2.7 ROUGE-1 and ROUGE-2 on the test set and up to +4 on the validation set, for both ROUGE-1 and ROUGE-2.

Back-summarization models – without pre-training – also improve the baseline, by a larger margin: +2.98 R1, +2.39 R2 on the test set, and up to +5.34 R1 and +4.80 R2 on the validation set.

Interestingly, mixing both approaches consistently outperforms other models, with large improvement on both the test set (+5.71 R1 and +4.64 R2) and the validation set (+6.40 R1 and +5.16 R2).

In addition to the ROUGE score, we also pay attention to how extractive the models are. As discussed in Sect. 6.1, we notice that the models are biased towards extractive summarization since they generate predictions with higher copy% on both reference sets (respectively 75.85 copy% and 55.38 copy% for test and valid). Even though we paid attention to this extractivity bias during the *back-summarization* evaluation, we still find *backsum* models to increase copy% more than self-supervised models.

Table 2. Scores on the `public_meetings` test set.

Model	Training steps	ROUGE score (F) $(R1, R2, RL)$	Copy % $R1$
Baseline	8 k	52.31/34.00/49.70	79.36
SelfSup	4 k (pre-trained 50 k)	55.08/36.76/52.43	83.72
Backsum	6 k	55.29/36.39/52.89	89.24
Both	6 k (pre-trained 50 k)	**58.02/38.64/55.56**	90.77

Table 3. Scores on the validation set.

Model	Training steps	ROUGE score (F) $(R1, R2, RL)$	Copy % $R1$
Baseline	6 k	33.83/15.86/31.05	74.25
SelfSup	4 k (pre-trained 50 k)	37.94/19.16/34.86	78.61
Backsum	10 k	39.17/20.06/36.19	86.96
Both	6 k (pre-trained 50 k)	**40.23/21.02/37.26**	88.03

7 Discussion

Copy% on Different Datasets. Running evaluations on two datasets makes it possible to compare how models behave when confronted with different kinds of examples. In particular, there is a 20% difference in terms of copy rate between evaluation sets (`public_meetings` and valid). On the other hand, for any given model, there is a difference of less than 5 copy%.

This suggests that models have a hard time adapting in order to generate more abstractive predictions when they should. In other words, models are bad at predicting the expected copy%.

ROUGE and Copy%. Our results suggest that better models, in terms of the ROUGE score also have a higher copy%. This poses the question of which metric we really want to optimize. On the one hand, we want to maximize ROUGE, on the other we want our prediction to be abstractive i.e. stick to the reference copy rate. These two objectives currently seem to be in opposition, making it hard to choose the most relevant model. For example, on the test set (Table 2) Backsum performs similarly to SelfSup with respect to ROUGE but has a much higher copy%, which would probably be penalized by human evaluators.

8 Conclusion

We presented a French meeting summarization task consisting in predicting exhaustive reports from meeting transcriptions. In order to avoid constraints from data annotation, we explore two approaches that take advantage of the large amount of available unaligned reports: (i) self-supervised pre-training on reports only; (ii) back-summarization to generate synthetic transcription for unaligned reports. Both approaches exhibit encouraging results, clearly outperforming the previous baseline trained on aligned data. Interestingly, combining the two techniques leads to an even greater improvement compared the baseline: +6/+5/+6 ROUGE-1/2/L.

References

1. Cho, K., et al.: Learning phrase representations using RNN encoder-decoder for statistical machine translation. In: Proceedings of the Conference EMNLP 2014 - 2014 Conference on Empirical Methods in Natural Language Processing, pp. 1724–1734 (2014)
2. Devlin, J., Chang, M.W., Lee, K., Toutanova, K.: BERT: pre-training of deep bidirectional transformers for language understanding. In: Proceedings of the 2019 Conference of the North American Chapter of the Association for Computational Linguistics: Human Language Technologies, vol. 1 (Long and Short Papers), pp. 4171–4186. Association for Computational Linguistics, Minneapolis, October 2019
3. Edunov, S., Ott, M., Auli, M., Grangier, D.: Understanding back-translation at scale. In: Proceedings of the 2018 Conference on Empirical Methods in Natural Language Processing, EMNLP 2018, pp. 489–500. Association for Computational Linguistics, August 2018
4. Gehrmann, S., Deng, Y., Rush, A.M.: Bottom-up abstractive summarization. In: Proceedings of EMNLP (2018)
5. Hernandez, F., Nguyen, V., Ghannay, S., Tomashenko, N., Estève, Y.: TED-LIUM 3: twice as much data and corpus repartition for experiments on speaker adaptation. In: Karpov, A., Jokisch, O., Potapova, R. (eds.) SPECOM 2018. LNCS (LNAI), vol. 11096, pp. 198–208. Springer, Cham (2018). https://doi.org/10.1007/978-3-319-99579-3_21
6. Klein, G., Kim, Y., Deng, Y., Senellart, J., Rush, A.M.: OpenNMT: open-source toolkit for neural machine translation. In: Proceedings of System Demonstrations ACL 2017 - 55th Annual Meeting of the Association for Computational Linguistics, pp. 67–72 (2017)
7. Lewis, M., et al.: BART: denoising sequence-to-sequence pre-training for natural language generation, translation, and comprehension. Technical report, October 2019
8. Lin, C.Y.: Rouge: a package for automatic evaluation of summaries. In: Proceedings of the Workshop on Text Summarization Branches Out (WAS 2004), pp. 25–26 (2004)
9. Manning, C.D., Surdeanu, M., Bauer, J., Finkel, J., Bethard, S.J., McClosky, D.: The Stanford CoreNLP natural language processing toolkit. In: Association for Computational Linguistics (ACL) System Demonstrations, pp. 55–60 (2014)
10. Mikolov, T., Chen, K., Corrado, G., Dean, J.: Efficient estimation of word representations in vector space, vol. 1, pp. 1–12, January 2013
11. Nallapati, R., et al.: Abstractive text summarization using sequence-to-sequence RNNs and beyond. In: Proceedings of CoNLL (2016)

12. Paulus, R., Xiong, C., Socher, R.: A deep reinforced model for abstractive summarization. In: 6th International Conference on Learning Representations, ICLR 2018 - Conference Track Proceedings, May 2017
13. Poncelas, A., Shterionov, D., Way, A., De Buy Wenniger, G.M., Passban, P.: Investigating backtranslation in neural machine translation. In: EAMT 2018 - Proceedings of the 21st Annual Conference of the European Association for Machine Translation, pp. 249–258 (2018)
14. Povey, D., et al.: Semi-orthogonal low-rank matrix factorization for deep neural networks. In: Proceedings of the Annual Conference of the International Speech Communication Association, INTERSPEECH, vol. 2018, pp. 3743–3747 September (2018)
15. Povey, D., et al.: The Kaldi speech recognition toolkit. In: IEEE Signal Processing Society, pp. 1–4 (2011)
16. Radford, A., Wu, J., Child, R., Luan, D., Amodei, D., Sutskever, I.: Language models are unsupervised multitask learners. Technical report, OpenAI (2019)
17. Rush, A.M., Chopra, S., Weston, J.: A neural attention model for abstractive sentence summarization. In: Proceedings of the Conference on Empirical Methods in Natural Language Processing (EMNLP), pp. 379–389, September 2015
18. Sak, H., Senior, A., Beaufays, F.: Long short-term memory based recurrent neural network architectures for large vocabulary speech recognition, February 2014
19. See, A., Liu, P.J., Manning, C.D.: Get to the point: summarization with pointer-generator networks. In: ACL 2017 - 55th Annual Meeting of the Association for Computational Linguistics, Proceedings of the Conference (Long Papers), pp. 1073–1083. Association for Computational Linguistics (2017). ISBN 9781945626753
20. Sennrich, R., Haddow, B., Birch, A.: Improving neural machine translation models with monolingual data. In: 54th Annual Meeting of the Association for Computational Linguistics, ACL 2016 - Long Papers, vol. 1, pp. 86–96 (2016)
21. Sennrich, R., Haddow, B., Birch, A.: Neural machine translation of rare words with subword units. In: Proceedings of the 54th Annual Meeting of the Association for Computational Linguistics (vol. 1: Long Papers), pp. 1715–1725. Association for Computational Linguistics, Berlin (2016)
22. Sutskever, I., Vinyals, O., Le, Q.V.: Sequence to sequence learning with neural networks. In: Advances in Neural Information Processing Systems (NIPS), pp. 3104–3112, September 2014
23. Tardy, P., Janiszek, D., Estève, Y., Nguyen, V.: Align then summarize: automatic alignment methods for summarization corpus creation. In: Proceedings of the 12th Conference on Language Resources and Evaluation (LREC 2020), pp. 6718–6724 (2020)
24. Vaswani, A., et al.: Attention is all you need. In: Guyon, I., et al. (eds.) Advances in Neural Information Processing Systems, vol. 30, pp. 5998–6008. Curran Associates, Inc., June 2017
25. Ziegler, Z.M., Melas-Kyriazi, L., Gehrmann, S., Rush, A.M.: Encoder-agnostic adaptation for conditional language generation. arXiv preprint arXiv:1908.06938 (2019)

Uncertainty of Phone Voicing
and Its Impact on Speech Synthesis

Daniel Tihelka[1(✉)] [iD], Zdeněk Hanzlíček[1] [iD], and Markéta Jůzová[2] [iD]

[1] New Technologies for the Information Society, University of West Bohemia,
Pilsen, Czech Republic
{dtihelka,zhanzlic}@ntis.zcu.cz

[2] Department of Cybernetics, Faculty of Applied Sciences,
University of West Bohemia, Pilsen, Czech Republic
juzova@kky.zcu.cz

Abstract. While unit selection speech synthesis is not at the centre of research nowadays, it shows its strengths in deployments where fast fixes and tuning possibilities are required. The key part of this method is target and concatenation costs, usually consisting of features manually designed. When there is a flaw in a feature design, the selection may behave in an unexpected way, not necessarily causing a bad quality speech output. One of such features in our systems was the requirement on the match between expected and real units voicing. Due to the flexibility of the method, we were able to narrow the behaviour of the selection algorithm without worsening the quality of synthesised speech.

Keywords: Speech synthesis · Unit selection · Target cost · Phone voicing

1 Introduction

In the past few years, the use of deep neural networks for speech synthesis become widely attractive [2,6,8,13,23,25]. Although the DNN can achieve very natural-sounding speech output, it still requires rather powerful hardware to run on. Also, since it models the speech in such a way that the model is somehow "spread" through the network weights, it is virtually impossible to make an ad-hoc "fix" when something goes wrong.

On the other hand, the unit selection approach suffers from occasional unnatural artefacts [10], causing speech perception annoyances on the otherwise very natural-sounding speech. Since it uses "raw" speech data, it closely mimics the voice style of the original speaker and thus it is not flexible in changing speaking style and/or other characteristics. On the other hand, when an artefact is perceived in the synthesised speech, the identification of its cause is rather straightforward [21] and it can be fixed much more easily. This is one of the factors why the deployment of unit selection is still considered in commercial applications, where fast fixes are desirable.

© Springer Nature Switzerland AG 2020
A. Karpov and R. Potapova (Eds.): SPECOM 2020, LNAI 12335, pp. 581–591, 2020.
https://doi.org/10.1007/978-3-030-60276-5_56

In the present paper we are going to show a case study of such an artefact fix in order to illustrate the flexibility of this synthesis method. And although we illustrate the problem on our particular feature handling, a linguistic/ phonological attributes are almost always used in many other systems, despite the fact that the actual features are rarely revealed in the research papers (possibly due to language dependency).

2   Costs in Unit Selection

The key part of the unit selection algorithm is the computation of target and join costs [1,4,5,22]. It is also expected that when synthesising a phrase being recorded in the corpus, the sequence of units from this phrase will be selected. It is simply due to the fact that the concatenation cost $CC(c_{i-1}, c_i) = 0$ for unit candidates c_{i-1} and c_i neighbouring in the speech corpus [18,20]. Similarly, the target cost $TC(t_i, c_i) = 0$ since the features in target specification t_i must be the same as features of the candidate c_i (the target feature generator used when building unit selection database is the same as that used when synthesising input not seen before).

In some of deployments of our English version of TTS system ARTIC [19] we had reports that despite the output sounding natural, it differs from the original when synthesising a phrase from speech corpus. Closer analysis revealed that there is one feature in target cost preventing the $TC(t_i, c_i) = 0$ requirement.

2.1   Voicing Mismatch Feature

The target cost features used in our TTS system ARTIC describe prosody on deep-level [14,15], so called IFF – independent feature formulation [16]. There is one special feature, called *voiced penalty*, introduced originally to prevent the selection of units with incorrect boundaries placement in the process of automatic speech segmentation [3,11]. It checks the expected and real voicing based on phonetic properties of a unit and F_0 computed from the unit signal:

$$TC_v(t_i, c_i) = \begin{cases} 0 & \Leftrightarrow V(t_i) == V(c_i) \\ 1 & \text{otherwise} \end{cases} \tag{1}$$

where

$$V(t_i) = \begin{cases} 1 & \text{for voiced phones} \\ 0 & \text{for unvoiced phones} \end{cases} \tag{2}$$

and

$$V(c_i) = \begin{cases} 1 & \Leftrightarrow F_0(c_i) > 0 \\ 0 & \Leftrightarrow F_0(c_i) \leq 0 \end{cases} \tag{3}$$

It penalises the selection of a candidate if it should be voiced but there is no F_0 detected at it, or the other way round, if it should be unvoiced and there is F_0 detected. Let us note that this example is for phones, but our system works with diphones, where the features are supposed to be stable enough [9]. Thus, there are two independent checks of TC_v, for the left phone and for the right phone respectively. The region in which the F_0 is analysed is 5 pitch periods long, centred around the diphone boundary [20].

Let us emphasise that this feature is not a hard-stopper for the selection – in such a case the affected unit candidates could be removed from the inventory a-priori. Instead, it rather penalises the selection of such candidates, but they can still be used for synthesis if there is no better candidate available (as regards the other target features and concatenation cost).

2.2 Voicing Mismatch Origins

Although the phones are strictly categorised into voiced and unvoiced in theory [9], there are a surprisingly large number of voicing mismatches in phone centres (diphone boundary) where the signal should be stable enough. In two of our corpora (Jan and Kateřina, see [19]) we examined, the 1.37% of 633, 387 and 2.45% of 557, 556 phones contain voicing mismatch as defined by Eq. 1.

Looking at the individual phones in the corpora, in Fig. 1 we show the relative number of candidates with voicing mismatch in the middle of phones. It can be seen that the majority of mismatches are for phones [P\] (in SAMPA notation [24]) for both voices, and for [Z] and [d_z] depending on the speaker. As noted before, all the statistics are related to the centres of the phones.

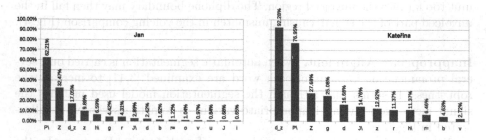

Fig. 1. Voicing mismatch occurrences (in % relative to unit count) for both examined voices as occurring in the speech corpus recordings. Only phones with mismatch >0.5% are presented.

The deeper analysis of the individual cases revealed various, but not the only, categories we have encountered:

GCI Detection Failure. Since the F_0 value is computed from glottal closure instants (pitch-marks), either from a glottal [7] or speech signals [12], the ability to reliably determine voicing parts is crucial in these algorithms. In Fig. 2 there is rather nice signal for clearly voiced [h\] phone, but an inability to detect GCI (pitch-marks) by [12] caused the middle of the phone to be considered as unvoiced.

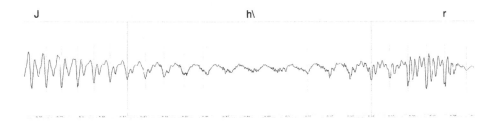

Fig. 2. Signal of [h\] from recording [... alespoJ h\ rubi: ...] with clearly visible voicing structure but without GCI detected. The black vertical lines are automatically detected phone boundaries, the red vertical lines are GCI instants assigned. (Color figure online)

Devoicing. Especially in the case of paired consonants, but not only there, a devoicing may occur under some conditions [9], causing a temporal stop in GCI detection and thus no F_0 assignment, as illustrated in Fig. 3. The devoicing and GCI detection failures were the reasons of voicing mismatch in the majority of cases for units with the highest mismatch score.

Inappropriate Segmentation. When a voiced unit neighbours with unvoiced, the process of automatic segmentation [3,11] may place the boundary of a voiced unit too far into the unvoiced region. The diphone boundary may then fall in the unvoiced part of the signal, causing mismatch in the voicing comparison (Fig. 4).

Inappropriate Alignment. When automatic segmentation is carried out, several pronunciation variants of each word are examined [3,11] to increase the robustness. It seems that although the segmentation model used matches the signal more precisely, an inappropriate variant is sometimes chosen, as illustrated in Fig. 5.

Naturaly, there is often the combination of such factors. For example, the significant amount of mismatches for [d_z] and [R] is caused by the GCI detection failure.

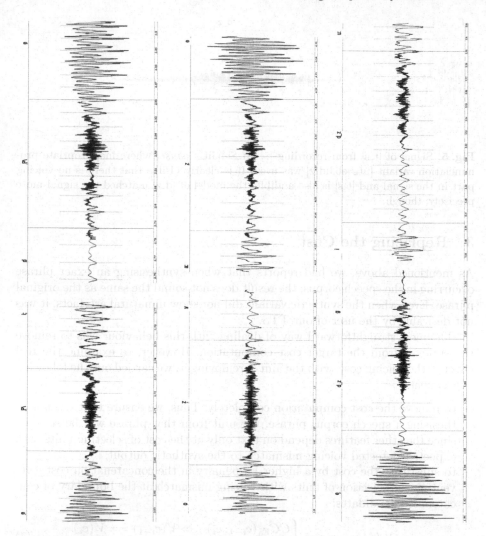

Fig. 3. Signal of phones [P\,z,d_z] with reduced voicing and thus without GCI detected at the left side. The right side shows the same phone with clear voicing structure, and thus with GCIs detected correctly.

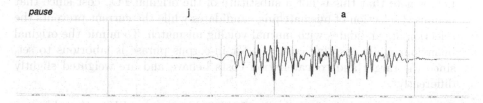

Fig. 4. Signal of [j] from recording with the left boundary placed incorrectly too far into the preceding pause. The black vertical lines are automatically detected phone boundaries, the red vertical lines are GCI instants assigned. (Color figure online)

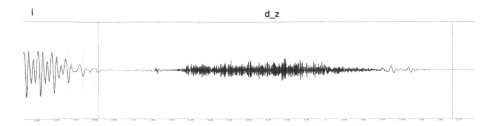

Fig. 5. Signal of [t̠s] from recording [. . . ut̠SebJit̠s *pause*] where inappropriate pronunciation variant [ut̠SebJid̠z] was used. It is clearly visible that there is no voicing part in the signal and [t̠s] is also audible; the model of [d̠z] matched the signal more precisely, though.

3   Replacing the Cost

As mentioned above, we had reports that when synthesising an exact phrase occurring in the speech corpus, the result does not sound the same as the original phrase. Even when the synthetic variant did not show unnatural artefacts, it was not desirable by the user of our TTS.

The most straightforward way of dealing with this behaviour was to remove the sub-cost from the target cost computation. However, to examine the real effect of the voicing cost, with the aim of removing it, we carried out the following experiments:

- to remove the cost computation completely. Thus, we ensure that when synthesising a speech corpus phrase, the unit from that phrase will be selected since the other features depend on text only at the cost of selecting units with expected/detected voicing mismatch to the synthetic output;
- to substitute the cost by a higher F_0 penalty in the concatenation cost, preventing the selection of units with voicing mismatch at the boundary of concatenated candidates:

$$CC'_{F_0}(c_{i-1}, c_i) = \begin{cases} CC_{F_0}(c_{i-1}, c_i) & \Leftrightarrow V(c_{i-1}) == V(c_i) \\ 1000 & \text{otherwise} \end{cases} \qquad (4)$$

where $CC_{F_0}(c_{i-1}, c_i)$ is the original F_0 cost computation as described in [20]. Let us note that this is not a substitute of the original TC_v cost since that penalised selection of mismatching candidates while the current prevents the selection of candidates with mutual voicing mismatch. To mimic the original behaviour while ensuring cost = 0 for in-corpus phrase is laborious to set, since the target and concatenation costs behave and are weighted slightly differently.

Then, we have synthesised nearly $150,000$ sentences and logged each usage of unit where the voicing mismatch occurred. In the following text, the *baseline* denotes the original implementation of target cost computation, taking the voicing mismatch into account and trying to avoid it, although it still does not have to

be avoided when there is no better candidate available (i.e. candidate with voicing mismatch $TC_v(t_i, c_i) > 0$ will be preferred to candidate with $TC_v(t_i, c_i) = 0$
if the first has better match of the other target features than the latter). The
no-VC will denote the version when the voicing match-mismatch in not examined at all (though, it is still logged), and F_0-*VC* will denote the selection with
modified concatenation cost, as defined by Eq. 4. Let us emphasise, that both
modifications ensure the selection of the original phrase in the required case.

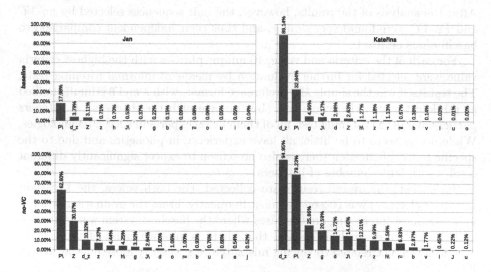

Fig. 6. Voicing mismatch occurrences (in % relative to unit count) for both examined
voices as occurred in the synthesised output of the individual baseline and *no-VC*
system (F_0-*VC* system is omitted since it looks very close to *no-VC*). Only the first
15 phones are presented.

It can be seen in Fig. 6 that the voicing mismatch is reduced in the *baseline*
system, as expected. On the contrary, the mismatches in *no-VC* and F_0-*VC* are
roughly the same. It means that the original $CC_{F_0}(c_{i-1}, c_i)$ was capable enough
in preventing concatenation boundary voicing mismatches.

4   Evaluation of Quality Impact

Both the proposed cost computation modifications ensure the selection of the
whole phrases from the corpus, when they appear at the TTS input. However,
it needs to be answered if, and how much, the modifications affect the overall
quality of the generated speech, while the expectation is that omitting voicing
mismatch evaluation will not perform worse than the baseline system.

To do so, we examined the logs collected during the synthesis of the $150,000$
phrases, following the methodology described in [17]. The difference criteria
$\delta(a, b)$ were defined as:

1. the number of different candidates in the selected sequence;
2. the number of expected/detected voicing mismatches (as defined by Eq. 1)

and both criteria were originally expected to be evaluated for combinations

1. baseline × no voicing cost (*no-VC*)
2. baseline × voicing cost moved to F_0 concatenation sub-cost (*F_0-VC*)
3. *no-VC* × *F_0-VC*

After the analysis of the results, however, the unit sequences selected by *no-VC* and *F_0-VC* were found very similar, and thus their independent comparison to baseline was omitted.

For each of the criteria and voice, 10 unique phrases with the highest criteria value were selected for further evaluation by means of informal listening tests. The test itself was the simple ABX preference format, with A and B stimuli shuffled at random through the whole test (but not through the listeners). 6 listeners participated in the listening test, all of them being experts in speech technologies. While 6 may seem to be little, all have experience in phonetics and due to the specific test configuration there is also no reason to expect significantly different results with larger number of listeners.

In Table 1, the overall results are collected. For both voices, the X variant (i.e. no preference) was chosen the most frequently. From the evaluation point of view, the most interesting are the cases where the baseline system was evaluated as better. Further analysis showed that there is another cause of the quality deterioration, not related to voicing mismatch (e.g. slightly more fluctuations in F_0).

Table 1. The results of ABX listening test. The numbers represent the count of preferences given to the corresponding system, the *total* is the sum through evaluation of sentences with the highest differences in candidate sequence and voicing mismatches.

	Candidates diff. no.			Voice mismatch no.			Candidates diff. no.		
	baseline	none	*no-VC*	baseline	none	*no-VC*	*no-VC*	none	*F_0-VC*
Jan	13	30	17	11	25	24	14	41	5
Kateřina	14	36	10	14	33	13	9	32	19
Total	all collected								
	52	124	64						

To test the statistical significance of the result, we have carried out a sign test with the null and alternative hypothesis:

H0: *The outputs of the both systems are perceived as equally good*
H1: *The output of one system sounds better*

The null hypothesis testing the same quality was chosen intentionally, as we need to check whether or not omitting the mismatch check will have a negative impact on the quality.

The sign test proved (at significance level $\alpha = 0.05$) that the version not considering the voicing mismatch is better for voice Jan (p-value $= 0.0042$) and that both version are of the same quality for voice Kateřina (p-value $= 0.1053$). Taking all the results together, the test proved that not considering voicing mismatch does not decrease the quality of the output, i.e. both systems are perceived as equally good (p-value $= 0.3502$).

5   Conclusion

We have identified the reason of the suspicious behaviour reports and narrowed it by removing the counterproductive voicing mismatch evaluation from unit selection cost computations. Using the listening tests designed to check a "worst-case" scenario, and knowing that the new system behaviour does not affect the quality of synthetic speech in any negative way, we can use *no-VC* in out TTS system now.

Despite the fact that the DNN-based speech synthesis is naturally moving to the centre of speech research, the relative ability to identify problems and tune and fix the behaviour of TTS system in relatively straightforward way remains one of the strengths of the unit selection approach.

Let us also emphasize that the observations we point out are cross-language (we have found similar issues in English and Russian voices), so the results can not only be extended to other speech synthesizers but also to other fields where phone voicing needs to be considered; at least as a caution that there may be such an uncerntainty.

Acknowledgments. This research was supported by the Technology Agency of the Czech Republic (project No. TH02010307), and by the grant of the University of West Bohemia, (project No. SGS-2019-027).

References

1. Železný, M., Krňoul, Z., Císař, P., Matoušek, J.: Design, implementation and evaluation of the Czech realistic audio-visual speech synthesis. Sig. Process. **12**, 3657–3673 (2006)
2. Hanzlíček, Z., Vít, J., Tihelka, D.: WaveNet-based speech synthesis applied to Czech: a comparison with the traditional synthesis methods. In: Sojka, P., Horák, A., Kopeček, I., Pala, K. (eds.) TSD 2018. LNCS (LNAI), vol. 11107, pp. 445–452. Springer, Cham (2018). https://doi.org/10.1007/978-3-030-00794-2_48
3. Hanzlíček, Z., Vít, J., Tihelka, D.: LSTM-based speech segmentation for TTS synthesis. In: Ekštein, K. (ed.) TSD 2019. LNCS (LNAI), vol. 11697, pp. 361–372. Springer, Cham (2019). https://doi.org/10.1007/978-3-030-27947-9_31

4. Hunt, A.J., Black, A.W.: Unit selection in a concatenative speech synthesis system using a large speech database. In: ICASSP 1996, Proceedings of International Conference on Acoustics, Speech, and Signal Processing, IEEE, Atlanta, Georgia, vol. 1, pp. 373–376 (1996)
5. Kala, J., Matoušek, J.: Very fast unit selection using Viterbi search with zero-concatenation-cost chains. In: ICASSP 2014, Proceedings of International Conference on Acoustics, Speech, and Signal Processing, IEEE, Florence, Italy, pp. 2569–2573 (2014)
6. Kalchbrenner, N., et al.: Efficient neural audio synthesis. arXiv preprint arXiv:1802.08435 (2018)
7. Legát, M., Matoušek, J., Tihelka, D.: A robust multi-phase pitch-mark detection algorithm. In: Interspeech, vol. 2007, pp. 1641–1644 (2007)
8. Lorenzo-Trueba, J., et al.: Towards achieving robust universal neural vocoding, pp. 181–185 (2019)
9. Machač, P., Skarnitzl, R.: Principles of Phonetic Segmentation. Epocha, Prague (2013)
10. Matoušek, J., Legát, M.: Is unit selection aware of audible artifacts? In: SSW 2013, Proceedings of the 8th Speech Synthesis Workshop, ISCA, Barcelona, Spain, pp. 267–271 (2013)
11. Matoušek, J., Romportl, J.: Automatic pitch-synchronous phonetic segmentation. In: INTERSPEECH 2008, Proceedings of 9th Annual Conference of International Speech Communication Association, ISCA, Brisbane, Australia, pp. 1626–1629 (2008)
12. Matoušek, J., Tihelka, D.: Using extreme gradient boosting to detect glottal closure instants in speech signal. In: IEEE International Conference on Acoustics, Speech and Signal Processing (ICASSP), Brighton, Great Britain, pp. 6515–6519 (2019)
13. van den Oord, A., et al.: WaveNet: a generative model for raw audio. arXiv preprint arXiv:1609.03499 (2016)
14. Romportl, J.: Structural data-driven prosody model for TTS synthesis. In: Proceedings of the Speech Prosody 2006 Conference, pp. 549–552. TUDpress, Dresden (2006)
15. Romportl, J., Matoušek, J.: Formal prosodic structures and their application in NLP. In: Matoušek, V., Mautner, P., Pavelka, T. (eds.) TSD 2005. LNCS (LNAI), vol. 3658, pp. 371–378. Springer, Heidelberg (2005). https://doi.org/10.1007/11551874_48
16. Taylor, P.: Text-to-Speech Synthesis, 1st edn. Cambridge University Press, New York (2009)
17. Tihelka, D., Grůber, M., Hanzlíček, Z.: Robust methodology for TTS enhancement evaluation. In: Habernal, I., Matoušek, V. (eds.) TSD 2013. LNCS (LNAI), vol. 8082, pp. 442–449. Springer, Heidelberg (2013). https://doi.org/10.1007/978-3-642-40585-3_56
18. Tihelka, D., Hanzlíček, Z., Jůzová, M., Matoušek, J.: First steps towards hybrid speech synthesis in Czech TTS system ARTIC. In: Karpov, A., Jokisch, O., Potapova, R. (eds.) SPECOM 2018. LNCS (LNAI), vol. 11096, pp. 676–686. Springer, Cham (2018). https://doi.org/10.1007/978-3-319-99579-3_69
19. Tihelka, D., Hanzlíček, Z., Jůzová, M., Vít, J., Matoušek, J., Grůber, M.: Current state of text-to-speech system ARTIC: a decade of research on the field of speech technologies. In: Sojka, P., Horák, A., Kopeček, I., Pala, K. (eds.) TSD 2018. LNCS (LNAI), vol. 11107, pp. 369–378. Springer, Cham (2018). https://doi.org/10.1007/978-3-030-00794-2_40

20. Tihelka, D., Matoušek, J., Hanzlíček, Z.: Modelling F0 dynamics in unit selection based speech synthesis. In: Sojka, P., Horák, A., Kopeček, I., Pala, K. (eds.) TSD 2014. LNCS (LNAI), vol. 8655, pp. 457–464. Springer, Cham (2014). https://doi.org/10.1007/978-3-319-10816-2_55

21. Tihelka, D., Matoušek, J., Kala, J.: Quality deterioration factors in unit selection speech synthesis. In: Matoušek, V., Mautner, P. (eds.) TSD 2007. LNCS (LNAI), vol. 4629, pp. 508–515. Springer, Heidelberg (2007). https://doi.org/10.1007/978-3-540-74628-7_66

22. Tihelka, D., Romportl, J.: Exploring automatic similarity measures for unit selection tuning. In: INTERSPEECH 2009, Proceedings of 10th Annual Conference of International Speech Communication Association, ISCA, Brighton, Great Britain, pp. 736–739 (2009)

23. Vít, J., Hanzlíček, Z., Matoušek, J.: Czech speech synthesis with generative neural vocoder. In: Ekštein, K. (ed.) TSD 2019. LNCS (LNAI), vol. 11697, pp. 307–315. Springer, Cham (2019). https://doi.org/10.1007/978-3-030-27947-9_26

24. Wells, J.C.: SAMPA computer readable phonetic alphabet. In: Gibbon, D., Moore, R., Winski, R. (eds.) Handbook of Standards and Resources for Spoken Language Systems. Mouton de Gruyter, Berlin and New York (1997)

25. Wu, Z., Watts, O., King, S.: Merlin: an open source neural network speech synthesis system. In: 9th ISCA Speech Synthesis Workshop (2016), pp. 218–223, September 2016

Grappling with Web Technologies: The Problems of Remote Speech Recording

Daniel Tihelka[1]([✉])[iD], Markéta Jůzová[2][iD], and Jakub Vít[2][iD]

[1] New Technologies for the Information Society,
University of West Bohemia, Pilsen, Czech Republic
dtihelka@ntis.zcu.cz
[2] Department of Cybernetics, Faculty of Applied Sciences,
University of West Bohemia, Pilsen, Czech Republic
{juzova,jvit}@kky.zcu.cz

Abstract. Modern web browsers are becoming operating systems of their own kind, allowing unified access to the underlying hardware. The sound device can thus be used by web-based communication systems, such a Google meet, Zoom and others. This attracts the idea of using such capabilities to record a speech synthesis corpus through the web, with there being cases of use where it is really beneficial – for example, the building of personalised speech synthesis. The present paper shows that although it may appear easy, there are some dark corners to take care of.

Keywords: Speech corpus · Speech recording · Speech synthesis · Web browser · Dropout

1 Introduction

In the field of speech synthesis, the traditional way of building a new voice is to record a (semi-)professional speaker in a highly controlled environment, i.e. high quality recording studio (unechoic, sound-proof) using a professional recording software [3,9]. Nevertheless, it is not suitable for large-scale recording of ordinary (inexperienced) speakers whose aim is to create their own voice [8], although primarily intended for people in a threat of voice loss [4,6].

The usual procedure of recording these people in a studio would be time and resource consuming, significantly limited by the distance from the studio, and even uncomfortable for many of such speakers; in addition, there is sometimes simply not enough time before the surgery [10,14]. To mitigate these inconveniences, we have first developed a recording application running on a speaker's local computer and monitoring the process of recording and the critical parameters of its quality instead of an audio engineer. Still, the installation of the application requires some level of computer skills.

At the present time, computer applications are being shifted from local machines to the Internet with its cloud technologies. It has been even more

© Springer Nature Switzerland AG 2020
A. Karpov and R. Potapova (Eds.): SPECOM 2020, LNAI 12335, pp. 592–602, 2020.
https://doi.org/10.1007/978-3-030-60276-5_57

notable in the recent years with web browsers becoming a kind of operating system in their own right. Therefore, the idea of building a cloud-based recording application is very attractive – after all, many online communication technologies use these interfaces and capabilities. There is no need to install any application locally, as only a web browser is required, with modern web browsers enabling developers to access sound systems in a unified way anyway and any improvements and fixes in the application being managed in the cloud only.

On the other hand, these interfaces were originally designed primarily for communication purposes, so the web-based speech corpus recording has its own unexpected specifics which must be taken into the consideration.

2   Speech Corpus Recording

Since the start of the development of the ARTIC [13] TTS system, we were focused on professional voices. We have been using SoundForge professional software, extended through scripting interface with simple GUI application displaying the texts to the speaker, controlling the recording session and checking the base parameters of the recordings, such appropriate loudness, minimum length of initial and final pauses, good quality of EGG signal (when recorded), and so on [3,7]. When the SoundForge is connected with a professional FireWire sound card, the ASIO device drivers ensure direct connection to sound hardware, bypassing the intermediary layers in Windows sound drivers.

Let us now consider a web-based page for the recording. There are plethora of combinations of microphones, computers, sound cards and acoustic environments in which the recording may be carried out. The only way of ensuring the coverage of a wide range of devices is the use of an abstract OS API by the browser (WASAPI on Windows 10). However, with it being a higher-level API, all the layers through which the sound must pass (and which ASIO drivers strive to avoid) cause problems with latencies, especially on devices with lower computational power. While this is not a big problem when on-line communication is being carried out, the signal starts to behave strangely when recording the sound. It is all the more apparent the more the device is loaded either with other tasks or with real-time recording visualisation, as shown in Fig. 5 or presented in [12]. In addition, the parameters of the recordings may differ from what is prescribed by the application – for example, we have found that the sampling rate of the sound received differs in some cases, most likely based on the native format of the sound device being used.

2.1   Dropout Types

When recording, the sound samples are stored in the device driver within buffer(s), subsequently passed through all the Windows sound system layers to the application for further processing. There is a trade-off between buffer length and latency (the difference in time when the sound is digitised and when the samples appear in the application); the shorter buffers are more suitable for real-time visualisation of the recording in form of STFT bars or waveform, as illustrated in Fig. 5. However, the shorter buffers are sensitive to the overall system performance, since there are more frequent calls from the Windows sound system to the application (and thus the ASIO type of drivers provide direct access to sound device buffers to mitigate this). The key aspect is thus the speed of buffer consumption. If the application is not fast enough, which happens when it is occupied with other tasks such as a visualisation or any other load, the buffers pending are dropped, not reaching the application. Following that, the *dropout* occurs. In reality, however, the buffers may not only be missed, but surprisingly malformed in various ways.

Here is the description of the dropout types we have encountered so far:

Missing Buffer. In the majority of cases, a speech sample buffer is simply missing in the recording. This kind of dropout is illustrated in Fig. 1.

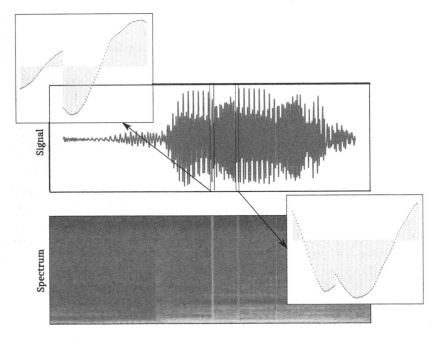

Fig. 1. The illustration of the most usual dropouts, where a recording buffer is missing. It is not visible in the signal unless zoomed in, but it is visible in the spectrum and clearly audible in the recording.

Repeated Buffer. Sometimes, the same speech buffer is repeated twice, or even multiple times, replacing what has been said in the affected part. Usually, the same buffer is repeated right after itself, as shown in Fig. 2 (upper), but we have also found a case where the frame is separated by a good signal, as illustrated in Fig. 2 (lower).

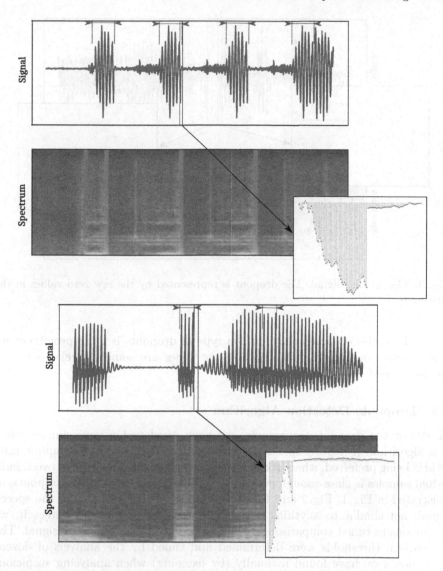

Fig. 2. The multiple repetitions of the signal buffer. In the upper example, the segment marked by red color is repeated 4×, and even a wider segment with phone [i] is repeated 3×. In the lower instance, the segment is taken from the good signal far behind it. (Color figure online)

Zero Buffer. Instead of buffer repetition, for one particular speaker the buffer (or subsequent buffers) is missing completely, replaced by raw zero values as shown in Fig. 3.

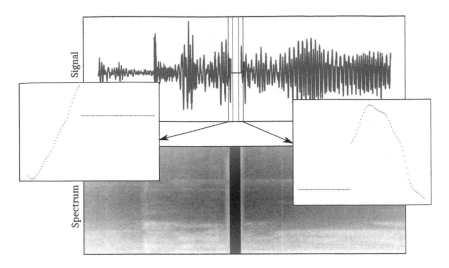

Fig. 3. The missing signal. The dropout is represented by the raw zero values in the missing part.

We have also observed that a given type of dropouts is more prominent for a particular speaker. It indicates that the types are somehow related to the hardware used.

2.2   Dropouts Detection Algorithm

To detect the dropout occurrences, we have developed a very simple detection algorithm. It expects speech recorded with at least 44.1 kHz sampling rate, 48 kHz being preferred, where it can be expected that the difference between individual samples is close enough even in the case of noise. Looking at dropouts, as illustrated in Fig. 1, Fig. 2 and Fig. 3, there is clear discontinuity in the speech signal, not similar to anything else in the "normal" signal. To detect it, we use manually tuned comparison of the 2nd-order differences in the signal. The comparison thresholds were determined and tuned by the analysis of dozens of dropouts we have found manually (by listening) when analysing suspicious segmentation outputs [5].

To describe the algorithm, let us have a signal samples s (in 48 kHz sampling rate). The first step is to compute the second-order difference S of the signal s

$$S = f''(s) \tag{1}$$

Choose the 2 ms frame f and obtain the absolute value of the signal S within the frame

$$F = |S[f]| \tag{2}$$

Find the maximum value in the frame

$$s^* = \max(F) \tag{3}$$

and choose the 5 samples sub-frame F^* centred around the s^* sample and the remaining samples in the frame F^- not containing the sub-frame

$$F^* = F[s^* - 2, s^* + 2], \ F^- = F\backslash F^* \tag{4}$$

compute mean and standard deviation in both sub-frames F^* and F^-:

$$\mu^* = \mu(F^*), \ \sigma^* = \sigma(F^*), \ \mu^- = \mu(F^-), \ \sigma^- = \sigma(F^-) \tag{5}$$

and maximum value in the F^- :

$$s^- = max(F^-) \tag{6}$$

Now we can compare the values and decide whether or not the dropout occurred in the frame $s[f]$. Note that the values were set heuristically, based on the analysis of the dropouts seen:

$$\mu^* > 14 \, \mu^- \quad \wedge \quad \sigma^* > 13 \, \sigma^- \quad \wedge \quad (s^* > 0.08 \quad \vee \quad s^* > 35 \, s^-)$$
$$\Rightarrow \text{dropout detected}$$

After the analysis, shift the frame by its halved size and continue by the next frame

$$f = f + f/2 \tag{7}$$

The behaviour of the algorithm is illustrated in Fig. 4. Note that the algorithm must work on the original signal as captured from the sound card. Even when the values are tuned for 48 kHz sampling rate, they also work sufficiently well for the 44.1 kHz sampling rate. For lower sapling rates, it has not been tested and is unlikely to work reliably. It is also not wise to re-sample the recorded signal to 48 kHz, as it will smooth the discontinuities in it.

Despite the algorithm being very simple, it behaves surprisingly well. We periodically examine recordings carried out through the course of the voice conservation project [6], where a wide range of recording device combinations are used, and the dropouts detected by the algorithm are then confirmed by their manual examination.

3   Analysis of Recordings

As already mentioned, in project [6] we record test users in order to tune the recording process. We have had 31 users so far, with 21, 055 prompts recorded in total, whereby a wide range of recording devices and environments have been used – we make no requirements for devices to be used, only base requirements on acoustic background.

Let us have a look at the dependency of dropouts on devices used for the recordings and the level of visual effects on the web page through the recording runs. The results are taken from the analysis of 18 recording session of 6 test users, and are presented in Table 1. The *visualisation level* of type *waveform*

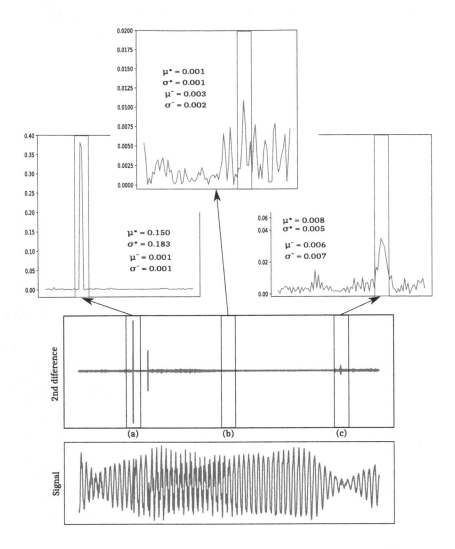

Fig. 4. The illustration of dropouts-detection algorithm. The region (a) represents frame F with dropout occurrence (there are two consecutive dropouts in the signal, both detected correctly); the regions (b) and (c) are frames without dropouts. The red rectangles in frames represent F^* signals with μ^*, μ^-, σ^* and σ^- values shown. (Color figure online)

is shown in Fig. 5, where a real-time changing waveform is displayed as the recording continues. On the contrary, the type *text* only displays "Recording..." text without additional graphical effects, as shown in Fig. 6. Both screenshots are from application created by our project partner, employing free JavaScript library *Web Audio Recorder* [11] with the buffer size set to 4,096 samples. The

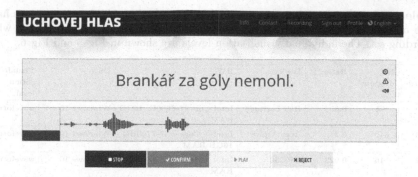

Fig. 5. The real-time visualisation of the recorded waveform. This requires small buffers passed to the application to minimise the latency, but it is also computationally demanding, consuming non-negligible additional resources. In Table 1 it represents visualisation type *waveform*.

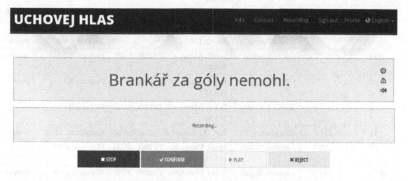

Fig. 6. The same application as in Fig. 5, but without real-time visualisation. Instead, only a frame with "Recording..." text is shown, lowering the dropout occurrences significantly. In Table 1 it represents the visualisation type *text*.

last *visualisation level* type *none* is from our minimal recording web interface, using the direct access to the sound recording interface [1]. Contrary to the previous application, it also has auto gain and noise reduction effects switched off.

It can bee seen that lowering the level of visualisation and increasing computation power has direct positive effect on the number of dropout occurrences. However, there are still cases when even no level of visualisation leads to some dropouts in the recordings, which may be caused by non-deterministic behaviour of JavaScript.

Table 1. The number of dropouts detected in the recordings when using various hardware and visualisation effects. The *ratio* is the proportion of dropouts to the whole recording set. The individual visualisation levels are shown in Fig. 5 and Fig. 6.

No. recordings	No. dropouts	Ratio	Device	HW	OS	Vizualization level
1600	5	0.003	Dell Latitude	Intel Core i7-6600U, 16GB RAM	Windows 10	waveform
1613	340	0.2	Acer Aspire	Intel Core i7-7700HQ, 16GB RAM	Windows 10	waveform
702	16	0.023	HP 250	Intel Celeron N3060, 4GB RAM	Windows 10	waveform
1614	655	0.41	Lenovo ThinkPad	Intel Core i5-5300U, 8GB RAM	Windows 10	waveform
2370	303	0.0128	Lenovo ThinkPad	Intel Core i5-5300U, 8GB RAM	Windows 10	waveform
400	8	0.020	HP EliteBook	Intel Core i5-7200U, 16GB RAM	Fedora 32	waveform
100	1465	14.65	BlackBerry	Key2 LE	Android 8.1	waveform
101	0		BlackBerry	Key2 LE	Android 8.1	text
100	0		BlackBerry	Key2 LE	Android 8.1	none
101	228	2.257	GPD Pocket	Intel Atom x7-Z8750, 8GB RAM	Fedora 32	waveform
101	0		GPD Pocket	Intel Atom x7-Z8750, 8GB RAM	Fedora 32	text
101	0		GPD Pocket	Intel Atom x7-Z8750, 8GB RAM	Fedora 32	none
100	11	0.11	Raspberry Pi	RPi 4, model B, 4GB RAM	Raspbian 10.4	waveform
200	37	0.185	Lenovo N500	Intel Pentium T3200, 2GB RAM	Windows 10	none
3395	4043	1.191	Lenovo N500	Intel Pentium T3200, 2GB RAM	Windows 10	waveform
879	1240	1.411	Lenovo N500	Intel Pentium T3200, 2GB RAM	Windows 10	waveform
101	0		Lenovo N500	Intel Pentium T3200, 2GB RAM	Windows 10	text
200	28	0.140	Lenovo N500	Intel Pentium T3200, 2GB RAM	Windows 10	none

4   Conclusion

We have pointed out the difficulties that must be taken into account when building a web application through which speech corpora are recorded. We recommend, in general, switching off all the integrated sound processing effects, avoiding as many real-time visualisations and animations as possible and using a rather more powerful computer in order to mitigate the fact that JavaScript has non-deterministic behaviour (garbage connection, interpreter lock, etc.). When impossible to guarantee a minimal hardware configuration the recording runs on, but not only then, as is the case in [2,6,8], it is also good to have a server–side check of the recordings [3], which the proposed, albeit simple, dropout-detection algorithm can be part of. The results of such checks may either tune the recording configuration (buffers size, look&feel) or instruct a speaker to adjust his/her recording conditions (close unnecessary applications, or to use a better device).

Acknowledgments. This research was supported by the Technology Agency of the Czech Republic (project No. TH02010307), and by the grant of the University of West Bohemia, (project No. SGS-2019-027).

References

1. Web audio concepts and usage. https://developer.mozilla.org/en-US/docs/Web/API/Web_Audio_API. Accessed 15 June 2020
2. Conkie, A., Okken, T., Kim, Y.J., Fabbrizio, G.D.: Building text-to-speech voices in the cloud. In: LREC 2012, ELRA, Istanbul, Turkey, pp. 3317–3321 (2012)
3. Grůber, M., Legát, M., Tihelka, D.: Corpus recording and checking on the recorded data. In: The 1st Young Researchers Conference on Applied Sciences, Západočeská univerzita, Plzeň, pp. 174–179 (2007)
4. Hanzlíček, Z., Romportl, J., Matoušek, J.: Voice conservation: towards creating a speech-aid system for total laryngectomees. In: Kelemen, J., Romportl, J., Žáčková, E. (eds.) Beyond Artificial Intelligence: Contemplations, Expectations, Applications, Topics in Intelligent Engineering and Informatics, vol. 4, pp. 203–212. Springer, Heidelberg (2012). https://doi.org/10.1007/978-3-642-34422-0_14
5. Hanzlíček, Z., Vít, J., Tihelka, D.: LSTM-based speech segmentation for TTS synthesis. In: Ekštein, K. (ed.) TSD 2019. LNCS (LNAI), vol. 11697, pp. 361–372. Springer, Cham (2019). https://doi.org/10.1007/978-3-030-27947-9_31
6. Jůzová, M., Tihelka, D., Matoušek, J., Hanzlíček, Z.: Voice conservation and TTS system for people facing total laryngectomy. In: Interspeech 2017, pp. 3425–3426 (2017)
7. Legát, M., Grůber, M., Matoušek, J.: The issue of checking the volume consistency of speech corpus during recording. In: The 1st Young Researchers Conference on Applied Sciences, Západočeská univerzita, Plzeň, pp. 206–211 (2007)
8. Malfrère, F., et al.: My-own-voice: a web service that allows you to create a text-to-speech voice from your own voice. In: Interspeech 2016, 17th Annual Conference of the International Speech Communication Association, San Francisco, CA, USA, pp. 1968–1969 (2016)
9. Matoušek, J., Tihelka, D., Romportl, J.: Building of a speech corpus optimised for unit selection TTS synthesis. In: LREC 2008, Proceedings of 6th International Conference on Language Resources and Evaluation, ELRA, Marrakech, Morocco, pp. 1296–1299 (2008)
10. Mertl, J., Žáčková, E., Řepová, B.: Quality of life of patients after total laryngectomy: the struggle against stigmatization and social exclusion using speech synthesis. Disabil. Rehabil. Assist. Technol. **13**(4), 342–352 (2018)
11. Miyane, Y.: Web audio recorder JavaScript library. https://github.com/higuma/web-audio-recorder-js. Accessed 15 June 2020
12. Stanislav, P., Šmídl, L., Švec, J.: An automatic training tool for air traffic control training. In: Interspeech 2016, 17th Annual Conference of the International Speech Communication Association, San Francisco, CA, USA, 8–12 September 2016, pp. 782–783 (2016)

13. Tihelka, D., Hanzlíček, Z., Jůzová, M., Vít, J., Matoušek, J., Grůber, M.: Current state of text-to-speech system ARTIC: a decade of research on the field of speech technologies. In: Sojka, P., Horák, A., Kopeček, I., Pala, K. (eds.) TSD 2018. LNCS (LNAI), vol. 11107, pp. 369–378. Springer, Cham (2018). https://doi.org/10.1007/978-3-030-00794-2_40

14. Řepová, B., Zábrodský, M., Plzák, J., Kalfert, D., Matoušek, J., Betka, J.: Text-to-speech synthesis as an alternative communication means after total laryngectomy. Biomedical Papers of the Medical Faculty of the University Palacky (2020)

Robust Noisy Speech Parameterization Using Convolutional Neural Networks

Ryhor Vashkevich[✉] and Elias Azarov

Belarusian State University of Informatics and Radioelectronics, Minsk, Belarus
ryhorv@gmail.com, azarov@bsuir.by

Abstract. A new neural network approach to speech characteristic features extraction is proposed. The features provide separable information about the pitch of a speech signal and amplitude spectral envelope. The features can compactly describe a speech signal and allows effectively training machine learning models that use the proposed features as an input. The experimental application to the voice activity detection (VAD) problem shows the advantage of the proposed features over the features that are widely used: melspectrogram, amplitude spectrum and MFCC. Experimental results show that a VAD based on proposed features is more accurate while using a simpler architecture and much less number of training parameters. The performance has been tested on noisy speech with different SNR ratios and the results show that the proposed features are more robust to noise. The experimental results indicate that the proposed VAD model outperforms the VAD from the WebRTC framework.

Keywords: Speech characteristic features · VAD · Speech parametrization

1 Introduction

Parametric speech representation is used in many digital speech processing tasks (speech synthesis, recognition, conversion, noise reduction, etc.). There are several classical speech representations such as amplitude spectrum, mel-frequency cepstral coefficients (MFCCs), linear predictive coding (LPC,) special parametric models, etc. Modern speech processing systems are based on artificial neural networks and as an input use a raw waveform or the classical representations of speech. For example, the WaveNet vocoder [1] uses a melspectrogram as input features, Deep Speech 2 [2] uses log amplitude spectrum, Wave2Letter [3] uses raw waveform.

The main motivation of using a parametric speech representation is to improve performance of the neural network. Models that use raw data as input features are usually more complex than models that use parametric data as an input. Speech parametrization by acoustic part of neural networks as a rule is not as effective in terms of computational resources. That is why the development of new neural network structures for acoustic modeling is required.

On the other hand, the classical approaches to the parametric presentation of speech have disadvantages that require the use of complex neural network models for subsequent processing. The main disadvantage is the redundancy due to the high variability of the fundamental frequency and the amplitude envelope of the spectrum of the

A. Karpov and R. Potapova (Eds.): SPECOM 2020, LNAI 12335, pp. 603–612, 2020.
https://doi.org/10.1007/978-3-030-60276-5_58

speech signal. Using the amplitude spectrum, it is required to model the joint distribution of all parameters, which leads to the need to use a large training dataset and a complex model with a large number of trainable parameters [1–4]. Existing studies in the field of parametric modeling of speech [5] for various applications indicate that the statistical relationship between the fundamental frequency and the amplitude envelope of the spectrum of the speech signal is not strong. In vocoder models, tone and envelope are described separately, which makes it possible to compactly describe the signal. Applying this concept to neural networks, it is possible to obtain an effective parametric representation of speech by separating the joint distribution of the fundamental frequency and the amplitude envelope.

The main goal of this work is to propose neural network structures for acoustic modeling that perform a separate parametric description of the fundamental frequency and amplitude envelope of the spectrum of the speech signal (pitch- inclusive and envelope-inclusive representations respectively).

The second problem considered in the paper is the robustness of the representation to noise. The fundamental frequency extraction methods are sensitive to noise, making it difficult to separate the pitch and the envelope. Pitch extraction errors lead to an incorrect envelope estimation and this makes the classic approach inapplicable. So we conducted a number of experiments to estimate overall robustness of the proposed approach to additive noise. As a practical application, the problem of voice activity detection (VAD) in noisy condition is considered. This particular problem is chosen as a very indicative to the resistance of the proposed features to noise.

VAD is an important stage of almost any speech processing task. It segments audio on voiced and unvoiced segments and these segments use in further stages of the entire speech processing pipeline. Thus, it is possible to formulate two main requirements for VAD models. On the one hand, the model should provide high accuracy and on the other hand, it should be computationally effective. The first requirement means that the model should be robust to variety species of noise with large range of signal-to-noise ratios (SNR). The second requirement means that the model should consume as little computational resources as possible and take minimum time to provide correct results.

Most publications propose approaches based on machine learning methods that provide high accuracy, but at the same time are computationally expensive. To reduce the computational cost, alternative approaches are considered. Including an analytical model for voice activity detection which based on the analysis of formants of a harmonic signal [6]. The disadvantage of this method includes the assumption that the speech signal always has a harmonic structure. As an alternative the paper [7] proposes to analyze the spectrogram of an audio signal, dividing it into two parts. It is assumed that information about the speech signal is always contained in the lower frequency band, and noise is in the higher. The disadvantage of this approach is sensitivity to low-frequency noise.

Approaches based on machine learning methods are generally more robust to noise, since they can take into account different species of noise and a large variation of the human voice. One of the simplest examples of the use of machine learning is proposed in [8], where the support-vector machines are used to classify input characteristic features of an audio signal on two classes – voice and noise. Many approaches use artificial neural networks (NN) which are a more powerful classification tool [9–14].

The high computational complexity of NN-based methods is associated with many parameters. In particular, the convolutional layers can model frequency variations of the input signal and the LSTM layers can take into account its temporal variations as proposed in [9]. This model has approximately 100,000 trainable parameters. The publications [10, 11] use only fully connected layers with many trainable parameters. A model from [11], for example, has 1 billion trainable parameters and it requires a large training dataset and tends to overfit. Approaches [10, 12] propose first of all to denoise an input signal using the first few layers of a model and further layers use as a classifier of clean audio.

The publications [9, 13, 14] are considered the fact that a speech signal is a long-time signal and it is possible to extract temporal features from this signal using recurrent layers.

In most of speech processing tasks, including VAD, MFCCs are used as the main characteristic features of a speech signal. The MFCC features was successfully applied to the VAD task in [8, 10, 11]. Papers [6, 7] use speech features based in the spectrum of the signal, taking into account the fact that the speech signal in most cases has a harmonic structure.

Recently, end-to-end approaches to building deep learning models are widespread [9]. The main idea of such approaches is that the characteristic features has not extracting from the input data. Instead of this the raw input data feed to a deep neural network model. However, such models are hard to train, since a large amount of data and complex models are required.

The general idea of the paper is to use special spectrum convolutional features that effectively describe the harmonic structure of the speech signal as described in the next section. The proposed features are compared with other frequently used features (amplitude spectrum, melspectrogram, MFCCs) it he context of a VAD task. Additionally, the proposed VAD model with the spectrum convolutional features is compared with a VAD model from the WebRTC framework [15].

2 Spectral Convolutional Features

The spectral convolutional features are obtained from the raw audio signal as described below.

A spectrum of the speech signal has a lot of vocal components (80% to 95% of the total duration of the speech signal) as shown in Fig. 1. All harmonics of the speech spectrum have frequencies that are multiples of the fundamental frequency ($F0$).

The fundamental frequency of a speech signal for most speech processing task has a value from $F0_{min}= 50$ Hz to $F0_{max}= 450$ Hz. Based on this range N hypothetical values of $F0_i$, is obtained by quantization of the frequency range $F0_{min}...F0_{max}$ on N values along log scale as shown in Fig. 2.

The hypothetical values $F0_i$ determine what possible value can the fundamental frequency of the input signal take. The frequency interval is quantized using the logarithmic scale to ensure equal resolution of the analysis in different frequency ranges of the fundamental tone.

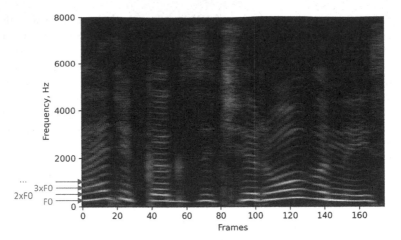

Fig. 1. Amplitude spectrum of the speech signal.

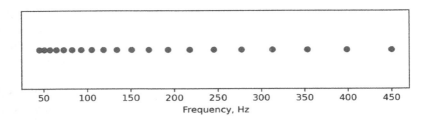

Fig. 2. Log-scale quantized frequency range.

For each value of $F0_i$ the M harmonics are selected from the input spectrum. The indices I_i of all M harmonics of the spectrum with the frequency resolution f_r for the given $F0_i$ are calculated as:

$$I_i^j = round\left(\frac{1+j}{2} \cdot \frac{F0_i}{f_r} + 1\right), j = 0\ldots(M-1).$$

The frequency resolution f_r of the spectrum is determined as:

$$f_r = \frac{f_s}{N_{fft}},$$

where f_s is the sample rate, N_{fft} is the size of the Fast Fourier transform.

To get the final features vector X for the given signal s it is necessary to compute spectrum S and then select from the spectrum only those components that correspond the calculated indices I_i^j:

$$S = \log_{10} |FFT(s)|,$$

$$X(i,j) = S(I_i^j), \quad i = 0 \ldots (N-1), \quad j = 0 \ldots (M-1).$$

Thus, for each input frame, the features matrix X is formed. The matrix has shape $N \times M$ and consists of N vectors of M components. Each vector hypothetically describes the harmonic structure of the speech signal. The features extraction process is show in Fig. 3.

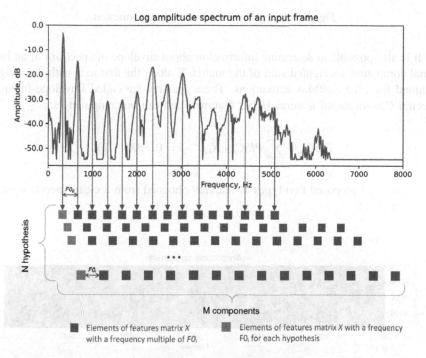

Fig. 3. Features extraction.

At the next step, a convolution layer with K filters of size M is applied to features matrix X along the second axis with a hop size one along the first axis and ReLU activation function. The activations of the first layer have size $N \times 1 \times K$. These activations then use as an input to a second convolution layer with one filter and kernel size K applied along the third axis. The SoftMax activation function is then applied along the first axis of the activations of the second convolution layer. The result of the SoftMax function has size $N \times 1 \times 1$. The final spectral convolutional features are obtained squeezing the last two dimensions of the result activations. The whole features extraction process is shown in Fig. 4.

The obtained features contain information about pitch probability distribution for each frame of the input speech signal as well as information about its vocalization. We called it Pitch-Inclusive spectral Convolutional features (*PISC* features).

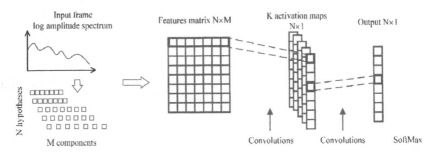

Fig. 4. Spectral Convolutional features extraction.

It is also possible to determine information about envelope of spectrum of an input signal computing a weighted sum of the matrix X along the first axis with the weights obtained from the SoftMax activations. These features are called Envelope-Inclusive Spectral Convolutional features (*EISC* features) and can be computed as:

$$EISC_j = \sum_{i=0}^{N-1} PISC_i * X_{i,j}, \quad j = 0 \ldots (M-1).$$

An example of proposed two types of features obtained from a clean speech signal is shown in Fig. 5.

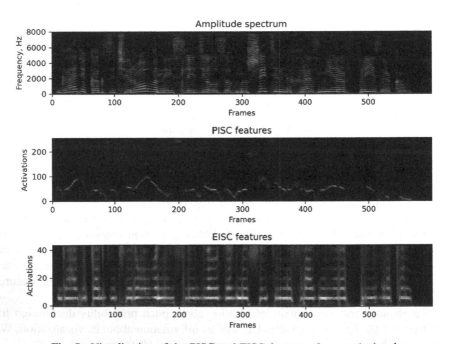

Fig. 5. Visualization of the PISC and EISC features of a speech signal.

Analyzing the example of the PISC features it is easy to determine which frames contain voice signal and which frames contain only noise (voiced frames has large values of one of its activations). Moreover, an activation with maximum value of each frame strongly correlates with the pitch of the input signal, an amplitude spectrum of which is shown above the PISC features.

The PISC features contain information about the pitch of a speech signal. The accuracy of the representation of the pitch information depends on the value of parameter N. To obtain the more accurate features it is necessary to use more hypothetical values of $F0_i$ to compute matrix X (parameter N should have large value). But large values of the parameter N require large computational costs to obtain these features.

The EISC features can be used as a compact representation of the spectrum envelope (44 bins along frequency axis instead of 512 bins of the full spectrum). They contain information about a spectrum envelope from the full available frequency range. The compact representation of the spectrum envelope is achieved by eliminating excessive information about the pitch of the signal by applying spectral convolutions.

3 Experiments and Results

3.1 Experimental Setup

Recurrent Neural Network Model for Robust VAD. To test the proposed features a simple recurrent neural network with a single GRU cell is used. The model consists of three trainable layers. The first layer is a dense layer with 16 units and ReLU activation functions. It projects frame-level features into a hidden representation. The activations of the dense layer feed a GRU cell with 16 units. The GRU layer output feed an output dense layer with a single unit and sigmoid activation. A dropout regularization with rate 0.2 is applied after the first dense and GRU layers.

The convolutional layers described in the previous chapter are trained together with the main model without any additional losses. The number N of hypothetical values of F0i is 256 and the number of harmonics M is 44.

A binary cross entropy is used as a loss function. To train the model the Adam [16] optimizer with learning rate 10^{-3} is used. The model is trained during 1000 epochs.

Training and Test Data. A private dataset with 11 h of noisy speech is used to train the model. Additional 2 h of noisy speech is used to evaluate the model. Clean records from The Noisy Speech Database [17] are used to evaluate the model's robustness to various types of noise. These records are corrupted by adding noise records from the Musan dataset [18] with specific SNR ratio. All the records have a sampling rate of 16 kHz. A STFT with FFT size 1024 and hop size 100 is used to obtain a amplitude spectrum from a waveform.

3.2 Experimental Results

General Performance. To evaluate the quality of the proposed features, the performance of the described VAD model with various input features was evaluated. As alternative characteristic features were used:

- 512 bin log amplitude spectrum;
- 80 channel melspectrogram;
- 20 MFCC coefficients.

The area under the curve (AUC) score is used as an objective metric to evaluate model performance. The model is evaluated on 2 h of noisy speech and the results is presented in Table 1.

Table 1. VAD results of a model with different input characteristic features.

Features type	AUC score
Magnitude spectrum	0.8117
Melspectrogram	0.8527
MFCC	0.8137
PISC	0.9422
EISC	0.9460

The results from Table 1 show that the model with the proposed characteristic features can better solve the VAD task.

The F1 score is also computed for proposed model and the VAD model from the WebRTC framework. The results are presented in Table 2.

Table 2. Results of comparing the proposed model with a WebRTC model.

Model	F1 score
Proposed with PISC features	0.9651
Proposed with EISC features	0.9642
WebRTC	0.8351

Noise Robustness. To test noise robustness clean records from The Noisy Speech Database are used. The records are corrupted by adding noise records from the Musan dataset with SNR ratio −20 dB, −10 dB, 0 dB and random SNR value from a range −20 dB...0 dB. The results the evaluation of different characteristics features are presented in Table 3.

The results presented in Table 3 indicate that the proposed features are more robust to noise than their widely used alternatives. Moreover, the performance of model with the proposed features almost unchanged depending on the SNR ratio.

Table 3. Experimental results in different noisy conditions.

Features type	−20 dB	−10 dB	0 dB	Random SNR
	AUC scores			
FFT	0.8356	0.8585	0.8827	0.8713
Melspectrogram	0.9018	0.9214	0.9318	0.9251
MFCC	0.8410	0.8651	0.8947	0.8835
PASC	0.9434	0.9427	0.9450	0.9447
EASC	0.9426	0.9439	0.9451	0.9440

4   Conclusion

A new approach to extracting of characteristic features from a noisy speech signal is proposed. The proposed features can separately represent the pitch and spectral envelope of a speech signal. The paper also investigate a simple VAD model. The performance of this model is tested in various noise conditions and using various alternative characteristic features, such as melspectrogram, amplitude spectrum, MFCC. Experimental results show that the proposed features help the model to solve the task much better, due to the compact representation of the speech signal, as well as provide robustness to noise. The proposed approach outperforms the popular open model from the WebRTC framework.

References

1. van den Oord, A., et al.: WaveNet: a generative model for raw audio. http://arxiv.org/abs/1609.03499. Accessed 26 Nov 2019
2. Amodei, D., et al.: Deep speech 2: end-to-end speech recognition in English and mandarin. http://arxiv.org/abs/1512.02595. Accessed 13 Jan 2020
3. Collobert, R., Puhrsch, C., Synnaeve, G.: Wav2Letter: an end-to-end convnet-based speech recognition system. http://arxiv.org/abs/1609.03193. Accessed 15 Jan 2020
4. Prenger, R., Valle, R., Catanzaro, B.: Waveglow: a flow-based generative network for speech synthesis. In: ICASSP 2019 - 2019 IEEE International Conference on Acoustics, Speech and Signal Processing (ICASSP), Brighton, United Kingdom, pp. 3617–3621 (2019). https://doi.org/10.1109/icassp.2019.8683143
5. Spanias, A., Painter, T., Atti, V.: Audio Signal Processing and Coding. Wiley, Hoboken (2007)
6. Yoo, I.-C., Lim, H., Yook, D.: Formant-based robust voice activity detection. Trans. Audio Speech Lang. Process. **23**(12), 2238–2245 (2015). https://doi.org/10.1109/TASLP.2015.2476762
7. Pang, J.: Spectrum energy based voice activity detection. In: The 7th IEEE Annual Computing and Communication Workshop and Conference (CCWC), Las Vegas, USA, pp. 1–5 (2017). https://doi.org/10.1109/CCWC.2017.7868454
8. Kinnunen, T., et al.: Voice activity detection using MFCC features and support vector machine. In: The 12th International Conference on Speech and Computer (SPECOM07), Moscow, Russia, vol. 2, pp. 556–561 (2007)

9. Zazo, R., Sainath, T.N., Simko, G., Parada, C.: Feature learning with raw-waveform CLDNNs for voice activity detection. In: 17th Annual Conference of the International Speech Communication Association, San Francisco, USA, pp. 3668–3672 (2016). https://doi.org/10.21437/interspeech.2016-268

10. Zhang, X., Wu, J.: Denoising deep neural networks based voice activity detection. In: International Conference on Acoustics Speech and Signal Processing, Vancouver, Canada, pp. 853–857 (2013). https://doi.org/10.1109/ICASSP.2013.6637769

11. Ryant, N., Liberman, M., Yuan, J.: Speech activity detection on youtube using deep neural networks. In: 14th Annual Conference of the International Speech Communication Association Lyon, France, pp. 728–731 (2013)

12. Wang, Q., et al.: A universal VAD based on jointly trained deep neural networks. In: 16th Annual Conference of the International Speech Communication Association, Dresden, Germany, pp. 2282–2286 (2015)

13. Hughes, T., Mierle, K.: Recurrent neural networks for voice activity detection. In: International Conference on Acoustics, Speech and Signal Processing, Vancouver, Canada, pp. 7378–7382 (2013). https://doi.org/10.1109/ICASSP.2013.6639096

14. Eyben, F., Weninger, F., Squartini, S., Schuller, B.: Real-life voice activity detection with LSTM recurrent neural networks and an application to Hollywood movies. In: International Conference on Acoustics, Speech and Signal Processing, Vancouver, Canada, pp. 483–487 (2013). https://doi.org/10.1109/ICASSP.2013.6637694

15. Python interface to the WebRTC Voice Activity Detector. https://github.com/wiseman/py-webrtcvad

16. Kingma, D.P., Ba, J.: Adam: a method for stochastic optimization. arXiv preprint arXiv:1412.6980 (2014)

17. Valentini-Botinhao, C.: Noisy speech database for training speech enhancement algorithms and TTS models [sound]. University of Edinburgh. School of Informatics. Centre for Speech Technology Research (CSTR) (2017). https://doi.org/10.7488/ds/2117

18. Snyder, D., Chen, G., Povey, D.: Musan: a music, speech, and noise corpus. arXiv preprint arXiv:1510.08484 (2015)

More than Words: Cross-Linguistic Exploration of Parkinson's Disease Identification from Speech

Vass Verkhodanova[1,2]([✉]), Dominika Trčková[3], Matt Coler[1], and Wander Lowie[2,3]

[1] Campus Fryslân, University of Groningen, Leeuwarden, The Netherlands
{v.verkhodanova,m.coler}@rug.nl
[2] Research School of Behavioural and Cognitive Neurosciences, University of Groningen, Groningen, The Netherlands
w.m.lowie@rug.nl
[3] Center for Language and Cognition Groningen, University of Groningen, Groningen, The Netherlands
d.trckova@student.rug.nl

Abstract. This study investigates the effect of listeners' first language and expertise on their perception of speech produced by people with Parkinson's Disease (PD). We compared assessment scores and identification accuracy of expert and non-expert Czech and Dutch listeners on two tasks: perception of speech healthiness and recognition of sentence type intonation. We collected speech data from 30 Dutch speakers diagnosed with PD and 30 Dutch speaking healthy controls. Short phrases from intonation tasks and spontaneous monologues were used as stimuli in an online perception experiment. 40 people (20 expert and 20 non-expert listeners) participated. Results show that both expertise and language familiarity are important factors in perception of speech of people with PD, however differences in identification accuracy depend on the task type. In recognition of PD speech, there are prominent acoustic cues that trigger perception of "unhealthiness" in the non-expert listeners, while experience with phonetics may lead to a different focus in their perception of such cues. In intonation accuracy recognition, both Czech and Dutch expert groups outperform both non-expert groups, indicating the added value of phonetic experience for the prosodic task. Yet, there is a clear benefit for having Dutch as the first language for both tasks, as Dutch listeners performed more accurately in both healthiness and sentence type recognition tasks.

Keywords: Parkinson's disease · Hypokinetic dysarthria · Prosody · Expert knowledge · Speech perception · Speech recognition

1 Introduction

Parkinson's Disease (PD) is a complex neurodegenerative disease that is characterized by a range of both motor and non-motor symptoms. A speech disorder

© Springer Nature Switzerland AG 2020
A. Karpov and R. Potapova (Eds.): SPECOM 2020, LNAI 12335, pp. 613–623, 2020.
https://doi.org/10.1007/978-3-030-60276-5_59

referred to as hypokinetic dysarthria (HD) is among them. It has been argued, that up to 90% of people diagnosed with PD develop HD [1], leading to a range of changes in speech, including monotonous voice, hypophonia, and "slurred" articulation. HD also affects patients' mental well-being, as the speech changes were reported to have a negative impact not only on their social-linguistic competence in interactions with others [13], but also their concept of self-identity and roles [11,14].

The most widely-used criteria for analysis and description of HD are the deviant speech dimensions, developed in late 60s, according to which prosody is the most perceptually affected cluster in HD speech as judged by speech language therapists (SLTs) [2]. Since then, although the knowledge of perceptual and acoustic parameters of HD speech has grown and automated technology to aid the diagnosis is in development [6,12], many issues are still unresolved [10] and the diagnosis still lies in the hands of clinicians. However, clinicians' perceptive abilities are varied [4,5], suggesting that the matter of the audience is an important factor. Studies on the perception of dysarthric speech are scarce, especially those exploring expertise as a differentiating factor among listeners. One such study focused on the assessments of dysarthric speech intelligibility [18] by dysarthric speakers, SLTs, and non-expert listeners. They reported mixed results: there were no significant differences between the three groups, though intra-rater reliability was lower and significant for the expert group of listeners. More recently Smith et al. [15] demonstrated that there was no significant difference between intelligibility assessments of speech affected by HD when judged by expert or non-expert listeners. However, less is known about the effect of listener's first language (L1) or their expertise on the perception of healthiness or prosody in HD speech.

In this study we investigate the topic of HD perception by non-expert listeners and experts in phonetics. We explore whether a different expertise with speech sounds could be beneficial for recognition of both sentence type in HD speech and HD itself. Furthermore, we address the question of effect of listeners' L1, exploring to what extent the verbal content plays a role in HD recognition. Thus, this study aims to answer the following research question: do listeners perceive dysarthric speech of people with PD differently on the basis of their L1 and expertise?

2   Methods

To investigate whether level of expertise and L1 affect the ability to distinguish speech of people with PD from healthy speakers, and the ability to correctly identify sentence type from voice only, we conducted an online perception experiment with four different groups of participants. We recruited participants with Czech and Dutch as their first languages to compare perception of listeners whose L1 are from two different language families. Their results were subjected to cross-comparisons in subsequent analyses.

2.1 Listeners

In total, 40 listeners participated in the experiment. The listeners include native Dutch and native Czech listeners with both education in speech sciences such as phonetics and phonology as well as listeners with the a lack of such specialist education. The subjects were divided into groups based on their L1 and expertise: Czech non-expert (CN), Dutch non-expert (DN), Czech expert (CP), and Dutch expert listeners (DP). The demographics of the groups were relatively similar, except for the Dutch experts, who reported higher age and this group included more male than female participants than the other three groups. Demographics of the groups are are summarized in Table 1. Every participant from the expert groups had a profession in the field of speech sciences (all Dutch experts) or had at least obtained an undergraduate degree in the Phonetics and Phonology (the case of a few Czech experts). The non-expert groups held degrees and professions from various fields except for speech sciences. All but one subject reported normal hearing – one non-expert participant reported a minor unspecified hearing problem, but as their answers did not significantly differ from the other subjects in the same group, we kept the subject's responses to preserve equal group sizes.

Table 1. Summary of listeners group demographics. Age is given in years.

	Czech expert	Czech non-expert	Dutch expert	Dutch non-expert	All
n	10	10	10	10	40
Mean age	30.7	31.7	40.9	29.7	33.3
SD age	10.3	9.4	14.2	11.5	11.4

2.2 Speech Data Collection

Speech recordings were collected from 60 Dutch native speakers, 30 speakers diagnosed with PD and 30 healthy control speakers (HC). None of these individuals exhibited cognitive impairments as assessed with the Minimal Mental State Examination (MMSE). The HC speakers had no history of neurological disorders. The demographics for both groups can be seen in Table 2. The recording sessions took place in quiet rooms at the University Medical Centre Groningen, Medical Centre Leeuwarden or at participants' homes with the TASCAM-DR100 recorder and an external Senheiser e865 microphone placed at around a 40 cm distance from a participant. Recording procedure followed the test protocol that included a range of speech tasks [7].

All participants gave their written informed consent to the speech tasks and the recording procedure. The collection and analysis of the material was approved by the Medical Ethics Committee of the University Medical Center Groningen.

Table 2. Summary of speaker group demographics. Age and duration of disease are given in years.

Group	Gender	Age (SD)	Duration of disease (SD)	MMSE Mean(SD)
PD	F: n = 16 M: n = 14	58.1 (10.49) 65.7 (8.82)	10.4 (6.16) 11.0 (7.83)	28.3 (1.49) 28.0 (1.27)
HC	F: n = 8 M: n = 22	66.4 (7.71) 67.9 (7.47)	–	29.2 (1.13) 28.9 (1.07)

2.3 Stimuli

To address two main topics of the study, we created two sets of stimuli. The first set was created from a free speech task – the best possible representative of naturally occurring speech – and was elicited during a conversational interview with open-ended questions on a familiar topics such as hobbies, daily routines, family or prior jobs. An exception had to be made in a case of one speaker, as their free speech interview task was unavailable due to technical reasons, and we used their free speech picture description task instead. The second set of stimuli was from a prosody elicitation task focusing on sentence typing, as previous research has shown that speakers with HD have problems in both realization of and conveying the question intonation [9,17]. Stimuli for the second set consisted of syntactically identical phrases differing in question or statement intonation. All the speech samples were intensity normalized.

2.4 Procedure

Participants of the perception experiment completed a recognition task where they listened to the stimuli in two blocks corresponding to the two topics of interest in the study: recognition of HD from speech and recognition of sentence type produced by speakers with HD. In the first block, participants had to assess the healthiness of phrases by answering the question about speech healthiness and they had to pick one option from a list ("Did this voice sound healthy to you? – 1) Rather healthy, 2) Rather unhealthy, 3) I don't know") and then specify how confident they were ("1) Rather sure, 2) Rather unsure"). The second block had a similar setting, where listeners were asked to identify the sentence type of the phrase ("Did this phrase sound as question or statement? – 1) Question, 2) Statement, 3) I don't know") and specify how confident they were. The experiment was built in javascript with jsPsych library [3] and set up online by means of JATOS tool for online experimentation [8].

For every block, the procedure consisted of a short practice session of two stimuli and a main session. For the main sessions of the first and second blocks there were 60 and 120 (60 statements and 60 questions) stimuli that were randomized within their corresponding block for each participant. Each speech sample could be heard and responded to only once.

3   Results

We explored the results of two listening tasks by measuring relative unhealthiness scores, and statement and question recognition scores per speaker. These scores were calculated based on the numerical values (0 and 1) which represented answers from both listening tasks. The relative unhealthiness score was calculated by dividing the number of answers identifying a speaker as unhealthy by the total number of answers for that speaker; statement recognition scores and question recognition were calculated by dividing the number of correctly identified sentence type of one speaker by the total number of answers for that speaker. Figure 1 shows the extent to which each speaker in the first listening task was perceived by each group of listeners, suggesting that PD speakers were generally perceived as more unhealthy than HC speakers, who were mostly found healthy by all the participants. This trend is apparent in all listener groups, but with Czech group rating more HC speakers as unhealthy than the Dutch group.

The comparison between question recognition accuracy and statement recognition per speaker (Fig 2) clearly shows that statement recognition was a simpler task for all the participants, as the scores, and therefore the success rate for each speaker, are much higher than in question recognition. This is especially the case with Dutch experts, who correctly identified a statement in more than 40 out of 60 speakers in total. Similar to relative unhealthiness scores (Fig 1), Dutch listeners in general were more successful than Czech listeners.

To analyze the effect of expertise and L1 on accuracy of listener's recognition of HD speech and sentence types, we applied bootstrap resampling (R=1000) to estimate the classification accuracy between PD and HC groups as well as between sentence types expressed by statistics comparing distributions of answers: $\|mean(HC_{scores}) - mean(PD_{scores})\|/mean(SD(HC_{scores}), SD(PD_{scores}))$.

Thus, we measured the participants' ability to distinguish a speaker with PD from a HC speaker as a bootstrapped accuracy of recognition based on their answers in healthiness assessment task. A similar procedure was applied for the participants' ability to correctly recognize sentence type which was based on participants' answers in the sentence type task. Accuracy measures for both tasks were calculated with and without confidence used as a weight. As Levene's test did show a significantly different variance of the accuracy scores, we chose the non-parametric Welch's F test instead of one-way ANOVA followed up with Tukey post-hoc tests.

Results on HD Recognition from Voice. Overall, Dutch listeners (M(DN) = 1.6, M(DP) = 1.4) were more accurate in recognizing an unhealthy voice from a healthy voice than Czech listeners (M(CN) = 1.2, M(CP) = 1.0), and non-expert listeners (M(DN) = 1.6, M(CN) = 1.2) were more accurate than

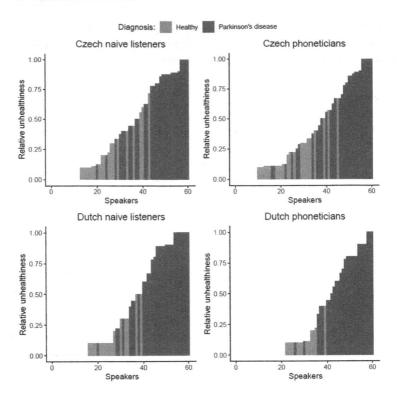

Fig. 1. Judgements of heathiness/unhealthuness presetned as relative rating for each speaker in an ordered fashion - from minimum (healthy) to maximum (unhealthy) for every listener group

expert listeners (M(DP) = 1.4, M(CP) = 1.0) in their respective languages. Specifically, Dutch non-expert listeners (M = 1.6) were the most accurate group, followed by Dutch experts (M = 1.4) and Czech non-expert listeners (M = 1.2); Czech experts (M = 1.0) were the least accurate group in unhealthiness recognition. The analysis of the confidence-weighted scores of unhealthiness recognition accuracy yielded very similar results to the ones without introduced weights. The only difference was that confidence appears to have boosted the answers of all four groups, the most for Dutch experts, whose mean grew from 1.4 to 1.6 and whose standard deviation decreased by 0.1 (while the standard deviations of the other groups remained the same).

We conducted Welch's F test to determine whether the differences in unhealthiness recognition accuracy scores were significantly related to the four groups of listeners. The test yielded significant differences, $F(3, 3996) = 3700.25$, $p < 0.001$, and an effect size $r = 0.74$. The Tukey post-hoc test showed that all the differences between all four groups were significant, $p < 0.001$, for all possible combinations. Welch's F test with confidence-weighted scores showed that the differences in weighed scores were also significant, $F(3, 3996) = 3396.93$,

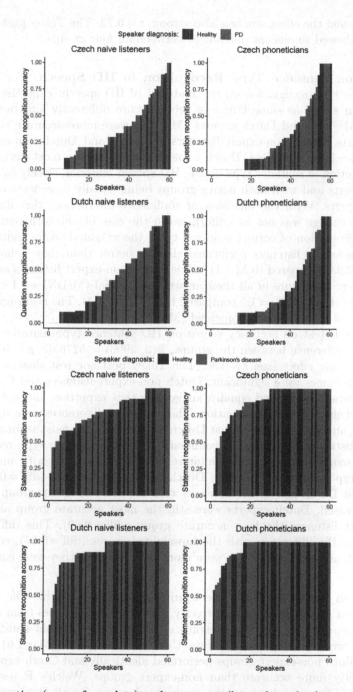

Fig. 2. Question (upper four plots) and statement (lower four plots) accuracy scores presented as relative rating for each speaker in an ordered fashion - from minimum (question) to maximum (statement) for every listener group

$p < 0.001$, and the effect size was also strong, $r = 0.72$. The Tukey post-hoc test similarly showed significant differences between all four groups.

Results on Sentence Type Recognition in HD Speech. For the task of sentence type recognition on the material of HD speech, expertise appears to have an effect, as same language groups score differently and both Czech experts ($M = 1.4$) and Dutch experts ($M = 1.6$) were more accurate than their counterparts, Czech non-expert listeners ($M = 1.0$) and Dutch non-expert listeners ($M = 1.0$). Therefore, Dutch experts performed the most accurately in the recognition of sentence type on the HD speech material, closely followed by Czech experts and with both native groups being equally inaccurate compared to the experts. With the inclusion of confidence interaction, the direction of the scores change was not as uniform as in the case of unhealthiness recognition. In recognition of correct sentence type, the weighted scores indicate that Czech non-expert listeners performed slightly better than they believed they did ($M = 0.93$, compared to $M = 1$), while Dutch non-expert listeners and Czech experts were more sure in all their answers in general ($M(DN) = 1.1$ compared to $M(DN) = 1.0$, $M(CP) = 1.5$ compared to $M(CP) = 1.4$). The accuracy and the confidence of Dutch experts coincided ($M = 16$).

Welch's F test on accuracy scores on HD sentence type material revealed significant difference between the groups, $F(3, 3996) = 5276.91$, $p < 0.001$, and the very strong effect size, $r = 0.80$. The Tukey post-hoc test showed that not all the differences were significant: Dutch non-expert listeners and Czech non-expert listeners performed equally, suggesting that expertise, and not the language, had effect in this combination. The remaining differences were significant, $p < 0.001$, and they confirmed that Dutch experts were the most accurate group. Same statistical calculations on confidence-weighted sentence type recognition accuracy scores demonstrated similar results. The difference with unweighted sentence type scores in Czech and Dutch non-expert listeners (diff = 0.19) suggested that Dutch non-expert listeners were the more confident group, as their score increased. Dutch experts were still the most accurate group and Czech non-expert listeners the least accurate group (diff = 0.69). This difference is larger than the difference with the unweighted scores (diff = 0.57), confirming that Czech non-expert listeners were more accurate than they were confident.

Results on Sentence Type Recognition in HC Speech. The measures on ability of all the listeners to correctly identify sentence types from a healthy voice (HC speech) showed that the only clear difference was that Dutch experts ($M = 2.2$) were much more accurate than Czech experts ($M = 1.6$) in their answers. Both non-expert groups performed identically and Czech experts were only slightly more accurate than non-expert groups. Welch's F test showed significant differences between the unweighted scores for all four groups, $F(3, 3996) = 5015.92$, $p < 0.001$, and a large effect size, $r = 0.79$. The Tukey post-hoc test showed that the effect is mostly due to the high accuracy score of Dutch experts, as group comparisons indicated high differences between them and other groups (diff(DE−CE) = 0.60, diff(DE−DN, DE−CN) = 0.68, $p < 0.001$). These results demonstrate that both familiarity with the language and phonetic expertise are important for accurate discrimination between sentence type in a healthy Dutch voice. Statistical calculations on confidence-weighted sentence type recognition accuracy scores demonstrated similar results.

4   Discussion

In this study we investigated how listeners perceive dysarthric speech of speakers with PD differently on the basis of familiarity with speakers' language and expertise. We indeed found that both language familiarity and expertise in speech sciences has a significant effect on the perception of HD speech. This effect was different for two speech perception tasks.

Overall, PD patients are perceived as much more unhealthy than healthy speakers, which is in line with described pathological symptoms due to prosodic insufficiency (first described by Darley et al. [2]). Statistical analysis revealed that the differences between four groups of listeners regarding their ability to accurately distinguish a dysarthric from a healthy voice were significant, demonstrating that both expertise and language affect the ability to recognize PD from a Dutch voice. Surprisingly, it was not experts who were able to accurately identify a PD patient from voice, which was expected due to their progressively trained perceptual awareness of dysphonic qualities [18]. Instead, both non-expert groups were significantly more accurate. The superiority of the non-expert groups in discriminating between healthy and PD speakers confirms the conclusions in [15,18], where authors found no differences between non-expert and expert groups of listeners, and of Van Borsel and Eeckhout [16], in whose study non-expert listeners rated stuttering speakers harsher than speech therapists. The finding that non-expert listeners are more sensitive to HD speech is perhaps due to a simplistic interpretation of the concept of healthiness on the part of the non-experts relative to the experts, who might assess healthiness with various strategies based on the specifics of their expertise. A similar conclusion was also reached by Walshe et al. [18]. The fact that Czech experts scored significantly worse than Dutch expert on HD recognition task may be due to different amounts of active experience in the field: while the Dutch expert sample consisted of university professors and assistant professors only, the Czech sample contained, besides university professors, also listeners who obtained an undergraduate degree in Phonetics and Phonology, but not all of whom were practicing professionals.

For the sentence type recognition task both language and expertise proved to be important factors. Both groups of expert were clearly more accurate than both non-expert groups, showing that expertise is a key factor for the ability to correctly identify sentence type in unhealthy voice. The leading role of phonetic expertise in sentence type identification in HD thus goes against Walshe et al. [18], Van Borsel & Eeckhout [16], and Smith et al. [15], in which studies experts performed identically, or even worse than non-expert listeners. This confirms that task type is an important factor [9]. It is interesting to see that in sentence type identification in the voices of healthy speakers, Dutch experts were the most accurate, while three other groups performed on a same level of accuracy. It appears that both having Dutch as L1 and phonetic expertise are beneficial to process sentence types in a healthy Dutch voice correctly. Such a difference with results on HD speech is surprising, as intonation – a primary prosodic manifestation of sentence type – is usually found to be rather problematic for PD patients when compared to healthy population. However, it is clear that recognition of healthy sentence types was easier. One possible explanation for this is that familiarity with the language of the stimuli had more crucial role amongst the experts. Presumably, Dutch native expert participants are more sensitive to language-specific speech disturbances than non-native expert participants. These findings highlight the need for more research into the language specific and language universal apects of HD.

5   Conclusion

With this study we demonstrated that both expertise and familiarity with the speakers' language are important factors in HD perception, and that differences in identification accuracy further depend on task type. There are acoustic cues that trigger recognition of PD speech as "unhealthy" in the non-expert listeners, while experience with phonetics may lead to a shifted focus of such ques. Both Czech and Dutch expert groups outperform non-expert groups in sentence type recognition, indicating the added value of phonetic experience for the prosodic task. The finding that non-expert listeners are more sensitive to HD speech based on healthiness perception has an implication of importance of their instincts regarding changes in speech, which holds a potential to contributing to an early detection of the disease.

Acknowledgments.. We thank Lea Busweiler, research assistant, for the help in the data collection. We also thank all the speakers and listeners who volunteered to participate in our study.

References

1. Brabenec, L., Mekyska, J., Galaz, Z., Rektorova, I.: Speech disorders in Parkinson's disease: early diagnostics and effects of medication and brain stimulation. J. Neural Transm. **124**(3), 303–334 (2017)
2. Darley, F.L., Aronson, A.E., Brown, J.R.: Clusters of deviant speech dimensions in the dysarthrias. J. Speech Hear. Res. **12**(3), 462–496 (1969)
3. De Leeuw, J.R.: jsPsych: a Javascript library for creating behavioral experiments in a web browser. Behav. Res. Methods **47**(1), 1–12 (2015)
4. Fonville, S., Van Der Worp, H., Maat, P., Aldenhoven, M., Algra, A., Van Gijn, J.: Accuracy and inter-observer variation in the classification of dysarthria from speech recordings. J. Neurol. **255**(10), 1545–1548 (2008)

5. Van der Graaff, M., et al.: Clinical identification of dysarthria types among neurologists, residents in neurology and speech therapists. Eur. Neurol. **61**(5), 295–300 (2009)
6. Hazan, H., Hilu, D., Manevitz, L., Ramig, L.O., Sapir, S.: Early diagnosis of Parkinson's disease via machine learning on speech data. In: 2012 IEEE 27th Convention of Electrical and Electronics Engineers in Israel, pp. 1–4. IEEE (2012)
7. den Hollander, J., Coler, M., Verkhodanova, V., Timmermans, S.: Test handleiding dysarthria project (2017)
8. Lange, K., Kühn, S., Filevich, E.: Just another tool for online studies (JATOS): an easy solution for setup and management of web servers supporting online studies. PloS one **10**(6), 14 p. (2015)
9. Martens, H., Van Nuffelen, G., Cras, P., Pickut, B., De Letter, M., De Bodt, M.: Assessment of prosodic communicative efficiency in Parkinson's disease as judged by professional listeners. Parkinson's Dis.**2011**, 10 p. (2011)
10. Mekyska, J., Smékal, Z., Košt'álová, M., Mračková, M., Skutilová, S., Rektorová, I.: Motorické aspekty poruch řeči u parkinsonovy nemoci a jejich hodnocení. Česká a slovenská neurologie a neurochirurgie (6), 662–668 (2011)
11. Miller, N., Noble, E., Jones, D., Allcock, L., Burn, D.J.: How do I sound to me? Perceived changes in communication in Parkinson's disease. Clin. Rehabil. **22**(1), 14–22 (2008)
12. Orozco-Arroyave, J.R., et al.: Towards an automatic monitoring of the neurological state of Parkinson's patients from speech. In: 2016 IEEE International Conference on Acoustics, Speech and Signal Processing (ICASSP), pp. 6490–6494. IEEE (2016)
13. Pell, M.D., Cheang, H.S., Leonard, C.L.: The impact of Parkinson's disease on vocal-prosodic communication from the perspective of listeners. Brain Lang. **97**(2), 123–134 (2006)
14. Shadden, B.B., Hagstrom, F., Koski, P.R.: Neurogenic Communication Disorders: Life Stories and the Narrative Self. Plural Publishing Inc., San Diego (2008)
15. Smith, C.H., et al.: Rating the intelligibility of dysarthic speech amongst people with Parkinson's disease: a comparison of trained and untrained listeners. Clin. Linguist. Phonetics **33**(10–11), 1063–1070 (2019)
16. Van Borsel, J., Eeckhout, H.: The speech naturalness of people who stutter speaking under delayed auditory feedback as perceived by different groups of listeners. J. Fluen. Disord. **33**(3), 241–251 (2008)
17. Verkhodanova, V., Timmermans, S., Coler, M., Jonkers, R., de Jong, B., Lowie, W.: How dysarthric prosody impacts naïve listeners' recognition. In: Salah, A.A., Karpov, A., Potapova, R. (eds.) SPECOM 2019. LNCS (LNAI), vol. 11658, pp. 510–519. Springer, Cham (2019). https://doi.org/10.1007/978-3-030-26061-3_52
18. Walshe, M., Miller, N., Leahy, M., Murray, A.: Intelligibility of dysarthric speech: perceptions of speakers and listeners. Int. J. Lang. Commun. Disord. **43**(6), 633–648 (2008)

Phonological Length of L2 Czech Speakers' Vowels in Ambiguous Contexts as Perceived by L1 Listeners

Jitka Veroňková$^{(\boxtimes)}$ and Tomáš Bořil

Institute of Phonetics, Charles University, Prague, Czech Republic
{jitka.veronkova,tomas.boril}@ff.cuni.cz

Abstract. The paper focuses on the vowel length of non-native speakers' Czech and their perception by native speakers. Due to its phonological status, the length of vowels in Czech is an important sound feature. Its improper realization can result in communication breakdown. From the Czech read speech of 8 Russian and Ukrainian female speakers, a perception test was created: 78 items consisting of 5 pairs and 1 triad of the same sentences that differed only in the target word. These two-syllable words were distinguished by a combination of short/long vowels, e.g., /lanu/ – /la:nu/, /la:nu/ – /la:nu:/. The L1 Czech listeners rated the degree of the foreign accent of the items and intelligibility of the target words. The agreement of listeners with the speakers' intent is evaluated, and the types of substitutions are analyzed in particular with respect to the combination of short/long vowel and the position of a stressed/unstressed syllable. The vowel duration and the formants of i-vowels were measured. In the second perception experiment, durations of both vowels in /lanu/ were manipulated in the range from 60 ms to 240 ms and native Czech speakers rated their perceived length.

Keywords: Czech as L2 · Vowel length · Duration · Formants · Intelligibility · Foreign accent · Perception

1 Introduction

One of the foundations for successful communication in a foreign language is an appropriate acquisition of target language sound features. Vowel length belongs among the segmental phenomena that cause difficulty in production and perception for speakers of Czech as L2 [1, 2].

Even languages from one language group may differ in the treatment of vowel length. Slavic languages, including Czech, may serve as an example. For instance, Croatian uses length as one of its distinctive features, while Polish vowel system contains just short vowels, unlike, e.g., Russian and Ukrainian, where variability in vowel quantity is present; however, it does not have phonological status; it is a feature governed by the word stress (similarly to, e.g., English or German) [3].

© Springer Nature Switzerland AG 2020
A. Karpov and R. Potapova (Eds.): SPECOM 2020, LNAI 12335, pp. 624–635, 2020.
https://doi.org/10.1007/978-3-030-60276-5_60

1.1 Vowel Length in Czech

In Czech, the vowel length is phonological. Two grades – short and long – are discriminated, and in the original vocabulary or grammatical endings, the long vowel is consistently marked in orthography. Vowel length distinguishes either lexemes, e.g., *dráha* (track; lane) – *drahá* (expensive; dear), or grammatical forms, e.g., *pracovat* (to work): *pracuji* ([I] work) – *pracují* ([they] work). Its incorrect pronunciation may result in difficulties understanding the speech or even change the meaning completely. Vowel length is fully independent of word stress in Czech, which is fixed on the first syllable of a word. The stressed syllable can be long or short with long syllables being not limited to a specific position within a word. Individual words may contain more long vowels or no long vowels at all.

In Czech, phonological length is acoustically manifested mainly in vowel duration; the latest data shows long/short duration ratio of 1.6 to 1.8 [4]. I-vowels are an exception with their vowel quality also contributing to the differentiation between short and long variants (short [ɪ] is more open than long [iː]), especially in the Bohemian part of the Czech Republic [5–7] and the ratio is lower (around 1.3) [4]. According to recent studies, spectral cue can contribute to the distinction between short and long variants in case of high back vowels as well [7].

Vowel duration is influenced by a variety of factors, such as the vocal quality itself (high vowels tend to be shorter than low ones), word stress (this factor is not applicable to Czech in such an extent), the phonetic neighborhood, position in the sentence, speech rate, etc., cf. [8–10]. Openness/closeness of syllable was recognized as an important factor for Czech [11].

1.2 Vowel Length for L2 Czech Speakers

L2 speakers can use different vowel length patterns because of negative transfer from their mother tongue or other foreign languages to Czech as the target language [12]. L1 listeners repeatedly report errors in vowel length in the speech of Czech L2 speakers as disturbing and contributing to foreign accent. In [13], the native Czech listeners evaluated the acceptability of L2 Czech speakers' speech (L1 Russian) and then specified the phonetic features that had influenced their evaluation. Approximately two-thirds of the comments referred to the segmental level, with one-third of that number to vowels. As for vowels, twice as many comments were about their length.

As follows from perception analyses, a widespread error of L2 Czech speakers is mixing of length and word stress, especially improper lengthening of the canonically short stressed syllable [2, 14] (L1 Russian); the difficulty for L2 Czech speakers lies in estimating adequate vowel duration – the canonically long vowels are often realized as half-long [2] (L1 Russian); the pronunciation of two adjacent long vowels is difficult [15] (L1 Polish). Errors in the production of vowel length are attested for advanced speakers as well [2, 13, 15].

The focus of this paper is the vowel length of L2 Czech speakers with Russian and Ukrainian mother tongues as perceived by L1 Czech listeners. Four novel aspects compared to the previous studies are presented. Unlike previous studies, this paper seeks to limit the role of content when perceiving the length, therefore an ambiguous context was chosen (experiment 1) or the stimuli were tested without context (experiment 2); measurement of vowel duration and partially spectral analysis was performed; vowel duration in the stimuli was manipulated (experiment 2); in both experiments, perception tests were administered.

Four main dimensions according to which L2 speech may be characterized are attested: foreign accent, comprehensibility, intelligibility and fluency [16]. The perceptual task in our experiments corresponds to intelligibility, defined here as "the extent to which a speaker's message is actually understood by a listener" [17]. This concept is not identical with the overall understanding of the content [18]. Apart from intelligibility, the degree of foreign accent was also examined. Testing both of these dimensions is another contribution of this paper.

We would like to thank anonymous reviewers for their helpful comments on earlier versions of this paper.

2 Method

2.1 Experiment 1

Material. For the purpose of broader research concerning Czech as L2, a database of L2 Czech speakers' recordings was acquired. Recordings were taken in a sound-treated and sound-proofed room (AKG C 4500 B-BC microphone, sample rate 48 kHz, 16-bit depth). From the collection of database texts, 13 sentences focusing on vowel duration in ambiguous context were selected (see Table 1): 5 pairs and 1 triad of the same sentences differed only in the target word. These target words were distinguished by a combination of short (S)/long (L) vowels. The target words were disyllabic, and their structure was CVCV, so all the vowels occurred in open syllables to eliminate the factor of syllable openness/closeness. No ambiguity in their correct pronunciation in standard Czech has been attested. All the remaining words in the sentence (and even the preceding sentence) contained only short vowels so that the linguistic surrounding does not affect the S/L pattern of the target word pronunciation. The target word was placed in the middle of the sentence in order to reduce the potential influence of the phrase juncture. In the text for recording, the sentences were mixed among others to mask the target sound phenomena.

Speakers. A group of speakers consisted of 8 female speakers; L1: 4 Russians, 3 Ukrainians, 1 Ukrainian/Russian. Age: 18–37; Czech proficiency: level B1–C1 according to CEFR [19] (students of Czech study programme, of a non-linguistic programme and of Czech language courses at Charles University, Prague).

Perceptual Test. The 13 sentences (see Table 1) formed the basis of the perception experiment. Some of their realizations were excluded because of slips of tongue. Finally, six realizations of each sentence were used in the experiment, i.e., 78 items in total (13 sentences, 6 realizations each; 9–10 items of each speaker), as a compromise between a sufficient number of items and a reasonable length for listeners.

Two sections of the experiment were created and, in each section, the same set of the 78 items was tested. In the first section, the listeners determined the degree of foreign accent on a 7-point symmetrical scale for each item (1 – no foreign accent, 4 – middle foreign accent, 7 – completely foreign). In the second section, they determined on the same set of 78 items, which version they heard: On the screen, they saw four variants, i.e., S/L combinations of a target word written in a phonological transcription, e.g., /viru/, /viru:/, /vi:ru/, /vi:ru:/. Phonological transcription was used intentionally instead of orthography to support the focus of listeners on the segmental intelligibility task, not on the content of the sentence. [18] For the same reason, all four versions of S/L combinations were used even if they do not appear in Czech (e.g., /lanu:/) or if the form does not fit lexically or grammatically in the carrier sentence.

Table 1. List of sentences in perception experiment.

(1) *Tyhle* **valy** */* **vály** *ze dřeva budily pozornost.* (SS–LS)
These wooden *pastry boards / mounds* attracted attention.
(2) *Podle něj se chlapci* **myli** */* **mýli** *poměrně často.* (SS–LL)
In his view, boys *washed themselves / are wrong* quite often.
(3) *K jednomu* **lanu** */* **lánu** *poslali dobropis.* (SS–LS)
To one *rope / large field*, they sent a credit note.
(4) *Sklizeň* **lánu** */* **lánů** *trvala až do večera.* (LS–LL)
Harvesting *the large field / fields* lasted until the evening.
(5) *Text o vlastnostech* **viru** */* **virů** */* **výrů** *publikovala v časopise.* (SS–SL–LL)
In a journal, she published a text about attributes of *a virus / viruses / eagle owls*
(6) *Uprostřed* **víru** */* **vírů** *plavaly rybky.* (LS–LL)
Fish swam in the middle of *a swirl / swirls*.
Note: 1) *ů* is just a graphic symbol, the sound is the same as with *ú* /u:/. 2) In Czech, there is no difference between *i/y*, *í/ý* pronunciation (unlike [i]/[ɨ] in Polish, Russian and Ukrainian).

The experiment was designed using Praat MFC (multiple forced-choice) environment [21]. Each section was preceded by a training part containing 5 items. Within each section, items were played in a random order, divided by a short break with a piece of music after every 20 items. In the first step, the accent section was conducted, and after a break, the intelligibility section followed. It was possible to replay each item up to three times.

Listeners. A group of listeners comprised 13 native listeners, students of the Phonetic programme (Faculty of Arts, Charles University, Prague). The experiment was conducted individually or in a small group; the listeners used headphones.

Acoustic Analysis. Perception analysis was supplemented by acoustic measurement of vowel duration. The items were manually labelled according to the rules set out in [20] using Praat software [21]. Normalized vowel durations were obtained as follows: For each target word, articulation rate was measured in syll/s. A mean articulation rate of each speaker was calculated from these values. Normalized vowel duration was computed by multiplication of the real duration by the mean articulation rate of the speaker and divided by total mean articulation rate of all speakers together. This procedure makes it possible to compare vowel duration among speakers.

As stated in 1.1, the real duration of i-vowels may not be the only or the main cue of perceived vowel length for native Czech. For these vowels, mean F1 and F2 formant frequencies in the middle third of their duration were analyzed using Burg method, looking for five formants in the range of maximum frequency of 5500 Hz in Praat [21]. These values were manually corrected according to visual inspection of spectrogram and audio listening to avoid nasal formants mismatch and other possible errors of the automatic algorithm.

2.2   Experiment 2

To evaluate the effect of vowel duration on perceived length, a perception experiment consisting of words /lanu/ with manipulated vowel durations was created. Both vowels were manipulated to five durations (60, 85, 120, 170, and 240 ms) yielding 25 combinations (five /a/-durations x five /u/-durations). The manipulations were performed in Praat [21] using the pitch-synchronized overlap-and-add method. Words by two speakers from the material of experiment 1 were used, thus the total number of items-of-interest was 50. Additional 18 non-manipulated two-syllable words with mixed short and long vowels were included into the set as distractors. The perception experiment was evaluated by 17 native Czech listeners; items introduced by a short desensitization beep sound were played in random order for each subject in Praat MFC environment [21] with 4 possible answers combining short/long choice for both vowels.

3   Results

3.1   Assessment of Foreign Accent and Intelligibility

Concerning the question of accent, listeners used all the points from the scale including the extreme points in certain utterances. From all the single judgements, a mean value of individual speaker's accent was calculated. This mean accentedness ranged from 2 (very weak accent, 2 speakers) to 5 (strong accent, 4 speakers), i.e., nobody sounded accentless or with a very strong accent or even completely foreign.

Speakers also differed in terms of intelligibility of target words. The frequency of the listener marking the target word in accordance with the speaker's intent was counted, and, in case of disagreement, the types of patterns involved were analyzed (for the relation between accent and intelligibility see Fig. 1). All the listeners' judgments are grouped by the accent mark, and for each group, a mean value of intelligibility is evaluated. With a higher degree of accent, the level of intelligibility decreases.

However, the relationship is not straightforward, as the assessment concerning single speakers shows. There was one speaker with a very high intelligibility (91%) and a very weak accent; the assessment of the two parameters is thus in accordance. On the contrary, another speaker with a very weak accent showed a discrepancy. Intelligibility of this speaker was much lower, around 50%. Three more speakers achieved similar intelligibility, but their accent was more perceptible, ranging from weak to strong. Concerning three speakers (all with a strong accent), the listeners identified around a third of target words correctly. This finding is in accordance with [16:1–6] that intelligibility and accentedness are partially independent.

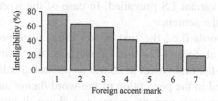

Fig. 1. The relation between foreign accent and intelligibility on the base of all the judgements.

3.2   Assessment of Individual Target Words

Regarding the segments, 49.2% of vowels were perceived in accordance with the speakers' intent, and 50.8% incorrectly, of which 42.6% were originally short and 57.4% were originally long. Regarding the target words, 47.3% of perceived variants corresponded to the canonical form, while 52.7% differed. Tables 2a)–d) show the perception of target words ordered according to their original S/L pattern.

Table 2 Intelligibility of target words: (a) SS, (b) LS, (c) LL, (d) SL. Columns: perceived variants – number of assessments. Bold numbers with asterisk: correct match, bold without asterisk: mismatch that fits into the sentence, plain: mismatch that does not fit into the sentence (semantic or grammar problem or the word does not exist).

(a)	SS	SL	LS	LL
valy /vali/	**56***	1	20	1
myli /mili/	**45***	5	24	4
lanu /lanu/	**30***	3	44	1
viru /viru/	**11***	12	48	7
Sum	142	21	136	13
%	45.5	6.7	43.6	4.2

(b)	SS	SL	LS	LL
vály /vaːli/	29	0	**48***	1
lánu /laːnu/	15	0	**63***	0
lánu /laːnu/	9	2	**61***	6
víru /viːru/	2	5	**54***	17
Sum	55	7	226	24
%	17.6	2.2	72.4	7.7

(c)	SS	SL	LS	LL
mýlí /miːliː/	30	5	37	**6***
lánů /laːnuː/	8	2	46	**22***
výrů /viːruː/	8	20	22	**28***
vírů /viːruː/	0	6	33	**39***
Sum	46	33	138	95
%	14.7	10.6	44.2	30.4

(d)	SS	SL	LS	LL
virů /viruː/	2	**17***	41	18
%	2.6	21.8	52.6	23.1

Among the realizations of the original SS words (i.e., both short vowels), two patterns prevailed. The SS pattern (following the original form), and the LS pattern (the first, i.e., stressed, syllable long). In total, both variants are perceived equally (with 45.5% for the former and 43.6% for the latter). Concerning single items, two words – /vali/ and /mili/ – were identified with greater certainty, as opposed to /lanu/ and /viru/, in which the incorrect variant LS prevailed. In case of the word /mili/, the variant LS does not even fit into the sentence.

The original LS words (i.e., the first vowel long) were predominantly perceived following the canonical form (72.4%). Concerning the word /vaːli/, the listeners evaluated the length of the first vowel in 25.6% as phonologically short (SS). The same SS pattern also occurred in the perception of the word /laːnu/ and the LL pattern in the word /viːru/. All the incorrect variants mentioned above fit into the sentence.

Comparing the original LL words (i.e., both vowels long) to the SS and LS words mentioned above, the layout of the perceived variants was somewhat more varied. In total, the highest score was achieved in the case of the LS pattern (i.e., with the second vowel short), although the predominance was not that apparent (44.2%). This LS pattern was also found in the word /miːliː/, although it did not fit into the sentence. The second most frequently used pattern was the agreement with the speaker's intent (LL pattern), however, its value was not so considerable (30.4%). Additionally, in the perception of the word /miːliː/, the LL pattern occurred in small numbers; apart from the LS pattern, this word was perceived as SS, i.e., with both vowels short. The short vowel in the first syllable also occurred in the word /viːruː/ with the orthographic form *výrů*.

The SL pattern was tested on just one target word. The dominant variant perceived is the LS pattern (52.6%). Compared to this one, the SL pattern, which follows the speaker's intent, did not reach even half the LS size of the agreement, similarly to the LL pattern.

The most frequent pattern among the variants perceived in the disagreement with the speaker's intent was the LS pattern (59.0%). The exchanges occurred even in the items assessed by some listeners as accentless: SS>LS (lengthening of the first stressed syllable): *lanu>lánu* (62%), *viru>víru* (56%) – 5 judgements indicating the item as accentless; LL>LS (shortening of the last syllable, difficulties pronouncing two adjacent long vowels): *lánů>lánu* (59%), *mýlí>míli* (47%), *vírů>víru* (42%) – 12 judgements indicating the item as accentless; SL>LS (shift of a long vowel to the first, stressed syllable): *virů>víru* (53%); 4 judgements.

These findings are consistent with Gersamia's data [14], who analyzed the realization of three-syllable words with different S/L patterns in Czech read texts of L1 Russian speakers. The most common patterns used by the speakers were LSS and SLS patterns, i.e. patterns with only one long vowel that was mostly associated with word stress (in the SLS pattern, speakers predominantly produced word stress on the second syllable). The patterns containing two adjacent syllables were produced by speakers very rarely.

A question may be asked whether the distribution of patterns (Table 2) is not affected by the frequency of single variants in the Czech corpus. In the spoken corpus [22], which is less extensive, the absolute frequency of the variants involved is very low, so we chose the written corpus [23]. Due to the extensive word homophony and homonymy of endings within the lexeme, it is not easy to obtain comparable data. However, it is possible to make the following assertions:

Of the words /vali/ and /va:li/, which meet the grammatical and lexical requirements of the SS carrier sentence, the word /vali/ is significantly more frequent; here (Table 2a), therefore, the perception of the correct SS pattern could be supported by frequency. In case of the words /lanu/ and /la:nu/ (dative case that fits into the SS carrier sentence), the SS word /lanu/ is more frequent, but the LS pattern prevails in the perception of the SS word /lanu/ (Table 2a), unlike in the previous case.

The SS word /mili/ is also common (*myli* – instances per million (imp) 0.49), but the LL word /mi:li:/ (in written form *mýlí*) has the highest frequency of all words used in the experiment (*mýlí* – imp 3.29). [23] However, the perception of this LL word is very low (see Table 2a and 2c). We believe that the vowel quality plays a role especially in the first syllable (see Sect. 3.3). The LS pattern /mi:li/ was perceived in a significant volume instead of the SS pattern; this LS word exists in Czech, but does not fit into the SS carrier sentence. This observation is consistent with the assumption that L2 Czech speakers (L1 Russian and Ukrainian) tend to lengthen a short stressed syllable. (This assessment is also a proof that the listeners performed the task as expected, i.e., that they focused on the intelligibility of the target words, rather than on the content of the whole sentence).

In addition to the last word /viru/, all the other three words fit into the SS carrier sentence (see Table 2a). Regarding the frequency, apart from the LL word /vi:ru:/ with lower (*vírů*) and very low (*výrů*) frequency, the frequency of other words (SS, LS, SL) suitable for carrier sentences is higher and comparable (imp 2.01–2.53) [23]. However, in the perception of the SS carrier sentence, a clear preference for the LS pattern can be observed (see also Table 2b and 2d). The quality of i-vowels may influence perception here as well.

From the above observations, we believe that the frequency of variants in the corpus does not play a large role in our experiment.

3.3 Objective Measurements

Figures 2a and 2b depict distribution of normalized vowel durations in target words with respect to the vowel length according to the original text, i.e., the L2 speakers' intent. The measured data is divided into (a) the so called non-i-vowels (i.e., all the vowels except i-vowels) and (b) i-vowels, as considerable differences between these groups were expected (see 1.2). Additionally, within i-vowels, the data is split according to the letter *i/í* or *y/ý* used in text because, as opposed to L1 Czech speakers, differences in L2 Czech speakers' pronunciation may be expected. The position of vowels within a word (first syllable, second syllable) is distinguished.

Figures 2c, 2d depict distribution of normalized vowel durations and their division into S/L groups as perceived by L1 Czech listeners. The data consists of all the listeners' single judgements. The non-i-vowels/i-vowels division and the position within a word are considered.

The distributions of duration with respect to the two aforementioned aspects, i.e., the L2 speakers' intent and L1 Czech listeners' perception, are compared. Distributions of non- i-vowel duration in the first syllable (Fig. 2a and Fig. 2c, top) indicate two visible peaks distinguishing S and L vowels in both groups, i.e., L2 speakers and L1 listeners; however the overlap of S and L vowels is lower in L1 Czech listeners (Fig. 2c). Second syllable vowel duration as perceived by L1 Czech listeners (Fig. 2c, bottom) features a similar pattern, although with a greater overlap; the peak of short vowels is shifted to lower durations by comparison with the first syllable, i.e., the duration of short vowels in the second syllable is lower than in the first one. The S and L durations in the production of L2 speakers (Fig. 2a, bottom) overlap substantially.

In the distribution of i-vowel (i, í) duration (see Fig. 2b), there are visible overlaps in L2 Czech speakers, however, one obvious peak indicating S vowels is present, both in the first and second syllables, opposite to L vowels appearing especially in the second syllable. On the contrary, the distribution of the duration of y/ý-words contains two clearly distinguished peaks for both S and L vowels in the first syllable, although its extension does not cover the entire range, i.e., their duration is lower. The originally S vowel y is extended as a band over the entire range. The distribution of i-sound in the second syllable, as perceived by L1 Czech listeners, is also outlined as a band with only some traces of hills. Regarding the i-sound in the first syllable, peaks distinguishing S and L vowels are clearly recognizable, but they are close together with a large overlap, as opposed to the first syllable of non- i-sound in L1 Czech perception.

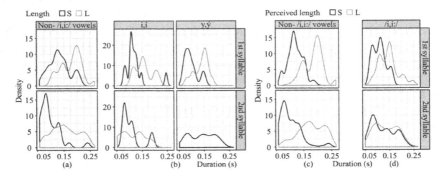

Fig. 2. (a), (b) Distribution of normalized vowel durations realised by L2 speakers split into S/L groups as written in the original text (i-vowels distinguished by the letter used in text). (c), (d) Distribution of normalized vowel durations split into S/L groups as perceived by L1 Czech listeners (without additional division according to the perception experiment setup).

The duration values of short/long vowels overlap is in accordance with the finding [2] that the canonically long vowels are often realized as half-long in Czech of L1 Russian speakers. L1 Czech listeners identified the half-length in both cases: as lengthening of a canonically short vowel and as the insufficient length of a canonically long vowel. For native Czechs, the clearer perceptual distinction between short and long vowels is in line with the findings of [8], who found that listeners with a mother tongue in which the vowel length is phonological have a more categorical perception of this phenomenon than speakers with a mother tongue containing long vowels but without a phonological status (similar finding in [1]).

For each speaker, the differences of F1 and F2 between short /i/ and long /i:/ were evaluated. With respect to the vowel length in the text, L2 speakers' mean difference of F2 (/i:/ minus /i/) is 116.8 ± 114 Hz (significance level α = 0.05), and difference of F1 is −37.7 ± 27.8 Hz (i.e., for /i:/, F2 is increased and F1 is decreased according to /i/, with one boundary of confidence interval very close to 0 Hz in both F1 and F2).

Differences of L1 Czech listeners' split between the short and long perception of i-vowels are as follows: L2 speakers' mean difference of F2 (/i:/ minus /i/) is 270.0 ± 115.7 Hz, and difference of F1 is −47.9 ± 34.1 Hz. The relatively frequent shift of the original *mýlí>myli* or *výrů>virů* can be explained by the possible insufficient quality of the first vowel regardless of its duration.

Figure 3 depicts durations of manipulated vowels in perception experiment 2 identified as short or long by L1 listeners. The 50% intervals of values for short and long vowels are clearly different and, moreover, they do not approach each other at all. This again supports the categorical perception mentioned above. As for the first and second syllables, the 50% intervals show no difference; the mean values for long vowels are only shifted further from each other. This corresponds to the fact that in Czech difference in length between stressed and unstressed syllables does not exist. These findings are confirmed also by the duration ratios: the mean ratio of the phonologically same vowels regarding their length is identical and is equal to 1; the long:short vowels mean ratio is identical regardless of their order within a word (SL and LS patterns) and is equal to 2.

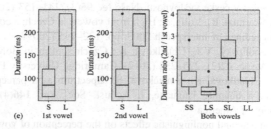

Fig. 3. Durations and duration ratios of short and long vowels in perception experiment 2 as evaluated by L1 Czech listeners.

4 Conclusions

In this paper, the perception of vowel length in L1 Czech listeners based on L2 Czech speech (L1 Russian and Ukrainian) was examined. In ambiguous contexts, the disyllabic target words with controlled S/L pattern (four variants) were miscomprehended in 52.7% assessments. The confusion appeared even in the items evaluated as accentless. The LS pattern showed the greatest agreement between L2 speakers' intent and L1 listeners (72.4%); the lowest agreement was achieved in SL and LL patterns. It was the LS pattern that was the most frequent pattern indicated within the judgements not corresponding to the L2 speakers' intent (59.0%). This most preferable LS pattern is in accordance with the generally observed tendency of Russian and Ukrainian speakers to produce a long vowel in the stressed syllable, even in the position of canonically short vowel.

The common confusion of short and long vowels in L2 speakers was confirmed by duration measurement as well; the values showed considerable overlap. The data indicates that categorical perception is typical of Czech speakers and the duration of vowels is an essential factor in determining phonological length; however, the perception of vowel length may be influenced by other features, such as quality of vowels.

Acknowledgements. This research was supported by the Czech Science Foundation project No. 18-18300S "Phonetic properties of Czech in non-native and native speakers' communication".

References

1. Janota, P., Palková, Z.: Testing perceptive and productive skills in language learning. In: Romportl, M., Janota, P. (eds.) Acta Universitatis Carolinae, Philologica 3, Phonetica Pragensia V, pp. 15–27. Charles University, Prague (1976)
2. Ramasheuskaya, K.: Specifika češtiny ruských studentů (se zaměřením na vybrané fonetické a morfosyntaktické jevy). Ph.D. thesis. Charles University, Prague (2014)
3. Sawicka, I. (ed.): Komparacja systemów i funkcjonowania współczesnych języków słowiańskich 2. Fonetyka/ Fonologia. Uniwersytet Opolski, Opole (2007)
4. Skarnitzl, R.: Dvojí i v české výslovnosti. Naše řeč **95**(3), 141–153 (2012)
5. Podlipský, V.J., Skarnitzl, R., Volín, J.: High front vowels in Czech: a contrast in quantity or quality?. In: Proceedings of Interspeech 2009, pp. 132–135 (2009)
6. Šimáčková, Š., Podlipský, V.J., Chládková, K.: Czech spoken in Bohemia and Moravia. J. Int. Phonetic Assoc. **42**(2), 225–232 (2012). Cambridge University Press, Cambridge
7. Podlipský, V.J., Chládková, K., Šimáčková, Š.: Spectrum as a perceptual cue to vowel length in Czech, a quantity language. J. Acoust. Soc. Am. **146**(4), 352–357 (2019). Acoustical Society of America
8. Keating, P.: Linguistic and nonlinguistic effects on the perception of vowel duration. UCLA Working Papers in Phonetics, vol. 60, pp. 20–40 (1985)
9. van Santen, J.P.H.: Contextual effects on vowel duration. Speech Commun. **11**(6), 513–546 (1992)
10. Rosner, B.S., Pickering, J.B.: Vowel perception and production. Oxford Psychology Series 23. Oxford University Press, Oxford (1994)
11. Janota, P., Jančák, P.: An investigation of Czech vowel quantity by means of listening tests. In: Romportl, M., Janota, P. (eds.) Acta Universitatis Carolinae, Philologica 1, Phonetica Pragensia II., pp. 31–68. Charles University, Prague (1970)
12. Escudero, P., Boersma, P.: Bridging the gap between L2 speech perception research and phonological theory. Stud. Second Lang. Acquisition **26**, 551–585 (2004)
13. Romaševská, K., Veroňková, J.: How Czech speech of Russian-speaking learners is perceived by native speakers of Czech and its correlation with age factors and language competence. In: Besters-Dilger, J., Gladkova, H. (eds.) Second Language Acquisition in Complex Linguistic Enviroments, Russian Native Speakers Acquiring Standard and Non-Standard Varieties of German and Czech, pp. 147–176. Peter Lang, Frankfurt am Main (2016)

14. Gersamia, G.: Kvantita tříslabičných slov v českých projevech nerodilých mluvčích. Na základě nahrávek rusky mluvících respondentů. Bachelors thesis, Charles University, Prague (2017)
15. Veroňková, J., Bořil, T., Palková, Z., Poukarová, P.: Délka českých samohlásek u polských mluvčích v taktech s různou strukturou kvantity. In: Bogoczová, I. a kol. AREA SLAVICA 3. (Jazyk na hranici – hranice v jazyku), pp. 51–61. Ostravská univerzita, Ostrava (2020)
16. Derwing, T.M., Munro, M.J.: Pronunciation fundamentals. Evidence-based Perspectives for L2 Teaching and Research. John Benjamins Publishing Company, Amsterdam (2015)
17. Munro, M.J., Derwing, T.M.: Foreign accent, comprehensibility, and intelligibility in the speech of second language learners. Lang. Learn. 49(1), 73–97 (1995)
18. Thomson, R.: Measurement of accentedness, intelligibility, and comprehensibility. In: Kang, O., Ginther, A. (eds.) Assessment in Second Language Pronunciation, pp. 11–29 (2018)
19. Common European Framework of Reference for Languages: Learning, teaching, assessment (CEFR). https://www.coe.int/en/web/common-european-framework-reference-languages. Accessed 10 June 2020
20. Machač, P., Skarnitzl, R.: Principles of Phonetic Segmentation. Nakladatelství Epocha, Praha (2009)
21. Boersma, P., Weenink, D.: Praat: doing phonetics by computer [Computer program], version 6.0.25 (2019)
22. Kopřivová, M., et al.: ORAL: korpus neformální mluvené češtiny, version 1 (2. 6. 2017). Ústav Českého národního korpusu FF UK, Praha (2017). http://www.korpus.cz
23. Křen, M., et al.: SYN2015: reprezentativní korpus psané češtiny. Ústav Českého národního korpusu FF UK, Praha (2015). http://www.korpus.cz

Learning an Unsupervised and Interpretable Representation of Emotion from Speech

Siwei Wang[(✉)], Catherine Soladié, and Renaud Séguier

FAST Research Team, CentraleSupélec, IETR, 6164 Rennes, France
{siwei.wang,catherine.soladie,renaud.seguier}@centralesupelec.fr

Abstract. One of the severe obstacles to naturalistic human affective computing is that emotions are complex constructs with fuzzy boundaries and substantial individual variations. Thus, an important issue to be considered in emotion analysis is generating a person-specific representation of emotion in an unsupervised manner. This paper presents a fully unsupervised method combining autoencoder with Principle Component Analysis to build an emotion representation from speech signals. As each person has a different way of expressing emotions, this method is applied to the subject level. We also investigate the relevancy of such a representation. Experiments on Emo-DB, IEMOCAP, and SEMAINE database show that the proposed representation of emotion is invariant among subjects and similar to the representation built by psychologists, especially on the arousal dimension.

Keywords: Unsupervised learning · Representation learning · Speech emotion analysis · Data-driven · Person-specific

1 Introduction

Affective computing on speech signals is commonly confronted in various scientific disciplines such as neuroscience, psychology, and cognitive sciences [17]. In terms of basic emotion categories, the interpretation of affective nonverbal expressions has been extremely accurate in specific tasks, e.g., speech emotion recognition [13,20,24]. In daily interactions, however, people exhibit non-basic, subtle, and rather sophisticated affective states like depression, instead of those basic emotions labeled in common databases [15]. It indicates that a single label (or any small number of discrete classes) may not reflect the authenticity and the complexity of the affective states [18]. Therefore, an alternative way to emotion analysis is to propose a suitable emotion representation that can not only interpret basic emotions but also express a large variety of emotional states and intensities. Such a representation should fulfill three requirements: *person-specific*, *data-driven*, and *explainable*.

First, the need for person-specific models stems from individual differences in emotions. Human emotions are a highly subjective phenomenon [25], which

© Springer Nature Switzerland AG 2020
A. Karpov and R. Potapova (Eds.): SPECOM 2020, LNAI 12335, pp. 636–645, 2020.
https://doi.org/10.1007/978-3-030-60276-5_61

can be influenced by personality, culture, biological sex, and genotype. The way to express happiness, for instance, usually varies between human beings, which might be reflected in the intensity of laughter or other manifestations.

Second, collecting data with labels is expensive and time-consuming [3]. Moreover, many researchers doubt the reliability of the ground truth provided by accessible databases in the field of emotion analysis [5]. Indeed, there is disunity in labeling on the variations of emotion among human beings. Taking the SEMAINE database as an example, the mean correlation coefficient between two annotators used for the ground truth labeling can only reach 0.45.

Third, such representation should be explainable. Most models of machine learning and artificial intelligence are typically regarded as black boxes, e.g., it is unclear if each neuron in hidden layers is provided with practical meaning [19]. Thus, this highlights the need for mappings between a data-driven representation and the existing emotion representations, like emotion categories and emotion dimensions, which benefits us to understand the outcomes from data simply.

Our experiments in this paper build on prior work that corresponds to these requirements. For person-specific emotion analysis, Soladié et al. [22] applied Principle Component Analysis (PCA) on several facial expressions to build a representation for emotions. They showed that the organization of expressions is invariant, even if the morphology and the way the expressions are performed are slightly different between different subjects. Although this work has been performed only for facial expression understanding, it is a good way to retain the emotion organization for other research fields, such as speech emotion analysis. Autoencoders (AE) [9] have been extensively used in the literature to learn unsupervised data-driven representations for speech emotion recognition. For instance, [7, 12] utilized autoencoder and its variants to generate the representation of affect from speech. These methods are likely to learn more meaningful, controllable, and discriminative features, but none of them analyzes the distribution of emotion from those representations, which requires to keep a stable representation. To solve this, Wang et al. [23] proposed an Organization-Controlled AutoEncoder (OCAE) to keep the organization of the latent space each time the process is launched on the same database. Yet, this study focuses on one database, but it does not extend to the possible invariance of this representation among different databases or different subjects.

In this paper, we analyze the link between speech signals and emotions from learning a person-specific latent representation in an unsupervised manner. We utilize OCAE to have an accurate and stable representation. Our first contribution is applying this method on the subject level to build a person-specific latent space that takes into account the specificities of each subject. Our second contribution concerns the different analyses of these latent representations. We perform two different analyses. The first one aims at showing the invariance of these data-driven representations among speakers. This invariance between different speakers can bring an extension to unknown subjects. To confirm this, we define an order similarity index computed by the Spearman rank correlation coefficient. The second analysis concerns the link between the representations and the existing emotion model built by psychologists. More precisely, we utilize the organization and the order similarity indexes and the Pearson correlation coefficient to verify the match between our representation and the psychological emotions.

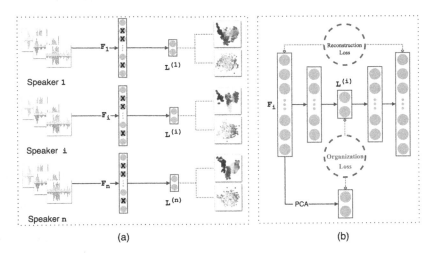

Fig. 1. (a) is the framework overview, which is made of 3 steps: data pre-processing, representation learning, and evaluation. For each speaker i, we obtain a person-specific latent space $L^{(i)}$ using OCAE. This latent space can be mapped with categorical or dimensional emotions, which are shown in different colors for evaluation. (b) focuses on the structure of OCAE [23] which combines a reconstruction loss and an organization loss for an accurate and stable representation learning. (Color figure online)

2 Proposed Method

Our system is made of three steps, as shown in Fig. 1. First, we pre-process the speech signals, including feature extraction, feature selection, and normalization to obtain relevant features. The detail of data pre-processing will be introduced in Sect. 3.2. Then, we learn a person-specific representation using OCAE (Sect. 2.1). Finally, we measure the performance of the data-driven representation. We map our representation to emotion categories and arousal for measuring the similarity of the representations between speakers (Sect. 2.2).

2.1 Organization-Controlled AutoEncoder

Autoencoder [9] is known to be efficient in dimension reduction in an unsupervised manner. It sets the target values to be close to the input, i.e., the reconstruction loss J_R in Fig. 1b. However, there are certain drawbacks. Several factors, such as the initialization, training epochs or the number of hidden layers, can lead to entirely different organizations of the latent space even on the same data set. A significant advantage of PCA [16], which is another unsupervised method for dimensionality reduction, is the invariance of the result as long as inputs are the same. Yet, an autoencoder with a 2-dimensional latent space could produce a better visualization of the data than PCA [9]. Thus, combining autoencoder with PCA can conserve the consistency of organization brought by PCA, as well as take advantage of having better discrimination from autoencoder. It has been proved in [23] that OCAE can provide a stable organization of the representation each time training is launched on the same data set with an accurate representation as AE alone. We add a constraint to the objective through the result of PCA, which matches the red circle in Fig. 1. This constraint is performed by adding an organization loss J_O:

$$J_O\left(\Theta\right) = \left\|L^{(i)} - p(F_i)\right\|^2 \tag{1}$$

where Θ is the parameter set of the encoder and the decoder, including the weight and bias of each hidden layer. For speaker i, the input of the encoder is the feature vector F_i and $L^{(i)}$ is the 2-dimensional representation, i.e., the output of the encoder. And $p(F_i)$ is the result of PCA on F_i, which has the same dimension as the latent space $L^{(i)}$. We add the complementary loss function to the objective function of autoencoder, i.e., $\alpha \cdot J_R + \beta \cdot J_O$, where α and β are the weight coefficients of the two losses. In this paper, we propose to apply OCAE separately for each subject to build a person-specific latent representation. Similarly, we suppose that it can keep the same organization of the latent spaces for different speakers. For the ith speaker, we obtain the latent space $L^{(i)}$ which can be used for visualization and mapping with other emotion representations. In our case, the dimension of $L^{(i)}$ is set to 2.

2.2   Similarity Between Person-Specific Representations

We construct a mapping to the existing representations to measure the performance of the data-driven emotion representation provided by OCAE. As an example of the mapping from categories to dimensions, emotion categories can be located in emotion dimension space as long as the coordinates of the emotion category have been determined. Then, the organization of emotions can be defined as the relative location of the emotion categories in emotion dimension space. We want to investigate the interpretability of the unsupervised 2-dimensional representation built by OCAE. For that purpose, we propose to study the organization of the learned latent space according to two higher-level characteristics: (i) emotion categories (angry, happy, etc.) and (ii) arousal (continuous label of activation). Thus, these person-specific representations can be visualized with the categorical and arousal labels, as shown in Fig. 1a, where each point represents one sample, and the color indicates the label. To measure the similarity of emotion organization of the representations between different subjects, we propose to compare the organization and the order of emotions located in these person-specific representations. Moreover, we propose to explore the relationship between our data-driven representation and the existing emotion representations based on psychology to understand how emotions are organized in our representation.

Assuming that a database with n speakers on m emotion states has N audio clips, we can obtain n latent spaces by OCAE applied to the data of each subject separately. For the ith speaker, each audio clip k gives one point $P_k^{(i)}$ on the latent space $L^{(i)}$ and is labeled with emotion $E_k^{(i)} = e \in \{1, 2, ..., m\}$. We compute the mean value $\mu_e^{(i)}$ of the representation labelled with emotion e for the ith speaker:

$$\mu_e^{(i)} = \frac{\sum_k \left(P_k^{(i)} | E_k^{(i)} = e \right)}{N_e^{(i)}} \tag{2}$$

where $N_e^{(i)}$ is the number of audio clips on e emotion state of ith speaker. Based on the relative location of each emotion's mean value on the latent space, we utilize two metrics to measure the similarity of the organization or the order between different speakers.

Organization Similarity Between Speakers. We utilize the presence/absence of edges between the Delaunay tessellations of two latent spaces to compare the similarity of organizations between the different subjects. Based on the mean value of emotions $\mu^{(i)}$ and $\mu^{(j)}$ on our 2-dimensional representations of subject i and j, we can get an m-Delaunay tessellations, which has a set of edges $D^{(i)}$ and $D^{(j)}$. The organization similarity index $S_{org}(i, j)$ is defined by

$$S_{org}(i, j) = \frac{2 \cdot \left| D^{(i)} \cap D^{(j)} \right|}{\left| D^{(i)} + D^{(j)} \right|} \tag{3}$$

where $|D^{(i)}|$ is the number of edges of the m-Delaunay tessellation of ith subject and $|D^{(i)} \cap D^{(j)}|$ is the number of edges in common of the m-Delaunay tessellations of both subjects i and j. Given m points, we can obtain an m-Delaynay tessellation and then compute the organization similarity index in random order as the baseline. Moreover, we computed the similarity index of 'one-switch': if we switch two points randomly, the organizations will be different. So 'one-switch' can be considered as a dividing line to distinguish similar organizations from different organizations.

Order Similarity of Emotion Between Speakers. Spearman's correlation coefficient ρ can be used to measure statistical dependence between the rankings of two variables, which is defined as the Pearson correlation coefficient between the rank variables [4]. To compare the person-specific latent spaces, we compute the emotion rankings $r^{(i)}$ and $r^{(j)}$ converted by $\mu^{(i)}$ and $\mu^{(j)}$ in one particular direction. Then, the rank correlation ρ_{ij} between the ith and the jth speaker can be computed by

$$\rho_{ij} = 1 - \frac{6 \sum_e d_{ij}^2 (e)}{n (n^2 - 1)} \tag{4}$$

where $d_{ij}(e) = r^{(i)}(e) - r^{(j)}(e)$ denotes the difference between two rankings on emotion e of the ith and the jth speaker. Thus, we can measure the similarity of emotion order among subjects by defining the mean value of the absolute order similarity index S_{ord}.

$$S_{ord} = \frac{\sum_i \sum_{j \neq i} |\rho_{ij}|}{m (m - 1)} \tag{5}$$

Also, we can explore the similarity of order between our representation of emotions and the existing model based on the psychology using the order similarity index defined in Eq. (5). Russell's circumplex model [18] is one classical mapping from emotion categories to Arousal-Valence representation based on psychology. We compute the order similarity index between the rankings derived by our representation and the arousal axis of Russell's circumplex model.

3   Experiments

3.1   Databases

- *Berlin Database of Emotional Speech* (Emo-DB) [1] consists of 535 utterances spread over ten subjects. It has categorical labels of 7 basic emotions. This database is chosen because it contains short utterances spoken in an exaggerated acting way. It will allow easy analysis of the organization of the proposed unsupervised representation in terms of simple emotion categories.
- The USC IEMOCAP database [2] consists of 7383 utterances spread over ten subjects. It has both categorical labels of 10 emotions and dimensional labels of arousal, valence, and dominance. This database is chosen because of the large number of audio clips and the diversity of the labels.
- The SEMAINE database [14] consists of 141 conversations spread over 15 subjects. It has dimensional labels of arousal, valence, power, and expectation. We chose this dataset because of the high quality of the speech recordings and because it contains spontaneous emotions, to a certain extent.

3.2   Experimental Setting

Indeed, it is still an open problem to determine the appropriate window length for emotion analysis [8]. Fortunately, a speech segment at least longer than 250 ms has been shown to contain sufficient emotional information [11]. For speech segmentation, we tried six different window lengths of 0.25, 0.5, 1, 2, 2.5, and 3 seconds to eliminate the effect of window size. Moreover, we select the sliding widow of 2-second length with a 100 ms shift. For feature extraction, we used OpenSMILE [6] to extract the original feature set. The selection of the original LLDs and statistical functionals is based on default feature sets (6530 in total) [21]. After feature selection, the number of the final feature subset is 286.

Normalization as a part of data pre-processing is critical, especially for unsupervised learning [10]. In particular, the values of each feature are of different orders of magnitude. The process of normalization can be divided into two stages, where two common normalization methods are adopted: min-max normalization and z-score normalization on subject and feature level, respectively.

For the network, there are four fully-connected hidden layers with random initialization (286-128-64-16-2). The activation functions of hidden layers are ReLUs except for the bottleneck layer (sigmoid). α and β in the final objective function are used for balancing two loss functions: reconstruction loss and organization loss. When training epoch = 1, we set $\alpha = 0$ and $\beta = 1$ to constrain the organization of emotions on the latent space. And then, we enable α increase by 0.01 every epoch until 100 epoch, and β is the opposite. In fact, organization loss only occurs in the early stage of the training.

3.3   The Invariance of Representations Among Speakers

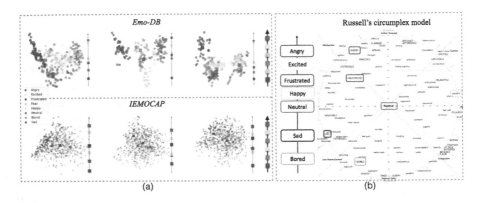

Fig. 2. The comparison of the organization of emotions between our person-specific representations and Russel's circumplex model. (a) shows the six latent spaces from six speakers on Emo-DB and IEMOCAP. On the right of each latent space, there is a projection of the mean values of each emotion to the x-axis, but we plot it on the vertical direction in order to match to arousal axis of the psychological model. The order of emotions from the first component of our representation accords with the order of emotions from the arousal of Russel's circumplex model

The advantage of person-specific representations is the ability to maintain diversity among subjects. However, the need for discovering the invariance between these representations is the basis for extending to different subjects. Due to the unbalance of these databases, six emotions are selected in our experiments (sad, bored, neutral, fear, happy, anger in Emo-DB and sad, neutral, happy, frustrated, excited, anger in IEMOCAP). Figure 2a shows the latent spaces from six subjects and the projection of the mean values of emotions to the x-axis on the right of each latent space. Yet, the projection is shown in the vertical direction to match the psychological model (Fig. 2b). We observe that the OCAE unsupervised representation clusters among the emotion categories, especially on Emo-DB. Moreover, the organization of emotions on the latent spaces among speakers is invariant. More precisely, the invariant emotion order of the first component among speakers matches the order of arousal provided by Russell's circumplex model.

For quantitative analysis, Table 1 presents the mean organization and order similarity indexes of 2D representations between different speakers on Emo-DB and IEMOCAP. The baseline is computed randomly, and the 'one-switch' is computed by switch two points randomly. The values of the baseline and 'one-switch' from Emo-DB and IEMOCAP are the same due to the number of emotions on both databases. The 'one-switch' can be regarded as a dividing line to differentiate the similar or the different organization of emotions. The organization and the order similarity indexes of OCAE are both higher than the value of 'one-switch'; the indexes of AE are the opposite. This result confirms that the emotions are organized in the same way for quite all the subjects on both databases. Yet, no visible evidence was found on the second component. Note that the direction of the x-axis might be opposite between different representations, so the alignment among speakers needs a few labels.

Table 1. The similarity indexes of 2D representations between different speakers on Emo-DB and IEMOCAP. We utilize the value of 'one-switch' to divide similar or different organization or order of emotions. OCAE is better than AE for keeping invariant organization of emotions on the latent spaces between different speakers.

	Organization similarity index		Order similarity index	
	[HTML]000000 Emo-DB	IEMOCAP	Emo-DB	IEMOCAP
Baseline	0.66		0.37	
One-switch	0.81		0.66	
AE	0.76	0.74	0.55	0.59
OCAE	**0.87**	**0.83**	**0.90**	**0.92**

3.4 The First Component Mapping to Arousal

Figure 2b shows Russel's circumplex model and the projection of the emotions used in the applied databases on the arousal axis. We compare the order of this projection with the order derived from the first component of our representation on Emo-DB and IEMOCAP. We found that the order derived from our representation mostly accords with the psychological representation of the arousal axis. Swap between 'sad' and 'bored' on Emo-DB may be due to the exaggerated acting way, which leads to all emotions in high-intensity states.

To confirm the observation, we can derive the rankings from the first component of our representation and the order of emotions located in Russell's circumplex model on the arousal axis. The order similarity indexes on Emo-DB and

IEMOCAP are 0.88 and 0.91, respectively. We can also directly compute the Pearson correlation coefficient (CC) to compare our representation with arousal on the IEMOCAP and the SEMAINE databases. The CCs on SEMAINE and IEMOCAP are 0.33 and 0.63, respectively, which also proves that the emotions on our representation are organized along the direction of arousal. A possible explanation for the poor result on SEMAINE is the disagreement on the variations of emotion among annotators. The mean correlation coefficient between the two annotators used for the ground truth labeling can only reach 0.45.

4   Conclusion

In this paper, we applied a fully unsupervised person-specific manifold method (OCAE) for speech emotion representation creation. Through the similarity index, we verified that the organization of our representation is invariant for emotions among speakers. Thanks to this invariance, it provides feasibility to extend to emotion-related tasks for person-specific models. Moreover, it brings an extension to unknown subjects. Our method can produce a meaningful latent space, especially on the first component, which carries arousal information. These results are valid in both acted and natural emotions evaluated on Emo-DB, IEMOCAP, and SEMAINE database. We underline that this method is fully unsupervised, which can be an alternative to the challenge of labeling in terms of the emotion that is something difficult and not especially 'safe'.

Acknowledgments.. Thanks to China Scholarship Council (CSC) and the French government funding program ANR REFLET No. ANR-17-CE19-0020-01 for funding.

References

1. Burkhardt, F., Paeschke, A., Rolfes, M., Sendlmeier, W.F., Weiss, B.: A database of German emotional speech. In: Ninth European Conference on Speech Communication and Technology (2005)
2. Busso, C., et al.: IEMOCAP: interactive emotional dyadic motion capture database. Lang. Resour. Eval. **42**(4), 335 (2008)
3. Caron, M., Bojanowski, P., Joulin, A., Douze, M.: Deep clustering for unsupervised learning of visual features. In: Proceedings of the European Conference on Computer Vision (ECCV), pp. 132–149 (2018)
4. Daniel, W.W.: Applied Nonparametric Statistics. Houghton Mifflin, Boston (1978)
5. Eskimez, S.E., Duan, Z., Heinzelman, W.: Unsupervised learning approach to feature analysis for automatic speech emotion recognition. In: IEEE International Conference on Acoustics, Speech and Signal Processing (ICASSP), pp. 5099–5103 (2018)
6. Eyben, F., Weninger, F., Gross, F., Schuller, B.: Recent developments in openS-MILE, the munich open-source multimedia feature extractor. In: Proceedings of the 21st ACM international conference on Multimedia, pp. 835–838 (2013)
7. Ghosh, S., Laksana, E., Morency, L.P., Scherer, S.: Representation learning for speech emotion recognition. In: Interspeech, pp. 3603–3607 (2016)

8. Han, K., Yu, D., Tashev, I.: Speech emotion recognition using deep neural network and extreme learning machine. In: Fifteenth Annual Conference of the International Speech Communication Association (2014)

9. Hinton, G.E., Salakhutdinov, R.R.: Reducing the dimensionality of data with neural networks. Science **313**(5786), 504–507 (2006)

10. Kaya, H., Karpov, A.A., Salah, A.A.: Fisher vectors with cascaded normalization for paralinguistic analysis. In: Sixteenth Annual Conference of the International Speech Communication Association (2015)

11. Kim, Y., Provost, E.M.: Emotion classification via utterance-level dynamics: a pattern-based approach to characterizing affective expressions. In: IEEE International Conference on Acoustics, Speech and Signal Processing (ICASSP), pp. 3677–3681 (2013)

12. Latif, S., Rana, R., Qadir, J., Epps, J.: Variational autoencoders for learning latent representations of speech emotion: a preliminary study. In: Interspeech, International Speech Communication Association (ISCA), pp. 3107–3111 (2018)

13. Lotfian, R., Busso, C.: Curriculum learning for speech emotion recognition from crowdsourced labels. IEEE/ACM Trans. Audio, Speech Lang. Process. **27**(4), 815–826 (2019)

14. McKeown, G., Valstar, M., Cowie, R., Pantic, M., Schroder, M.: The SEMAINE database: annotated multimodal records of emotionally colored conversations between a person and a limited agent. IEEE Trans. Affect. Comput. **3**(1), 5–17 (2011)

15. Op't Eynde, P., De Corte, E., Verschaffel, L.: Accepting emotional complexity: a socio-constructivist perspective on the role of emotions in the mathematics classroom. Educ. Stud. Math. **63**(2), 193–207 (2006)

16. Pearson, K.: LIII on lines and planes of closest fit to systems of points in space. London Edinb. Dublin Philos. Mag. J. Sci. **2**(11), 559–572 (1901)

17. Poria, S., Cambria, E., Bajpai, R., Hussain, A.: A review of affective computing: from unimodal analysis to multimodal fusion. Inform. Fusion **37**, 98–125 (2017)

18. Russell, J.A.: A circumplex model of affect. J. Pers. Soc. Psychol. **39**(6), 1161 (1980)

19. Samek, W., Binder, A., Montavon, G., Lapuschkin, S., Müller, K.R.: Evaluating the visualization of what a deep neural network has learned. IEEE Trans. Neural Netw. Learn. Syst. **28**(11), 2660–2673 (2016)

20. Schuller, B., Rigoll, G., Lang, M.: Hidden Markov model-based speech emotion recognition. In: IEEE International Conference on Acoustics, Speech and Signal Processing (ICASSP), vol. 2, pp. II-1 (2003)

21. Schuller, B., Steidl, S., Batliner, A.: The interspeech 2009 emotion challenge. In: Tenth Annual Conference of the International Speech Communication Association (2009)

22. Soladié, C., Stoiber, N., Séguier, R.: Invariant representation of facial expressions for blended expression recognition on unknown subjects. Comput. Vis. Image Underst. **117**(11), 1598–1609 (2013)

23. Wang, S., Soladié, C., Séguier, R.: OCAE: Organization-controlled autoencoder for unsupervised speech emotion analysis. In: 5th International Conference on Frontiers of Signal Processing (ICFSP), pp. 72–76. IEEE (2019)

24. Wu, S., Falk, T.H., Chan, W.Y.: Automatic speech emotion recognition using modulation spectral features. Speech Commun. **53**(5), 768–785 (2011)

25. Zhao, S., Ding, G., Han, J., Gao, Y.: Personality-aware personalized emotion recognition from physiological signals. In: IJCAI, pp. 1660–1667 (2018)

Synchronized Forward-Backward Transformer for End-to-End Speech Recognition

Tobias Watzel$^{(\boxtimes)}$ (iD), Ludwig Kürzinger(iD), Lujun Li(iD), and Gerhard Rigoll(iD)

Chair of Human-Machine Communication,
Technical University of Munich, Munich, Germany
{tobias.watzel,ludwig.kuerzinger,lujun.li,rigoll}@tum.de

Abstract. Recently, various approaches utilize transformer networks, which apply a new concept of self-attention, in end-to-end speech recognition. These approaches mainly focus on the self-attention mechanism to improve the performance of transformer models. In our work, we demonstrate the benefit of adding a second transformer network during the training phase, which is optimized on time-reversed target labels. This new transformer receives a future context, which is usually not available for standard transformer networks. We have access to future context information, which we integrate into the standard transformer network by proposing two novel synchronization terms. Since we only require the newly added transformer network during training, we are not changing the complexity of the final network and only adding training time. We evaluate our approach on the publicly available dataset TEDLIUMv2, where we achieve relative improvements of 9.8% for the dev and 6.5% on the test set, respectively, if we employ synchronization terms with euclidean metrics.

Keywords: Speech recognition · Transformer · Forward-backward transformer · Regularization · Synchronization

1 Introduction

Nowadays, sequence-to-sequence (Seq2Seq) models are popular candidates for automatic speech recognition (ASR) systems, as they have multiple advantages compared to traditional ASR systems. One significant advantage is that Seq2Seq models merge several speech modules into a single end-to-end system. In the past, these modules were hand-crafted and required specific knowledge, whereas the end-to-end systems do not require any knowledge and can be directly trained to predict, e.g., characters (chars).

Seq2Seq model approaches can be roughly divided into four categories: connectionist temporal classification (CTC) [8,9], transducer [7,10,19], attention [1,3,5,14,15,25], and transformer [20,21,24,26] approaches.

© Springer Nature Switzerland AG 2020
A. Karpov and R. Potapova (Eds.): SPECOM 2020, LNAI 12335, pp. 646–656, 2020.
https://doi.org/10.1007/978-3-030-60276-5_62

For CTC, a recurrent neural network (RNN) learns an alignment between the input speech features and a given transcript. This approach does not require any labeled data since it is based on a loss function, which applies a forward-backward algorithm to calculate a total probability for all possible alignments [8]. The forward-backward algorithm is not ideal due to the high correlation of consecutive speech features as it assumes conditional interdependence of these features. The conditional independence can be relaxed by utilizing the concept of transducers [10]. Therefore, another RNN is added to the model, which is trained to create a function between all previous features and the output of the network.

For attention models, the transducer approach is extended by an attention network [1]. The first RNN, the encoder, transforms the speech features into a robust high-dimensional space, which encodes essential aspects of the speech signal. The attention networks learn to create attention weights based on these high-level features and a second decoder RNN. These weights are scaling the high-level features of the encoder and are getting summed up to a glimpse, which is utilized in the decoder.

All these approaches are heavily relying on RNNs, which are computationally expensive. Transformers replace RNNs with simple feed-forward networks and apply self-attention (SA) layers [26]. The self-attention is defined in a way that all keys, values, and queries are coming from the same location: the output of the previous encoder layer. Therefore, the encoder layers have access to all positions of the previous layer.

In our work, we propose a novel way to add a second transformer network in the training phase. This right-to-left (R2L) transformer is only utilized during training and is removed later in decoding. It supports the traditional left-to-right (L2R) transformer and is trained on time-reversed labels, i.e., receives future context information. To synchronize the transformer outputs during optimization, we add two novel regularization terms depending on the selected target labels. Currently, most of the state-of-the-art models for the English language are either trained on char or byte pair encoding (BPE) units as target labels, where BPE units seem superior if enough data is available [4,25].

Recent studies have already demonstrated, at least for attention models, that adding a second R2L decoder is beneficial to improve the overall performance [16, 28,29]. In [16], a novel forward-backward algorithm is proposed that is applied during decoding to improve the beam search. However, they did not put much emphasis on the training scheme and applied a fixed weight for the loss of both decoders, which favored the L2R decoder with a higher weight. [29] employed the idea of a dual decoder setup and proposed a synchronization for the outputs of the L2R and R2L decoder in the domain of text-to-speech. They utilized a L2 regularization since they assumed equal sequence lengths between both decoders, which is valid for Mandarin or Japanese. However, the English language, can also be encoded with BPE units, which generate unequal sequences for a normal and a time-reversed transcript. Due to the fact, that there have not been any works to solve these issues for attention models, the authors in [28] proposed two novel

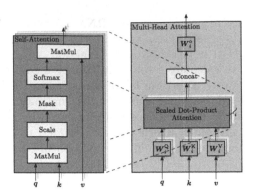

Fig. 1. Self-attention module of the transformer network proposed in [26], which is utilized in the multi-head attention module.

regularization terms for models trained on char or BPE units, respectively. For normal chars, we added a L2 regularizer term inspired by [29], since sequences generated by L2R and R2L decoders have the same lengths. For BPE units, we integrated another term based on a soft version of the dynamic time warping (DTW) algorithm to synchronize the decoder outputs. Both regularizing terms were solely added during training and improved the overall performance of a standard attention model.

So far, there have been no works for transformer models, which apply the concept of a dual transformer setup. Since it is unclear if a standard transformer model can also benefit from a second transformer network during training, our contributions can be summed up as follows:

– We demonstrate the advantage of employing a second transformer during optimization.
– Depending on the utilized target labels, we propose synchronization terms for char and BPE units to synchronize the output of the transformers.
– We compare the euclidean and cosine distances as distance measurement of the transformer outputs.

2 Proposed Method

2.1 Transformer Network

The standard transformer network consists of two modules: stacked encoder networks and stacked decoder networks [26]. Let $X = (x_1, \cdots, x_t, \cdots, x_T)$ be the input speech features and $y = (y_1, \cdots, y_k, \cdots, y_K)$ the target labels. The fundamental idea of transformer networks is to heavily rely on a SA modules, which are only dependent on the previous layer [26]. As in Fig. 1 depicted, transformer networks apply only feed-forward networks and avoid RNNs. The usage

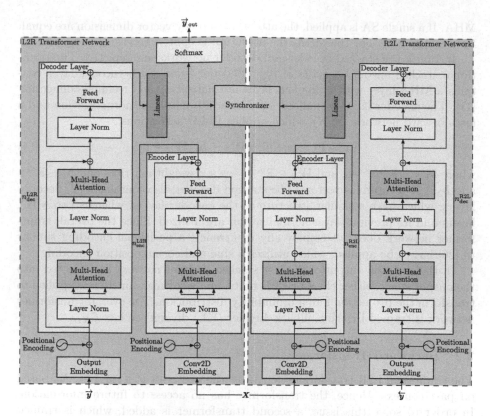

Fig. 2. Dual transformer setup with L2R and R2L transformer networks. Both networks are synchronized with a synchronizer.

of SA has the advantage that recurrences in the overall network structure are discarded, which are computationally expensive. The SA is defined as:

$$\text{SelfAttention}(\boldsymbol{q}, \boldsymbol{k}, \boldsymbol{v}) = \text{Softmax}(\frac{\boldsymbol{q}\boldsymbol{k}^T}{\sqrt{d_k}})\boldsymbol{v}, \tag{1}$$

where $\boldsymbol{q}$, $\boldsymbol{k}$, and $\boldsymbol{v}$ are the query, key, and value vectors, respectively, and d_k is the dimension of the key vector. In order to allow access to information in different positions, a multi-head attention (MHA) is utilized. The MHA network has multiple heads $\boldsymbol{h}_i$, where each head is built up by a SA network:

$$\text{MultiHeadAttention}(\boldsymbol{q}, \boldsymbol{k}, \boldsymbol{v}) = \text{Concat}(\boldsymbol{h}_1, \cdots, \boldsymbol{h}_i, \cdots, \boldsymbol{h}_{N_\text{h}})\boldsymbol{W}_i^\text{o}, \tag{2}$$

where

$$\boldsymbol{h}_i = \text{SelfAttention}(\boldsymbol{q}\boldsymbol{W}_i^Q, \boldsymbol{k}\boldsymbol{W}_i^K, \boldsymbol{v}\boldsymbol{W}_i^V). \tag{3}$$

Here, $\boldsymbol{W}_i^Q \in \mathbb{R}^{d_\text{att} \times d_k}$, $\boldsymbol{W}_i^K \in \mathbb{R}^{d_\text{att} \times d_k}$ and $\boldsymbol{W}_i^V \in \mathbb{R}^{d_\text{att} \times d_k}$ are projection matrices, d_att is the dimension of the SA module and N_h is the number of heads in the

MHA. If a single SA is applied, the attention and key vector dimension are equal. Hence, $d_{att} = d_k$. For the MHA, the attention is split, whereas $d_k = d_{att}/N_h$.

In Fig. 2, the stacked encoder and decoder networks are depicted. The traditional stacked encoder consists of n_{enc}^{L2R} encoder layer. Each encoder layer contains a MHA layer with a previous layer normalization and residual connection [11]. The encoder layer is finalized with an additional layer normalization and a feed-forward network (FFN), which applies ReLU activations:

$$FFN(x) = \max(0, \boldsymbol{x}\boldsymbol{W}_1 + \boldsymbol{b}_1)\boldsymbol{W}_2 + \boldsymbol{b}_2, \tag{4}$$

where $\boldsymbol{W}_1 \in \mathbb{R}^{d_{att} \times d_{ffn}}$, and $\boldsymbol{W}_2 \in \mathbb{R}^{d_{ffn} \times d_{att}}$ are weight matrices and $\boldsymbol{b}_1 \in \mathbb{R}^{d_{att}}$ and $\boldsymbol{b}_2 \in \mathbb{R}^{d_{att}}$ are bias vectors. The stacked decoder also contains n_{dec}^{L2R} decoder layers, which have a similar structure as the encoder layers, whereas two MHA layers and its previous layer normalization are utilized. Since the overall transformer network does not employ any recurrency, a positional encoding similar to [26] has to be applied, which utilizes a sine and cosine position encoding.

In the end, the output of the last stacked decoder network is projected into the output space, either chars or BPE units, and a softmax function is applied to transform the projection into a probability distribution. The overall transformer network is trained by the Kullback–Leibler (KL) divergence loss $\mathcal{L}_{KL}$.

2.2 Backward Transformer

The traditional L2R transformer generates the posterior $p(\overrightarrow{y}_k|y_{1:k-1}, \boldsymbol{X})$ based on past context. Hence, the transformer has no access to future information. In order to solve this issue, a second transformer is added, which is trained on time-reversed target labels. This R2L transformer generates the posterior $p(\overleftarrow{y}_l|y_{L:l+1}, \boldsymbol{X})$, which is dependent on future context. In an ideal case, both posteriors are equal, since both transformer networks receive the same amount of information:

$$p(\overrightarrow{\boldsymbol{y}}|\boldsymbol{X}, \overrightarrow{\theta}) = p(\overleftarrow{\boldsymbol{y}}|\boldsymbol{X}, \overleftarrow{\theta}), \tag{5}$$

where $\overrightarrow{\theta}$ are the weights of the L2R and $\overleftarrow{\theta}$ are the weights of the R2L transformer network. However, this is not valid, as the L2R and R2L transformers depend on a different context, i.e., past and future context. As a result, the training criterion is distinct, and the predictions of the transformer networks differ.

2.3 Synchronizing Transformer Outputs of Equal Sequence Length

Chars as target labels produce equal sequence length $K = L$, since there is no length difference between an utterance in regular or time-reversed order. The outputs of the L2R and R2L transformers are synchronized by extending the KL loss to the total loss $\tilde{\mathcal{L}}$:

$$\begin{aligned} \tilde{\mathcal{L}} = &\ \alpha[\beta\mathcal{L}_{CTC}(\overrightarrow{\theta}) + (1-\beta)\mathcal{L}_{KL}(\overrightarrow{\theta})] \\ &+ (1-\alpha)[\beta\mathcal{L}_{CTC}(\overleftarrow{\theta}) + (1-\beta)\mathcal{L}_{KL}(\overleftarrow{\theta})] + \lambda\Omega(\overrightarrow{\theta}, \overleftarrow{\theta}), \end{aligned} \tag{6}$$

where α define the weighting between the L2R and R2L transformer losses, β weights the internal losses $\mathcal{L}_{KL}$ and $\mathcal{L}_{CTC}$[12] in each transformer and λ is a scalar to scale the synchronization term $\Omega(\overrightarrow{\theta}, \overleftarrow{\theta})$:

$$\Omega(\overrightarrow{\theta}, \overleftarrow{\theta}) = \frac{1}{K} \sum_{k=0}^{K} d(\overrightarrow{y}_k, \overleftarrow{y}_k), \tag{7}$$

with $d(\cdot, \cdot)$ representing a distance function. Note that these distance functions can only be applied, as both sequences are equal.

2.4 Synchronizing Transformer Outputs of Unequal Sequence Length

If BPE units as target labels are applied, sequences of unequal lengths are created, since the encoding for ordinary and time-reversed utterances differs. The sequences can be synchronized by a soft version of the DTW algorithm [6], which has already been proposed in [28]:

$$\min\nolimits^{\gamma}\{a_1, \cdots, a_n\} := \begin{cases} \min_{i \leq n} a_i & \gamma = 0 \\ -\gamma \log \sum_{i=1}^{n} e^{a_i/\gamma} & \gamma > 0. \end{cases} \tag{8}$$

The scalar γ adjusts the softening of the algorithm, e.g., $\gamma = 0$ returns the typical DTW algorithm. By utilizing Eq. 8, a synchronization loss for unequal sequence lengths is defined:

$$\Omega(\overrightarrow{\theta}, \overleftarrow{\theta}) = \min\nolimits^{\gamma}\{\langle A, \Delta(\overrightarrow{y}, \overleftarrow{y})\rangle, A \in \mathcal{A}_{k,l}\}. \tag{9}$$

Here, the alignment matrix A is a binary matrix from a set $\mathcal{A}_{k,l} \subset \{0,1\}^{k,l}$. The set consists of matrices that define paths from $(1,1)$ to (k,l). These paths are restricted, as only $\downarrow$, $\rightarrow$ and $\searrow$ moves are allowed. In order to measure the distance between two vectors of unequal lengths, the distance function $\Delta(\overrightarrow{y}, \overleftarrow{y}) := [d(\overrightarrow{y}_k, \overleftarrow{y}_l)]$ is introduced, which can be any arbitrary distance function $d(\cdot, \cdot)$. This inner product of the alignment matrix A and the distance function $\Delta(\overrightarrow{y}, \overleftarrow{y})$ results in an alignment cost, containing all possible alignments for $\overrightarrow{y}$ and $\overleftarrow{y}$.

2.5 Distance Functions

There are several ways to define distance functions $d(\cdot, \cdot)$ to measure the distance between two vectors. Since it is unclear, which distance metric is suitable for the synchronization terms, two popular metrics are applied: The euclidean distance d_{euc} and the cosine angle d_{cos}.

The euclidean distance d_{euc} between two vectors of the same length v_1 and v_2 is defined as the norm of the difference of these vectors:

$$d_{euc}(v_1, v_2) = ||v_1 - v_2||, \tag{10}$$

where $|| \cdot ||$ represents the norm of a vector. The distance can also be replaced by a cosine distance $d_{\cos}$ between two vectors. Inspired by [2], the angle of two vectors is:

$$d_{\cos}(\boldsymbol{v}_1, \boldsymbol{v}_2) = \frac{1}{\pi} \arccos(\frac{\langle \boldsymbol{v}_1, \boldsymbol{v}_2 \rangle}{||\boldsymbol{v}_1|| \cdot ||\boldsymbol{v}_2||}), \qquad (11)$$

where $\langle \cdot, \cdot \rangle$ defines the inner product of two vectors.

3 Experiments

We evaluate our approach on the publicly available dataset TEDLIUMv2 [18], which contains nearly 200 h of training data and a corresponding 150 k lexicon. The dataset is already split into train, dev, and test set.

In Fig. 2, we illustrate our proposed architecture. The transformer models are implemented in the ESPnet toolkit [27], which is popular for training Seq2Seq models. First, we extract 83-dimensional feature vectors utilizing Kaldi [17]. The resulting feature vector consists of 80-dimensional log Mel and its pitch features. Next, we create the corresponding chars and BPE units as target labels. In all experiments, we set the number of BPE units to 100.

After creating the dataset, we apply an individual Conv2D front-end for each L2R and R2L transformer. The front-end is built-up by two conv2D layers with ReLU activations. Each conv2D layer applies $d_{att} = 256$ filters, a 3×3 kernel, and a stride length of two. The network is finished by a linear layer of size $d_{att} = 256$ and the position encoding proposed in [26].

The encoder network of both transformers has $n_{enc}^{L2R} = n_{enc}^{R2L} = 12$ stacked encoder layers with 2048 units. The decoder of the transformer networks employs the same number of units. However, they only consist of $n_{dec}^{L2R} = n_{dec}^{R2L} = 6$ stacked decoder layers. The dimension of the SA in each encoder/decoder layer is set to $d_{att} = 256$, with a total of four heads in every MHA.

We separately pretrain both L2R and R2L transformer beforehand, initialize the dual transformer setup with all pretrained weights and add the synchronization term from Eq. 7. For all experiments, we apply the same training setup with a batch-size of 32. The transformers are trained by Adam optimizer [13] with a warmup of 25.000 steps and a learning-rate of 5.0. After the warmup phase, the learning-rate is reduced based on the function described in [26]. To avoid over-fitting, we employ residual dropout [26] and standard dropout [22] with a rate of 0.1 in every encoder/decoder layer. Moreover, the target labels are smoothed by a label smoothing with a value of 0.1 [23]. For guiding the training of the transformer network, we extend the internal KL loss by adding a CTC [8] network and its loss and set $\beta = 0.3$.

In the dual transformer setup, we set $\alpha = 0.9$ to mainly focus the training on the L2R transformer. The scalar λ for scaling the synchronization is dependent on the utilized target labels and the distance functions. In our experiments, we select a value of $\lambda = 1.0$ for chars with a euclidean distance function and $\lambda = 15.0$ with a cosine distance function. For BPE units, where we apply the soft-DTW algorithm, we scale the synchronization term by $\lambda = 10^{-3}$ for the euclidean

Table 1. Evaluation of our approach on TEDLIUMv2 [18] with the resulting WERs for all five setups.

TEDLIUMv2 [18]				
	chars		BPE	
Methods	Dev	Test	Dev	Test
Forward (L2R)	18.28	16.27	16.08	14.13
Backward (R2L)	23.12	17.75	24.33	19.39
Dual with cos sync	17.85	15.31	15.14	13.62
Dual with euc sync	**16.49**	**15.25**	**14.49**	**13.19**

distance function, whereas λ is set to 10^{-1} for the cosine distance. The softening parameter of the soft-DTW loss is set to $\gamma = 1.0$.

The synchronization terms and the R2L transformer are only utilized during the training phase. Later in the decoding phase, both modules are discarded, and a single L2R transformer is evaluated. As a result, we do not increase the number of parameters or adding complexity to the final model.

3.1 Benchmark Details

We evaluate our approach on four different experiment setups:

1. *Forward*: A standard L2R transformer is training on regular target labels and serves as our baseline.
2. *Backward*: A R2L transformer is optimized on time-reversed target labels.
3. *Dual with euc sync*: The same setup as in *Dual no sync*, however, a synchronization term from Eq. 10 is added to the total loss, which employs a euclidean distance function.
4. *Dual with cos sync*: This setup corresponds to the setup *Dual with euc sync* and replaces the euclidean distance with the cosine distance function from Eq. 11.

In the decoding phase, we always apply a shallow fusion between the transformer outputs and their corresponding CTC outputs, where we weigh the output of the CTC network with 0.3. Furthermore, we perform a beam search with a beam size of 10 and do not employ any language model.

3.2 Results

In Table 1, we depict all the results of the different setups. For the setups *Forward* and *Backward*, we notice a significant difference between the performance of both transformers. The R2L transformer returns a higher WER for chars and BPE units than the standard L2R transformer.

The R2L transformer in the *Backward* setup cannot cope with the performance of the L2R transformer. The drop in the WER is even worse if BPE units are applied. A reason for the minor performance of the R2L transformer could be the amount of data, which has already been stated in [28]. It seems that networks trained on time-reversed target labels require more data to equalize with networks trained on standard target labels. In English, sentences usually begin with similar structures (e.g., conjunctions) and are finished with a variety of different words, which makes it challenging for the transformer network to learn a robust alignment between input features and output chars or BPE units. Even though the R2L transformer returns minor results, it encodes valuable information of the time-reversed label sequence. As mentioned above, the information is usually not available for the standard L2R transformer since it cannot obtain future target labels.

In order to utilize the information of the R2L transformer, we add our synchronization terms from Eq. 7 and Eq. 9 in the *Dual with cos sync* and *Dual with euc sync* setups, hence, create a connection between the L2R and R2L transformers. The transformer networks have to minimize the synchronization terms to reduce the distance of their outputs, which allows them to retrieve valuable information from each other. The proposed synchronization terms employ a distance metric, which can be freely chosen. Since it not clear, which of the most popular distance functions, i.e., the euclidean and the cosine metric, performs superior, we evaluate our model on both distance metrics. As presented in Table 1, both metrics improve the performance of the L2R transformer, if it is synchronized with the R2L transformer during training. The overall best result can be observed utilizing an euclidean metric, even though the difference of the WERs is minor for chars in the test set. For chars, we achieve a relative improvement of 9.8% WER on the dev and 6.3% WER on the test set, whereas for BPE units, we retrieve a relative WER improvement of 9.9% on the dev and 6.7% on the test set.

The superior results for the *Dual euc sync* demonstrate, that our synchronization terms help to add valuable information from future target labels into the traditional L2R transformer. The L2R transformer exploits this information to emit better output labels. We observe no benefit from applying a cosine metric in the synchronization terms, which perform worse compared to the euclidean metric.

4   Conclusion

In our work, we presented a novel approach to synchronize the outputs of L2R and R2L transformer networks. The proposed synchronization terms are simple to integrate into existing transformer models and only slightly increase the training time, whereas the complexity of the final model is not increased. Our synchronization returns superior results compared to standard transformer models and is applicable for chars as well as BPE units as target labels. For future work, we want a deep integration of both decoders into the training scheme to improve the standard transformer model.

References

1. Bahdanau, D., Cho, K., Bengio, Y.: Neural machine translation by jointly learning to align and translate. arXiv preprint arXiv:1409.0473 (2014)
2. Cer, D., et al.: Universal sentence encoder. arXiv preprint arXiv:1803.11175 (2018)
3. Chan, W., Jaitly, N., Le, Q., Vinyals, O.: Listen, attend and spell: a neural network for large vocabulary conversational speech recognition. In: 2016 IEEE International Conference on Acoustics, Speech and Signal Processing (ICASSP), pp. 4960–4964. IEEE (2016)
4. Chiu, C.C., et al.: State-of-the-art speech recognition with sequence-to-sequence models. In: 2018 IEEE International Conference on Acoustics, Speech and Signal Processing (ICASSP), pp. 4774–4778. IEEE (2018)
5. Chorowski, J., Bahdanau, D., Cho, K., Bengio, Y.: End-to-end continuous speech recognition using attention-based recurrent nn: first results. arXiv preprint arXiv:1412.1602 (2014)
6. Cuturi, M., Blondel, M.: Soft-DTW: a differentiable loss function for time-series. In: Proceedings of the 34th International Conference on Machine Learning-Volume 70, pp. 894–903. JMLR.org (2017)
7. Graves, A.: Sequence transduction with recurrent neural networks. arXiv preprint arXiv:1211.3711 (2012)
8. Graves, A., Fernández, S., Gomez, F., Schmidhuber, J.: Connectionist temporal classification: labelling unsegmented sequence data with recurrent neural networks. In: Proceedings of the 23rd International Conference on Machine Learning, pp. 369–376. ACM (2006)
9. Graves, A., Jaitly, N.: Towards end-to-end speech recognition with recurrent neural networks. In: International Conference on Machine Learning, pp. 1764–1772 (2014)
10. Graves, A., Mohamed, A.r., Hinton, G.: Speech recognition with deep recurrent neural networks. In: 2013 IEEE International Conference on Acoustics, Speech and Signal Processing, pp. 6645–6649. IEEE (2013)
11. Han, Y.S., Yoo, J., Ye, J.C.: Deep residual learning for compressed sensing ct reconstruction via persistent homology analysis. arXiv preprint arXiv:1611.06391 (2016)
12. Hori, T., Watanabe, S., Zhang, Y., Chan, W.: Advances in joint ctc-attention based end-to-end speech recognition with a deep cnn encoder and rnn-lm. arXiv preprint arXiv:1706.02737 (2017)
13. Kingma, D.P., Ba, J.: Adam: a method for stochastic optimization. arXiv preprint arXiv:1412.6980 (2014)
14. Lu, L., Zhang, X., Renais, S.: On training the recurrent neural network encoder-decoder for large vocabulary end-to-end speech recognition. In: 2016 IEEE International Conference on Acoustics, Speech and Signal Processing (ICASSP), pp. 5060–5064. IEEE (2016)
15. Luong, M.T., Pham, H., Manning, C.D.: Effective approaches to attention-based neural machine translation. arXiv preprint arXiv:1508.04025 (2015)
16. Mimura, M., Sakai, S., Kawahara, T.: Forward-backward attention decoder. In: Interspeech, pp. 2232–2236 (2018)
17. Povey, D., et al.: The kaldi speech recognition toolkit. In: IEEE 2011 Workshop on Automatic Speech Recognition and Understanding. No. CONF, IEEE Signal Processing Society (2011)
18. Rousseau, A., Deléglise, P., Esteve, Y.: Enhancing the TED-LIUM corpus with selected data for language modeling and more TED talks. In: LREC, pp. 3935–3939 (2014)

19. Sak, H., Shannon, M., Rao, K., Beaufays, F.: Recurrent neural aligner: an encoder-decoder neural network model for sequence to sequence mapping. In: Interspeech, pp. 1298–1302 (2017)
20. Salazar, J., Kirchhoff, K., Huang, Z.: Self-attention networks for connectionist temporal classification in speech recognition. In: ICASSP 2019–2019 IEEE International Conference on Acoustics, Speech and Signal Processing (ICASSP), pp. 7115–7119. IEEE (2019)
21. Sperber, M., Niehues, J., Neubig, G., Stüker, S., Waibel, A.: Self-attentional acoustic models. arXiv preprint arXiv:1803.09519 (2018)
22. Srivastava, N., Hinton, G., Krizhevsky, A., Sutskever, I., Salakhutdinov, R.: Dropout: a simple way to prevent neural networks from overfitting. J. Mach. Learn. Res. **15**(1), 1929–1958 (2014)
23. Szegedy, C., Vanhoucke, V., Ioffe, S., Shlens, J., Wojna, Z.: Rethinking the inception architecture for computer vision. In: Proceedings of the IEEE Conference on Computer Vision and Pattern Recognition, pp. 2818–2826 (2016)
24. Tian, Z., Yi, J., Tao, J., Bai, Y., Wen, Z.: Self-attention transducers for end-to-end speech recognition. Proc. Interspeech **2019**, 4395–4399 (2019)
25. Tüske, Z., Audhkhasi, K., Saon, G.: Advancing sequence-to-sequence based speech recognition. Proc. Interspeech **2019**, 3780–3784 (2019)
26. Vaswani, A., et al.: Attention is all you need. In: Advances in Neural Information Processing Systems, pp. 5998–6008 (2017)
27. Watanabe, S., et al.: Espnet: end-to-end speech processing toolkit. arXiv preprint arXiv:1804.00015 (2018)
28. Watzel, T., Kürzinger, L., Li, L., Rigoll, G.: Regularized forward-backward decoder for attention models. arXiv preprint arXiv:2006.08506 (2020)
29. Zheng, Y., et al.: Forward-backward decoding for regularizing end-to-end TTS. arXiv preprint arXiv:1907.09006 (2019)

KazNLP: A Pipeline for Automated Processing of Texts Written in Kazakh Language

Zhandos Yessenbayev[✉], Zhanibek Kozhirbayev, and Aibek Makazhanov

National Laboratory Astana, Nur-Sultan, Kazakhstan
{zhyessenbayev,zhanibek.kozhirbayev,aibek.makazhanov}@nu.edu.kz

Abstract. We present the current results of our ongoing work on develop-ing tools and algorithms for processing Kazakh language in the framework of KazNLP project. The project is motivated by the need in accessible, easy to use, cross-platform, and well-documented auto-mated text processing tools for Kazakh, particularly user generated text, which includes transliteration, code switching, and other artifacts of language-specific raw data that needs pre-processing. Thus, apart from a basic tokenization-tagging-parsing pipeline, and downstream applica-tions such as named entity recognition and spell checking, KazNLP offers pre-processing tools such as text normalization and language identifica-tion. All of the KazNLP tools are released under the Creative Commons license. Since the detailed description of the methods and algorithms that were used in KazNLP are published or to be published in various venues, reference to which is given in the corresponding sections, this work provides just an overview of the tools and their performance level.

Keywords: Kazakh language · Natural language processing · Corpus linguistics · Computational linguistics · Programming tools

1 Introduction

Kazakh language is the official language of the Republic of Kazakhstan, spoken by over 12 million people. From the NLP point of view, Kazakh is an interesting research object that presents several challenges as an agglutinative language with complex morphology and a relatively free word order. However, NLP research on this language is rather scarce, while demand in automated processing of Kazakh text is rising. For example, consider automatic spellchecking, one of the clas-sic NLP tasks. To our knowledge there are only two research papers [12,20] that discuss spell-checking for Kazakh. Moreover, the method described in [12] achieved only 83% accuracy, which was enough to outperform MS Word spell-checking system. Thus, while Kazakh spellcheckers are far from being perfect, there is almost no research in this direction. This is not surprising, because to achieve high performance on many NLP problems for languages with complex

A. Karpov and R. Potapova (Eds.): SPECOM 2020, LNAI 12335, pp. 657–666, 2020.
https://doi.org/10.1007/978-3-030-60276-5_63

morphology, one needs the very basic tools such as morphological analyzers and taggers. To be fair, we admit that such tools exist for Kazakh, but as far as we can tell, they are limited in access [7], ease of use [1,31], or the degree of readiness ("raw" experimental systems) [16]. Moreover, existing tools are documented either poorly or not at all and have not been evaluated on user generated text, such as internet reviews and message boards.

To address these issues, we have developed KazNLP, an open source Python library for processing Kazakh texts. KazNLP consists of the following modules, that can be used in isolation or in a pipeline: 1) initial normalization tool; 2) sentence-word tokenizer; 3) language identification tool; 4) morphological analyzer; 5) morphological tagger; 6) syntactic parser; 7) spelling checking and correction tool; 8) secondary normalization tool; 9) named entity recognizer. The set of tools is chosen to strike a balance between research and engineering applications, and will, hopefully, be extended in the future. To facilitate efficiency and performance trade-off, where possible, we have developed both statistical and neural implementations. Importantly, we avoided any hard-coded rules, to make the tools adaptable to the announced shift of the Kazakh alphabet from Cyrillic to Latin. The code is released under the *CC-SA-BY* license (Creative Commons Attribution 4.0), which allows any use, including commercial.

In what follows, we describe KazNLP in more detail. Specifically, Sect. 2 describes the datasets and corpora used in the development of KazNLP. Section 3 provides an overview of KazNLP components and corresponding NLP tasks. Section 4 briefly describes the software package and the demo web service. Finally, we draw conclusions and discuss future work in Sect. 5.

2   Text Datasets and Corpora

The development of KazNLP would not be possible without language resources described below. All the datasets are available either online or upon request.

2.1   Kazakh Language Corpus

The Kazakh Language Corpus (KLC) [17] was collected to aid linguistics, computational linguistics, and NLP research for Kazakh. It contains more than 135 million words in more than 400 thousand documents classified by genres into the following five sections: 1) *literary*; 2) *official*; 3) *scientific*; 4) *publicistic*; 5) *informal language*. KLC also has a portion of data annotated for syntax and morphology. It should be noted that initially the syntactic tagset comprised a compact set of syntactic categories, which were later improved during the development of Kazakh Dependency Treebank [13] described next.

2.2   Kazakh Dependency Treebank

Another important linguistic resource that lays the grounds of the current work is the Kazakh Dependency Treebank [13,14]. Following the best practices of the

Universal Dependencies (UD) guidelines [23] for consistent annotation of grammar, we have relabeled a part of KLC with lexical, morphological, and syntactic annotation that can be used by computer scientists working on language processing problems and by linguists likewise. The treebank contains about 61 K sentences and 934.7 K tokens (of those 772.8 K alphanumeric), stored in the UD-native CoNLL-U format. In addition, we annotated all the proper nouns of the corpus with such labels as a person's name, location, organization and others.

2.3   Other Datasets

To perform experiments on text normalization, language identification, and senti-ment analysis, we used a data set of user generated text collected from the three of the most popular Kazakhstani online news outlets, namely nur.kz, zakon.kz, and tengrinews.kz [21]. The noisy data was corrected and labeled semi-automatically resulting in total of 27,236 comments annotated for language and sentiment. Positive and negative comments amounted to 5995 (22%) and 7409 (27.2%), respectively, with the rest being neutral. In terms of languages, data set contains 63% of Kazakh texts, 34.4% of Russian texts and the rest is mixed comments (i.e. code-switched between Kazakh and Russian).

3   KazNLP Components

3.1   Initial Normalization Module

User generated content (UGC) generally refers to any type of content created by Internet users. UGC as a text is notoriously difficult to process due to prompt introduction of neologisms, peculiar spelling, code-switching or transliteration. All of this increases lexical variety, thereby aggravating the most prominent problems of NLP, such as out-of-vocabulary lexica and data sparseness. It has been shown [6] that certain preprocessing, known as lexical normalization or simply normalization, is required for them to work properly. Kazakhstani segment of Internet is not except from noisy UGC and the following cases are the usual suspects in wreaking the "spelling mayhem":

- *spontaneous transliteration*, e.g. Kazakh word "" can be spelled in three additional ways: "", "", and "biz";
- use of *homoglyphs*, e.g. Cyrillic letter "" (U+0456) can be replaced with Latin homoglyph "i" (U+0069);
- *code switching* – use of Russian words and expressions in Kazakh text and vice versa;
- *word transformations*, e.g. "","" instead of "" (great), or seg-mentation of words, e.g. " "or "____".

We have implemented a module for initial normalization of UGC. However, unlike with lexical normalization [6], we do not attempt to detect ill-formed words and recover their standard spelling. All that we really care about at this point is to provide an intermediate representation of the input UGC that will not necessarily match its lexically normalized version, but will be less sparse. Thus, we aim at improving performance of downstream applications by reducing vocabulary size (effectively, parameter space) and OOV rate. To this end, initial normalization does two things: (i) converts the input into a common script; (ii) recovers word transformations and does various minor replacements. Our simple *rule-based* approach (using regular expressions and dictionary) amounts to successive application of three straightforward procedures: (i) homoglyph resolution - substitutes Latin letters with Cyrillic counterparts (ii) common script transliteration - substitutes similar Latin and Cyrillic letters with common letter (iii) replacement and transformation - applies regular expressions to correct ill-formed word forms. It is difficult to perform a direct evaluation of this algorithm, but an extrinsic evaluation can be found in our previous work [21].

3.2   Sentence-Word Tokenizer

Sentence segmentation is a problem of segmenting text into sentences for further processing; and tokenization is a problem of segmenting text into chunks that for a certain task constitute basic units of operation (e.g. words, digits, etc.). At first glance, the problems might seem trivial, but this is not always the case, as many standard punctuation symbols might be used in different contexts (abbreviations, emoji, etc.), and the meaning of "tokens" and "sentences" may vary depending on a task at hand. Thus, to solve sentence and token segmentation problems one cannot blindly segment texts at the occurrences of certain symbols, and has to resort to a more sophisticated approach.

We are aware of ready to use tools that can be adapted to Kazakh, such as Punkt in NLTK [3], Elephant [4], and Apache OpenNLP [26]. However, they mostly use hand-crafted features. The only free-distributed tokenizer for Kazakh is based on the finite state transducer and is built into the morphological analyzer [31], which is not always convenient and necessary. Therefore, we decided to implement our own module for sentence splitting and word tokenization, which treats the problem as a single sequence labeling task.

The current version of KazNLP includes only HMM-based implementation of the module, but we have also experimented with an LSTM-based approach. This approach utilizes character embeddings, i.e. represent characters as vectors in a multidimensional continuous space. We evaluated our method on three typologically distant languages, namely Kazakh, English, and Italian. Experi-ments show that the performance on F1-measure achieves 95.60% on sentence and 99.61% word segmentation, respectively, for Kazakh language which is better than that of popular systems like Punkt and the Elephant. An interested reader is referred to our work in [30].

3.3   Language Identification Module

Language identification (LID) is the task of determining a language in which a given piece of information is presented be it text, audio or video. Early works on LID for texts have achieved near perfect accuracy when applied to small sets of monolingual texts, languages and domains, which led to a popular misconception that LID task was solved. The authors of [2] argue that LID is far from being solved, and show that as number of target languages grows and documents become shorter, performance of the standard methods drops. Another challenge that the authors mention, but do not address, is the fact that input text may be multilingual. Indeed, the results of the *Workshop on Computational Approaches to Code Switching* show that word-level LID on multilingual input is significantly harder than classical LID, i.e. document-level on monolingual input.

In KazNLP, we aim to provide tools for processing real world data, including noisy UGC, i.e. tweets, comments, Internet forum dialogs, etc. Apart from noise, there is a certain amount of Kazakh-Russian code-switching, which allows us to experiment with multilingual word-level LID. Following a common strategy of dealing with noisy UGC, prior to LID, we perform normalization as described above. For LID, we trained and applied character-based LSTM neural networks. The best performance of 99.73% was achieved for 150-character long sequences. Our results suggest improvement over the state-of-the-art for Kazakh language based on popular LangID [11] or Bayesian approach [9].

3.4 Morphological Analyzer and Tagger

Morphological analysis is the task of identifying the constituents of a word (root and morphemes), while morphological tagging is the task of identifying the most suitable analysis in the context of a sentence. Thus, this problems are closely related. A common approach for the agglutinative languages like Kazakh is to develop a finite state transducers to identify all possible morphological parses of a word [31], and on top of that apply a morphological disambiguation model based on statistical or other machine learning approaches [16,29].

In this module, we addressed the morphological analysis and tagging problem as a single sequence-to-sequence modeling task similar to neural machine transla-tion. In particular, we applied a character-level encoder-decoder neural models with attention using bi-directional LSTM network implemented in Open-NMT [8]. For comparison, we experimented with language independent packages like UDPi-pe [27] and Lemming [19]. In both tasks, our model outperformed the latter models, achieving the accuracy on morphological analysis of 87.8% and on mor-phological tagging of 96.8%.

3.5 Syntactic Parser

In computational linguistics and NLP, syntactic parsing usually amounts to determining a parse tree of a given sentence, according to a predefined grammar formalism, e.g. constituency, dependency, etc. grammars. In KazNLP, we adhere to dependency formalism and develop a dependency parser. Commonly used approaches to tackle this problem fall into two main classes: graph-based (MST-Parser [18]) and transition-based (Malt-parser [24]) parsing models. As an initial version, we have implemented a graph-based parser using data-driven statistical approach to compute weights of the search graph [32]. Thus, the goal is to find a minimum spanning tree in the given weighted directed graph. The performance of the parser in terms of unlabeled attachment score was 61.08, which is barely comparable to the state-of-the-art neural-based models. Nonetheless, the tool can be used in the domains with limited amount of data, where the advanced models are not applicable.

3.6   Spelling Correction Module

The spelling correction can be divided into two tasks: word recognition and error correction. For languages with a fairly straightforward morphology recognition may be reduced to a trivial dictionary look up. Correction is done through generating a list of possible suggestions: usually words within some minimal edit distance to a misspelled word. For agglutinative languages, such as Kazakh, even recognizing misspelled words becomes challenging as a single root may produce hundreds of word forms. It is practically infeasible to construct a dictionary with all possible word forms included: apart from being gigantic such a dictionary would be all but verifiable. For the same reason the correction task becomes challenging as well.

To address this problem, we followed the approach presented by Oflazer and Güzey [25]. In particular, we used a mixture of lexicon-based and generative approach by keeping a lexicon of roots and generating word forms from that lexicon on the fly for lookup and correct suggestions. To rank a list of suggestions, we use a Bayesian argument that combines error and source models. For our error model we employ a noisy channel-based approach proposed by Church and Gale [6]. Our source model is built upon the theoretical aspects that were used for morphological disambiguation in [5]. For the purpose of comparison we experimented with built-in Kazakh spelling packages in OpenOffice and Microsoft Office 2010. We show that although our method is more accurate than the open source and commercial analogues, achieving the overall accuracy of 83% in generating correct suggestions, the generation of those suggestions still needs improvement in terms of pruning and better ranking [12].

3.7   Secondary Normalization Module

In addition to the initial normalization and spelling correction modules, we implemented another tool to normalize and correct texts - secondary normalization module. The secondary normalization is designed to directly convert UGC to the standard language overcoming the drawbacks of two former stages, namely, incompleteness of rule-based approach. In this task, we applied a statistical machine translation strategy to convert from noisy texts (source language) to lexically correct texts (target language) given the corresponding parallel data set. The performance of this approach was 21,67 in terms of BLEU metrics, which is considered as a moderate result [22]. It should be noted, that this tool still can be used after initial normalization and spelling correction, as it also able to correctly deal with phrases.

3.8 Named Entity Recognizer

Named entity recognition (NER) is considered one of the important NLP task. It is a problem of recognizing real world objects found in a sentence, such as geographical location, person's name, organization, and others. There are several approaches based on manually created grammar rules, statistical models, and machine learning to solve the NER problem.

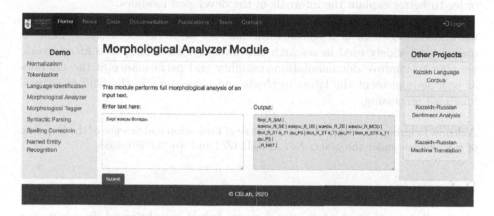

Fig. 1. A screenshot demonstrating a work of the morphological analyzer module.

Previously we have implemented a model for recognition of named entities in Kazakh language based on conditional random field (CRF) [28]. However, in our latest setup, we use a hybrid approach combining a bidirectional LSTM neural network and a CRF models. The main idea is to feed the features determined by CRF as input to LSTM network, thus, replacing the linear scoring by non-linear neural network scoring. The performance of this model in terms of F1-measure is 88%. For more details see our work [10, 28].

4 Toolkit and Web Service

This project aims at building free/open source language processing tools for Kazakh. The proposed set of tools is designed to tackle a wide range of NLP problems that can be divided into pre-processing, core processing, and application-level routines. Each NLP task is implemented as a separate programming module in *Python 3* programming language and released under *CC-SA-BY* license. The current version of source code of KazNLP and the documentation are available on Github repository [15]. KazNLP can be installed and run on all platforms including Linux, OS X, and Windows, where Python is supported.

In addition to the packages, we have developed a web service that can be tested by an interested user. The web service is accessible online via link http://kazcorpus.kz/kaznlp/. Figure 1 shows a screenshot demonstrating a work of the morphological analyzer module.

5   Conclusion and Future Work

In this paper we presented a new package KazNLP which implements the core natural language processing tasks for automated processing of texts in Kazakh language. Rather than emphasizing the design and usage of the package, for which the reader is referred to the corresponding documentation in the repository, we focused on the NLP tasks and outlined the way we tackled them in order to better explain the internals of the developed modules.

Although KazNLP is a unique and holistic tool oriented for Kazakh language, we understand there is a considerable amount of work necessary to make it mature and widely used in research and industry. Therefore, as a future work, we plan to improve documentation, usability and performance of the modules as well as implement the latest methods and algorithms in the field of natural language processing.

Acknowledgments. Supported by Ministry of Education and Science of the Republic of Kazakhstan under the grants No. AP05134272 and No. AP08053085.

References

1. Assylbekov, Z., et al.: A free/open-source hybrid morphological disambiguation tool for kazakh (2016)
2. Baldwin, T., Lui, M.: Language identification: the long and the short of the matter. In: Human Language Technologies: The 2010 Annual Conference of the North American Chapter of the Association for Computational Linguistics, pp. 229–237. Association for Computational Linguistics, Los Angeles, California (2010)
3. Bird, S., Loper, E.: NLTK: the natural language toolkit. In: Proceedings of the ACL Interactive Poster and Demonstration Sessions, pp. 214–217. Association for Computational Linguistics (2004)
4. Evang, K., Basile, V., Chrupała, G., Bos, J.: Elephant: sequence labeling for word and sentence segmentation. In: Proceedings of the 2013 Conference on Empirical Methods in Natural Language Processing, pp. 1422–1426. Association for Computational Linguistics (2013)
5. Hakkani-Tür, D.Z., Oflazer, K., Tür, G.: Statistical morphological disambiguation for agglutinative languages. In: Proceedings of the 18th Conference on Computational Linguistics - Volume 1, pp. 285–291. Association for Computational Linguistics (2000)
6. Han, B., Baldwin, T.: Lexical normalisation of short text messages: makn sens a #twitter. In: Proceedings of the 49th Annual Meeting of the Association for Computational Linguistics: Human Language Technologies, pp. 368–378. Association for Computational Linguistics (2011)
7. Kessikbayeva, G., Cicekli, I.: Rule based morphological analyzer of kazakh language. In: Proceedings of the 2014 Joint Meeting of SIGMORPHON and SIGFSM, pp. 46–54. Association for Computational Linguistics (2014)
8. Klein, G., Kim, Y., Deng, Y., Senellart, J., Rush, A.: OpenNMT: open-source toolkit for neural machine translation. In: Proceedings of ACL 2017, System Demonstrations, pp. 67–72. Association for Computational Linguistics, Vancouver, Canada July 2017 https://www.aclweb.org/anthology/P17-4012

9. Kozhirbayev, Z., Yessenbayev, Z., Makazhanov, A.: Document and word-level language identification for noisy user generated text. In: 2018 IEEE 12th International Conference on Application of Information and Communication Technologies (AICT), pp. 1–4 (2018)
10. Kozhirbayev, Z., Yessenbayev, Z.: Named entity recognition for the kazakh language. Journal of Mathematics, Mechanics and Computer Science, 106(2) (2020) (Submitted)
11. Lui, M., Baldwin, T.: Langid.py: an off-the-shelf language identification tool. In: Proceedings of the ACL 2012 System Demonstrations, pp. 25–30. Association for Computational Linguistics (2012)
12. Cheng, V., Li, C.H.: Combining supervised and semi-supervised classifier for personalized spam filtering. In: Zhou, Z.H., Li, H., Yang, Q. (eds.) PAKDD 2007. LNCS (LNAI), vol. 4426, pp. 449–456. Springer, Heidelberg (2007). https://doi.org/10.1007/978-3-540-71701-0_45
13. Makazhanov, A., Sultangazina, A., Makhambetov, O., Yessenbayev, Z.: Syntactic annotation of kazakh: following the universal dependencies guidelines. a report. In: 3rd International Conference on Turkic Languages Processing, (TurkLang 2015), pp. 338–350 (2015)
14. Makazhanov, A., Yessenbayev, Z.: NLA-NU Kazakh Dependency Treebank. https://github.com/nlacslab/kazdet. Accessed 10 June 2020
15. Makazhanov, A., Yessenbayev, Z., Kozhirbayev, Z.: KazNLP: NLP Tools for Kazakh Language. https://github.com/nlacslab/kaznlp. Accessed 10 June 2020
16. Makhambetov, O., Makazhanov, A., Sabyrgaliyev, I., Yessenbayev, Z.: Data-driven morphological analysis and disambiguation for kazakh. In: Gelbukh, A. (ed.) CICLing 2015. LNCS, vol. 9041, pp. 151–163. Springer, Cham (2015). https://doi.org/10.1007/978-3-319-18111-0_12
17. Makhambetov, O., Makazhanov, A., Yessenbayev, Z., Matkarimov, B., Sabyrgaliyev, I., Sharafudinov, A.: Assembling the kazakh language corpus. In: Proceedings of the 2013 Conference on Empirical Methods in Natural Language Processing, pp. 1022–1031. Association for Computational Linguistics (2013)
18. McDonald, R., Pereira, F.: Online learning of approximate dependency parsing algorithms. In: 11th Conference of the European Chapter of the Association for Computational Linguistics. Association for Computational Linguistics (2006)
19. Müller, T., Cotterell, R., Fraser, A., Schütze, H.: Joint lemmatization and morphological tagging with lemming. In: Proceedings of the 2015 Conference on Empirical Methods in Natural Language Processing, pp. 2268–2274. Association for Computational Linguistics (2015)
20. Mussayeva, A.: Kazakh language spelling with hunspell in openoffice.org. Tech. rep., The University of Nottingham (2008)
21. Myrzakhmetov, B., Yessenbayev, Z., Makazhanov, A.: Initial normalization of user generated content: case study in a multilingual setting. In: 2018 IEEE 12th International Conference on Application of Information and Communication Technologies (AICT), pp. 1–4 (2018)
22. Myzakhmetov, B., Yessenbayev, Z.: Normalization of noisy user comments in kazakh language using statistical machine translation approach. Bulletin of the Eurasian National University (2020) (Submitted)
23. Nivre, J.: Universal Dependencies. https://universaldependencies.org/. Accessed 10 June 2020

24. Nivre, J., Hall, J., Nilsson, J.: MaltParser: a data-driven parser-generator for dependency parsing. In: Proceedings of the Fifth International Conference on Language Resources and Evaluation (LREC'06). European Language Resources Association (ELRA) (2006)

25. Oflazer, K., Guzey, C.: Spelling correction in agglutinative languages. In: Fourth Conference on Applied Natural Language Processing, pp. 194–195. Association for Computational Linguistics (1994)

26. OpenNLP: The Apache OpenNLP Library. https://opennlp.apache.org/. Accessed 13 June 2020

27. Straka, M., Hajič, J., Straková, J.: UDPipe: trainable pipeline for processing CoNLL-u files performing tokenization, morphological analysis, POS tagging and parsing. In: Proceedings of the Tenth International Conference on Language Resources and Evaluation (LREC 2016), pp. 4290–4297. European Language Resources Association (ELRA) (2016)

28. Tolegen, G., Toleu, A.: Named entity recognition for kazakh using conditional random fields. In: The 4-th International Conference on Computer Processing of Turkic Languages (TurkLang 2016) (2016)

29. Toleu, A., Tolegen, G., Makazhanov, A.: Character-aware neural morphological disambiguation. In: Proceedings of the 55th Annual Meeting of the Association for Computational Linguistics (Volume 2: Short Papers), pp. 666–671. Association for Computational Linguistics (2017)

30. Toleu, A., Tolegen, G., Makazhanov, A.: Character-based deep learning models for token and sentence segmentation. In: The 5-th International Conference on Computer Processing of Turkic Languages (TurkLang 2017) (2017)

31. Washington, J., Salimzyanov, I., Tyers, F.: Finite-state morphological transducers for three kypchak languages. In: Proceedings of the Ninth International Conference on Language Resources and Evaluation (LREC 2014), pp. 3378–3385. European Language Resources Association (ELRA) (2014)

32. Yessenbayev, Z., Kozhirbayev, Z.: Data-driven dependency parsing for kazakh. Bulletin of the Eurasian National University (2020) (Submitted)

Diarization Based on Identification
with X-Vectors

Zbyněk Zajíc(✉) ⓘ, Josef V. Psutka ⓘ, and Luděk Müller ⓘ

University of West Bohemia, Faculty of Applied Sciences, NTIS - New Technologies
for the Information Society and Department of Cybernetics,
Univerzitní 8, 301 00 Plzeň, Czech Republic
{zzajic,psutka_j,muller}@ntis.zcu.cz

Abstract. In this paper, we describe a diarization of mono channel
telephone recordings from The Language Consulting Center providing
the Czech language consultancy service. In our proposed approach to a
diarization, we use information about the known identity of one speaker
(the language counsellor) acquired from the text transcription at the
beginning of the conversation. In the state-of-the-art diarization based
on the x-vectors clustering, we replace the clustering step by the identification of each segment of the recording against the counsellor's identity x-vector and the general x-vector model that represents the client.
Our proposed diarization without resegmentation step can be used as an
online approach. Because of the uniqueness of our data, we compare our
results with the Kaldi diarization as the baseline system.

Keywords: Diarization · Identification · X-vector · Automatic speech
recognition

1 Introduction

The main goal of the project "Access to a Linguistically Structured Database
of Enquiries from the Language Consulting Center"[1] is to publish data from
The Language Consulting Center (LCC) of the Czech Language Institute of the
Academy of Sciences of the Czech Republic. This institute provides a unique
language consultancy service in the matters of the Czech language mostly via a
telephone line open to public calls.

The Automatic Speech Recognizer (ASR) and the language processing methods (like topic detection, keyword spotting, etc.) are being designed to describe
the speech data to allow their better accessibility, more in [23,24].

Part of the telephone calls from the LCC, almost 8k recordings, were recorded
on the analog telephone line (8kHz, μ-law resolution) during 2013–2016 time
period and stored only in mono channel - language counsellor and client are
mixed in one channel. The diarization of these data can improve the ASR results

---

[1] https://starfos.tacr.cz/en/project/DG16P02B009.

© Springer Nature Switzerland AG 2020
A. Karpov and R. Potapova (Eds.): SPECOM 2020, LNAI 12335, pp. 667–678, 2020.
https://doi.org/10.1007/978-3-030-60276-5_64

using an adaptation of the acoustic and language model. The topic identification method has also shown the differences in using only an answer from language counsellor instead of the whole conversation to categorize the topic in the recordings [24]. Therefore, our goal is to apply a method for the Speaker Diarization (SD) to separate the question of the LCC's client from the answer of the language counsellor.

The most common approach to the SD consists of the segmentation of an input signal, followed by the merging of the segments into the clusters corresponding to individual speakers [10,12,15]. Alternatively, the segmentation and the clustering step can be combined into a single iterative process [9,17].

In this paper, we expand the state-of-the-art off-line SD approach [4,22] based on the x-vector [18] representation of the speech segments applied for the First and Second DIHARD Challenge [11], by exploiting the known information about the identity of one speaker (the language counsellor) a similar way as in work [1,6].

Our previous paper [23] used the information about counsellor's identity only for the resegmentation step to initialize the speaker's Gaussian Mixture Models (GMM). In this paper, we use the x-vector identity models instead of GMM and replace the clustering step of diarization by x-vector identification process. As a resegmentation step in this paper, the Variational Bayes (VB) inference [3,13] is preferred instead of an early approach based on the Viterbi algorithm [16].

2   Baseline Diarization System

As our baseline system, we have chosen the Kaldi with the recipe for the diarization[2] [14]. A diagram of the system is shown in Fig. 1.

After the Mel Frequency Cepstral Coefficients (MFCCs) feature extraction, the constant window segmentation is applied to generate segments of test recording. Consequently, the x-vector [18] is extracted for each of these segments using Time Delay Neural Network (TDNN). The x-vectors are then clustered using Agglomerative Hierarchical Clustering (AHC) method with the stopping threshold set for two clusters. The Probabilistic Linear Discriminant Analysis (PLDA) model [8] is used to compute the similarity between the segments. The x-vectors are whitened before the PLDA computation by subtracting the mean and transforming by the Linear Discriminant Analysis (LDA) matrix. The gross scaling given by the segmentation with the constant window can be refined by featurewise resegmentation using VB method.

3   Diarization System Based on Identification

Our proposed system for SD is based also on comparison the speech segments represented by x-vectors. Instead of clustering x-vectors [14,20], we have decided to incorporate known information about the speakers in the recordings. The main

---

[2] https://github.com/kaldi-asr/kaldi/tree/master/egs/callhome_diarization/v2.

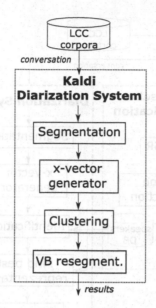

Fig. 1. Diagram of the Kaldi diarization process.

contribution to our problem is the knowledge of the identity of one part of the conversation, the language counsellor. In contrast with the identity of the client (where there is a potentially infinite number of the speakers), the list of the language counsellors answering the language queries is limited and known. A diagram of our proposed diarization system is shown in Fig. 2.

This section further describes the main steps of our proposed diarization process. The feature extraction, segmentation and x-vector computation are the same as in Kaldi system, described in Sect. 2.

3.1   Feature Extraction

As a feature vector, we use MFCC with a Hamming window of length 25 ms with 10 ms shift. Then FFT and the power spectrum are computed. There are 23 triangular overlapping bins whose centres are equally spaced in the mel-frequency domain (in a range from 20 Hz to 7,4 kHz), the log of the energies is computed and the cosine transformation is taken, keeping 23 coefficients. Cepstral Mean Normalization (CMN) is applied to compensate for channel variations. This module also performs an energy-based voice activity detection (VAD), with every frame being labelled as speech or silence based on a threshold (= 5.5).

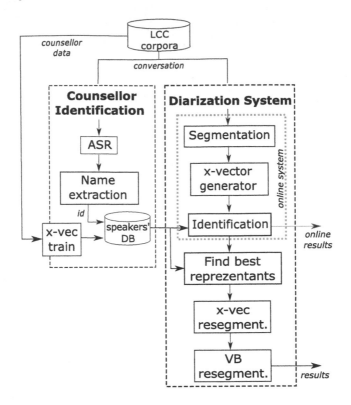

Fig. 2. Diagram of the diarization based on identification.

3.2 Segmentation

The segmentation provides chunks of speech between important non-speech events. To minimize the presence of more than one speaker in a segment, the longer segments are split into intervals of max. 1.5 s, with 0.75 s overlaps. This "blind" constant window segmentation instead of the segmentation based on the Speaker Change Detection (SCD) [7,19,21] is used, mainly because of the advantage of x-vectors' ability to represent short segments and the application of the feature-wise resegmentation step on the end of the diarization.

3.3 Segment Description

Each segment is described by the x-vector. A TDNN (with three time-delayed and two fully connected layers) is used as an x-vector extractor, and x-vectors are extracted from the affine component of the second-to-last layer with dimension 512. The LDA is used as the whitening transformation with mean subtraction.

3.4 Segment Identification

The LCC's telephone recordings generally contain only two speakers (the counsellor and the client). Instead of the clustering process (see Sect. 2), we have decided to use the client's and the counsellor's identity x-vector model to decide if the segment belongs to the client or counsellor.

As a distance measure, we use the PLDA model, with the between-class dimension equal to the feature dimension, for calculating the similarity of two x-vectors. The acquiring of the counsellor's x-vector is described in Sect. 5. As the unknown client's x-vector the universal male and female x-vectors are used. The segment is assigned to the identity (counsellor or client) with highest PLDA similarity, as shown in Algorithm 1.

Algorithm 1: Segment identification using maximal PLDA similarity to the counsellor's $\mathbf{c}$, male $\mathbf{m}$ and female $\mathbf{f}$ x-vector model.

Data:
test x-vectors: $\mathbf{X}^{test} = [\mathbf{x}_1^{test}, \mathbf{x}_2^{test}, ..., \mathbf{x}_T^{test}]$,
identity x-vectors: $[\mathbf{c}, \mathbf{m}, \mathbf{f}]$
Result:
labels: $\mathbf{Y}^* = [y_1^*, y_2^*, ..., y_T^*]$,
PLDA similarity: $\mathbf{Sim} = [sim_1, sim_2, ..., sim_T]$

for $i = 1, 2, ..., T$ **do**
- find label:
  $$y_i^* = argmax(plda(\mathbf{x}_i^{test}, [\mathbf{c}, \mathbf{m}, \mathbf{f}]))$$
- compute similarity (used in Alg. 2 for resegmentation):
  $$sim_i = plda(\mathbf{x}_i^{test}, \mathbf{c}) - max(plda(\mathbf{x}_i^{test}, [\mathbf{m}, \mathbf{f}]))$$

3.5 Resegmentation Based on Refined X-Vectors

Because of the missing info about the client and also the imprecise counsellor model (more in Sect. 5), the previous step brings some faulty decisions about the identity of the segments. Therefore, we choose only the representative segments with sufficient similarity between the segment and the identity models (these with the bigger difference in similarity to these identity models).

From these representative segments, new identity models are created by x-vectors averaging ($ver0$), as shown in Algorithm 2, and the whole recording is resegmented according to these two new identity x-vectors.

Algorithm 2: Computation of the new x-vector models for counsellor $\mathbf{c}^{new}$ and client $\mathbf{mf}^{new}$ from the segments with the maximal PLDA similarity. These models are then used for resegmentation by the same approach as in Alg. 1.

Data:
test x-vectors: $\mathbf{X}^{test} = [\mathbf{x}_1^{test}, \mathbf{x}_2^{test}, ..., \mathbf{x}_T^{test}]$,
plda similarity: $\mathbf{Sim} = [sim_1, sim_2, ..., sim_T]$
Result:
new identity x-vectors: $[\mathbf{c}^{new}, \mathbf{mf}^{new}]$

Compute:
- find thresholds:
$$Th^c = mean(sim_i) \quad \text{for} \quad \forall i \quad \text{where} \quad sim_i > 0$$
$$Th^{mf} = mean(sim_i) \quad \text{for} \quad \forall i \quad \text{where} \quad sim_i < 0$$
- compute new identity models:
$$\mathbf{c}^{new} = mean(\mathbf{x}_i^{test}) \quad \text{for} \quad \forall i \quad \text{where} \quad sim_i > Th^c$$
$$\mathbf{mf}^{new} = mean(\mathbf{x}_i^{test}) \quad \text{for} \quad \forall i \quad \text{where} \quad sim_i < Th^{mf}$$

Note: the original counsellor's x-vector $\mathbf{c}$ is relatively accurate, therefore, we can
- add $\mathbf{c}$ to the set of nearest x-vectors to get more data for the new counsellor's model $\mathbf{c}^{new}$ (*ver*1).
- or left only the original counsellor's x-vector $\mathbf{c}$, $\mathbf{c}^{new} = \mathbf{c}$ (*ver*2).

3.6   VB Resegmentation

After the final clustering, the VB resegmentation [3,13] can be used to refine the diarization feature-wise. For the initialization of the VB, the labelling from the previous step is used. In this work, we used implementation of VB available in Kaldi[3].

4   Online Diarization Base on Identification

From the nature of this approach, the first part of the system (excluding both resegmentation steps) can be used as online diarization, see Fig. 2. Because of the replacing the clustering step by the identification of the segment against identity models (counsellor, male and female client), the diarization decision about the segment can be done immediately after obtaining this segment (1.5s or 0,75s respectively for other than the first one).

In our task, where only one of the speaker in the recording is known (counsellor), the online diarization by identification against universal male and female x-vectors is inaccurate, see the results in Sect. 6.2.

---

[3] https://github.com/kaldi-asr/kaldi/tree/master/egs/callhome_diarization/v1/ diarization.

5   Counsellor Identification

The identity of the counsellor in the recording is obtained from the transcription (done by the annotators for the sake of this paper). The whole list of counsellors employed by the LCC is known and they were instructed to introduce themselves to the phone with their name and organization (LCC). For this reason, the identification task is reduced to find one name from the list appearing at the beginning of the transcription.

For each counsellor, the identity x-vector with 512 components was extracted from his/her data obtained from manually transcribed and force-aligned mono recordings, the oldest ones. These recordings were divided into the counsellor and client data according to the transcription also. For the other side in the conversation, the LCC's client, we extracted the only general x-vector model for a female and a male client. As we have described in the previous paper [23], these transcriptions were not flawless, especially in the case of the overlapped speech.

This whole process is represented by the left branch of the diagram in Fig. 2.

6   Experiments

This section describes our experiments on 715 recordings, a small part of the mono channel data from the LCC which contains a manual transcription.

6.1   Training Data

The following LDC corpora were used as training data for VB, TDNN and PLDA: NIST 2004,05,06 (LDC2006S44, LDC2011S01, LDC2011S04, LDC2011 S09, LDC2011S10, LDC2012S01, LDC2011S05, LDC2011S08), SWBD2 Phase2,3 (LDC99S79, LDC200-2S06) and SWBD Cellular1,2 (LDC2001S13, LDC2004S07). Additionally, data augmentation (additive noise, music, babble and reverberation) was applied to this data.

The list of counsellors consisted of 9 speakers (3 males and 6 females). We excluded the first 10 recordings (sorted by date of acquisition) for each counsellor to represent his/her identity, the rest of the data was used for testing. The counsellor's x-vector is generated on 1 to 10 recordings to investigate the influence of the amount of data for identification. In Fig. 3, the mean and standard deviation of training data duration across all counsellors is depicted in seconds.

To representing the client's model, the general male and female x-vector models were generated from the client side of all 715 recordings (210 males and 505 females).

The hyper-parameters for the diarization process (length of the segmentation window, whitening transformation, etc.) were tuned on a related task with CallHome corpus [2], so development set was not needed.

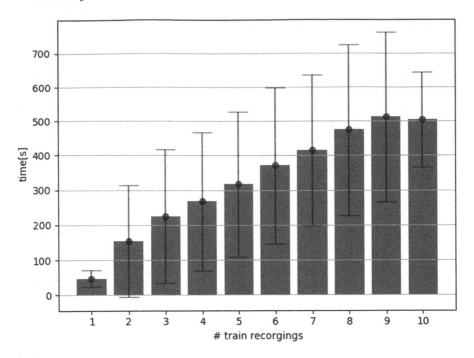

Fig. 3. Mean and standard deviation of train recordings duration across each counsellor.

6.2 Results

Figure 4 presents the results in terms of Diarization Error Rate (DER) [5] on test recordings for our proposed diarization system based on identification with a different number of recordings to represent the counsellor's models. The results without resegmentation steps can be considered as an online system (SD_id_online). The results of three versions to the x-vector resegmentation are labelled as $SD\_id\_ver0 - 2$ (see the note in Sect. 3.5). For comparison to the baseline, the Kaldi system result is also depicted (SD_Kaldi).

Table 1 shows results of our proposed diarization system based on identification with 10 recordings to represent the counsellor's models by x-vectors for three versions of x-vector resegmentation ($SD\_id\_ver0 - 2$), compared with Kaldi system SD_Kaldi. The results are stated with and without VB resegmentation. The pure identification step without any resegmentation is labelled as SD_id_online.

7 Discussion

Our proposed approach based on identification ($SD\_id\_ver0 - 2$) improved the Kaldi results in both settings - without and with VB resegmentation, see Table 1. From Fig. 4, we can see that 3–5 training recordings (approx. 200–300 s) for the identity's x-vectors is sufficient to outperform the baseline system.

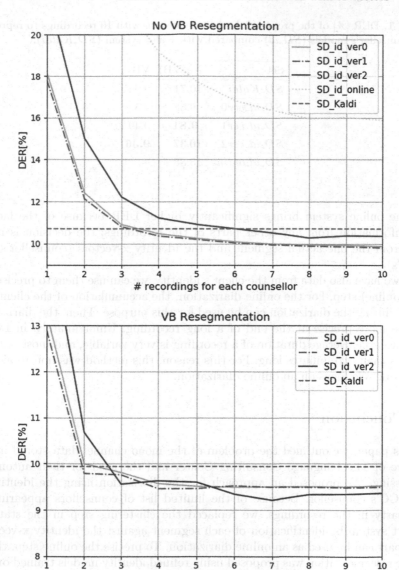

Fig. 4. Graph of DER development depending on the number of training sentences with and without VB resegmentation. Kaldi system works without identification model, therefore its results are invariant to the number of recordings used for this model.

System SD_id_ver0 and SD_id_ver1 have very similar results, adding the original counsellor's model (SD_id_ver1) brings minimal improvement. It can be explained by the relatively precise previous diarisation step in case of counsellor's data.

Table 1. DER [%] of the proposed diarization system with 10 recordings to represent the counsellor's models (*SD_id*) compared with Kaldi system (*SD_Kaldi*).

Set	No(VB)	VB
SD_Kaldi	10.71	9.94
SD_id_ver0	9.88	9.52
SD_id_ver1	**9.81**	9.49
SD_id_ver2	10.37	**9.36**
SD_id_online	15.88	–

The online system brings significantly higher DER because of the lack of any information from any other parts of the recordings, the decision is made only from the one actual segment and the identity x-vectors (counsellor's and client's).

If we have also data from the client's identity, we can use them to precise the first (online) step. For the online diarization, the accumulation of the client's x-vector during the diarization can be used for this purpose. Then, the diarization will be more precise at the end of a long recording. Unfortunately, in LCC's database, the average duration of a recording is very variable, and most of them are less than two minutes long. For this reason, this method was not applied to this work. We focus on an offline diarization.

8 Conclusion

In this paper, we outlined the problem of the mono channel data stored in the archive of The Language Consulting Center and the need for their automatic processing. We proposed an approach for the diarization using the identity of the LCC's counsellor. Because of the limited list of counsellors appearing as one party in the recordings, we replaced the clustering step in the state-of-the-art system by identification of each segment against the identity x-vectors. This part can be used as an online diarization. To precise the online step, the x-vector resegmentation was proposed using refined identity models trained on the representative segments from the previous step. For the comparison, we applied a Kaldi recipe for the diarization.

Acknowledgements. The work described herein has been supported by the Ministry of Education, Youth and Sports of the Czech Republic, Project No. LM2018101 LINDAT/ CLARIAH-CZ and the Ministry of Culture Czech Republic, project No. DG16P02B009. Access to computing and storage facilities owned by parties and projects contributing to the National Grid Infrastructure MetaCentrum, provided under the program "Projects of Large Research, Development, and Innovations Infrastructures" (CESNET LM20150-42), is greatly appreciated.

References

1. Campr, P., Kunešová, M., Vaněk, J., Čech, J., Psutka, J.: Audio-Video Speaker diarization for unsupervised speaker and face model creation. In: Sojka, P., Horák, A., Kopeček, I., Pala, K. (eds.) TSD 2014. LNCS (LNAI), vol. 8655, pp. 465–472. Springer, Cham (2014). https://doi.org/10.1007/978-3-319-10816-2_56
2. Canavan, A., Graff, D., Zipperlen, G.: CALLHOME American English Speech, LDC97S42. In: LDC Catalog. Philadelphia: Linguistic Data Consortium (1997)
3. Diez, M., Burget, L., Matejka, P.: Speaker diarization based on bayesian HMM with eigenvoice priors. In: Odyssey - Speaker and Language Recognition Workshop, pp. 147–154 (2018)
4. Diez, M., et al.: BUT system for DIHARD speech diarization challenge 2018. In: Intespeech, pp. 2798–2802. Hyderabad (2018)
5. Fiscus, J.G., Radde, N., Garofolo, J.S., Le, A., Ajot, J., Laprun, C.: The rich transcription 2006 spring meeting recognition evaluation. Mach. Learn. Multimodal Interact. **4299**, 309–322 (2006)
6. Geiger, J.T., Wallhoff, F., Rigoll, G.: GMM-UBM based open-set online speaker diarization. In: Interspeech, pp. 2330–2333. Makuhari (2010)
7. Hrúz, M., Zajíc, Z.: Convolutional neural network for speaker change detection in telephone speaker diarization system. In: ICASSP, pp. 4945–4949. IEEE, New Orleans (2017)
8. Ioffe, S.: Probabilistic linear discriminant analysis. Lecture Notes in Computer Science, pp. 531-542 (2006)
9. Kenny, P., Reynolds, D., Castaldo, F.: Diarization of telephone conversations using factor analysis. IEEE J. Sel. Top. Sign. Process. **4**(6), 1059–1070 (2010)
10. Rouvier, M., Dupuy, G., Gay, P., Khoury, E., Merlin, T., Meignier, S.: An open-source state-of-the-art toolbox for broadcast news diarization. In: Interspeech, pp. 1477–1481. Lyon (2013)
11. Ryant, N., et al.: The second DIHARD diarization challenge: dataset, task, and baselines. In: INTERSPEECH. Gratz (2019)
12. Sell, G., Garcia-Romero, D.: Speaker diarization with PLDA I-vector scoring and unsupervised calibration. In: IEEE Spoken Language Technology Workshop, pp. 413–417. South Lake Tahoe (2014)
13. Sell, G., Garcia-Romero, D.: Diarization resegmentation in the factor analysis subspace. In: ICASSP, pp. 4794–4798. IEEE, April 2015
14. Sell, G., et al.: Diarization is hard : some experiences and lessons learned for the JHU team in the inaugural DIHARD challenge. In: Interspeech, pp. 2808–2812. Hyderabad (2018)
15. Senoussaoui, M., Kenny, P., Stafylakis, T., Dumouchel, P.: A study of the cosine distance-based mean shift for telephone speech diarization. Audio Speech Lang. Process. **22**(1), 217–227 (2014)
16. Shum, S., Dehak, N., Chuangsuwanich, E., Reynolds, D., Glass, J.: Exploiting intra-conversation variability for speaker diarization. In: Interspeech, pp. 945–948. Florence (2011)
17. Shum, S.H., Dehak, N., Dehak, R., Glass, J.R.: Unsupervised methods for speaker diarization: an integrated and iterative approach. Audio Speech Lang. Process. **21**(10), 2015–2028 (2013)
18. Snyder, D., Garcia-Romero, D., Sell, G., Povey, D., Khudanpur, S.: X-vectors: robust DNN embeddings for speaker recognition. In: ICASSP, pp. 5329–5333 (2018)

19. Wang, R., Gu, M., Li, L., Xu, M., Zheng, T.F.: Speaker segmentation using deep speaker vectors for fast speaker change scenarios. In: ICASSP, pp. 5420–5424. IEEE, New Orleans (2017)
20. Zajíc, Z., Kunešová, M., Radová, V.: Investigation of segmentation in i-vector based speaker diarization of telephone speech. In: Ronzhin, A., Potapova, R., Németh, G. (eds.) SPECOM 2016. LNCS (LNAI), vol. 9811, pp. 411–418. Springer, Cham (2016). https://doi.org/10.1007/978-3-319-43958-7_49
21. Zajíc, Z., Hrúz, M., Müller, L.: Speaker diarization using convolutional neural network for statistics accumulation refinement. In: Interpeech, pp. 3562–3566. Stockholm (2017)
22. Zajíc, Z., Kunešová, M., Hrúz, M., Vaněk, J.: UWB-NTIS speaker diarization system for the DIHARD II 2019 challenge. In: Interspeech submitted. Graz (2019)
23. Zajíc, Z., Psutka, J.V., Zajícová, L., Müller, L., Salajka, P.: Diarization of the language consulting center telephone calls. In: Salah, A.A., Karpov, A., Potapova, R. (eds.) SPECOM 2019. LNCS (LNAI), vol. 11658, pp. 549–558. Springer, Cham (2019). https://doi.org/10.1007/978-3-030-26061-3_56
24. Zajic, Z., et al.: First insight into the processing of the language consulting center Data. In: Karpov, A., Jokisch, O., Potapova, R. (eds.) SPECOM 2018. LNCS (LNAI), vol. 11096, pp. 778–787. Springer, Cham (2018). https://doi.org/10.1007/978-3-319-99579-3_79

Different Approaches in Cross-Language Similar Documents Retrieval in the Legal Domain

Vladimir Zhebel$^{(\boxtimes)}$ [iD], Denis Zubarev [iD], and Ilya Sochenkov [iD]

Federal Research Center Computer Science and Control of the Russian Academy of Sciences, 44-2 Vavilov Str, Moscow 119333, Russia
{zhebel,zubarev,sochenkov}@isa.ru

Abstract. The problem of cross-lingual information retrieval in the legal domain is up-to-date, because of the need of studying the best international practices to improve legislation. One of the possible solutions is thematically similar document retrieval. However, there is an important task to transfer between languages. The paper describes different approaches to solve this problem: from classical mediator-based methods to modern procedures of distributive semantics. As a test collection, we have used the UN digital library. The combination of the extended translation model and BM25 ranking function demonstrates the best results.

Keywords: Cross-Lingual document retrieval · Distributional semantics · Information retrieval in the legal domain

1 Introduction

In a condition of ongoing global progress – especially in the technological field – and the birth of new entities, there is a need to supplement and adjust existing legislation to determine the legal status of new concepts. So, for example, one of the tasks set in the Decree of the President of the Russian Federation dated 05.05.2018 "On national goals and strategic tasks for the development of the Russian Federation for the period up to 2024" is the creation of a system for legal regulation for the digital economy, based on a flexible approach in each area, and also the introduction of civil turnover based on digital technologies. It would be impossible to solve this problem without studying best international practices.

However, the volume of existing legal information is so high – not to mention the speed of its growth – that it is not possible to analyze the world experience manually, especially with the fact that in different countries the legislation is written in their language, and legal terms will vary among themselves.

As a consequence, solving the problem of cross-lingual information retrieval is of the current interest. Since the generation of correct search query requires the researcher to have a sufficiently deep understanding of the peculiarities of legal vocabulary in the language of the documents under consideration, the possibility of searching documents

© Springer Nature Switzerland AG 2020
A. Karpov and R. Potapova (Eds.): SPECOM 2020, LNAI 12335, pp. 679–686, 2020.
https://doi.org/10.1007/978-3-030-60276-5_65

for thematically similar texts on a given reference document or its fragment is the most interesting task.

This paper describes various approaches to find thematically similar documents written in different languages. For documents similarity calculation we use model "bag of words" with cosine and hamming measures.

2   Experimental Dataset

For quality evaluation of the considered methods and data source for model training, it is necessary to choose a suitable parallel corpus of documents. We decided to use the United Nations digital library texts in Russian and English as such corpora: this dataset is an excellent aligned parallel set of legal documents. We collected about 33 thousand pairs of documents.

3   Usage of Mediator

As noted in [1], one of the common effective approaches is to use a mediator collection – a parallel corpus of documents in two languages – to transfer between languages.

3.1   Explicit Semantic Analysis

Firstly, let us consider the CL-ESA method described in [2]. In this model, documents are presented as a weighted vector of concepts determined by Wikipedia articles.

We selected about 800 thousand pairs of aligned articles in Russian and English (articles that identified as comparable across languages by the Wikipedia community) as a concept mediator.

As a result, for each examined document D, the weight of concept C is defined as the cosine of the angle between the Top-M keywords of document D and the corresponding keywords of the article attached to the concept:

$$\frac{\sum_{w_i \in D} v_i c_i}{\sqrt{\sum_{w_i \in D} v_i^2} \sqrt{\sum_{w_i \in D} c_i^2}},$$

where v_i is the $tf * idf$ measure of the word w_i in the text D, and c_i is $tf * idf$ measure of the word w_i in concept-attached article.

For precomputing concept vectors, we used 200 top keywords (with weight >0.05) of a document to compute weights of concepts. We kept the maximum 1200 concepts with the largest weight per document. Since we build vectors of Wikipedia articles using Wikipedia articles as concepts, we excluded a concept that represents the same article from the vectors.

3.2 Simplified Model

Now let us simplify the model described above: instead of building a vector space of concepts, we will use only the transitivity of the document similarity function. In other words, let us have a monolingual function for searching for similar documents $F(d)$. Also, there is a set of parallel documents $\{A_i, A_i^*\}$.

Then, to find documents similar to the given Russian-language document $d \in D_{rus}$ in English collection, one can use the following algorithm:

1. Find the documents similar to d in mediator collection A.
2. Take the documents from A^* that correspond to the documents founded in step 1.
3. Search for similar documents obtained in step 2.

The described approach allows us to use any mediator collection on an existing search index base without the building a new vector space for each pick.

As a method for monolingual search for similar documents, we will use the approach described in [3], which is perfect for processing large amounts of data.

Also, we use Wikipedia as a mediator collection, as well as our set of documents from UN digital library. During the search, we exclude the analyzed pair of documents from the mediator.

4 Approaches Based on Word Embeddings

The ideas of distributive semantics, firstly proposed in [4], play a special role in natural language processing. The essence of this approach is building such word vector space that similar words will have close (for example, in terms of cosine distance).

4.1 Extended Translation Models

One of the possible solutions for the cross-lingual document retrieval is the development of the conceptual mediator based on word embeddings as concepts. Such a method was proposed in [5] for some probabilistic search models – for example, the widespread BM25 ranking function [6]. In this case, the authors define an extended tf and df measures for the word t in document d as follows:

$$\widehat{tf}_{t,d} = tf_{t,d} + \sum_{t' \in R(t)} P_T(t|t')tf_d(t');$$
$$\widehat{df}_t = |\{d \in D : t \in T_d \vee \exists t' \in R(t), t' \in T_d\}|,$$

where $P_T(t|t')$ is the probability of translation, and $R(t)$ is the set of related terms to a given one.

So, other components are defined as follows:

- document length $\hat{L}_d = \sum_{t \in \hat{T}_d} \widehat{tf}_d(t)$;
- average document length $\widehat{avgdl} = \frac{1}{|D|} \sum_{d \in D} \hat{L}_d$.

So, for BM25 we get the following expression:

$$BM25_{ET}(q,d) = \sum_{t \in \hat{T}_d \cap T_q} \frac{(k_1 + 1)\overline{\widehat{tf}_d(t)}}{k_1 + \overline{\widehat{tf}_d(t)}} \frac{(k_2 + 1)tf_q(t)}{k_2 + tf_q(t)} \log \frac{|D| + 0.5}{\widehat{df}_t + 0.5}$$

with

$$\overline{\widehat{tf}_d(t)} = \frac{\widehat{tf}_d(t)}{\hat{B}(d)}, \hat{B}(d) = (1 - b) + b \frac{\hat{L}(d)}{\widehat{avgdl}}.$$

4.2 Retrieval Based Approach

Another interesting approach is building a bilingual vector space described in [7]. The main idea is to build one pseudo-bilingual sentence instead of two monolingual ones during the model training. For example, for the sentences "мама мыла раму" and "mother washed a beautiful frame", the result will be "мама mother мыла washed раму beautiful frame".

We have implemented such a model in [8], as the dataset we have used pairs of parallel sentences from the following corpora [9]:

- News Commentary;
- TED Talks 2013;
- MultiUN (first 2 million sentences);
- Wiki;
- JW300;
- QED;
- Tatoeba.

Moreover, we used sentences from the parallel Yandex corpora.

All parallel sentences are preprocessed: we split each sentence into tokens, lemmatize tokens and parse texts. We use AOT for the Russian language and Udpipe for English language. Besides, we removed words with non-important part of speech: conjunction, pronoun, preposition, etc., and common stop-words.

After that, we filter out all pairs that have difference in length more than ten words. We use syntactic phrases up to four words to enrich the vocabulary. We take only phrases that are common for the corpus, e.g. has more than ten occurrences. We duplicate a sentence multiple times if there are some overlapping phrases. For example, from the sentence with the phrase "Russian presidential election …" we generate three variations with different phrases:

- "Russian_presidental_election …";
- "Russian_election presidential_election …";
- "Russian presidential election …".

Finally, we have assembled a corpus of more than 5.1 million sentences (more than 10 million sentences with phrases variations). The dictionary size was around 680 000 words/phrases.

For search we use a special implementation of the inverted index [3]: we have indexed words and syntactic phrases (up to 4 words) with weight (e.g. TF-IDF) that represents the association of this word with a document.

$$TF_D(w_i) = \log_{len(D)+1}(Cnt(w_i) + 1),$$

for word w_i from a document D.

$$IDF = \max\left(0, \log_{10} \frac{N - w_{cnt} + 0.5}{w_{cnt} + 0.5}\right),$$

where N is total amount of documents in collection.

So, at query time, we take top keywords (phrases) from the given document, map them to other language keywords by using cross-lingual embeddings (described above). Then we retrieve corresponding documents from the index and merge them to weighted vectors. After that, we compare the target vectors with all other ones by cosine and Hamming measures.

4.3   Topic Comparison (top2vec)

Furthermore, there is an exciting approach of comparing documents by their topics: for each document from the dataset we retrieve keywords by $tf * idf$ measure. Then we build vectors for each keyword by using bilingual space described previously. So, every document can be represented as an averaged vector of its keywords. To compare such vectors, we use cosine measure.

4.4   LASER Model

To compare our results with other models, we decided to take the universal multilingual model LASER (Language-Agnostic Sentence Representations), introduced in [10]. This system uses single, language agnostic BiLSTM encoder to build sentence embeddings, which is coupled with an auxiliary decoder and trained on parallel corpora. This model was trained for open data for 93 languages.

To train this model authors used the combination of the following publicly available parallel corpora:

- Europarl: 21 European languages;
- United Nations: first two million sentences in Arabic, Russian and Chinese;
- OpenSubtitles2018: a parallel corpus of movie subtitles in 57 languages;
- Global Voices: news stories from the Global Voices website (38 languages);
- Tanzil: Quran translations in 42 languages;
- Tatoeba.

As a result, any sentence is translated into a vector of dimension 1024 in a single multilingual space.

As a search model, we use approximate nearest neighbor for vector space. We use Faiss IVFFlatIndex [11] for indexing document embeddings.

We use this model in two ways described below.

Sentence Comparison. In this case, it is proposed to build vectors for each sentence in each document, and then search for the most similar sentences by cosine measure. As a result, the ranking of documents is based on the number of matching offers. At the same time, we excluded from the search offers containing less than ten characters (document numbers, section names, etc.).

Keywords Comparison. Since the data volumes are large enough, and the dimension of the vectors in this model is quite big, we decided to try a topic comparison model with LASER embedding space.

5   Results

To compare the quality of the methods described above, we used standard metrics: precision, recall and MAP (mean average precision).

For each pair of 33 thousand documents in our corpora, we searched for English-language materials corresponding to a given Russian-language one. The results are collected in Table 1.

Table 1. Metrics per each size group

Method	P@1	Rec@5	Rec@10	Rec@20	Rec@150	MAP@150
CL-ESA	0.13	0.24	0.3	0.36	0.48	0.186
CL-ESA, simpl., Wiki mediator	0.0016	0.004	0.0075	0.01	0.036	0.0037
CL-ESA, simpl., UN mediator	0.19	0.29	0.365	0.427	0.535	0.246
ETM+BM25	0.78	0.87	0.89	0.9	0.91	0.82
RBA, cos	0.58	0.73	0.77	0.8	0.83	0.645
RBA, ham	0.62	0.75	0.79	0.81	0.83	0.678
Top2Vec	0.7	0.81	0.83	0.85	0.89	0.749
LASER, keywords	0.05	0.118	0.15	0.19	0.33	0.088
LASER, sentences	0.05	0.27	0.415	0.55	0.767	0.16

We use the following abbreviation:

- CL-ESA – cross-lingual explicit semantic analysis;
- ETM – extended translation model;
- RBA – retrieval-based approach;
- LASER – Language-Agnostic Sentence Representations;
- Cos – cosine distance measure;
- Ham – Hamming distance measure.

It should be noted that the highest results were achieved using the extended translation model in combination with the BM25 ranking function. And all word embedding models show better results than mediator-only model.

Besides, using the CL-ESA methods and the simplified version using different mediators demonstrates the importance of choosing a collection for mediator: even simplified model achieves better results when it uses a collection that is thematically relevant to the documents from the search dataset.

6 Conclusion

We have compared different approaches to conducting a cross-lingual search for similar documents in the legal domain. The best results were demonstrated by the combination of the ETM and BM25 ranking function.

However, the task of similar documents retrieval is the only one of the stages of a search study in a comparative analysis of the legal domain. This stage can be used to prepare a list of candidate documents for further study. We suppose the next step of the research is to analyze and develop approaches to identify the semantically similar legal language in documents in different languages.

Acknowledgements. This study was funded by RFBR according to the research projects №. 18-29-16172 and №. 18-29-16022.

References

1. Zhebel, V., Kreskin, A., Sochenkov, I.: Cross-lingual document analysis in legal domain. Trudy Instituta sistemnogo analiza rossiyskoy akademii nauk **70**(1), 24–29 (2020). https://doi.org/10.14357/20790279200103
2. Potthast, M., Barrón-Cedeño, A., Stein, B., Rosso, P.: Cross-language plagiarism detection. Lang. Res. Eval. **45**(1), 45–62 (2011)
3. Sochenkov, I.V., Zubarev, D.V., Tikhomirov, I.A.: Exploratory patent search. Inform. Appl. **12**(1), 89–94 (2018)
4. Mikolov, T., Chen, K., Corrado G., Dean, J.: Efficient estimation of word representations in vector space. In: ICLR Workshop (2013)
5. Rekabsaz, N., Lupu, M., Hanbury, A., Zuccon, G.: Generalizing translation models in the probabilistic relevance framework. In: Proceedings of CIKM (2016)
6. Robertson, S.E., et al.: Okapi at TREC-3.0. In: Proceedings of the Third Text REtrieval Conference (TREC 1994), Gaithersburg, USA, November 1994
7. Vulić, I., Moens, M.F.: Bilingual word embeddings from non-parallel document-aligned data applied to bilingual lexicon induction. In: Proceedings of the 53rd Annual Meeting of the Association for Computational Linguistics and the 7th International Joint Conference on Natural Language Processing, vol. 2, pp. 719–725 (2015)
8. Zubarev, D.V., Sochenkov, I.V.: Cross-lingual similar document retrieval methods. Proc. Inst. Syst. Prog. **31**(5), 127–136 (2019). https://doi.org/10.15514/ISPRAS-2019-31(5)-9
9. Tiedemann, J.: Parallel data, tools and interfaces in OPUS. In: Proceedings of the Language Resources and Evaluation (LREC), pp. 2214–2218 (2012)

10. Artetxe, M., Schwenk, H.: Massively Multilingual Sentence Embeddings for Zero-Shot Cross-Lingual Transfer and Beyond. Trans. Assoc. Comput. Linguist. **7**, 597–610 (2019)
11. Johnson, J., Douze, M., Jégou, H.: Billion-scale similarity search with GPUs. arXiv:1702.08734 (2017)

Author Index

Printed in the United States
By Bookmasters